T0181125

Lecture Notes in Computer Science 13232

More information about this series at https://link.springer.com/bookseries/558

Stan Sclaroff · Cosimo Distante · Marco Leo ·
Giovanni M. Farinella · Federico Tombari (Eds.)

Image Analysis and Processing – ICIAP 2022

21st International Conference
Lecce, Italy, May 23–27, 2022
Proceedings, Part II

Springer

Editors
Stan Sclaroff
Boston University
Boston, MA, USA

Cosimo Distante
National Research Council
Lecce, Italy

Marco Leo
National Research Council
Lecce, Italy

Giovanni M. Farinella
University of Catania
Catania, Italy

Federico Tombari
Technische Universität München
Garching, Germany

ISSN 0302-9743 ISSN 1611-3349 (electronic)
Lecture Notes in Computer Science
ISBN 978-3-031-06429-6 ISBN 978-3-031-06430-2 (eBook)
https://doi.org/10.1007/978-3-031-06430-2

This Springer imprint is published by the registered company Springer Nature Switzerland AG
The registered company address is: Gewerbestrasse 11, 6330 Cham, Switzerland

Preface

The International Conference on Image Analysis and Processing (ICIAP) is an established biennial scientific meeting promoted by the Italian Association for Computer Vision, Pattern Recognition and Machine Learning (CVPL - formerly GIRPR), which is the Italian IAPR Member Society, and covers topics related to theoretical and experimental areas of Computer Vision, Image Processing, Pattern Recognition, and Machine Learning with emphasis on theoretical aspects and applications. The 21st International Conference on Image Analysis and Processing (ICIAP 2022) was held in Lecce, Italy, during May 23–27, 2022 (postponed from 2021 due to the COVID-19 pandemic), in the magnificent venue of the Teatro Apollo, (http://www.iciap2021.org). It was organized by the Lecce Institute of Applied Sciences and Intelligent Systems, an Institute of CNR, the National Research Council of Italy. ICIAP 2022 was sponsored by NVIDIA, E4 Computer Engineering s.p.a., and ImageS s.p.a., and was endorsed by the Apulia Region and the Province of Lecce.

The conference covered both established and recent scientific trends, with particular emphasis on Video Analysis and Understanding, Pattern Recognition and Machine Learning, Deep Learning, Multiview Geometry and 3D Computer Vision, Image Analysis, Detection and Recognition, Multimedia, Biomedical and Assistive Technology, Digital Forensics and Biometrics, Image Processing for Cultural Heritage, and Robot Vision.

The ICIAP 2022 main conference received 297 paper submissions from all over the world, including Austria, Azerbaijan, Bangladesh, Belgium, Brazil, Canada, China, the Czech Republic, the UK, Finland, France, Germany, Greece, Hungary, India, Ireland, Japan, Latvia, Lebanon, Mongolia, New Zealand, the Netherlands, Italy, Pakistan, Peru, Poland, Portugal, Russia, Syria, Spain, South Korea, Sri Lanka, Luxembourg, South Africa, Sweden, Switzerland, Turkey, the United Arab Emirates, and the USA. Two rounds of submissions were introduced, independent from each other, in order to improve the quality of papers. Paper selection was carried out by 25 expert researchers who acted as area chairs, together with the International Program Committee and an expert team of reviewers. The rigorous peer-review selection process, carried out by three distinct reviewers for each submission, ultimately led to the selection of 162 high-quality manuscripts, with an overall acceptance rate of 54%.

The main conference program included 45 oral presentations, 145 posters, and five invited talks. The invited talks were presented by leading experts in computer vision and pattern recognition: Larry S. Davis, University of Maryland and Amazon (USA), Roberto Cipolla, University of Cambridge (UK), Dima Damen, University of Bristol (UK), and Laura Leal-Taixe, Technische Universität München (Germany).

ICIAP 2022 also included 11 tutorials and hosted 16 workshops, seven competitions, and a special session on topics of great relevance with respect to the state of the art. The tutorial and workshop organizers came from both industry and academia.

Several awards were presented during the ICIAP 2022 conference. The Best Student paper award was supported by the MDPI Journal of Imaging, and several other prizes were conferred under the Platinum level sponsorship provided by NVIDIA.

The success of ICIAP 2022 is credited to the contribution of many people. Special thanks should be given to the area chairs, who did a truly outstanding job. We wish to thank the reviewers for the immense amount of hard work and professionalism that went into making ICIAP 2022 a successful meeting. Our thanks also go to the Organizing Committee for their unstinting dedication, advice, and support.

We hope that ICIAP 2022 has helped to build a piece of the future, a future where technologies can allow people to live comfortably, healthily, and in peace.

May 2022

Cosimo Distante
Stan Sclaroff
Giovanni Maria Farinella
Marco Leo
Federico Tombari

Organization

General Chairs

Cosimo Distante — National Research Council, Italy
Stan Sclaroff — Boston University, USA

Technical Program Chairs

Giovanni Maria Farinella — University of Catania, Italy
Marco Leo — National Research Council, Italy
Federico Tombari — Google and TUM, Germany

Area Chairs

Lamberto Ballan — University of Padua, Italy
Francois Bremond — Inria, France
Simone Calderara — University of Modena and Reggio Emilia, Italy
Modesto Castrillon Santana — University of Las Palmas de Gran Canaria, Spain
Marco Cristani — University of Verona, Italy
Luigi Di Stefano — University of Bologna, Italy
Sergio Escalera — University of Barcelona, Spain
Luiz Marcos Garcia Goncalves — UFRN, Brazil
Javier Ortega Garcia — Universidad Autonoma de Madrid, Spain
Costantino Grana — University of Modena and Reggio Emilia, Italy
Tal Hassner — Facebook AML and Open University of Israel, Israel
Gian Luca Marcialis — University of Cagliari, Italy
Christian Micheloni — University of Udine, Italy
Fausto Milletarì — NVIDIA, USA
Vittorio Murino — Italian Institute of Technology, Italy
Vishal Patel — Johns Hopkins University, USA
Marcello Pelillo — Università Ca' Foscari Venice, Italy
Federico Pernici — University of Florence, Italy
Andrea Prati — University of Parma, Italy
Justus Piater — University of Innsbruck, Austria
Elisa Ricci — University of Trento, Italy
Alessia Saggese — University of Salerno, Italy
Roberto Scopigno — National Research Council, Italy

Filippo Stanco   University of Catania, Italy
Mario Vento   University of Salerno, Italy

Workshop Chairs

Emanuele Frontoni   Università Politecnica delle Marche, Italy
Pier Luigi Mazzeo   National Research Council, Italy

Publication Chair

Pierluigi Carcagni   National Research Council, Italy

Publicity Chairs

Marco Del Coco   National Research Council, Italy
Antonino Furnari   University of Catania, Italy

Finance and Registration Chairs

Maria Grazia Distante   National Research Council, Italy
Paolo Spagnolo   National Research Council, Italy

Web Chair

Arturo Argentieri   National Research Council, Italy

Tutorial Chairs

Alessio Del Bue   Italian Institute of Technology, Italy
Lorenzo Seidenari   University of Florence, Italy

Special Session Chairs

Marco La Cascia   University of Palermo, Italy
Nichi Martinel   University of Udine, Italy

Industrial Chairs

Ettore Stella   National Research Council, Italy
Giuseppe Celeste   National Research Council, Italy
Fabio Galasso   Sapienza University of Rome, Italy

North Africa Liaison Chair

Dorra Sellami   University of Sfax, Tunisia

Oceania Liaison Chair

Wei Qi Yan   Auckland University of Technology, New Zealand

North America Liaison Chair

Larry S. Davis   University of Maryland, USA

Asia Liaison Chair

Wei Shi Zheng   Sun Yat-sen University, China

Latin America Liaison Chair

Luiz Marcos Garcia Goncalves   UFRN, Brazil

Invited Speakers

Larry S. Davis   University of Maryland and Amazon, USA
Roberto Cipolla   University of Cambridge,UK
Dima Aldamen   University of Bristol, UK
Laura Leal-Taixe   Technische Universität München, Germany

Steering Committee

Virginio Cantoni   University of Pavia, Italy
Luigi Pietro Cordella   University of Napoli Federico II, Italy
Rita Cucchiara   University of Modena and Reggio Emilia, Italy
Alberto Del Bimbo   University of Firenze, Italy
Marco Ferretti   University of Pavia, Italy
Fabio Roli   University of Cagliari, Italy
Gabriella Sanniti di Baja   National Research Council, Italy

Endorsing Institutions

International Association for Pattern Recognition (IAPR)
Italian Association for Computer Vision, Pattern Recognition and Machine Learning
(CVPL)
Springer

Institutional Patronage

Institute of Applied Sciences and Intelligent Systems (ISASI)
National Research Council of Italy (CNR)
Provincia di Lecce
Regione Puglia

Contents – Part II

Digital Forensics and Biometrics

Image Analysis, Detection and Recognition

Multiview Geometry and 3D Computer Vision

Towards Reconstruction of 3D Shapes in a Realistic Environment

Mohammad Zohaib[1,2]([∅])(iD), Matteo Taiana[1](iD), and Alessio Del Bue[1](iD)

[1] Pattern Analysis and Computer Vision (PAVIS), Italian Institute of Technology, Genoa, Italy
{mohammad.zohaib,matteo.taiana,alessio.delbue}@iit.it
[2] Department of Marine, Electrical, Electronic and Telecommunications Engineering, University of Genoa, Genoa, Italy

Abstract. This paper presents an end-to-end approach for single-view 3D object reconstruction in a realistic environment. Most of the existing reconstruction approaches are trained on synthetic data and they fail when evaluated on real images. On the other hand, some of the methods require pre-processing in order to separate an object from the background. In contrast, the proposed approach learns to compute stable features for an object by reducing the influence of image background. This is achieved by feeding two images simultaneously to the model; synthetic with white background and its realistic variant with a natural background. The encoder extracts the common features from both images and hence separates features of the object from features of the background. The extracted features allow the model to predict an accurate 3D object surface from a real image. The approach is evaluated for both real images of the Pix3D dataset and realistic images rendered from the ShapeNet dataset. The results are compared with state-of-the-art approaches in order to highlight the significance of the proposed approach. Our approach achieves an increase in reconstruction accuracy of approximately 6.1% points in F_1 score with respect to Mesh R-CNN on the Pix3D dataset.

Keywords: 3D reconstruction · Single-view image · Realistic background

1 Introduction

Recent advancement in computer vision and deep learning has brought 3D object reconstruction to a level useful for a variety of applications including Augmented Reality (AR), autonomous driving, robotics applications and game development. Reconstruction approaches in the literature rely on data from one or more input modalities: from simple images [1–5], depth images [6,7], to point clouds [8–10]. The field of 3D object reconstruction from 2D images is evolving quickly, with part of the field focused on estimating very accurate 3D shapes in simplified settings, i.e., using images with a blank background as input, while another part strives to estimate 3D object shapes in the wild.

Supplementary Information The online version contains supplementary material available at https://doi.org/10.1007/978-3-031-06430-2_1.

S. Sclaroff et al. (Eds.): ICIAP 2022, LNCS 13232, pp. 3–14, 2022.
https://doi.org/10.1007/978-3-031-06430-2_1

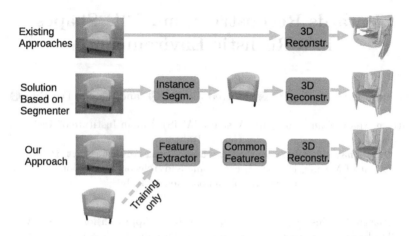

Fig. 1. Overview of the presented approach. Reconstruction systems with high accuracy do not perform well when applied to natural images directly (top). An instance segmentation algorithm can be considered as a simple solution for removing the background (middle). In comparison, the proposed approach reconstructs accurate 3D shapes by estimating common features for realistic and white background images (bottom).

Some of the most recent efforts in 3D object reconstruction from images have been focused on solving a simplified problem: reconstructing objects from synthetic images which present no background clutter and perfect foreground/ background segmentation [2,3]. Data for training and evaluating such systems is collected by rendering 3D object models on a white background. A commonly used source for the 3D models is the ShapeNet dataset [11]. Object reconstruction systems trained on this kind of data achieve high reconstruction accuracy, however their performance drops dramatically for natural images. On the other hand, some of the current research approaches focus on reconstructing 3D objects from natural images [4]. These systems have the advantage of being applicable to real-world images, but this comes at the cost of lower reconstruction accuracy.

In comparison, we propose an end-to-end approach that estimates the 3D shape of an object from a real image without requiring any pre-processing module. The approach learns to generate features from an image with a background that are similar to the features that would be generated from the same image without a background. These features allow the model to estimate a comparatively accurate object shape from a real image. During training, both versions of the image are fed to an encoder in parallel that extracts common features. Whereas during inference, only the real images are used. A sketch of the proposed approach is illustrated in Fig. 1. The main contributions are as follows:

- The proposed approach reconstructs 3D shape in an end-to-end manner from natural images even in the presence of background
- Unlike the existing approaches, our approach extracts features from realistic images that are closer to those extracted from similar synthetic images

– The presented approach does not require any segmentation approach; it separates object features from features of the background
– Our results are comparable to those obtained using a combination of state-of-the-art (SOTA) methods for segmentation and 3D reconstruction from background-less synthetic images

The remainder of the paper is organized as follows: Sect. 2 reviews techniques from the literature, Sect. 3 describes the proposed method and Sect. 4 presents the experimental details. Results are compared with SOTA techniques in Sect. 5. Finally, conclusions are summarised in Sect. 6.

2   Related Work

Estimating 3D object shapes from images is a relevant problem that has actively been tackled by the scientific community. As a result, several approaches have been proposed in the last few years.

Mescheder *et al.* [2] presented ONet, a system for 3D reconstruction based on a continuous 3D occupancy function. It discretizes a volumetric space continuously by evaluating occupancy probability. Deng *et al.* presented CvxNet [3], an approach for geometry representation based on primitive decomposition. It models the shape of one object as the union of a small set of convex components. The number of convexes needs to be specified at training time, which negatively affects the flexibility of the reconstruction system. BSP-Net [12], a similar approach proposed by Chen *et al.*, generates 3D meshes via binary space partitioning, relieving the need for specifying the number of components.

The above mentioned approaches focus on solving the 3D object reconstruction problem in a simplified setting: the input images are obtained by projecting 3D object models onto images with a white background. In an effort towards extending the capabilities of reconstruction algorithms to natural images, Wu *et al.* [13] substituted the white background in their input data with randomised real images. A decisive step towards object reconstruction in the wild was taken with the disclosure of the Pix3D benchmark [14], a dataset composed of natural images for which the poses of the visible objects have been accurately estimated, exploiting a combination of human labeling and automatic optimization. Pix2Vox++ [15] exploits multiple views of one object to generate multiple 3D shapes. It reports results on synthetic images and natural images from Pix3D, restricted to the 'chair' category. Other similar works [16–19] also learn to produce object 3D shapes in the presence of background. However, their aim is to estimate shapes in the form of either a point cloud or a voxel grid. Both representations have their own limitations; point clouds lack connectivity information while memory requirements grow cubically with resolution in voxel grids.

On the other hand, SIST [20], a self-supervised approach for image to 3D shape translation, uses an implicit field decoder to estimate continues object's 3D surface. However, SIST was not tested on natural images but on Pix3D images whose background had been painted white, exploiting ground truth segmentation information. Mesh R-CNN [4] separates itself from the rest of the field because it was designed to work on natural images. As impressive as its performance is, Mesh R-CNN achieves a lower reconstruction accuracy with respect to the other existing approaches e.g., ONet and CvxNet. Salvi *et al.* presented a method that improves ONet by introducing self-attention module in encoder [21]. Whereas the decoder is the same as ONet. Although they have tested the approach on natural images taken from online product dataset [22], they have not presented any quantitative results. Secondly, it is obvious since ONet is trained on synthetic data, it performs poorly on images with a background at inference.

In comparison to the reported approaches, the proposed approach estimates object 3D shapes (in mesh form) from natural images. It is trained to learn object features from realistic images that contain objects and random real backgrounds. Furthermore, we also present a setup that enables the existing synthetic approaches to perform well on natural images. We deal with the requirement imposed by such algorithms of having input images with a white background by combining them with an instance segmentation algorithm.

3   Methodology

Given a single realistic image, goal of the research is to estimate an accurate 3D shape. In this regard, an end-to-end approach is proposed that learns to extract object features from realistic images – containing a synthetic object with a background. These features allow the approach to decode comparatively accurate 3D shapes from natural images. The designed model is inspired from Occupancy Network (ONet) [2]. It is based on two main modules: feature extractor and 3D shape predictor. The first module extracts object features in the presence of a background, whereas the second module estimates a 3D object boundary in the form of an occupancy function. The process is described in detail in the following subsections. The architecture of the proposed method is illustrated in Fig. 2.

3.1   Feature Extractor

The module extracts object features that are required for 3D shape reconstruction. It is based on an encoder that takes two images, a synthetic image along with its variant with added background, in parallel fashion and produces two feature vectors. The vectors are compared in order to instruct the encoder to produce common features for both images. In the beginning, the encoder produces different features; however, after some learning iterations, it starts extracting the same features. We calculate Mean Absolute Error (MAE) for features comparison as

$$\mathcal{L}_{enc} = \frac{1}{M} \sum_{m=1}^{M} |f_m^{(S)} - f_m^{(R)}|, \tag{1}$$

where $f_m^{(S)}$ and $f_m^{(R)}$ denote features extracted from synthetic and realistic image. m is an element of a feature vector of size M. The loss $\mathcal{L}_{enc}$ is only used to update the weights of the encoder. The process of $\mathcal{L}_{enc}$ computation and the weights update is highlighted in the left part of the Fig. 2.

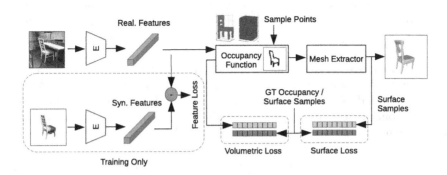

Fig. 2. Architecture of the proposed approach. During training, an image with and without background is fed to the encoder in parallel, producing two feature vectors; synthetic and realistic. The vectors are compared in order to enable the encoder to extract common features from both image versions. The shape predictor module based on an occupancy function uses the feature vector (coming from a realistic image) for object boundary estimation. The boundary is evaluated by computing volumetric and surface loss. At inference time, only the real image is fed to the system.

3.2 Shape Predictor

The shape predictor estimates an object's 3D shape by utilizing an occupancy function in a similar way as defined in ONet [2,21]. The function $\mathcal{F} : R^3 \rightarrow [0;1]$ evaluates N uniform sample points in space for estimating occupancy probability for each point. Where 0 and 1 represent if the samples are outside or inside the object boundary, respectively. It is a combination of 5 fully-connected ResNet blocks. Each block contains a pair of Conditional Batch-Normalization (CBN), ReLU activation functions, and a fully connected layer. The output of the last block is downsampled to a 1D vector and passed through a sigmoid activation function in order to obtain estimated probabilities. For a detailed visualization of the occupancy function see Fig. 2 of [21]. The shape predictor takes common features extracted by the feature extractor and corresponding GT uniform 3D sample points. It first computes 256D features for every point using a fully connected network and passes them along with common features to the occupancy function. The function estimates volumetric occupancy for every point with respect to the object's boundary. The estimated occupancies indicate whether the points belong to the object or its surroundings. These occupancies are compared with corresponding GT values. We use Binary Cross-Entropy (BCE) Loss for comparison as

$$\mathcal{O} = \mathcal{F}(f_{com}, P_i) \mid i : 1 \, to \, N$$
$$\mathcal{L}_{vol} = \mathcal{L}_{BCE}(\mathcal{O}, \mathcal{Q}), \tag{2}$$

where f_{com} represents common features, P_i are the N 3D points for evaluation and the corresponding output of occupancy function for all the points is depicted by vector $\mathcal{O}$. The volumetric loss $\mathcal{L}_{vol}$ is a BCE loss between predicted ($\mathcal{O}$) and label ($\mathcal{Q}$) occupancies. The loss improves the geometry of the predicted shape by reasoning on the 3D volumetric space. The mesh extractor module produces meshes in a two-step process; by applying Multiresolution IsoSurface Extraction (MISE) [2] that utilizes occupancy function in order to achieve the required resolution and by using Marching Cubes algorithm [23] for final mesh extraction.

Additionally, in order to improve the surface of the predicted shape, a surface loss is introduced. Since comparing meshes is a complex and resource consuming task, random points are sampled for predicted (X) and GT (Y) mesh. These sample points are used to calculate Chamfer L_1 Distance as

$$\mathcal{L}_{surf} = \frac{1}{n_X} \sum_{x \in X} \min_{y \in Y} |x - y| + \frac{1}{n_Y} \sum_{y \in Y} \min_{x \in X} |y - x|, \tag{3}$$

where x and y represent a single sample from set X and Y, respectively. n_X and n_Y show the total number of samples in each set, which is the same in our case. The overall shape prediction loss can be defined as

$$\mathcal{L}_{overall} = \mathcal{L}_{vol} + \mathcal{L}_{surf}. \tag{4}$$

Unlike encoder loss, the shape prediction loss ($\mathcal{L}_{overall}$) contributes in updating the weights of the whole model.

3.3   Inference

At inference, the lower part of the feature extractor is removed from the model as highlighted in Fig. 2. Real images are fed to the encoder for 2D feature extraction. The extracted features are then used by the second module for producing occupancy probabilities for all the sample points in a 3D volumetric space.

4   Experimental Details

In this section, we describe the experimental details we used to assess the performance of the proposed approach in comparison with SOTA. The proposed approach is validated using two types of data; real from Pix3D and rendered from ShapeNet in the presence of background.

4.1   Implementation Details

Unlike [21], in our approach we use ResNet18 [24] encoder (without self-attention module) with pre-trained weights for ImageNet dataset [25]. The last layers are modified to obtain a 256D feature vector. We train the model by using synthetic and realistic images that are rendered from ShapeNet objects on white and random backgrounds, respectively. We set the learning rate and weight decay to 1e-5. The rest of the hyperparameters are set to the same values as used in [2].

4.2   Datasets

Mainly three datasets are used in the experiments. For training, we use synthetic images rendered by Choy *et al.* [26] and realistic images that are generated by adding random backgrounds to the synthetic images. The background images are taken from SUN dataset [27]. For evaluation on real images we use the Pix3D dataset [14]. We select only the object categories in Pix3D for which ONet, Mesh R-CNN, and YOLACT [28] were trained, i.e. tables, chairs, and sofas. For evaluation on some other categories (e.g., car, airplane, bench) real images are taken from PASCAL [29] and COCO dataset [30]. The details of the selected test samples from the Pix3D dataset based on a comparison of YOLACT and Mask R-CNN [31] segmentation are provided in the supplementary material.

4.3   Performance Measurement

We use the Chamfer L_1 distance and F_1 score for evaluation. The Chamfer distance computes the average distance between predicted and ground truth meshes with the help of surface sample points. We select 100k sample points randomly from the surface of the predicted and the ground truth mesh and use a KD-tree to associate each point with its nearest neighbor from the other mesh. The final value for the Chamfer distance (as depicted in Eq. 3) is an average of all the absolute L_1 distances measured in both directions: from the estimate to the ground truth and vice versa. The F_1 score describes how accurate a prediction is, given the tolerance factor τ (surface thickness) and is computed as the harmonic mean between precision and recall. Precision is defined as the fraction of points that have a distance below τ when going from the estimated mesh to the ground truth mesh. A recall is defined as the fraction of points that have a distance below τ when going in the opposite direction. We chose the value of 0.001 for τ from the ones present in the literature because it is the one that better highlights the difference in performance in our experiments. Better performances correspond to lower values of the Chamfer L_1 distance and higher values of the F_1 score. We report F_1 score values as percentages of its range $[0, 1]$. Furthermore, for a fair comparison, the estimated shapes are resized and aligned using Iterative Closest Point (ICP). It is due to the fact that Mesh R-CNN produces estimates whose pose is dependent on the viewpoint of the input image, while other approaches reconstruct shapes in a canonical pose and of a specific size.

4.4   Baselines

We compare our results with SOTA baselines including Mesh R-CNN, CvxNet, and ONet. We use a pre-trained model of Mesh R-CNN. Whereas, CvxNet and ONet are trained on the realistic dataset. It is because the model trained on white background images could never perform well on natural images. Additionally, we also present a setup that enables CvxNet and ONet that are trained on the synthetic dataset to produce 3D shapes in the presence of background. The setup integrates an instance segmenter before the 3D reconstruction module.

The segmenter processes natural images, partitioning the pixels into two groups: foreground and background. The foreground pixels represent segmented objects. We separate an object of interest and use the segmentation information in order to paint the background pixels with a uniform white color. In order to make it identical to a synthetic sample, we center the object and apply padding by considering the dimensions of the images used during training. The resulting image is then processed by the reconstruction module, which outputs an estimate of the 3D shape of the object. The proposed system is presented graphically in Fig. 3.

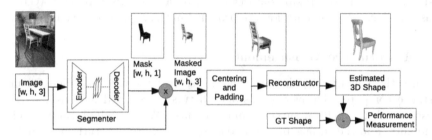

Fig. 3. Setup to execute CvxNet and ONet on real images. The approaches can produce good results for real images if the input image is processed appropriately; separating an object by applying an instance segmentation algorithm, pasting it on the center of the white image, and padding the image in order to make it similar to synthetic.

We chose YOLACT [28] as the instance segmentation algorithm for the system. For differentiating from the original versions CvxNet and ONet that are trained on the realistic dataset, we are using the letter M (mask) with them. So, the CvxNet-M and ONet-M show the versions when the segmenter is integrated with them. A bar chart comparing the performance of the approaches with and without the presented setup is given in supplementary material.

All the approaches are tested on the Pix3D dataset. However, for testing generalization, an experiment is conducted by splitting the pix3D dataset into two parts; white background and color background images. Due to limited space, extra results are provided in the supplementary material.

5   Results and Analysis

In this section, we discuss the results of the baseline approaches and compare them with the results of our approach. CvxNet, ONet, Mesh R-CNN, and the proposed approach are tested using natural images without any pre-processing. However, CvxNet-M and ONet-M compute a masked version of the original image then reconstruct the 3D shape. Qualitative results are illustrated in Fig. 4. Where input real RGB images, their masked versions and the expected 3D shapes for every category are shown in the first three columns, respectively. The next columns illustrate reconstructions by the baselines and the proposed approach (last column). The results of the Cvxnet-M and ONet-M are more accurate than those of CvxNet and ONet. That is due to the fact that they use the segmented foreground part of an image. Reconstructions of the Mesh R-CNN are not complete. In many scenarios, the self-occluded regions are not accurately reconstructed. In comparison, the presented approach outperforms by estimating sharp and smoother surface without requiring any processing on the images.

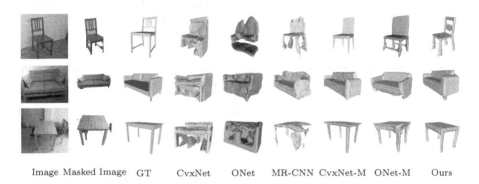

Image Masked Image   GT   CvxNet   ONet   MR-CNN CvxNet-M ONet-M   Ours

Fig. 4. Qualitative comparison of the proposed approach (ours) with the baselines. The masked images are obtained by removing the background, centering the object and padding. CvxNet-M and ONet-M use masked images, whereas the rest of the approaches i.e., CvxNet, ONet, Mesh R-CNN (MR-CNN), and ours use natural images.

A quantitative analysis highlighting the performance of the approaches on the Pix3D dataset is illustrated in Table 1. The CvxNet-M and ONet-M perform well for both metrics in comparison with CvxNet and ONet. This validates that feeding masked images to a 3D reconstruction approach that is trained on synthetic images is beneficial. Second, CvxNet-M and ONet-M show better results than Mesh R-CNN, with the exception of the table class. The problem with the table class originates at the segmentation stage: when the segmentation is not accurate, the 3D reconstruction suffers. In comparison, our approach performs overall well on both metrics. However, for the chair category, ONet-M outperforms ours on Chamfer distance. That is because most of the images in the chair category contain multiple objects of interest and at various positions. In ONet-M, the segmenter selects only one object with the highest score and hence the reconstructor produces a good shape. Whereas, in our approach, the encoder considers all the prominent objects in an image for gathering the features.

For testing on realistic images, we train CvxNet, ONet, and our approach on the realistic dataset. The results are presented in the Table 2. Although the CvxNet and ONet are trained on the realistic dataset, they still could not perform well on the test set. That is due to the fact that they consider background for features learning, which is variant. On the other hand, our approach performs well as it is trained to extract object features in the presence of the background.

Table 1. Quantitative comparison between the baselines and the approach on the Pix3D dataset. Our approach achieves better reconstruction accuracy on a scale of F_1 score in all the cases, while ONet-M retains an advantage for the masked version of the chair category for Chamfer L_1 distance. The best values are highlighted in bold.

Category	CvxNet	ONet	Mesh R-CNN	CvxNet-M	ONet-M	Ours
F_1 Score (%), τ=0.001 ↑/Chamfer L_1 Distance ↓						
Chair	35.43/2.73	34.21/2.45	37.63/1.99	46.88/1.91	46.24/**1.54**	**47.16**/1.82
Sofa	41.99/1.94	42.45/1.91	53.61/1.76	58.35/1.68	53.73/1.75	**61.58/1.63**
Table	33.15/5.07	28.79/5.01	48.12/2.41	44.98/3.72	45.19/2.79	**48.91/2.14**
Average	36.86/3.25	35.15/3.12	46.45/2.05	50.07/2.44	48.39/2.03	**52.55/1.86**

Table 2. Quantitative results for the realistic dataset. The performance of our approach is comparatively better than CvxNet and ONet on both metrics.

Category	Chair	Sofa	Table	Airplane	Bench	Phone	Display	Vessel	Car	Avg.
F_1 Score (%), τ=0.001 ↑										
CvxNet	49.13	62.39	49.87	71.21	54.23	62.93	47.10	53.20	50.83	55.65
ONet	53.21	68.02	62.32	58.76	60.72	66.78	40.82	47.20	64.33	58.02
Ours	58.32	71.23	65.93	73.30	62.43	68.11	52.21	54.74	66.32	63.22
Chamfer L_1 Distance ↓										
CvxNet	1.83	0.87	1.82	1.01	1.32	1.19	1.95	0.92	0.87	1.31
ONet	1.52	0.91	0.83	1.22	1.58	1.27	2.06	1.06	0.92	1.26
Ours	1.48	0.78	0.71	0.93	1.21	1.03	1.89	0.87	0.83	1.08

6   Conclusions

The objective of this paper is to reconstruct 3D shapes from real images. In this regard, an end-to-end approach is proposed that strives to extract object features from a real image by reducing the influence of the image background. During training, a synthetic image is fed to the encoder with its realistic version. The encoder extracts common features from both images that represent features of the object. The extracted features are used by the model to estimate object 3D shape. During inference, we test on real images. The proposed approach outperforms SOTA approaches which are validated by conducting a series of experiments. Furthermore, a system is designed that enables CvxNet and ONet to extract accurate 3D shapes from real images. That system entails segmenting the object of interest from the input images and removing the image background before passing them to the reconstruction algorithms.

References

1. Wang, Z., Isler, V., Lee, D.D.: Surface HOF: surface reconstruction from a single image using higher order function networks. In 2020 IEEE International Conference on Image Processing (ICIP), pp. 2666–2670 (2020)

2. Mescheder, L., Oechsle, M., Niemeyer, M., Nowozin, S., Geiger, A.: Occupancy networks: learning 3D reconstruction in function space. In: Proceedings of the IEEE/CVF Conference on Computer Vision and Pattern Recognition, pp. 4460–4470 (2019)

3. Deng, B., Genova, K., Yazdani, S., Bouaziz, S., Hinton, G., Tagliasacchi, A.: CVXNET: learnable convex decomposition. In: Proceedings of the IEEE/CVF Conference on Computer Vision and Pattern Recognition, pp. 31–44 (2020)

4. Gkioxari, G., Malik, J., Johnson, J.: Mesh R-CNN. In: Proceedings of the IEEE/CVF International Conference on Computer Vision, pp. 9785–9795 (2019)

5. Spezialetti, R., Tan, D.J., Tonioni, A., Tateno, K., Tombari, F.: A divide et Impera approach for 3D shape reconstruction from multiple views. In: 2020 International Conference on 3D Vision (3DV), pp. 160–170 (2020)

6. Yang, B., Rosa, S., Markham, A., Trigoni, N., Wen, H.: Dense 3D object reconstruction from a single depth view. IEEE Trans. Pattern Anal. Mach. Intell. **41**(12), 2820–2834 (2018)

7. Gupta, K., Jabbireddy, S., Shah, K., Shrivastava, A., Zwicker, M.: Improved modeling of 3D shapes with multi-view depth maps. In: International Conference on 3D Vision (3DV), pp. 71–80 (2020)

8. Gu, J., et al.: Weakly-supervised 3D shape completion in the wild. In: Vedaldi, A., Bischof, H., Brox, T., Frahm, J.-M. (eds.) ECCV 2020. LNCS, vol. 12350, pp. 283–299. Springer, Cham (2020). https://doi.org/10.1007/978-3-030-58558-7_17

9. Yin, K., Chen, Z., Chaudhuri, S., Fisher, M., Kim, V.G., Zhang, H.: COALESCE: component assembly by learning to synthesize connections. In: 2020 International Conference on 3D Vision (3DV), pp. 61–70 (2020)

10. Lombardi, S., Oswald, M.R., Pollefeys, M.: Scalable point cloud-based reconstruction with local implicit functions. In: 2020 International Conference on 3D Vision (3DV), pp. 997–1007 (2020)

11. Chang, A.X., et al.: ShapeNet: an information-rich 3D model repository. arXiv:1512.03012 (2015)

12. Chen, Z., Tagliasacchi, A., Zhang, H.: BSP-NET: generating compact meshes via binary space partitioning. In: Proceedings of the IEEE/CVF Conference on Computer Vision and Pattern Recognition, pp. 45–54 (2020)

13. Wu, J., Wang, Y., Xue, T., Sun, X., Freeman, W.T., Tenenbaum, J.B.: MarrNet: 3D shape reconstruction via 2.5 D sketches. In: 31st Conference on Neural Information Processing Systems (2017)

14. Sun, X., Wu, J., Zhang, X., Zhang, Z., Zhang, C., Xue, T., Tenenbaum, J.B., Freeman, W.T.: Pix3D: dataset and methods for single-image 3D shape modeling. In: Proceedings of the IEEE Conference on Computer Vision and Pattern Recognition, pp. 2974–2983 (2018)

15. Xie, H., Yao, H., Zhang, S., Zhou, S., Sun, W.: Pix2Vox++: multi-scale context-aware 3D object reconstruction from single and multiple images. Int. J. Comput. Vision **128**(12), 2919–2935 (2020)

16. Girdhar, R., Fouhey, D.F., Rodriguez, M., Gupta, A.: Learning a predictable and generative vector representation for objects. In: Leibe, B., Matas, J., Sebe, N., Welling, M. (eds.) ECCV 2016. LNCS, vol. 9910, pp. 484–499. Springer, Cham (2016). https://doi.org/10.1007/978-3-319-46466-4_29

17. Feng, Q., Luo, Y., Luo, K., Yang, Y.: Look, cast and mold: learning 3D shape manifold from single-view synthetic data. arXiv preprint arXiv:2103.04789 (2021)

18. Pinheiro, P.O., Rostamzadeh, N., Ahn, S.: Domain-adaptive single-view 3D reconstruction. In: IEEE International Conference on Computer Vision, pp. 7638–7647 (2019)

19. Yang, S., Xu, M., Xie, H., Perry, S., Xia, J.: Single-view 3D object reconstruction from shape priors in memory. arXiv preprint arXiv:2003.03711, pp. 3152–3161 (2020)
20. Kaya, B., Timofte, R.: Self-supervised 2D image to 3D shape translation with disentangled representations. In: 2020 International Conference on 3D Vision (3DV), pp. 1039–1048 (2020)
21. Salvi, A., Gavenski, N., Pooch, E., Tasoniero, F., Barros, R.: Attention-based 3D object reconstruction from a single image. In: International Joint Conference on Neural Networks (IJCNN), pp. 1–8 (2020)
22. Oh Song, H., Xiang, Y., Jegelka, S., Savarese, S.: Deep metric learning via lifted structured feature embedding. In: IEEE Conference on Computer Vision and Pattern Recognition, pp. 4004–4012 (2016)
23. Lorensen, W.E., Cline, H.E.: Marching cubes: a high resolution 3D surface construction algorithm. ACM SIGGRAPH Comput. Graph. **21**(4), 163–169 (1987)
24. He, K., Zhang, X., Ren, S., Sun, J.: Deep residual learning for image recognition. In: Proceedings of the IEEE Conference on Computer Vision and Pattern Recognition, pp. 770–778 (2016)
25. Deng, J., Dong, W., Socher, R., Li, L.J., Li, K., Fei-Fei, L.: ImageNet: a large-scale hierarchical image database. In: IEEE Conference on Computer Vision and Pattern Recognition, pp. 248–255 (2009)
26. Choy, C.B., Xu, D., Gwak, J.Y., Chen, K., Savarese, S.: 3D-R2N2: a unified approach for single and multi-view 3D object reconstruction. In: Leibe, B., Matas, J., Sebe, N., Welling, M. (eds.) ECCV 2016. LNCS, vol. 9912, pp. 628–644. Springer, Cham (2016). https://doi.org/10.1007/978-3-319-46484-8_38
27. Xiao, J., Hays, J., Ehinger, K.A., Oliva, A., Torralba, A.: Sun database: large-scale scene recognition from abbey to zoo. In: IEEE Computer Society Conference on Computer Vision and Pattern Recognition, pp. 3485–3492 (2010)
28. Bolya, D., Zhou, C., Xiao, F., Lee, Y.J.: YOLACT: real-time instance segmentation. In: Proceedings of the IEEE/CVF International Conference on Computer Vision, pp. 9157–9166 (2019)
29. Everingham, M., Eslami, S.A., Van Gool, L., Williams, C.K., Winn, J., Zisserman, A.: The pascal visual object classes challenge: a retrospective. Int. J. Comput. Vis. **111**(1), 98–136 (2015)
30. Lin, T.-Y., et al.: Microsoft COCO: common objects in context. In: Fleet, D., Pajdla, T., Schiele, B., Tuytelaars, T. (eds.) ECCV 2014. LNCS, vol. 8693, pp. 740–755. Springer, Cham (2014). https://doi.org/10.1007/978-3-319-10602-1_48
31. He, K., Gkioxari, G., Dollár, P., Girshick, R.: Mask R-CNN. In: Proceedings of the IEEE International Conference on Computer Vision, pp. 2961–2969 (2017)

SVP-Classifier: Single-View Point Cloud Data Classifier with Multi-view Hallucination

Seyed Saber Mohammadi[1,2](✉) ⓘ, Yiming Wang[2,3] ⓘ, Matteo Taiana[2] ⓘ, Pietro Morerio[2] ⓘ, and Alessio Del Bue[2] ⓘ

[1] Department of Marine, Electrical, Electronic and Telecommunications Engineering, University of Genoa, Genoa, Italy

[2] Pattern Analysis and Computer Vision (PAVIS), Italian Institute of Technology, Genoa, Italy
{seyed.mohammadi,yiming.Wang,matteo.taiana,pietro.morerio, alessio.delbue}@iit.it

[3] Deep Visual Learning (DVL), Fondazione Bruno Kessler, Trento, Italy

Abstract. We address single-view 3D shape classification with partial Point Cloud Data (PCD) inputs. Conventional PCD classifiers achieve the best performance when trained and evaluated with complete 3D object scans. However, they all experience a performance drop when trained and evaluated on partial single-view PCD. We propose a Single-View PCD Classifier (SVP-Classifier), which first hallucinates the features of other viewpoints covering the unseen part of the object with a Conditional Variational Auto-Encoder (CVAE). It then aggregates the hallucinated multi-view features with a multi-level Graph Convolutional Network (GCN) to form a global shape representation that helps to improve the single-view PCD classification performance. With experiments on the single-view PCDs generated from ModelNet40 and ScanObjectNN, we prove that the proposed SVP-Classifier outperforms the best single-view PCD-based methods, after they have been retrained on single-view PCDs, thus reducing the gap between single-view methods and methods that employ complete PCDs. Code and datasets are available: https://github.com/IIT-PAVIS/SVP-Classifier.

Keywords: Multi-view feature hallucination · 3D object classification · Partial point cloud

1 Introduction

Object recognition using 3D data has recently gained increasing research attention because of its vital role in many real-world applications, such as robotic manipulation and navigation. Although several types of 3D data representation exist, Point Cloud Data (PCD) is mostly preferred for its simplicity and availability from a wide range of sensors, e.g., Lidars, RGB-D and stereo cameras. Thanks to the introduction of the PointNet [10] and PointNet++ [12] architectures, direct analysis on unordered PCDs can be performed using deep models [5]. Existing state-of-the-art method [7,9,12,16,19] are mostly tested on the pre-processed and complete 3D scan of an

Supplementary Information The online version contains supplementary material available at https://doi.org/10.1007/978-3-031-06430-2_2.

object, and they experience a drastic performance drop when evaluated on partial PCDs captured from a single viewpoint [15], thus reducing classification accuracy. Retraining the models with partial PCDs can greatly reduce the gap, yet the performance is still about 3–5% lower than that of the models operating on the full PCDs.

However, for operational scenarios, a single-view 3D shape classifier is more desirable, as obtaining a full 3D object scan requires a feasible camera path that can perform a complete 360° trajectory around the object. Such trajectory requires more energy consumption and it might not be viable for autonomous agents as their motion can be constrained by a specific path or a set of views. When multiple single-view PCDs are available, multi-view aggregation has demonstrated advantageous performance with Graph Neural Networks [8].

Inspired by the observation that using multi-view PCDs is beneficial for classification performance, we propose a novel method called *SVP-Classifier* that given only a single-view PCD, hallucinates the multi-view feature representing the unseen part of the object to boost the performance for shape classification. Note that single-view PCD represents real-world settings with challenges introduced by occlusions. As it can be visualised in Fig. 1, We first hallucinate the multi-view feature by feeding a single-view PCD into a Conditional Variational Auto-Encoder (CVAE), where the condition is given as the relative camera pose transformation from the input view to the output view. As training a generative network is a notoriously challenging task, we propose to guide and regularise the CVAE learning with an auxiliary task of partial PCD completion. We then exploit a multi-level GCN to aggregate the shape features from the generated multi-view shape features, to further exploit the relations among multiple views in the feature space, and eventually improve the classification performance.

To summarise, our main contributions are listed as: (1) we propose a multi-view feature hallucination network employing a CVAE to generate the features of the missing viewpoints given a single-view PCD. (2) We propose to exploit PCD completion as an auxiliary task to regularise the learning of the CVAE. (3) We improve the state-of-the-art single-view accuracy by 0.6% and 1.1% on synthetic and real data, respectively.

2   Related Work

We first cover previous works addressing the shape classification problem with PCD input data. We then review the task of dataset generation employing a generative model, i.e., a Variational Auto-Encoder.

Shape Classification with 3D Object Scans. There are a wide range of approaches addressing the shape classification task with various 3D input representations. Voxel-based CNN [5] shows promising performance but at the cost of high memory usage. PointNet [10] was proposed to directly process the unordered PCD with a neural network for shape analysis. PointNet is designed to describe the global shape without a particular focus on encoding local structures, thus having limitations in the recognition of fine-grained patterns. The follow-up work, PointNet++ [12], addresses those limitations by processing PCDs sampled in a metric space in a hierarchical fashion. Dynamic Graph CNN (DGCNN) [16] further proposes an architecture with a novel

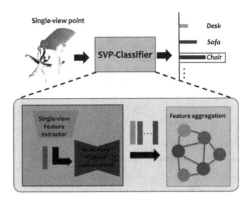

Fig. 1. We propose *SVP-Classifier* to classify objects based on a single-view PCD. The proposed classifier first extracts the single-view shape features and then passes them to a multi-view feature hallucination network to generate multi-view shape features. The generated multi-view shape features are then passed to a multi-level Graph Convolutional Network to form a global aggregated shape feature for the classification.

operation, named EdgeConv, to dynamically capture local geometric structure through edge features that describe the relationships among a point and its neighbours. Differently, PointView-GCN [8] shows a significant improvement in classification accuracy by proposing a network with multi-level GCNs to aggregate the shape features from partial PCDs captured from multiple views. However, the network proposed in [8] has a access to multi-view PCDs, while, in our case, only a single-view PCD is used as input. Instead, in this work, we focus on the more challenging case where an incomplete PCD from a single viewpoint is available as this is the typical setting when acquiring 3D data in practical scenarios.

Data Generation with Variational Auto-Encoders. Data generation is the task of generating new realistic data given a training sample. The Variational Auto-Encoder (VAE) [1] is one of the most powerful tools for generating high dimensional structured data. VAEs have an encoder-decoder structure, and apply regularisation of the latent space. They process the input by initially encoding it as a distribution over the latent embedding. The input for the decoder is then sampled from the distribution, employing randomly generated noise. In the usual case of such an Auto-Encoder [4] being used to produce the reconstruction of the input data on its output, the VAE is trained by minimising reconstruction error. Conditional VAEs [13] were introduced so that the decoded output can be conditioned on supplemental information respect to that processed by the encoder. However, most of the mentioned methods are commonly used to reconstruct the input data as the output of the network. Differently, we propose to use a CVAE to hallucinate multi-view feature data given the single-view features.

3 SVP-Classifier

Given the single-view PCD, *SVP-Classifier* first hallucinates the missing multi-view features, then aggregates the hallucinated shape features using a multi-level GCN to

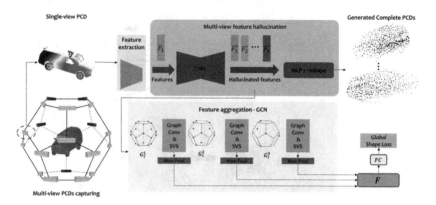

Fig. 2. The architecture of *SVP-Classifier* consists of three modules: the feature extraction, the multi-view feature hallucination and the feature aggregation. During feature hallucination, the CVAE takes the features extracted from a single-view PCD as input and hallucinates the feature vectors for the missing views. Each hallucinated vector is used to generate a completed version of the input PCD. During feature aggregation, the GCN processes the input feature vector together with the hallucinated ones. Each level of the GCN updates the nodes by aggregating the neighbourhood information and then forms a sub-view-graph via selective view-sampling (SVS). Finally, the output of the GCN is processed to estimate the object class.

improve classification accuracy. Figure 2 shows the overall architecture of the proposed network in the training phase, which consists of three modules: the feature extractor, the feature hallucination module, which contains the PCD completion head, and the feature aggregation module. We will briefly introduce each module in the following.

Feature Extractor. Given the multiple single-view PCDs, we separately extract the shape feature F_i from each single-view PCD corresponding to the ith view using the backbone network from PCD-based sota methods for shape recognition [12].

Feature Hallucination. We apply the CVAE to hallucinate the feature $F'_{n|i}$ for the missing n^{th} view, describing a PCD that is not visible from the ith view.

Feature Aggregation. The proposed network for feature aggregation consists of multiple levels of GCNs. At each level j, we perform graph convolution on the input graph G^j to update the node feature F_i^j followed by a selective view sampling to form a smaller graph G^{j+1} with the most discriminative views selected from G^j [8]. The updated features in the new graph G^{j+1} are then fed as input to the next level of convolution. At the input level, each node v_i^1 in the view graph, G^1, is represented by the node feature F_i^1 which is either the input single view feature F_i or the hallucinated shape feature $F'_{n|i}$ given the feature of the input view F_i and the relative view transform from the ith view to the nth view.

Section 3.1 describes the CVAE in detail, while in Sect. 3.2 we describe the workflow of the feature aggregation network. Finally, in Sect. 3.3 we define the loss functions used for training each network.

3.1 CVAE Architecture

As shown in Fig. 3, the proposed CVAE network utilises two encoders with shared weights and one decoder, at training time. One of the encoders takes as input the shape feature F_i extracted from one single-view PCD (among multiple single-view data), while the other encoder is fed with the desired view feature F_n, the one we aim to hallucinate. We condition our network with the relative camera pose transformation between the input and the desired view. The output of the decoder is the estimated feature vector for the desired view: F'_n. At test time, the desired view feature F_n is not available, so the corresponding encoder is not used and the feature vector is filled with random values instead. We explain the workflow of CVAE in detail in the remainder of this section.

Let $p(\mathbf{z})$ be the latent variable distribution which is a centered isotropic multivariate Gaussian with diagonal covariance, where $p(\mathbf{z}) = \mathcal{N}(0, \mathbf{I})$ and $q(\mathbf{z}|F) = \mathcal{N}(\boldsymbol{\mu}(F), \text{diag}(\boldsymbol{\sigma}^2(F))$ where $F = [F_i, F_n]$ and F is a concatenation of both input F_i and desired view feature F_n. The encoder outputs two vectors, the mean $\boldsymbol{\mu}$ and the standard deviation $\boldsymbol{\sigma}$, and it can be defined as: $(\boldsymbol{\mu}, \boldsymbol{\sigma}) = Enc([F, T]; \theta)$, where the encoder $Enc(\cdot)$ with related parameters θ takes as the input the extracted shape feature F that is concatenated (or conditioned) with the relative camera pose transformation T between F_i and F_n. This will output $\boldsymbol{\mu}(F), \boldsymbol{\sigma}(F) \in \mathbb{R}^d$ with d dimension of the latent space.

We can sample $\mathbf{z}$ from $q(\mathbf{z}|F)$ with the reparameterization trick [6] by first sampling a random vector $\mathbf{u}$ from a unit Gaussian $\mathcal{N}(0, \mathbf{I})$, and then multiplying it by the standard deviation $\boldsymbol{\sigma}(F)$ and adding the mean $\boldsymbol{\mu}(F)$: $\mathbf{z} = \boldsymbol{\mu}(F) + \boldsymbol{\sigma}(F) \odot \mathbf{u}$, where $\odot$ is the element-wise product. At test time, we do not have access to the desired view feature F_n vector, so we sample the latent variable z_n randomly from a Gaussian distribution. Finally, the decoder takes the latent space z as input and outputs the reconstructed desired view feature $F'_n = D(Z; \psi)$, where $D(\cdot)$ is the decoder with related parameter ψ.

The feature vector F'_n that has been hallucinated by the CVAE is then fed to a Fully Connected (FC) layer with $3 \cdot p$ output size, where p is the number of points. Furthermore, we reshape the output of the FC layer to obtain a $p \times 3$ matrix that represents the completed PCD. We employ this auxiliary task in order to improve the quality of the feature hallucinated by the CVAE.

3.2 Graph Convolutions

We refer to the features of all nodes in a graph G^j as $\mathbf{F}^j = \{F_i^j\}_{i \in N^j}$. Furthermore, we define the notion of neighbourhoods among nodes: node v_p^j is considered a neighbour of node v_i^j if v_p^j is within the k-Nearest Neighbours (k-NN) of v_i^j with $k = 4$, which are defined by its viewpoint position. The binary adjacency matrix $\mathbf{A}^j$ encodes the neighbourhoods of G^j. With the input view-graph G^j at j^{th} level constructed with all features $\mathbf{F}^j$, we perform three main steps: $a)$ local graph convolution, $b)$ non-local message passing, and $c)$ selective view sampling (SVS).

First, the local graph convolution updates the feature of each node v_i^j by considering its neighbouring nodes as: $\mathbf{F}^j = f\left(\mathbf{A}^j \mathbf{F}^j \mathbf{W}^j; \alpha^j\right)$, where $f(\cdot)$ represents the LeakyReLU operation with related parameters α^j, while $\mathbf{W}^j$ is the weight matrix.

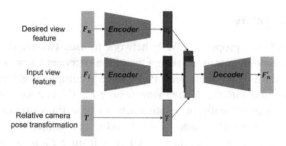

Fig. 3. Architecture of the proposed CVAE, with shared encoder weights. The encoding of the input view feature, that of the desired view feature, and the relative pose transformation between the two camera poses are concatenated and fed to the decoder in order to produce the hallucinated feature vector F_n'. Note that during test we do not have access to the desired view feature, so the latent embedding variable z_n is sampled from a random distribution instead.

Then, we update features $\mathbf{F}^j$ considering the long-range relationships among all nodes in the view-graph G^j with non-local message passing. Each node v_i^j first updates the state of its edge with any other node v_p^j via a message $m_{i,p}^j$ that is computed as: $m_{i,p}^j = \mathcal{R}\left(F_i^j, F_p^j; \beta^j\right)_{i,p \in N^j}$, where $\mathcal{R}(\cdot)$ is the relation function among a pair of views with the related parameters β^j. We then update the feature of node v_i^j as a function of its current state and of all the received pairwise messages as: $F_i^j = \mathcal{C}\left(F_i^j, \sum_{p=1, p \neq i}^{N^j} m_{i,p}^j; \gamma^j\right)$, where $\mathcal{C}(\cdot)$ is a combination function with related parameters γ^j.

Finally, we utilise selective view sampling (SVS) to select the most descriptive views among each neighbourhood. We subsample the view-graph G^j using Farthest Point Sampling (FPS) [12]. For each subsampled node the encoding of each of its nearest neighbour is passed through a fully connected layer followed by a softmax function for class prediction. We select the node with the highest response for the correct class after the softmax function to form the graph of next level. This updated view-graph G^{j+1}, with its associated updated features $\mathbf{F}^{j+1}$, is the output of a level of the GCN and is fed as input to the following level, until the last level of graph convolution is reached.

After each level of graph convolution, max-pooling performs on the updated features $\mathbf{F}^j$ to obtain a global shape feature F_{global}^j. When the GCN finalises data processing, the final global shape feature F_{global} is computed as the concatenation of the features pooled at all levels. This final feature is then processed by a fully connected layer and by the softmax function to compute the estimated class for the input object.

3.3   Network Training

Training *SVP-Classifier* implies to first train the CVAE network and then train the GCN.

Training the CVAE. The training loss L_{cvae} is the weighted sum of the Kullback-Leibler divergence loss (L_{kl}) [3], the Mean Squared Error loss (L_{mse}) [13], and the Chamfer Distance loss (L_{cd}) [18], as defined in Eq. 1:

$$L_{cvae} = \eta \cdot \left(L_{mse}(F_n, F_n') + \zeta \cdot L_{kl}(\mu, \sigma) \right) + \lambda \cdot \left(L_{cd}(PCD_{gt}, PCD_{gen}) \right), \quad (1)$$

where F_n is the ground-truth feature vector of the desired view, and F_n' is the reconstructed feature vector for the same view. L_{kl} is defined to regularise the organisation of the latent space by forcing the distributions which are computed by the encoder to be as close as possible to normal distributions. The Chamfer Distance loss, L_{cd}, measures the difference between the ground-truth complete point cloud PCD_{gt} and the reconstructed point cloud PCD_{gen}. Coefficients η, ζ and λ are used for weighting the contributions of the different loss functions.

Training the GCN. The GCN is trained using a loss function containing global shape loss

$$L_{GCN} = L_{global}^{CE} \left(\mathcal{S}(F_{global}), y \right) \quad (2)$$

where L_{global}^{CE} is the cross-entropy loss, $\mathcal{S}$ is a classifier composed of a fully connected layer followed by the softmax function, and y is the ground-truth shape category.

4   Experiments

To validate our design choices for *SVP-Classifier*, we conduct extensive experiments in terms of the architecture and of the loss functions used to train the CVAE. Additionally, we compare our proposed method against the sota methods for 3D shape classification. We evaluate the performance of different methods on single-view datasets [8] that are generated from two benchmark datasets: ModelNet40 [14] and ScanObjectNN [2]. Note that as performance metric, we use the overall accuracy [11].

4.1   Implementation Details

The input point clouds have a variable number of points so, as a first step, we resample each input PCD to 1024 points using Farthest Point Sampling (FPS). Then, each PCD is processed using a PCD-based backbone, to compute a 512-dimensional feature vector. The feature vector is then processed by the CVAE to perform the feature hallucination for the missing views. The encoder of the CVAE contains four FC layers. The same encoder is used to compute the embedding from the input view features and from the desired view features. The size of such embedding is 40. The embedding for the relative pose between the virtual sensor that acquired the input view and that for the desired view consist simply of the linearisation of a 4×4 pose matrix. The three embeddings are concatenated to produce the input data for the decoder, which consists of four FC layers and which is used to hallucinate the features of the missing views. The hallucinated missing view features are then fed to the GCN for feature aggregation and object

Table 1. Comparison with sota methods on ModelNet40/Model-D and ScanObjectNN/Scan-D. Best results with single-view PCD inputs are highlighted in **bold**. Methods marked with * are evaluated with full 3D scans of the objects. The proposed *SVP-Classifier*, using PointNet++ as backbone for encoding the input PCD, achieves the highest classification accuracy among the single-view methods.

Method	Input	Views	Overall-accuracy	
			ModelNet40	ScanObjectNN
SpiderCNN* [17]	full PCD	–	92.4	79.5
PointNet* [10]	full PCD	–	89.2	79.2
PointNet++* [12]	full PCD	–	90.7	84.3
DGCNN* [16]	full PCD	–	92.9	86.2
PointView-GCN	*multi-view PCD*	20	93.3	85.9
PointNet++	single-view PCD	1	87.5	78.6
DGCNN	single-view PCD	1	86.9	78.0
Ours (PointNet++)	single-view PCD	1	**88.1**	**79.7**

classification. The trainable parameters of the GCN are: $W^j, \alpha^j, \beta^j, \gamma^j$. As an auxiliary task, each hallucinated feature vector is used to estimate the complete object point cloud as it would be seen from the desired view. For this, the output of the CVAE is fed to one FC layer with 3072 output neurons. Finally, that data is reshaped to 1024×3, which represents the 3D coordinates of the points of the reconstructed complete PCD.

We utilise the Adam optimiser to train the networks with a learning rate of 0.0001 and weight decay set to 0.09. Our models are trained for 100 epochs, either on Model-D or on Scan-D. The values selected for weighting the different losses contributing to train the CVAE were: $\eta = 0.5$, $\zeta = 0.0001$, and $\lambda = 0.5$. We verified empirically that the training is robust to small variations around these values, always leading to similar final classification accuracy.

4.2   Evaluation of *SVP-Classifier*

Establishing a Baseline on Single-View PCDs. PCD-based object classifiers are commonly evaluated on the full point cloud of an object, but we are interested in comparing their performance with that of *SVP-Classifier* in similar conditions: when training and evaluating on single-view PCDs. Because of this, we trained and tested two models from the sota on single-view data: PointNet++ [12] and DGCNN [16]. Results for the ModelNet40/Model-D and ScanObjectNN/Scan-D datasets are listed in Table 1.

We observe that using single-view input data, leads to a moderate but consistent drop in classification performance. For ModelNet40, the performance of PointNet++ drops by 3.2%, while that of DGCNN drops by 6.0%. For ScanObjectNN, using single-view input leads to a drop in performance of 5.7% for PointNet++, and 8.2% for DGCNN. The higher drops incurred when using ScanObjectNN are likely due to the higher complexity of that data, which is derived from scans or real objects, rather than from digitally generating point clouds from synthetic models. It can be seen that using *SVP-Classifier* leads to an improvement of classification performance of 0.6% on Model-D and of 1.1% on Scan-D, respect to simply using PointNet++. A gap still remains to be filled respect to the approaches that employ a full PCD as input and with the one that used 20 partial views at once. For more details, please refer to the supplementary material.

4.3 Ablation Studies

In this section, we report the results of the ablation studies that have been performed with the single-view PCD datasets to assess the impact of various design choices on the classification performance. In particular, we compared the impact of different architectures for the module which hallucinates the features from the missing views, and the impact of using different combinations of loss functions.

Feature Hallucination Network. We considered four increasingly complex architectures to perform feature hallucination: (1) **VAE**: In the simplest architecture, we employed a Variational AutoEncoder for feature hallucination. The VAE takes as input the encoded features for the input view and exploits its nondeterministic nature to produce hallucinated features for the remaining 19 views. (2) **CVAE**: Considering that the relative pose between the input view and the view that we want to hallucinate contains useful information for the hallucination process, we performed the hallucination with a Conditional VAE, using the relative pose as the conditioning factor. (3) **CVAE-SE**: For this architecture, we enriched the embedding of the CVAE by concatenating the embedding described at the previous point, with the ground-truth features for the desired view, processed by the same encoder used for those of the input view. This enrichment only happens at training time, while at test time, there exist no desired view feature and the encoder for it, and we randomly sample the latent variable from a Gaussian distribution. We call this architecture "CVAE with Shared Encoder" (CVAE-SE). (4) **CVAE-SE-PG**: For this last architecture, we added one more head to our network, and tasked it with reconstructing the full object PCD, given the features hallucinated for each of the 19 views. This auxiliary task was intended to improve the hallucination performance of the entire model. We call this architecture CVAE-SE-PG, where PG stands for PCD Generation.

Loss Functions for Feature Hallucination. We defined three loss functions for driving the learning of the feature hallucination module: (1) **Mean Squared Error (mse)**, L_{mse}: This loss function, often addressed as the "reconstruction term" in the context of VAEs, is computed between the values of the features hallucinated by the VAE/CVAE and those of the ground-truth features for the corresponding views. (2) **Kullback-Leibler divergence**, L_{KL}: This loss function, often addressed as the "regularisation" term in the context of VAEs, is computed on the latent embedding. (3) **Chamfer Distance**, L_{CD}: The Chamfer Distance is commonly used in the context of PCD-based shape reconstruction. In our network it is applied to measure the difference between the full ground-truth point cloud and the completed PCD we estimate starting from a vector of hallucinated features. The results of the experiments on the architecture of

Table 2. Effects of different feature hallucination network architectures on the classification performance. The network employing the CVAE with the Shared Encoder and the head for PCD Generation achieves the best results (highlighted in bold).

Method	Acc. Model-D	Acc. Scan-D
PointNet++, baseline	87.5	78.6
SVP-Classifier (VAE)	87.4	78.4
SVP-Classifier (CVAE)	87.7	78.9
SVP-Classifier (CVAE-SE)	87.8	79.0
SVP-Classifier (CVAE-SE-PG)	**88.1**	**79.7**

Table 3. Effects of different loss functions on the classification performance. Combining L_{mse}, L_{KL} and L_{CD} leads to the best results (highlighted in bold).

Method	CVAE Losses	Acc. Model-D	Acc. Scan-D
PointNet++, baseline	N.A	*87.5*	*78.6*
SVP-Classifier (CVAE-SE)	L_{mse}	87.2	78.4
SVP-Classifier (CVAE-SE)	$L_{mse} + L_{KL}$	87.7	79.3
SVP-Classifier (CVAE-SE-PG)	$L_{mse} + L_{KL} + L_{CD}$	**88.1**	**79.7**

the feature hallucination network are listed in Table 2. The performance of the fine-tuned PointNet++ on single-view input is listed as baseline. Employing a simple VAE for feature hallucination is slightly detrimental respect to the baseline, while using a CVAE (which allows one to condition the hallucination on the relative transformation between cameras) leads to a small improvement. Adding the encoding of the PCD from the desired view (at training time) to the inputs of the autoencoder leads to another small improvement (CVAE-SE). Finally, adding one auxiliary head to reconstruct PCD's from the hallucinated features (CVAE-SE-PG) leads to the best classification accuracy, with improvements over the baseline of 0.6% and 1.1% on Model-D and Scan-D, respectively. It should be noted that all the versions of *SVP-Classifier* mentioned in the table were trained with a weighted combination of L_{mse} and L_{KL}, except for CVAE-SE-PG, which also employed the L_{CD}.

The results on the ablation on using different loss function are listed in Table 3. Training the feature hallucination network simply with L_{mse} leads to a decrease in classification accuracy for the whole network, while training it with a weighted combination of L_{mse} and L_{KL}, as is usual for VAEs, is beneficial. The extra supervision provided by adding the loss component on PCD reconstruction (L_{CD}) to the weighted sum leads to the best performance, with improvements over the baseline of 0.6% and 1.1% on Model-D and Scan-D, respectively.

5   Conclusion

The task of classifying objects given a single-view PCD is of practical relevance, as the full PCDs of objects are seldom available in real settings (autonomous driving, robotic agents). With *SVP-Classifier*, we contribute to the improvement of single-view PCD classification. *SVP-Classifier* hallucinates information on the object of interest, as it would be seen from other points of view. Such information is then aggregated using a GCN, leading to improved classification accuracy over the state-of-the-art. Experiments on synthetic and real data confirm the positive effect of *SVP-Classifier*. While the accuracy of the methods that process full PCDs remains out of reach, we will analyse limitations of the proposed method for future research.

References

1. An, J., Cho, S.: Variational autoencoder based anomaly detection using reconstruction probability. Spec. Lect. IE **2**(1), 1–18 (2015)
2. Angelina Uy, M., Pham, Q.H., Hua, B.S., Thanh Nguyen, D., Yeung, S.K.: Revisiting point cloud classification: a new benchmark dataset and classification model on real-world data. arXiv, arXiv-1908 (2019)
3. Bengio, Y., Courville, A., Vincent, P.: Representation learning: a review and new perspectives. IEEE Trans. Pattern Anal. Mach. Intell. **35**(8), 1798–1828 (2013)
4. Dong, G., Liao, G., Liu, H., Kuang, G.: A review of the autoencoder and its variants: a comparative perspective from target recognition in synthetic-aperture radar images. IEEE Geosci. Remote Sens. Mag. **6**(3), 44–68 (2018)
5. Ioannidou, A., Chatzilari, E., Nikolopoulos, S., Kompatsiaris, I.: Deep learning advances in computer vision with 3D data: a survey. ACM Comput. Surv. (CSUR) **50**(2), 1–38 (2017)
6. Kingma, D., Welling, M.: Auto-encoding variational Bayes. In: ICLR 2014 (2014)
7. Li, Y., Bu, R., Sun, M., Wu, W., Di, X., Chen, B.: PointCNN: convolution on x-transformed points. Adv. Neural Inf. Process. Syst. **31**, 820–830 (2018)
8. Mohammadi, S.S., Wang, Y., Del Bue, A.: PointView-GCN: 3D shape classification with multi-view point clouds. In: 2021 IEEE International Conference on Image Processing (ICIP), pp. 3103–3107. IEEE (2021)
9. Qi, C.R., Liu, W., Wu, C., Su, H., Guibas, L.J.: Frustum PointNets for 3D object detection from RGB-D data. In: Proceedings of the IEEE Conference on Computer Vision and Pattern Recognition, pp. 918–927 (2018)
10. Qi, C.R., Su, H., Mo, K., Guibas, L.J.: PointNet: deep learning on point sets for 3D classification and segmentation. In: Proceedings of the IEEE Conference on Computer Vision and Pattern Recognition, pp. 652–660 (2017)
11. Qi, C.R., Su, H., Nießner, M., Dai, A., Yan, M., Guibas, L.J.: Volumetric and multi-view CNNs for object classification on 3D data. In: Proceedings of the IEEE Conference on Computer Vision and Pattern Recognition, pp. 5648–5656 (2016)
12. Qi, C.R., Yi, L., Su, H., Guibas, L.J.: PointNet++: deep hierarchical feature learning on point sets in a metric space. In: Advances in Neural Information Processing Systems, pp. 5099–5108 (2017)
13. Sohn, K., Lee, H., Yan, X.: Learning structured output representation using deep conditional generative models. Adv. Neural Inf. Process. Syst. **28**, 3483–3491 (2015)
14. Vishwanath, K.V., Gupta, D., Vahdat, A., Yocum, K.: ModelNet: towards a datacenter emulation environment. In: Proceedings of the IEEE Ninth International Conference on Peer-to-Peer Computing, pp. 81–82. IEEE (2009)

15. Wang, Y., Carletti, M., Setti, F., Cristani, M., Bue, A.D.: Active 3D classification of multiple objects in cluttered scenes. In: Proceedings of the IEEE/CVF International Conference on Computer Vision Workshops, pp. 2602–2610 (2019)
16. Wang, Y., Sun, Y., Liu, Z., Sarma, S.E., Bronstein, M.M., Solomon, J.M.: Dynamic graph CNN for learning on point clouds. ACM Trans. Graph. (TOG) **38**(5), 1–12 (2019)
17. Xu, Y., Fan, T., Xu, M., Zeng, L., Qiao, Yu.: SpiderCNN: deep learning on point sets with parameterized convolutional filters. In: Ferrari, V., Hebert, M., Sminchisescu, C., Weiss, Y. (eds.) ECCV 2018. LNCS, vol. 11212, pp. 90–105. Springer, Cham (2018). https://doi.org/10.1007/978-3-030-01237-3_6
18. Yuniarti, A., Suciati, N.: A review of deep learning techniques for 3D reconstruction of 2D images. In: 2019 12th International Conference on Information & Communication Technology and System (ICTS), pp. 327–331. IEEE (2019)
19. Zhao, Y., Birdal, T., Deng, H., Tombari, F.: 3D point capsule networks. In: Proceedings of the IEEE Conference on Computer Vision and Pattern Recognition, pp. 1009–1018 (2019)

3D Key-Points Estimation
from Single-View RGB Images

Mohammad Zohaib[1,2]($\boxtimes$)(ID), Matteo Taiana[1](ID), Milind Gajanan Padalkar[1](ID),
and Alessio Del Bue[1](ID)

[1] Pattern Analysis and Computer Vision (PAVIS), Italian Institute of Technology,
Genoa, Italy
{mohammad.zohaib,matteo.taiana,milind.padalkar,alessio.delbue}@iit.it
[2] Department of Marine, Electrical, Electronic and Telecommunications Engineering,
University of Genoa, Genoa, Italy

Abstract. The paper presents an end-to-end approach that leverages
images for estimating an ordered list of 3D key-points. Most of the existing methods either use point clouds or multiple RGB/depth images to estimate 3D key-points, whereas the proposed approach requires only a single-
view RGB image. It is based on three steps: extracting latent codes, computing pixel-wise features, and estimating 3D key-points. It also computes
a confidence score of every key-point that enables it to predict a different
number of key-points based on an object's shape. Therefore, unlike existing approaches, the network can be trained to address several categories
at once. For evaluation, we first estimate 3D key-points for two views of
an object and then use them for finding a relative pose between the views.
The results show that the average angular distance error of our approach
(6.39°) is 8.01° lower than that of KP-Net (14.40°) [1].

Keywords: 3D Key-points · Single-view RGB images · Pose
estimation

1 Introduction

Finding objects' position and pose in images is a necessary step for solving higher
level tasks such as object search, manipulation, navigation in cluttered environments, path planning, and human-robot interaction. Recent research reveals that
estimating object location and pose can be improved significantly with the help
of 3D key-points. It is due to the fact that they provide information about
Points of Interest (PoI) and are invariant to transformation i.e., rotations, scale,
etc. [1–6]. Moreover, they also contain semantic information, which is helpful
while reasoning on correspondences between points in two shapes [3,7,8].

Most of the recent studies use 3D key-points for various human related applications including joint detection, motion capturing, pose estimation, etc., which

Supplementary Information The online version contains supplementary material
available at https://doi.org/10.1007/978-3-031-06430-2_3.

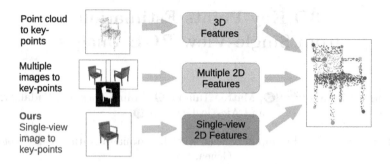

Fig. 1. Comparison with the other paradigms. Some existing methods use point clouds (top) or multiple images representing different views of an object (middle) as inputs and compute 2D/3D features for key-points estimation. In comparison, the proposed approach considers a single-view RGB image, extracts object 2D features, and use them for estimating 3D key-points (bottom).

deal with a single category (human) and a fixed number of key-points [9–17]. On the other hand, key-points are also used in applications related to rigid objects (i.e. car, chair, etc.), where an object's structure and the number of key-points may vary depending on the category. To simplify the problem, the existing approaches train their network separately for every category for a fixed number of key-points. In the literature, most of the works including [7, 18–20] compute 3D key-points from 3D point clouds. On the other hand, a method proposed in [1] uses RGB images. However, it estimates 3D key-points in the form of 2D pixels and associated depth values. In comparison, we present an approach that uses a single-view RGB image and estimates an ordered list of key-points in 3D space. An overview of our approach is shown in Fig. 1, highlighting the key difference with the existing approaches. Our main contributions are as follows;

– The proposed approach estimates key-points from a single-view RGB image.
– Unlike the existing approaches, the proposed approach estimates the confidence score for every key-point that allows it to predict the different number of key-points based on the object's shape.
– The estimated key-points provide order-wise semantic information that is independent of the object's view.

The remainder of the paper is organized as follows: Sect. 2 presents a literature review, Sect. 3 describes the proposed system, and Sect. 4 discusses the experiments we performed and compares the results with state-of-the-art (SOTA) techniques. Finally, conclusions are summarised in Sect. 5.

2   Related Work

Key-points provide an object's structural information, which can be utilized in geometric reasoning. The popularity stems from the fact that they require minimum processing resources and are easy to handle in comparison to complete

3D point clouds of an object or 2D pixels of an image. Moreover, in some cases, they also contain semantic information by ensuring their unique order.

In literature, most of the approaches use point clouds for estimating 3D key-points. Adamczyk et al. present an evolutionary algorithm for the selection of visual key-points from unordered sets to address the classification problem [21]. An approach that estimates category-specific 3D key-points in an unsupervised way is presented in [22]. It considers linear symmetric shapes and produces order-wise correspondences with consistent semantics. Jakab et al. [23] use key-points for aligning two shapes. Their network takes two shapes and finds the key-points for shape deformation from a set of randomly sampled surface points. Chen et al. [24] present an unsupervised approach that computes key-points from the object's point cloud to represent good abstraction and approximation of the input 3D shape. This approach encodes local features using PointNet++ and uses them for producing a set of ordered 3D key-points. You et al. [27] present a method that uses geodesic consistency loss for producing dense semantic embeddings. Sun et al. present an unsupervised approach for object parts decomposition using key-points estimation [31]. Their network is trained to compute K semantic correspondence key-points by feeding two randomly rotated versions of an object in the form of point clouds. The SK-Net proposed in [25] generates random spatial key-points in 3D space and converges them to an object's point cloud by learning geometric features. Unlike the other approaches, the spatial key-points are not a part of the object's point cloud.

Other works have focused on depth images and/or RGB. Georgakis et al. present an approach that uses RGB and depth images to compute object 3D pose by matching predicted key-points to the corresponding CAD model [26]. He et al. present a point-wise 3D key-points voting network that uses key-points for calculating object pose in six Degrees of Freedom (6DoF) [28]. They train the network using RGBD images and predict key-points used by the least-squares fitting method for estimating the object's 6D pose. Another RGBD images based approach is presented in [29] that uses estimated 3D key-points for tracking an object's pose. Although their network does not require 3D shapes during training, they are however, required test objects that are relatively similar to those used as training samples [30]. Barabanau et al. [32] present an approach that estimates 2D key-points from an RGB image and transforms these key-points to a 3D model of an object taken from a predefined set of 5 CAD templates only. The intrinsic camera parameters are known. Lu et al. present an approach that uses key-points for finding the pose of a robotic arm [33]. Initially, key-points are sampled on the kinematic chain and are filtered in order to select optimal ones using RANSAC [34]. A similar approach that finds semantic correspondence between two images using both appearance and geometry reasoning by incorporating 2D key-points is presented by Han et al. in [35]. The approach uses these semantic correspondences for producing a warped version of two images. Suwajanakorn et al. compute 3D key-points in an unsupervised way by using two views of an object in different poses and knowledge of the object category [1]. During inference, they use single-view RGB images. However, their estimates are in the form of 2D pixels and depths. In comparison, we present a supervised approach to estimate key-points in 3D space from a single-view RGB image.

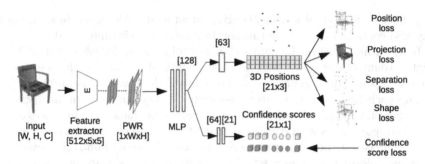

Fig. 2. The proposed architecture. An RGB image is fed to a feature extractor to produce object features that are up-sampled in order to achieve a Pixel-wise Representation (PWR). Finally, a Multilayer Perceptron (MLP) is added that uses PWR for estimating 21 key-points in 3D space along with confidence scores.

3   Methodology

Given an RGB image, our work aims at estimating an ordered list of 3D key-points that are semantically and geometrically consistent across different instances of an object category. For this, an end-to-end approach is proposed that extracts an object's features from an image, computes Pixel-wise Representation (PWR) and uses the representation for estimation of 3D key-points along with confidence scores. The architecture of the approach is illustrated in Fig. 2.

The presented approach is based on three modules. The first module (feature extractor) takes an RGB image as input and produces feature vectors. These extracted features are converted to PWR in the second module. The PWR has the same width and height as the input image. However, instead of representing the RGB value, every pixel represents a feature for the corresponding pixel of the input image. The third module contains a Multilayer Perceptron (MLP) based on four linear layers. The PWR features are flattened to a 1D tensor before feeding to the MLP. The MLP uses them for estimating the 21 key-points. For every key-point, a position in 3D space $[x, y, z]$ and a confidence score (from 0 to 1) is computed. The confidence score reflects how confident the network is that the key-point exists for the object. If such a value is greater than 0.5, it means that the predicted key-point exists for the object, and it is considered as a valid key-point. Otherwise, it is discarded. In this way, the network separates an object's valid key-points from the predicted 21 key-points. So, the total number of valid key-points could be different for different shapes of objects.

3.1   Training Loss

The network is trained separately for every category (as followed in literature) as well as jointly for all the categories. We found that the results with both methods are approx. the same. So, we report the results of a network trained jointly for all the categories. The network minimizes five losses: 3D position loss, 2D projection loss, separation loss, shape consistency loss and the confidence score loss. The **3D position loss** ($\mathcal{L}_{pos}$) measures how accurate the 3D position corresponding to the predicted key-point is, w.r.t. the ground truth. For this we compute Mean Square Error (MSE) between 3D positions of predicted $\mathcal{P} = \{p_i | i = 1, ..., N_p\}$ and ground truth $\mathcal{Q} = \{q_i | i = 1, ..., N_q\}$ key-points as:

$$\mathcal{L}_{pos} = \frac{1}{N_q} \sum_{i=1}^{N_q} \left\| p_i - q_i \right\|_2^2, \tag{1}$$

where N_q is the total number of ground truth key-points, which could be up to 21. We skip the extra predicted key-points that are not valid for an object (w.r.t. ground truth). In order to predict a more accurate position of the key-points, we also compute loss in 2D space. To do so, both the valid estimated and ground truth key-points are transformed from 3D to 2D pixel coordinates using the known transformation T (camera intrinsic and extrinsic) [1]. The **2D projection loss** ($\mathcal{L}_{proj}$) is computed by taking the Mean Absolute Error (MAE) between estimated and ground truth 2D pixels as:

$$p_i = [x_{p_i}, y_{p_i}, z_{p_i}]^\top, \quad q_i = [\bar{x}_{q_i}, \bar{y}_{q_i}, \bar{z}_{q_i}]^\top$$
$$[u_i, v_i]^\top = T(p_i), \quad [\bar{u}_i, \bar{v}_i]^\top = T(q_i)$$
$$\mathcal{L}_{Proj} = \frac{1}{3N_q} \sum_{i=1}^{N_q} \left\| [u_i, v_i]^\top - [\bar{u}_i, \bar{v}_i]^\top \right\|_1, \tag{2}$$

where $1/3$ is a scaling factor to balance the effect of the projection loss. We consider a **separation loss** ($\mathcal{L}_{sep}$) that ensures that no more than one key-point can exist at the same 3D location. The loss penalizes two predicted key-points that are closer than hyperparameter δ^2 (0.05). It is calculated as:

$$\mathcal{L}_{sep} = \frac{1}{N_q^2} \sum_{i=1}^{N_q} \sum_{j \neq i}^{N_q} \max(0, \delta^2 - \left\| p_i - p_j \right\|_2^2), \tag{3}$$

where, p_i and p_j represent the ith and j^{th} predicted key-point, respectively.

The point clouds based approaches predict key-points from the input point clouds and hence they do not consider object's surrounding. In contrast, an image based approach can predict 3D key-points in object's surroundings. To prevent this issue, a **shape consistency loss** ($\mathcal{L}_{shape}$) is included that forces the network to predict key-points closer to the surface of the object. It can be described as:

$$d_i = \left\| p_i - kNN(p_i, \mathcal{PC}) \right\|_2,$$
$$\mathcal{L}_{shape} = \frac{min(1, M)}{max(1, M)} \sum_{i=1}^{N_q} d_i, \quad \text{if } d_i > \gamma, \tag{4}$$

where, $kNN()$ is a function that finds the nearest neighbor of a valid predicted key-point p_i from a point cloud $\mathcal{PC}$ of an object which is available during the training time. The d_i is the distance of a key-point p_i from its nearest neighbor point while M is a count of considered distances that are greater than γ (0.05). $\mathcal{L}_{shape}$ is an average of the considered distances.

During training, we use ground truth information for the separation of valid key-points (from the predicted 21 key-points) for calculating the losses. However, during inference, to identify these valid key-points without ground truth information, a confidence score is required for every predicted key-point. For that, we compute a **confidence score loss** ($\mathcal{L}_{conf}$) by comparing predicted scores ($\mathcal{C}_\mathcal{P} = \{cp_i \,|\, i = 1, ..., N\}$) with ground truth scores ($\mathcal{C}_\mathcal{Q} = \{cq_i \,|\, i = 1, ..., N\}$) as;

$$\mathcal{L}_{conf} = \frac{1}{N} \sum_{i=1}^{N} \|cp_i - cq_i\|, \tag{5}$$

where cp_i could range from 0 to 1, whereas, cq_i could be either 1 or 0. The 1 and 0 represent if the key-point exists (is valid) for the object or not, respectively. N is the total number of predicted key-points, which is 21. Moreover, we pad zeros at the end of the ground truth score vector if it contains less than N scores. Where the zeros represent invalid key-points. The overall loss can be defined as:

$$\mathcal{L}_{overall} = \mathcal{L}_{pos} + \mathcal{L}_{proj} + \mathcal{L}_{sep} + \mathcal{L}_{shape} + \mathcal{L}_{conf}. \tag{6}$$

3.2   Inference

During inference, the network predicts 21 3D key-points along with their confidence scores from a single image. All the key-points having a confidence score greater than 0.5 are selected as valid key-points. The rest of the key-points are discarded. For better visualization, the predicted valid key-points are illustrated on the original point cloud of the object (i.e. Fig. 3).

3.3   Implementation Details

The feature extractor module is based on ResNet-18 that is pre-trained on ImageNet dataset [36]. We discard its last two layers to extract features of dimensions $512 \times 5 \times 5$. The network is implemented in PyTorch and trained with Adam optimizer. The learning rate is 10^{-3}, and the batch size is 512.

4   Experiments

The section presents an arrangement of the dataset, explains metrics selected for performance evaluation and compares the results with SOTA approaches.

4.1   Dataset

As extensively evaluated in previous approaches, we use KeypointNet dataset [3] to analyse the performance of our approach. It contains 8329 3D models, corresponding point clouds, and 83231 key-points of 16 object categories. However, it does not contain images along with camera parameters that are required in our experiments. We render (RGB/RGBA) images in 24 different views by placing the object's 3D model at the origin of the reference frame and the virtual cameras at different locations, pointed towards the origin. During training, original point clouds along with the ground truth key-points are transformed w.r.t. the objects view in the input images. It allows the proposed approach to estimate key-points in different poses. We use the data split provided by KeypointNet.

4.2 Performance Measurement

We compare our results with those of KP-Net [1]. Unlike the existing point cloud based methods [3,22–25], their approach (in inference) uses single image and estimates 3D key-points (pixel $[u, v]$ and depth $[d]$). It estimates 3D key-points for two views of an object. The key-points are then used for finding a pose (rotation matrix) between the object views. The estimated pose is compared with the ground truth pose by computing an angular distance error.

We follow the same procedure and estimate key-points for two views (A and B) of an object using our approach. However, for evaluation, we use these key-points in two different methods. The first method is exactly the same as KP-Net, where we compute relative rotation matrix ($\bar{R}$) between object views using Procrustes analysis and then calculate the angular distance error (E_{rot_mat}) between computed and ground truth relative rotation matrix (R) as:

$$E_{rot\_mat} = 2\,arcsin\left(\frac{1}{2\sqrt{2}}\,||\bar{R} - R||_F\right). \tag{7}$$

As a second evaluation, we transform the estimated key-points of view A ($\mathcal{A} = \{a_i | i = 1, ..., N\}$) using the predicted ($\bar{R}$) and the ground truth ($R$) rotation matrix and call them $\mathcal{A}_p = \{ap_i | i = 1, ..., N\}$ and $\mathcal{A}_q = \{aq_i | i = 1, ..., N\}$, respectively. Generally, both the key-points $\mathcal{A}_p$ and $\mathcal{A}_q$ should lie on the same positions as the key-points of view B (see Fig. 3). Every key-point ap_i/aq_i of $\mathcal{A}_p/\mathcal{A}_q$ is considered as a vector from the origin ($\mathbf{ap}_i, \mathbf{aq}_i$). An angular distance error (E_{3D_pos}) between $\mathcal{A}_p$ and $\mathcal{A}_q$ is computed using vector dot product as:

$$E_{3D\_pos} = \frac{1}{N}\sum_{i=1}^{N} arccos\left(\frac{\mathbf{ap}_i \cdot \mathbf{aq}_i}{|\mathbf{ap}_i|\,|\mathbf{aq}_i|}\right), \tag{8}$$

where N is total number of estimated valid key-points. For a fair comparison with the KP-Net, we consider the first evaluation. Nevertheless, for validation on other categories, results from both evaluations are presented.

4.3 Results and Analysis

The proposed approach is evaluated using white background images of 13 different categories – 10 more than the KP-Net [1]. Two views of an object are passed

Table 1. Error in pose estimation between two views of an object. Angular distance error is computed in degrees between; (1) estimated and ground truth rotation matrices (Eq. 7) and (2) 3D positions (Eq. 8) of the predicted key-points in two views. MSE is computed between predicted and ground truth key-points.

Category	Error b/w Rot. matrices		Error b/w key-points 3D positions		MSE b/w predicted & ground truth key-points	
	Mean	Median	Mean	Median	Mean	STD
Airplane	6.581	3.145	5.963	2.565	0.006	0.017
Car	6.761	2.980	5.316	2.456	0.008	0.040
Chair	13.562	5.017	11.247	4.566	0.015	0.049
Table	23.919	3.635	18.079	2.975	0.053	0.159
Vessel	14.652	4.392	11.655	3.478	0.026	0.075
Bed	28.598	12.422	25.332	9.049	0.094	0.163
Cap	16.904	8.193	13.634	6.261	0.031	0.063
Helmet	26.947	16.058	23.504	15.243	0.062	0.076
Knife	25.330	13.006	20.599	12.490	0.008	0.006
Motorcycle	9.467	3.226	6.490	2.507	0.011	0.045
Guitar	19.559	5.289	7.247	2.926	0.003	0.006
Mug	18.470	9.135	10.320	5.942	0.026	0.056
Bottle	17.118	14.854	14.674	12.013	0.023	0.027
Average	16.962	7.690	12.822	6.190	0.028	0.060

to the network for estimating 3D key-points for every view. The Procrustes analysis is used that utilizes the estimated key-points to compute a pose (rotation matrix) between the views. The error in the estimated pose is computed using both the evaluations (Eqs. 7 and 8). The results are depicted in Table 1. The last two columns present the mean and Standard Deviation (STD) of the MSE between predicted and ground truth 3D key-points. The error is comparatively high for some categories. It is due to the structural variation (single/bunk beds, tables), different key-points for similar object shapes (helmet, knife, etc.), and differences in center of rotation and the center of mass of the object (i.e., mug). Qualitative results are given in supplementary material (**Sup. Fig.** 1).

To compare our results with the KP-Net, we consider the same three categories (cars, airplanes, and chairs) as reported in [1]. For the evaluation, we compute the angular distance error between the estimated and ground truth pose of two views of an object (see Eq. 7). The error is depicted in Table 2. In [1], the results are presented for four different versions; (1) supervised KP-Net that learns from ground truth 2D pixels and corresponding depths, (2) supervised KP-Net with a pretrained Orientation Network (O-Net) that provides an object's orientation information, (3) KP-Net (unsupervised) with O-Net, and (4) KP-Net without O-Net. It is reported in [1] that the KP-Net without O-Net

Table 2. Error in pose estimation between two views of the same object. Mean and median angular distance errors are calculated (in degrees) between ground truth rotation and the rotation computed by Procrustes estimates between predicted key-points of the two views. Results of the baselines (first four rows) are the same as reported in [1]. All the results are produced for transparent images.

Method	Car		Airplane		Chair	
	Mean	Median	Mean	Median	Mean	Median
Supervised KP-Net	16.268	5.583	18.350	7.168	21.882	8.771
Supervised KP-Net with O-Net	13.961	4.475	17.800	6.802	20.502	8.261
KP-Net with O-Net	13.500	4.418	18.561	6.407	14.238	5.607
KP-Net	11.310	3.372	17.330	5.721	14.572	5.420
Ours	**5.190**	**2.073**	**3.257**	**2.053**	**10.732**	**4.096**

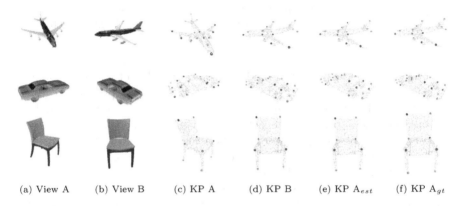

(a) View A   (b) View B   (c) KP A   (d) KP B   (e) KP A$_{est}$ (f) KP A$_{gt}$

Fig. 3. Computing pose between two views (a) and (b) of an object. The corresponding estimated key-points are shown on the original point clouds in (c) and (d). The key-points of view A (c) are transformed to view B using estimated and ground truth rotation matrix as illustrated in (e) and (f), respectively.

performs overall well. The first four rows of the Table 2 present results of the KP-Net versions. In comparison to those, our results are more accurate.

Qualitative results are illustrated in Fig. 3. Columns (a) and (b) show two views of the same object. The corresponding estimated key-points are presented in columns (c) and (d), respectively. Finally, the key-points (and point clouds) of view A after transformation using estimated (A_{est}) and the ground truth (A_{gt}) rotation are illustrated in (e) and (f). It can be visualized that the pose of the transformed key-points (e and f) is the same as the pose of key-points of view B (d). The experiment highlights that; (1) the estimated key-points can be used for computing a pose between two views, (2) the key-points are in semantical order, which is independent of the object view, and (3) the network can predict key-point of the occluded part of the object (i.e., back legs of the chair). Furthermore, we present another experiment that highlights the significance of the confidence

Table 3. Comparison of the key-points predicted as valid by our network based on confidence scores (Pred.) with the key-points selected using ground truths (GT). The pose estimation error in two views of an object is approx. the same in both the cases; either the Pred. or GT key-points are used. Mean and SE of the pose error (calculated in both the methods using (a) rotation matrices (Eq. 7) and (b) key-points 3D positions (Eq. 8)) is given for RGB and RGBA images.

| Category | Metric | Error b/w rotation matrices | | | | Error b/w 3D positions | | | |
| | | RGB | | RGBA | | RGB | | RGBA | |
		Pred.	GT	Pred.	GT	Pred.	GT	Pred.	GT
Airplane	Mean	6.581	6.552	3.257	3.267	5.963	5.974	2.805	2.797
	SE	0.195	0.194	0.075	0.076	0.003	0.003	0.001	0.001
Car	Mean	6.761	6.764	5.190	5.187	5.316	5.334	4.040	4.057
	SE	0.318	0.318	0.277	0.280	0.004	0.004	0.004	0.004
Chair	Mean	13.56	13.56	10.73	10.71	11.25	11.25	7.53	7.54
	SE	0.340	0.340	0.330	0.327	0.004	0.004	0.006	0.004

score. We compare the predicted valid key-points (to whom the network assigns confidence greater than 0.5) with the key-points known to be present because of ground truths. The results are approx. the same in both cases, which validates that the confidence score helps the network in classifying the valid key-points for every object. The results are given in Table 3 that show mean angular distance error and the Standard Error (SE) which is calculated as $\sigma/\sqrt{n}$, where σ is the standard deviation of n angular distance errors. Additional results, including evaluation for realistic images, can be found in the supplementary material.

In a nutshell, it can be inferred that if a network could not estimate the confidence scores, it should predict fixed numbers of key-points as followed by the existing approaches. Otherwise, It may not be possible for the network to separate valid key-points from the total predicted N (21) key-points. Moreover, the confidence score allows jointly training a network for several categories with a different number of key-points. Otherwise, either the network can be trained for a single category, or the total key-points should be fixed for all the categories.

5   Conclusions

The paper presents an end-to-end solution for 3D key-points estimation from a single-view RGB image. The proposed approach extracts object features from an image, computes pixel-wise features by upsampling, and uses them for estimating 3D key-points along with confidence scores that reflect validity of the key-points. It enables the network to predict a different number of key-points based on the object shape. The key-points are estimated in an ordered semantic list, which increases its significance. Moreover, the network can be trained together for all the classes. The approach is evaluated by computing the pose between two views of an object. Our results are more accurate than those reported by KP-Net [1].

References

1. Suwajanakorn, S., Snavely, N., Tompson, J., Norouzi, M.: Discovery of latent 3D keypoints via end-to-end geometric reasoning. In: NeurIPS (2018)
2. Spezialetti, R., Salti, S. and Di Stefano, L.: Performance evaluation of 3D descriptors paired with learned keypoint detectors. AI **2**(2), pp. 229–243 (2021)
3. You, Y., et al.: KeypointNet: a large-scale 3D keypoint dataset aggregated from numerous human annotations. In: CVPR, pp. 13647–13656 (2020)
4. Bisio, I., Haleem, H., Garibotto, C., Lavagetto, F., Sciarrone, A.: Performance evaluation and analysis of drone-based vehicle detection techniques from deep learning perspective. IEEE Internet Things J. **14**(8) (2021)
5. Shu, Z., et al.: Detecting 3D points of interest using projective neural networks. IEEE Trans. Multimed. (2021)
6. Lin, Y., Chen, L., Huang, H., Ma, C., Han, X., Cui, S.: Beyond farthest point sampling in point-wise analysis. arXiv preprint arXiv:2107.04291 (2021)
7. Zheng, Z., Yu, T., Dai, Q., Liu, Y.: Deep implicit templates for 3D shape representation. In: CVPR, pp. 1429–1439 (2021)
8. Zhao, W., Zhang, S., Guan, Z., Zhao, W., Peng, J., Fan, J.: Learning deep network for detecting 3D object keypoints and 6D poses. In: CVPR, pp. 14134–14142 (2020)
9. Liu, L., Yang, L., Chen, W., Gao, X.: Dual-view 3D human pose estimation without camera parameters for action recognition. IET Image Processing. (2021)
10. Tang, R., Wang, L., Guo, Z.: A multi-task neural network for action recognition with 3D key-points. In: ICPR, pp. 3899–3906 (2021)
11. Paoletti, G., Cavazza, J., Beyan, C., Del Bue, A.: Unsupervised human action recognition with skeletal graph Laplacian and self-supervised viewpoints invariance. BMVC (2021)
12. Yuan, Y., Wei, S.E., Simon, T., Kitani, K., Saragih, J.: SimPoE: simulated character control for 3D human pose estimation. In: CVPR, pp. 7159–7169 (2021)
13. Wandt, B., Rudolph, M., Zell, P., Rhodin, H., Rosenhahn, B.: CanonPose: self-supervised monocular 3D human pose estimation in the wild. In: CVPR, pp. 13294–13304 (2021)
14. Zhang, C., Zhan, F., Chang, Y.: Deep monocular 3D human pose estimation via cascaded dimension-lifting. arXiv preprint arXiv:2104.03520 (2021)
15. Wan, C., Probst, T., Gool, L.V., Yao, A.: Self-supervised 3D hand pose estimation through training by fitting. In: CVPR, pp. 10853–10862 (2019)
16. Li, Y., Torralba, A., Anandkumar, A., Fox, D., Garg, A.: Causal discovery in physical systems from videos. arXiv preprint arXiv:2007.00631 (2020)
17. Paoletti, G., Cavazza, J., Beyan, C. and Del Bue, A.: Subspace clustering for action recognition with covariance representations and temporal pruning. In: ICPR, pp. 6035–6042 (2021)
18. Shi, R., Xue, Z., You, Y., Lu, C.: Skeleton merger: an unsupervised aligned keypoint detector. In: CVPR, pp. 43–52 (2021)
19. You, Y., Liu, W., Li, Y.L., Wang, W., Lu, C.: UKPGAN: unsupervised keypoint GANeration. arXiv preprint arXiv:2011.11974 (2020)
20. Bojanić, D., Bartol, K., Petković, T., Pribanić, T.: A review of rigid 3D registration methods. In: 13th International Scientific-Professional Symposium Textile Science and Economy, pp. 286–296 (2020)
21. Adamczyk, D., Hula, J.: Keypoints selection using evolutionary algorithms. In: ITAT, pp. 186–191 (2020)

22. Fernandez-Labrador, C., Chhatkuli, A., Paudel, D.P., Guerrero, J.J., Demonceaux, C., Gool, L.V.: Unsupervised learning of category-specific symmetric 3D keypoints from point sets. In: Vedaldi, A., Bischof, H., Brox, T., Frahm, J.-M. (eds.) ECCV 2020. LNCS, vol. 12370, pp. 546–563. Springer, Cham (2020). https://doi.org/10.1007/978-3-030-58595-2_33

23. Jakab, T., Tucker, R., Makadia, A., Wu, J., Snavely, N., Kanazawa, A.: Keypoint-Deformer: unsupervised 3D keypoint discovery for shape control. In: CVPR, pp. 12783–12792 (2021)

24. Chen, N., et al.: Unsupervised learning of intrinsic structural representation points. In: CVPR, pp. 9121–9130 (2020)

25. Wu, W., Zhang, Y., Wang, D., Lei, Y.: SK-Net: deep learning on point cloud via end-to-end discovery of spatial keypoints. AAAI **34**(04), 6422–6429 (2020)

26. Georgakis, G., Karanam, S., Wu, Z., Kosecka, J.: Learning local RGB-to-CAD correspondences for object pose estimation. In: ICCV, pp. 8967–8976 (2019)

27. You, Y., et al.: Fine-grained object semantic understanding from correspondences. arXiv preprint arXiv:1912.12577 (2019)

28. He, Y., Sun, W., Huang, H., Liu, J., Fan, H., Sun, J.: PVN3D: a deep point-wise 3D keypoints voting network for 6DoF pose estimation. In: CVPR, pp. 11632–11641 (2020)

29. Wang, C., et al.: 6-PACK: category-level 6D pose tracker with anchor-based keypoints. In: ICRA, pp. 10059–10066 (2020)

30. Devgon, S., Ichnowski, J., Balakrishna, A., Zhang, H., Goldberg, K.: Orienting novel 3D objects using self-supervised learning of rotation transforms. In: IEEE 16th International Conference on Automation Science and Engineering (CASE), pp. 1453–1460 (2020)

31. Sun, W., et al.: Canonical capsules: unsupervised capsules in canonical pose. arXiv preprint arXiv:2012.04718 (2020)

32. Barabanau, I., Artemov, A., Burnaev, E., Murashkin, V.: Monocular 3D object detection via geometric reasoning on keypoints. arXiv preprint arXiv:1905.05618 (2019)

33. Lu, J., Richter, F., Yip, M.: Robust keypoint detection and pose estimation of robot manipulators with self-occlusions via sim-to-real transfer. arXiv preprint arXiv:2010.08054 (2020)

34. Fischler, M.A., Bolles, R.C.: Random sample consensus: a paradigm for model fitting with applications to image analysis and automated cartography. Commun. ACM **24**(6), 381–395 (1981)

35. Han, K., et al.: SCNET: learning semantic correspondence. In: ICCV, pp. 1831–1840 (2017)

36. Deng, J., Dong, W., Socher, R., Li, L.J., Li, K., Fei-Fei, L.: ImageNet: a large-scale hierarchical image database. In: CVPR, pp. 248–255 (2009)

Online Panoptic 3D Reconstruction as a Linear Assignment Problem

Leevi Raivio[(✉)] and Esa Rahtu

Tampere University, Korkeakoulunkatu 7, 33720 Tampere, Finland
{leevi.raivio,esa.rahtu}@tuni.fi

Abstract. Real-time holistic scene understanding would allow machines to interpret their surrounding in a much more detailed manner than is currently possible. While panoptic image segmentation methods have brought image segmentation closer to this goal, this information has to be described relative to the 3D environment for the machine to be able to utilise it effectively. In this paper, we investigate methods for sequentially reconstructing static environments from panoptic image segmentations in 3D. We specifically target real-time operation: the algorithm must process data strictly online and be able to run at relatively fast frame rates. Additionally, the method should be scalable for environments large enough for practical applications. By applying a simple but powerful data-association algorithm, we outperform earlier similar works when operating purely online. Our method is also capable of reaching frame-rates high enough for real-time applications and is scalable to larger environments as well. Source code and further demonstrations are released to the public at: https://tutvision.github.io/Online-Panoptic-3D/

Keywords: 3D reconstruction · Panoptic segmentation · Real time

1 Introduction

Panoptic segmentation [17] is a recent computer vision topic, which combines the tasks of semantic segmentation – assign a class label to each pixel – and instance segmentation – detect, classify and segment each object instance. The objective is to segment and classify both *stuff* – amorphous, unquantifiable areas of the image like floor, buildings and roads – and *things* – quantifiable objects. Many potential applications – *e.g.* in the fields of context-aware augmented reality [21], autonomous driving [13] and robotics [11] – require rich semantic information in real time from the environment. For instance, real-time semantic knowledge of objects around a robot would allow it to interact with it's environment at a higher level of autonomy and utilise the information for more robust localisation.

While panoptic image segmentation has been researched quite extensively [3,13,16,20,23], panoptic 3D reconstruction has not been studied as much. Segmented images alone are often not sufficient: in many applications the segments 3D location relative to the actor needs to be known as well. The segmentation of 3D reconstructions has gained quite a lot of attention recently [1,6,24], but many

© The Author(s), under exclusive license to Springer Nature Switzerland AG 2022
S. Sclaroff et al. (Eds.): ICIAP 2022, LNCS 13232, pp. 39–50, 2022.
https://doi.org/10.1007/978-3-031-06430-2_4

(a) Original image   (b) Segmented image   (c) 3D reconstruction

Fig. 1. An example of online panoptic 3D reconstruction. RGB-D Images (a) are segmented (b) and integrated sequentially into a panoptic 3D reconstruction (c)

works assume offline processing – *i.e.* that all the data is available simultaneously rather than sequentially – which rules out many real-time applications.

Contributions and Scope. This work is focused on online panoptic 3D reconstruction of static scenes from sequences of RGB-D images or point clouds. Segmentation of completed 3D reconstructions and other ways of achieving the same result offline are considered out of scope. Since the focus is on what happens after panoptic segmentation, the related segmentation methods are only discussed briefly. The main contributions in this article are as follow:

- We revisit the method introduced with PanopticFusion [21], explore some unanswered aspects and provide an open-source implementation with the updated method,
- formulate online panoptic 3D data-association as a Linear Assignment Problem (LAP), separate from segmentation and reconstruction,
- provide a simple yet effective baseline for real-time operation, with TSDF integration and optimal LAP solution in relation to association likelihood.

Organisation of the Rest of the Article. Background and a brief introduction to the applied methods are provided in Sect. 2. Section 3 introduces our proposed method, which is evaluated in Sect. 4. Finally, Sect. 5 concludes the paper, adding some final remarks on the system's applications and future possibilities.

2   Related Work

PanopticFusion [21] is the first work to propose online integration of panoptic image segmentations to a 3D reconstruction. They integrate point clouds generated from segmented images to a TSDF voxel volume [5,22] by greedily matching detected segments with the reconstruction and regulating each voxel's corresponding instance with a weighting function. Semantic labels are inferred in a bayesian manner based on confidence scores provided by the segmentation model. They also apply a Conditional Random Field (CRF) to regularise the reconstruction, improving results significantly. Voxblox++ [10] – introduced later the same year – is a similar approach that also integrates segmented RGB-D images into a TSDF volume. It leverages geometric segmentation of depth images to improve instance segmentation accuracy. Both geometric and semantic segments are used to compute a pair-wise weight, which is used to greedily match them with segments in the reconstruction. Because of the geometric segmentation, the method allows segmentation of objects with no known semantic class in addition to objects recognised by the instance segmentation model.

Recently, [11] built upon the idea of Voxblox++. They apply Voxblox++ for 3D instance integration, with two small but effective modifications: the pair-wise weight is replaced by a triplet weight that also takes semantic labels into account in the fusion, and – in addition to geometric segments – instance segments are fused if they overlap by a significant amount. The article introduces a method for searching and aligning CAD models to reconstructed objects based on geometry and semantic class, as well as geometrical and physical rules. With the CAD models, a contact graph and interactive virtual scene are reconstructed to allow a robot to simulate its interaction with the environment. SceneGraphFusion [26] is another approach that forms a scene graph online from a stream of RGB-D images, but unlike the above-mentioned approach, it generates the graph with a deep neural network, after which the panoptic labels for geometrically segmented portions of the 3D reconstruction are produced a side product.

Panoptic-MOPE [12] is another recent approach, which integrates sequences of RGB-D images into a surfel reconstruction. Unlike other mentioned approaches – which assume the camera pose either known or estimated elsewhere – it also tracks camera movements based on geometric-, appearance- and semantic cues. The method also applies a novel RGB-D panoptic segmentation model. Although it is only tested on room-sized environments, the authors claim it could be scaled to larger environments as well.

3 Methods

Figure 2 depicts a flow diagram of our panoptic 3D reconstruction pipeline. First, instance IDs and semantic classes are acquired from RGB images with a panoptic segmentation model. The IDs and classes are combined with depth information to form a panoptic point cloud, which is transformed to a global coordinate frame with camera pose information and quantised to a voxel grid. Voxel clusters corresponding to detections are matched with ones found in the current reconstruction. Afterwards, the new voxels are integrated into the volume. If there are no matches for some of the detected segments, new targets are generated accordingly.

Similar to [10,21] and [11], we apply the popular Voxblox ROS library [22] to produce a TSDF voxel volume online. Our panoptic tracking system is built alongside the TSDF integrator, separate from the segmentation model and the

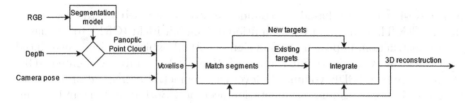

Fig. 2. A flow diagram of the reconstruction pipeline

integrator itself. Although our implementation isn't highly optimised, some parallelisation has been utilised to reach competitive speeds. We apply EfficientPS [20] for segmentation because of its efficient design and impressive performance on benchmark datasets.

3.1   Instance Association as a Linear Assignment Problem

The most essential part of integrating 2D panoptic segmentations to a 3D reconstruction is the association of detected segments with both temporally and spatially consistent labels. While classes determined as *stuff* – amorphous, unquantifiable regions like sky, road or floor – can always simply be associated with the same persistent ID, the association of *things* – quantifiable objects – is not as trivial. The output of an image segmentation model is only consistent within the single frame: the same instance label is guaranteed to point to the same object only in the image it originates from. On the other hand, while video segmentation methods like [15] can be applied to maintain temporal consistency of instance labels in the current camera view, they can not necessarily hold spatial consistency throughout the sequence: if an object is first perceived in the sequence, then lost, and later seen again, it will be assigned a new instance label unless it can be re-identified as the same object. A static object instance can be tracked in 3D, however, by associating detected instances with the existing portion of the 3D reconstruction even if it leaves the current camera view.

The data-association task of matching *thing* detections with the reconstruction can be formulated as a Linear Assignment Problem (LAP). Assume we have n detected segments S_d, each a set of 3D points associated with an instance label and the segmentation model's confidence scores corresponding to semantic classes. On the other hand, let's assume a set of m tracking targets S_t in the 3D reconstruction formed thus far with similar properties. Since there are no overlaps between the segments in panoptic segmentation and the model's objective is to segment each object as a whole, we can make the common assumption that a detection can only originate from a single target and only one target should be associated with a detection. [4,7,25] With these assumptions, the problem can be formulated into a $n \times m$ matrix, where each row corresponds to a detection, each column corresponds to a target and cell values are the likelihoods of the detections originating from each target. An optimal solution to the problem is then achieved by matching each row with a single column in a way that maximises the sum of likelihoods across all matches.

Because the amount of possible objects in the reconstructed scene is unknown beforehand, a mechanism for generating new target segments is required. With an LAP such a mechanism is quite simple to implement: if an optimal detection-target match has a likelihood below a certain threshold, the detection is assumed to be a new object and it is associated with a new instance label. Since the number of detections and targets might not be equal, a number of temporary dummy IDs are generated to form a square matrix. The likelihood of a detection or a target – depending on which set is smaller – matching these dummies is set to zero. Thus, whenever a detection is matched with one – because there are no more real targets available – it will be assigned a new ID. On the other hand, dummy detections matched with any target – when there were less detections than targets – can be ignored.

The problem can be solved optimally in $\mathcal{O}(n^3)$ time with the Hungarian Algorithm [18]. Variants of the method have been applied in many multi-object tracking works. While faster, approximate, ways to solve the problem have also been invented [4,7,21], the optimal solution can be found fast enough in our case since only objects in the current camera view have to be considered. Both PanopticFusion [21] and Voxblox++ [10] apply a greedy algorithm to the task, which is fast, but will not guarantee an optimal solution unless the likelihood threshold is greater than 0.5 [17].

3.2   Integrating Panoptic Labels into a Voxel Grid

Because we represent the environment with a voxel volume, each target object is a set of voxels. Voxel size is assumed uniform across the whole scene. On the other hand – while detections are originally point clouds – they are transformed into sets of voxels in the same grid to make them consistent with the targets, and to reduce the number of points used to represent them. We employ a simple yet effective weighting strategy inspired by PanopticFusion [21] to determine a voxel's instance ID. For each timestep t and voxel v in the camera view the weight is computed as

$$W(t,v) = \begin{cases} W(t-1,v) + w(t,v) & l_d(t,v) = l_\tau(t,v), \\ W(t-1,v) - w(t,v) & l_d(t,v) \neq l_\tau(t,v) \wedge W(t-1,v) \geq w(t,v) \\ w_t(v,d) & l_d(t,v) \neq l_\tau(t,v) \wedge W(t-1,v) < w(t,v) \end{cases}$$
(1)

where $l_\tau(t,v) \in L$ is the persistent ID of a target instance currently associated with the voxel, $l_d(t,v) \in L$ a persistent ID of a matched detection segment and $w(t,v)$ the TSDF weight of the voxel. If a weight is reduced significantly – i.e. when $l_d(t,v) \neq l_\tau(t,v) \wedge W(t-1,v) < w(t,v)$, the last case above – the voxel's persistent instance ID is reset as $l_\tau(t,v) := l_d(t,v)$. This way, only one ID has to be stored, and one does not need to keep track of all the instances associated with each voxel.

With an approach similar to [21], each time a detected segment is matched with a target, the target's confidence scores are integrated in a probabilistic manner. However, because confidence is hard to define in panoptic segmentation, and EfficientPS therefore does not output confidence scores, each detection's confidence is assumed to be one for the detected class and zero for others. Therefore, the semantic class of a target ID $l \in L$ at timestep t is determined by

$$class(t, l) = \begin{cases} argmax_c(n_c^l/n^l) & max_c(n_c^l/n^l) > \theta \\ void & max_c(n_c^l/n^l) \leq \theta \end{cases} \qquad (2)$$

where n_c^l is the number of times the target has been associated with the semantic class c, while n^l total number of associations for the target across all classes. If the score is lower than a given threshold θ, it will be assigned to the *void* semantic class, representing objects that the segmentation model does not recognise.

Imprecise borders between segments in the image can cause problems when back-projecting them to 3D. For instance, if an object's segmentation surpasses the actual object's borders, a part of the detection can appear behind the object *e.g.* on a wall. An accurate segmentation model can alleviate this issue, but nonetheless effective outlier rejection can make the system more robust to segmentation errors. Both confidence intervals on Gaussian probability distributions and clustering with the DBScan algorithm [8] were found to be effective. However, when the voxel weighting approach introduced in [21] was employed, the accuracy gains from both approaches were almost insignificant. Therefore, we believe that with a sufficiently precise segmentation model, simply introducing voxel weights as described above is enough. With noisier detections, however, one could benefit from the more complex approaches as well.

3.3   Association Likelihood Estimation

The likelihood of a detection matching a target can be evaluated in many ways. PanopticFusion [21] applies the Intersection over Union (IoU) metric, popular in both evaluation of object detection and image segmentation works [9,17], as well as estimating overlap in object tracking methods [25]. On the other hand, Voxblox++ [10] and it's recent follower [11] simply count intersecting voxels. Applying statistical distance metrics – *e.g.* the Mahalanobis distance [19] – have also been proposed, increasing accuracy when tracking dynamic objects [4,7].

We found IoU over visible segments to work best in our case. It is computed by dividing the intersection – in this case, number of intersecting voxels between two segments – by union – the total number of voxels between the two. Only parts of the target segments seen in the current camera view are considered, thus objects being only partially visible should not affect the metric as much. The IoU scores are normalised across detection-target pairs to estimate a probability distribution. To avoid setting a fixed threshold for generating new targets, we instead chose to set the threshold as $1/n$, where n is the number of possible matches. In our tests, this strategy seemed to provide better results than any single fixed threshold. The association algorithm's precision and recall could be further tuned by multiplying the threshold with some constant, however we found that to not be necessary in our case.

The same method could be applied with other likelihood metrics as well. Bhattacharyya distance [2] – a divergence metric between two probability distributions – was also considered as a likelihood metric. By representing voxel clusters as multivariate Gaussian distributions, we could also take into account the object's shape and size in addition to overlap. However, we found the metric to be less consistent than IoU with the current system and dataset, most likely because many objects are only partially visible.

4   Evaluation

We evaluate the systems performance on the ScanNet dataset [6]. Since PanopticFusion [21] is the only similar approach on the dataset, we only report its results as a comparison. While [10] and [11] are not evaluated on the dataset, their data-association method is quite similar to the one in PanopticFusion, thus their performance can be roughly estimated with its results as well.

In all of the tests below, only every 10th frame of the RGB-D video feed is processed. The original frame-rate of ScanNet data is 30 Hz, thus with the new rate we are required to process each frame in 333 ms or less to run the algorithm in real time. Processing more frames than this does not seem to increase quality of the results much. On the other hand, the number of points in the panoptic point cloud has a huge effect the amount of computation required, therefore the point cloud resolution is reduced from the original 640×480 points after segmentation, multiplier depending on the voxel size used. The authors of PanopticFusion did not mention which one of Voxblox's TSDF integrators they applied in the article, thus we apply the 'fast' integrator.

To compare them to the ground truth, the results have to be labelled in the original meshes, therefore all our results are transformed to ScanNet evaluation meshes via an approximate nearest neighbour search implemented in Faiss [14]. Panoptic Quality is computed similarly to the 2D metric [17], however IoU is computed over mesh vertices instead of pixels. Panoptic image segmentations are inferred separately, and are read from disk during operation so that they can be re-used in all of the tests. All our tests were run on Intel Core i7-8665u CPU at 1.90 GHz.

4.1   Data

All training and evaluation in this manuscript are performed on the ScanNet [6] dataset. It contains 2.5 million RGB-D video frames and ground-truth poses from different indoor scenes. Each scene has both 2D and 3D ground-truth annotations with two stuff classes ('wall' and 'floor'), as well as 18 thing classes of objects commonly found in indoor environments. A test set with hidden ground-truth is provided for evaluation of semantic segmentation and object segmentation both

Table 1. Average 3D Panoptic Quality on open ScanNet validation set. Mean Panoptic Quality (PQ), Segmentation Quality (SQ) and Recognition Quality (RQ) are reported over all classes, as well as over *things* and *stuff* separately.

	all			stuff			things			frame rate (Hz)[2]
	PQ	SQ	RQ	PQ	SQ	RQ	PQ	SQ	RQ	
Results reported in PanopticFusion[1] [21]										
offline, with CRF	33.5	73.0	45.3	58.4	70.7	80.9	30.8	73.3	41.3	–
online, without CRF	29.7	**71.2**	41.1	**56.7**	69.5	**79.6**	26.7	**71.4**	36.8	4.30
Results with our method										
Greedy, 2.4 cm voxels	29.6	66.0	41.4	51.1	69.5	71.4	27.2	65.6	38.0	0.32
Greedy, 5 cm voxels	20.0	54.1	28.9	46.3	64.6	68.9	17.1	53.0	24.5	2.66
Greedy, 10 cm voxels	28.9	57.0	42.4	47.3	64.7	70.3	26.9	56.1	39.3	9.83
Hungarian, 2.4 cm voxels	33.5	65.4	47.6	53.4	**70.0**	74.3	31.3	64.9	44.6	0.33
Hungarian, 5 cm voxels	**34.0**	68.0	**47.8**	52.4	69.0	74.3	**31.9**	67.9	**44.5**	3.53
Hungarian, 10 cm voxels	31.7	62.3	47.0	47.2	66.5	68.8	30.0	61.9	44.5	**11.63**

[1] Open validation set of the PanopticFusion paper contains different ScanNet scenes than the validation set of this work.

[2] PanopticFusion and our method were evaluated on different hardware, thus framerates should not be directly compared between them. They are nevertheless reported to clarify the effect of reducing image and voxel-grid resolutions.

in 2D and 3D, but since no panoptic ground-truth is available for the hidden set, a part of the training data is applied as an open test set in this work.

A subset of 25 000 images – frames sub-sampled from video sequences approximately every 100 frames – are provided with 2D ground truth for training image segmentation models, which are used to train the panoptic segmentation model. A part of the training data – consisting of roughly five percent of the images in the 2D training set – is applied as a validation set. The randomly picked validation scenes are separated from training scenes: locations present in validation dataset are not found in training data. The same scenes found in 2D validation set are also used in evaluating 3D performance. However, since multiple scenes are sometimes captured from the same locations and the reconstructions are usually quite similar, only one instance of each location is stored for 3D evaluation. Therefore, 3D evaluation is performed on 35 randomly picked unique indoor scenes not found in training data.

4.2 Panoptic Quality on ScanNet with an Open Validation Set

Table 1 compares our method to PanopticFusion on the Panoptic Quality (PQ) metric [17], which is a combination of Segmentation Quality (SQ) and Recognition Quality (RQ). The results have been reported as an average over all classes, as well as over stuff and thing classes separately. The results of the original

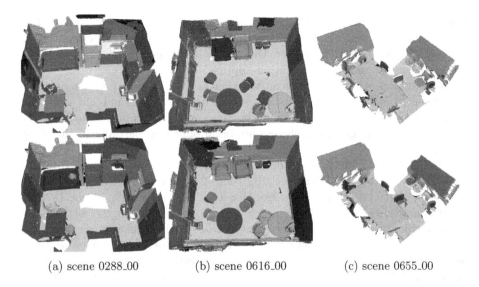

(a) scene 0288_00   (b) scene 0616_00   (c) scene 0655_00

Fig. 3. Qualitative examples of typical scenes in the ScanNet dataset. Top: panoptic reconstructions with our method, bottom: ground truth. Colors of 'thing' instances can be different from ground truth because they are randomised.

PanopticFusion [21] have been reported both with and without Conditional Random Field (CRF) regularisation. Because the test data and panoptic segmentation model are not the same as reported in [21], we have also re-implemented the Greedy data-association algorithm of PanopticFusion: detection segments are processed in descending order based on size, by choosing the target corresponding to highest IoU. IoU threshold for the greedy algorithm is set to 0.25 as in the original work. The results on the test closest to the original Panoptic-Fusion results – Greedy algorithm with 2.4 cm voxels – seem to be quite close to the results reported in the original paper.

The visual quality of the reconstruction varies mostly based on the resolution of the voxel grid: the smaller the voxels are, the more detailed the meshes become. Voxel size seems to affect computational requirements the most as well. While image resolution could be reduced up to 0.1 times the original after segmentation with voxels larger than 5 cm without affecting the quality of the outcome much, with smaller voxels there is a significant loss of quality. When image resolution is only reduced to half of the original, the quality is much better, but computational requirements grow respectively. In tests with 5 cm and 10 cm voxels point cloud resolution was reduced to 0.1 times the original, and in tests with 2.4 cm voxels the resolution was reduced to 0.5 times the original.

While our method seems to be roughly on par with the version of Panoptic-Fusion applying CRF, the one without CRF seems to perform significantly worse than ours on PQ. This is further supported by the tests on our re-implementation of their method. While both PanopticFusion configurations perform better on 'stuff' classes and Segmentation Quality (SQ) than ours, they both seem to fall behind on 'thing' classes and Recognition Quality (RQ). This implies that while our method's segmentation seems to be slightly worse, our data association strategy seems to be performing better. Differences in SQ most likely originate from differences in the image segmentation approaches.

Some examples of panoptic reconstruction on typical scenes found in ScanNet are presented in Fig. 3. Object instances seem to have been found and separated from one another correctly most of the time, and segmentations look pretty accurate, even though larger objects are often only seen partially in the video sequences.

Volumetric integration slows down a lot when voxel size is decreased taking on average 46, 106 or 1188 ms for 10, 5 and 2.4 cm voxels respectively. Data-association with the Hungarian algorithm took on average 47, 177 or 1875 ms for the same sizes, while greedy matching took on average 53, 247 or 1949 ms. Most of this time is spent on computing the IoUs, which takes more time with larger amounts of voxels. Compared to the relations of these timings reported in the PanopticFusion paper, our association algorithm is somewhat slower than the original. Our implementation of the greedy algorithm is a bit slower than the Hungarian method, most likely because the segments need to be sorted every iteration. We can reach reconstruction quality similar to the version of PanopticFusion with CRF regularisation even when decimating point clouds, reducing voxel grid resolution and processing less frames. While the CRF is only run every 10 s, it is reported to take around 4.5 s and to slow down along increasing reconstruction size. For a real-time application to get the benefits of the regularisation, one would need to tolerate long delays during operation, thus we see it more useful as a post-processing tool for cleaning the reconstruction offline.

5   Conclusion

In this work, we revisited the idea of sequentially integrating panoptic image segmentation to 3D reconstruction introduced in [21]. We formulated the task as a Linear Assignment Problem and studied a way of solving it optimally fast enough for real-time applications. Our method seems to outperform earlier works when operating strictly online.

In the future, we aim to research real-time applications that benefit from the more sophisticated scene understanding that panoptic reconstruction offers. Other possible research topics also include extending this method for tracking dynamic objects in 3D simultaneously and applying the method for different sensor modalities. Because input data is only required to be segmented point clouds and pose information, the system could also be adapted for LiDARs and multi-camera setups relatively easily. These would provide increased real-time situational awareness compared to a single camera, albeit at the cost of increased computational requirements. Because the method only processes data seen in the current camera view, it should also be scalable to larger environments as well.

References

1. Armeni, I., Sener, O., Zamir, A.R., Jiang, H., Brilakis, I., Fischer, M., Savarese, S.: 3d semantic parsing of large-scale indoor spaces. In: IEEE/CVF Conference on Computer Vision and Pattern Recognition (CVPR) (2016)
2. Bhattacharyya, A.: On a measure of divergence between two multinomial populations. Sankhyā Indian J. Stat. **7**(4), 401–406 (1946)
3. Cheng, B., et al.: Panoptic-DeepLab: a simple, strong, and fast baseline for bottom-up panoptic segmentation. In: IEEE/CVF Conference on Computer Vision and Pattern Recognition (CVPR) (2020)
4. Chiu, H., Prioletti, A., Li, J., Bohg, J.: Probabilistic 3d multi-object tracking for autonomous driving. arXiv preprint (2020)
5. Curless, B., Levoy, M.: A volumetric method for building complex models from range images. In: ACM SIGGRAPH (1996)
6. Dai, A., Chang, A.X., Savva, M., Halber, M., Funkhouser, T., Nießner, M.: Scan-Net: richly-annotated 3d reconstructions of indoor scenes. In: IEEE/CVF Conference on Computer Vision and Pattern Recognition (CVPR) (2017)
7. Dao, M.-Q., Frémont, V.: A two-stage data association approach for 3d multi-object tracking. Sensors **21**(9), 2894 (2021)
8. Ester, M., Kriegel, H.P., Sander, J., Xu, X.: A density-based algorithm for discovering clusters in large spatial databases with noise. In: International Conference on Knowledge Discovery and Data Mining, pp. 226–231 (1996)
9. Everingham, M., Gool, L.V., Williams, C.K.I., Winn, J.M., Zisserman, A.: The pascal visual object classes (VOC) challenge. Int. J. Comput. Vis. **88**, 303–338 (2009)
10. Grinvald, M.: Volumetric instance-aware semantic mapping and 3d object discovery. IEEE Robot. Autom. Lett. **4**(3), 3037–3044 (2019)
11. Han, M., et al.: Reconstructing interactive 3d scenes by panoptic mapping and cad model alignments. In: IEEE International Conference on Robotics and Automation (ICRA) (2021)
12. Hoang, D.C., Lilienthal, A.J., Stoyanov, T.: Panoptic 3d mapping and object pose estimation using adaptively weighted semantic information. IEEE Robot. Autom. Lett. **5**(2), 1962–1969 (2020)
13. Hou, R., et al.: Real-time panoptic segmentation from dense detections. In: IEEE/CVF Conference on Computer Vision and Pattern Recognition (CVPR) (2020)
14. Johnson, J., Douze, M., Jégou, H.: Billion-scale similarity search with GPUs. arXiv preprint arXiv:1702.08734 (2017)
15. Kim, D., Woo, S., Lee, J.Y., Kweon, I.S.: Video panoptic segmentation. In: IEEE/CVF Conference on Computer Vision and Pattern Recognition (CVPR) (2020)
16. Kirillov, A., Girshick, R., He, K., Dollar, P.: Panoptic feature pyramid networks. In: IEEE/CVF Conference on Computer Vision and Pattern Recognition (CVPR) (2019)
17. Kirillov, A., He, K., Girshick, R., Rother, C., Dollar, P.: Panoptic segmentation. In: IEEE/CVF Conference on Computer Vision and Pattern Recognition (CVPR), June 2019 (2019)
18. Kuhn, H.W.: The Hungarian method for the assignment problem. Nav. Res. Logist. Q. **2**(1–2), 83–97 (1955)

19. Reprint of: Mahalanobis, P.C. (1936) "On the Generalised Distance in Statistics". Sankhya A. **80**(1), 1–7 (2019). https://doi.org/10.1007/s13171-019-00164-5
20. Mohan, R., Valada, A.: EfficientPS: efficient panoptic segmentation. Int. J. Comput. Vis. **129**(5), 1551–1579 (2021)
21. Narita, G., Seno, T., Ishikawa, T., Kaji, Y.: PanopticFusion: online volumetric semantic mapping at the level of stuff and things. In: IEEE/RSJ International Conference on Intelligent Robots and Systems (IROS), pp. 4205–4212 (2019)
22. Oleynikova, H., Taylor, Z., Fehr, M., Siegwart, R., Nieto, J.: Voxblox: Incremental 3d Euclidean signed distance fields for on-board MAV planning. In: IEEE/RSJ International Conference on Intelligent Robots and Systems (IROS) (2017)
23. Porzi, L., Rota Bulò, S., Colovic, A., Kontschieder, P.: Seamless scene segmentation. In: IEEE/CVF Conference on Computer Vision and Pattern Recognition (CVPR) (2019)
24. Roynard, X., Deschaud, J.-E., Goulette, F.: Paris-Lille-3D: a large and high-quality ground-truth urban point cloud dataset for automatic segmentation and classification. Int. J. Robot. Res. **37**(6), 545–557 (2018)
25. Weng, X., Wang, J., Held, D., Kitani, K.: 3D Multi-object tracking: A baseline and new evaluation metrics. In: 2020 IEEE/RSJ Int. Conf. Intell. Robots Syst. (IROS), 10359–10366 (2020). https://doi.org/10.1109/IROS45743.2020.9341164
26. Wu, S.C., Wald, J., Tateno, K., Navab, N., Tombari, F.: SceneGraphFusion: incremental 3d scene graph prediction from RGB-D sequences. In: IEEE/CVF Conference on Computer Vision and Pattern Recognition (CVPR), pp. 7515–7525 (2021)

Temporal Up-Sampling of LIDAR Measurements Based on a Mono Camera

Zoltan Rozsa[1,2(✉)] and Tamas Sziranyi[1,2]

[1] Machine Perception Research Laboratory of Institute for Computer Science and Control (SZTAKI), Eötvös Loránd Research Network (ELKH), Kende u. 13-17, 1111 Budapest, Hungary
{zoltan.rozsa,tamas.sziranyi}@sztaki.hu
[2] Faculty of Transportation Engineering and Vehicle Engineering, Budapest University of Technology and Economics (BME-KJK), Műegyetem rkp. 3, 1111 Budapest, Hungary

Abstract. Most of the 3D LIDAR sensors used in autonomous driving have significantly lower frame rates than modern cameras equipped to the same vehicle. This paper proposes a solution to virtually increase the frame rate of the LIDARs utilizing a mono camera, making possible the monitoring of dynamic objects with fast movement in the environment. First, dynamic object candidates are detected and tracked in the camera frames. Next, LIDAR points corresponding to these objects are identified. Then, virtual camera poses can be calculated by back projecting these points to the camera and tracking them. Finally, from the virtual camera poses, the object movement (transformation matrix transforming the object between frames) can be calculated (knowing the real camera poses) to the time moment, which does not have a corresponding LIDAR measurement. Static objects (rigid with the scene) can also be transformed to this time movement if the real camera poses are known. The proposed method has been tested in the Argoverse dataset, and it has outperformed earlier methods with a similar purpose.

Keywords: LIDAR-camera fusion · 3D geometry · Trajectory estimation

1 Introduction

3D LIDARs are essential elements of most sensor systems assembled for autonomous transportation to provide accurate depth information. These LIDARs have a typical frame rate range of 5–20 Hz (e.g., Velodyne[1] - one of the most popular manufacturers - LIDARs are adjustable in this range too). In the case of the adjustable rotation rate, the angular (horizontal) resolution decreases as the rotation rate increases. In many cases, we would like to have as much detail from the environment as possible. For example, to detect distant objects in 3D, it is indispensable to have a high horizontal resolution. When we

---

[1] https://velodynelidar.com.

© The Author(s), under exclusive license to Springer Nature Switzerland AG 2022
S. Sclaroff et al. (Eds.): ICIAP 2022, LNCS 13232, pp. 51–64, 2022.
https://doi.org/10.1007/978-3-031-06430-2_5

reconstruct a LIDAR's inter-frame point-cloud, it can enhance the 3D detection and tracking capabilities of the whole system.

An appropriately up-sampled higher frame rate may help to monitor dynamic objects more frequently. In this way, we can detect their direction changes and react to them in time, avoiding hazardous events. Besides, one can estimate a dynamic object's velocity or its next pose if two earlier poses are known (with three preceding poses, acceleration too). However, we cannot do that when we detect a vehicle for the first time (there is no previous information about its positions). Vehicles on highways can move with $130\,km/h = 36\,m/s$ (or even faster), which implies that with 5 Hz LIDAR measurement rate, the vehicle can move about 7.2 m between two frames (and being the first detection, there is no way to tell its speed and predict its movement). In this case, we would like to know its next pose as soon as possible to execute other estimations, so it is necessary to increase LIDARs frame rate beyond the limit.

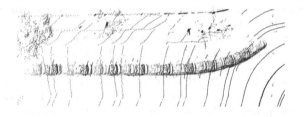

Fig. 1. Upsampled object trajectory. Red point clouds show real measurements, green ones are the virtual ones generated by the proposed method. All the points are transformed to the coordinate system of the first LIDAR frame. Environment points of this frame are colored black. For better visualization purposes not every frame is illustrated. (Color figure online)

In addition to LIDAR, another vital element of autonomous driving sensor systems is the camera. In most of the cases, the equipped cameras measure at least 30 FPS (e.g., [2, 22]) from them, much more frequent information is available than from the LIDARs (specific cameras can operate with the multiple of this FPS value). However, despite the high frame rate, LIDARs are also necessary to get accurate 3D information, that is why we propose a solution based on sensor fusion.

In this paper, we propose a method based on the cameras' much more frequent information to generate virtual LIDAR measurements (focusing on dynamic objects) between real ones. A resulted dynamic object trajectory is illustrated Fig. 1, while a whole point cloud in Fig. 4.

Contribution: The main contribution of the paper is the proposal of a novel methodology to temporally up-sample LIDAR point clouds based on camera. Our advantages compared to other currently available point cloud prediction methods (Sect. 2.2):

- To do virtual point cloud generation, only one previous LIDAR measurement is required (previous methods uses 5 generally).
- As the proposed methodology is based on geometry, no learning is required.
- It does not depend on the LIDAR resolution or the characteristics of the point cloud.
- The method can process higher range in real-time than others.
- This real-time run of the whole pipeline already includes object detection, tracking and fusion. These are requirements for most of the higher-level processing algorithms like change detection, SLAM, online-fusion, etc. [1,3].

In order to support our statements evaluation is presented about the proposed method's performance on large public database (Sect. 5).

2   Related Works

In the first part of this section, LIDAR data point up-sampling is discussed, in the second part, methods that can serve a similar purpose (as the current proposal) are evaluated.

2.1   Up-Sampling of LIDAR Data Points and Interpolation

The resolution of currently available LIDAR sensors is much behind modern cameras in terms of temporal and spatial frequencies.

While measurements acquired by cameras have millions of pixels, data points of today's LIDARs are in the range of several hundreds of thousands. Thus spatial up-sampling depth maps from LIDAR measurements (to match RGB images) have a relatively long history, and there are mature methods for that. There are conventional methods like bilateral filtering [14] or semantic-based upsampling [19], but nowadays, neural network-based methods [7,26] provide state of the art performance for depth completion task, which is a more general term for this problem. A benchmark for depth map completion is also available in the KITTI dataset [21].

The ratio of the frequency of LIDAR measurements (frequency of complete 360° rotation in case of rotating LIDARs, cc. "circular frames") and camera measurements (briefly discussed in Sect. 1) are also small. Despite this, there are no available method in the literature to solve the temporal up-sampling to the best of our knowledge.

The closest researches maybe those which are related to the temporal interpolation of LIDAR clouds. The goal of [11] and [10] to solve the frequency mismatching (synchronization) problem of cameras and LIDARs. In practice (if the synchronization is not solved), the camera frame closest to the LIDAR frame timestamp is considered accurate enough (see more in Sect. 3.1). These interpolation methods cannot be compared to our proposal as they need a 'future' (not available during online processing) LIDAR frame to generate the actual one. In fact, they can be used as a preprocessing step to our method if the data has a synchronization problem.

2.2 Point Cloud Prediction

There are methods in the literature that theoretically can be capable of substituting the one proposed in this paper for virtual LIDAR point cloud generation. These are methods designed for point cloud prediction. Methods like [4,23,24] typically deep-learning-based ones and uses about 5 previous frames to forecast the next point cloud frames. However, there are several drawbacks of relying on these methods:

- Several previous frames are necessary. (If an object appears only in the last frame, it is highly likely to forecast its points as static or with high errors.)
- They use only previous information, cannot deal with high accelerations or direction changes. They can never reach (theoretically) the accuracy of methods like ours based on geometry and actual information for generating virtual measurements.
- Training is required before inference, and separate training for LIDAR point clouds with different characteristics or resolution.
- Prediction works (quasi) real-time only in close range, despite the need for as soon as possible detection [17].

3 Preliminaries

Before going to details about the proposed method, preliminaries have to be discussed.

3.1 LIDAR-Camera Synchronization

As we said previously, in practice, in most cases, the camera frame with the closest timestamp to the LIDAR frame is considered as the corresponding camera frame. The correctness of this assumption depends on the rotation frequency of the rotating LIDAR as data points in the LIDAR frame have different timestamps in reality. In this way, ego-movement can corrupt our measurement, as the first measured point can be measured from a significantly different viewpoint than the last one. Nevertheless, using the correct equipment (for high-frequency localization) and methods like [6], all the data points of one frame can be de-skewed (transformed to the same coordinate system). Also, other dynamic objects could cause a problem, but if we talk about points of the same vehicle, as they are close to each other, the time difference between the measurements of their data points will be negligible. The remaining error is caused by the frequency mismatching of the camera and LIDAR. However, assuming a camera with a sampling time of 0.03 s (about 30 FPS framerate), we will get that our LIDAR frame timestamp can differ from the nearest camera frame time step only by about 0.015 s (even less with a higher camera framerate). These are the reasons, using the closest camera frame is accepted in practice. As this accuracy is available as ground truth, we also aim to achieve this quality with our estimation.

3.2   Time Notations

In the paper, we differentiate only two notations of time moments. T_{t-1} will indicate the reference timestamp, and T_t the current timestamp (with index $t \in \mathbb{N}^+$ indicating the tth measurement). As their interpretation may not obvious, we summarize it in case of the different sensors:

T_{t-1} Timestamp:

- *Camera:* In case of camera, we refer T_{t-1} as the timestamp of the camera frame corresponding to the last available complete LIDAR frame.
- *LIDAR:* In the case of LIDAR, T_{t-1} means the timestamp of the last available complete circular frame of the sensor. Using the most basic assumption, it is considered to be the same as T_{t-1} of the camera. (See more about the neglected time differences of camera's and LIDAR's T_{t-1} in Sect. 3.1)

T_t Timestamp:

- *Camera:* In the most simple case, T_t is the timestamp of the next available camera frame after T_{t-1}. If one does not want to use the highest available up-sampling rate with the proposed method, then this can be different; see explanation of Eq. 1 for more information.
- *LIDAR:* Exactly the same virtual timestamp as T_t in case of camera. At this time moment, there is no complete circular LIDAR frame measurement available. That is why we generate a virtual measurement $P_{t,v}$ to this time moment.

The time difference between the two time moments:

$$\Delta t = T_t - T_{t-1} = \frac{n}{f_C} \tag{1}$$

It means that the elapsed time is the multiplication of $n \in \mathbb{N}^+$ and the inverse of the camera's sampling frequency f_C. n can be chosen based on our need and computing power. The ratio of camera and lidar sampling frequency limits its maximum: $n < \frac{f_C}{f_L}$. $n = 1$ results in the highest possible up-sampling, with the highest computation need. $n = \frac{f_C}{f_L}$ would result in no up-sampling; this value is used in the quantitative evaluation part as in this case, ground truth measurement is provided for the estimation.

In the following, we introduce our method on the basis of T_{t-1} and T_t for virtual point cloud generation for the time moment T_t. If we repeat these steps for any t and $t-1$ it results in an up-sampled LIDAR measurement, as generally $\frac{f_C}{f_L} > 1$ and so there is no available complete circular LIDAR measurement at T_t.

4   The Proposed Method

From here, we describe the proposed method in detail. The flowchart of the system is illustrated in Fig. 2. The five steps of our virtual frame generation (temporal up-sampling):

1. Detecting dynamic object candidates (T_{t-1} time moment)
2. Determine corresponding LIDAR data points of detected objects and project back to image (T_{t-1} time moment)
3. Tracking dynamic objects and object points on camera image (to T_t time moment)
4. Compute virtual camera motion for the tracked points for each tracked object
5. Transform LIDAR data points (acquired at) T_{t-1} to T_t time moment using the estimated transformation matrices

4.1   Detecting Dynamic Object Candidates on RGB Images

In our experiments, we used Yolo_v2 [16] for vehicle detection. Example detections can be seen in Fig. 2. (Naturally, this can be substituted with any other detector.) In the tests, we detected only vehicles (but of course, it can be extended with any category), because of the following reasons. The two most common traffic participants and dynamic object candidates on roads are vehicles and pedestrians. Pedestrians walking with about $5\,km/h = 1.4\,m/s$ means that at the lowest LIDAR measurement rate, 5 Hz, they can move about 0.28 m (up-scaling their movement to 10 Hz would mean half of that - 0.14 m). This 10 cm range (even smaller in case of higher LIDAR measurement frequency) is similar to the LIDARs accuracy, so their points can be well estimated as static in this small time duration. In contrast to vehicles, cars can change direction and move significant amount during this time (see described example in Sect. 1).

We refer vehicles as dynamic object candidates. The reason is that for each vehicle, we calculate the transformation, which maps it from the previous LIDAR frame to the current (virtual) one. In the case of parking vehicles, it would not be necessary as they are rigid with the scene. However, we cannot tell in advance that a moving vehicle will not brake and stop or a parking vehicle will not depart. Also, it is possible that, we do not have previous information about the vehicle being static or dynamic.

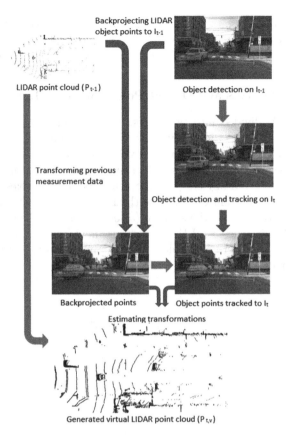

Fig. 2. Overview of the proposed method. Images and point clouds illustrate different states of the process. Arrows indicate processes, the red ones correspond to processes through data with different measurement times. (Color figure online)

4.2 Backprojecting LIDAR Object Points to Image

First, the intrinsic calibration [25] and LIDAR-camera calibration [27] should be done.

In this subsection, we deal with measurements recorded at T_{t-1} timestamp. We have a point cloud from the LIDAR sensor P_{t-1} and an RGB image I_{t-1} acquired by the camera. The elapsed time between T_{t-1} and T_t is assumed to be the sampling rate of the camera to get the highest possible up-sampling (Eq. 1).

Knowing the intrinsic matrix $\mathbf{K}$ and the transformation matrix between the coordinate system of the LIDAR and the camera $\mathbf{T}_{L,C}$, we can project 3D LIDAR data points to the 2D image. LIDAR data points that are in front of the camera and project into the image border can be selected and we deal with only them. In a second step, we can select the corresponding LIDAR points for each detected object which projection are in the 2D bounding box of the object. These points can belong to multiple objects, so Euclidean cluster extraction [18] is executed on the 3D pairs of the points, and only the points of the biggest cluster are used. Illustration of 2D points back-projected from the LIDAR for each object can be seen in Fig. 2 in the left-hand side of the last but one row.

4.3 Tracking Objects and Object Points on RGB Images

This subsection is divided into two parts. The first part explains object-level tracking; the second part deals with point-level tracking. From here, we are talking about processing which should be executed in time moment T_t. Naturally, the detection part of our system uses the same detector on each frame.

To track the detected objects (based on their bounding boxes) from the last camera frame to the current one, we use the Hungarian algorithm [12].

Each object from the previous frame (also identified in the current one) has points assigned from the last LIDAR measurement (where both LIDAR and camera measurements were available). These points are tracked to the current frame with Kanade-Lucas-Tomasi (KLT) algorithm [8].

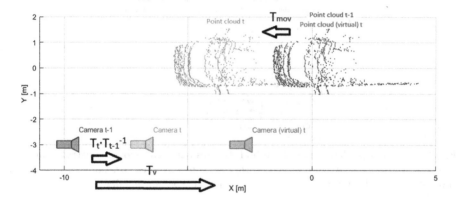

Fig. 3. Illustration of the calculation of the position of dynamic object point clouds (and transformations) when only camera measurement is available in case of a simple case of forward ego moving and object moving in the opposing lanes. Camera positions and points of a car are illustrated in a global coordinate system and colored with blue at timestep T_{t-1} and red in T_t. $\mathbf{T}_{t-1}$ and $\mathbf{T}_t$ (transformations between the coordinate systems of the real cameras and the global coordinate system) are known. We search for $\mathbf{T}_{mov}$ of the illustrated car (i^{th} object), describing its movement in the global coordinate system. In order to that, we assume that the car is being static and we estimate a transformation matrix between the real camera at timestep T_{t-1} and a virtual camera (colored with green) at T_t, $\mathbf{T}_v = \mathbf{T}_{t,i} \cdot \mathbf{T}_{t-1,i}^{-1}$ based on our image points on I_t (tracked from I_{t-1}). From this data, $\mathbf{T}_{mov}$ can be calculated. One can observe that, the object points have the same coordinate in the local coordinate system of the virtual camera (constructed by assuming the illustrated car being static) and the real one at T_t as these are the same. (Color figure online)

4.4   Estimate 3D Transformations

As a prerequisite for this step, we need to know the coordinate transformation between the static scene of T_{t-1} and T_t (the extrinsic camera parameters of I_{t-1} and I_t). This can be calculated from the sensor poses of the two measurements (in our experiments, we used the GPS-based poses provided by the Argoverse dataset [2], or alternatively, it can be determined solely from the images, for example, with methods like ORB-SLAM [13]. Thus, we have homogeneous transformation matrices $\mathbf{T}_{t-1} = [\mathbf{R}_{t-1}|\vec{t}_{t-1}]$ and $\mathbf{T}_t = [\mathbf{R}_t|\vec{t}_t]$ transforming to camera coordinate systems of T_{t-1} and T_t from a global coordinate system, with $\mathbf{R}$-s and $\vec{t}$-s being rotation matrices and translation vectors respectively. (If we talk about epipolar geometry, most of the cases, the coordinate system of the first camera is defined as global coordinate system and only $\mathbf{T}_t$ is estimated, but we keep using this general terminology in the following.)

For each tracked object we can estimate a virtual camera pose based on their 3D-2D (3D in P_{t-1} and 2D in I_t) point pairs (Subsect. 4.3). We evaluated point pairs in our experiments at least 50 matches (this filter very far and unstable tracks). We used the method of [5] with MLESAC [20] robust estimator to solve the Perspective-n-Point problem. Illustration of the transformations (and a simple case for object and camera movements) is visible in Fig. 3. With the estimation we get transformation matrices $\mathbf{T}_{D,i}$ between the LIDAR (at T_{t-1}) and virtual cameras (az T_t) corresponding to i^{th} dynamic object. Transformations between (the real) camera coordinate system (at T_{t-1}) and coordinate system of the ith virtual camera:

$$\mathbf{T}_{t,i} = \mathbf{T}_{L,C} \cdot \mathbf{T}_{D,i} \cdot \mathbf{T}_{L,C}^{-1} \tag{2}$$

Finally, if we would like to determine the transformation matrix corresponding to the real movement ($\mathbf{T}_{mov,i}$) of the ith 3D object in the global coordinate system, we can do it by rearranging the equation (knowing the facts, we have one $\mathbf{K}$, and coordinate systems of virtual and real cameras are the same in both cases of T_{t-1} and T_t):

$$\mathbf{T}_t \cdot \mathbf{T}_{mov,i} \cdot \mathbf{T}_{t-1}^{-1} = \mathbf{T}_{t,i} \cdot \mathbf{T}_{t-1,i}^{-1} \tag{3}$$

4.5   Transformation of LIDAR Data Points (Acquired at) T_{t-1} to T_t Time Moment Using the Estimated Transformation Matrices

In this step, we differentiate two types of point clouds. The first type of point cloud corresponds to the static scene. All the projections that were outside of the bounding boxes of dynamic object candidates fall into this category. They can be transformed to the virtual estimation of T_t timestamp by the following matrix:

$$\mathbf{T}_S = \mathbf{T}_{L,C}^{-1} \cdot \mathbf{T}_t \cdot \mathbf{T}_{t-1}^{-1} \cdot \mathbf{T}_{L,C} \tag{4}$$

For the remaining points, different transformation matrices are used based on which dynamic object candidates they correspond to:

$$\mathbf{T}_{D,i} = \mathbf{T}_{L,C}^{-1} \cdot \mathbf{T}_t \cdot \mathbf{T}_{mov,i} \cdot \mathbf{T}_{t-1}^{-1} \cdot \mathbf{T}_{L,C} \tag{5}$$

Using $\mathbf{T}_S$ and $\mathbf{T}_{D,i}$-s for the appropriate points, we can generate $P_{t,v}$, our virtual LIDAR point cloud in time moment T_t (where we originally did not have LIDAR measurement). We can determine these transformation matrices and transform the previous LIDAR measurement for all the camera measurements (without corresponding LIDAR), and this way, we temporally upsample our LIDAR measurements.

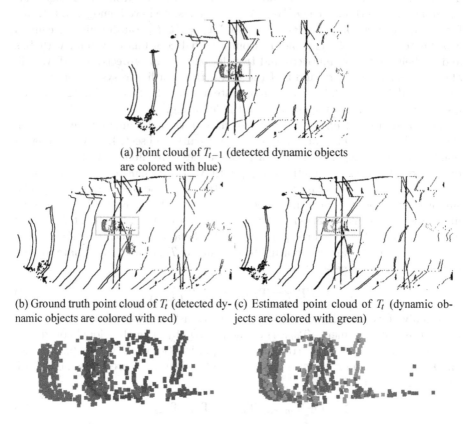

(a) Point cloud of T_{t-1} (detected dynamic objects are colored with blue)

(b) Ground truth point cloud of T_t (detected dy- (c) Estimated point cloud of T_t (dynamic ob-
namic objects are colored with red)   jects are colored with green)

(d) Yellow highlighted vehicle at T_{t-1} (blue) (e) Yellow highlighted vehicle's ground truth
and T_t (red)   (red) and estimated (green) position at T_t

Fig. 4. Illustration of point cloud estimation of the proposed method. (a) subfigure shows the last LIDAR frame (P_{t-1}), (b) shows the ground truth and (c) our estimation of the point cloud at T_t. Subfigure (d) highlights ground truth points of a vehicle at T_{t-1} and T_t. While (e) visualizes together the ground truth point cloud and our estimation. (Color figure online)

5   Results

Our method has been tested on the Argoverse dataset [2], which is a large public database for autonomous driving. We compare the accuracy of our proposed method with the latest methods, also generating virtual point clouds like [4] and [23]. The error measures are Chamfer Distance (CD), and Earth Movers Distance (EMD) which are also applied in the cited literature. CD measures the distance between the nearest neighbors of points in the predicted and the ground truth one in both directions, while EMD describes the similarity of the predicted point cloud to the ground truth one by calculating the cost of global matching problem.

The formula for Chamfer Distance between two point clouds:

$$CD = \frac{1}{N_t} \sum_{p_{t,v} \in P_{t,v}} \min_{p_t \in P_t} ||p_t - p_{t,v}||^2 + \frac{1}{N_t} \sum_{p_t \in P_t} \min_{p_{t,v} \in P_{t,v}} ||p_t - p_{t,v}||^2 \quad (6)$$

where $p_{t,v} \in \mathbb{R}^{N_t \times 3}$ and $p_t \in \mathbb{R}^{N_t \times 3}$ are the data points of the predicted (virtual) $P_{t,v}$ and ground truth P_t point clouds respectively.

The Earth Movers Distance between the two point clouds:

$$EMD = \min_{\phi: P_t \to P_{t,v}} \frac{1}{N_t} \sum_{p_t \in P_t} ||p_t - \phi(p_t)||^2 \quad (7)$$

where ϕ is bijection. EMD calculates the point-to-point mapping between two point cloud P_t and $P_{t,v}$ by finding global nearest neighbors.

The competitor methods require more than one previous LIDAR frame and also learning (see more comparison in Sect. 2.2). We used all the test sequences (4168 frames) of the Argoverse dataset (LIDAR and front facing center camera) for the evaluation. In our case, error metrics were calculated to dynamic object candidates for fair evaluation because the estimation covered that part of the point cloud (remaining parts are determined with prior knowledge). The results are summarized in Table 1.

Table 1. Quantitative evaluation of the proposed pipeline.

Methods	CD [m²]	EMD [m²]
Copy last input	0.5812	1.0667
PointNet++ · LSTM [15]	0.3826	1.0011
PointCNN · ConvLSTM [9]	0.3457	0.9659
[4]'s PointGRU	0.2994	0.9084
[4]'s PointLSTM	0.2966	0.8892
[4]'s PointRNN	0.2789	0.8964
[23]'s proposal	0.3010	0.9010
[23]'s alignment	0.5250	0.7710
Proposed	**0.1983**	**0.3007**

The deep learning methods in Table 1 used 5 previous LIDAR frames to estimate consecutive ones, while we use only one previous LIDAR frame. However, relying on (actual) camera information, another level of accuracy can be reached. As our virtual point cloud is generated from real measurements by transformation, the EMD value (similarity in point cloud characteristics) is much smaller than the competitors' results.

Processing Effort: The pipeline is implemented in the Matlab environment, configuration: Intel Core i7-4790K @ 4.00 GHz processor, 32 GB RAM, Nvidia GTX 1080 GPU, Windows 10 64 bit. The processing time is heavily dependent on the image resolution (but for specific system components - e.g., detection - smaller resolution can be used). By downscaling the Argoverse images of original resolution of 1920×1200 to 25% of its original size, we were able to keep all the important detection for the evaluation. The above methods uses only downsampled and cropped point clouds to ensure real-time running, (e.g. [23] reported about 5 FPS speed with n NVIDIA RTX2080Ti GPU for their system with for only 4096 data points). We measured 105 ms average running time for our whole system without that, meaning, we can temporally 5 Hz LIDAR measurements by using the current limited, research configuration.

[23] and all the other methods above (evaluated at [4]) cropped the frames $10 \, \text{m} \times 10 \, \text{m}$, using the point cloud range $[-5 \, \text{m}, 5 \, \text{m}]$ (in both x and y direction). As we used front facing camera, we evaluated objects with centers within the $[0 \, \text{m}, 10 \, \text{m}]$ range in case of x direction. Note that: by increasing the distance we got a harder estimation problem.

6   Conclusions

This paper presented a novel approach to temporally up-sample LIDAR measurements based on mono camera and epipolar geometry. We provide state-of-the-art accuracy and similarity using only two camera frames and the previous LIDAR frame. In the future, we plan to test other datasets, extend the method to estimate occluded objects and further improve the generated point cloud characteristics similarity to real measurements.

Acknowledgements. The research was supported by the Ministry of Innovation and Technology NRDI Office within the framework of the Autonomous Systems National Laboratory Program and by the Hungarian National Science Fundation (NKFIH OTKA) No. K139485.

References

1. Benedek, C., Majdik, A., Nagy, B., Rozsa, Z., Sziranyi, T.: Positioning and perception in LIDAR point clouds. Digit. Sig. Process. **119**, 103193 (2021)
2. Chang, M.F., et al.: Argoverse: 3D tracking and forecasting with rich maps. In: Proceedings of the IEEE/CVF Conference on Computer Vision and Pattern Recognition (CVPR), June 2019 (2019)
3. Debeunne, C., Vivet, D.: A review of visual-LiDAR fusion based simultaneous localization and mapping. Sensors **20**(7), 2068 (2020)

4. Fan, H., Yang, Y.: PointRNN: point recurrent neural network for moving point cloud processing. arXiv arXiv:1910.08287 (2019)
5. Gao, X.S., Hou, X.R., Tang, J., Cheng, H.F.: Complete solution classification for the perspective-three-point problem. IEEE Trans. Pattern Anal. Mach. Intell. **25**(8), 930–943 (2003)
6. He, L., Jin, Z., Gao, Z.: De-skewing lidar scan for refinement of local mapping. Sensors **20**, 1846 (2020)
7. Hu, M., Wang, S., Li, B., Ning, S., Fan, L., Gong, X.: Towards precise and efficient image guided depth completion (2021)
8. Kalal, Z., Mikolajczyk, K., Matas, J.: Forward-backward error: automatic detection of tracking failures. In: 20th International Conference on Pattern Recognition, pp. 2756–2759 (2010)
9. Li, Y., Bu, R., Sun, M., Wu, W., Di, X., Chen, B.: PointCNN: convolution on X-transformed points. In: Advances in Neural Information Processing Systems, vol. 31. Curran Associates, Inc. (2018)
10. Liu, H., Liao, K., Lin, C., Zhao, Y., Guo, Y.: Pseudo-LiDAR point cloud interpolation based on 3D motion representation and spatial supervision. IEEE Trans. Intell. Transp. Syst., 1–11 (2021)
11. Liu, H., Liao, K., Zhao, Y., Liu, M.: PLIN: a network for pseudo-LiDAR point cloud interpolation. Sensors **20**, 1573 (2020)
12. Miller, M.L., Stone, H.S., Cox, I.J., Cox, I.J.: Optimizing Murty's ranked assignment method. IEEE Trans. Aerosp. Electron. Syst. **33**, 851–862 (1997)
13. Mur-Artal, R., Tardós, J.D.: ORB-SLAM2: an open-source SLAM system for monocular, stereo and RGB-D cameras. IEEE Trans. Rob. **33**(5), 1255–1262 (2017)
14. Premebida, C., Garrote, L., Asvadi, A., Ribeiro, A., Nunes, U.: High-resolution LIDAR-based depth mapping using bilateral filter, November 2016, pp. 2469–2474 (2016)
15. Qi, C., Yi, L., Su, H., Guibas, L.: PointNet++: deep hierarchical feature learning on point sets in a metric space. In: NIPS (2017)
16. Redmon, J., Farhadi, A.: YOLO9000: better, faster, stronger. In: 2017 IEEE Conference on Computer Vision and Pattern Recognition (CVPR), pp. 6517–6525 (2017)
17. Rozsa, Z., Sziranyi, T.: Object detection from a few LIDAR scanning planes. IEEE Trans. Intell. Veh. **4**(4), 548–560 (2019)
18. Rusu, R.B.: Semantic 3D object maps for everyday manipulation in human living environments. KI - Künstliche Intelligenz **24**(4), 345–348 (2010)
19. Schneider, N., Schneider, L., Pinggera, P., Franke, U., Pollefeys, M., Stiller, C.: Semantically guided depth upsampling. In: Rosenhahn, B., Andres, B. (eds.) GCPR 2016. LNCS, vol. 9796, pp. 37–48. Springer, Cham (2016). https://doi.org/10.1007/978-3-319-45886-1_4
20. Torr, P.H.S., Zisserman, A.: MLESAC: a new robust estimator with application to estimating image geometry. Comput. Vis. Image Underst. **78**, 138–156 (2000)
21. Uhrig, J., Schneider, N., Schneider, L., Franke, U., Brox, T., Geiger, A.: Sparsity invariant CNNs. In: International Conference on 3D Vision (3DV) (2017)
22. Wang, P., Huang, X., Cheng, X., Zhou, D., Geng, Q., Yang, R.: The ApolloScape open dataset for autonomous driving and its application. IEEE Trans. Pattern Anal. Mach. Intell. **42**, 2702–2719 (2019)
23. Wencan, C., Ko, J.H.: Segmentation of points in the future: Joint segmentation and prediction of a point cloud. IEEE Access **9**, 52977–52986 (2021)

24. Weng, X., Wang, J., Levine, S., Kitani, K., Rhinehart, N.: Inverting the Pose Forecasting Pipeline with SPF2: Sequential Pointcloud Forecasting for Sequential Pose Forecasting. CoRL (2020)
25. Zhang, Z.: A flexible new technique for camera calibration. IEEE Trans. Pattern Anal. Mach. Intell. **22**(11), 1330–1334 (2000)
26. Zhao, S., Gong, M., Fu, H., Tao, D.: Adaptive context-aware multi-modal network for depth completion. IEEE Trans. Image Process. **30**, 5264–5276 (2021)
27. Zhou, L., Li, Z., Kaess, M.: Automatic extrinsic calibration of a camera and a 3D LiDAR using line and plane correspondences. In: 2018 IEEE/RSJ International Conference on Intelligent Robots and Systems (IROS), pp. 5562–5569 (2018)

StandardSim: A Synthetic Dataset for Retail Environments

Cristina Mata[1]([⊠]), Nick Locascio[2], Mohammed Azeem Sheikh[2],
Kenny Kihara[2], and Dan Fischetti[2]

[1] Stony Brook University, Stony Brook, NY 11790, USA
cfmata@cs.stonybrook.edu
[2] Standard Cognition, 965 Mission Street, San Francisco, CA 94103, USA
{nick,mohammed,kenny,dan}@standard.ai

Abstract. Autonomous checkout systems rely on visual and sensory inputs to carry out fine-grained scene understanding in retail environments. Retail environments present unique challenges compared to typical indoor scenes owing to the vast number of densely packed, unique yet similar objects. The problem becomes even more difficult when only RGB input is available, especially for data-hungry tasks such as instance segmentation. To address the lack of datasets for retail, we present StandardSim, a large-scale photorealistic synthetic dataset featuring annotations for semantic segmentation, instance segmentation, depth estimation, and object detection. Our dataset provides multiple views per scene, enabling multi-view representation learning. Further, we introduce a novel task central to autonomous checkout called change detection, requiring pixel-level classification of takes, puts and shifts in objects over time. We benchmark widely-used models for segmentation and depth estimation on our dataset, show that our test set constitutes a difficult benchmark compared to current smaller-scale datasets and that our training set provides models with crucial information for autonomous checkout tasks. Our code and data can be found at https://standard-ai. github.io/Standard-Sim/.

Keywords: Change detection · Monocular depth estimation

1 Introduction

Autonomous checkout is a fast-spreading technology poised to change the way customers shop in brick and mortar stores. It often relies on cameras and other sensors to build an understanding of a retail environment and make a final decision about what a shopper purchases. Computer vision plays a crucial role in understanding this data, especially in systems where cameras are the only sensors available. While vision-only autonomous checkout is relatively new, progress in the domain has not led to any benchmarks or new tasks. We hypothesize that understanding of retail environments not only requires data that is tailored to this domain, but also a new computer vision task that identifies changes in retail

© The Author(s), under exclusive license to Springer Nature Switzerland AG 2022
S. Sclaroff et al. (Eds.): ICIAP 2022, LNCS 13232, pp. 65–76, 2022.
https://doi.org/10.1007/978-3-031-06430-2_6

scenes over time. Thus, in this paper, we describe a new dataset StandardSim, as well as a new task that identifies changes in retail scenes over time.

While object detection in retail environments has been broached by [9], which introduced the dense object detection task and dataset SKU-110K in the retail setting, they do not provide semantic annotations of objects beyond bounding boxes. Other large-scale datasets have been created synthetically for indoor environments, but they do not transfer well to retail because objects are not as diverse or densely packed. Additionally, these datasets do not provide diverse viewpoints, often showing scenes from the perspective of a person navigating the scene, as opposed to from the ceiling or a corner. Thus, when models trained on these datasets are applied to real retail environments they have not learned information needed to identify small objects from ceiling camera perspectives.

The change detection task is meant to simulate a shopper's actions in a retail environment by providing a model with a pair of images of a scene, before and after a series of interactions with objects in the scene. After a shopper interacts with objects in a retail environment, objects may be taken, added or shifted around the scene. Each image pair in our dataset displays what a random interaction might look like and is annotated similarly to segmentation, where each pixel belongs to a take, put, shift or no change. Objects tend to be small and changes are sparse. Due to the task's similarity to segmentation, we adapt a popular state of the art segmentation model, Deeplabv3 [5], to set a benchmark for this task, and show that due to the sparse changes and small size of objects, StandardSim is a very difficult benchmark.

With these problems in mind, we introduce StandardSim, a large-scale synthetic dataset made from highly accurate store models and featuring annotations for depth estimation, object detection, instance segmentation and a novel task which we call change detection. This dataset consists of over 25,000 images from 2,134 unique scenes. Each scene includes views from multiple cameras, enabling multi-view reconstruction and the generation of shape estimation annotations. Compared to previous datasets, StandardSim provides annotations for more tasks and fills a void in the retail environment domain.

In addition to change detection we focus on monocular depth estimation because depth provides important cues about the movement of objects over time. We benchmark the state of the art monocular depth estimation model Dense Prediction Transformer [19], based on MiDaS [20], on StandardSim and compare its performance on other datasets. We find that it has a much higher error on our dataset, suggesting that StandardSim constitutes a difficult new benchmark for monocular depth estimation. StandardSim will be made accessible via URL post-publication.

2   Related Work

2.1   Datasets for Retail Environments

So far work on computer vision for retail environments has focused on object detection in densely packed scenes [9]. In these environments, traditional object detection solutions fail because the boxes are either inaccurate or overlap to a degree that makes them less useful. In order to address this, the authors of [9] released a new dataset, SKU-110K, that includes densely packed views of shelves that are annotated with bounding boxes around the objects (typically called stock-keeping units (SKUs) in the retail environment). For this dataset, the primary limitations are that the data consists of front facing views of the objects only and the objects are only annotated with bounding boxes. Our synthetic dataset does much to address some of these shortcomings, including providing a variety of shelf styles (such as hanging items), pixel level masks for each object in the image, and different, skewed camera perspectives for each object.

2.2   Monocular Depth Estimation

Another shortcoming of retail datasets is the lack of depth information, which is key to both detect the layout of stores and to detect potential changes. Our dataset includes depth maps for each image, as well as multiple camera views to allow for reconstruction through multiple view points. Current state of the art in monocular depth estimation includes zero shot approaches [20] trained on multiple datasets to adapt well to various environments. Other approaches combine semantic segmentation and instance information with RGB images to obtain monocular depth maps [22,28] and using 360 views combined with edge and corner information to obtain good indoor depth maps [12]. Combining instance segmentation, semantic segmentation, and monocular depth estimation has also been explored [8]. Finally, domain adaptation from synth-to-real is becoming increasingly important [29], since many real world applications require accurate depth maps with no easy way to measure ground truth.

2.3   Change Detection

In addition to the work on retail environments, we also look at a new task, that of visual change detection. Visual change detection has been explored for outdoor urban scenes [26], which is similar to the task definition we provide, except that our scenes are indoors with labels tailored to retail settings. Other modalities for change detection, such as a purely classification based approach [14], exist in the urban change detection area. However, not many datasets exist. In addition, a deep literature exists for change detection in remote sensing applications [23], along with many open source datasets.

In addition to outdoor and remote sensing tasks, the new work ChangeSim [17] highlights change detection in an indoor setting. Several differences exist between this work and our task. Our cameras are fixed for a given set of changes, and our images are paired, with a well-defined before and after image. Our dataset also consists of retail environments rather than industrial environments. Finally we have multiple camera views per change, allowing for the potential implementation of multi-camera change detection. However, we do share many

Fig. 1. From left to right, an image sample with corresponding random matting on the shelves, depth annotation, and surface normal annotation.

similarities. Both tasks involve change detection using semantic masks in challenging environments, in which background characteristics such as lighting are in flux.

2.4 Synthetic Indoor Datasets

Besides ChangeSim [17], other photorealistic synthetic datasets exist. Two prominent datasets are Hypersim [21] and SceneNet [15]. Both focus heavily on several important aspects of scene understanding, with a focus on semantic segmentation, object detection, and depth estimation. These datasets are often composed of highly detailed indoor environments, but do not include retail environments. They have several pieces in common with our dataset, including varied lighting and different textures available for each scene, as well as detailed semantic labels for each object in the image.

3   StandardSim

3.1   Overview

Our dataset consists of 2,134 unique scenes of retail environments based on three real store layouts. Each scene consists of a set of actions in which several items may be taken, put or shifted. There are also scenes in which no changes occurred, which reflects real-world retail environments. Our dataset is focused on the change detection task which aims to discern what changes occur in a scene at a pixel level, so we structure a sample as a pair of before and after images. In total our dataset contains 6,359 such samples. We also add random matting to the shelves, resulting in two distinct mattes for each sample. Thus we have 4 images per sample from 2 different mattes, leading to 25,436 images in total. Figure 2 shows views of the empty stores in good lighting so that all details are viewable. It features diverse shelving and lighting configurations, as well as reflective surfaces from glass.

We chose a 70/15/15 training, validation and testing split. We randomly assigned scenes while keeping the distribution of samples across stores as even as possible, that is, we have 1/3 of the samples from each split come from each store. Additionally, there is an average of 3 camera views per scene to provide data for multi-view tasks.

Table 1. Comparison of StandardSim to other datasets. The columns refer to the type of annotation provided: from left to right these are Depth, Object Detection, Instance Segmentation and Change Detection annotations. The Domain refers to the type of locations displayed in the dataset.

	Depth	Object Det.	Instance	Change Det.	Domain	# Images
ChangeSim	✓	✗	✗	✓	Industrial	130,000
SKU-110K	✗	✓	✗	✗	Retail	11,762
HyperSim	✓	✓	✓	✗	Indoor	74,619
SceneNet	✓	✓	✓	✗	Indoor	5M
StandardSim	✓	✓	✓	✓	Retail	25,436

Fig. 2. Store layouts used in the dataset. Best viewed in color and zoomed in.

In Table 1 we show that compared to other datasets for retail and change detection, our dataset features more images and annotations for more tasks. Other large-scale synthetic datasets may provide more images but they do not provide change detection annotations and do not structure their samples for change detection. Our dataset provides scenes more suitable for autonomous checkout.

3.2   Dataset Creation

We utilize Blender [7] for the core of our data generation pipeline. Our data generation pipeline utilizes Blender's python interface to dynamically modify the store, the products, and the camera placements. We utilize Blender's cycles render engine to produce photo-realistic RGB, RGB with randomized textures, RGB with blank textures, z-depth, segmentation masks, and surface normals. Each image render takes approximately 20 s utilizing 2 GPUs (Tesla V100 16 GB). With the publication of this paper the generation and rendering pipelines will be made available and open-source.

Each store model is a to-scale replica of a real retail store. To create these replica assets, we perform a 3d scan of the store using a lidar-based Matterport [2] device, and utilize an in-house asset creator to model the store in Blender with the 3d scan as a guide. This ensures highly accurate dimensions and structure of the 3d model, while also producing high quality textures and meshes required for photo-realistic rendering from many viewpoints. We model the store and indicate which shelves meshes are shelves, but leave all shelves empty.

For our SKU models and textures we purchased a variety of assets from the online marketplace Turbosquid [1]. We build a collection we call SkuBank of

456 assets with high quality meshes and textures as .obj, .mtl, and .png files. Due to the licensing restrictions of these acquired assets, they are not part of our release. However, our data generation pipeline is flexible to any SkuBank built in this way, so substituting other objects can be done in a plug-and-play fashion. We place the objects from SKUBank on the store shelves in an automated, random fashion that aims to replicate the densely packed distribution seen in retail environments.

Once a store's product placement is complete, we focus on generating random, but valid extrinsic and intrinsic camera parameters. We supply these camera parameters in standard opencv matrix format [3]. Intrinsic parameters are fixed other than the field-of-view of the camera which we sample from the range [50, 70]. Extrinsic camera placement in world coordinates follows a simple algorithm, CamPlace. The camera height (z coordinate) is placed on the ceiling with some random height adjustments. The camera (x, y) global coordinates are determined by sampling from a 5×5 m plane around a random, front-facing sku. To filter out degenerate camera positions, we utilize Blender's built-in raytracing to ensure that the object in question is visible from the camera. This ensures that the camera is placed within the store, in free space, and avoids issues around the camera spawning inside of objects which may block its view.

In order to generate data for our change detection task, we must perturb the items in a realistic and principled way. Our change creation algorithm, SkuRemove, selects a random 2–10% of product sections, and then selects a random (1–6) number of products items to remove from that section. When a customer takes a product from a section, nearby products often shift around it due to their proximity. We model this directly by randomly shifting nearby SKUs with a higher probability than the random shifting applied to all SKUs.

4 Change Detection Task

We formulate the change detection task in the context of autonomous checkout, where it is crucial to know how objects have moved or shifted over time to put together a shopper's receipt. In our problem formulation, our input is a pair of images, (X_{t_1}, X_{t_2}), where $t_1 < t_2$ and t_1, t_2 are distinct time stamps. Our goal is to output a pixel-wise labeling $Y = \{y\}_1^N$ where N is the number of pixels in the frame and $y \in \{\text{no change, take, put, shift}\}$. These classes represent the basic actions we are interested in, and in other applications such as industrial ones have been extended to include rotation [17], a more specific semantic label than shift.

We follow the COCO-style annotation formatting [13] to provide annotations enabling change detection, semantic and instance segmentation, object detection and depth estimation. We separate image indexing from annotations for ease of use.

Table 2. Class breakdown (% IOU)

Class	No change	Put	Take	Shift
Val	98.30	24.46	19.30	2.56
Test	96.45	25.04	19.85	2.83

5 Experiments

5.1 Change Detection

Change detection requires a pixel-level understanding of where changes occur in a scene, leading us to choose a popular semantic segmentation model, Deeplabv3 [5], to benchmark on our dataset. We chose the Resnet50 [11] backbone as our architecture and found this provides a good trade-off of accuracy versus computational efficiency. We load COCO [13] pretrained weights into the encoder and finetune the entire network on our training set until convergence. We use a batch size of 16, the SGD optimizer with an initial learning rate of 0.0004 and cosine decay. We apply random crops and resizing to height 360 and width 512. All our experiments are conducted using PyTorch [18].

We include several data augmentations specific to the change detection task. First we randomly modify change detection pairs by flipping the change order, that is, we switch the before and after images and keep the same label. Second, we randomly apply left-right flipping to both images and labels. We also randomly add noise to pixel values to simulate real camera images. In addition, the images include changes in lighting that might occur in real camera feeds.

During training we measure the mean Jaccard Index or Intersection over Union (IOU) on the validation set and stop training when this metric plateaus at 1950 epochs. In our initial experiments, the model experienced mode collapse predicting all background, but we found that adding loss weights alleviates this issue. Our final model achieves 36.15% IOU on the validation set and 36.04% IOU on our test set, indicating that change detection is a very difficult task similar to fine parts segmentation. A breakdown of IOU by class can be found in Table 2, which shows that the model performs best on the put class and struggles most on the shift class. These results also suggest that our test set is unbiased relative to our validation set due to the similarity in results between the two. Figure 3 shows how the model performs qualitatively on our dataset, showing good results on even small objects.

Additionally, we qualitatively show how a model trained on StandardSim performs on a real change detection example in Fig. 4. The left and middle images correspond to before and after frames, and items are put on the shelf on the bottom right corner. The model's output is highlighted in the rightmost image, and shows that the synthetic to real domain transfer is achievable.

5.2 Depth Estimation

Many datasets for monocular depth estimation exist and have been used to train the state of the art monocular depth estimation method MiDaS [20]. The first

Fig. 3. Qualitative results of our change detection model. The leftmost column shows the initial scene, the second column shows the scene after changes. The third column shows the model's predictions and the fourth column shows the ground truth label. Green signifies puts, red are takes and purple are shifts. Best viewed in color and zoomed in. (Color figure online)

Fig. 4. Three frames from a real scene of a retail environment where changes occur in the bottom right corner of the scene. The left and middle images show the before and after frames. On the right the output from a model trained on StandardSim is overlaid on top.

version of MiDaS took advantage of 3D movies from which it reconstructed depth from neighboring frames. Their code provides an inference script and trained weights for their models, but no training script or implementation of their loss functions. The next iteration of their model builds off of recent developments in transformers for vision, and they develop Dense Prediction Transformers [19] for several dense prediction tasks including monocular depth estimation. We benchmark their DPT-Large model that was pretrained on their MIX 6 meta-dataset and keep their model parameters unchanged. We find that on our test set it achieves an absolute relative error of 83.59, and absolute difference of 10.03 and a square relative error of 625.80.

We note that in some of our scenes, there are areas of the camera in that are blocked by a lighting fixture, and this results in a ground truth depth of 0 m. To prevent NANs in the metric calculations we masked out the pixels where ground truth is 0, which amounted to ignoring 0.089% of the pixels. This is a negligible amount and unlikely to sway the overall metrics.

Table 3 compares the model's performance on our test set versus other datasets and shows that its performance is considerably worse on our dataset in absolute relative error. We include some qualitative results in Fig. 5 on our dataset and on the NYUv2 dataset [16], consisting of indoor samples, as well as on an outdoor sample. The model performs well on NYUv2 and on our dataset

Table 3. Absolute Relative Error (ARE) of DPT-Large on multiple datasets

	Sintel [4]	TUM [24]	Kitti [25]	NYUv2 [16]	Ours (test)
ARE	0.27	9.97	8.46	8.43	83.59

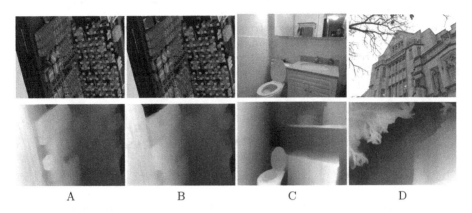

A   B   C   D

Fig. 5. Comparing DPT-Large results on different datasets. The top row shows the original image and the bottom row shows the output. Columns A and B show a change image pair. The model is able to identify general outlines of boxes well but fails to distinguish small changes in depth between small objects. Column C shows an indoor scene from the NYUv2 dataset and column D shows an outdoor scene. The model struggles with reflective surfaces and fine structures.

except that it cannot distinguish small changes in depth between small objects. This is also apparent from the depth estimation of the tree branches in the rightmost column. Further, when we reconstruct the depth estimate from a 3D perspective in Fig. 6 on our dataset, we find that corners and areas between shelves are inaccurate. This suggests that our dataset may serve as a useful augmentation to their MIX 6 to improve estimation of small and distant objects, especially since our ground truth is accurate.

6   Discussion

Top-Down cameras present difficulties for models trained on existing datasets. We parameterize the angle of tilt toward the ground as alpha, where 0 means pointed at the horizon, 90 means pointed straight down, and −90 means pointed straight up. Self-driving datasets are gathered from dashboard videos like Kitti have an alpha of 0. Datasets gathered from movies and disparity maps have limited alpha range as well. Our dataset has an alpha range of [30, 90].

Additionally, many datasets are gathered in outdoor settings, which means that the pixels near the top are the sky and have a filler infinite value. Our dataset generally has ceilings in the upper part of the image rather than a sky which can present difficulties for models trained on outside datasets.

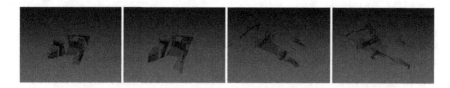

Fig. 6. Two depth estimation reconstruction examples. The leftmost column shows the ground truth depth while the second column from the left shows the DPT-Large's prediction reconstruction. The rightmost two images show the ground truth and prediction from a different perspective. The spaces between shelves indicate that corners are challenging for depth estimation.

Fig. 7. Our dataset offers multiple views of each scene. Here we have the same scene rendered from three different cameras.

One specific task that we have not explored in our dataset, but which is possible with the data provided, is that of multiview 3D reconstruction [10]. Since each scene has 3 different views along with associated intrinsic and extrinsic calibrations for the cameras, it is possible to use either classical computer vision methods [10] or deep learning methods to reconstruct the full 3D scene [27]. This has potential applications to many of the problems discussed above, allowing for 3D object detection and instance segmentation [6] of items if the views are detailed enough. 3D reconstructions can also provide information about store layout changes.

Labels for instance segmentation can be generated using the polygons provided from in the annotation files. The relevant functions are provided in the repository, along with some simple examples to set up for the instance segmentation task. Detailed item labels can be generated, as well as larger item group labels, such as labels for a group of candy bars.

7   Conclusion

In this paper we analyze why progress in autonomous checkout systems has been slow even though major advances continue to be made in computer vision. We conclude that this lag in progress is due to a lack of datasets for retail, and address this by presenting StandardSim, a large-scale synthetic open dataset with annotations for a variety of computer vision tasks. We also introduce the change detection task for retail, where pixel-wise labels that identify semantic changes in the environment for image pairs are tailored to shopper interactions with shelves. Finally, we benchmark state of the art models for change detection and monocular depth estimation on our dataset and compare their performance to that on other datasets. We conclude that StandardSim's retail environment domain is unique and challenging compared to other datasets, and identify further applications where it can be used.

References

1. 3D Models for Professionals: TurboSquid. https://www.turbosquid.com
2. Pro2 3D Camera. https://matterport.com/cameras/pro2-3D-camera
3. Bradski, G.: The OpenCV library. Dr. Dobb's J. Softw. Tools **120**, 122–125 (2000)
4. Butler, D.J., Wulff, J., Stanley, G.B., Black, M.J.: A naturalistic open source movie for optical flow evaluation. In: Fitzgibbon, A., Lazebnik, S., Perona, P., Sato, Y., Schmid, C. (eds.) ECCV 2012. LNCS, vol. 7577, pp. 611–625. Springer, Heidelberg (2012). https://doi.org/10.1007/978-3-642-33783-3_44
5. Chen, L.-C., Zhu, Y., Papandreou, G., Schroff, F., Adam, H.: Encoder-decoder with atrous separable convolution for semantic image segmentation. In: Ferrari, V., Hebert, M., Sminchisescu, C., Weiss, Y. (eds.) ECCV 2018. LNCS, vol. 11211, pp. 833–851. Springer, Cham (2018). https://doi.org/10.1007/978-3-030-01234-2_49
6. Chen, X., Ma, H., Wan, J., Li, B., Xia, T.: Multi-view 3d object detection network for autonomous driving. In: IEEE CVPR, vol. 1, p. 3 (2017)
7. Community, B.O.: Blender - a 3D modelling and rendering package. Blender Foundation, Stichting Blender Foundation, Amsterdam (2018). http://www.blender.org
8. Goel, K., Srinivasan, P., Tariq, S., Philbin, J.: QuadroNet: multi-task learning for real-time semantic depth aware instance segmentation. In: Proceedings of the IEEE/CVF Winter Conference on Applications of Computer Vision (WACV), January 2021, pp. 315–324 (2021)
9. Goldman, E., Herzig, R., Eisenschtat, A., Goldberger, J., Hassner, T.: Precise detection in densely packed scenes. In: Proceedings of the Conference on Computer Vision and Pattern Recognition (CVPR) (2019)
10. Hartley, R.I., Zisserman, A.: Multiple View Geometry in Computer Vision, 2nd edn. Cambridge University Press (2004). ISBN 0521540518
11. He, K., Zhang, X., Ren, S., Sun, J.: Deep residual learning for image recognition. arXiv preprint arXiv:1512.03385 (2015)
12. Jin, L., et al.: Geometric structure based and regularized depth estimation from 360 indoor imagery. In: Proceedings of the IEEE/CVF Conference on Computer Vision and Pattern Recognition (CVPR), June 2020 (2020)
13. Lin, T.Y., et al.: Microsoft COCO: common objects in context (2014). http://arxiv.org/abs/1405.0312
14. Liu, X., Lathrop Jr., R.G.: Urban change detection based on an artificial neural network. Int. J. Remote Sens. **23**(12), 2513–2518 (2002)
15. McCormac, J., Handa, A., Leutenegger, S., J. Davison, A.: SceneNet RGB-D: can 5M synthetic images beat generic imagenet pre-training on indoor segmentation? (2017)
16. Silberman, N., Hoiem, D., Kohli, P., Fergus, R.: Indoor segmentation and support inference from RGBD images. In: Fitzgibbon, A., Lazebnik, S., Perona, P., Sato, Y., Schmid, C. (eds.) ECCV 2012. LNCS, vol. 7576, pp. 746–760. Springer, Heidelberg (2012). https://doi.org/10.1007/978-3-642-33715-4_54

17. Park, J.M., Jang, J., Yoo, S.M., Lee, S.K., Kim, U., Kim, J.H.: ChangeSim: towards end-to-end online scene change detection in industrial indoor environments. In: 2021 IEEE/RSJ International Conference on Intelligent Robots and Systems (IROS). IEEE (2021). https://arxiv.org/abs/2103.05368
18. Paszke, A., et al.: PyTorch: an imperative style, high-performance deep learning library. In: Wallach, H., Larochelle, H., Beygelzimer, A., d'Alche Buc, F., Fox, E., Garnett, R. (eds.) Advances in Neural Information Processing Systems 32, pp. 8024–8035. Curran Associates, Inc. (2019), http://papers.neurips.cc/paper/9015-pytorch-an-imperative-style-high-performance-deep-learning-library.pdf
19. Ranftl, R., Bochkovskiy, A., Koltun, V.: Vision transformers for dense prediction. arXiv preprint (2021)
20. Ranftl, R., Lasinger, K., Hafner, D., Schindler, K., Koltun, V.: Towards robust monocular depth estimation: mixing datasets for zero-shot cross-dataset transfer. IEEE Trans. Pattern Anal. Mach. Intell. (TPAMI) **44**, 1623–1637 (2020)
21. Roberts, M., et al.: Hypersim: a photorealistic synthetic dataset for holistic indoor scene understanding (2021)
22. Saeedan, F., Roth, S.: Boosting monocular depth with panoptic segmentation maps. In: Proceedings of the IEEE/CVF Winter Conference on Applications of Computer Vision (WACV), January 2021, pp. 3853–3862 (2021)
23. Shi, W., Zhang, M., Zhang, R., Chen, S., Zhan, Z.: Change detection based on artificial intelligence: state-of-the-art and challenges. Remote Sens. **12**(10) (2020). https://doi.org/10.3390/rs12101688. https://www.mdpi.com/2072-4292/12/10/1688
24. Sturm, J., Engelhard, N., Endres, F., Burgard, W., Cremers, D.: A benchmark for the evaluation of RGB-D SLAM systems. In: Proceedings of the International Conference on Intelligent Robot Systems (IROS), October 2012 (2012)
25. Uhrig, J., Schneider, N., Schneider, L., Franke, U., Brox, T., Geiger, A.: Sparsity invariant CNNs. In: International Conference on 3D Vision (3DV) (2017)
26. Varghese, A., Gubbi, J., Ramaswamy, A., Balamuralidhar, P.: ChangeNet: a deep learning architecture for visual change detection. In: Leal-Taixé, L., Roth, S. (eds.) ECCV 2018. LNCS, vol. 11130, pp. 129–145. Springer, Cham (2019). https://doi.org/10.1007/978-3-030-11012-3_10
27. Wang, K., Shen, S.: MVDepthNet: real-time multiview depth estimation neural network. In: International Conference on 3D Vision (3DV), September 2018 (2018)
28. Wang, L., Zhang, J., Wang, O., Lin, Z., Lu, H.: SDC-depth: semantic divide-and-conquer network for monocular depth estimation. In: Proceedings of the IEEE/CVF Conference on Computer Vision and Pattern Recognition (CVPR), June 2020 (2020)
29. Zhao, Y., Kong, S., Shin, D., Fowlkes, C.: Domain decluttering: simplifying images to mitigate synthetic-real domain shift and improve depth estimation. In: Proceedings of the IEEE/CVF Conference on Computer Vision and Pattern Recognition (CVPR), June 2020 (2020)

A Strong Geometric Baseline for Cross-View Matching of Multi-person 3D Pose Estimation from Multi-view Images

Sam Dehaeck[(⊠)] , Corentin Domken , Abdellatif Bey-Temsamani ,
and Gabriel Abedrabbo

Flanders Make, DecisionS/ProductionS/CodesignS, Oude Diestersebaan 133,
3920 Lommel, Belgium
sam.dehaeck@flandersmake.be

Abstract. Reconstructing the 3D pose from multiple people viewed from several cameras is a complex problem. After the 2D human pose estimation is performed in each image separately (e.g. by a deep learning algorithm), the next step is matching the individuals on the different images. In the present contribution, a novel non-deep learning and parameter-free algorithm is developed to this end, which is only based on geometrical cues (i.e. reprojection errors). This is shown to achieve perfect precision and recall figures on a challenging 102-person benchmark.

Keywords: Pose estimation · Multi-person 3d pose · Multi-view geometry · Cross-view matching

1 Introduction

A correct localization of all joints of one or multiple humans in a room, is a powerful enabler for multiple end goals. Using a few cameras to monitor a room, one could, for instance, try to enforce a minimum distance of 1.5 m between people for social distancing. Another interesting application is the accurate measurement of joint angles to enable an ergonomic assessment of a working situation.

Recreating the 3D pose of multiple people as seen from a few cameras, is one of those computer vision problems which has been revolutionized by the developments in deep learning. In essence, the problem can be decomposed into the following sub-problems: 1) Joint detection 2) Skeleton Assembly 3) Cross-view Matching 4) Triangulation 5) Temporal Tracking. The order of these steps can be changed or even combined in a single neural network or algorithm. Steps 1 and 2 are solved nowadays by human pose estimation networks, such as [2,5,6]. The output of such networks are the 2D coordinates of the found joints grouped into multiple persons. Step 3 consists of matching the same person in all views. When 2 or more viewpoints are available for a joint, its 3D coordinates can be triangulated in step 4. Finally, for application purposes, it often is also required

S. Sclaroff et al. (Eds.): ICIAP 2022, LNCS 13232, pp. 77–88, 2022.
https://doi.org/10.1007/978-3-031-06430-2_7

to follow a given person over time. In this work, we will elaborate a parameter-free algorithm capable of solving the cross-view matching problem with 100% precision and recall on a challenging 102-person dataset. Our algorithm is not specific to human pose detection and thus can work for the matching of any set of keypoints.

2   Related Work

As our study will focus on the cross-view matching problem, we will limit our literature study to solutions that have addressed this topic for a multi-view multi-person setting. Early work, focused on combining cross-view matching and triangulation in a 3D pictorial structure model [4]. Belagiannis et al. [1] extended the 3D pictorial model to multiple people and reduced the 3D state space which needs to be discretized. To this end, all possible pairings of humans and views are triangulated and only the space this occupies is discretised. For a low number of people present in a scene, this reduction in space can be substantial. In order to eliminate the false pairings, penalty terms connected to joint confidence, reprojection error, visibility in multiple views and temporal consistency are formulated, next to the regular constraints inherent to a pictorial model. However, all these error terms need to be weighted and this is learnt for each setup separately. This step introduces a setup-dependency in the system, which requires a hand-labelled scene with multiple people. Our technique does not require such manual labelling. In addition, their approach remains quite costly and scales to the power 6 with the discretisation of 3D space.

Another new promising approach is being investigated in which cross-view matching is tackled together with the 2D pose estimation in an end-to-end learnable approach (e.g. [8,10,12]). Here, an 'unprojection' layer is created which accumulates all the features (or heatmaps) from the individual views after deprojection and a '3D' network then pinpoints the joint locations directly in 3D space (or even in 4D-space as tracking in time is included in [10]). While clearly this seems like the golden target, there is a risk that training could be setup specific, especially when taking temporal evolution into account. In addition, while having relevant multi-view training data is perhaps possible (in a limited amount) for tracking humans, this is likely not the case when applying 3D pose estimation to other, more industrially relevant, data.

Instead, our approach will handle the cross-view matching step separately, starting from an off-the-shelf human pose detection algorithm [6]. Solving the cross-view matching problem based on only the extracted joint coordinates and joint groupings, has also been studied in literature before. In [7], the authors performed the association between the 2D poses, based on a cost calculation consisting of geometric and appearance (or re-ID) terms. In [3], the authors perform a greedy algorithm in which intersections are upgraded with the best-fitting pose correspondences incrementally (so starting from pairs and ending with N-view correspondences). The merge candidates are first sorted based on a cost which is proportional to a geometric reprojection term and inversely proportional to the amount of common joints and the amount of views already contained in the candidate. Cycle consistency is also checked and a threshold is used to validate the merge. In [11], a similar greedy algorithm is proposed based on upgrading the best pairs to the maximum amount of views possible. The cost is only dependent on the average reprojection error between the new view and the already present views (thus not inversely proportional to the amount of joints or combination order). In [13], the assignment problem is solved on the three levels at the same time; grouping joints into skeletons per image, cross-view matching and temporal tracking. The per-frame parsing cost follows from the used human pose detection algorithm. For cross-view matching, a geometric cost was calculated per joint based on epipolar distance (line-to-line distance), whereas for temporal tracking, the line-to-point distance was calculated with respect to the 3D triangulated skeletons determined in the previous time-step. Combining these inherently differently scaled costs, leads to empirically defined weighing factors. Nevertheless, state-of-the-art results were obtained in several benchmarks.

Our work is closely related to [3,7,11,13]. However, as opposed to [7], our work does not include a re-ID network to perform matching based on clothing similarity. We will demonstrate that our simple formulation of the geometric cost is sufficient to obtain perfect precision scores on a synthetic dataset. This simplification allows to remove many ad-hoc parameters and allows to use the algorithm e.g. in sports settings or blue-collar workfloors. Compared to [3] and [11], our algorithm is not greedy and actually builds assignments from N-view correspondences down to lower orders (so reverse order). We will demonstrate how starting from pairs is prone to yielding wrong assignments, which supposedly would undermine this algorithm in more difficult testcases. Finally, compared to [13], our algorithm does not require any empirical parameters, whereas [13] requires them to balance the different cost functions. Note how such factors would inevitably be setup-dependent (e.g. the temporal tracking cost should clearly scale with the fps of the video). In addition, the canonical part on which they build their cost calculations is a limb, i.e. a combination of two joints. As we will show in our work, the discriminative power for the geometric cost decreases dramatically when the amount of keypoints considered decreases.

3   Material/Methods

3.1   Dataset

A synthetic dataset was created in Unity, in which a single person with a fixed pose is placed (and rotated) randomly in a 8×6 m room (102 positions recorded), see e.g. Fig. 1. For this dataset, 4 cameras are used. In order to test only the cross-view matching algorithm independently from the 2D human pose detection problem, we will also analyze the correct assignment of all 102 detected skeletons in the different views. Figure 1, overlays all the (Unity) poses from above, showing that this dataset is more difficult than could be achieved in a real situation due to the big overlap.

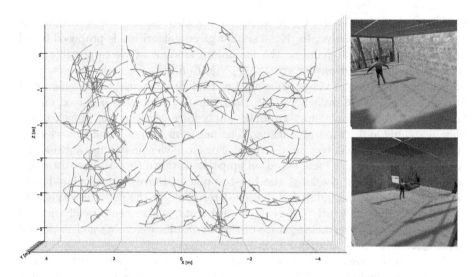

Fig. 1. Sample images from the Unity dataset and all skeletons combined in top-view

3.2 Human Pose Detection

HRNET. One of the human pose detection networks that is used in this work is Higher HRNET [6]. The codebase found online[1] was used.

TRTPOSE. This pose detection network does principally the same as HRNET, however, it is capable of performing in real-time on embedded Nvidia-Jetson boards [2,5]. The codebase used can be found online[2].

Average Head. In order to demonstrate that having multiple keypoints is preferable for having a good assignment, we will also reduce the found keypoints using HRNET to a single keypoint corresponding to the averaged 'head' location. This means that the locations of the first 5 keypoints (L-R ear, L-R eye and nose) are averaged to a single value.

3.3 Cross-View Matching Algorithm

As mentioned, the basis for our cross-view matching algorithm will be purely geo-metrical and is based on reprojection errors. As described in [9], the Sampson distance[3] is a better error metric to characterise the reprojection error than certain others. Conceptually, what is being captured in this metric, is the distance between the detected points in two views, taking into account their calibration.

---

[1] https://github.com/HRNet/HigherHRNet-Human-Pose-Estimation.
[2] https://github.com/NVIDIA-AI-IOT/trt_pose.
[3] The version implemented in OpenCV is used.

The lower this metric, the more confident one could be about the two joints forming a correct match.

Of course, from this simple metric defined per joint and per camera pair, there are many different ways in which to compute a per-body error or a multi-view error. Clearly, it is in this step, that our contribution differs mostly from other works in literature. One possibility is to extend the notion of a fundamental matrix to a system of 3 (or more) cameras (see [9]), and error metrics could be calculated for a three-way pairing for instance. However, in the developed method, everything is based on Sampson distances calculated for pairs of cameras. The flow of the algorithm is as follows for finding 4-way matches (quartets):

1. Calculate the Sampson error averaged out over all common joints for all possible person-view combinations and store this in a hashlist (with the camera+person index as hash, e.g. 'A1-B4' denoting person #1 in viewpoint A with person #4 in viewpoint B). If no common joints are found between two views, then try to calculate the Sampson error based on the averaged head-keypoint (details given in Sect. 3.2). If the head-keypoint cannot be calculated as well, then don't register this combination.
2. Knowing the amount of people visible in each view, we can create the full list of quartets that can be created (using e.g. the numpy.meshgrid method). Notice that the length of this list scales as the amount of people to the power 4 (for triplets it is to the power 3).
3. For each combination in this list, create the list of person-view pairs that can be constructed from it and look up those costs in the hash list. For instance, when examining the quartet A1-B4-C2-D3, we need to look up the pairs A1-B4, A1-C2, A1-D3, B4-C2, B4-D3 and C2-D3. Once the cost of all six of them are obtained, average them out and store this in a new array. If a pair cost is not available in the hash list (due to not having common joints), then set the cost to a value that can easily be filtered out in the next step (e.g. NaN). Note that by looking up all the possible pairs, we take care of the cycle-consistency requirement.
4. Sort the obtained cost list and try to perform the assignment from low to high errors. An assignment of a particular combination is only successful if all person-views it holds are unassigned so far. If, for instance, an assignment fails because one person-view has been previously taken, the other, previously untaken person-views are still switched to 'unavailable' for the remainder. The underlying reason being that any follow-up assignment is considered a 'second choice', which creates a reasonable doubt to its trustworthiness.

Now, from this basis, there are two strategies that can be formulated:

– The **optimal** strategy that avoids false positives as much as possible, will start by calculating the highest combination orders first and then continues with the lower orders (i.e. top-down). So, for a 4-camera setup, this starts by assigning all possible quartets. The remaining people-views that are not assigned, are then examined for valid triplets (i.e. steps 2–4). Pairs are avoided, for reasons which will be explained in Sect. 4.2.

– The **greedy** strategy that maximizes speed, works bottom-up. However, in contrary to literature where assignments start at the pair-level [3, 7], we will only start at the triplet-level. For the valid triplets, extra views to add are searched for in the remaining person-views. The list of possible quartets for a given triplet are then sorted in costs (c.f. step 3) and the smallest one is chosen if the cost falls below a given threshold. This completion step can be repeated multiple times, each time creating a higher order combination. However, note that the starting order needs to be larger than half the amount of cameras. Another important difference with the optimal strategy is that the greedy algorithm requires a single parameter, where the optimal one needs none. However, this parameter can be determined from looking at the distribution of good triplet costs (e.g. equal to the largest assigned cost) and hence can be deduced for instance from a video with a single person walking the room.

Remark that the fallback to the averaged head keypoint is important so as to increase the possibility of forming higher order intersections. Indeed, a quartet should have all 6 of its pairs return valid Sampson distances. Dropping one pair because there are no common joints is quite drastic and falling back to a single keypoint for one of the person-view pairs is preferable.

4   Results

4.1   Detection Precision

The first topic we will investigate is the Sampson error per joint so as to have a good indication of the precision of the detection. In Fig. 2(a), the histogram is shown of the Sampson distance calculated for all keypoint pairs for all camera combinations and images of the Unity Dataset. All distributions appear to be half-normal, with the head keypoint having the narrowest distribution. This is also logical as it is the average of a few keypoints picked from the HRNET distribution. The TRTPOSE detector clearly has the largest errors. The origin of this difference is elucidated in Fig. 2(b). On the x-axis of this figure, the size of the rescaled image is indicated before inference. Rescaling is a common way to increase inference speed with (hopefully) a minimal accuracy impact. TRT-POSE works nominally at 224×224 pixel images, whereas HRNET operates at 512×512 pixels (or at least pretrained weights are available for this resolution). The dashed lines, which are plotted versus the left axis, indicate the 95% percentile of the Sampson error distribution for all joints. We note that the two detection algorithms are actually very similar in their precision and hence, the effect we notice in (a) is entirely due to the rescaled resolution difference (TRT-POSE calculated at 224×224 versus HRNET at 512×512). Summarizing, the Sampson error for two joints can easily reach multiple pixels and scales inversely with the rescaled image size.

The second topic we wish to investigate is whether there are other important performance differences between the two human pose detection algorithms. The solid lines in Fig. 2(b), which are plotted against the right axis, indicate the average amount of joints detected per person (in all 4×102 images). The solid black line indicates that HRNET has a large zone where the resolution only has a moderate impact on the amount of detected keypoints, from 352 pixels onwards. Below this resolution, performance degrades quickly. TRTPOSE has a more or less similar working region, but its plateau is noticeably lower than that of HRNET and less flat. This indicates that it recognizes less joints (or less people as the average is calculated) and it is also more resolution dependent. Note, of course, that it is likely that the particular value of 352 pixels is dependent on the scene configuration and the average person size (around 100 pixels 'high'). Another sidenote is that the precise behaviour in Fig. 2(b) above a size of 512 pixels could be influenced by the fact that these are upscaled images (original image resolution is 512×512) and as such, resolutions above 512 present 'empty magnification', which can hinder a further useful decrease in the errors.

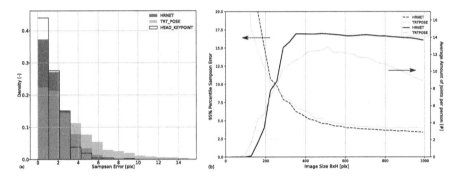

Fig. 2. (a) Comparison of the Sampson distance histogram for all keypoint-pairs (extracted from all image pairs in the Unity Dataset) for the different keypoint detectors considered. (b) 95%-percentile and average amount of keypoints per person versus rescaled image size for the different detection algorithms.

4.2 Cross-View Matching

For demonstrating the precision and recall of our assignment algorithm under challenging conditions, we will reinterpret the Unity dataset as consisting of a scene with 102 people imaged from 4 different cameras. As the starting point will be the raw 2D human pose detections obtained in single-person views, obstruction or difficulties in joint grouping due to large overlaps are avoided. In essence, this decouples the performance of the cross-view matching algorithm from the performance of the multi-person 2D detector, while still taking into account a realistic localisation precision.

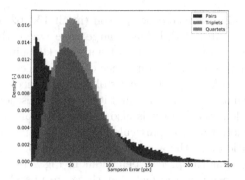

Fig. 3. Density histogram of Sampson error for bad combinations compared for pairs, triplets and quartets.

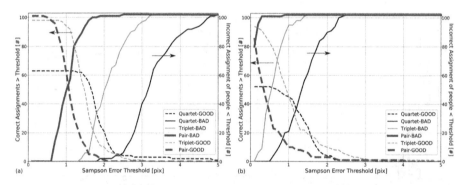

Fig. 4. (a) False negative and false positive assignments for a given Sampson error threshold for a full keypoint detection with HRNET (b) Same plot as (a) but for a single head keypoint.

Before describing the performance of our specific assignment algorithm, we will first show the behaviour of our error metric for bad associations. In Fig. 3, the density distribution of the Sampson per-person error over all 'bad' pairs, triplets and quartets is shown. This indicates that the pair distribution resembles a log-normal distribution with a low probability of having zero error and it only gradually declines to a value of 250 pixels (to be compared to the image size of 512 pixels). Now, as the mean Sampson error for a triplet A1B4C2 is calculated as the average of the pairs A1B4, B4C2 & A1C2, we can simply see the distribution for the triplets as being the average of three random points taken from the pairs distribution. Now, as the law of large numbers states, the formed distribution will approach the average of the underlying distribution more and more as we average out more points. As a quartet is the average of 6 pairs, we can see why the standard distribution is even smaller here. The beneficial effect that this 'sharpening' of the distribution leads to is that the probability of having a bad quartet with a small averaged error becomes very small. This is simply because the probability that all 6 view pairs in a bad quartet have a low Sampson error is very small. The behaviour of the metric for good associations is not shown here, but a similar 'sharpening' around the average in Fig. 2(a) is visible. This implies that a zero total cost becomes unlikely on the one hand, on the other hand the 99% percentile also decreases.

From this result, one can intuitively understand that, as the amount of cameras is increased, the good and bad association distributions will stop overlapping, allowing for a simple separation into good and bad matches. Now, let us try and quantify this amount of overlap versus the amount of cameras. In Fig. 4(a), we simulate the performance of a simple threshold based separation. For a given threshold of Sampson distance (cf. the x-axis), the dashed lines are plotted versus the left y-axis and correspond to the amount of good assignment combinations that fall above this threshold (i.e. False Negatives). The solid lines, on the other hand, are plotted against the right axis and these correspond to the amount of people present in combinations that have an averaged error below the threshold. While this represents the false positives in some measure, it is expressed here in the amount of people tied up incorrectly in order to have a similar scale for pairs, triplets and quartets. Now, as can be clearly noted, for quartet combinations, a threshold value of 2.5 pix would lead to a very low false negative count and also a low false positives count. For triplets, the situation is slightly worse but for pairs, the crossover happens at almost 50%! One more thing to note in the graph concerns the fact that only 62 people can be reconstructed from the quartets (although 102 are imaged). This is due to the fact that the zone in the room where people are visible from all cameras is smaller than the area chosen to randomly distribute the people. We do not believe that this difference in total count between triplets and quartets modifies the observation that quartets have more discriminative power than triplets. Relating this to results in literature, we note that [8] and others also note that using more than 4 cameras leads to only minimal changes in final performance.

The next result we will highlight is the advantage of including multiple keypoints. In Fig. 4(b), the false-positive/negative plot is repeated for the case of only using the averaged head keypoint. This shows that the threshold for avoiding false positives has decreased a lot, making it almost impossible to avoid them for all combination orders. Comparing this to Fig. 4(a), it is clear that having multiple keypoints is instrumental in pushing the error for incorrect assignments to larger values and creating a useable discriminative power.

Now, as these plots have demonstrated, a threshold based approach could work for the HRNET-based keypoint detector and 4 camera views. However, determining the correct threshold in practice is difficult and therefore a parameter-free assignment policy was chosen, which is explained in more detail in Sect. 3.3. The advantage of this approach over a threshold-based technique is that it can deal with an overlapping FP and FN curve.

To test our algorithm, we have quantified the assignment precision and recall on the dataset and tested the influence of basing it on HRNET, Head or TRT-POSE detections (based on 512×512 rescaling). As can be seen in Table 1(b), using only a single keypoint leads to dramatic precision and recall figures. This confirms the expectations from Fig. 4(b), where it was highlighted that good and bad combinations are not separated properly with respect to the mean Sampson error. Our HRNET algorithm based on quartets reaches 100% and 95% recall, which is excellent. Using only triplets, precision and recall stay equally good. However, there are 92 people assigned correctly with triplets, versus only 60 based on quartets. As mentioned this is due to the increased coverage reached when three out of four views is sufficient. Another suspicion that is confirmed, is that a pair of cameras has a significantly lower discriminative power than triplets and quartets. In this case, we already have 20% false positives and a recall of only 77%. The last 2 columns quantify the performance of the 'optimal' and 'greedy' strategy (i.e. top-down or bottom-up). The 'greedy' strategy is a simple copy of the numbers for the triplets. This is understandable as in this algorithm, only extra views are searched for to improve the triangulation accuracy (and completeness), but it is not the goal to find extra people. The 'optimal' strategy on the other hand starts from the quartets and then assigns extra people as all possible triplets from the remaining people-views are examined. This leads to a perfect 100–100% score.

In table (c), the performance of the TRTPOSE algorithm is shown calculated on 512×512 images. On first glance, the precision and recall performance is very competitive to the HRNET network. However, there is a subtle difference. The total coverage between the two networks is smaller for TRT-512 (i.e. true positives + false negatives). Looking at the 'optimal' strategy performance, we note that 83 people were successfully identified out of a total of 91 present in the dataset for TRTPOSE. With HRNET, we identified all 98 people. This is 15% more, which can be important in the field. This effect is tied to the plot in Fig. 2(b), where we noted that TRTPOSE detects, on average, up to 2 joints less than HRNET per person. This can lead to two views not sharing a common

Table 1. Precision and Recall of the assignment algorithm for (a) HRNET full keypoint detection (b) Head keypoint detection (c) TRT-POSE full keypoint detection based on 512×512 images.

	HRNET	Quartets	Triplets	Pairs	Optimal Str	Greedy Str
(a)	Precision	$60/60 = 100\%$	$92/92 = 100\%$	$79/96 = 82\%$	$98/98 = 100\%$	$92/92 = 100\%$
	Recall	$60/63 = 95\%$	$92/98 = 94\%$	$79/102 = 77\%$	$98/98 = 100\%$	$92/98 = 94\%$

	Head-KP	Quartets	Triplets	Pairs	Optimal Str	Greedy Str
(b)	Precision	$21/28 = 75\%$	$15/31 = 48\%$	$14/85 = 16\%$	$35/52 = 67\%$	$15/31 = 48\%$
	Recall	$21/52 = 40\%$	$15/94 = 16\%$	$14/102 = 14\%$	$35/94 = 37\%$	$15/94 = 16\%$

	TRT-512	Quartets	Triplets	Pairs	Optimal Str	Greedy Str
(c)	Precision	$53/54 = 98\%$	$74/76 = 97\%$	$44/77 = 57\%$	$83/85 = 98\%$	$74/76 = 97\%$
	Recall	$53/54 = 98\%$	$74/91 = 81\%$	$44/102 = 43\%$	$83/91 = 91\%$	$74/91 = 81\%$

joint and therefore not being present in the data for the assignment algorithm to pick up.

Finally, with respect to performance aspects of both algorithms; the greedy approach is based on triplets and hence scales as the amount of people to the power 3. The optimal strategy scales with a power 4. For matching a 10-person scene, a single-threaded Python implementation requires 0.4 s for the optimal strategy and 0.14 s for the greedy strategy, making both approaches real-time.

5   Conclusion

In this paper we have developed a cross-view matching algorithm, based on geometric Sampson errors only, which reaches 100% precision and recall in a synthetic dataset containing 102 people in a single room. This proves that knowing the position of the keypoints in at least 3 views is enough to solve the cross-view matching problem and that no appearance information nor temporal tracking is necessary. When the number of detected keypoints decreases or only 2 unobstructed views are available, the reliability decreases, as was also demonstrated in the paper.

Acknowledgements. We would like to acknowledge the help of E. Kikken for the many interesting discussions. We also thank E. Hage for the construction of the Unity dataset. This research received funding from the Flanders Make '2018-134-Ergo-Eyehand-CONV-ICON' and '2018-148-Pils-CONV-SBO' projects and from the Flemish Government under the "Onderzoeksprogramma Artificiele Intelligentie (AI) Vlaanderen" program.

References

1. Belagiannis, V., Amin, S., Andriluka, M., Schiele, B., Navab, N., Ilic, S.: 3D pictorial structures revisited: multiple human pose estimation. IEEE Trans. Pattern Anal. Mach. Intell. **38**(10), 1929–1942 (2016). https://doi.org/10.1109/TPAMI.2015.2509986

2. Xiao, B., Wu, H., Wei, Y.: Simple baselines for human pose estimation and tracking. In: Ferrari, V., Hebert, M., Sminchisescu, C., Weiss, Y. (eds.) ECCV 2018. LNCS, vol. 11210, pp. 472–487. Springer, Cham (2018). https://doi.org/10.1007/978-3-030-01231-1_29

3. Bridgeman, L., Volino, M., Guillemaut, J.: Multi-person 3d pose estimation and tracking in sports. In: Conference on Computer Vision and Pattern Recognition (CVPR) (2019). https://doi.org/10.1109/cvprw.2019.00304

4. Burenius, M., Sullivan, J., Carlsson, S.: 3d pictorial structures for multiple view articulated pose estimation. In: Computer Vision and Pattern Recognition (CVPR) (2013). https://doi.org/10.1109/cvpr.2013.464

5. Cao, Z., Simon, T., Wei, S., Sheikh, Y.: Realtime multi-person 2d pose estimation using part affinity fields. In: IEEE Conference on Computer Vision and Pattern Recognition (2017). https://doi.org/10.1109/cvpr.2017.143

6. Cheng, B., Xiao, B., Wang, J., Shi, H., Huang, T., Zhang, L.: HigherHRNet: scale-aware representation learning for bottom-up human pose estimation. arXiv arxiv:1908.10357v3 (2019). https://doi.org/10.1109/cvpr42600.2020.00543

7. Dong, J., Jiang, W., Huang, Q., Bao, H., Zhou, X.: Fast and robust multi-person 3d pose estimation from multiple views. In: Conference on Computer Vision and Pattern Recognition Workshops (CVPR) (2019). https://doi.org/10.1109/cvpr.2019.00798

8. Elmi, A., Mazzini, D., Tortella, P.: Light3DPose: real-time multi-person 3d pose estimation from multiple views. arXiv arXiv:2004.02688v1 (2020). https://doi.org/10.1109/icpr48806.2021.9412652

9. Hartley, R., Zisserman, A.: Multiple View Geometry in Computer Vision. Cambridge University Press (2004). https://doi.org/10.1017/cbo9780511811685

10. Reddy, N., Guigues, L., Pishchulin, L., Eledath, J., Narasimhan, S.: TesseTrack: end-to-end learnable multi-person articulated 3d pose tracking. In: Conference on Computer Vision and Pattern Recognition (CVPR) (2021). https://doi.org/10.1109/cvpr46437.2021.01494

11. Tanke, J., Gall, J.: Iterative greedy matching for 3d human pose tracking from multiple views. In: German Conference on Pattern Recognition (2019). https://doi.org/10.1007/978-3-030-33676-9_38

12. Tu, H., Wang, C., Zeng, W.: VoxelPose: towards multi-camera 3d human pose estimation in wild environment. arXiv arXiv:2004.06239v4 (2020). https://doi.org/10.1007/978-3-030-58452-8_12

13. Zhang, Y., And, L., Yu, T., Li, X., Li, K., Liu, Y.: 4d association graph for real-time multi-person motion capture using multiple video cameras. In: CVPR (2020). https://doi.org/10.1109/cvpr42600.2020.00140

Multi-view 3D Objects Localization from Street-Level Scenes

Javed Ahmad[1,2]([✉]) [iD], Matteo Toso[1] [iD], Matteo Taiana[1] [iD], Stuart James[1] [iD], and Alessio Del Bue[1] [iD]

[1] Pattern Analysis and Computer Vision (PAVIS),
Istituto Italiano di Tecnologia (IIT), Genova, Italy
{javed.ahmad,matteo.toso,matteo.taiana,stuart.james,
alessio.delbue}@iit.it
[2] Department of Electrical, Electronic and Telecommunications Engineering,
and Naval Architecture (DITEN), Universita degli Studi di Genova, Genoa, Italy

Abstract. This paper presents a method to localize street-level objects in 3D from images of an urban area. Our method processes 3D sparse point clouds reconstructed from multi-view images and leverages 2D instance segmentation to find all objects within the scene and to generate for each object the corresponding cluster of 3D points and matched 2D detections. The proposed approach is robust to changes in image sizes, viewpoint changes, and changes in the object's appearance across different views. We validate our approach on challenging street-level crowd-sourced images from the Mapillary platform, showing a significant improvement in the mean average precision of object localization for the available Mapillary annotations. These results showcase our method's effectiveness in localizing objects in 3D, which could potentially be used in applications such as high-definition map generation of urban environments. The code is publicly available (https://github.com/IIT-PAVIS/Multi-view-3D-Objects-Localization-from-Street-level-Scenes).

Keywords: Street-level objects · Point-cloud · Mapillary · Localization

1 Introduction

The detection of common objects within an urban area from a set of noisy crowd-sourced images has a wide range of interesting applications, such as making better maps, creating a safer traffic environment, visualizing places and stories, and the municipal management [2,18]. Nowadays, for many cities there is an abundance of publicly available, crowd-sourced images [2,3] captured with consumer cameras, and modern state-of-the-art 2D object detectors [5] can easily detect common objects (e.g. traffic signs, benches, ...). At the same time, these images

Supplementary Information The online version contains supplementary material available at https://doi.org/10.1007/978-3-031-06430-2_8.

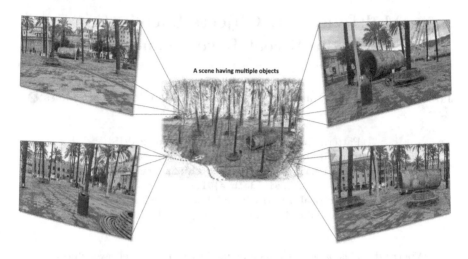

A scene having multiple objects

Fig. 1. A scene from Porto Antico (Genoa). We show how detections of common urban objects can be matched across different views. An example of how our method works. The images used in this figure are from Mapillary (https://www.mapillary.com/).

can be used to create sparse 3D points by employing structure from motion (SfM) pipelines [27]. We aim to leverage these two sources of information, combining SfM sparse point clouds with instance segmentation to accurately localize the objects present in a scene, and to match 2D instances of the same object across views, as shown in Fig. 1.

While some prior works have attempted a similar process, they focus on a specific class of objects either trees or traffic signs, exploit temporal constraints by using a set of sequential frames, or require limited changes in viewpoint across images, meaning they have limited applicability. These works exploit existing GPS measurements to localise objects, as the GPS localisation is known to be inaccurate[12–14, 16, 18, 19, 29, 33]. The task of localizing and matching from crowd-sourced images faces many challenges: for example, the images are captured in different illumination conditions (even day/night), using different devices (different camera models and intrinsics, different resolution), and of different levels of quality (blurred images). Moreover, urban settings present challenges of their own, like the presence of multiple objects of the same class in the same scene, high variability in object scale, and a lot of occlusions due to moving objects (cars, pedestrians). Our method uses the 3D sparse points instead of GPS position while localizing objects in a scene, which handles these challenges robustly.

Recently, a massive amount of street-level views are available everywhere by Mapillary [2] (a company that provides services based on crowd-sourced images). We use our method to process from small to large-scale scenes based on street-level views and to localize static objects given image-level instance segmentation provided by Mapillary. We choose static street-level objects, for example, traffic signs, traffic lights, street lights, benches, manholes, trash cans, poles, etc., whose 2D detections are provided by Mapillary.

Our approach follows three main steps, (i) pre-processing, (ii) object localization via clustering, (iii) 2D detections matching. In the preprocessing step, we select all Mapillary images available for a given area, generate sparse point clouds using Colmap [27], and extract all instances belonging to the chosen object classes. In the second step, we select only the 3D point cloud elements whose projection on the 2D images lies within the selected instances. Applying density-based clustering on the scene at hand, we obtain 3D clusters representing the location of objects in the scene. In the last step, we use these clusters to match the 2D instances associated to the same 3D object.

Our contributions are the following:

- Formation of a multi-view 3D objects localization pipeline that uses 3D sparse reconstruction by colmap [27] and instance segmentation available with images to localize objects in a given scene.
- Evaluation of the proposed method on the challenging Mapillary street-level scenes formed using Mapillary Python SDK [1] and comparison against the Mapillary method of objects derived from multiple detections in multiple images.
- Manual 3D object annotation of 2 scenes. Since there is no method available that detects enough classes of street-level objects in SfM sparse point clouds, hence, we annotate two Mapillary scenes (small and large) in Blender [6].

2 Related Work

Our work is related to different research domains in computer vision, such as street objects detection and localization, structure from motion with objects, and 3D object detection in SfM point cloud. Briefly, we highlight the most relevant research in this section.

Street Objects Localization and Matching: A related topic is finding the refined geo-location of objects within an area using multi-view detections. For example, Wegner et al. in [7,29] propose a method based on the conditional random field to geo-locate trees detected in image sequences for surveying purposes. Zhang et al. [34] geo-locate poles using modified brute-force line-of-bearing approach. Other works instead match the objects back in images after refined geo-locations [8,15,18,36]. Localizing objects from small to large scenes containing the same class of multiple objects is a challenging task, hence, we localize the objects by finding enough 3D points in the scene. It provides robustness while matching back the 2D detections as well.

Structure from Motion with Objects: Another related field of work is structure from motion with objects (SfMO); particularly deep neural network-based approaches expect a large amount of training and testing data, and our method is a good fit for this. For example, instead of key points matches in the structure from motion, the use of object representation is more robust to face challenges such as handling noise, different illumination changes, or views having less co-visibility. Initially, [9] proposed SfMO and shows simultaneous recovery of camera poses and 3D occupancy of objects as ellipsoids using 2D detections and matched feature points. Later works [11,20,30] follow similar approaches to estimate cuboids or ellipsoids using only 2D bounding boxes produced by typical object detection approaches such as [17,25,26].

3D Object Detection: The field most closely linked to our work is 3D object detections, and we highlight here the main contributions. Ahmed et al. in [4] proposes 3D object detection in point clouds using density based clustering. The PointNet [23], PointNet++ [24], VoteNet [22], H3DNet [35] are famous for their feature-learning based approaches and ImVoteNet [21] proposes a multi-modal approach i.e. lifting features from 2D to 3D to boost 3D detection. CenterPoint [31] proposes a center-based representation for 3D object detection and tracking. Pseudo-LiDAR [28,32] based methods produce a virtual point cloud from RGB images with noisy stereo depth estimates. In contrast to the methods mentioned above, our work uses SfM created sparse point clouds to localize objects in the scene.

3   Method

In this section, we describe each step of our method. In the first step, we create sparse point clouds from the multi-views images and extract the desired instances from those images. In the second step, we find objects' locations in 3D, in other words, the task of objects localization by clustering 3D points in the scene. In the last step, we use localized objects to perform 2D matching. Figure 2 shows our complete proposed approach step by step and the subsequent sections contain the detail of each step.

Pre-processing: We consider only the static objects available in the multiple images such as benches, street lights, poles, traffic signs, or manholes. Mapillary [2] provides instance segmentation for each image, or can also be acquired by an off-the-shelf 2D object detector, e.g. (YOLOv5 [5]). We extract the relevant instance segmentations as polygons $\mathcal{G}$ for all images, and compute their 2D bounding boxes $\mathcal{B}$ by looking for the smallest axis-aligned bounding box enclosing all vertices. We then use Colmap [27] to generate a sparse point-cloud reconstruction of the scene; we use Colmap as it is the gold standard for large scene SfM. The 3D points, camera poses, and 3D-2D association computed by colmap are accurate enough for the next step of our purposed method.

Find Objects in 3D: For the given scene $\mathcal{S}$, we have a set of 3D points $\mathcal{P} = \{P_i \mid i = 1, \ldots, n\}$, a set of images $\mathcal{I} = \{I_j \mid j = 1, \ldots, m\},]$ and a set of 3D to 2D associations $\mathcal{A} = \{A_{ij} \mid ij = 1, \ldots, n \times m\}$ representing where 3D point i is observed on image j. Moreover, we have $\mathcal{G}$ or $\mathcal{B}$ as well. We then want to segment the scene in a set of semantically meaningful objects $\mathcal{O} = \{O_k \mid k = 1, \ldots, r\}$.

We start by selecting scene elements whose 2D projection onto the image plane via operator Π lie within the available boundaries of polygons i.e. for each

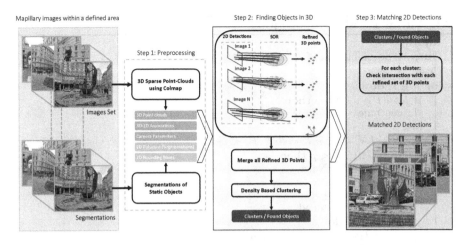

Fig. 2. Method: given multiple images of a scene with instance segmentation, generate a sparse reconstruction via SfM and select the desired instance (Step1). Leverage 3D-2D association, select the 3D points within the boundary of each instance, apply SOR to remove isolated points. Apply DBSCAN [10] on the scene at hand and get 3D object locations (Step 2). Use clusters to matched the 2D detection (Step 3). The images used in this figure are from Mapillary.

image j we select the only 3D points from P whose $A_{ij} = \Pi P_i$ is with in the selected polygons.

$$\hat{\mathcal{P}} = \{P_i : P_i \in \mathcal{P} \ \& \ \Pi P_i \in \mathcal{G}\}. \tag{1}$$

While selecting the 3D points (associated 3D points to a polygon) from $\mathcal{P}$, we also apply statistical outlier removal (SOR) to remove far/noisy 3D points selected because of image segmentation errors. We then represent as $\hat{\mathcal{P}}$ the union of all refined 3D points from all polygons. This allows to significantly reduce the size of the point cloud, keeping only the points relevant for the task at hand. Despite this first screening process, the point cloud still has some noisy 3D points. Moreover, we do not know *a priori* the total number r of objects present in the scene. Therefore, we use an unsupervised scanning method: the density-based scanning approach DBSCAN [10], for which we provide pseudocode in Algorithm 1. This algorithm depends on two hyperparameters: the minimum density of 3D points per object cluster ρ, and a nearest-neighbor maximal distance ϵ. Since the locations wherever objects are present in $\hat{\mathcal{P}}$ are highly denser than the rest part, we keep the default hyperparameter values of DBSCAN that works best to cluster 3D points as an object in our case. This process results in an initial cloud segmentation, that we then refine and get $\mathcal{O}$ after discarding objects that are either not associated with any 2D detection (it happens when a cluster formed on noisy points) or if any cluster has less than 2 points.

Matching 2D Detections: For each object in $\mathcal{O}$, we check the intersection with 3D point sets associated with polygons. If overlap occurs at-least 50%,

---

Algorithm 1: DBSCAN algorithm

Input: $\mathcal{P}$, ρ, ϵ
Output: Segmentation of $\mathcal{P}$
cluster id $= 0$;
for $P \in \mathcal{P}$ **do**
  if $label(P)$ *not None* **then**
  | continue
  else
    $\mathcal{N} = \{N \mid N \in \mathcal{P}\ \&\ dist(N,P) < \epsilon\}$
    if $density(\mathcal{N}) < \rho$ **then**
      | $label(P) =$Noise
      | continue
    cluster id $+= 1$
    $label(P) =$cluster id
    for $N \in \mathcal{N}$ **do**
      if $label(N)$ *is Noise* **then**
      | $label(N) =$ cluster id
      $\mathcal{M} = \{M \mid M \in \mathcal{P}\ \&\ dist(M,N) < \epsilon\}$
      if $density(\mathcal{M}) \geq \rho$ **then**
      | $\mathcal{N} += \mathcal{M}$

---

we consider both as the same object and assign O_k object's label to the corresponding polygon. We then consider matched all instance segmentation polygons associated to the same 3D point cluster.

4 Experiments

We check how our method performs computationally and qualitatively by running on several small and large-scale scenes, based on Mapillary street-level images from various European cities (Barcelona, Berlin, Lisbon, Paris, Genoa, Trondheim, Vienna). We choose these areas based on a higher density of images in the region and use the latest release of Mapillary Python SDK [1] to download images and metadata. Figure 3 shows an example of an area of a 50-m radius and an image from that area showing instance segmentation. Figure 4a shows the $\hat{\mathcal{P}}$ sparse scene just before clustering, and the Fig. 4b shows the DBSCAN-based sparse point cloud clustering. Figure 5 contains some examples of localized objects that appeared into multiple 2D detections, and it shows how objects with very high differences in scale, viewpoint, or even texture are still matched as the same object. Qualitatively, it shows promising results considering the challenges we are concerned about. In the case when objects are very close and due to considerable noisy points, DBSCAN clusters them as one object, it becomes so hard to distinguish in the matching step, as shown in Fig. 6.

We randomly selected 22 scenes (small and large) from the cities mentioned above and observe the processing time (except colmap) between 30 to 200 s depending on the size of the scene. We run our proposed pipeline on a single NVIDIA GeForce GTX 1180 GPU. Extensive visualization of each step while

Fig. 3. Left: An area of radius 50 m in Porto Antico, Genoa. Magenta dots represent the images locations. Right: An image from scene showing selected instances. The image in this figure is taken from Mapillary. (Color figure online)

(a)   (b)

Fig. 4. A scene from Porto Antico, Genova. (a) represent 3D scene after refined point-cloud, (b) represent 3D clusters as objects after employing DBSCAN.

processing a scene and further details about the considered scenes is available in the supplementary material.

Evaluation and Comparison: We evaluate our method on the scenes covering enough object classes either in a small or in a large scene. To do so, we first must produce ground-truth (GT) 3D object detection. We do this by manual annotation using Blender [6] to draw 3D bounding boxes in the sparse 3D reconstruction of Porto Antico, Genoa, and Marxergasse, Vienna. Note that we annotate only the instances and give the class labels available in Mapillary, and these objects have been localized by Mapillary baseline [1] as well, where only 2D matches are available. Our total numbers of 3D GTs and detected clusters are 52 and 82, 133 and 203, in Porto Antico and Marxergasse, respectively. First, we check the 3D intersection over union (IOU) between the GT 3D bounding boxes and the bounding boxes of our clusters. From this, we compute true positives (TP), false positives (FP), and false negatives (FN) rations. We finally

(a) Vienna. Examples of localized advertisement(19), streetlight(26) and traffic signs(62)(3)

(b) Genoa Porto Antico. Some examples of localized bench (8) and manhole (32)

(c) Genoa Largo XII Ottobre. Some example of localized advertisement (10) and support pole (34)

Fig. 5. Three different scenes to show how our method forms very well in localization in the scene that instance can be matched even with changed texture and size, or different light conditions, or even challenging camera viewpoints. The images used in this figure are from Mapillary.

Fig. 6. It is an example when our method fails to label the object correctly. It happens for the instances very close to each where their 3D points sufficiently overlap. The images used in this figure are from Mapillary.

Fig. 7. In the first and second row, correctly projected GT 3D boxes (in green color) and matched 2D detection (in blue color) after localized objects by our method has been plotted together. The third row shows an unintended localized object, this happens because of the wrong image instance's label. The images used in this figure are from Mapillary. (Color figure online)

compute the precision and recall, as a measure of how correct the algorithm is when localizing objects in 3D.

We then project these 3D annotations on the images, to provide 2D GT instances for each object. Figure 7 shows examples of projected GT boxes and 2D boxes of localized objects by our method together, also an example of when an unintended object can appear in evaluation. Having obtained GT 2D annotations, we proceed to evaluate our model's performances, where we measure the mean average precision (mAP) on the 2D matches using 2D GT instances. We compare against Mapillary's baseline of objects derived from multiple detections in multiple images [1] (the object localization method used by Mapillary). This comparison with numerical values reported in Table 1, showed a substantial mean average precision improvement, both at class level and overall. It shows how our method performs better than the latest available localization method.

Table 1. Performance evaluation and comparison with Mapillary's method of objects derived from multiple detections in multiple images [1]

	Genoa (Porto Antico)			Vienna (Marxergasse)		
	Ours (3D)	Ours (2D)	Mapillary (2D)	Ours (3D)	Ours (2D)	Mapillary (2D)
	(prec, rec)	mAP	mAP	(prec, rec)	mAP	mAP
Overall	0.57, 0.90	**0.54**	0.23	0.36, 0.55	**0.61**	0.17
Classes						
Benches	1.00, 1.00	0.60	0.31	–	–	–
Street lights	0.70, 1.00	1.00	0.33	0.30, 1.00	0.91	0.05
Trash cans	0.50, 1.00	0.60	0.05	–	–	–
Manholes	0.30, 0.40	0.30	0.43	–	–	–
Catch-basin	0.50, 0.80	0.40	0.05	–	–	–
Traffic signs	0.44, 0.66	0.40	0.39	0.43, 0.50	0.6	0.45
Poles	0.35, 0.87	0.47	0.05	0.10, 0.40	0.51	0.05
Traffic lights	–	–	–	0.31, 0.33	0.42	0.17

5   Conclusions and Future Directions

Exploring a scene to localize and match objects identified via instance segmentation given crowd-sourced images is a very challenging task. The crowd-sourced images are acquired using various devices (i.e., different camera models, type of lenses, intrinsics, user's proficiency) over a large time scale (i.e. different levels of illumination, different times of day, different seasons, human-made changes), which make recognizing the same instance across multiple images difficult. We show how leveraging SfM 3D information helps in localizing objects in 3D. Our method pre-processes a sparse point cloud, obtained from standard SfM techniques, to get a reduced 3D scene element where DBSCAN clusters the 3D points belonging to objects. We use these clusters to match the 2D instance detections, and experimentally validate on two urban scenes the accuracy of our results, comparing them against one baseline approach.

The main limitations of our approach are the following. It works offline, which means it depends on sparse reconstruction by Colmap [27]. It does not care about the appearance of an object while matching the 2D detection, which is good as well because we want to localize even if an object changes its appearance. This leads to the jump of the label to a very nearby object in some cases, such as very close and small signs on one pole.

We anticipate that this method will be a baseline in generating a large amount of ground-truth data for deep learning models, developing the cities, advancing urban planning, or autonomous driving.

Acknowledgements. This project has received funding from the European Union's Horizon 2020 research and innovation programme under grant agreement No 870743.

References

1. Mapillary Python SDK, mapillary api v4. https://github.com/mapillary/mapillary-python-sdk. Accessed 15 Dec 2021
2. Mapillary, the street-level imagery platform that scales and automates mapping. https://www.mapillary.com/. Accessed 15 Jul 2021
3. OpenStreetMap. openstreetmap. Accessed 15 Jul 2021
4. Ahmed, S.M., Chew, C.M.: Density-based clustering for 3d object detection in point clouds. In: Proceedings of the IEEE/CVF Conference on Computer Vision and Pattern Recognition, pp. 10608–10617 (2020)
5. Jocher, G., et al.: ultralytics/yolov5: v5.0 - YOLOv5-P6 1280 models, AWS, Supervise.ly and YouTube integrations (April 2021). https://doi.org/10.5281/zenodo.4679653
6. Blender Online Community: Blender - a 3D modelling and rendering package. Blender Foundation, Blender Institute, Amsterdam (2016). http://www.blender.org
7. Branson, S., Wegner, J.D., Hall, D., Lang, N., Schindler, K., Perona, P.: From Google Maps to a fine-grained catalog of street trees. ISPRS J. Photogramm. Remote. Sens. **135**, 13–30 (2018). https://doi.org/10.1016/j.isprsjprs.2017.11.008
8. Chen, X., Ma, H., Wan, J., Li, B., Xia, T.: Multi-view 3d object detection network for autonomous driving. In: Proceedings of the IEEE CVPR, pp. 1907–1915 (2017)
9. Crocco, M., Rubino, C., Del Bue, A.: Structure from motion with objects. In: Proceedings of the IEEE Conference on Computer Vision and Pattern Recognition, pp. 4141–4149 (2016)
10. Ester, M., Kriegel, H.P., Sander, J., Xu, X., et al.: A density-based algorithm for discovering clusters in large spatial databases with noise. In: KDD, vol. 96, pp. 226–231 (1996)
11. Gay, P., Rubino, C., Bansal, V., Del Bue, A.: Probabilistic structure from motion with objects (PSfMO). In: Proceedings of the IEEE International Conference on Computer Vision, pp. 3075–3084 (2017)
12. Hanif, M.S., Ahmad, S., Khurshid, K.: On the improvement of foreground-background model-based object tracker. IET Comput. Vis. **11**(6), 488–496 (2017)
13. Hebbalaguppe, R., Garg, G., Hassan, E., Ghosh, H., Verma, A.: Telecom inventory management via object recognition and localisation on Google Street View images. In: 2017 IEEE WACV, pp. 725–733. IEEE (2017)
14. Krylov, V.A., Kenny, E., Dahyot, R.: Automatic discovery and geotagging of objects from street view imagery. Remote Sens. **10**(5), 661 (2018)
15. Ku, J., Mozifian, M., Lee, J., Harakeh, A., Waslander, S.L.: Joint 3d proposal generation and object detection from view aggregation. In: 2018 IEEE/RSJ International Conference on Intelligent Robots and Systems (IROS), pp. 1–8. IEEE (2018)
16. Liu, C.J., Ulicny, M., Manzke, M., Dahyot, R.: Context aware object geotagging. arXiv preprint arXiv:2108.06302 (2021)
17. Liu, W.: SSD: single shot multibox detector. In: Leibe, B., Matas, J., Sebe, N., Welling, M. (eds.) ECCV 2016. LNCS, vol. 9905, pp. 21–37. Springer, Cham (2016). https://doi.org/10.1007/978-3-319-46448-0_2

18. Nassar, A.S., D'Aronco, S., Lefèvre, S., Wegner, J.D.: GeoGraph: graph-based multi-view object detection with geometric cues end-to-end. In: Vedaldi, A., Bischof, H., Brox, T., Frahm, J.-M. (eds.) ECCV 2020. LNCS, vol. 12352, pp. 488–504. Springer, Cham (2020). https://doi.org/10.1007/978-3-030-58571-6_29
19. Nassar, A.S., Lefèvre, S., Wegner, J.D.: Simultaneous multi-view instance detection with learned geometric soft-constraints. In: Proceedings of the IEEE/CVF International Conference on Computer Vision, pp. 6559–6568 (2019)
20. Nicholson, L., Milford, M., Sünderhauf, N.: QuadricSLAM: dual quadrics from object detections as landmarks in object-oriented slam. IEEE Robot. Autom. Lett. **4**(1), 1–8 (2018)
21. Qi, C.R., Chen, X., Litany, O., Guibas, L.J.: ImVoteNet: boosting 3d object detection in point clouds with image votes. In: Proceedings of the IEEE/CVF CVPR, pp. 4404–4413 (2020)
22. Qi, C.R., Litany, O., He, K., Guibas, L.J.: Deep Hough voting for 3d object detection in point clouds. In: Proceedings of the IEEE/CVF ICCV, pp. 9277–9286 (2019)
23. Qi, C.R., Su, H., Mo, K., Guibas, L.J.: PointNet: deep learning on point sets for 3d classification and segmentation. In: Proceedings of the IEEE CVPR, pp. 652–660 (2017)
24. Qi, C.R., Yi, L., Su, H., Guibas, L.J.: PointNet++: deep hierarchical feature learning on point sets in a metric space. arXiv preprint arXiv:1706.02413 (2017)
25. Redmon, J., Farhadi, A.: YOLOv3: an incremental improvement. arXiv preprint arXiv:1804.02767 (2018)
26. Ren, S., He, K., Girshick, R., Sun, J.: Faster R-CNN: towards real-time object detection with region proposal networks. Adv. Neural. Inf. Process. Syst. **28**, 91–99 (2015)
27. Schonberger, J.L., Frahm, J.M.: Structure-from-motion revisited. In: Proceedings of the IEEE Conference on Computer Vision and Pattern Recognition, pp. 4104–4113 (2016)
28. Wang, Y., Chao, W.L., Garg, D., Hariharan, B., Campbell, M., Weinberger, K.Q.: Pseudo-lidar from visual depth estimation: Bridging the gap in 3d object detection for autonomous driving. In: Proceedings of the IEEE/CVF Conference on CVPR, pp. 8445–8453 (2019)
29. Wegner, J.D., Branson, S., Hall, D., Schindler, K., Perona, P.: Cataloging public objects using aerial and street-level images-urban trees. In: Proceedings of the IEEE Conference on Computer Vision and Pattern Recognition, pp. 6014–6023 (2016)
30. Yang, S., Scherer, S.: CubeSLAM: monocular 3d object detection and slam without prior models. arXiv preprint arXiv:1806.00557 (2018)
31. Yin, T., Zhou, X., Krahenbuhl, P.: Center-based 3d object detection and tracking. In: Proceedings of the IEEE/CVF Conference on CVPR, pp. 11784–11793 (2021)
32. You, Y., et al.: Pseudo-LiDAR++: accurate depth for 3d object detection in autonomous driving. arXiv preprint arXiv:1906.06310 (2019)
33. Zhang, C., Fan, H., Li, W.: Automated detecting and placing road objects from street-level images. Comput. Urban Sci. **1**(1), 1–18 (2021). https://doi.org/10.1007/s43762-021-00019-6
34. Zhang, W., Witharana, C., Li, W., Zhang, C., Li, X., Parent, J.: Using deep learning to identify utility poles with crossarms and estimate their locations from Google Street View images. Sensors **18**(8), 2484 (2018)

35. Zhang, Z., Sun, B., Yang, H., Huang, Q.: H3DNet: 3D object detection using hybrid geometric primitives. In: Vedaldi, A., Bischof, H., Brox, T., Frahm, J.-M. (eds.) ECCV 2020. LNCS, vol. 12357, pp. 311–329. Springer, Cham (2020). https://doi.org/10.1007/978-3-030-58610-2_19

36. Zhao, J., Zhang, X.N., Gao, H., Yin, J., Zhou, M., Tan, C.: Object detection based on hierarchical multi-view proposal network for autonomous driving. In: 2018 International Joint Conference on Neural Networks (IJCNN), pp. 1–6. IEEE (2018)

Improving Multi-View Stereo
via Super-Resolution

Eugenio Lomurno[✉] , Andrea Romanoni , and Matteo Matteucci

Politecnico di Milano, Milano, Italy
{eugenio.lomurno,andrea.romanoni,matteo.matteucci}@polimi.it

Abstract. Today, Multi-View Stereo techniques can reconstruct robust and detailed 3D models, especially when starting from high-resolution images. However, there are cases in which the resolution of input images is relatively low, for instance, when dealing with old photos or when hardware constrains the amount of data acquired. This paper shows how increasing the resolution of such input images through Super-Resolution techniques reflects in quality improvements of the reconstructed 3D models. We show that applying a Super-Resolution step before recovering the depth maps leads to a better 3D model both in the case of patchmatch and deep learning Multi-View Stereo algorithms. In detail, the use of Super-Resolution improves the average f1 score of reconstructed models. It turns out to be particularly effective in the case of scenes rich in texture, such as outdoor landscapes.

Keywords: Multi-View Stereo · Super-Resolution · Single-Image Super-Resolution · 3D reconstruction

1 Introduction

Recovering the 3D model of a scene captured by images is a relevant problem in a wide variety of scenarios, e.g., city mapping, archaeological heritage preservation, autonomous driving, and robot localization. In the Computer Vision community, this task goes under the name of Multi-View Stereo (MVS), and it aims to reconstruct 3D models as accurately and completely as possible.

Currently, the most successful workflow to perform such reconstructions starts from a Structure from Motion algorithm that estimates camera parameters such as their positions and orientations [18]. Then, it follows the depth maps estimation step, for which the most common approaches rely on patchmatch [1] or deep learning [24] techniques. The former approaches lead to very accurate results, while the latter produce more complete models, even if they still suffer scalability issues. As the last step, depth maps are projected on 3D space and fused together, obtaining a dense point cloud.

Under controlled scenarios, in which the hardware adopted to collect the images is not subject to particular constraints, it is relatively easy to acquire high-resolution images and obtain a high-quality reconstruction of the scene.

S. Sclaroff et al. (Eds.): ICIAP 2022, LNCS 13232, pp. 102–113, 2022.
https://doi.org/10.1007/978-3-031-06430-2_9

However, in several cases, the input of an MVS method consists of low-resolution images. For instance, when power consumption constrains the hardware, e.g., with drones or telescopes, or when processing images taken in low-resolution such as with old photos.

In these cases, the recovered 3D model most likely lacks details or is incomplete, regardless of the adopted MVS algorithm. We claim that algorithmically increasing input images resolution can overcome this issue by enhancing their information content and quality. This is possible via Super-Resolution techniques that have recently reached impressive performance in many application fields despite the possibility of generating some artefacts.

This paper shows the benefits for MVS pipelines of upscaling low-resolution images through Single-Image Super-Resolution (SISR) techniques. In particular, we test SISR contribution over COLMAP [19] and CasMVSNet [6] MVS pipelines, and validate it over ETH3D Low-resolution many-view [20] and Tanks and Temples [10] benchmarks. Perceptive and numerical results demonstrate that SISR improves the quality of the dense point clouds produced by MVS algorithms by effectively balancing the increased amount of generated points and their position in the space.

2   Related Work

In the literature, there are some attempts to exploit Super-Resolution (SR) with the goal of improving the quality of 3D reconstructions. For instance, Goldlücke et al. [5] proposed a variational method to improve 3D models appearance by estimating textures with SR techniques. More recently, Li et al. [13] proposed a novel model-based SR method that better exploits geometric features to enrich the texture applied to a 3D model.

Other approaches exploiting SR in the 3D reconstruction realm aims to increase depth maps resolution. Lei et al. [12] relied on bilinear interpolation of multiple depth maps to increase the resolution of a single depth map. The authors in [27] and [21] used high-resolution RGB images to guide a deep learning model to increase depth maps resolution.

Differently from previous works, we aim at improving models geometry instead of their texture appearance, by applying SR directly on input images. We argue that SR can improve the reconstruction from low-resolution images, and different stages of a 3D reconstruction pipeline could benefit from the availability of SR images, e.g., camera calibration and mesh refinement. Surprisingly, to the best of our knowledge, no paper has ever analyzed if and to what extent MVS 3D reconstruction pipelines can benefit from input images enhanced through SR.

2.1 Single-Image Super-Resolution

Single-Image Super-Resolution (SISR) aims at recovering a high-resolution image from a single low-resolution image [23]. In the last few years, we have seen how modern deep learning pipelines overtook non-learning-based algorithms, such as nearest-neighbours and bicubic interpolation.

As first attempt, Dong *et al.* [4] proposed a convolutional neural network to map low- to high-resolution images. This network architecture has been extended with a combination of new layers, and skip-connections by Kim *et al.* [9]. Subsequently, other methods have exploited different combinations of residual and dense connections [15,26].

Recent works show that networks with novel feedback mechanisms further improve the quality of the SR images. For instance, Li *et al.* [14] combine a feedback block with curriculum learning. In their most recent work, Haris *et al.* released the Deep Back-Projection Network architecture [7]. The idea is to generate numerous degraded and high-resolution hypothesis images that the network uses to improve the output result. In the last revision, the authors have implemented dense connections, adversarial loss and recurrent layers, making the entire architecture more scalable and performing. Chen *et al.* [3] proposed a self-supervised encoding network based on their implicit neural representation technique to learn continuous mappings for super-resolution.

2.2 Multi-View Stereo

MVS aims at recovering a dense 3D representation of a scene perceived by a set of calibrated cameras. The main step adopted by the most successful MVS methods is depth maps estimation, i.e., the process of computing the depth of each pixel belonging to each image. Once computed, these maps are fused into a dense point cloud or a volumetric representation.

The most performing depth estimation approaches are based on the patch-match algorithm [1], which relies on the idea of choosing for each pixel a random guess of the depth and then propagating the most likely estimates to its neighbourhood. The work proposed by Schönberger *et al.* [19], named COLMAP, can be considered the cornerstone of modern patchmatch-based algorithms. It is a robust framework able to process high-quality images and jointly estimate pixel-wise camera visibility, as well depth and normal maps for each view. Since this method heavily relies on the Bilateral NCC Photometric-Consistency, it often fails in recovering areas with low texture. Recently, to compensate for this, TAPA-MVS [17] proposed to explicitly handle textureless regions by propagating in a planar-wise fashion the valid depth estimates to neighbouring textureless areas. Kuhn *et al.* [11] extended this method with a hierarchical approach improving the robustness of the estimation process.

Another family of MVS algorithms relies on deep learning. DeepMVS [8] and MVSNet [24] were the first approaches proposing an effective MVS pipeline based on DNNs. For each camera, both the approaches build a cost volume by projecting nearby images on planes at different depths, then they classify [8] or regress [24] the best depth for each pixel. Yao *et al.* [25] introduced an RNN to regularize the cost volume, while Luo *et al.* [16] built a model to learn how to aggregate the cost to compute a more robust depth estimate. MVS-CRF [22], finally, refines the MVSNet estimate through Markov Random Field, and Point-MVSNet [2] through a graph-based neural architecture. The huge limitation of learning-based approaches relies on their computational complexity. Usually, it is not feasible to handle high-resolution images as both memory and time costs grow cubically as the volume resolution increases, causing a limitation on the accuracy and completeness of the reconstructed models. The best attempt to handle this problem is the work of Xiaodong *et al.* [6], named CasMVSNet, in which they applied a coarse-to-fine approach that considerably improves the scalability of MVSNet-based methods.

3   Methods

Our work aims to provide an overview of the effects of Single-Image Super-Resolution (SISR) when applied in the head of Multi-View Stereo (MVS) algorithms. In order to generalize this phenomenon, we chose two different SISR algorithms to conduct our ablation studies. In particular, we identified the bicubic interpolation algorithm as a candidate algorithm for a more traditional SISR approach, while Deep Back-Projection Network (DBPN) by Haris *et al.* [7] to investigate the effects of a more recent pipeline based on deep learning. It is known in the literature that SR techniques are often prone to artifact generation, especially with the growth of the upscaling factor. In order to exploit both the SISR algorithms at the best of their performance, we fixed it to 2. Concerning DBPN we exploited the best set of weights provided by the authors for this upscaling factor, i.e., "DBPN-RES-MR64-3".

Regarding the MVS pipelines, we chose two different approaches based on two different technologies. The first one is COLMAP [19], in which depth maps estimation is heavily based on patchmatch. Moreover, it turns out to be one of the most efficient algorithms of its family due to its parallel maps computation. The second one is CasMVSNet, the deep learning architecture based on MVS-Net developed by Xiaodong *et al.* [6]. With the addition of a new cost volume technique, built upon a feature pyramid encoding geometry, it learns to estimate the depth space of the scene at gradually finer scales. In detail, it narrows the disparity range for every stage thanks to an iterative prediction made from the previous stages. It then gradually increases the cost volume resolution to obtain accurate outputs. This technique allowed us to exploit a deep learning approach, even with datasets composed of numerous images and limited hardware capabilities. For this algorithm, like for DBPN, we used the pre-trained model provided by the authors in order to facilitate the experiments reproducibility.

Table 1. F1 scores on the ETH3D low-resolution multi-view train set with COLMAP. We compare scores starting from low-resolution images against the ones obtained from bicubic interpolation and Deep Back-Projection Network Super-Resolution

τ (cm)	Overall			Indoor			Outdoor		
	Low-res	Bicubic	DBPN	Low-res	Bicubic	DBPN	Low-res	Bicubic	DBPN
1	35.80	39.85	**40.00**	40.68	**43.68**	43.42	32.55	37.29	**37.71**
2	53.41	54.83	**54.98**	56.15	**57.37**	57.09	51.59	53.14	**53.58**
5	72.16	72.58	**72.64**	**74.35**	74.05	73.62	70.70	71.59	**71.99**
10	81.83	**82.17**	82.13	**83.97**	83.24	82.69	80.40	81.46	**81.76**
20	88.98	**89.20**	89.14	**91.50**	90.55	90.11	87.30	88.30	**88.50**
50	95.29	**95.70**	**95.70**	**97.33**	97.19	97.15	93.93	94.71	**94.75**

Fig. 1. COLMAP disparity maps of a sample from storage_room_2 dataset (Indoor). On the left side, the low-resolution estimation, on the right side, the one enhanced via Deep Back-Projection Network.

4   Ablation Study

In order to calibrate the chosen algorithms, we have conducted an ablation study over the training set of the ETH3D Low-resolution many-view benchmark [20], which is composed of five gray-scale datasets split in indoor (two) and outdoor (three). For this benchmark, the accuracy and completeness metrics are available to compute the goodness of the reconstruction with respect to the ground truth. In order to take both into account, we relied our analysis on their harmonic mean, i.e., the f1 score. This section's experiments have been executed on an Intel(R) Xeon(R) CPU E5-2630 v4 @ 2.20 GHz with an Nvidia GTX 1080Ti GPU.

For each dataset, we computed its enhanced twins via bicubic interpolation and DBPN. Then, we applied COLMAP and CasMVSNet pipelines on all these datasets to obtain the dense point clouds. Finally, we evaluated the results with respect to the ground truths. We compare, with different distance tolerances, the performance of each dataset.

According to the literature, and after some preliminary trials, we noticed that COLMAP achieves poor performance in indoor datasets despite their resolution. This issue is related to its patchmatch algorithm heavily based on the Bilateral Normalized Cross-Correlation Photometric-Consistency, which is translated into

Table 2. F1 scores on the ETH3D low-resolution multi-view train set with CasMVS-Net. We compare scores starting from low-resolution images against the ones obtained from bicubic interpolation and Deep Back-Projection Network Super-Resolution

τ (cm)	Overall			Indoor			Outdoor		
	Low-res	Bicubic	DBPN	Low-res	Bicubic	DBPN	Low-res	Bicubic	DBPN
1	38.28	39.58	**39.59**	37.24	38.43	**38.46**	38.97	40.34	**40.35**
2	49.00	49.65	**49.66**	48.08	48.07	**48.09**	49.61	50.70	**50.71**
5	60.58	**61.17**	61.15	**61.10**	60.49	60.40	60.25	61.63	**61.65**
10	67.59	**68.43**	68.37	**69.39**	69.25	69.10	66.39	**67.89**	**67.89**
20	74.00	**74.93**	74.84	77.33	**77.44**	77.25	71.79	**73.26**	73.23
50	82.60	83.59	**83.61**	87.14	**87.31**	**87.31**	79.56	81.10	**81.14**

artifacts and poor estimates in textureless regions, e.g., monochromatic and reflective surfaces. In order to cope with this unwanted behavior, we modified COLMAP parameters when dealing with Indoor scenes by increasing the robustness of the depth estimates and thus trading-off with a higher computational cost. Specifically, we reduce the minimum NCC threshold and increase the window radius by 2. Then we filter the resulting depth maps with a speckle filter algorithm before fusing them. For this widely used filter, we have chosen a max depth range equal to 5 and set the maximum speckle size to 1% of the depth map dimension. These parameters have been tuned to maximize the average performance over both low-resolution datasets and their versions enhanced via SR.

Table 1 shows the results of this first set of experiments computed via COLMAP. On average, both SR techniques lead to noticeable improvements in the quality of the reconstructed models. This improvement tends to be more evident when tolerances are computed concerning little points neighborhoods. It is also observable a different behavior between indoor and outdoor scenes: in the first case, we observe good synergies between SR and the MVS algorithms concerning small tolerances, while the effects turn out to degrade the performance by considering more flexible evaluation criteria. As can be seen in the disparity map example displayed in Fig. 1, on the one hand, the effect of DBPN is translated into less noisy depth estimations, on the other, it erases the small number of points belonging to the textureless regions such as the inner part of the bricks or doors.

In the second case, the advantages of applying SISR are spread with evidence among the tolerances, reaching remarkable f1 score improvements (+5.16% for $\tau = 1$ cm). An example of the benefits can be visually appreciated in the disparity maps comparison in Fig. 2. In this case, it is evident how COLMAP is able to exploit the increased amount of input information, on the one hand, to identify pieces of bushes in the foreground, on the other, to better perceive the image depth and produce a more detailed disparity map.

In Table 2 are instead summarized the results obtained by carrying out the same set of experiments with CasMVSNet as MSV algorithm. In general, the behavior of this algorithm when its input is enhanced with SR is coherent with the one demonstrated by COLMAP. In fact, also in this case, it is evident a

Fig. 2. COLMAP disparity maps of a sample from forest dataset (Outdoor). On the left side, the low-resolution estimation, on the right side, the one enhanced via Deep Back-Projection Network.

Table 3. F1, accuracy and completeness scores over ETH3D low-resolution multi-view benchmark. We compare the presented models grouped in many subsets with a tolerance $\tau = 1\,\mathrm{cm}$

Model	Overall			Indoor			Outdoor		
	F1	Acc	Comp	F1	Acc	Comp	F1	Acc	Comp
COLMAP	36.6	**40.7**	33.8	34.4	**38.1**	31.7	38.0	**42.4**	35.1
COLMAP (DBPN)	**40.6**	37.2	**45.8**	**36.5**	35.7	**38.2**	**43.3**	38.2	**50.8**
CasMVSNet	36.8	**42.4**	34.3	27.8	36.0	23.7	42.7	**46.8**	41.3
CasMVSNet (DBPN)	**37.7**	41.4	**36.9**	**28.9**	**37.4**	**24.7**	**43.6**	44.0	**45.0**

generalized f1 score improvement, especially while considering the most restrictive tolerances.

We can thus argue that, in general, both patchmach and deep learning MSV algorithms demonstrate on average to benefit from the presence of richer input information. Moreover, from the results of this ablation study is possible to assert that this is true regardless of the SR algorithm chosen.

5   Experiments and Results

After having calibrated the algorithms, we evaluated their performance in a broader set of experiments. For this purpose, we executed two independent validation runs, both computed on an Intel(R) Xeon(R) CPU E5-2630 v4 @ 2.20 GHz with an Nvidia GTX 1080Ti.

5.1   Evaluation over ETH3D Benchmark

We evaluated the proposed approach over the whole ETH3D Low-resolution many-view benchmark [20], which is composed of ten gray-scale datasets split into indoor (four) and outdoor (six).

We used DBPN as SR algorithm for these experiments and COLMAP and CasMVSNet for the MVS pipeline. From Table 3 it is possible to summarize that,

Table 4. F1, precision and recall scores over Tanks and Temples train benchmark. We compare the downscaled version (low-res) with the SISR approaches (bicubic and DBPN) and with a theoretical high-resolution version (high-res). All the reconstructions are computed via COLMAP.

Model	Overall			Indoor			Outdoor		
	F1	Prec	Rec	F1	Prec	Rec	F1	Prec	Rec
Low-res (1x)	9.56	20.40	6.47	8.93	18.97	6.24	10.02	21.47	6.63
Bicubic (2x)	31.44	37.96	27.91	26.24	36.39	21.86	35.35	39.14	32.46
DBPN (2x)	**32.94**	**40.09**	**29.04**	**26.85**	**36.39**	**22.15**	**37.5**	**41.74**	**34.27**
High-res (2x)	37.65	45.28	33.19	32.03	42.63	26.57	41.87	47.27	38.15

concerning the most significant evaluation criterion, i.e., $\tau = 1$ cm, the trade-off represented by the f1 score is strongly raised up from the completeness metric. Like in the ablation study, we can observe a general benefit brought about by the SR usage, both in the case of patchmatch and deep learning MVS pipelines.

From these results and the visual comparison of reconstruction details is shown in Fig. 3, we argue that the improvement is related to the increased amount of reconstructed points allowed by the increased amount of pixels in input. In fact, the completeness boost means that there are many more points close to ground truth points regarding the low-resolution case. On the other hand, there are also many points that are far away from the ground truth, and this is translated into an accuracy drop.

Given the obtained performance, we can argue that this trade-off brings positive outcomes that are perceptively translated into denser reconstructions and characterized by a reduction of texture holes.

5.2 Evaluation over Tanks and Temples Benchmark

Finally, we have conducted a further experiment over the Tanks and Temples train benchmark [10], composed of 7 RGB high-resolution datasets split into indoor (three) and outdoor (four).

Given its images good quality, we performed an evaluation test to estimate the goodness of the SISR approaches with respect to theoretical high-resolution data. In detail, we downscaled every dataset by a factor of 4, considering it as the low-resolution benchmark. We did the same but with a factor of 2, considering this as the high-resolution benchmark. Finally, starting from the low-resolution datasets, we computed with bicubic interpolation and DBPN algorithms their twins enhanced via SR with upscaling factor 2, and we compared all the results. For these experiments, we have used COLMAP as MVS pipeline.

For this benchmark, the precision and recall metrics are available to compute the goodness of the reconstruction with respect to the ground truth. In order to compare these results with the ones of the previous experiments, these metrics can be intended from the reconstructed points perspective (if ETH3D benchmark

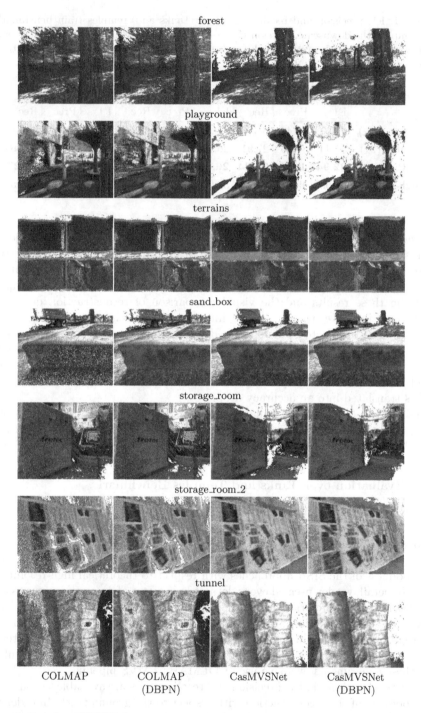

Fig. 3. Details of ETH3D low-res many-view benchmark 3D reconstructions. We compare the same view for each proposed pipeline in both low-resolution and enhanced via DBPN versions.

completeness can be assumed as the accuracy of the reconstructed points with respect to the nearest point of the ground truth, given a fixed tolerance, now we are dealing with its precision and its recall). Also in this case, in order to take both into account, we relied our analysis mainly on the f1 score. This time, the tolerance is fixed and different for each dataset according to the authors evaluation rules.

The results of this experiment are displayed in Table 4. In this scenario, it is evident how the SISR turns out to be very effective for every metric taken into account. The reason behind such a remarkable improvement has to be addressed to the shallow resolution of the low-res benchmark, which made very difficult for COLMAP to estimate good depths and subsequently fuse them into accurate dense point clouds. The most interesting result relies on comparing reconstructions from the benchmark enhanced via DBPN and the high-resolution one. In fact, despite there is still a performance gap between them, this gap is much lower than the one with the scores obtained starting from the low-resolution benchmark.

Summarizing this result with the ones obtained from previous analysis, we can conclude that SISR algorithms can help MVS techniques to produce better and denser reconstructions when put on top of the pipeline. Despite the presence of artifacts, the results are qualitatively closer to high-resolution theoretical reconstructions, and this approach can be applied both on MVS algorithms based on patchmatch and deep learning.

6   Conclusions

In this paper, we presented a study on how to improve 3D reconstruction starting from low-resolution images through the use of Single-Image Super-Resolution techniques, demonstrating Super-Resolution effectiveness for Multi-View Stereo algorithms based on both patchmatch and deep learning. Moreover, we have demonstrated the existence of a strong correlation between starting images and 3D models qualities and that an increased amount of input information provided by Super-Resolution is effectively translated into more robust and dense representations in the 3D space by Multi-View Stereo pipelines. Despite the Super-Resolution algorithm chosen, we have shown how the 3D models obtained results to benefit from the Single-Image Super-Resolution improvement of the input images the more they do not have a starting high-resolution.

References

1. Bleyer, M., Rhemann, C., Rother, C.: PatchMatch stereo-stereo matching with slanted support windows. In: BMVC, vol. 11, pp. 1–11 (2011)
2. Chen, R., Han, S., Xu, J., Su, H.: Point-based multi-view stereo network. In: Proceedings of the IEEE International Conference on Computer Vision, pp. 1538–1547 (2019)

3. Chen, Y., Liu, S., Wang, X.: Learning continuous image representation with local implicit image function. In: Proceedings of the IEEE/CVF Conference on Computer Vision and Pattern Recognition, pp. 8628–8638 (2021)
4. Dong, C., Loy, C.C., He, K., Tang, X.: Image super-resolution using deep convolutional networks. IEEE Trans. Pattern Anal. Mach. Intell. **38**(2), 295–307 (2015)
5. Goldlücke, B., Aubry, M., Kolev, K., Cremers, D.: A super-resolution framework for high-accuracy multiview reconstruction. Int. J. Comput. Vis. **106**(2), 172–191 (2014)
6. Gu, X., Fan, Z., Zhu, S., Dai, Z., Tan, F., Tan, P.: Cascade cost volume for high-resolution multi-view stereo and stereo matching. arXiv preprint arXiv:1912.06378 (2019)
7. Haris, M., Shakhnarovich, G., Ukita, N.: Deep back-projection networks for super-resolution. In: Proceedings of the IEEE Conference on Computer Vision and Pattern Recognition, pp. 1664–1673 (2018)
8. Huang, P.H., Matzen, K., Kopf, J., Ahuja, N., Huang, J.B.: DeepMVS: learning multi-view stereopsis. In: Proceedings of the IEEE Conference on Computer Vision and Pattern Recognition, pp. 2821–2830 (2018)
9. Kim, J., Kwon Lee, J., Mu Lee, K.: Accurate image super-resolution using very deep convolutional networks. In: Proceedings of the IEEE Conference on Computer Vision and Pattern Recognition, pp. 1646–1654 (2016)
10. Knapitsch, A., Park, J., Zhou, Q.Y., Koltun, V.: Tanks and temples: benchmarking large-scale scene reconstruction. ACM Trans. Graph. (ToG) **36**(4), 1–13 (2017)
11. Kuhn, A., Lin, S., Erdler, O.: Plane completion and filtering for multi-view stereo reconstruction. In: Fink, G.A., Frintrop, S., Jiang, X. (eds.) DAGM GCPR 2019. LNCS, vol. 11824, pp. 18–32. Springer, Cham (2019). https://doi.org/10.1007/978-3-030-33676-9_2
12. Lei, J., Li, L., Yue, H., Wu, F., Ling, N., Hou, C.: Depth map super-resolution considering view synthesis quality. IEEE Trans. Image Process. **26**(4), 1732–1745 (2017)
13. Li, Y., Tsiminaki, V., Timofte, R., Pollefeys, M., Gool, L.V.: 3d appearance super-resolution with deep learning. In: Proceedings of the IEEE Conference on Computer Vision and Pattern Recognition, pp. 9671–9680 (2019)
14. Li, Z., Yang, J., Liu, Z., Yang, X., Jeon, G., Wu, W.: Feedback network for image super-resolution. In: Proceedings of the IEEE Conference on Computer Vision and Pattern Recognition, pp. 3867–3876 (2019)
15. Lim, B., Son, S., Kim, H., Nah, S., Mu Lee, K.: Enhanced deep residual networks for single image super-resolution. In: Proceedings of the IEEE Conference on Computer Vision and Pattern Recognition Workshops, pp. 136–144 (2017)
16. Luo, K., Guan, T., Ju, L., Huang, H., Luo, Y.: P-MVSNet: learning patch-wise matching confidence aggregation for multi-view stereo. In: Proceedings of the IEEE International Conference on Computer Vision, pp. 10452–10461 (2019)
17. Romanoni, A., Matteucci, M.: TAPA-MVS: textureless-aware patchmatch multi-view stereo. In: Proceedings of the IEEE International Conference on Computer Vision, pp. 10413–10422 (2019)
18. Schonberger, J.L., Frahm, J.M.: Structure-from-motion revisited. In: Proceedings of the IEEE Conference on Computer Vision and Pattern Recognition, pp. 4104–4113 (2016)
19. Schönberger, J.L., Zheng, E., Frahm, J.-M., Pollefeys, M.: Pixelwise view selection for unstructured multi-view stereo. In: Leibe, B., Matas, J., Sebe, N., Welling, M. (eds.) ECCV 2016. LNCS, vol. 9907, pp. 501–518. Springer, Cham (2016). https://doi.org/10.1007/978-3-319-46487-9_31

20. Schops, T., et al.: A multi-view stereo benchmark with high-resolution images and multi-camera videos. In: Proceedings of the IEEE Conference on Computer Vision and Pattern Recognition, pp. 3260–3269 (2017)
21. Voynov, O., et al.: Perceptual deep depth super-resolution. In: Proceedings of the IEEE International Conference on Computer Vision, pp. 5653–5663 (2019)
22. Xue, Y., et al.: MVSCRF: learning multi-view stereo with conditional random fields. In: Proceedings of the IEEE International Conference on Computer Vision, pp. 4312–4321 (2019)
23. Yang, W., Zhang, X., Tian, Y., Wang, W., Xue, J.H., Liao, Q.: Deep learning for single image super-resolution: a brief review. IEEE Trans. Multimedia 21(12), 3106–3121 (2019)
24. Yao, Y., Luo, Z., Li, S., Fang, T., Quan, L.: MVSNet: depth inference for unstructured multi-view stereo. In: Ferrari, V., Hebert, M., Sminchisescu, C., Weiss, Y. (eds.) ECCV 2018. LNCS, vol. 11212, pp. 785–801. Springer, Cham (2018). https://doi.org/10.1007/978-3-030-01237-3_47
25. Yao, Y., Luo, Z., Li, S., Shen, T., Fang, T., Quan, L.: Recurrent MVSNet for high-resolution multi-view stereo depth inference. In: Proceedings of the IEEE Conference on Computer Vision and Pattern Recognition, pp. 5525–5534 (2019)
26. Zhang, Y., Tian, Y., Kong, Y., Zhong, B., Fu, Y.: Residual dense network for image super-resolution. In: Proceedings of the IEEE Conference on Computer Vision and Pattern Recognition, pp. 2472–2481 (2018)
27. Zuo, Y., Wu, Q., Fang, Y., An, P., Huang, L., Chen, Z.: Multi-scale frequency reconstruction for guided depth map super-resolution via deep residual network. IEEE Trans. Circ. Syst. Video Technol. 30(2), 297–306 (2019)

Efficient View Clustering and Selection
for City-Scale 3D Reconstruction

Marco Orsingher[1,2]([✉]), Paolo Zani[2], Paolo Medici[2], and Massimo Bertozzi[1]

[1] Università degli Studi di Parma Dipartimento di Ingegneria e Architettura,
Parma, Italy
`marco.orsingher@unipr.it`
[2] Vislab Srl - Ambarella Inc., Parma, Italy

Abstract. Image datasets have been steadily growing in size, harming the feasibility and efficiency of large-scale 3D reconstruction methods. In this paper, a novel approach for scaling Multi-View Stereo (MVS) algorithms up to arbitrarily large collections of images is proposed. Specifically, the problem of reconstructing the 3D model of an entire city is targeted, starting from a set of videos acquired by a moving vehicle equipped with several high-resolution cameras. Initially, the presented method exploits an approximately uniform distribution of poses and geometry and builds a set of overlapping clusters. Then, an Integer Linear Programming (ILP) problem is formulated for each cluster to select an optimal subset of views that guarantees both visibility and matchability. Finally, local point clouds for each cluster are separately computed and merged. Since clustering is independent from pairwise visibility information, the proposed algorithm runs faster than existing literature and allows for a massive parallelization. Extensive testing on urban data are discussed to show the effectiveness and the scalability of this approach.

Keywords: Large-scale 3D reconstruction · Scalable multi-view stereo

1 Introduction

Nowadays, arbitrarily large datasets of high-resolution images have become available thanks to almost unlimited sources of data, such as videos acquired by autonomous vehicles with multi-camera rigs and Internet pictures uploaded by millions of users on Social Media. While this large amount of data can be used to effectively reconstruct a precise 3D view of the world, the increasing size and high redundancy introduce both feasibility and efficiency challenges.

A traditional image-based 3D reconstruction pipeline is composed by two main steps: Structure From Motion (SFM) and Multi-View Stereo (MVS). SFM algorithms produce camera poses and sparse point clouds with raw images as input [13]. Then, MVS is used to generate depth and normal maps for each image, which can be converted either to dense point clouds or meshed surfaces [14].

To deal with such challenges, view selection algorithms have been proposed in order to select a representative subset of input images [9,10]. Moreover, each

S. Sclaroff et al. (Eds.): ICIAP 2022, LNCS 13232, pp. 114–124, 2022.
https://doi.org/10.1007/978-3-031-06430-2_10

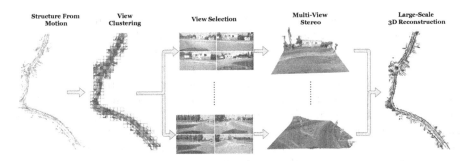

Structure From Motion **View Clustering** **View Selection** **Multi-View Stereo** **Large-Scale 3D Reconstruction**

Fig. 1. Overview of the proposed approach.

image shares common visibility information only with a local group of neighboring views, therefore the whole dataset can be divided in partially overlapping clusters to be processed simultaneously and independently [5,8,11,16]. Several approaches have been developed to reconstruct large buildings using community photo collections [1,5,16]. These works rely on unique and recurrent architectonic features that belong to the building of interest, when computing shared visibility information among images. On the other hand, urban scenarios exhibit a peculiar set of challenges and they have rarely been targeted in detail [2]. First of all, the presence of large portions of textureless surfaces, such as roads, vegetation or sky, makes it difficult to extract reliable keypoints during the SFM phase. Secondly, moving vehicles provide an unprecedented amount of data. In principle, this should enable 3D reconstruction to scale up from single buildings to entire cities. However, existing algorithms are limited to significantly smaller datasets and therefore impractical in the considered setting.

In this paper, a novel framework for efficient view clustering and selection for images of urban scenarios is presented. The proposed approach is specifically devoted to city-scale 3D reconstruction and designed to solve the aforementioned issues, assuming an arbitrarily large dataset of images acquired by a moving vehicle. The method overview is shown in Fig. 1: starting from the output of a SFM module, the view clustering algorithm (Sect. 3) builds uniform clusters in the observed world, while the view selection step (Sect. 4) computes the optimal subset of views in parallel for each cluster. Then, each cluster is reconstructed separately with any MVS method of choice and partial results are merged to get the full 3D model of the scene.

2 Related Work

As previously discussed, large-scale 3D reconstruction poses scalability issues, which mainly arise in literature in the context of architectural datasets collected from unstructured Internet images. Furukawa *et al.* [5] propose to perform global view selection on the whole set of images to remove redundancy and to build a visibility graph with the remaining cameras. These similarities are collected in a matrix that represents the adjacency matrix of the visibility graph. Then, an optimization procedure is applied to iteratively divide the graph into clusters using normalized-cuts, enforcing a size constraint, and add cameras back if a coverage constraint is violated. This process is repeated until convergence.

Several other approaches are based on this visibility graph formulation. Ladikos *et al.* [8] apply spectral graph theory on the similarity matrix and use mean shift to select the number of clusters, while Mauro *et al.* [11] employ the game theoretic model of dominant sets to find regular overlapping clusters. In a subsequent work [9], Mauro *et al.* propose to place selection after clustering and formulate an ILP problem with cameras as binary variables. The goal is to select the minimum number of views for each cluster, such that coverage and matchability are guaranteed. The same approach is adopted in this work. However, their formulation requires to execute the expensive Bron-Kerbosch algorithm [3] for each keypoint in the cluster. Since neighboring keypoints are likely to share the same camera subgraph, a more efficient alternative is provided in Sect. 4.

Furthermore, all the methods presented so far cluster images according on their relative visibility information and without considering the 3D structure of the scene. Zhang *et al.* [16] suggest to perform joint camera clustering and surface segmentation, which are formulated as a constrained energy minimization problem. Similarly, the clustering algorithm proposed in this work operates directly in 3D by exploiting the approximately uniform distribution of poses and geometry as produced by a moving vehicle.

It must be underlined that the naive solution of clustering images using temporal information from videos [2] is not ideal, since it does not consider multi-camera settings, where optimal viewpoints might be acquired by different sensors at very different time instants, and it fails when certain regions of the world are observed multiple times, since new clusters would be created each time, despite belonging to the same region.

For these reasons, existing works designed for architectural Internet datasets fail to tackle efficiently the unique set of challenges posed by large-scale urban 3D reconstruction. Therefore, the proposed approach makes the following contributions: (i) a novel framework that targets specifically urban scenarios and data acquired by a moving vehicle; (ii) a clustering algorithm (Sect. 3) that is independent from pairwise relationships between cameras; (iii) a more efficient formulation of the ILP problem for view selection (Sect. 4) with respect to [9].

3   View Clustering

In the considered setup, images are acquired by a moving vehicle with a sensor suite of N_{cams} cameras with framerate f and processed by a SFM module to obtain intrinsic and extrinsic parameters, as well as a set of sparse keypoints, for each image. These data are the input to the view clustering algorithm, which starts by building a 2D grid on the (x, y) plane with block size (x_b, y_b) and overlap $d_{overlap}$, given the input range of the sparse keypoints. This process is shown in Fig. 2 (left): since the majority of clusters is empty and several blocks

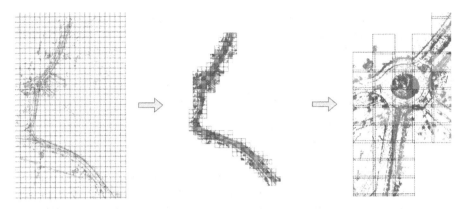

Fig. 2. View clustering example: the raw 2D grid built from SFM (left), the full sequence clustered and filtered (center) and overlapping clusters in more detail (right). Each cluster is represented with its borders and in a different color.

contain noisy keypoints very distant from the vehicle trajectory, a filtering step is required. To this end, each keypoint is assigned to the corresponding block simply by projecting it onto the (x, y) plane and empty clusters without associated points are removed. In this way, the sparse input point cloud is divided into partially overlapping sets of keypoints which can be processed independently and in parallel.

At this point, cameras must be associated to clusters. Intuitively, the assignment should be done based on the number and the quality of keypoints seen by each camera for each cluster. However, in urban scenarios with large textureless areas, the SFM module produces typically very few keypoints. Moreover, they are mostly concentrated in small textured patches of the world. Therefore, the 3D information of each non-empty cluster is augmented by sampling uniform points with resolution r within the boundaries of the corresponding block. This allows cameras to be assigned in clusters within their field of view even if SFM did not extract any keypoint at that specific location.

Then, each set of points is projected onto each camera lying within a given distance from its centroid and all the cameras that see at least one point in the cluster are associated to it. Finally, clusters with a few number of cameras are iteratively merged with their neighbors, until a minimum target set of views is reached for each cluster. This lower bound is set to 10 views in experiments, in order to provide enough information for 3D reconstruction. The output of the view clustering algorithm is a set of C independent clusters of cameras with common visibility, shown in Fig. 2 (center) for a full sequence and in Fig. 2 (right) in a greater detail.

The key novelty of the proposed method is that it does not require the computation of complex pairwise relationships between cameras for the whole dataset, differently from previous literature [5,9]. The use of the full similarity matrix quadratically scales as $O(N^2)$ and becomes quickly impractical in urban scenarios where up to $N = 60 \times f \times N_{cams}$ images are acquired every minute. Using the presented approach, a camera can be associated to at most K neighboring clusters and this redundancy is independent from the size of the scene. In the worst case scenario, each cluster has $N_c = K\frac{N}{C}$ cameras associated. Even in this unlikely case, building C separate and smaller visibility graphs is still significantly cheaper, requiring $O\left(\sum_{c=1}^{C} N_c^2\right)$ operations:

$$\sum_{c=1}^{C} N_c^2 = \frac{K^2 N^2}{C} < N^2 \iff K < \sqrt{C} \tag{1}$$

In practical situations, a camera is associated at most to a cluster and its immediate neighbors (i.e. $K \leq 8$ by design), while the number of clusters quickly grows to several hundreds or thousands as the vehicle moves. This allows the approach to scale up to arbitrarily large scenarios, since the improvement gap grows as the dataset size increases. Furthermore, each block can be independently processed, thus boosting up the performance thanks to the high degree of parallelization that can be obtained.

4   View Selection

At the end of the view clustering algorithm, every camera is added to all clusters where even a single associated point is present. Since this is likely to produce a highly redundant set of cameras for each block, a view selection algorithm is needed in order to choose the optimal subset of views for each cluster. In this context, *optimal* means the smallest subset of cameras that guarantees that each point in the cluster is seen by at least N_{vis} cameras and each camera have at least N_{match} other cameras to be successfully matched with.

Two cameras are considered to be *matchable* if they see a sufficient number of common points. This differs from previous literature, where the typical similarity measure is the average Gaussian-weighted triangulation angle between the two camera centers and the common keypoints [5,9]. In urban scenarios, with very sparse features, this is not a reliable metric and it would require tuning the Gaussian parameters for each cluster, due to the high variability and sparsity of urban keypoints. Note that as the sampling resolution $r \to 0$, the number of common points is essentially a measure of the intersection between the two camera frustum and the cluster itself.

Therefore, an Integer Linear Programming (ILP) problem is formulated for each cluster, with binary variables $x_i \in \{0, 1\}$ representing cameras:

$$\min \sum_{i=1}^{N_c} x_i$$
$$\text{s.t.} \sum_{i=1}^{N_c} x_i \geq N_{min} \tag{2}$$
$$\mathbf{A}_i^\top \mathbf{x} \geq 0 \qquad \forall i = 1, \ldots, N_c$$
$$\mathbf{B}_j^\top \mathbf{x} \geq N_{vis} \qquad \forall j = 1, \ldots, P_c$$

The first constraint requires each camera in the cluster to have at least N_{match} matchable cameras. A linear formulation can be derived from the similarity matrix $\mathbf{S}_c$ of the considered cluster as follows. Let $\tilde{\mathbf{S}}_c = \mathbf{S}_c > 0$ a binary matrix with $\tilde{s}_{c,ij} = 1$ if cameras i and j share common keypoints, 0 otherwise. The constraint vectors $\mathbf{A}_i$ are computed as the rows of the matrix $\mathbf{A} = \tilde{\mathbf{S}}_c - N_{match} \cdot \mathbf{I}$, being $\mathbf{I}$ the identity matrix of size $N_c \times N_c$. This formulation effectively activates the constraint only for the selected cameras, ignoring the others.

The visibility constraint vectors $\mathbf{B}_j$ for each point in the cluster are composed by binary coefficients $b_{ij} = 1$ if the point j is visible in camera i, 0 otherwise. This is the first improvement with respect to the ILP formulation in [9], where the coverage constraint requires that each point must be seen by at least one clique in the visibility subgraph associated to that point. This formulation implies that the Bron-Kerbosch algorithm for finding maximal cliques needs to be executed separately for each point. However, such requirement is both inefficient and redundant, since the same set of cameras is likely to see multiple points.

As a second important difference, the efficiency of the algorithm is boosted further by providing a good initial guess to the ILP solver. A minimum number of cameras N_{min} must be selected for each cluster, in order to guarantee enough information for the reconstruction. This threshold is adaptively selected according the cluster size and clamped between boundary values N_{low} and N_{high}. Then, cameras are sorted by the number of visible points and force the solver to select the N_{min} views with the best visibility.

5   Results

5.1   Experimental Setup

The proposed framework has been implemented in C++ and tested on a consumer Intel Core i5-5300U 2.30 GHz CPU. The code relies on the open-source library OR-Tools [12] for solving the ILP problem during view selection. The approach is independent from the specific choice of both the SFM module producing the required input and the MVS software generating the dense reconstruction. Camera poses and sparse keypoints are obtained by a custom implementation of SFM with bundle adjustment, while the state-of-the-art MVS algorithm proposed in [15] has been chosen, with the implementation provided by the authors.

To the best of our knowledge, the only two available large-scale urban datasets with a multi-camera setting are nuScenes [4] and the recently released DDAD [6]. However, they contain short sequences (20 s and 5–10 s, respectively) at a low framerate (12 Hz and 10 Hz, respectively), making it difficult to evaluate the proposed approach on those data. Therefore, the algorithm is validated on custom sequences acquired in the city of Parma, Italy, representing diverse real-world situations. The vehicle is equipped with $N_{cams} = 7$ cameras with resolution 3840×1920 and framerate $f = 30$ Hz.

The following set of parameters has been used for experiments: block size $(x_b, y_b) = (20, 20)$ m, with $d_{overlap} = 2$ m; sampling resolution $r = 1$ m; $N_{vis} = N_{match} = 2$, $N_{low} = 10$ and $N_{high} = 30$ for view selection.

Table 1. Quantitative results for each dataset and each stage of the pipeline.

Dataset	Seq. 1	Seq. 2	Seq. 3	Seq. 4	Seq. 5
# views (N)	5131	6782	8477	9912	11956
# keypoints (P)	389621	321696	349616	372434	393246
# clusters (C)	156	173	200	261	324
N after clustering	27103	32757	39014	49758	58364
K after clustering	5.28	4.83	4.60	5.02	4.88
Avg. N_c after clustering	173.75	189.34	195.07	190.64	180.13
$t_{clustering}$ (s)	22.35	28.52	32.7	37.32	40.89
N after selection	4951	5842	6266	8411	10138
K after selection	0.96	0.86	0.74	0.84	0.85
Avg. N_c after selection	31.73	33.76	31.33	32.22	31.29
$t_{selection}$ (s)	50.54	76.13	92.4	124.85	144.02
t_{tot} (s)	72.89	104.65	125.1	162.17	184.91

5.2 Quantitative Evaluation

Table 1 provides a quantitative description of the input sequences and the algorithm results for each stage of the pipeline. The first thing to note is that urban data contain relatively few keypoints (P), when compared to architectural image sets with similar size [5]. This justifies the choices of sampling additional points for each cluster and avoiding similarity measures based on the triangulation angle. Secondly, the clustering phase produces extremely redundant data (N after clustering), with each view assigned to approximately 5 different clusters (K after clustering), on average. Most of the clustering time ($t_{clustering}$) is spent for the camera association procedure, since several hundreds of views must be tested for each cluster. When processing extremely large sequences with millions of images, this phase can be naturally parallelized since each cluster is independent from the others, after the sampling step. Moreover, the optimization algorithm during view selection exploits the clustering redundancy to automatically remove useless views. The number of output views (N after selection) is consistently lower than the input for each sequence, even considering that some cameras covering the overlap between two clusters are assigned to both. Finally, the efficiency condition imposed in Eq. (1) ($K < \sqrt{C}$) is satisfied for each sequence by a large margin and this gap increases with the dataset size.

Direct quantitative comparison with state-of-the-art approaches is difficult, since they all target a substantially different scenario [5,16], where the assumption of uniformly distributed poses is violated. In terms of the running time as a function of the input number of images, Fig. 3 shows that the proposed method is much faster and scales linearly with the dataset size, while [5,16] both require several hours to process a few thousands of images. This demonstrates that they are not suited for urban applications, where thousands of images are acquired every minute by the vehicle. The considered ILP baseline [9] shares the same

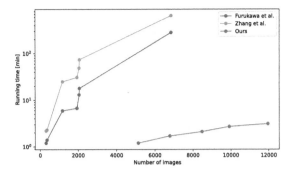

Fig. 3. Running time comparison between proposed method and state-of-the-art solutions [5,16]. The Y axis is in log-scale for better visualization.

issue and numerical performances are available only for datasets smaller by an order of magnitude. The reported runtime for 706 images divided into 36 clusters is around 3 min. However, the Bron-Kerbosch algorithm scales exponentially as $O(3^{\frac{n}{3}})$ with the number of cameras n. In the collected sequences, an instance of such algorithm with $n \approx 10^2$ would be executed for each keypoint of each cluster, making the approach impractical even for very short scenes. On the other hand, the proposed framework can approximately process 4000 images per minute.

5.3 Qualitative Evaluation

Figure 4 provides a visualization of clustered cameras, before and after view selection, in order to show how redundancy is exploited and reduced by the optimization algorithm. Three main situations arise in urban data: (i) when all the views lie along a straight line outside the cluster boundaries (Fig. 4a, left), the ILP solver selects a well-distributed subset of cameras (Fig. 4b, left); (ii) when the vehicle trajectory intersects the cluster (Fig. 4a, center), two disjoint sets of cameras are selected (Fig. 4b, center); (iii) for complex trajectories such as roundabouts (Fig. 4a, right), the framework generalizes well by selecting cameras from diverse viewpoints (Fig. 4b, right).

Furthermore, Fig. 5 shows that the proposed method effectively cluster images based on shared visual content, which allows to produce a detailed 3D reconstruction of the world. Each point cloud has been cropped at the corresponding cluster boundaries and it is stored in this way for the subsequent global fusion step. Only qualitative evaluation of the resulting reconstruction is provided, since performances depend solely on the choice of MVS algorithm. The goal of the presented approach is to show that 3D reconstruction can be achieved in very large-scale scenarios where processing all the images in a single batch is not practically feasible. For a detailed comparison of state-of-the-art MVS algorithms for large-scale outdoor scenes, the reader can refer to [7].

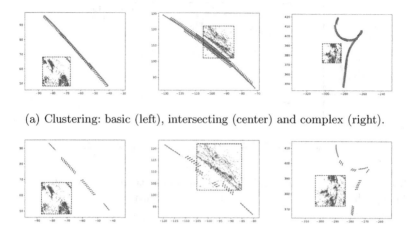

(a) Clustering: basic (left), intersecting (center) and complex (right).

(b) Selection: basic (left), intersecting (center) and complex (right).

Fig. 4. From clustering to selection: each cluster has black dashed borders, keypoints (blue) and cameras with viewing direction (red). (Color figure online)

Fig. 5. Clusters of images (left) and corresponding 3D point cloud (right).

Finally, a shorter dataset with $N = 1000$ images is considered, in order to have a relatively small instance of the problem where full reconstruction in a single batch is still possible. The proposed algorithm selects a subset of 786 views (21.4% less than N), divided into $C = 26$ clusters. Then, MVS is executed separately for each cluster and results are combined into a global 3D model. Figure 6 shows a comparison of the point clouds obtained by considering the whole set of images at once (left) and by merging multiple clusters (right). From a quantitative point of view, the full point cloud contains 3.08×10^6 points, which is approximately 7.8% more than the 2.84×10^6 points belonging to the result of the presented framework. This variation is lower than the difference between the input data size and it is mostly concentrated in ambiguous regions, such as the rotary center with identical trees, where MVS benefits from higher redundancy. Furthermore, increasing the N_{min} parameter always reduces this gap.

While ground truth for numerical evaluation in 3D is not available, it can be seen that the proposed *divide-and-conquer* approach maintains a good reconstruction quality, while being able to scale up to entire cities, where batch reconstruction is not an option.

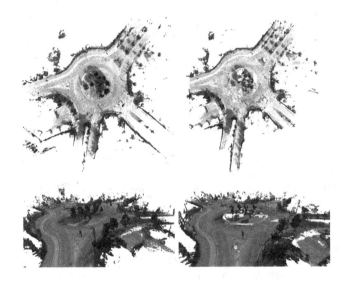

Fig. 6. Qualitative point cloud comparison between batch reconstruction (left) and merged local clusters computed with the proposed algorithm (right).

6   Conclusion

In this work, a method for enabling city-scale 3D reconstruction from images acquired by a moving vehicle has been proposed. The algorithm builds a set of partially overlapping clusters over the vehicle trajectory, then it selects the optimal subset of views to compute a 3D point cloud, independently for each cluster. All the local point clouds are then fused together in order to obtain a full 3D model of the sequence. The presented framework focuses on efficiency, by clustering images independently from their pairwise similarities and providing a novel formulation for the view selection step. These contributions reduce the processing time with respect to state-of-the-art methods, not designed for the urban scenario, allowing the algorithm to scale up to arbitrarily large datasets.

References

1. Agarwal, S., et al.: Building Rome in a day. Commun. ACM **54**(10), 105–112 (2011)
2. Akbarzadeh, A., et al.: Towards urban 3D reconstruction from video. In: Third International Symposium on 3D Data Processing, Visualization, and Transmission (3DPVT 2006), pp. 1–8. IEEE (2006)
3. Bron, C., Kerbosch, J.: Algorithm 457: finding all cliques of an undirected graph. Commun. ACM **16**(9), 575–577 (1973)
4. Caesar, H., et al.: nuScenes: a multimodal dataset for autonomous driving. arXiv preprint arXiv:1903.11027 (2019)
5. Furukawa, Y., Curless, B., Seitz, S.M., Szeliski, R.: Towards internet-scale multiview stereo. In: 2010 IEEE Computer Society Conference on Computer Vision and Pattern Recognition, pp. 1434–1441. IEEE (2010)
6. Guizilini, V., Ambrus, R., Pillai, S., Raventos, A., Gaidon, A.: 3D packing for self-supervised monocular depth estimation. In: IEEE Conference on Computer Vision and Pattern Recognition (CVPR) (2020)
7. Knapitsch, A., Park, J., Zhou, Q.Y., Koltun, V.: Tanks and temples: benchmarking large-scale scene reconstruction. ACM Trans. Graph. **36**(4), 1–13 (2017)
8. Ladikos, A., Ilic, S., Navab, N.: Spectral camera clustering. In: 2009 IEEE 12th International Conference on Computer Vision Workshops, ICCV Workshops, pp. 2080–2086. IEEE (2009)
9. Mauro, M., Riemenschneider, H., Signoroni, A., Leonardi, R., Van Gool, L.: An integer linear programming model for view selection on overlapping camera clusters. In: 2014 2nd International Conference on 3D Vision, vol. 1, pp. 464–471. IEEE (2014)
10. Mauro, M., Riemenschneider, H., Signoroni, A., Leonardi, R., Van Gool, L.: A unified framework for content-aware view selection and planning through view importance. Proc. BMVC **2014**, 1–11 (2014)
11. Mauro, M., Riemenschneider, H., Van Gool, L., Leonardi, R.: Overlapping camera clustering through dominant sets for scalable 3D reconstruction. Proc. BMVC **2013**(2013), 1–11 (2013)
12. Perron, L., Furnon, V.: Or-tools. https://developers.google.com/optimization/
13. Schonberger, J.L., Frahm, J.M.: Structure-from-motion revisited. In: Proceedings of the IEEE Conference on Computer Vision and Pattern Recognition, pp. 4104–4113 (2016)
14. Seitz, S.M., Curless, B., Diebel, J., Scharstein, D., Szeliski, R.: A comparison and evaluation of multi-view stereo reconstruction algorithms. In: 2006 IEEE Computer Society Conference on Computer Vision and Pattern Recognition (CVPR 2006), vol. 1, pp. 519–528. IEEE (2006)
15. Xu, Q., Tao, W.: Planar prior assisted patchmatch multi-view stereo. In: Proceedings of the AAAI Conference on Artificial Intelligence, vol. 34, pp. 12516–12523 (2020)
16. Zhang, R., Li, S., Fang, T., Zhu, S., Quan, L.: Joint camera clustering and surface segmentation for large-scale multi-view stereo. In: Proceedings of the IEEE International Conference on Computer Vision, pp. 2084–2092 (2015)

Digital Forensics and Biometrics

U Can't (re)Touch This – A Deep Learning Approach for Detecting Image Retouching

Daniel Aumayr$^{(\boxtimes)}$ and Pascal Schöttle🆔

Digital Business and Software Engineering, Management Center Innsbruck, Innsbruck, Austria
d.aumayr@mci4me.at, pascal.schoettle@mci.edu

Abstract. Mobile applications make it easier than ever to edit images and share these images on the Internet. Even if this image manipulation is not a criminal offense in most cases, retouched images can have a number of negative effects.

The goal of this work is to provide a solution for an automatic recognition of image retouching. For this, we adapt a well known convolutional neural network (CNN) from [5] to the domain of RGB images and create a data set to train it. Specifically, we process 1 000 images with nine different filters using Snapseed, an image editing app. Then, we use these 10 000 images in two different experiments to train the adapted CNN. The first experiment compares each single filter with the original images, while the second experiment tries to distinguish all ten classes at once.

Overall, the first experiment achieves an accuracy of over 99% for most classes, and the second experiments a total accuracy of roughly 88%. Finally, we analyze, why some of the classes have lower accuracies.

Keywords: Image forensics · Image retouching · Deep learning · Convolutional neural network

1 Introduction

Regular use of Facebook, Instagram, and other social media has become the norm in recent years. Excessive use or chasing after strangers can have a negative impact on the psychological well being of young people. When consuming images of other users, some of which have been heavily edited, they compare themselves with an image that does not correspond to reality [18].

Governments are also getting involved in this issue. Norway sees the spread of pictures in which bodies or body parts are retouched as a reason for body dysmorphic disorder - excessive concentration on slight or imagined effects in the appearance [26]. The Norwegian government has therefore changed the marketing law and "influencers" have to flag edited photos in paid posts on social media such as Instagram.[1] Also researchers addressed raising concerns about the dependability of visual information on social media platforms [21].

---

[1] https://www.bbc.com/news/newsbeat-57721080 [Last accessed: April 16, 2022].

© The Author(s), under exclusive license to Springer Nature Switzerland AG 2022
S. Sclaroff et al. (Eds.): ICIAP 2022, LNCS 13232, pp. 127–138, 2022.
https://doi.org/10.1007/978-3-031-06430-2_11

The detection of image modifications traditionally belongs to the field of image forensics. Here, the main focus is on several forgery techniques such as copy-move, image splicing, and re-sampling. These techniques are increasingly being used for illegal applications. Post-processing or retouching of images is classified as relatively harmless and therefore generally has a lower priority [20].

Furthermore, there is an increasing number of different applications for image processing aside from (semi-)professional image editing software such as Adobe Photoshop or Lightroom. Mobile applications such as Instagram,[2] Snapseed,[3] or Pixlr[4] become more and more popular and have not yet been examined as closely by the image forensics community [23,24].

In recent years, thanks to technical developments, deep learning approaches have found their way into image forensics. In comparison to classical methods, the deep learning algorithms have a good performance and have been able to establish themselves in the field of image forensics [5,7,29].

The remainder of this paper is organized as follows: Sect. 2 introduces the necessary theoretical foundation and covers the current state of the art on manipulation detection. Section 3 presents the design of our experiments. Specifically, we introduce the data set, the architecture of the neural network and the different experiments, before we present the experimental results in Sect. 4. Section 5 concludes the paper and discusses further research.

2 Related Work

Manipulating images has never been easier, therefore the field of multimedia forensics becomes more relevant these days. Layman can edit media data with editing software, without deep background knowledge. Multimedia forensics tries to authenticate digital sensor data, by developing tools to discover manipulations or to deduce knowledge about the source device [8].

Hereby, multimedia forensics belongs to the broader research area of multimedia security that also includes digital watermarking and steganography. Digital watermarking modifies a given multimedia object by adding information about the properties or content of the media data, with the aim of confirming its authenticity [10]. In steganography and steganalysis you try to hide respectively identify information in media data. The goal of steganography is to conceal the existence of a hidden message by preserving the image properties. Steganalysis unveils hidden information without having access to the original image [15].

So, while digital watermarking and steganography aim at modifying images to achieve their purpose, image forensics deals with the detection of image modifications. E.g., by analyzing the intrinsic fingerprints of devices camera-based or editing-based clues can be identified and a conclusion about the authenticity can be drawn [25]. Those methods need prior knowledge of the used device and are not as suitable for real-world problems [27].

---

[2] https://www.instagram.com/.

[3] https://snapseed.online.

[4] https://pixlr.com/.

So-called one-class methods need a large set of untampered data from the camera of interest for the analysis. Manufacturing imperfections create small deviations in the sensor elements and create noise-like patterns called photo-response non-uniformity (PRNU) noise. The PRNU-forgery detection is very powerful and can detect all kinds of forgery attacks but needs prior knowledge. If the given noise is missing in an area, a manipulation has likely happened. The first approach for PRNU-based forgery detection was proposed in [17]. Newer methods further developed the approach and use deep learning with noiseprints instead of PRNU [12]. Blind methods have been introduced, which rely only on the given data and try to detect anomalies in order to infer tampering. Blind methods cover lens distortion [30], color filter array artifacts [14], noise level and noise pattern [11], compression artifacts [22] and editing artifacts [2,3].

All blind methods look for divergent traces which are left behind by different image processing, therefore researchers extract the features of those traces to detect a specific image manipulation. Researchers can successfully detect image manipulations such as resampling, resizing, median filtering, contrast enhancement, or multiple JPEG compression by extracting features related to tampering traces and then identifying the used manipulation technique [7].

In recent years, convolutional neural networks (CNN) have found their way into the field of multimedia forensics, and researchers introduced new methods for identifying different manipulation methods [4,19].

A universally applicable approach to recognize image manipulations was presented by Bayar and Stamm in [5]. The aim of their approach is to modify a CNN in such a way that it focuses on the properties of the image manipulations instead of the content of an image. This is implemented by introducing a new convolutional layer, the so-called constrained convolutional layer, that recognizes the manipulations without assigning a priority to the individual processing methods [5,7]. The experiments carried out in their works consider five different editing operations: median filtering, gaussian blurring, additive white gaussian noise, resampling using bilinear interpolation, and JPEG compression. Both, single manipulation and multiple manipulation detection for these five filters with a data set of 60 000 respectively 132 000 256 × 256 grayscale images were examined. Their results showed an accuracy of over 99% in most cases [7].

3   Experimental Setup

In this section, we describe the data set for our experiments, elucidate the CNN architecture, which is inspired by [5,7], and detail our experimental set-up and evaluation metrics.

3.1   Data Set

As we are not aware of publicly accessible data sets with color images that were post-processed using a mobile application, we decided to create a new data set.[5]

---

[5] https://github.com/DanielAumayr/U-Can-t-re-touch-this.

We decided to use 1 000 landscape and architecture images from the RAISE data set [13]. Originally, we planned to apply common Instagram filters to the selected images, as Instagram is undoubtedly the most popular mobile image editing application. However, because of the algorithm introduced by Instagram in 2018, which recognizes automated postings and bans the relevant accounts, an alternative approach had to be found [1] and thus we decided for using Snapseed and the following processing pipeline:

First, we used scikit-image (skimage) [28] to crop the original images from RAISE to a square format, resize them to 256 × 256 pixel and then save them in JPEG format with quality factor 100.

Since Snapseed is a mobile application, the Android emulator Bluestacks[6] is used to edit the photos with Snapseed. In Snapseed, we processed the 1 000 images with nine different filters, creating a data set of 10 000 images in total.

The filters we used to create our retouched images (see Fig. 1) are numbered as follows:

01 Original – original image
02 Smooth – blur
03 Pop – higher contrast
04 Accentuate – higher contrast and more intensive colours
05 Morning – bottom part of image has warm colouring
06 DramaDark – higher contrast, image will be darker
07 Vintage – Vintage Look, brown-filter
08 Retrolux8 – image gets a used look, like old pictures
09 Crunch3 – image gets a texture and dark vignette
10 Grain104 – grain is added to image

3.2 Architecture

Bayar and Stamm [5,7] presented a model in their work that is able to recognize processing in grayscale images. We use their CNN architecture but adapt the first layer such that it can process RGB color images.

The most important property of the suggested CNN architecture is its first convolutional layer, termed constrained convolutional layer. When classic CNNs such as ImageNet [16] are used to detect manipulation, they learn the content of the image and usually not the fingerprints of a manipulation. To prevent this, the constrained convolutional layer is supposed to suppress the content of the image and the CNN adaptively learns the local pixel relationships and is thus able to detect traces left behind by image manipulation attempts [5,7,9]. This happens consciously in the first layer, as this takes over the forensic activity.

To do this, the 5 × 5-filters in the constrained convolutional layer are adaptively learned but have to fulfill the following constraints [7]:

$$\begin{cases} w_k^{(1)}(0,0) & = -1, \\ \sum_{m,n \neq 0} w_k^{(1)}(m,n) & = 1, \end{cases} \tag{1}$$

---

[6] https://www.bluestacks.com/de/index.html.

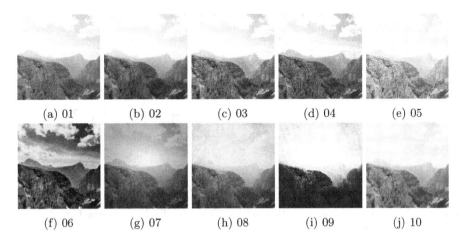

Fig. 1. Exemplary images: first row from left to right filters 01–05, second row from left to right filters 06–10 (best viewed in color)

where the superscript $^{(1)}$ indicates the first layer, and the subscript $_k$ the k-th convolutional filter in this layer. The $(0,0)$ represents the central point of the filter and (m,n) the remaining positions of the 5×5-filter [5,7]. In other words, this means that after updating the filter values, e.g., with stochastic gradient descent (SGD), the central value of the filter is set to -1 and the surrounding coefficients are normalized so that their sum is 1 again. This normalization is implemented by dividing each individual weight by the sum of all weights excluding the central value. The filters $w_k^{(1)}$ are updated in every iteration.

Overall, the CNN, as depicted in Fig. 2, consists of five convolutional layers (of which the first is the constrained convolutional layer as described above), three max-pooling layers, one average-pooling layer, and three fully-connected layers. As the extraction of the features in the constrained convolutional layer is very unstable and can easily be destroyed by pooling or activation [6], the result of the first layer will be passed on unchanged to the second convolutional layer [7]. Conceptually, Bayar and Stamm [7] divide the CNN into four blocks: Prediction Error Feature Extraction, Hierarchical Feature Extraction, Cross Feature Maps Learning, and Classification.

Note that in contrast to the CNN from [7], we adapted the first layer to the dimensions $256 \times 256 \times 3$ and the second layer to the dimensions $128 \times 128 \times 96$.

3.3 Experiments Design

Similar to the approach in [7], we designed two different experiments to assess the performance of our adapted CNN. In the first experiment, we trained nine different CNNs with binary output, one for each of the nine filters (cf. Sect. 3.1) separately. This means, that each of this nine CNNs is trained on a combination of original (unfiltered) images and the counterparts all retouched by one specific

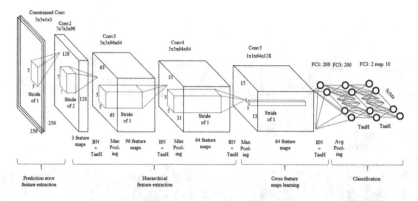

Fig. 2. CNN Architecture, BN: Batch-Normalization; TanH: Tangens hyperbolicus; SoMa: Soft-max; adapted from [7]

filter. These CNNs are then supposed to differentiate untouched from retouched images in the inference phase.

In the second experiment, we train the CNN with images from all ten classes at the same time and in the inference phase, the CNN is supposed to determine not only if an image was retouched but also which of the 9 filters was used.

All experiments are carried out on an Intel Core i7-9700K CPU with 3.60 GHz. The GPU used is a NVIDIA GeForce RTX 2070 with 8 GB of display memory and 16 GB of shared memory and DDR4-3000 RAM with 32 GB memory. The experiment is implemented in Jupyter Notebook 6.1.4 with TensorFlow 2.4.1.

The images are divided into a training data set and a test data set with a division of 80% to 20%. The images are saved as an array with $256 \times 256 \times 3$ with values between 0 and 255 of the data type Float32. The batch size is set to 64 images per batch [7] for training and testing.

We use Stochastic Gradient Descent for training with the following parameters: In the first experiment (two classes) we use a momentum of 0.95, a decay of 10^{-6} and learning rate of 10^{-3}. The training runs until a loss (on the training data) of less than or equal to $5*10^{-5}$ has been reached or 600 epochs have been trained. In the second experiment (ten classes) we use a momentum of 0.95, a decay of 10^{-7} and learning rate of 10^{-3}. The training runs until a loss of less than or equal to $5*10^{-5}$ has been achieved on the training data set or 800 epochs have been run through.

3.4   Evaluation

After completing the training, we measure true positive (TP), true negative (TN), false positive (FP), and false negative (FN) rates on the test data. Then, we calculate the Accuracy (2), Recall (3), Precision (4), and F1-Score (5).

Table 1. Mean accuracy (and standard deviation) of all nine classifiers in the binary comparison and the number of epochs trained

Filter	Accuracy (std)	# of Epochs
Smooth (02)	74.40% (2.705)	464
Accentuate (03)	99.60% (0.300)	334
Pop (04)	99.10% (0.464)	273
Morning (05)	99.35% (0.255)	162
DramaDark (06)	99.30% (0.458)	113
Vintage (07)	96.65% (0.903)	161
Retrolux (08)	99.60% (0.255)	150
Crunch (09)	99.40% (0.490)	158
Grain (10)	99.25% (0.570)	217

$$ACC = {}^{TP+TN}\!/_{TP+TN+FN+FP} \tag{2}$$

$$recall = {}^{TP}\!/_{TP+FN} \tag{3}$$

$$precision = {}^{TP}\!/_{TP+FP} \tag{4}$$

$$F1 - Score = {}^{2}\!/_{precsion^{-1}+recall^{-1}} \tag{5}$$

Furthermore, for the experiment with the ten classes, we additionally report confusion matrices. All experiments are carried out five times and we report mean values and standard deviations of these five runs.

4 Results

In this section, we present the results of the experiments laid out above. Furthermore we conduct an additional analysis, where we take a closer look into the misclassification of certain filters.

4.1 Experiments with 2 Classes

In the first experiment, we compare the performance of all nine trained CNNs, as described above. Table 1 shows the collected results, where each row compares the respectively filtered images with non-filtered ones.

As can be seen in Table 1, the CNN achieved an accuracy of over 99% with seven of the nine filters and acceptable standard deviations. When training with the Vintage filter (07), an accuracy of 96.65% and a standard deviation of approx. 0.90 was achieved on the test data set. The only filter that has an accuracy of less than 95% is the Smooth filter (02), where the CNN only achieves an accuracy of 74.40% with a standard deviation of approx. 2.71.

Furthermore, we can see in Table 1 that none of the nine CNNs needed the full 600 epochs to reach the early stopping criterion.

Table 2. Accuracy multi-class classification

Run (initialization value)	# of Epochs	Accuracy
1 (24)	800	87.90%
2 (69)	800	88.55%
3 (420)	800	89.40%
4 (8)	800	87.75%
5 (42)	800	88.20%
Mean	800	88.36% (0.59)

Table 3. Confusion matrix with absolute values

Filter	01	02	03	04	05	06	07	08	09	10
Original (01)	164	41	1	1	0	0	1	0	1	1
Smooth (02)	37	156	0	0	0	0	6	0	0	1
Accentuate (03)	0	0	144	56	1	5	0	0	0	0
Pop (04)	1	0	57	138	2	0	0	0	0	0
Morning (05)	0	0	2	5	190	0	0	0	0	0
DramaDark (06)	0	0	7	0	0	193	0	0	0	0
Vintage (07)	0	3	0	0	0	0	191	0	0	0
Retrolux (08)	0	0	0	0	0	0	1	193	0	0
Crunch (09)	0	0	0	0	0	0	0	0	210	0
Grain (10)	0	0	0	0	0	0	0	0	0	189

4.2 Experiment with 10 Classes

In the second experiment, we conducted a multi-class classification. The original images and all nine edited variants were fed into the CNN at the same time. At inference, the CNN is supposed to assign the correct filter to the tested images.

Table 2 shows the accuracies of all 5 splits of the training data. As can be seen, the CNN achieves an average accuracy of 88.36% with a standard deviation of 0.59. No training run was interrupted prematurely.

Table 3 shows a confusion matrix of the predictions of the trained model. The data shown is the arithmetic mean of the five training runs. The columns represent the predictions, the rows represent the ground truth labels and thus, the diagonal shows the correctly assigned images. As can be seen, the filters Morning (05), DramaDark (06), Vintage (07), Retrolux (08), Crunch (09), and Grain (10) are predicted very well. These 6 filters have a recall and precision between 95% and 99%, as can be seen in Table 4.

The remaining four filters are confused more often and also possess significantly lower recall, precision, and F1 scores. But, a closer look at Table 3 reveals that the filters Original (01) and Smooth (02) are almost exclusively confused among each other and the same holds for the two filters Accentuate (03) and Pop (04).

Table 4. Recall, Precision, and F1 scores for multi-class experiment

Filter	Recall	Precision	F1-score (std)
Original	78.55%	80.98%	79.53% (0.026)
Smooth	77.89%	78.08%	77.76% (0.018)
Accentuate	69.79%	68.39%	69.01% (0.017)
Pop	69.76%	68.89%	69.18% (0.022)
Morning	95.97%	98.43%	97.18% (0.004)
DramaDark	96.34%	97.31%	96.82% (0.008)
Vintage	97.93%	96.12%	97.01% (0.012)
Retrolux	99.46%	99.80%	99.63% (0.003)
Crunch	99.72%	99.62%	99.66% (0.003)
Grain	99.76%	98.87%	99.31% (0.001)

In summary, Table 4 also shows the F1 scores (and standard deviations) of all filters. The F1 score shows a similar picture as the recall and precision. The filters Morning, DramaDark, Vintage, Retrolux, Crunch and Grain have F1 scores between 96% and 99% with a standard deviation between 0.001 and 0.012. Original and Smooth have F1 scores of 79.53% and 77.76% with standard deviations of 0.026 and 0.018, respectively. Accentuate and Pop have the worst F1 scores of 69.01% and 69.18% with standard deviations of 0.017, resp. 0.022.

4.3 Similarity Evaluation Miss-Classified Images

As seen in Subsects. 4.1 and 4.2 the classes Original (01) and Smooth (02) as well as the classes Accentuate (03) and Pop (04) are confused more often than the other classes. As we do not know the exact workings of the Snapseed filters we used, we assume that these two "clusters" stem from the fact that the filters work in a similar way. To assess if this holds true, we conducted another experiment. Inspired by the workings of filters in general and the constrained convolutional layer in our CNN, we did the following: For each of the four classes mentioned above, we select all misclassified images. For each of these images, we consider all 10 versions. Then, we calculate filtered versions of all of these images by calculating the distance of central pixels to their 5×5 neighborhood. Finally, in Fig. 3, we report the Euclidean distances of all misclassified images from the classes Original (01) and Smooth (02) (top row) and the classes Accentuate (03) and Pop (04) (bottom row) to all other classes. It can clearly be seen that these classes are closest to each other thus explaining the respective misclassifications.

As we have shown that and why these misclassifications happen almost exclusively within these two "clusters", we could subsume the filters Original (01) and Smooth (02) in one class (depicted orange in the confusion matrix in Table 3) and also subsume the filters Accentuate (03) and Pop (04) as one class (depicted turquoise in the confusion matrix in Table 3). These two "superclasses" would

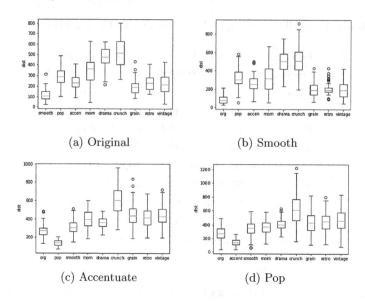

(a) Original

(b) Smooth

(c) Accentuate

(d) Pop

Fig. 3. Comparison Original (a), Smooth (b), Accentuate (c), Pop (d)

then have recall, precision, and F1 scores of 97.27%, 98,47%, and 97,87% (std: 0.007) for the superclass Original/Smooth and 97.75%, 96.10%, and 96.92% (std: 0.006) for the superclass Accentuate/Pop, respectively. Thereby, the performance is at par with that of the other classes.

5   Conclusion

In this paper, we introduced a data set with images processed with a mobile application and proposed a CNN architecture slightly adapted from Bayar and Stamm [7] to perform an image manipulation detection on RGB images. The proposed CNN is able to suppress the content of the RGB images and learn the image manipulation detection features without processing the images beforehand.

We have shown through a series of experiments, that our CNN can both detect single manipulations as well as differentiate between several manipulations.

We further analyzed the mismatch between the classes Original and Smooth as well as Accentuate and Pop in more detail and were able to show that these classes are very similar regarding the relation of the central pixel to its surrounding pixels in a 5 × 5 filter, thus explaining the misclassifications. Furthermore, subsuming the mismatched classes in two "superclasses" yields high accuracies again.

The sometimes poorer results in comparison to the original work by Bayar and Stamm [7] might be explained by the smaller data set. We plan to assess the effect of the size of the training set on classifier performance in future work.

Nonetheless, especially our multi-class approach shows that the automatic classification of retouched images is feasible. Thus, in response to recent government initiatives, social networking services like Instagram could automatically flag user's photos as very likely edited edited with a specific filter and thus taking their part in the responsibility to protect their users.

Acknowledgements. The second authors is supported by the Austrian Science Fund (FWF) under grant no. I 4057-N31 ("Game Over Eva(sion)").

References

1. Agung, N.F.A., Darma, G.: Opportunities and challenges of Instagram algorithm in improving competitive advantage. Int. J. Innov. Sci. Res. Technol. **4**(1), 743–747 (2019)
2. Amerini, I., Ballan, L., Caldelli, R., Del Bimbo, A., Del Tongo, L., Serra, G.: Copy-move forgery detection and localization by means of robust clustering with J-Linkage. Signal Process. Image Commun. **28**(6), 659–669 (2013)
3. Amerini, I., Ballan, L., Caldelli, R., Del Bimbo, A., Serra, G.: A SIFT-based forensic method for copy-move attack detection and transformation recovery. IEEE Trans. Inf. Forensics Secur. **6**(3), 1099–1110 (2011)
4. Amerini, I., Uricchio, T., Ballan, L., Caldelli, R.: Localization of JPEG double compression through multi-domain convolutional neural networks. In: 2017 IEEE Conference on Computer Vision and Pattern Recognition Workshops (CVPRW), pp. 1865–1871. IEEE (2017)
5. Bayar, B., Stamm, M.C.: A deep learning approach to universal image manipulation detection using a new convolutional layer. In: Proceedings of the 4th ACM Workshop on Information Hiding and Multimedia Security, pp. 5–10 (2016)
6. Bayar, B., Stamm, M.C.: Design principles of convolutional neural networks for multimedia forensics. Electron. Imaging **2017**(7), 77–86 (2017)
7. Bayar, B., Stamm, M.C.: Constrained convolutional neural networks: a new approach towards general purpose image manipulation detection. IEEE Trans. Inf. Forensics Secur. **13**(11), 2691–2706 (2018)
8. Böhme, R., Freiling, F.C., Gloe, T., Kirchner, M.: Multimedia forensics is not computer forensics. In: Geradts, Z.J.M.H., Franke, K.Y., Veenman, C.J. (eds.) IWCF 2009. LNCS, vol. 5718, pp. 90–103. Springer, Heidelberg (2009). https://doi.org/10.1007/978-3-642-03521-0_9
9. Chen, J., Kang, X., Liu, Y., Wang, Z.J.: Median filtering forensics based on convolutional neural networks. IEEE Signal Process. Lett. **22**(11), 1849–1853 (2015)
10. Cox, I., Miller, M., Bloom, J., Fridrich, J., Kalker, T.: Digital Watermarking and Steganography. Morgan kaufmann (2007)
11. Cozzolino, D., Poggi, G., Verdoliva, L.: Splicebuster: A new blind image splicing detector. In: 2015 IEEE International Workshop on Information Forensics and Security (WIFS), pp. 1–6. IEEE (2015)
12. Cozzolino, D., Verdoliva, L.: Camera-based image forgery localization using convolutional neural networks. In: 2018 26th European Signal Processing Conference (EUSIPCO), pp. 1372–1376. IEEE (2018)
13. Dang-Nguyen, D.T., Pasquini, C., Conotter, V., Boato, G.: RAISE: a raw images dataset for digital image forensics. In: Proceedings of the 6th ACM Multimedia Systems Conference, pp. 219–224 (2015)

14. Ferrara, P., Bianchi, T., De Rosa, A., Piva, A.: Image forgery localization via fine-grained analysis of CFA artifacts. IEEE Trans. Inf. Forensics Secur. **7**(5), 1566–1577 (2012)
15. Kirchner, M., Böhme, R.: Hiding traces of resampling in digital images. IEEE Trans. Inf. Forensics Secur. **3**(4), 582–592 (2008)
16. Krizhevsky, A., Sutskever, I., Hinton, G.E.: ImageNet classification with deep convolutional neural networks. Commun. ACM **60**(6), 84–90 (2017)
17. Lukáš, J., Fridrich, J., Goljan, M.: Detecting digital image forgeries using sensor pattern noise. In: Security, Steganography, and Watermarking of Multimedia Contents VIII, vol. 6072, p. 60720Y. International Society for Optics and Photonics (2006)
18. Lup, K., Trub, L., Rosenthal, L.: Instagram# Instasad?: exploring associations among Instagram use, depressive symptoms, negative social comparison, and strangers followed. Cyberpsychol. Behav. Soc. Netw. **18**(5), 247–252 (2015)
19. Marra, F., Gragnaniello, D., Verdoliva, L., Poggi, G.: A full-image full-resolution end-to-end-trainable CNN framework for image forgery detection. IEEE Access **8**, 133488–133502 (2020)
20. Meena, K.B., Tyagi, V.: Image forgery detection: survey and future directions. In: Shukla, R.K., Agrawal, J., Sharma, S., Singh Tomer, G. (eds.) Data, Engineering and Applications, pp. 163–194. Springer, Singapore (2019). https://doi.org/10.1007/978-981-13-6351-1_14
21. Pasquini, C., Amerini, I., Boato, G.: Media forensics on social media platforms: a survey. EURASIP J. Inf. Secur. **2021**(1), 1–19 (2021). https://doi.org/10.1186/s13635-021-00117-2
22. Pasquini, C., Boato, G., Pérez-González, F.: Statistical detection of JPEG traces in digital images in uncompressed formats. IEEE Trans. Inf. Forensics Secur. **12**(12), 2890–2905 (2017)
23. Piva, A.: An overview on image forensics. International Scholarly Research Notices 2013 (2013)
24. Shashidhar, T., Ramesh, K.: Reviewing the effectivity factor in existing techniques of image forensics. Int. J. Electr. Comput. Eng. (IJECE) **7**(6), 3558–3569 (2017)
25. Swaminathan, A., Wu, M., Liu, K.R.: Digital image forensics via intrinsic fingerprints. IEEE Trans. Inf. Forensics Secur. **3**(1), 101–117 (2008)
26. Lahousen, T., Linder, D., Gieler, T., Gieler, U.: Der Hautarzt **68**(12), 973–979 (2017). https://doi.org/10.1007/s00105-017-4064-7
27. Verdoliva, L.: Media forensics and DeepFakes: an overview. IEEE J. Sel. Top. Signal Process. **14**(5), 910–932 (2020)
28. Van der Walt, S., et al.: scikit-image: image processing in Python. PeerJ **2**, e453 (2014)
29. Yang, P., Baracchi, D., Ni, R., Zhao, Y., Argenti, F., Piva, A.: A survey of deep learning-based source image forensics. J. Imaging **6**(3), 9 (2020)
30. Yerushalmy, I., Hel-Or, H.: Digital image forgery detection based on lens and sensor aberration. Int. J. Comput. Vision **92**(1), 71–91 (2011)

A Robust and Efficient Overhead People Counting System for Retail Applications

Antonio Greco[1]([✉]), Alessia Saggese[1], and Bruno Vento[2]

[1] Department of Information Engineering, Electrical Engineering and Applied Mathematics (DIEM), University of Salerno, Salerno, Italy
{agreco,asaggese}@unisa.it
[2] A.I. Tech srl, Salerno, Italy
bruno.vento@aitech.vision

Abstract. Counting people with overhead cameras is a feature that is increasingly required by the retail market, for surveillance and business intelligence. However, despite the great advances in modern neural networks, it is far from simple to train effective systems in all possible real-world scenarios. The main problem is that the publicly available datasets are not sufficiently representative or are not annotated for this purpose. To this aim, in this paper we demonstrate that a considerable effort in the collection of heterogeneous data in real scenarios, producing a new dataset of about 30,000 images, allowed to realize an effective and efficient people counting system, able to process at least 10 FPS on board of three different types of smart cameras. In addition, the F-Score higher than 95% over the test set demonstrates the effectiveness of the proposed people counting system and its suitability for real applications.

Keywords: People counting · People detection · Smart camera

1 Introduction

Counting the people entering and exiting a gate has become a very important and not negligible feature for both business intelligence and safety purposes [10]. For example, the knowledge of the number of people inside a building may allow to close the entrance if the number exceeds the maximum allowed in the building for safety purposes [20]. Viceversa, the awareness of the number of people inside the shopping mall or inside a specific selling area in a certain day or week in retail environments, could allow to evaluate peak hours [3]. This is why, in the last years, we have assisted to an increasing interest towards the design and the development of people counting systems [16]. According to a study conducted by Research and Market[1] in 2020, the overall people counting system market is expected to grow from USD 818 million in 2020 to USD 1,333 billion by 2025, with an impressive CAGR of 10.31%.

A trivial and cheap solution for the people counting problem is the usage of existing surveillance systems [5]. Anyway, the performance achievable with a

---

[1] https://www.researchandmarkets.com.

S. Sclaroff et al. (Eds.): ICIAP 2022, LNCS 13232, pp. 139–150, 2022.
https://doi.org/10.1007/978-3-031-06430-2_12

standard surveillance camera, which frames the people frontally, is limited due to occlusions between people, caused by the view of the camera that do not allow to count all of them, especially in crowded environments. Thus, the only reliable solution is the introduction of overhead cameras and algorithms able to analyze the people from the top, instead of evaluating the full body [14].

Starting from this consideration, a first attempt in this direction has been made in [15]: the authors introduce a head detection method based on background subtraction and foreground mask extraction for blob detection, followed by the Hough transform for the identification of the circles associated to the head. Then, a tracking based on optical flow has been proposed. This algorithm typically fails in case of camouflage and shadows, due to errors in the search of the circles. In order to address this issue, in [2] the authors use a depth camera instead of a traditional RGB one. In addition, in [6] RGB and depth images are combined for improving the performance of the detection based on traditional background subtraction and foreground mask extraction. Instead of using a tracking algorithm, the authors introduce a *virtual sensor* based on integral operations that is resilient to errors in the foreground mask and very efficient. In [4] a radar sensor has been used to solve the problems related to the presence of camouflage and shadows; a detector based on a Recurrent Neural Network (RNN), able to simultaneously evaluate both spatial and temporal information, was applied to these data. Even if very promising, sensors different from RGB cameras are very expensive and also prevent the possibility to revamp already existing installations, where traditional RGB cameras are already positioned.

More recently, various convolutional neural networks (CNNs) have been proposed for overhead people counting. One of the first attempts in this direction has been proposed in [1], where a Single Shot multi-box Detector has been considered. Even if the idea is very promising, the dataset used by the authors is composed by only 500 images acquired in only two environments, one indoor and one outdoor. Thus, the achieved performance, about 95%, can not be considered enough representative of the different conditions that we can find in real situations. In [11] the authors combine a SSD detector with a SORT based tracking, training their system on a dataset composed of 1,500 images, where they achieved 83% of accuracy. Even if the performance is not very high if compared with the previously mentioned experiment and also can not be a warranty of the generalization capability of the system given the limited size of the dataset, the interesting point was the integration of the system over a Raspberry Pi, thus confirming the possibility to run such a category of algorithms over embedded systems with limited resources.

From the analysis of the recent scientific literature on this topic, two main critical points emerge. On one hand, there is no experimental evidence that the problem can be considered solved and that the systems are robust enough in different real conditions and ready for being deployed in real systems. On the other hand, the dataset can be considered an additional open issue with overhead people counting systems. Indeed, we can mention plenty of datasets suited for frontal people detection, such as MOTP [7], PETS [9] or CrowdHuman [19] just to mention a few of them, but only a few are available for overhead view people counting.

Starting from this assumption, in this paper we propose a system combining a CNN, optimized for running on board of smart cameras, with an overlap tracking for head view people counting. We experimentally demonstrated that the datasets already available are not sufficient and representative of the real world; thus, we acquired our new dataset in 12 different scenarios by using various camera sensors, obtaining the biggest dataset nowadays available, composed of more than 30,000 annotated images. Please note that the biggest annotated dataset for standard people detection, namely CrowdHuman, consists of *only* 15,000 images.

The achieved F-Score, higher than 95% on average, confirms the generalization capability and the effectiveness of the proposed solution in the wild. In addition, the system is able to run directly on board of smart cameras, with a processing frame rate higher than 10 fps, confirming that it is ready for the market.

2 System Description

The architecture of the proposed people counting system is depicted in Fig. 1. The application is installed on the smart camera and is executed directly on board, with the possibility to exploit the embedded hardware accelerators (GPU or TPU). The processing pipeline includes people detection (see Sect. 2), people tracking (see Sect. 2) and people counting (see Sect. 2).

People Detection. The people detection module receives in input the image I_t acquired by the camera at time t and returns a list of bounding boxes $p = \{p_1, ..., p_{|p|}\}$, where $|p|$ is the number of people detected in I_t. The generic detected person p_i is represented as a quadruple (x, y, w, h), where x and y are the abscissa and ordinate of the top-left point, while w and h are the width and height of the bounding box surrounding the person. We implement the people detection module with a convolutional neural network (CNN) suitably designed to comply with the limited memory and processing resources available on smart cameras. To this aim, we adapt to our purpose a CNN commonly used for embedded devices, namely MobileNetv2-SSD [13,18]. The backbone is based on a streamlined architecture that uses depth-wise separable convolutions to build lightweight CNNs. The basic idea is to replace a fully convolutional operator with a factorized version that splits convolution into two separate layers, one for filtering (depth-wise convolution) and one for combining (point-wise convolution). This factorization has the effect of drastically reducing the number of operations and the model size. To save additional processing time and memory space, we set the network input size to 512×288; in this way, we maintain an aspect ratio of 16:9 and obtain an excellent trade-off between accuracy and computational complexity. Hereinafter, we will refer to this CNN with the name *PDNet*, namely the abbreviation of People Detection Network.

People Tracking. The people tracking module receives the people detected at time t $\{p_1, ..., p_{|p|}\}$ and associate them to the people tracked at time $t - 1$,

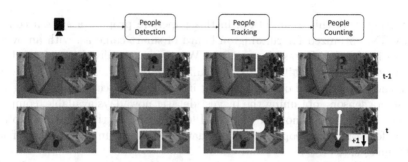

Fig. 1. Architecture of the proposed system. The application, running on board of a smart camera, performs real-time people detection, tracking and counting.

to produce a new set of tracked people $P = \{P_1, ..., P_{|P|}\}$, where $|P|$ is the number of people tracked at time t. The goal is to assign a unique identifier to each person, in order to collect the trajectories of the people and evaluate the direction of the passage. We developed this module as proposed in [8], namely with a one-to-one overlap tracking algorithm, based on a similarity matrix which takes into account the position of the bounding boxes; with an overhead camera, whose installation avoid strong occlusions between people, this is an effective and efficient solution.

People Counting. The people counting module takes as input a certain number of virtual counting lines (typically two, one for each direction), characterized by a crossing direction, and the trajectories traveled by people; as outputs, it generates a notification for each person who has crossed one of the lines at time t. The crossing is verified by checking if the end point of the trajectory is on the opposite side of the virtual line with respect to the starting point. The notification, labeled with the timestamp t, contains metadata related to the bounding box, the id and the virtual line crossed by the person.

3   Dataset

Top-View Multi-Person Tracking (TVMPT). The Top-View Multi-Person Tracking (TVMPT) dataset [17] includes 4,000 images taken from 5 overhead different cameras installed in a canteen, where people are in a queue or sitting at the table. The dataset is challenging, as evident in Fig. 3a, since there are people very close to each other and the camera is positioned at twice the height of typical people counting installations, so people appear quite small. The images, whose resolution is 1280×960, are annotated with the bounding boxes surrounding each person, so we have used them for training and testing our *PDNet*.

MIVIA People Counting (MIVIA PC). The MIVIA People Counting dataset [6] consists of videos 640×480 collected at a frame rate of 30 fps with an overhead camera installed at a height of 3.2 m, framing people passing left to right or bottom to top and viceversa (see Fig. 3b). This dataset does not provide the bounding boxes for people detection, so we manually annotated almost 13,000 images to use it for training and testing our *PDNet*.

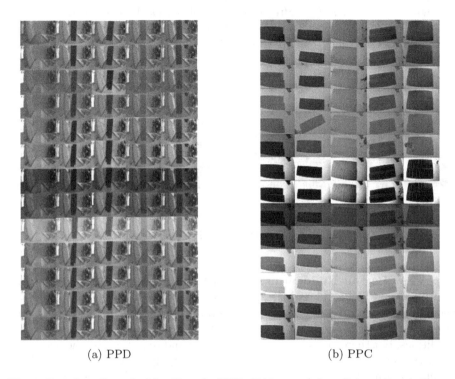

(a) PPD   (b) PPC

Fig. 2. Samples collected with a Hanwha XND-6010 camera from Private People Detection (a) and with an Axis P3227-LV camera from Private People Counting (b). In the columns we report the 5 different considered backgrounds. In (a) the rows contain the original images (first row), modified with different sharpness (32), contrast (1, 25, 75, 100), gamma (1.0, 0.75, 0.20) and color level (1, 25, 75, 100) values, while in (b) the rows contain the images modified with 2 different sharpness values (0, 100), 4 saturation values (0, 25, 75, 100), 4 contrast values (0, 25, 75, 100) and 4 brightness values (0, 25, 75, 100).

Private People Detection (PPD). This dataset includes images that we collected in 11 real scenarios, with different cameras installed at heights varying from 2.5 to 6 m in very heterogeneous case studies (supermarket, shopping center, canteen, office). In addition, to include in the dataset several background variations, we collected additional images with a Hanwha XND-6010 camera installed at a height of 3 m by changing the carpet on the floor and sharpness, contrast, gamma and color level values with respect to the original ones (50, 50, 0.5, 50); as shown in Fig. 2a we used 1 sharpness value (32), 4 contrast values

(1, 25, 75, 100), 3 gamma values (1.0, 0.75, 0.20) and 4 color level values (1, 25, 75, 100). The images are collected in uncontrolled conditions, so there are passages one at a time, side by side and in a row and various people very close to each other. The labeling procedure was very onerous. We annotated 4,527 images in a semi-automatic way, i.e. we used the bounding boxes obtained from a previously trained *PDNet* after having double-checked the correctness of the results. Then, we manually annotated other 4,035 images, in order to include in the dataset as many examples on which the previous detector was unable to detect people. Finally, we selected 4,352 background images, to be included in the training set to also provide samples without people and reduce the false positive rate.

   (a) TVMPT   (b) MIVIA PC   (c) TVPR - L2R   (d) TVPR - R2L

Fig. 3. Samples from TVMPT, MIVIA People Counting and TVPR datasets.

Top View Person Re-identification (TVPR). The Top View Person Re-identification (TVPR) dataset [12] includes 23 videos in which 100 different people pass under an overhead camera, installed at around 4 m, left to right and viceversa (see Fig. 3c and 3d). The 640 × 480 images are acquired in controlled conditions, but there are changes in illumination and reflections on the floor that make the detection challenging in some cases. We found this dataset useful for people counting, but it is not annotated for this purpose. Therefore, we defined two virtual crossing lines, one for each direction, and manually annotated the time instant and the direction of each passage.

Private People Counting (PPC). Since the TVPR dataset is recorded in a single scenario and includes a limited number of passages, we collected additional videos for increasing the representativeness of the test set for people counting. The acquisition procedure is similar to the one adopted for the last scenario of the Private Detection dataset, as shown in Fig. 2b, but in this case we used an Axis P3227-LV camera and 2 sharpness values (0, 100), 4 saturation values (0, 25, 75, 100), 4 contrast values (0, 25, 75, 100) and 4 brightness values (0, 25, 75, 100), initially set to 50. We collected 4 videos, which last about 3 h in total, with the camera installed at different heights: 2.25 m, 2.50 m, 2.85 m and 3.15 m. Hereinafter, we will refer to them with the names PPC-2.25, PPC-2.50, PPC-2.85 and PPC-3.15. Also in this case, there are uncontrolled passages one at a time, side by side and in a row. We manually annotated for these videos the time instant and the direction of each passage, for a total of 1,795 passages.

Final Dataset. The fusion of the standard datasets, namely TVMPT+MIVIA, includes 14,838 training images and 2,251 test images, with 68.095 and 13,857 people samples, respectively. The PPD, that we completely collected and annotated from scratch, consists of 12,060 training images and 854 test images, with 9,252 and 1,037 people samples, respectively.

4   Experimental Analysis

Experimental Framework. For our experiments, we trained two versions of $PDNet$: $PDNet_b$ was trained by using only data from standard datasets, namely the training set of TVMPT+MIVIA; $PDNet_f$ was trained, instead, by combining both the training sets of TVMPT+MIVIA and PPD. This choice allows to compare the results obtained by using only standard data with the ones achieved by extending, with great attention and effort, the existing datasets with additional samples collected in real scenarios. Both $PDNet_b$ and $PDNet_f$ have been trained with the same procedure, initializing the weights to the ones learned from COCO (the version available in Tensorflow Object Detection API) and using batches with 16 images and the Root Mean Square Propagation (RMSProp) optimizer. The initial learning rate has been set to 0.004 and scaled of a factor 0.95 every 500,000 steps. As data augmentation policies, we randomly applied crop, horizontal flip, brightness and contrast; the pseudo-random procedure ensures that at most two of these policies are applied to the same image. The training has been stopped when the performance on the validation set did not improve for 5 consecutive steps. Finally, $PDNet_b$ reached convergence after 707,883 steps, while $PDNet_f$ required 1,345,466 steps, being the training set more complex and representative.

As typically done in the literature, we evaluate the detection and counting performance of our method in terms of Recall (R), Precision (P) and F-Score (F):

$$R = \frac{TP}{TP + FN} \tag{1}$$

$$P = \frac{TP}{TP + FP} \tag{2}$$

$$F = \frac{2 * P * R}{P + R} \tag{3}$$

where TP, FN and FP are the number of true positives, false negatives and true positives, respectively.

We have to define for detection and counting two different concepts of true positive. In case of detection, we adopt the intersection over union (IoU) and the Pascal Criterion. In particular, we consider a detected person a true positive if the IoU between the bounding box b_i and an element in the groundtruth g_i is greater than 0.5:

$$IoU = \frac{b_i \cap g_i}{b_i \cup g_i} \geq 0.5 \tag{4}$$

otherwise it is a false positive. All the groundtruth elements not associated with a detected person are false negatives.

In case of counting, we consider a detected passage as a true positive if in the groundtruth there is a passage in the same direction within a window of 1 s; otherwise it is a false positive. All the groundtruth passages not counted by the system are considered false negatives.

Results. The detection results achieved by the two versions of $PDNet$ are reported in Table 1. We can note that the performance of the two neural networks is quite similar on a standard dataset, namely the TVMPT+MIVIA test set, being the F-Score obtained by $PDNet_b$ equal to 0.926, while $PDNet_f$ achieved 0.936. On the other hand, we notice a substantial difference on the challenging PPD test set, over which $PDNet_b$ does not go beyond a F-Score of 0.538, while $PDNet_f$ obtains a remarkable 0.980.

This result is the clear proof that a training with standard datasets, such as TVMPT and MIVIA, is not sufficient for giving to the CNN capability to generalize over new scenarios and backgrounds. In fact, by observing the errors, we noticed that $PDNet_b$ was not able to detect people in scenarios with certain backgrounds (especially with shades of red) or low contrast between the color of the people and the floor, while $PDNet_f$ is significantly more robust and accurate, as demonstrated by the quantitative results. Although the annotation procedure of PPD has been very strenuous, the new dataset allowed to train an effective CNN for people detection from top-view cameras; so much better that $PDNet_f$ outperformed $PDNet_b$ even on the TVMPT+MIVIA test set.

The proposed system, including $PDNet_f$, demonstrates its effectiveness over the TVPR and PC datasets, as shown in Table 2. In fact, the F-Score achieved in all the scenarios is between 0.909 (over PPC-2.25) and 0.974 (over PPC-2.50). We can note that an installation between 2.50 and 3.15 m (F-Score equal to 0.974, 0.957 and 0.961) allows to maximize the performance, while going under 2.25 is harder because people become only partially visible under the camera. In TVPR the camera is even higher, but we attribute the slight drop in performance (F-Score equal to 0.927) not only to this aspect but also to the presence of reflections and to a background with a particular texture.

The proposed system has been also tested on three different smart cameras equipped with GPU, namely Panasonic WV-X2251L and Hanwha PNV-A6081R equipped with a GPU and Axis Q1615 Mk III TPU provided. The processing frame rate depends on the computational resources available on the specific cameras, but it varies between 10 and 24 fps, fast enough to count people passing quickly under an overhead camera. Therefore, the proposed system is ready for being used in real plants in terms of accuracy and processing time.

Table 1. People detection results in terms of Recall (R), Precision (P) and F-Score (F) achieved by $PDNet_b$ and $PDNet_f$.

Test set	Model	R	P	F
TVMPT+MIVIA	$PDNet_b$	0.894	0.960	0.926
	$PDNet_f$	0.916	0.956	0.936
PPD	$PDNet_b$	0.436	0.603	0.538
	$PDNet_f$	0.990	0.972	0.980

Table 2. People counting results in terms of Recall (R), Precision (P) and F-Score (F) achieved by the proposed system over the TVPR and PPC datasets.

Test set	TP	FN	FP	R	P	F
TVPR	165	13	13	0.927	0.927	0.927
PPC-2.25	75	8	7	0.904	0.915	0.909
PPC-2.50	112	4	2	0.966	0.982	0.974
PPC-2.85	1110	69	30	0.941	0.974	0.957
PPC-3.15	391	26	6	0.938	0.985	0.961

Discussion. The training with standard datasets, such as TVMPT and MIVIA, which however required considerable effort to produce the missing annotations for our purpose, does not allow to achieve acceptable performance in real and different scenarios. This is the clear insight we deduce from the results, being the F-Score obtained by $PDNet_b$ equal to 0.926 on the test set of standard datasets and 0.538 in the different scenarios collected in PPD. On the other hand, the additional considerable effort done for labelling PPD allowed us to train $PDNet_f$, a CNN that has been shown to generalize well under several conditions (F-Score equal to 0.936 and 0.980 over TVMPT+MIVIA and PPD test sets, respectively). The same CNN, integrated in an application performing people counting through detection and tracking for determining a passage as line crossing in a certain direction, demonstrated to be effective and ready for a reliable overhead people counting system; indeed, the F-Score is always greater than 0.900, with a peak of 0.974 and an average around 0.950. We can conclude that we have achieved the goal and that the effort to collect a truly representative dataset, significantly extending the standard datasets, has been considerable.

The quantitative analysis we performed gives an idea of the accuracy of the proposed system in very heterogeneous conditions in terms of camera height, background and behavior of the people passing under the camera. However, in real installations it is possible to fix some operating constraints that may allow to maximize the performance of the system. Therefore, we also carried out a qualitative analysis of the errors, in order to define the optimal conditions for using the people counting system. Before going ahead, a first clear insight can be deduced from the quantitative results in Table 2: the optimal installation height seems to be between 2.50 and 3.15 m, while further reducing the height (e.g. PPC-2.25) may negatively affect the performance of the system, since part of the person is not inside the field of view of the camera.

From the qualitative analysis, we noticed remarkable performance in very variable conditions, with passages in a single line, side by side, parallel and crossed in opposite directions; however, we found possible errors that may happen in particular operating conditions, whose examples are reported in Fig. 4. In more detail, we observed that most of the times in which the head of the person is covered with a big object, as shown in Fig. 4a, the detection fails; we also noted that people with hats and helmets are correctly detected, so we consider this error reasonable. Another common error is the failure to detect a person partially outside the camera field of view (see Fig. 4b); this error is frequent when the head of the person is not visible, but it can be easily solved by correctly installing the camera. We have also noticed spurious false positives due to objects (trash bins, hangers, bags) present in the transit area, whose top-view appearance is similar to a person; these errors can be removed by ensuring that the transit area is empty. Less frequent errors occur when people pass very close to each other in line (see Fig. 4c) or side by side (see Fig. 4d), even if they carry together a load; in this case, sometimes the bounding boxes are merged and a single person is counted instead of two. This type of error can not be solved, but it happens not frequently and with people really close to each other.

(a) Covered   (b) Partially out   (c) Too close in line   (d) Too close side

Fig. 4. Examples of detection errors due to challenging operating conditions.

5   Conclusion

In this paper we have shown that it is possible to realize an effective and efficient people counting system from top-view cameras. On the other hand, we demonstrated that achieving this goal is anything but simple, since an enormous effort is necessary to collect and annotate a representative dataset to train CNN and carry out a comprehensive experimental analysis. Our effort produced a dataset with around 30,000 images, which allowed to achieve a F-Score higher than 0.95 in people detection and counting. The quantitative analysis, which also demonstrated the suitability of the system for smart cameras (at least 10 FPS), has been extended with a careful qualitative assessment, that provided us with useful insights for determining the expected performance of the system under specific operating conditions and for recommending installation constraints that maximize the accuracy of the system.

References

1. Ahmad, M., Ahmed, I., Ullah, K., Ahmad, M.: A deep neural network approach for top view people detection and counting. In: 2019 IEEE 10th Annual Ubiquitous Computing, Electronics Mobile Communication Conference (UEMCON), pp. 1082–1088 (2019). https://doi.org/10.1109/UEMCON47517.2019.8993109
2. Carletti, V., Del Pizzo, L., Percannella, G., Vento, M.: An efficient and effective method for people detection from top-view depth cameras. In: 2017 14th IEEE International Conference on Advanced Video and Signal Based Surveillance (AVSS), pp. 1–6 (2017). https://doi.org/10.1109/AVSS.2017.8078531
3. Castelo-Branco, F., Reis, J.L., Vieira, J.C., Cayolla, R.: Business intelligence and data mining to support sales in retail. In: Rocha, Á., Reis, J.L., Peter, M.K., Bogdanović, Z. (eds.) Marketing and Smart Technologies. SIST, vol. 167, pp. 406–419. Springer, Singapore (2020). https://doi.org/10.1007/978-981-15-1564-4_38
4. Choi, J.H., Kim, J.E., Jeong, N.H., Kim, K.T., Jin, S.H.: Accurate people counting based on radar: deep learning approach. In: 2020 IEEE Radar Conference (RadarConf20), pp. 1–5 (2020). https://doi.org/10.1109/RadarConf2043947.2020.9266496
5. Cruz, M., Keh, J.J., Deticio, R., Tan, C.V., Jose, J.A., Dadios, E.: A people counting system for use in CCTV cameras in retail. In: 2020 IEEE 12th International Conference on HNICEM, pp. 1–6. IEEE (2020)
6. Del Pizzo, L., Foggia, P., Greco, A., Percannella, G., Vento, M.: Counting people by RGB or depth overhead cameras. Pattern Recogn. Lett. **81**, 41–50 (2016)
7. Dendorfer, P., et al.: CVPR19 tracking and detection challenge: how crowded can it get? arXiv:1906.04567 [cs], June 2019
8. Di Lascio, R., Foggia, P., Percannella, G., Saggese, A., Vento, M.: A real time algorithm for people tracking using contextual reasoning. Comput. Vis. Image Underst. **117**(8), 892–908 (2013)
9. Ellis, A., Ferryman, J.: PETS 2010 and PETS2009 evaluation of results using individual ground truthed single views. In: 2010 7th IEEE International Conference on Advanced Video and Signal Based Surveillance, pp. 135–142. IEEE (2010)
10. Javare, P., Khetan, D., Kamerkar, C., Gupte, Y., Chachra, S., Joshi, U.: Using object detection and data analysis for developing customer insights in a retail setting. In: Proceedings of the 3rd International Conference on Advances in Science & Technology (ICAST) (2020)
11. Le, M.C., Le, M.H., Duong, M.T.: Vision-based people counting for attendance monitoring system. In: 2020 5th International Conference on Green Technology and Sustainable Development (GTSD), pp. 349–352 (2020). https://doi.org/10.1109/GTSD50082.2020.9303117
12. Liciotti, D., Paolanti, M., Frontoni, E., Mancini, A., Zingaretti, P.: Person reidentification dataset with RGB-D camera in a top-view configuration. In: Nasrollahi, K., et al. (eds.) FFER/VAAM -2016. LNCS, vol. 10165, pp. 1–11. Springer, Cham (2017). https://doi.org/10.1007/978-3-319-56687-0_1
13. Liu, W., et al.: SSD: single shot MultiBox detector. In: Leibe, B., Matas, J., Sebe, N., Welling, M. (eds.) ECCV 2016. LNCS, vol. 9905, pp. 21–37. Springer, Cham (2016). https://doi.org/10.1007/978-3-319-46448-0_2
14. Massa, L., Barbosa, A., Oliveira, K., Vieira, T.: LRCN-RetailNet: a recurrent neural network architecture for accurate people counting. Multimedia Tools Appl. **80**(4), 5517–5537 (2021)

15. Mukherjee, S., Saha, B., Jamal, I., Leclerc, R., Ray, N.: Anovel framework for automatic passenger counting. In: 2011 18th IEEE International Conference on Image Processing, pp. 2969–2972 (2011). https://doi.org/10.1109/ICIP.2011.6116284
16. Pazzaglia, G., et al.: People counting on low cost embedded hardware during the SARS-CoV-2 pandemic. In: Del Bimbo, A., et al. (eds.) ICPR 2021. LNCS, vol. 12662, pp. 521–533. Springer, Cham (2021). https://doi.org/10.1007/978-3-030-68790-8_41
17. Prodaiko, I.: Person re-identification in a top-view multi-camera environment. Master Thesis (2020)
18. Sandler, M., Howard, A., Zhu, M., Zhmoginov, A., Chen, L.C.: Inverted residuals and linear bottlenecks: mobile networks for classification, detection and segmentation. arXiv (2018)
19. Shao, S., et al.: CrowdHuman: a benchmark for detecting human in a crowd. CoRR arXiv:1805.00123 (2018)
20. Wang, Q., Gao, J., Lin, W., Li, X.: NWPU-crowd: a large-scale benchmark for crowd counting and localization. IEEE Trans. Pattern Anal. Mach. Intell. **43**(6), 2141–2149 (2020)

Deepfake Style Transfer Mixture: A First Forensic Ballistics Study on Synthetic Images

Luca Guarnera[1,2]([⊠]) [iD], Oliver Giudice[1,3] [iD], and Sebastiano Battiato[1,2] [iD]

[1] Department of Mathematics and Computer Science, University of Catania, Catania, Italy
luca.guarnera@unict.it, {giudice,battiato}@dmi.unict.it
[2] iCTLab Spinoff of University of Catania, Catania, Italy
[3] Applied Research Team, IT Department, Banca d'Italia, Rome, Italy

Abstract. Most recent style-transfer techniques based on generative architectures are able to obtain synthetic multimedia contents, or commonly called deepfakes, with almost no artifacts. Researchers already demonstrated that synthetic images contain patterns that can determine not only if it is a deepfake but also the generative architecture employed to create the image data itself. These traces can be exploited to study problems that have never been addressed in the context of deepfakes. To this aim, in this paper a first approach to investigate the image ballistics on deepfake images subject to style-transfer manipulations is proposed. Specifically, this paper describes a study on detecting how many times a digital image has been processed by a generative architecture for style transfer. Moreover, in order to address and study accurately forensic ballistics on deepfake images, some mathematical properties of style-transfer operations were investigated.

Keywords: Image ballistics · Deepfake · Multimedia forensics

1 Introduction

Advances in Deep learning algorithms and specifically the introduction of Generative Adversarial Networks (GAN) [11] architectures, enabled the creation and widespread of extremely refined techniques able to *manipulate* digital data, alter it or create contents from scratch. These algorithms achieved surprisingly realistic results leading to the birth of the Deepfake phenomenon: multimedia contents synthetically modified or created through machine learning techniques. Deepfakes, or also called synthetic multimedia content, could be employed to manipulate or even generate realistic places, animals, objects and human beings. In this paper only the most dangerous deepfake images of people's faces were taken into account. Specifically, manipulations on people's faces could be categorized into four main groups:

Fig. 1. Deepfake manipulation categories.

– **Entire face synthesis**: creates entire non-existent face images [18–20].
– **Attribute manipulation**: also known as *face editing* or *facial retouching*, modifies facial attributes [4,5] such as hair color, gender, age.
– **Identity swap**: replaces a person's face in a video with another person's face[1]).
– **Expression swap**: also known as *face reenactment* [28,29], modifies the person's facial expression.

Figure 1 shows an example of each kind of manipulations.

Unfortunately, these types of manipulations are often used for malicious purposes such as industrial espionage attacks and threats to individuals having their face placed in a porn video. To counteract the illicit use of this powerful technology, forensic researchers have created in the last years several deepfake detection algorithms able to solve the *Real* Vs *Deepfake* classification task. State-of-the-art methods have demonstrated that Convolutional Neural Networks (CNNs) [7,16,23,31,32] and analytical approaches [9,12,13,25] could be employed to extract a unique fingerprint on synthetic content to define not only whether the digital data are deepfakes but also to determine the GAN architecture that was used in the creation phase. These unique patterns can be extracted from deepfake images to analyze and study them from the perspective of the science of Image Ballistics [2,6,10,27], in order to reconstruct the history of the data itself (a study that has never been addressed in the world of deepfakes). A first basic approach is proposed in this paper by defining whether a deepfake images is subject to single or double deepfake manipulation operations. In our experiments, only the Style-Transfer category as deepfake manipulation type was taken into account. Several mathematical properties of style-transfer operations were also analysed and the obtained results could give extremely interesting hints to further study the Forensic Ballistics task on Deepfake images with techniques similar to double quantization detection [8].

The remainder of this paper is organized as follows: Sect. 2 presents some state-of-the-art methods of Deepfakes creation in order to understand the process of creating synthetic digital content. The proposed approach and experimental results are described in Sect. 3. In Sect. 4 some mathematical properties of the Style-Transfer operation are demonstrated. Finally, Sect. 5 concludes the paper.

---

[1] https://github.com/deepfakes/faceswap.

2   State of the Art

An overview on Media forensics with particular focus on Deepfakes has been proposed in [14,30,33]. Deepfakes are generally created by Generative Adversarial Networks (GANs) firstly introduced by Goodfellow et al. [11]. Authors proposed a new framework for estimating generative models via an adversarial mode in which two models train simultaneously: a generative model G, that captures the data distribution, and a discriminative model D, able to estimate the probability that a sample comes from the training data rather than from G. The training procedure for G is to maximize the probability of D making an error, resulting in a min-max two-player game.

The best results in the deepfake creation process were achieved by Style-GAN [19] and StyleGAN2 [20], two *entire face synthesis* algorithms. Style-GAN [19], while being able to create realistic pseudo-portraits of high quality, small artifacts could reveal the fakeness of the generated images. To correct these imperfections, Karras et al. made some improvements to the StyleGAN generator (including re-designed normalization, multi-resolution, and regularization methods) by proposing StyleGAN2 [20].

Face attribute manipulation use image-to-image translation techniques in order to address the style diversity [1,17,22,24,26]. The main limitation of these methods is that they are not scalable to the increasing number of domains (e.g., it is possible to train a model to change only the air color, only the facial expression, etc.). To solve this problem, Choi et al. proposed StarGAN [4], a framework that can perform image-to-image translations on multiple domains using a single model. Given a random label as input, such as hair color, facial expression, etc., StarGAN is able to perform an image-to-image translation operation. This framework has a limitation: it does not capture the multi-modal nature of data distribution. In other words, this architecture learns a deterministic mapping of each domain, i.e. given an image and a fixed label as input, the generator will produce the same output for each domain. This mainly depends on the fact that each domain is defined by a predetermined label. To solve this limitation, Choi et al. recently proposed StarGAN v2 [5]. The authors changed the domain label to domain-specific style code by introducing two modules: (a) a mapping network to learn how to transform random Gaussian noise into a style code, and (b) a style encoder to learn how to extract the style code from a given reference image.

He et al. [15] proposed a new technique called AttGAN in which an attribute classification constraint is applied to the generated image, in order to guarantee only the correct modifications of the desired attributes. Achieved results showed that AttGAN exceeds the state of the art on the realistic modification of facial attributes.

An interesting style transfer approach was proposed by Cho et al. [3], with a framework called "group-wise deep whitening-and coloring method" (GDWCT) for a better styling capacity. GDWCT has been compared with various cutting-edge methods in image translation and style transfer improving not only computational efficiency but also quality of generated images.

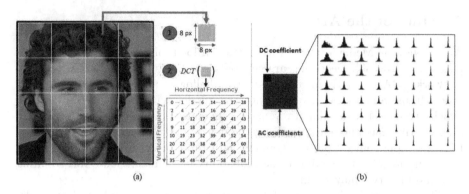

(a)   (b)

Fig. 2. (a-1) The input image I is divided into non-overlapping blocks of size 8×8. The DCT (a-2) is applied to each block and the DC/AC coefficients are obtained. (b) The DC histogram and the AC histograms are modeled, as demonstrated in [21], by considering all the coefficients of each block: e.g., the histogram of the AC coefficient of position 1 is obtained by modeling (following the Laplacian distribution) all the coefficients of position 1 of each block.

3   Forensic Ballistics on Deepfake Images

In order to define whether a multimedia content has undergone one or two deepfake manipulation operations belonging to the style-transfer category, we considered two approaches:

- A first analytical approach (Method 1) in which the DCT coefficients were analyzed to study whether there are discriminative frequency statistics.
- Given the potential of deep neural networks to solve different deepfake classification tasks, our second method (Method 2) relies on the use of a basic encoder (RESNET-18).

Method 1: Let I be a digital image. Following the JPEG pipeline, I is divided into non-overlapping blocks of size 8×8 (Fig. 2(a-1)). The Discrete Cosine Transform (DCT) is then applied to each block obtaining the 64 DCT coefficient. They are sorted into a zig-zag order starting from the top-left element to the bottom right (Fig. 2(a-2)). The DCT coefficient at position 0 is called DC and represents the average value of pixels in the block. All others coefficients namely AC, corresponds to specific bands of frequencies. By applying evidence reported in [21], the DC coefficient can be modelled with a Gaussian distribution while the AC coefficients were demonstrated to follow a zero-centred Laplacian distribution (Fig. 2(b)) described by:

$$P(x) = \frac{1}{2\beta} exp\left(\frac{-|x - \mu|}{\beta}\right) \tag{1}$$

with $\mu = 0$ and $\beta = \sigma/\sqrt{2}$ is the scale parameter where σ corresponds to the standard deviation of the AC coefficient distributions. The β values define the amplitude of the Laplacian graph that characterizes the AC parameters. It is possible to analyze all β values to determine if they can be discriminative to solve the proposed task. For this reason, the following feature vector was obtained for each involved image:

$$\vec{\beta} = \{\beta_1, \beta_2, \ldots, \beta_N\} \tag{2}$$

with $N = 63$. Note that β_1 through β_N represent only the β values extracted from the 63 AC coefficients: the DC coefficient is excluded because it represent, as described above, only the average value of the pixels in the block. The $\vec{\beta}$ feature vectors, extracted for each involved images, were used to train the following standard classifiers: k-NN (with $k = \{1, 3, 5, 7, 11, 13, 15\}$), SVM (with linear, poly, rbf and sigmoid kernels), Linear Discriminant Analysis (LDA), Decision-Tree, Random Forest, GBoost.

Method 2: The Pytorch implementation of RESNET-18 was used starting from the pre-trained model trained on ImageNet.[2] A fully-connected layer with an output size of 2, followed by a SoftMax, were added to the last layer of the Resnet-18 in order to be trained on the specific task. The network was set considering the following parameters: $batch-size = 30$; $learning-rate = 0.0001$; $number-of-classes = 2$; $epochs = 100$; $criterion = CrossEntropyLoss$; $optimizer = SGD$ with $momentum = 0.9$.

Creating Datasets for Both Methods: Let $S(s,t) = s \oplus t$ be a generic style transfer architecture. The function S takes as input a source s and a target t. The style transfer operation (which we defined with the symbol $\oplus$) is performed between s and t: attributes (such as hair color, facial expression) are captured by t to be transferred to s. So, for our task, we define two classes:

- *Deepfake-2* representing all images where a style transfer operation is applied once (the 2 in the Deepfake-2 class label refers to the input number: the source and the target). Thus, we collected a set of images $D_1 = S(s_1, t_1)$, where s_1 and t_1 represent two different dataset;
- *Deepfake-3* representing all images where a style transfer operation is applied twice. Thus, we have collected a set of images $D_2 = S(D_1, t_2)$, where D_1 is the source representing the output of the first style transfer operation and $t_2 \neq t_1$, $t_2 \neq s_1$.

3.1 Experiment Results

For our experiments, we collected images generated by StarGAN-V2 [5], because it is the best style transfer architecture capable of obtaining deepfake images with high quality detail and resolution. Figure 3 shows examples of StarGAN-V2 deepfake images related to the two classes. For the methods described above, 1200 images were considered as the train set and 200 images as the test set for each class. So in total, our train set consists of 2400 images and the test set consists of 400 images.

---

[2] https://pytorch.org/vision/stable/models.html.

Fig. 3. Examples of images generated by StarGAN-V2 engine [5]. (a) images in which a style transfer operation was applied once; (b) images in which a style transfer operation was applied twice.

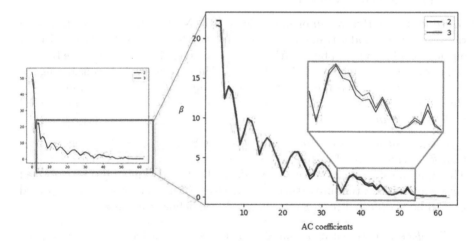

Fig. 4. The average β values for each i-th AC coefficient of the involved dataset are reported. Label "2" and "3" represents the trend of the AC coefficients related to the Deepfake-2 and Deepfake-3 class images. The abscissa axis are the 64 coefficients of the 8×8 block, while the ordinate axis are the average of the β values.

Regarding Method 1, Fig. 4 shows how the feature vector $\vec{\beta}$ extracted from all the involved images can to be discriminative. In detail, we plot the average β values for each i-th AC coefficient of the train set images. It can be seen that some β statistics (albeit slightly) appear to be discriminative in solving the proposed task. Table 1 shows the classification results obtained by training the standard classifiers (listed above) considering $\vec{\beta}$ feature vectors as input. As a first approach in forensic ballistics on deepfake images, we obtained at best a classification accuracy value of 81% using the Random Forest classifier.

Figure 5 shows the results obtained considering *Method 2*. We are able to distinguish the images given by the two classes reaching 92.75% as the best classification accuracy value (Loss value equal to 0.23). We are not surprised that we obtain better results than in Method 1. This happens because

Table 1. Classification results of the proposed analytical method.

Classifiers		Classes	Precision	Recall	F1-score	Accuracy (%)
k-NN	k = 3	Deepfake-2	0.78	0.69	0.73	74%
		Deepfake-3	0.72	0.79	0.75	
	k = 5	Deepfake-2	0.79	0.67	0.73	75%
		Deepfake-3	0.71	0.82	0.76	
	k = 7	Deepfake-2	0.78	0.66	0.71	73%
		Deepfake-3	0.70	0.81	0.75	
	k = 11	Deepfake-2	0.79	0.67	0.72	74%
		Deepfake-3	0.70	0.82	0.76	
	k = 13	Deepfake-2	0.79	0.65	0.71	74%
		Deepfake-3	0.70	0.82	0.75	
	k = 15	Deepfake-2	0.79	0.65	0.71	73%
		Deepfake-3	0.69	0.82	0.75	
SVM	Linear	Deepfake-2	0.78	0.65	0.71	73%
		Deepfake-3	0.69	0.81	0.74	
	Poly	Deepfake-2	0.71	0.41	0.52	62%
		Deepfake-3	0.58	0.83	0.68	
	rbf	Deepfake-2	0.69	0.41	0.51	61%
		Deepfake-3	0.57	0.81	0.67	
	Sigmoid	Deepfake-2	0.66	0.51	0.57	62%
		Deepfake-3	0.59	0.73	0.65	
LDA		Deepfake-2	0.77	0.67	0.72	73%
		Deepfake-3	0.70	0.79	0.75	
Decision-Tree		Deepfake-2	0.76	0.72	0.74	74%
		Deepfake-3	0.73	0.76	0.74	
Random Forest		Deepfake-2	0.85	0.76	0.80	81%
		Deepfake-3	0.78	0.86	0.82	
Gboost		Deepfake-2	0.82	0.67	0.74	76%
		Deepfake-3	0.72	0.85	0.78	

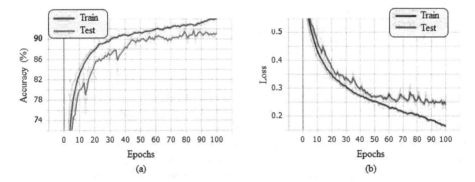

Fig. 5. Training accuracy (%) values (a) and loss values (b) were shown in the ordinate axis in (a) and (b) respectively, calculated over 100 epochs (x-axis) of Deepfake-2 and Deepfake-3 class images. The best accuracy was 92.75% with a loss value of 0.23.

neural architectures are optimized to infer to their highest degree those features in order to solve the proposed task. The binary classification task seems to be at first glance a simple task. This is incorrect since RESNET-18 is comparing images of people's faces for both classes and it is well understood that the task that we propose has a higher difficulty than the classic classification tasks (such as classifying with respect to gender). It is clear that the neural architecture is not focusing on the semantics of the synthetic data, but is able to capture those intrinsic features that allow us in our case to classify with respect to the number of style-transfer manipulations performed. It is important to note another key element in our classification task. As demonstrated by Guarnera et al. [13] each generative architecture leaves a unique fingerprint on the synthetic data. This is the main reason why deepfake detection methods achieve excellent classification results. Under this aspect, our classification task turns out to be more complicated, since we are analyzing images manipulated by the same architecture. Consequently we expect that the images belonging to the two classes contain almost the same fingerprint. Despite this, we manage to solve the proposed task very well. In general, an image manipulated via generative architectures will contain a first unique fingerprint. If the same synthetic content is subjected to a further style-transfer operation (considering the same or other GAN architectures) then the multimedia data will contain another unique fingerprint that we could somehow consider as a "combination" with the first trace left by the GAN engine.

Having demonstrated the high accuracy results obtained in the classification task, it may be useful to understand, for complete and accurate forensic ballistics analysis on deepfake images, whether mathematical properties of the style transfer operation are satisfied. This further analysis is reported in Sect. 4.

4   Mathematical Properties

Let A, B, and C be three input images. We want to demonstrate:

- **Property 1:** $A \oplus \phi = A$
- **Property 2:** $A \oplus B = B \oplus A$
- **Property 3:** $(A \oplus B) \oplus C = A \oplus (B \oplus C)$

where ϕ is the neutral element.

As regards the last two properties we calculated the *Structural Similarity Index Measure* (SSIM) scores and SSIM maps between the output images. In general we expect that all the properties listed above are not satisfied because even minimally, the GAN architecture will manipulate the source image. For this reason, we can also analyze the style transfer with respect to the RGB colors, since the source is also manipulated with the characteristics related to the color distribution of the target. For this reason, we compute two RGB color histograms $(H_1$ and $H_2)^3$ related to the two output deepfake images and compare them by

---

[3] e.g.: supposing we want to prove property 2, then H_1 and H_2 will be the RGB color histograms of the deepfake images obtained from $A \oplus B$ and $B \oplus A$ respectively.

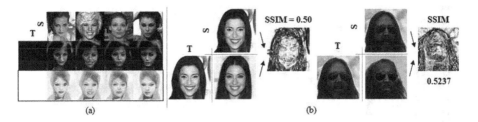

(a)   (b)

Fig. 6. Style transfer operations. S and T represent the source and the target, respectively. (a) The source are images of people's faces and the target are total black and total white images. (b) The source and target are the same images.

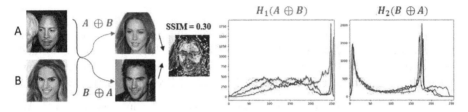

Fig. 7. Style transfer operations related to *Property 2*. $H(\cdot)$ represents the function that computes the RGB histogram.

using[4] (i) *Correlation(H_1, H_2)* with an output defined in the interval $[-1; 1]$ where 1 is perfect match and -1 is the worst; (ii) *Chi-Square(H_1, H_2)*: with an output defined in the interval $[0; +\infty[$ where 0 is perfect match and mismatch is unbounded; (iii) *Bhattacharyya distance(H_1, H_2)* with an output defined in the interval $[0; 1]$ where 0 is perfect match and 1 mismatch.

Demonstration 1: We want to try to figure out if there is a neutral element ϕ in the style-transfer operation. The neutral element will be that target image such that the style-transfer operation between the source and target gives as output the same source image. The neutral element should be a simple image, just a neutral image. Therefore we fix as targets: an all-white image and an all-black image. As shown in Fig. 6(a), the fixed targets seem not to satisfy property 1. The neutral element could be a target image with the same characteristics as the source. For this reason we try to check if it is the source image itself. From the results shown in Fig. 6(b), we can say that there is no neutral element because even in the case where target and source represent the same face, the architecture GAN will reconstruct equally the output image, going to transfer, even if equal, characteristics of the target in the source.

Demonstration 2: Figure 7 shows an example of the outputs obtained. The latter turn out to be visually different. Moreover, the 0.3009 score and the SSIM map show us that this property is not satisfied either. Finally, we check if the style transfer referred to RGB colors of Property 2 is satisfied. As shown in

---

[4] https://docs.opencv.org/3.4/d8/dc8/tutorial_histogram_comparison.html.

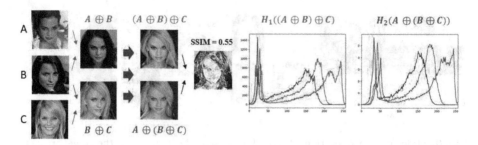

Fig. 8. Style transfer operations related to *Property 3*. $H(\cdot)$ represents the function that computes the RGB histogram.

Fig. 7, the distribution of colors is already visually different. We compare the histograms using the metrics listed above. The following values were obtained: *Correlation*: 0.47; *Chi-Square*: 699.03; *Bhattacharyya distance*: 0.56. This shows that even style transfer referring to RGB colors of Property 2 is not satisfied.

Demonstration 3: Figure 8 shows an example of the outputs obtained. We can observe that the resulting images are very similar to each other. The score 0.55002 and the SSIM map show us, however, that even this property does not seem to be fully satisfied. Let's check if the style transfer referred to RGB colors of Property 3 is satisfied. As shown in Fig. 8, the distribution of colors is visually very similar. The difference values calculated between the two RGB histograms are as follows: *Correlation*: 0.931; *Chi-Square*: 10.52; *Bhattacharyya distance*: 0.21. Considering a set of 1000 randomly selected images, we get on average a *Correlation*: 0.863 (with variance = 0.0034), *Chi-Square*: 14.255 (with variance = 31.2064) and *Bhattacharyya distance*: 0.262 (with variance = 0.0019). From the results obtained we can say that Property 3 is almost satisfied. So the latter can be rewritten in the following way: $(A \oplus B) \oplus C \simeq A \oplus (B \oplus C)$.

5   Conclusions

In this paper, we proposed a first approach to investigate forensic ballistics on deepfake images subjected to style transfer manipulations. In detail, we tried to define if a synthetic image has been subjected to one or two deepfake manipulation processes (performed using StarGAN-v2) by proposing two possible methods: one purely analytical and another based on a deep neural architecture. To better understand the type of style-transfer manipulation we also demonstrated several mathematical properties of the operation itself. From the excellent results obtained, this study represents a good starting point to create more sophisticated methods in the field of forensic ballistics of deepfake images. In future works we will address this issue considering also other different manipulations (e.g. face reenactment, face swap) and we will try to generalize as much as possible also on audio and video deepfakes.

References

1. Almahairi, A., Rajeshwar, S., Sordoni, A., Bachman, P., Courville, A.: Augmented CycleGAN: learning many-to-many mappings from unpaired data. In: International Conference on Machine Learning, pp. 195–204. PMLR (2018)
2. Battiato, S., Giudice, O., Paratore, A.: Multimedia forensics: discovering the history of multimedia contents. In: Proceedings of the 17th International Conference on Computer Systems and Technologies 2016, pp. 5–16 (2016)
3. Cho, W., Choi, S., Park, D.K., Shin, I., Choo, J.: Image-to-image translation via group-wise deep whitening-and-coloring transformation. In: Proceedings of the IEEE Conference on Computer Vision and Pattern Recognition, pp. 10639–10647 (2019)
4. Choi, Y., Choi, M., Kim, M., Ha, J.W., Kim, S., Choo, J.: StarGAN: unified generative adversarial networks for multi-domain image-to-image translation. In: Proceedings of the IEEE Conference on Computer Vision and Pattern Recognition, pp. 8789–8797 (2018)
5. Choi, Y., Uh, Y., Yoo, J., Ha, J.W.: StarGAN v2: diverse image synthesis for multiple domains. In: Proceedings of the IEEE/CVF Conference on Computer Vision and Pattern Recognition, pp. 8188–8197 (2020)
6. Farid, H.: Digital image ballistics from JPEG quantization: a followup study. Technical Report TR2008-638, Department of Computer Science, Dartmouth College (2008)
7. Gandhi, A., Jain, S.: Adversarial perturbations fool deepfake detectors. In: 2020 International Joint Conference on Neural Networks (IJCNN), pp. 1–8. IEEE (2020)
8. Giudice, O., Guarnera, F., Paratore, A., Battiato, S.: 1-D DCT domain analysis for JPEG double compression detection. In: Ricci, E., Rota Bulò, S., Snoek, C., Lanz, O., Messelodi, S., Sebe, N. (eds.) ICIAP 2019. LNCS, vol. 11752, pp. 716–726. Springer, Cham (2019). https://doi.org/10.1007/978-3-030-30645-8_65
9. Giudice, O., Guarnera, L., Battiato, S.: Fighting deepfakes by detecting GAN DCT anomalies. J. Imaging **7**(8), 128 (2021). https://doi.org/10.3390/jimaging7080128
10. Giudice, O., Paratore, A., Moltisanti, M., Battiato, S.: A classification engine for image ballistics of social data. In: Battiato, S., Gallo, G., Schettini, R., Stanco, F. (eds.) ICIAP 2017. LNCS, vol. 10485, pp. 625–636. Springer, Cham (2017). https://doi.org/10.1007/978-3-319-68548-9_57
11. Goodfellow, I., et al.: Generative adversarial nets. In: Advances in Neural Information Processing Systems, pp. 2672–2680 (2014)
12. Guarnera, L., Giudice, O., Battiato, S.: DeepFake detection by analyzing convolutional traces. In: Proceedings of the IEEE/CVF Conference on Computer Vision and Pattern Recognition Workshops, pp. 666–667 (2020)
13. Guarnera, L., Giudice, O., Battiato, S.: Fighting deepfake by exposing the convolutional traces on images. IEEE Access **8**, 165085–165098 (2020). https://doi.org/10.1109/ACCESS.2020.3023037
14. Guarnera, L., Giudice, O., Nastasi, C., Battiato, S.: Preliminary forensics analysis of DeepFake images. In: 2020 AEIT International Annual Conference (AEIT), pp. 1–6. IEEE (2020). https://doi.org/10.23919/AEIT50178.2020.9241108
15. He, Z., Zuo, W., Kan, M., Shan, S., Chen, X.: AttGAN: facial attribute editing by only changing what you want. IEEE Trans. Image Process. **28**(11), 5464–5478 (2019)

16. Hsu, C.C., Zhuang, Y.X., Lee, C.Y.: Deep fake image detection based on pairwise learning. Appl. Sci. **10**(1), 370 (2020). https://doi.org/10.3390/app10010370

17. Huang, X., Liu, M.-Y., Belongie, S., Kautz, J.: Multimodal unsupervised image-to-image translation. In: Ferrari, V., Hebert, M., Sminchisescu, C., Weiss, Y. (eds.) ECCV 2018. LNCS, vol. 11207, pp. 179–196. Springer, Cham (2018). https://doi.org/10.1007/978-3-030-01219-9_11

18. Karras, T., Aittala, M., Hellsten, J., Laine, S., Lehtinen, J., Aila, T.: Training generative adversarial networks with limited data. In: Larochelle, H., Ranzato, M., Hadsell, R., Balcan, M.F., Lin, H. (eds.) Advances in Neural Information Processing Systems, vol. 33, pp. 12104–12114. Curran Associates, Inc. (2020). https://proceedings.neurips.cc/paper/2020/file/8d30aa96e72440759f74bd2306c1fa3d-Paper.pdf

19. Karras, T., Laine, S., Aila, T.: A style-based generator architecture for generative adversarial networks. In: Proceedings of the IEEE Conference on Computer Vision and Pattern Recognition, pp. 4401–4410 (2019)

20. Karras, T., Laine, S., Aittala, M., Hellsten, J., Lehtinen, J., Aila, T.: Analyzing and improving the image quality of StyleGAN. In: Proceedings of the IEEE/CVF Conference on Computer Vision and Pattern Recognition, pp. 8110–8119 (2020)

21. Lam, E.Y., Goodman, J.W.: A mathematical analysis of the DCT coefficient distributions for images. IEEE Trans. Image Process. **9**(10), 1661–1666 (2000)

22. Lee, H.-Y., Tseng, H.-Y., Huang, J.-B., Singh, M., Yang, M.-H.: Diverse image-to-image translation via disentangled representations. In: Ferrari, V., Hebert, M., Sminchisescu, C., Weiss, Y. (eds.) ECCV 2018. LNCS, vol. 11205, pp. 36–52. Springer, Cham (2018). https://doi.org/10.1007/978-3-030-01246-5_3

23. Li, L., et al.: Face X-ray for more general face forgery detection. In: Proceedings of the IEEE/CVF Conference on Computer Vision and Pattern Recognition, pp. 5001–5010 (2020)

24. Mao, Q., Lee, H.Y., Tseng, H.Y., Ma, S., Yang, M.H.: Mode seeking generative adversarial networks for diverse image synthesis. In: Proceedings of the IEEE/CVF Conference on Computer Vision and Pattern Recognition, pp. 1429–1437 (2019)

25. McCloskey, S., Albright, M.: Detecting GAN-generated imagery using saturation cues. In: 2019 IEEE International Conference on Image Processing (ICIP), pp. 4584–4588 (2019). https://doi.org/10.1109/ICIP.2019.8803661

26. Na, S., Yoo, S., Choo, J.: MISO: mutual information loss with stochastic style representations for multimodal image-to-image translation. arXiv preprint arXiv:1902.03938 (2019)

27. Piva, A.: An overview on image forensics. International Scholarly Research Notices 2013 (2013)

28. Thies, J., Zollhöfer, M., Nießner, M.: Deferred neural rendering: image synthesis using neural textures. ACM Trans. Graph. (TOG) **38**(4), 1–12 (2019)

29. Thies, J., Zollhofer, M., Stamminger, M., Theobalt, C., Nießner, M.: Face2Face: real-time face capture and reenactment of RGB videos. In: Proceedings of the IEEE Conference on Computer Vision and Pattern Recognition, pp. 2387–2395 (2016)

30. Verdoliva, L.: Media forensics and DeepFakes: an overview. IEEE J. Sel. Top. Signal Process. **14**(5), 910–932 (2020)

31. Wang, S.Y., Wang, O., Zhang, R., Owens, A., Efros, A.A.: CNN-generated images are surprisingly easy to spot... for now. In: Proceedings of the IEEE/CVF Conference on Computer Vision and Pattern Recognition, pp. 8695–8704 (2020)

32. Xuan, X., Peng, B., Wang, W., Dong, J.: On the generalization of GAN image forensics. In: Sun, Z., He, R., Feng, J., Shan, S., Guo, Z. (eds.) CCBR 2019. LNCS, vol. 11818, pp. 134–141. Springer, Cham (2019). https://doi.org/10.1007/978-3-030-31456-9_15

33. Zhang, T.: Deepfake generation and detection, a survey. Multimedia Tools Appl. **81**, 6259–6276 (2021). https://doi.org/10.1007/s11042-021-11733-y

Experimental Results on Multi-modal Deepfake Detection

Sara Concas[1]([✉]), Jie Gao[2], Carlo Cuccu[1], Giulia Orrù[1], Xiaoyi Feng[2], Gian Luca Marcialis[1], Giovanni Puglisi[1], and Fabio Roli[3]

[1] University of Cagliari, Cagliari, Italy
{sara.concas90c,carlo.cuccu,giulia.orru,puglisi,marcialis}@unica.it
[2] Northwestern Polytechnical University, Xi'an, China
jie_gao@mail.nwpu.edu.cn, fengxiao@nwpu.edu.cn
[3] University of Genova, Genoa, Italy
fabio.roli@unige.it

Abstract. The advantages of deepfakes in many applications are counterbalanced by their malicious use, for example, in reply-attacks against a biometric system, identification evasion, and people harassment, when they are widespread in social networks and chatting platforms (cyberbullying) as recently documented in newspapers. Due to its "arms-race" nature, deepfake detection systems are often trained on a certain class of deepfakes and showed their limits on never-seen-before classes. In order to shed some light on this problem, we explore the benefits of a multi-modal deepfake detection system. We adopted simple fusion rules, which showed their effectiveness in many applications, for example, biometric recognition, to exploit the complementary of different individual classifiers, and derive some possible guidelines for the designer.

Keywords: Deepfake · Pattern recognition · Multiple classifiers

1 Introduction

To our knowledge, the first appearance of deepfake detection methods is related to [1], where deepfakes are presented as a technique for replacing the face of a targeted individual in a video with the face of another person. Born apparently for entertainment applications, deepfakes were rapidly viewed as possible threats in computer security, for example, face recognition [13], and reported in the first extensive survey on the topic [24]. Deepfakes can also be used in social networks as a new generation of harassments and cyberbullying attacks.[1]

The interest has rapidly grown in deepfakes, and a simple search on Google Scholar reports a number of 5,600 contributions between 2018 and 2022. We can roughly separate the contributions into two classes: the first ones present novel deepfake generation approaches, second ones run after the first ones. Since such attacks are generated from deep networks, deep networks are adopted to

---

[1] https://cyberbullying.org/deepfakes.

S. Sclaroff et al. (Eds.): ICIAP 2022, LNCS 13232, pp. 164–175, 2022.
https://doi.org/10.1007/978-3-031-06430-2_14

counteract, based on the incremental training on more and more data as they are available [1,20]. Therefore, due to its arms-race nature, each solution suffers from the drop of effectiveness when dealing with never-seen-before deepfakes, provided by novel approaches [2,5], similarly to other computer vision security fields such as fingerprint presentation attack detection [4]. Although many recent works proposed to test the novel architecture by adopting a cross-data set protocol, where the performance is also assessed on deepfakes generated by "unknown" methods, fewer contributions were conceived as being "intrinsically robust".

In our opinion, incremental training cannot be considered the definitive solution. Due to the large availability of deepfake detection methods, often trained on different sets of samples, it could be worthy to investigate their potential when combined together by multiple classifier systems approaches [3,18]. The outcome of the investigated fusion models can offer a feedback on how to approach the design of the individual deepfake detection methods. Worth noting, we do not fine-tune such methods to put ourselves in the realistic hypothesis that data are intrinsically limited in size and acquisition time.

Section 2 briefly reviews the state of the art on deepfake detection methods and highlights their potential complementarity. Section 3 describes the fusion models and experiments are carried out in Sect. 4 to point out when our claim is or is not confirmed. Conclusions are drawn in Sect. 5.

2   Related Works on Deepfake Detection

Deepfakes can be categorized into four types [24]: face synthesis, attribute manipulation and face identity or expression swap. The generation of these manipulated images/videos occurs mainly with deep-learning techniques such as 3D models [22], AutoEncoders and Generative Adversarial Networks (GAN). Based on realism and technology used for the realization, two generations of deepfake can be distinguished [23]. The first generation is characterized by deepfakes with low quality and visible artifacts [14]; the second generation is visually indistinguishable from the original videos by naked eyes [16].

Since the deepfake creation techniques increased in visual quality and dissemination, the need for reliable and robust deepfake detection methods has grown in importance in the last few years. Like the generation, deepfake detection has also evolved over the years, and different technologies are used depending on the deepfake to be analyzed. Detection methods can be classified into [26]:

- general network-based methods, that classify at the frame level with specially designed methods or transfer learning techniques [1];
- visual artefacts-based methods, based on the analysis of image quality and resolution and search for visual artifacts [15];
- biological signals-based methods, based on the analysis of biological signals as eye blinking frequency [14] and heart rate detection [11];
- camera fingerprints-based methods, that perform a search for specific traces due to generation via GAN [6];

– temporal consistency-based methods, which in addition to spatial information use the temporal information given by consecutive frames [9].

This distinction is made on the basis of the type of features used for the classification. In general, the classification can be performed with both shallow and deep-learning based techniques. In addition, some methods use attention mechanisms that allow detectors to "learn" which part of the input is the most relevant. Furthermore, some authors [3] propose an ensemble of classifiers trained in different ways to improve the detection performance.

In this work, we investigate if combining different detection algorithms can improve the system's generalization capabilities. In fact, many of the proposed state-of-the-art (SOTA) systems are specialized in one type of manipulation and fail to correctly classify never-seen-before deepfakes. The analysis of the related outcome may lead to some feedbacks on how to approach the countermeasure design. Thus, we selected three detectors representative of the SOTA.

The first one is a frame-by-frame binary classifier based on the XceptionNet model, a traditional CNN pretrained on Imagenet and transferred to the deepfake detection task by replacing the final fully-connected layer with two outputs [20]. The second detector is visual artefacts-based and uses the transfer learning technique. In particular, the well-known ResNet50 [10] model is fine-tuned using faces images with warping artifacts as negative samples. The third detector [3] is based on the ensemble of different EfficientNetB4 models obtained by adding an attention mechanism and using two different training techniques, one classical end-to-end strategy and a siamese strategy.

These three detectors are based on different assumptions and work well on different types of manipulations. Therefore, they may exhibit a different dynamic of outcomes, especially on never-seen-before deepfakes, which can be managed by an appropriate fusion rule.

3 Multi-modal Systems and Fusion in Deepfake Detection

Although the theory of multiple classifier systems is largely known as effective in many pattern recognition problems [19], we focus here on multi-modal biometric fusion. Fusion is a technique for improving a biometric system's performance by integrating diverse samples or biometric modalities (iris, fingerprint, face). This approach usually leads to increased robustness against intra- and inter-class variations [25], and its extension has proven to be effective over time [7].

Fusion can be carried out at different levels of the biometric system [17]: sensor level (the samples are combined into a single one), feature level (the extracted features are combined together to obtain the final feature set), score-level (the similarity or distance scores are combined to obtain a single score) and decision level (the scores are converted into decisions before combining them). Fusion techniques look also promising in unimodal biometric scenarios. For example, in [17], the authors tested different levels of fusion on a facial recognition system, identifying the score-level as the best of those analyzed.

Coming back to the deepfake detection problem, in [21], the authors propose a feature-level fusion model by considering both spatial domain features (the high-level semantic features of the image) and frequency domain features (the low-level texture features). Similarly, in [28] Zhao et al. point out that a fake sample contains at least three kinds of artifacts: microcosmic (statistical) features, mesoscopic features and macroscopic (semantic) features. They propose a multi-layer fusion neural network (MFNN) able to capture the artifacts at different levels through the use of specially designed shallow, middle and deep layers. The extracted features are then fused together before classification. In [27], the authors propose a model that includes RGB features and textural information extracted by neural networks and signal processing methods. The described feature-level fusion techniques attempted to construct a more effective detection ability by studying the artifact typology, thereby addressing the system design from the "cause" perspective. Deepfake detection, on the other hand, still performs well when evaluated on the same typology of fakes on which it was trained, implying that a system that combines different features representing different artifacts may not be able to identify never-before-seen typologies of fakes [2]. Therefore, we focus our attention on the "effect" point-of-view, namely, the posterior probability distributions of a fake given the input feature set $P(deepfake|x)$, and observe how these distributions change when combined under score-level fusion rules. Our goal is to determine which score-level fusion technique is suited for deepfakes and to offer insights into how to build an effective multi-modal system.

In particular, we chose three deepfake detectors [3,10,20] as reported in Sect. 2. Modelling the outcomes as "scores" or individual posterior probabilities, we applied five non-parametric rules: average, bayesian, product, maximum, and minimum [12]. We introduce here a common notation useful to explain the fusion rules analysed. Let's consider the fusion of N classifiers' scores. Let $P_i(\text{deepfake}|x)$ be the score of the i-th classifier on the same frame x, i.e. the estimated posterior probability that the analyzed frame contains a deepfake. According to this notation, we get rid of the fact that the original sample x may be processed to derive a specific feature set, the actual input to the i-th classifier. It is possible to compute the fusion scores of the N classifiers with the following formulas:

$$avg\_score = \frac{1}{N} \sum_{i=1}^{N} P_i(\text{deepfake}|x) \tag{1}$$

$$bayes\_score = \frac{\prod_{i=1}^{N} P_i(\text{deepfake}|x))}{\prod_{i=1}^{N} P_i(\text{deepfake}|x)) + \prod_{i=1}^{N} (1 - P_i(\text{deepfake}|x))} \tag{2}$$

$$prod\_score = \sqrt[N]{\prod_{i=1}^{N} P_i(\text{deepfake}|x)} \tag{3}$$

$$max\_score = max_i(P_i(\text{deepfake}|x)) \tag{4}$$

$$min\_score = min_i(P_i(\text{deepfake}|x)) \qquad (5)$$

4 Experimental Analysis and Results

The goal of this work is to determine if the score-level fusion of four deepfake models results in a system that can generalize by exploiting their complimentary and to determine which fusion rule is most suited for this purpose.

The models used are: XceptionNet [20] trained on FaceForensics++ excluding Neural Textures, the residual model ResNet50 [15] trained on a non-public dataset, and two models obtained by the ensemble of multiple models based on EfficientNet [3], one trained on FF++ and the other on DFDC.

For each of the mentioned detectors, we performed the tests using the pretrained models provided by the authors.

In the experimental protocol, the EfficientNet ensemble trained on FF++ was not tested on FF ++ because pretrained on data in common with our test set (Table 1).

Table 1. Pre-trained models used in experimental evaluation.

Model	Xception	Resnet50	Ensemble DFDC	Ensemble FF
Pretrained on	FF++ (without Neural Textures)	Non-public data	DFDC	FF++

4.1 Datasets

The following data sets were used to test the proposed methods:

FaceForensics++ dataset [20] contains both examples of 1st and 2nd generation deepfakes and a variety of different manipulation techniques. It includes four automated face manipulation methods applied to 1,000 original videos downloaded from YouTube. In particular: (1) **FaceSwap**: a graphics-based approach used to transfer the face region from a source video to a target video; (2) **Deepfakes**: a target face is replaced with that of a source video or image using autoencoders; (3) **Face2Face**: a facial reenactment method where the expressions of the source video are transferred to a target video while maintaining the target identity; (4) **Neural Textures**: a facial reenactment method. The original video data is used to learn a neural texture of the target.

Deepfake Detection Challenge Dataset (DFDC), a dataset created for the DeepFake Detection Challenge (DFDC) Kaggle competition [8]. It contains 100,000 total clips sourced from 3,426 paid actors, produced with several Deepfake, GAN-based, and non-learned methods.

We adopted the same split proposed in [20] for the FaceForensics++ dataset and the split proposed in [3] for the DFDC dataset. For each video of the datasets we used 32 frames (as in [3]).

4.2   Results

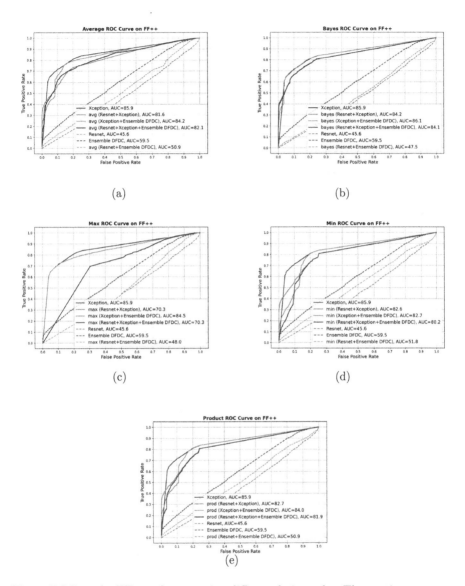

Fig. 1. ROC on the FF ++ dataset using different fusion rules. The continuous curves relate to intra-dataset tests, the dashed ones to cross-dataset tests.

In Fig. 1 the ROCs related to the FF++ dataset tests are shown for each fusion rule analyzed. In particular, the intra-dataset tests, for which at least one of the classifiers is trained on the same type of test data, are shown with a solid line, while the cross-dataset tests are shown with a dashed line. The difference

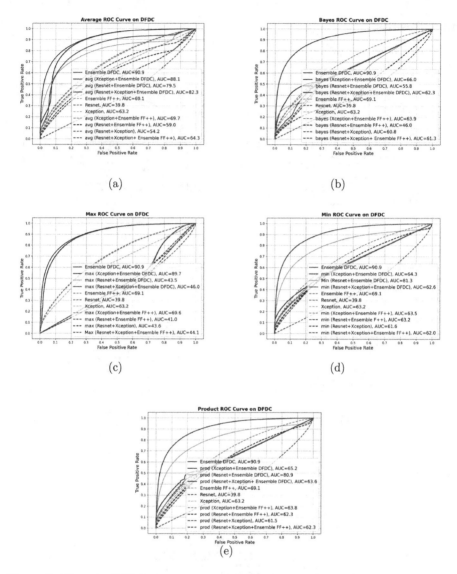

Fig. 2. ROC on the DFDC dataset using different fusion rules. The continuous curves relate to intra-dataset tests, the dashed ones to cross-dataset tests.

between the two types of tests is particularly evident from ROCs and confirms previous findings in the literature [2]. On the FF ++ dataset, among the intra-dataset models, Xception outperforms the others. Among those cross-datasets, Ensemble DFDC is the most performing. As regards the fusion rule, the Bayesian is better suited for intra-dataset comparisons and allows to increase the performance of the individual methods. The min rule, on the other hand, is the best for cross-dataset comparisons and is better than the least performing classifiers but worse than the most performing.

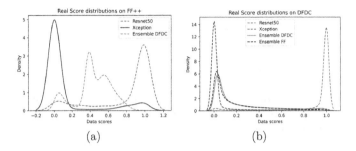

(a)   (b)

Fig. 3. Score (probability of belonging to the fake class) distributions of the real samples obtained from the three models. The continuous curves relate to intra-dataset tests, the dashed ones to cross-dataset tests.

A completely different analysis emerges from the tests on the DFDC dataset (Fig. 2), for which the average is particularly effective for both intra-dataset and cross-dataset tests. Also in this case the Ensemble models were the best performing, both in the intra-dataset and cross-dataset case, proving to be the most able to generalize. However, from the analysis of the two test datasets it is not possible to extract information on the best fusion rule. These variations show the need to act at the level of design and model training. Are these results useful to highlight good training practices and the choice of the most suitable score-level fusion rules? In fact, they are strongly influenced not only by the type of manipulation, but also by the characteristics of the original videos used to train the models. The answer to our question is suggested by the distributions of the real samples reported in Fig. 3. Paying attention to cross-dataset comparisons (dashed lines), the related distributions are highly chaotic due to the bias introduced in the models training by the specific types of acquisition sensors and image features. In fact, an unbiased model is able to generalize the representation of reals, as in the case of XceptionNet, and returns values close to zero even on unknown types of images. A biased model returns values close to zero on types of real images that it knows (intra-dataset) and very different score values for unknown types (cross-dataset). This is evident for the DFDC-trained ensemble, represented by the green curve in Fig. 3. In particular, models trained on FF++ are more robust than those trained on DFDC or proprietary data. Therefore, we may draw a possible suggestion for the designer: a good training step should therefore use various types of real video to eliminate or mitigate this bias. The real samples are indeed the "controllable" part of the deepfake detection problem and the never-seen-before effect should not occur. If we consider a model without bias at the level of real samples, we expect probabilities close to zero for them. We can therefore analyze the results of the fusion only from the point of view of deepfakes. In particular, the distributions of the deepfake samples for the single

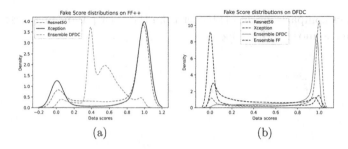

Fig. 4. Score (probability of belonging to the fake class) distributions of the fake samples obtained from the three models. The continuous curves relate to intra-dataset tests, the dashed ones to cross-dataset tests.

classifiers are shown in Fig. 4. The cross-dataset distributions are chaotic in this case too: however, this is expected because the never-seen-before attacks are the "uncontrollable" face of the story, and over time new deepfakes will be created, unknown to the detectors in use, thus also leading to a non-stationary process of detection. The distributions of the fake after the fusions are shown in Figs. 5 and 6, for the FF ++ and DFDC datasets respectively. The scores appear close to 1 only for the Bayesian and max rule fusions. The other fusion rules have different behaviors depending on the training and test data used.

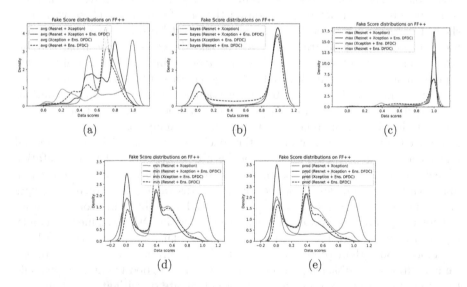

Fig. 5. Score (probability of belonging to the fake class) distributions of the FF++ fake samples obtained from the different fusions on the three models. The continuous curves relate to intra-dataset tests, the dashed ones to cross-dataset tests.

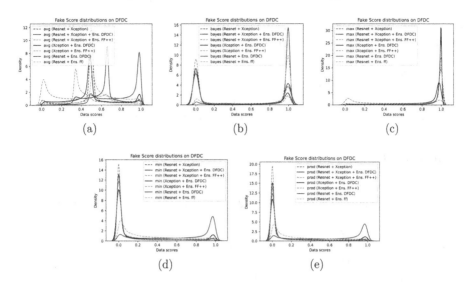

Fig. 6. Score (probability of belonging to the fake class) distributions of the DFDC fake samples obtained from the different fusions on the three models. The continuous curves relate to intra-dataset tests, the dashed ones to cross-dataset tests.

On the basis of what observed, we can deduce some useful information for a deepfake detectors designer:

– if the designer manages to obtain a model almost without bias on real samples, a max rule-based fusion (Figs. 5(c) and 6(c)) brings particular benefit in the detection of never-seen-before deepfake samples;
– bayesian fusion (Figs. 5(b) and 6(b)) helps separate real and fake distributions and only fails when both classes are biased;
– min and product (Figs. 5(d), 5(e), 6(d) and 6(e)) are not suitable for the realization of a multi-modal deepfake detector as they tend to lower the deepfake scores, which could already be very low for cross-dataset comparisons.

5 Conclusions

The cross-dataset evaluation of deepfake detection systems is an open issue due to the continuous evolution of fake generation techniques. In this paper, we explored the possibility of exploiting the complementarity of different models through score-level fusion techniques on three methods representative of the SOTA. Experiments have also highlighted some good practices underlying the development of an efficient multi-modal deep-fake detector. In particular, we highlighted the importance of reducing the bias introduced by different real videos and of selecting the fusion rule on the basis of the individual classifiers' real and fake distributions. This work is a preliminary analysis for the development of a multi-modal deep-fake detection system capable of classifying never-seen-before deepfakes. We believe that the drafted path is promising; thus, we will extend our experiments to further individual classifiers and fusion models, in order to draw more accurate guidelines for the designer.

Acknowledgment. This work is supported by the Italian Ministry of Education, University and Research (MIUR) within the PRIN2017 - BullyBuster - A framework for bullying and cyberbullying action detection by computer vision and artificial intelligence methods and algorithms (CUP: F74I19000370001).

References

1. Afchar, D., Nozick, V., Yamagishi, J., Echizen, I.: MesoNet: a compact facial video forgery detection network, pp. 1–7 (2018). https://doi.org/10.1109/WIFS.2018. 8630761
2. Bekci, B., Akhtar, Z., Ekenel, H.K.: Cross-dataset face manipulation detection. In: 2020 28th Signal Processing and Communications Applications Conference (SIU), pp. 1–4. IEEE (2020)
3. Bonettini, N., Cannas, E.D., Mandelli, S., Bondi, L., Bestagini, P., Tubaro, S.: Video face manipulation detection through ensemble of CNNs. In: 2020 25th International Conference on Pattern Recognition (ICPR), pp. 5012–5019 (2021). https://doi.org/10.1109/ICPR48806.2021.9412711
4. Chugh, T., Jain, A.K.: Fingerprint spoof detector generalization. IEEE Trans. Inf. Forensics Secur. **16**, 42–55 (2021). https://doi.org/10.1109/TIFS.2020.2990789
5. Cozzolino, D., Gragnaniello, D., Poggi, G., Verdoliva, L.: Towards universal GAN image detection. In: 2021 International Conference on Visual Communications and Image Processing (VCIP), pp. 1–5 (2021). https://doi.org/10.1109/VCIP53242. 2021.9675329
6. Cozzolino, D., Verdoliva, L.: Noiseprint: a CNN-based camera model fingerprint. IEEE Trans. Inf. Forensics Secur. **15**, 144–159 (2020). https://doi.org/10.1109/ TIFS.2019.2916364
7. Didaci, L., Marcialis, G.L., Roli, F.: Analysis of unsupervised template update in biometric recognition systems. Pattern Recogn. Lett. **37**, 151–160 (2014). https:// doi.org/10.1016/j.patrec.2013.05.021
8. Dolhansky, B., Howes, R., Pflaum, B., Baram, N., Ferrer, C.C.: The DeepFake Detection Challenge (DFDC) preview dataset (2019)
9. Güera, D., Delp, E.J.: Deepfake video detection using recurrent neural networks. In: 2018 15th IEEE International Conference on Advanced Video and Signal Based Surveillance (AVSS), pp. 1–6 (2018). https://doi.org/10.1109/AVSS.2018.8639163
10. He, K., Zhang, X., Ren, S., Sun, J.: Deep residual learning for image recognition. In: 2016 IEEE Conference on Computer Vision and Pattern Recognition (CVPR), pp. 770–778 (2016). https://doi.org/10.1109/CVPR.2016.90
11. Hernandez-Ortega, J., Tolosana, R., Fierrez, J., Morales, A.: DeepFakes detection based on heart rate estimation: single- and multi-frame, pp. 255–273. Springer, Cham (2022). https://doi.org/10.1007/978-3-030-87664-7_12
12. Jain, A., Nandakumar, K., Ross, A.: Score normalization in multimodal biometric systems. Pattern Recogn. **38**(12), 2270–2285 (2005). https://doi.org/10.1016/ j.patcog.2005.01.012
13. Korshunov, P., Marcel, S.: DeepFakes: a new threat to face recognition? Assessment and detection. arXiv preprint arXiv:1812.08685 (2018)

14. Li, Y., Chang, M.C., Lyu, S.: In ictu oculi: exposing AI created fake videos by detecting eye blinking. In: 2018 IEEE International Workshop on Information Forensics and Security (WIFS), pp. 1–7 (2018). https://doi.org/10.1109/WIFS.2018.8630787

15. Li, Y., Lyu, S.: Exposing DeepFake videos by detecting face warping artifacts. In: Proceedings of the IEEE/CVF Conference on Computer Vision and Pattern Recognition (CVPR) Workshops, June 2019

16. Li, Y., Yang, X., Sun, P., Qi, H., Lyu, S.: Celeb-DF: a large-scale challenging dataset for DeepFake forensics. In: Proceedings of the IEEE/CVF Conference on Computer Vision and Pattern Recognition (CVPR), June 2020

17. Punyani, P., Kumar, A.: Evaluation of fusion at different levels for face recognition. In: 2017 International Conference on Computing, Communication and Automation (ICCCA), pp. 1052–1055, May 2017. https://doi.org/10.1109/CCAA.2017.8229950

18. Rana, M.S., Sung, A.H.: DeepfakeStack: a deep ensemble-based learning technique for deepfake detection. In: 2020 7th IEEE CSCloud/2020 6th IEEE EdgeCom, pp. 70–75 (2020). https://doi.org/10.1109/CSCloud-EdgeCom49738.2020.00021

19. Roli, F., Giacinto, G., Vernazza, G.: Methods for designing multiple classifier systems. In: Kittler, J., Roli, F. (eds.) MCS 2001. LNCS, vol. 2096, pp. 78–87. Springer, Heidelberg (2001). https://doi.org/10.1007/3-540-48219-9_8

20. Rössler, A., Cozzolino, D., Verdoliva, L., Riess, C., Thies, J., Nießner, M.: Face-Forensics++: learning to detect manipulated facial images. In: ICCV 2019, pp. 1–11 (2019). https://doi.org/10.1109/ICCV.2019.00009

21. Sun, F., Zhang, N., Xu, P., Song, Z.: Deepfake detection method based on cross-domain fusion. Secur. Commun. Netw. **2021**, 2482942 (2021). https://doi.org/10.1155/2021/2482942

22. Thies, J., Zollhöfer, M., Stamminger, M., Theobalt, C., Nießner, M.: Face2Face: real-time face capture and reenactment of RGB videos. Commun. ACM **62**(1), 96–104 (2018). https://doi.org/10.1145/3292039

23. Tolosana, R., Romero-Tapiador, S., Fierrez, J., Vera-Rodriguez, R.: DeepFakes evolution: analysis of facial regions and fake detection performance. In: Del Bimbo, A., et al. (eds.) ICPR 2021. LNCS, vol. 12665, pp. 442–456. Springer, Cham (2021). https://doi.org/10.1007/978-3-030-68821-9_38

24. Tolosana, R., Vera-Rodriguez, R., Fierrez, J., Morales, A., Ortega-Garcia, J.: Deepfakes and beyond: a survey of face manipulation and fake detection. Inf. Fusion **64**, 131–148 (2020). https://doi.org/10.1016/j.inffus.2020.06.014

25. Vishi, K., Mavroeidis, V.: An evaluation of score level fusion approaches for fingerprint and finger-vein biometrics (2018)

26. Yu, P., Xia, Z., Fei, J., Lu, Y.: A survey on deepfake video detection. IET Biom. **10**(6), 607–624 (2021). https://doi.org/10.1049/bme2.12031

27. Zhao, L., Zhang, M., Ding, H., Cui, X.: MFF-Net: deepfake detection network based on multi-feature fusion. Entropy **23**(12), 1692 (2021). https://doi.org/10.3390/e23121692

28. Zhao, Z., Wang, P., Lu, W.: Multi-layer fusion neural network for deepfake detection. Int. J. Digit. Crime Forensics (IJDCF) **13**(4), 26–39 (2021). https://doi.org/10.4018/IJDCF.20210701.oa3

Fooling a Face Recognition System with a Marker-Free Label-Consistent Backdoor Attack

Nino Cauli[1]([✉]), Alessandro Ortis[2], and Sebastiano Battiato[2]

[1] Università degli Studi di Cagliari, via Università 40, 09124 Cagliari, Italy
nino.cauli@unica.it
[2] Università degli Studi di Catania, Piazza Università 2, 95131 Catania, Italy
{ortis,battiato}@dmi.unict.it

Abstract. Modern face recognition systems are mostly based on deep learning models. These models need a large amount of data and high computational power to be trained. Often, a feature extraction network is pretrained on large datasets, and a classifier is finetuned on a smaller private dataset to recognise the identities from the features. Unfortunately deep learning models are exposed to malicious attacks both during training and inference phases. In backdoor attacks, the dataset used for training is poisoned by the attacker. A network trained with the poisoned dataset performs normally with generic data, but misbehave with some specific trigger data. These attacks are particularly difficult to detect, since the misbehaviour occurs only with the trigger images. In these paper we present a novel marker-free backdoor attack for face recognition systems. We generate a label-consistent poisoned dataset, where the poisoned images matches their labels and are difficult to spot by a quick visual inspection. The poisoned dataset is used to attack an Inception Resnet v1. We show that the network finetuned on the poisoned dataset is successfully fooled, identifying one of the author as a specific target identity.

Keywords: Backdoor attack · Adversarial attack · Face recognition · Label-consistent · Deep learning

1 Introduction

Since 2012, when AlexNet won the ImageNet competition, Deep Learning (DL) models have become the *de facto* standard for image classification (IC) and, more recently, for face recognition (FR) [2,8,12,14,15]. Despite the incredible performances of DL for solving IC and FR problems, these approaches are exposed to malicious attacks both in their training and inference phases. Adversarial attacks are dangerous attacks performed at inference time. The goal of these attacks is to modify input images of DL models in order to change the classification results.

The work in this paper was founded by the project PON AIM1893589 promoting the attraction of researchers back to Italy.

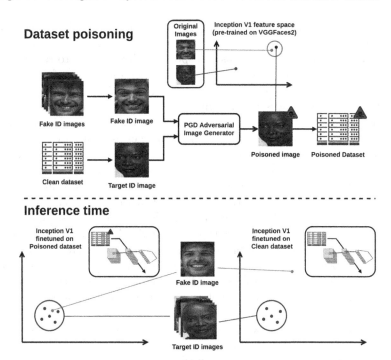

Fig. 1. The backdoor attack presented in this paper: on the top the dataset poisoning; on the bottom the different classification of two networks trained with the poisoned and clean dataset respectively

The changes on the images are small enough not to be spotted by a human [10]. Unfortunately, proper small pixel variations in the input space, can result in a substantial shift in the output feature space of a DL model, leading to misclassification. Although adversarial attacks can be very effective in misleading DL models for IC and FR, they present some drawbacks. In some cases it is not possible to feed the DL model with the digitally crafted adversarial image at inference time, because the input images are captured from physical cameras (e.g., live inference). In these scenarios, classical methods for generating adversarial images fail [19]. The robustness of the DL classifiers to adversarial attacks can be increased through adversarial training, where the model to defend is trained on adversarial attacks generated by the defender [6,18]. Moreover, although adversarial images are crafted to be undetectable by human eyes, they are often easy to be detected by an automated system [5,10,11].

A different approach consists in attacking the DL model during the training phase, poisoning the training dataset with modified entries. A network trained with a poisoned dataset performs normally on benign testing samples, but, for some specific inputs, it changes the prediction to a proper target label specified by the attacker. This type of approach belongs to the class of the so called backdoor attacks: when training the network on a poisoned dataset we are embedding a backdoor in the trained model, which is then triggered during inference time by some specific inputs (see [7] for a survey on backdoor attacks). Usually modified entries in a poisoned dataset are easy to spot by a human due to the mismatch between the poisoned image and its label. Recently researchers face the problem introducing label-consistent backdoor attacks [13,17,21]. In label-consistent attacks the poisoned images visually match their labels, making them difficult to spot by human eyes. Turner et al. [17] generate the poisoned dataset adding a small backdoor pattern in the corners of the images associated with the target label. After training, if the same pattern is added to the image at inference time, the image is classified as the target. In order to be label-consistent, the poisoned images are synthetically generated to be similar to the target class in pixel space, but far from the target class in the classifier prediction. Zhao et al. [21] apply this approach to attack video recognition models. Even if these solutions are label-consistent, a small backdoor pattern is still visible in the poisoned images. In [13], Saha et al. solve the problem hiding the backdoor pattern. They generate the poisoned images to match their labels in pixel space, and to be as close as possible in the classifier features space to random images with the backdoor pattern visible. This results in a label-consistent poisoned dataset, invisible to human eyes.

These label-consistent backdoor attacks are very promising, but they require to add a backdoor pattern to images at inference time in order to fool the classifier. Therefore, although the poisoned training dataset is hidden, a visual inspection of the classifier inputs can detect an attack. We claim that label-consistent backdoor attacks can be achieved without the use of backdoor patterns if applied to FR problems. Datasets for generic IC problems are very diverse, while FR datasets contain similar images of faces, with significant features (nose, mouth, eyes) always at the same locations [3,16]. These common features are easier to learn for adversarial image generators, making possible to create label-consistent poisoned datasets without the use of backdoor patterns.

In this paper we present a backdoor attack able to fool a FR system to recognise a chosen identity (our backdoor) as a target identity. During the attack the targeted system is trained on a label-consistent poisoned dataset generated using an approach similar to [13]: poisoned images are generated to match their clean labels in pixel space, while being close to a target class in the classifier features space. State of the art adversarial generators can be used to create the poisoned images. In our knowledge, the contributions of this paper are the following: 1) first study on label-consistent backdoor attacks to FR systems; 2) first implementation of a label-consistent backdoor attack without relying on backdoor patterns.

2   Proposed Method

In this paper we propose a backdoor strategy to attack a FR system based on Deep Neural Networks (DNN). The goal of the attack is to make the FR model mistake a fake identity for a specific target one. The attack consists in poisoning a dataset used to finetune the FR model: part of the images belonging to the target identity are replaced with poisoned adversarial images. The adversarial images are created to be visually similar to the target images (close in pixel space) and, at the same time, classified as a fake identity image (close in feature space) by the FR model pretrained on a generic dataset (see top of Fig. 1). The poisoned images in the dataset are difficult to spot by a quick human visual verification. The provider of the FR system will then use the poisoned dataset to finetune its model without noticing the poisoned entries. At inference time, the model finetuned with the poisoned dataset classifies both fake and target identity images as images of the same person (see bottom of Fig. 1). Before to describe the proposed backdoor attack in more details, it is useful to introduce few basic notions of Face Recognition and Adversarial Attacks.

The objective of a FR system is to recognise the identity of a person from an image of his/her face. A FR system is divided in three subsystems: a face detector $d()$, a feature extractor $f()$, and a classifier $c()$. First, the face must be located and cropped using a face detector. Second, the cropped images are used as input for the feature extractor. The most successful IC network architectures are used to implement the feature extractor: AlexNet for DeepFace [15]; Inception Resnet V1 for Facenet [14]; VGGNet for VGGface [12]; ResNet for SphereFace [8]; SENet for VGGface2 [2]. At the end, a classifier assigns an identity to the extracted features.

After training, the performance of face recognition models are evaluated with respect to the following tasks: **face verification** is the problem of verify if a pair of input images depict the same person or not; **face identification** is the problem of assigning the identity to a face in an image from a pool of testing identities. Face identification systems can be divided in two classes: **Closed-Set** and **Open-Set** identification. In Closed-Set identification, any subject presented to the identifier is known to be part of the pool of testing identities. In Open-Set identification, on the other hand, it is unknown whether the subject presented is contained in the system's identities set or not.

Despite the incredible performances of Deep Learning for solving FR problems, these models are exposed to adversarial attacks. The goal of these attacks is to modify the FR model inputs in order to change the classification results. The changes on the input are small enough to be difficult to spot by a visual inspection. An adversarial attack can have two distinct goals: make the FR model miss-classify the adversarial images (**obfuscation**, Eq. 1); make the FR model classify an adversarial image of a target class as an image of a specific different class (**replacement**, Eq. 2).

$$\mathbf{I}_{adv} = \underset{\|\mathbf{I}-\mathbf{I}_{tar}\|_p \leq \varepsilon}{\arg\max} \ \mathcal{L}(\mathbf{I}, \mathbf{I}_{tar}, \theta), \tag{1}$$

$$\mathbf{I}_{adv} = \underset{\|\mathbf{I}-\mathbf{I}_{tar}\|_p \leq \varepsilon}{\arg\min} \ \mathcal{L}(\mathbf{I}, \mathbf{I}_{fake}, \theta), \tag{2}$$

where $\mathbf{I}$, $\mathbf{I}_{tar}$, $\mathbf{I}_{fake}$, and $\mathbf{I}_{adv}$ are the image to attack, the target image, the fake ID image, and the adversarial image respectively, $\mathcal{L}()$ is a loss function between the predicted class of the adversarial image and the one of the target image, θ are the weights of the FR model, $\|\cdot\|_p$ is some l_p-norm, and ε is a small constant.

A straightforward way to generate the adversarial images $\mathbf{I}_{adv}$ is calculating the gradient of the loss function $\mathcal{L}()$ with respect to the pixel of the original image $\mathbf{I}$. In the Fast Sign Gradient Method (FGSM) the pixels of $\mathbf{I}_{adv}$ are modified in the direction of the sign of the gradient with respect to $\mathbf{I}$ in a single step (obfuscation):

$$\mathbf{I}_{adv} = \mathbf{I} + \varepsilon \text{sgn}(\nabla_{\mathbf{I}}\mathcal{L}(\mathbf{I}, \mathbf{I}_{tar}, \theta)), \tag{3}$$

where $\nabla_{\mathbf{I}}$ is the gradient with respect $\mathbf{I}$.

Better performances are achieved with iterative versions of the FGSM algorithm, like the Project Gradient Descend (PGD) method [9]. There are two borderline situations for adversarial attacks: in **white-box attacks** the attacker knows exactly the model to be attacked and the statistics of the training set; in **black-box attacks** the attacker does not have any information on the system to be attacked. White-box attacks are easier to perform, but black-box attacks reflect better what happens in real life situations.

In this paper we attack a FR model for face identification. Multi-task Cascaded Convolutional Networks (MTCNNT) [20] is used as face detector. The attacked feature extractor is Facenet [14], pretrained using VGGface2 [2] dataset. We use the PGD algorithm to generate the poisoned images. We are in a replacement scenario: we want to generate poisoned images visually similar to a target class, but classified as a different fake identity by the attacked FR model. The loss function that we minimise is the following:

$$\mathcal{L}(\mathbf{I}, \mathbf{I}_{fake}, \theta_f) = \|f(\mathbf{I}, \theta_f) - f(\mathbf{I}_{fake}, \theta_f)\|_2, \tag{4}$$

where θ_f are the weights of the feature extractor model $f()$ (Inception Resnet V1), and $\|\cdot\|_2$ is the l_2-norm.

3   Results

Let us assume that a company wants to use a FR model for some particular task (i.e. as identification system, to analyse surveillance cameras data or to access to protected data). The company acquires a pretrained model from a external provider and generate a small training set with the images of its employees in order to finetune the FR model. A malicious attacker able to get access to the pretrained FR model and able to alter the training set (white-box attack), can poison the data and make the security system identify her/him as an employee of the company. Using the label-consistent backdoor attack presented in this paper, the poisoned dataset will be difficult to spot.

The FR model chosen for the experiments is Inception Resnet V1, pretrained on VGGFace2. The model takes as input cropped RGB images of faces with dimensions $3 \times 160 \times 160$, and it gives as output an array of 512 features. The cropped images are generated using the MTCNNT algorithm. In real life scenario, the face recognition network would be finetuned on a small face dataset

Fig. 2. Samples of the cropped images used to generate the clean training set (10 females and 10 males)

(e.g. a dataset of the employees of a company). We generated the training set used for finetuning selecting 800 images of 20 random subjects form the VGGFace2 test set (10 women and 10 men, 40 images each). In order to test the finetuned network we created a test set of 400 different images of the same subjects (20 each). Figure 2 shows an example of one cropped image for each of the subjects. The clean training set was poisoned using 20 images of an author of this paper. 18 different images of the author were used to test the efficiency of the attack. The target subject was chosen to be as challenging as possible. In particular, the target has different gender and ethnicity than the author of this paper. Figure 3 shows an example of 3 poisoned images generated from 3 target and 3 fake ID images. To implement the FR model and the MTCNNT algorithm we used the Facenet pytorch repository[1]. For the finetuning, a fully connected layer 512×20 was attached to the output of Facenet. This last layer works as classifier, with its 20 outputs representing the probability of each identity in the training set. The finetuning was run for 8 epochs for each experiment, using Cross Entropy loss function and Adam optimiser (learning rate 0.001). The poisoned images were created using Foolbox, a Python library[2] with an implementation of the PGD algorithm, with $\varepsilon = 0.1$. All the experiments were run on a intel core i7-4720HQ, with 16GB of RAM and an nvidia GeForce GTX 960M graphic card. We assigned a label to each one of the 20 subjects in the training set (*woman0-9* and *man0-9*). The clean training set was then poisoned substituting 20 of the 40 images of the *woman8* subject with poisoned images. The poisoned images were created running the PGD algorithm, embedding a different author image in each of the 20 *woman8* attacked images (see Fig. 3). After being finetuned on the poisoned set, the FR model is expected to recognise the images of the author as images of *woman8* subject. We performed two different tests: in the first one, only the classifier (last layer weights) was finetuned; in the second, we finetuned all the weights of the network (inception resnet v1 and last layer weights). In both cases we are assuming that the attacker knows the FR model and its pretrained weights (white-box attack) and that all the subjects to be identified are present in the training set (closed-set identification). In order to evaluate the results, we also finetuned the networks on the clean training set.

---

[1] https://github.com/timesler/facenet-pytorch.
[2] https://foolbox.jonasrauber.de/.

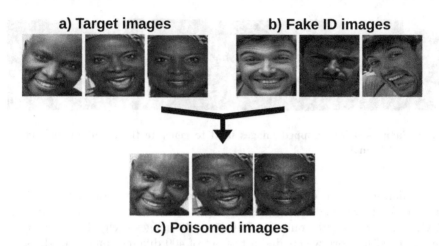

a) Target images   b) Fake ID images

c) Poisoned images

Fig. 3. Example of the poisoning process for 3 images: a) target images to be poisoned; b) fake identity images to be embedded; c) poison images generated by PGD algorithm

Table 1. Finetuning test results: accuracy on the 20 classes set and misclassification (author = *woman8*) ratio on the Fake ID set

F-tuning	Last layer		All layers	
Testset	20 class	Fake ID	20 class	Fake ID
Clean	0.9952	0.0 (0/18)	0.9519	0.0 (0/18)
Poisoned	0.9976	0.89 (16/18)	0.9663	0.0 (0/18)

All the networks were tested using a test set with 400 images of the 20 subjects, and a 18 images set with images of the author face (Fake ID set) as described in Sect. 3. For the first test set (20 classes) we calculate the accuracy of the finetuned models in predicting the correct class of each image. For the Fake ID set (author face images) we calculate the ratio of images classified as *woman8*.

Figure 4a and 4b are plots of the 512 output features of Facenet projected in a bidimensional space using t-distributed stochastic neighbor (t-SNE) algorithm. In order to improve the output of t-SNE, we first extracted the first 50 principal components using the principal component analysis (PCA) algorithm. The points represent the network prediction for each image in the test sets and for the poisoned images. The color of the dots represent the real class of the image (orange for the target class and blue for all the others), while the background color represent the predicted class from the network. We approximate the classification boundaries using a Voronoi tessellation of the space: we colored the Voronoi cell of each point with the color of the class predicted by the network for that image. Figure 4a shows the results for the networks finetuned only on the last layer. In this representation, the poisoned and target images are grouped in two separated and distinguishable sets (red plusses and orange crosses signs).

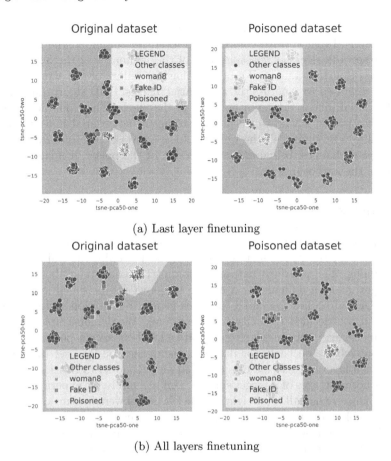

(a) Last layer finetuning

(b) All layers finetuning

Fig. 4. Plots of the feature layer output of the network projected in two dimensions using PCA followed by t-SNE algorithm. Identity replacement experiment finetuning only the classification layer (a) and all the layers (b). (Color figure online)

The clean network fails to classify the author images (green squares) as target images (the squares are on top of blue area). The poisoned network, on the other hand, classify almost all the author images as the target (orange area). The quantitative results shown in Table 1 confirm these findings. The first two columns show the results of the network finetuned both on the clean and on the poisoned set. The two networks have similar accuracy on the 20 class test set (first column), making the attack difficult to be spotted. On the other hand, the poisoned network classify 16 out of 18 author images as *woman*8, while the clean network 0 out of 18 (second column).

Figure 4b shows the results after finetuning all the weights. In this case the backdoor attack fails: both clean and poisoned networks rearrange the feature space bringing together the poisoned and target images (red plusses and orange crosses signs), and moving away the author ones (green squares). The results on accuracy on the 20 subjects set are inline with the ones obtained finetuning only the last layer (Table 1, third column). However, finetuning all the weights, the poisoned network does not classifies any of the 18 images of the author as the target class (fourth column). We believe that the failure of the attack is due to the algorithm used to generate the adversarial images. PGD algorithm generates the adversarial images adding a noise to all the pixels. This approach does not exploit the specific features of a face, resulting in adversarial images not able to generalise. Finetuning all the weights of the network, the organization of the feature space changes, bringing the projection of the poisoned images far from the one of the author images.

4 Conclusions

In this paper we proposed a preliminary work on label-consistent backdoor attacks to a FR system. We attacked an FR model poisoning the training set used for finetuning. The poisoned dataset maintains the consistency between labels and images, making difficult for a human to detect the poisoned images. Until now similar approaches were used to attack generic recognition systems. On the other hand, we decided to focus on FR systems. We demonstrated that, in a white-box scenario, it is possible to generate a label-consistent poisoned training set without relying on backdoor patterns. Using the poisoned set to finetune a classification layer, we successfully attacked an FR model to misclassify images of the author as images of a target subject. Unfortunately, we experienced a drop in performances when all the weights of the FR model were finetuned. We believe that the problem lies in the method used to poison the images, since the changes are uniformly distributed in the entire image. We expect to obtain better performances using an adversarial images generator that changes the images only in the areas corresponding to important facial features. Our next step will be to study the frequency domain to find typical frequencies and/or facial structures in the images, generating a poisoned dataset tailored on FR problems [1,3,4]. In this way we will relax the constraint on white-box attacks and we will perform experiments in a black-box scenario.

References

1. Balakrishnan, G., Xiong, Y., Xia, W., Perona, P.: Towards causal benchmarking of bias in face analysis algorithms. In: Vedaldi, A., Bischof, H., Brox, T., Frahm, J.-M. (eds.) ECCV 2020. LNCS, vol. 12363, pp. 547–563. Springer, Cham (2020). https://doi.org/10.1007/978-3-030-58523-5_32
2. Cao, Q., Shen, L., Xie, W., Parkhi, O.M., Zisserman, A.: VGGFace2: a dataset for recognising faces across pose and age. In: International Conference on Automatic Face and Gesture Recognition (2018)
3. Deb, D., Zhang, J., Jain, A.K.: ADVFaces: adversarial face synthesis. In: 2020 IEEE International Joint Conference on Biometrics (IJCB), pp. 1–10. IEEE (2019)
4. Giudice, O., Guarnera, L., Battiato, S.: Fighting deepfakes by detecting GAN DCT anomalies. J. Imaging 7(8), 128 (2021). https://doi.org/10.3390/jimaging7080128

5. Gong, Z., Wang, W., Ku, W.S.: Adversarial and clean data are not twins. arXiv preprint arXiv:1704.04960 (2017)
6. Goodfellow, I., Shlens, J., Szegedy, C.: Explaining and harnessing adversarial examples. In: 3rd International Conference on Learning Representations, ICLR 2015 - Conference Track Proceedings (2015)
7. Li, Y., Wu, B., Jiang, Y., Li, Z., Xia, S.T.: Backdoor learning: a survey. arXiv preprint arXiv:2007.08745 (2020)
8. Liu, W., Wen, Y., Yu, Z., Li, M., Raj, B., Song, L.: SphereFace: deep hypersphere embedding for face recognition. In: Proceedings of the IEEE Conference on Computer Vision and Pattern Recognition, pp. 212–220 (2017)
9. Madry, A., Makelov, A., Schmidt, L., Tsipras, D., Vladu, A.: Towards deep learning models resistant to adversarial attacks. In: International Conference on Learning Representations (2018)
10. Massoli, F.V., Carrara, F., Amato, G., Falchi, F.: Detection of face recognition adversarial attacks. Comput. Vis. Image Underst. **202**, 103103 (2021)
11. Papernot, N., McDaniel, P.: Deep k-nearest neighbors: towards confident, interpretable and robust deep learning. arXiv preprint arXiv:1803.04765 (2018)
12. Parkhi, O.M., Vedaldi, A., Zisserman, A.: Deep face recognition. In: Proceedings of the British Machine Vision Conference (BMVC), pp. 41.1–41.12, September 2015. https://doi.org/10.5244/C.29.41
13. Saha, A., Subramanya, A., Pirsiavash, H.: Hidden trigger backdoor attacks. In: Proceedings of the AAAI Conference on Artificial Intelligence, vol. 34, pp. 11957–11965 (2020)
14. Schroff, F., Kalenichenko, D., Philbin, J.: FaceNet: a unified embedding for face recognition and clustering. In: Proceedings of the IEEE Conference on Computer Vision and Pattern Recognition, pp. 815–823 (2015)
15. Taigman, Y., Yang, M., Ranzato, M., Wolf, L.: DeepFace: closing the gap to human-level performance in face verification. In: Proceedings of the IEEE Conference on Computer Vision and Pattern Recognition, pp. 1701–1708 (2014)
16. Tinsley, P., Czajka, A., Flynn, P.: This face does not exist... but it might be yours! identity leakage in generative models. In: Proceedings of the IEEE/CVF Winter Conference on Applications of Computer Vision, pp. 1320–1328 (2020)
17. Turner, A., Tsipras, D., Madry, A.: Label-consistent backdoor attacks. arXiv preprint arXiv:1912.02771 (2019)
18. Wang, B., et al.: Neural cleanse: identifying and mitigating backdoor attacks in neural networks. In: 2019 IEEE Symposium on Security and Privacy (SP), pp. 707–723 (2019). https://doi.org/10.1109/SP.2019.00031
19. Zhang, B., Tondi, B., Barni, M.: Adversarial examples for replay attacks against CNN-based face recognition with anti-spoofing capability. Comput. Vis. Image Underst. **197–198**, 102988 (2020)
20. Zhang, K., Zhang, Z., Li, Z., Qiao, Y.: Joint face detection and alignment using multitask cascaded convolutional networks. IEEE Signal Process. Lett. **23**(10), 1499–1503 (2016)
21. Zhao, S., Ma, X., Zheng, X., Bailey, J., Chen, J., Jiang, Y.G.: Clean-label backdoor attacks on video recognition models. In: Proceedings of the IEEE/CVF Conference on Computer Vision and Pattern Recognition, pp. 14443–14452 (2020)

DeepFakes Have No Heart: A Simple rPPG-Based Method to Reveal Fake Videos

Giuseppe Boccignone⬤, Sathya Bursic⬤, Vittorio Cuculo⬤,
Alessandro D'Amelio(✉)⬤, Giuliano Grossi⬤, Raffaella Lanzarotti⬤,
and Sabrina Patania⬤

PHuSe Lab - Dipartimento di Informatica, Università degli Studi di Milano,
20133 Milan, Italy
{giuseppe.boccignone,sathya.bursic,vittorio.cuculo,alessandro.damelio,
giuliano.grossi,raffaella.lanzarotti,sabrina.patania}@unimi.it

Abstract. We present a simple, yet general method to detect fake videos displaying human subjects, generated via Deep Learning techniques. The method relies on gauging the complexity of heart rate dynamics as derived from the facial video streams through remote photoplethysmography (rPPG). Features analyzed have a clear semantics as to such physiological behaviour. The approach is thus explainable both in terms of the underlying context model and the entailed computational steps. Most important, when compared to more complex state-of-the-art detection methods, results so far achieved give evidence of its capability to cope with datasets produced by different deep fake models.

Keywords: DeepFake detection · rPPG · Image forensics · Biological signals

1 Introduction

The term "DeepFake" (DF, a portmanteau of "deep learning" and "fake") refers to videos created by deep learning techniques, especially generative models such as variational auto-encoders and generative adversarial networks, aiming at producing a believable media [1]. Concerning human faces, four DF categories can be identified: re-enactment [2–4], swapping [5], editing [6,7], and synthesis [8].

DF techniques date back to 2017 (a celebrated synthesized version of Obama [9]). Since then, impressive improvements have been achieved, in terms both of realism [10], accessibility and reduction of source data required to realize DF [11]. Yet, very realistic DFs have fostered unethical and malicious applications, posing a series of threats for individuals (e.g. fake porn), organizations (blackmail to managers to stop sharing their compromising DFs), politicians (e.g. fake news to sabotage government leaders) [12]. Thus, efforts have been devoted to DF detection methods allowing to discriminate between real and forged videos [13,14]. Based on the artifacts such methods look for, [12] seven categories can be drawn up: *Blending*, spatial artifacts that appear when the generated content is

S. Sclaroff et al. (Eds.): ICIAP 2022, LNCS 13232, pp. 186–195, 2022.
https://doi.org/10.1007/978-3-031-06430-2_16

blended back to the frame; *Environment*, spatial artifacts that emerge comparing the fake face with the context of the rest of the frame; *Forensics*, artifacts corresponding to subtle features or patterns introduced in the video by the model; *Behaviour*, anomalies in mannerisms or other human behaviours introduced by the model; *Physiology*, physiological signals (e.g. heart beat, blood flow, breathing) disrupted by DF methods; *Synchronization*, temporal inconsistencies (e.g., between visemes and phonemes) introduced in fake videos; *Coherence*, disrupted temporal coherence.

Markedly, behavioural and physiological signals allow to characterize the original videos, resulting in general detectors, independent of the DF generative model, or the dataset adopted for training [15]. Behavioural artifacts involve inconsistencies in physical attributes. The main features investigated are facial attributes [16], head pose [17], facial expression [18], emotions [19,20], gaze tracking [21,22], and blink detection [23].

Physiological artifacts relate to the corruption of physiological signals, such as respiratory pattern [24,25] and heart rate variability [26,27], latently conveyed by original human videos. In particular, several methods have proven their effectiveness as fake detectors using heart rate estimation. In [28] the virtual heart rate is not explicitly computed, while motion-magnified spatial-temporal maps are derived to highlight the chrominance spatio-temporal signal. A dual-spatial-temporal attention network is adopted to alleviate the influence of various interferences such as head movement or illumination variations. In [29] a two stage network is proposed based on the conjecture that in DFs the real PPG signal is lost, while a rhythmic pattern persists, which is a mixture of PPG signals, depending on the adopted generative method. DeepFakesonPhys [30] consists of a convolutional attention network composed of two parallel CNNs to extract and combine spatio-temporal information from the video.

Overall, these approaches prove their effectiveness in the intra-method and inter-method experiments, yet we argue that they suffer from two cogent issues. The first is related to the lack of explicit assessment of the physiological information; indeed, adopting end-to-end approaches brings to consider a broader set of factors (e.g. appearance, texture, behaviour) besides the physiology itself. Clearly, this would potentially allow to reach higher levels of accuracy at the cost of hiding the actual contribution of the information coming from physiological signals alone, if not framed in a principled framework. Second, in this concerning realm, explainability of the method and results achieved should be a serious issue. To the best of our knowledge, the FakeCatcher method [31] is the only proposal in such direction. To characterize interconsistency, such method trains an SVM on 126-dimensional feature vectors computed from the rPPG-derived signal, extracted from three face regions of interest by two rPPG methods. To improve performance, a CNN classifier is trained on PPG maps. However, this way, the explainability is lost, though the system is still based on biological signals.

2   Proposed Method

Background and Rationale. Consider a video of a talking agent involved in some kind of interaction (dyadic, small group, giving a speech). This can be conceived as the observable measurement of the behaviour of the agent according to an agent-in-context (AIC) model [32]. In brief (cfr. Fig. 1), over time, the agent is influenced at the conceptual level by the social and environmental context, beliefs, memories and learning. At the perceptual level the agent takes into account, both exteroceptive sensations from the world and interoceptive sensations from the body. Accordingly, the agent regulates his/her body's visceral physiology and behavioural outflow.

It is out of the scope of this note to discuss in detail the AIC model. It will suffice to remark that level coordination over time stands on a principled generative/predictive framework [32,33], where forward (bottom-up, from periphery to cortex) and backward (top-down, feedback) signalling is synchronized and enforced to support embodied simulation [33]. Indeed, the measurable behavioural/physiological outflow - e.g., facial behaviour or heart rate variability (HRV) that are cogent for the work presented here - is the result of such simulation loop [32–34].

Clearly, our research hypothesis here is that out-of-context manipulation of (faking) one of the observable variables (markedly, the face) the overall coordination is disrupted.

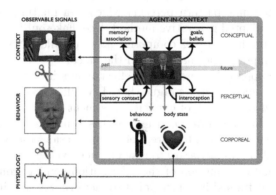

Fig. 1. Agent-in-context model. Right: over time, the agent exhibits a dynamics which involves coordination at different levels: conceptual, perceptual, and corporeal (behavioural/physiological) responses. Left: faking interventions on the observable face dynamics might disrupt (iconized by scissors) observable context/behaviour and behaviour/physiology coordination.

Gauging context/behaviour disruption (cfr. left-most side of Fig. 1), e.g., such as addressed in *Synchronization* and *Behaviour* approaches, is intriguing though complex. Thus, in this work we address behaviour/physiology disruption.

To such end, rPPG [35–38] analysis is a promising approach since HR behaviour and related vascular control can be straightforwardly assessed from the video signal itself. Indeed, rPPG is a suitable mean to sense capillary dilation and constriction related to heart beats. Dilation and constriction modulate the transmission or reflection of visible (or infra-red) light emitted to and detected from the skin. The amount of reflected light changes according to the blood volume and cardiac-synchronous variations (but see [39]).

Fake Detection via rPPG. The hidden physiological information is firstly estimated from RGB videos. To this end, the face of the (possibly) manipulated subject is detected and a set of 100 patches is automatically tracked on it.

The pixels color intensities within the i-th patch at time t $\{p_i^j(t)\}_{j=1}^{N_i}$ are averaged, thus resulting in 100 RGB traces. Denote $q_i(t)$ the RGB trace obtained from the i-th patch:

$$q_i(t) = \frac{1}{N_i} \sum_{j=1}^{N_i} p_i^j(t), \qquad i = 1, \ldots, 100. \tag{1}$$

The trace is then split into K overlapping time-windows, $q_i^k(t) = q_i(t)w(t - k\tau F_s)$, $k = 0, \ldots, K-1$, where F_s represents the video frame rate, τ is the amount of overlap and w is the rectangular window. The Blood Volume Pulse (BVP) signal is then estimated for each patch, at each time frame, using the POS rPPG method [39],

$$x_i^k(t) = POS\left[q_i^k(t)\right]. \tag{2}$$

This procedure is carried out by resorting to the pyVHR Python framework [26], which is suitable to scrutinize and interpret the processing steps outlined above.

Cogently, our goal here is to address signal features that have a clear semantics as to the physiological behavior. The proposed approach employs two sets of features as predictors of the presence of faking interventions. According to the AIC model, we would expect the BVP signals estimated from each patch to be disrupted. Disruption is here gauged in terms of complexity of the estimated BVP signals. Signal complexity can be addressed in many different ways; we refer to the ensemble of predictors as to *Intra-Patch BVP Complexity Measures*.

Further, it is reasonable to assume that a genuine face video would yield BVP estimates with a certain amount of coherence across the tracked patches, while exhibiting manifold behaviours in forged ones. Hence, besides individual patch-based BVP complexity, we quantify the amount of coherence between patches by means of *Inter-Patch BVP Coherence Measures*.

Intra-Patch BVP Complexity Measures. A simple and widely used measure of complexity of a time series is represented by the number of Zero-Crossings of the signal; that is the rate at which a time-series changes from positive to negative, or vice-versa. Additionally, the Hjorth Mobility and Complexity parameters

[40] are computed as indicators of the statistical properties of the signal in the time domain. In the frequency domain, the Shannon entropy of the Power Spectral Density of the estimated BVP is computed. Another common approach is to measure the entropy of a time series; more precisely, a BVP signal x can be defined as a family of time-indexed random variables $\{x(t)\}_{t \in \mathcal{T}}$. The *Entropy Rate* of a sequence of N random variables,

$$H(x) = \lim_{N \to \infty} \frac{1}{N} H\left(x(1), x(2), \dots x(N)\right) \tag{3}$$

describes the rate of growth of the entropy of the sequence with N. The analytical computation of the Entropy Rate of a time-series requires knowing the joint distribution of the random variables composing it; however, many efficient algorithms for its approximation from a data sample have been proposed, e.g., Approximate Entropy [41], Sample Entropy, Permutation Entropy [42], SVD Entropy [43].

Another powerful statistical index of complexity, is the fractal dimension (FD) of the waveform. A variety of algorithms are available for the computation of the FD. Here we consider four approaches that have been widely adopted for the analysis of biological signals [44], namely Detrended Fluctuation Analysis (DFA), Katz FD, Petrosian FD and Higuchi FD. All such quantities are averaged across the 100 patches in each time frame in order to yield a 12-D feature vector of Intra-Patch Complexity Measures.

Inter-Patch BVP Coherence Measures. The degree of consistency between the BVP estimates across the patches is quantified for a given time window both in the time and frequency domain.

The level of spectral concordance between the BVP signals estimated from the i-th and j-th patches in the k-th time frame, can be computed as their *magnitude-squared Coherence* function,

$$C(x_i^k, x_j^k) = \frac{|G(x_i^k, x_j^k)|^2}{G(x_i^k, x_i^k)G(x_j^k, x_j^k)},$$

where $G(x_i^k, x_j^k)$ is the *cross-spectral density* between the BVP estimates from the i-th and j-th patches, respectively; $G(x_i^k, x_i^k)$ and $G(x_j^k, x_j^k)$ are the *Power Spectral Densities* of the signals. The magnitude-squared Coherence function denotes the amount of similarity between the two signals at each frequency. Here we consider the average of the magnitude-squared Coherence function over frequencies as a scalar measure of spectral similarity between BVP estimates.

We also evaluate the similarity of the signals in the time domain by measuring the Mutual Information between the BVPs:

$$MI(x_i^k, x_j^k) = \int_{x_i^k} \int_{x_j^k} P(x_i^k, x_j^k) log\left(\frac{P(x_i^k, x_j^k)}{P(x_i^k)P(x_j^k)}\right) d_{x_i^k} d_{x_j^k}$$

Such quantities are computed for each possible pair (i, j) of patches and for each time frame k. The 8-D Inter-Patch BVP Coherence feature vector for the k-th time window is then obtained by computing the first four moments of the resulting distributions (mean, standard deviation, skewness, kurtosis) for both spectral similarity and Mutual Information.

Feature Selection and Classification. The procedure described above, yields two sets of measures that when joined produce a 20-D vector of features for every frame k of a given video. Sequential floating forward selection (SFFS) [45] is thus employed in order to select the best subset of features without sensible loss of information. According to SFFS, the following 6 features result to be the most informative for DF classification via rPPG signals: Zero Crossing Rate, Petrosian FD, SVD Entropy, Higuchi FD, Average Spectral similarity and Standard Deviation of Mutual Information. Remarkably, 4 features are Complexity related quantities on BVP signals, while 2 are between-patches coherence measures.

A Support Vector Machine (SVM) is eventually adopted for classification (real vs. fake). Training and classification are carried out at the time-frame level; video level predictions are obtained by picking the most frequently predicted label.

3   Results

We test the effectiveness of the proposed approach on the FaceForensics++ dataset [46] consisting of 1000 original video sequences that have been manipulated with five methods, namely DeepFakes [47], Face2Face [48], FaceShifter [49], FaceSwap [50] and NeuralTextures [51]. Figure 2 shows one example (via FaceShifter) of swapping faces from a source to a target subject. The proposed approach is tested on each DF algorithm outcome separately. Table 1 reports the 5-fold Nested Cross-Validation Accuracy results. As can be observed, regardless of the swapping method, we are able to discriminate real from fake videos with remarkable accuracy. The table shows the results obtained by a recent approach [31]; reported accuracy scores are comparable. Most important, we test the ability of our method to generalize across different face swapping algorithms. To

Fig. 2. Face swapping. Source (left), target (center) and swapped (right) example from the FaceForensics++ dataset

Table 1. 5-Fold nested cross validation accuracy for the video-level classification of Real *vs.* Fake videos for each face swapping algorithm

	DeepFakes	Face2Face	FaceShifter	FaceSwap	NeuralText.	Avg.
Our	90.68 ± 0.61	94.46 ± 0.03	98.88 ± 0.13	95.39 ± 0.07	87.57 ± 0.35	93.40%
[31]	**94.87%**	**96.37%**	-	**95.75%**	**89.12%**	**94.02%**

this end, the original videos belonging to the "Real" class are partitioned into train/test sets with 80/20%, while the "Fake" class is built by joining all the data from every swap algorithm except one. We train our SVM on this dataset and test the model on the left out data. As a consequence, the "Fake" class contains about 5 times the number of examples if compared to the "Real" one. In Table 2, the results of the cross-method evaluation are reported. Due to the class imbalance, besides accuracy scores we also report the F1 and AUC scores, too. Notably, the proposed approach outperforms the results obtained by [31], thus exhibiting better generalization abilities despite representing a much simpler and straightforward method.

Table 2. Cross-method results: testing on a swap algorithm while training on the others

	Metric	DeepFakes	Face2Face	FaceSh.	FaceSwap	NeuralText.	Avg.
Our	AUC	89.82%	88.97%	91.52%	88.59%	89.76%	89.75%
	F1	96.72%	96.70%	98.21%	97.44%	94.14%	96.64%
	Acc.	**94.53%**	94.48%	**96.98%**	95.66%	**90.61%**	**94.45%**
[31]	Acc.	93.75%	**95.25%**	-	**96.25%**	81.25%	91.62%

4   Conclusions

We have presented a simple method for DF detection. The features extracted from the rPPG-based signal have a clear semantics as to the heart rate physiological behaviour. The approach is thus explainable both in terms of the underlying context model and the entailed computational steps. Most important, when compared to more complex state-of-the-art detection methods, present results give evidence of its capability to cope with datasets produced by different DF models. Beyond the reported results, the agent-in-context model, which frames our approach, is suitable for paving the way to a seamless integration of analyses addressing different levels of the faked agent's behavior. The present investigation is to be intended as a *proof of concept*, hence further experiments on larger and more challenging datasets (e.g. [52,53]) are planned to be carried out in future works.

References

1. Lee, S.-H., Yun, G.-E., Lim, M.Y., Lee, Y.K.: A study on effective use of bpm information in deepfake detection. In: 2021 International Conference on Information and Communication Technology Convergence (ICTC), pp. 425–427. IEEE (2021)

2. Bansal, A., Ma, S., Ramanan, D., Sheikh, Y.: Recycle-GAN: unsupervised video retargeting. In: Proceedings of the European Conference on Computer Vision (ECCV), pp. 119–135 (2018)

3. Tran, L., Yin, X., Liu, X.: Representation learning by rotating your faces. IEEE Trans. Pattern Anal. Mach. Intell. **41**(12), 3007–3021 (2018)

4. Bursic, S., D'Amelio, A., Granato, M., Grossi, G., Lanzarotti, R.: A quantitative evaluation framework of video de-identification methods. In: 2020 25th International Conference on Pattern Recognition (ICPR), pp. 6089–6095. IEEE (2021)

5. Peng, B., Fan, H., Wang, W., Dong, J., Lyu, S.: A unified framework for high fidelity face swap and expression reenactment. In: IEEE Transactions on Circuits and Systems for Video Technology (2021)

6. Gupta, A., Khan, F.F., Mukhopadhyay, R., Namboodiri, V.P., Jawahar, C.: Intelligent video editing: incorporating modern talking face generation algorithms in a video editor. In: Proceedings of the Twelfth Indian Conference on Computer Vision, Graphics and Image Processing, pp. 1–9 (2021)

7. Ding, H., Sricharan, K., Chellappa, R.: Exprgan: facial expression editing with controllable expression intensity. In: Proceedings of the AAAI Conference on Artificial Intelligence, vol. 32 (2018)

8. Karras, T., Laine, S., Aittala, M., Hellsten, J., Lehtinen, J., Aila, T.: Analyzing and improving the image quality of stylegan. In: Proceedings of the IEEE/CVF Conference on Computer Vision and Pattern Recognition, pp. 8110–8119 (2020)

9. Suwajanakorn, S., Seitz, S.M., Kemelmacher-Shlizerman, I.: Synthesizing Obama: learning lip sync from audio. ACM Trans. Graph. **36**(4), 1–13 (2017)

10. Nirkin, Y., Keller, Y., Hassner, T.: FSGAN: subject agnostic face swapping and reenactment. In: Proceedings of the IEEE/CVF International Conference on Computer Vision, pp. 7184–7193 (2019)

11. Lattas, A., et al.: Avatarme: realistically renderable 3D facial reconstruction in-the-wild. In: Proceedings of the IEEE/CVF Conference on Computer Vision and Pattern Recognition, pp. 760–769 (2020)

12. Mirsky, Y., Lee, W.: The creation and detection of deepfakes: a survey. ACM Comput. Surv. **54**(1), 1–41 (2021)

13. Tolosana, R., Vera-Rodriguez, R., Fierrez, J., Morales, A., Ortega-Garcia, J.: Deepfakes and beyond: a survey of face manipulation and fake detection. Inf. Fusion **64**, 131–148 (2020)

14. Nguyen, T.T., Nguyen, C.M., Nguyen, D.T., Nguyen, D.T., Nahavandi, S.: Deep learning for deepfakes creation and detection: a survey. arXiv preprint arXiv:1909.11573 (2019)

15. Ciftci, U.A., Demir, I., Yin, L.: How do the hearts of deep fakes beat? deep fake source detection via interpreting residuals with biological signals. In: 2020 IEEE International Joint Conference on Biometrics (IJCB), pp. 1–10. IEEE (2020)

16. Matern, F., Riess, C., Stamminger, M.: Exploiting visual artifacts to expose deepfakes and face manipulations. In: 2019 IEEE Winter Applications of Computer Vision Workshops (WACVW), pp. 83–92. IEEE (2019)

17. Yang, X., Li, Y., Lyu, S.: Exposing deep fakes using inconsistent head poses. In: ICASSP 2019–2019 IEEE International Conference on Acoustics, Speech and Signal Processing (ICASSP), pp. 8261–8265. IEEE (2019)
18. Agarwal, S., Farid, H., Gu, Y., He, M., Nagano, K., Li, H.: Protecting world leaders against deep fakes. In: CVPR Workshops, vol. 1 (2019)
19. Mittal, T., Bhattacharya, U., Chandra, R., Bera, A., Manocha, D.: Emotions don't lie: an audio-visual deepfake detection method using affective cues. In: Proceedings of the 28th ACM International Conference on Multimedia, pp. 2823–2832 (2020)
20. Hosler, B., et al.: Do deepfakes feel emotions? a semantic approach to detecting deepfakes via emotional inconsistencies. In: Proceedings of the IEEE/CVF Conference on Computer Vision and Pattern Recognition, pp. 1013–1022 (2021)
21. Demir, I., Ciftci, U.A.: Where do deep fakes look? synthetic face detection via gaze tracking. In: ACM Symposium on Eye Tracking Research and Applications, pp. 1–11 (2021)
22. Cuculo, V., D'Amelio, A., Lanzarotti, R., Boccignone, G.: Personality gaze patterns unveiled via automatic relevance determination. In: Mazzara, M., Ober, I., Salaün, G. (eds.) STAF 2018. LNCS, vol. 11176, pp. 171–184. Springer, Cham (2018). https://doi.org/10.1007/978-3-030-04771-9_14
23. Jung, T., Kim, S., Kim, K.: Deepvision: deepfakes detection using human eye blinking pattern. IEEE Access 8, 83144–83154 (2020)
24. Prathosh, A., Praveena, P., Mestha, L.K., Bharadwaj, S.: Estimation of respiratory pattern from video using selective ensemble aggregation. IEEE Trans. Signal Process. 65(11), 2902–2916 (2017)
25. Chen, M., Zhu, Q., Zhang, H., Wu, M., Wang, Q.: Respiratory rate estimation from face videos. In: 2019 IEEE EMBS International Conference on Biomedical & Health Informatics (BHI), pp. 1–4. IEEE (2019)
26. Boccignone, G., Conte, D., Cuculo, V., D'Amelio, A., Grossi, G., Lanzarotti, R.: An open framework for remote-PPG methods and their assessment. IEEE Access 8, 216083–216103 (2020)
27. Rouast, P.V., Adam, M.T., Chiong, R., Cornforth, D., Lux, E.: Remote heart rate measurement using low-cost RGB face video: a technical literature review. Front. Comput. Sci. 12(5), 858–872 (2018)
28. Qi, H., et al.: DeepRhythm: exposing deepfakes with attentional visual heartbeat rhythms. In: Proceedings of the 28th ACM International Conference on Multimedia, pp. 4318–4327 (2020)
29. Liang, J., Deng, W.: Identifying rhythmic patterns for face forgery detection and categorization. In: 2021 IEEE International Joint Conference on Biometrics (IJCB), pp. 1–8 (2021)
30. Hernandez-Ortega, J., Tolosana, R., Fierrez, J., Morales, A.: DeepFakesON-Phys: Deepfakes detection based on heart rate estimation. arXiv preprint arXiv:2010.00400 (2020)
31. Ciftci, U.A., Demir, I., Yin, L.: FakeCatcher: detection of synthetic portrait videos using biological signals. IEEE Transactions on Pattern Analysis and Machine Intelligence (2020)
32. Koban, L., Gianaros, P.J., Kober, H., Wager, T.D.: The self in context: brain systems linking mental and physical health. Nat. Rev. Neurosci. 22(5), 309–322 (2021)
33. Hutchinson, J.B., Barrett, L.F.: The power of predictions: an emerging paradigm for psychological research. Curr. Direct. Psychol. Sci. 28(3), 280–291 (2019)

34. Boccignone, G., Conte, D., Cuculo, V., D'Amelio, A., Grossi, G., Lanzarotti, R.: Deep construction of an affective latent space via multimodal enactment. IEEE Trans. Cognit. Develop. Syst. **10**, 865–880 (2018)
35. Wieringa, F.P., Mastik, F., Steen, A.F.W.v.d.: Contactless multiple wavelength photoplethysmographic imaging: a first step toward "spo2 camera"technology". Ann. Biomed. Eng. **33**(8), 1034–1041 (2005)
36. Humphreys, K., Ward, T., Markham, C.: Noncontact simultaneous dual wavelength photoplethysmography: a further step toward noncontact pulse oximetry. Rev. Sci. Instrum. **78**(4), 044304 (2007)
37. Verkruysse, W., Svaasand, L.O., Nelson, J.S.: Remote plethysmographic imaging using ambient light. Opt. Express **16**(26), 21434–21445 (2008)
38. McDuff, D.J., Estepp, J.R., Piasecki, A.M., Blackford, E.B.: A survey of remote optical photoplethysmographic imaging methods. In: 2015 37th Annual International Conference of the IEEE Engineering in Medicine and Biology Society (EMBC), pp. 6398–6404. IEEE (2015)
39. Wang, W., den Brinker, A.C., Stuijk, S., De Haan, G.: Algorithmic principles of remote PPG. IEEE Trans. Biomed. Eng. **64**(7), 1479–1491 (2016)
40. Hjorth, B.: Eeg analysis based on time domain properties. Electroencephalogr. Clin. Neurophysiol. **29**(3), 306–310 (1970)
41. Pincus, S.M., Gladstone, I.M., Ehrenkranz, R.A.: A regularity statistic for medical data analysis. J. Clin. Monitor. **7**(4), 335–345 (1991)
42. Bandt, C., Pompe, B.: Permutation entropy: a natural complexity measure for time series. Phys. Rev. Lett. **88**(17), 174102 (2002)
43. Roberts, S.J., Penny, W., Rezek, I.: Temporal and spatial complexity measures for electroencephalogram based brain-computer interfacing. Med. Biol. Eng. Comput. **37**(1), 93–98 (1999)
44. Esteller, R., Vachtsevanos, G., Echauz, J., Litt, B.: A comparison of waveform fractal dimension algorithms. IEEE Trans. Circuits Syst. I: Fundam. Theory Appl. **48**(2), 177–183 (2001)
45. Pudil, P., Novovičová, J., Kittler, J.: Floating search methods in feature selection. Pattern Recogn. Lett. **15**(11), 1119–1125 (1994)
46. Rössler, A., Cozzolino, D., Verdoliva, L., Riess, C., Thies, J., Nießner, M.: Face-Forensics++: learning to detect manipulated facial images. In: International Conference on Computer Vision (ICCV) (2019)
47. Deepfakes. https://github.com/deepfakes/faceswap
48. Thies, J., Zollhofer, M., Stamminger, M., Theobalt, C., Nießner, M.: "Face2face: real-time face capture and reenactment of RGB videos. In: Proceedings of the IEEE Conference on Computer Vision and Pattern Recognition, pp. 2387–2395 (2016)
49. Li, L., Bao, J., Yang, H., Chen, D., Wen, F.: Faceshifter: towards high fidelity and occlusion aware face swapping. arXiv preprint arXiv:1912.13457 (2019)
50. Faceswap. https://github.com/MarekKowalski/FaceSwap/
51. Thies, J., Zollhöfer, M., Nießner, M.: Deferred neural rendering: image synthesis using neural textures. ACM Trans. Graph. **38**(4), 1–12 (2019)
52. Li, Y., Yang, X., Sun, P., Qi, H., Lyu, S.: Celeb-df: a large-scale challenging dataset for deepfake forensics. In: Proceedings of the IEEE/CVF Conference on Computer Vision and Pattern Recognition, pp. 3207–3216 (2020)
53. Dolhansky, B., et al.: The deepfake detection challenge (dfdc) dataset. arXiv preprint arXiv:2006.07397 (2020)

Image Analysis, Detection
and Recognition

Exact Affine Histogram Matching by Cumulants Transformation

Andrea Fusiello[✉]

DPIA, Università di Udine, Via Delle Scienze, 208 Udine, Italy
andrea.fusiello@uniud.it

Dedicated to the memory of Egle Poggi (1936–2022).

Abstract. This paper proposes an exact solution to histogram matching under affine transformation, i.e., it shows how to retrieve an unknown affine transformation between the colors of two images. The key is the use of the third central moment (skewness) in addition to covariance and mean. These three moments (a.k.a. cumulants) are sufficient to determine all the d.o.f. of an affine mapping of the RGB space.

Keywords: Color matching · Color transfer · Colour mapping

1 Introduction

Histogram matching is the transformation of the colors of one image (source) so that its histogram matches a specified target one (see Fig. 1). The well-known histogram equalization method is a special case in which the target histogram is a uniform distribution. It can be used to normalize two or more images, when they were acquired with different sensors (or from a sensor whose response changes over time), atmospheric conditions or global illumination. Affine transformations are particularly relevant for color histograms of images taken under varied illuminant conditions are related by an affinity [9].

Fig. 1. Source (left) and target (right) images. The goal is to transform the source so as to match the histogram of the target. Image from the UPenn dataset [19]

Histogram matching, a.k.a. color transfer, has been well studied in the literature, see [12] for a review. In their seminal paper, Reinhard et al. [13] match the mean and standard deviation of each axis separately after converting the source and target images into the decorrelated colour space $l\alpha\beta$ [15]. They write:

> While it appears that the mean and standard deviation alone suffice to produce practical results, the effect of including higher moments remains an interesting question.

This paper shows empirically that an affine transformation is completely and exactly determined by the first three *multivariate* moments.

In recent years, the problem of color transfer evolved into "style transfer" (see e.g., [6]) where algorithms based on spatial (local) color mappings can handle applications such as time-of-day [18], weather and season change [4], painterly stylization [1] and transfer of artistic edits [17].

Motivated by the reduction of seam artefacts in mosaicking from aerial images [16], in this work we aim at recovering the *exact* affine mapping between the colors of two images (assuming that such mapping exists). Previous work in histogram matching fail to reach this goal: Pitié et al. [11] obtain an *approximate* non parametric map; others [10,20] arbitrarily fix some degrees of freedom of a general affinity (this point will be clarified in Sect. 3), or restrict the solution to a subset of affine mappings [9,13].

2   Background

In single-channel (gray-scale) images, the brightness values can be considered as the samples of a univariate random variable X, whose probability density function (pdf) is the normalized histogram. For any function g, also $Y = g(X)$ is a random variable.

2.1   Q-Q Plots

Let F_X and F_Y be cumulative distribution function (CDF) of X and Y, respectively. The functions F_Y^{-1} and F_X^{-1} are the *quantile functions*, i.e. $F^{-1}(\alpha)$ is the $\alpha - th$ quantile of F. If g is monotone increasing, then $F_Y(y) = F_X(g^{-1}(y))$, and also $F_Y^{-1}(u) = g(F_X^{-1}(u))$. This means that the quantile functions transform according to g, hence the graph of g is described by the pairs $(F_Y^{-1}(\alpha), F_X^{-1}(\alpha))$ for all $\alpha \in [0,1]$. This graph, where the quantiles of two variables are plotted one against the other is called the *Q-Q plot*. So, for gray-scale images, histogram matching can accomplished by means of the Q-Q plot, that allows to find the exact transformation that makes the quantiles of X and Y to match. For example, if g is an affine map, the Q-Q plot will be a line, with a given (positive) slope and intercept.

In the case of color (RGB) images the underlying random variable is multivariate, the CDF is $\mathbb{R}^3 \to \mathbb{R}$ and the quantile function is not defined. The Q-Q plot trick can only be applied to each channel separately, but this would yield a very special affinity, whose linear part is represented by a (positive) diagonal matrix. Note that, since any matrix A can be decomposed with SVD as

$A = UDV^\top$, if one transform X with $V^\top$ and Y with $U^\top$, then the residual transformation is the diagonal D and can be recovered with three Q-Q plots. The problem is that these U and V are unknown. In fact, Pitié et al. [11] iteratively apply random rotations to the color space and solve using Q-Q plots on each channel separately.

2.2 Cumulants

Instead of exploiting the quantiles of the two probability distributions, let us turn our attention to its *moments*, in particular to the first raw moment – the mean, and the second and third central moments, a.k.a. variance and skewness. These three moments coincide with the first three *cumulants* of the probability distribution (see e.g. [8]), and so they will be collectively referred to in the following. The cumulants of a probability distribution provide an alternative description of the distribution to those given by its moments: any two probability distributions whose moments are identical will have identical cumulants, and vice versa. As a matter of fact, the first three cumulants are also moments, but fourth and higher-order cumulants are not.

Using the Kronecker product, the first three cumulants of X write:

- Average $\kappa_1(X) = \mathbb{E}[X] = \mu_X$
- Covariance $\kappa_2(X) = \mathbb{E}[(X - \mu_X) \otimes (X - \mu_X)]$
- Skewness $\kappa_3(X) = \mathbb{E}[(X - \mu_X) \otimes (X - \mu_X) \otimes (X - \mu_X)]$

The third cumulant of a d-dimensional random vector $X = (X_1, \dots, X_d)^\top$ is an $d^2 \times d$ matrix, containing at most $d(d+1)(d+2)/6$ distinct elements. This is analogous to $\kappa_2(X)$, that is $d \times d$ and has at most $d(d+1)/2$ distinct elements, being symmetric. An elimination matrix G_d $d(d+1)(d+2)/6 \times d^3$ can be defined [7] that extracts only the unique elements from $\mathrm{vec}\, \kappa_3(X)$.

Given a random sample from X of size N, the third sample cumulant is computed as follows:

$$\kappa_3(X) = \frac{1}{N} \sum_{i=1}^{N} (\mathbf{x}_i - \mathbf{m}) \otimes (\mathbf{x}_i - \mathbf{m})^\top (\mathbf{x}_i - \mathbf{m}) \tag{1}$$

where $\mathbf{x}_i \in \mathbb{R}^d$ is the i-th sample and $\mathbf{m} \in \mathbb{R}^d$ is the sample mean.

It is particularly useful to our goals to be able to determine how the cumulants change under a generic affine transformation $Y = AX + t$, where A is a $d \times d$ non-singular matrix and $\mathbf{t}$ is a $d \times 1$ vector:

$$\kappa_1(Y) = A\kappa_1(X) + \mathbf{t} \tag{2}$$

$$\kappa_2(Y) = A\kappa_2(X)A^\top \tag{3}$$

$$\kappa_3(Y) = (A \otimes A)\kappa_3(X)A^\top. \tag{4}$$

The first two relationships are well-known, while the third is proved in [3]. Using the "vec trick" [14], we get the following equivalent formulae:

$$\text{vec } \kappa_1(Y) = (I_3 \otimes A) \text{ vec } \kappa_2(X) + \mathbf{t} \tag{5}$$

$$\text{vec } \kappa_2(Y) = (A \otimes A) \text{ vec } \kappa_2(X) \tag{6}$$

$$\text{vec } \kappa_3(Y) = (A \otimes A \otimes A) \text{ vec } \kappa_3(X). \tag{7}$$

3 Method

Given two multivariate random variables X and Y we want to find the affine transformation that match the first three cumulants of Y with those of X.

The translation can be recovered from Eq. (2) once A has been determined, so let us concentrate on Eqs. (3) and (4) and on the linear transformation A.

Let us start with the second cumulant (covariance), and follow [10]. Since $\kappa_2(X)$ and $\kappa_2(Y)$ are positive definite, let $\Sigma_X \Sigma_X^\top = \kappa_2(X)$ and, likewise, $\Sigma_Y \Sigma_Y^\top = \kappa_2(Y)$ (see Lemma 1 in the Appendix). Substituting this into (3) yields:

$$\Sigma_Y \Sigma_Y^\top = (A\Sigma_X)(A\Sigma_X)^\top \tag{8}$$

from which a solution is derived as:

$$A = \Sigma_Y \Sigma_X^{-1}. \tag{9}$$

Since the factorization of a psd matrix as per Lemma 1 is not unique, we arbitrarily set

$$\Sigma_X = UD^{1/2}\text{diag}(1,1,\det(U)) \tag{10}$$

where

$$\kappa_2(X) = UDU^\top \tag{11}$$

is the spectral decomposition of $\kappa_2(X)$, with ordered eigenvalues. It is easy to see that $\Sigma_X \Sigma_X^\top = \kappa_2(X)$. The same applies to Σ_Y likewise.

Then, as observed by [10], the general solution, that matches the second order cumulants of X and Y, is given by:

$$A = \Sigma_Y Q \Sigma_X^{-1} \tag{12}$$

where $QQ^\top = I$. This leaves three d.o.f. in the choice of Q, that have been fixed according to different considerations, in the literature.

For example, in the principal components (PCA) method [9,20], a rotation is sought that align the principal axis of $\kappa_2(X)$ with those of $\kappa_2(Y)$, and Q is chosen among four possible matrices, namely:

$$\text{diag}(1, \ 1, \ 1), \quad \text{diag}(1,-1,-1), \quad \text{diag}(-1, \ 1,-1), \quad \text{diag}(-1,-1, \ 1). \tag{13}$$

In the Monge-Kantorovitch (MK) solution [10]:

$$A_{\text{MK}} = \hat{\Sigma}_X^{-1} \left(\hat{\Sigma}_X \kappa_2(Y) \hat{\Sigma}_X \right)^{\frac{1}{2}} \hat{\Sigma}_X^{-1} \tag{14}$$

where $\hat{\Sigma}_X = \kappa_2(X)^{\frac{1}{2}}$ and $\hat{\Sigma}_Y = \kappa_2(Y)^{\frac{1}{2}}$. This corresponds to

$$Q = \Sigma_Y^{-1} A_{\mathrm{MK}} \Sigma_X. \tag{15}$$

Keeping in mind that $\hat{\Sigma}_X = U D^{1/2} U^\top$ (and likewise for $\hat{\Sigma}_Y$) it can be proven that Q is indeed an orthogonal matrix.

None of the previous methods, however is able to exactly recover the affine transformation, when the two histograms are affinely related. Our method instead achieve exact recovery by exploiting the third order cumulant κ_3 in Eq. (7). In particular, we fix the remaining three d.o.f. by solving the non-linear system of 10 equations:

$$G_3 \kappa_3(Y) = G_3(A \otimes A \otimes A)\mathrm{vec}\,\kappa_3(X) \quad \text{with } A = \Sigma_Y Q \Sigma_X^{-1} \tag{16}$$

in the least squares sense with the Levenberg-Marquardt method.

Experiments reported in the next section show that in this way the affinity is always recovered, upon convergence of the numerical method.

If we restrict to A with positive determinant, then $Q \in SO(3)$ and so we can parametrize it with the Euler angles ω, ϕ, ϑ. This assumption can be relaxed at the price of searching for the minimum in $-SO(3)$ as well, since $O(3)$ has two connected components.

The Matlab code that implements our method is available on the web at http://www.diegm.uniud.it/fusiello/demo/ahm/

4   Experiments

In order to validate the method in a controlled setting, we applied random affine transformations (with positive determinant) to the reference image shown in Fig. 1 and then run our method to retrieve the affine transformation matrix. Convergence is only local, so we started the minimization in a neighborhood of the ground truth.

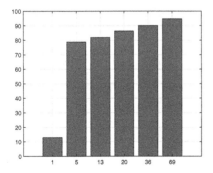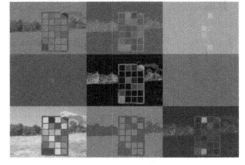

Fig. 2. Convergence rate as a function of the number of initial points (left); the first bar corresponds to the MK solution, while the second includes the four PCA solutions. Sample of images mapped according to the random affine transformation (right).

The error was measured as the Frobenius norm of the difference between the output matrix and the ground truth one. Upon convergence it was of the order of $1e-10$ (or less). To be more realistic we also started our method from multiple points by picking N regularly spaced samples in $SO(3)$, in addition to the MK solution and the four PCA solutions. These samples are the centers of an equal area partitioning of the unit sphere in $\mathbb{R}^4$ [5], which is a double cover of $SO(3)$ (antipodal points represent the same rotation) via the quaternion representation of rotations. Figure 2 shows the convergence rate as a function of N for 1000 random affine transformations. These experiments indicates empirically that the method is able to recover the correct affine transformation, up to numerical errors, when suitably initialized.

In the real experiments that follow the underlying transformations are more regular than the random ones (see Fig. 2 for a sample), although not necessarily affine, so we initialized instead the method with the MK solution, that provides a reasonable starting point, thanks to the minimization of the transportation cost. Our solution is nevertheless different from MK, as one can see in Fig. 3.

Source   Target

IDT   MK   Ours

Fig. 3. Comparison with results reported in [11] (IDT) and [10] (MK).

Source   Output   Checkerboard (target/output)

Fig. 4. Indoor scene with varying light temperature. The top image is the target, then from the second row onward images are: source, output and checkerboard comparison.

Source   Output   Checkerboard (target/output)

Fig. 5. Outdoor scene in different time of the day. The top image is the target, then from the second row onward images are: source, output and checkerboard comparison.

In our first real experiment we shoot eight pictures of an indoor scene with a Sony α6300 camera, while varying the temperature of the LED light source from 3200K to 5500K. The image at 5500K was chosen as the target one and all the others have been transformed with the affinity found by our algorithm to match the target histogram. The results are shown in Fig. 4.

A similar experiment has been conducted outdoor (Fig. 5). A series of nine photos have been taken during the day, one has been selected as target and all the others have been transformed to match its histogram.

Figure 6 shows the result of our method on a portion of the dataset that originally motivated this research [16], where the effect of the illumination change is particularly prominent.

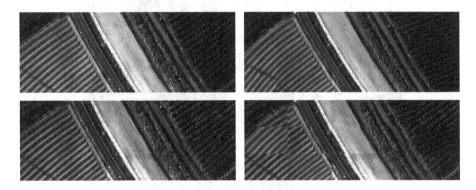

Fig. 6. Top row: two images of the same area under different illumination, source (left) and target (right) Bottom row: output (left), i.e., source image transformed to match the target, and checkerboard comparison (right) between target and output (a small misalignment is present).

In the last experiment we tried also to apply our algorithm to color transfer, although this was not our focus. As a global method, it achieved reasonable results on some of images used in [6] (Fig. 7), while it failed on more challenging cases where local operations (and a segmentation) are required, e.g. changing the color of the eyes, or lighting the windows of a building.

Source   Target   Output

Fig. 7. Some examples of color transfer on images from [6]. The output is the affine transform of the source to match the target palette.

5   Conclusions

In this paper we have shown how to retrieve an unknown affine transformation between the colors of two images (formally, between two probability distributions) thanks to the use of the third cumulant, which allows to fix the three d.o.f. that remain after matching mean and covariance. This study was motivated by image mosaicking, but can also find application in color transfer or in histogram normalization [9]. Future work will study the extension of this approach to more general classes of transformations.

Acknowledgements. Images shown in Figs. 4 and 5 are courtesy of Leonardo Fusiello, Fig. 6 are courtesy of Helica srl. The MATLAB implementation of the recursive zonal equal area sphere partitioning is due to P. Leopardi.

Appendix

Lemma 1 [2]. A real matrix M is positive semi-definite (psd) if and only if it can be decomposed as $M = \Sigma \Sigma^\top$.

The decomposition is not unique: if $M = \Sigma \Sigma^\top$ also $M = \Sigma Q Q^\top \Sigma^\top = (\Sigma Q)(\Sigma Q)^\top$ with $QQ^\top = I$. Requiring Σ to be psd as well makes the factorization unique. In this case Σ is referred to as the *square root* of M.

References

1. Gatys, L.A., Ecker, A.S., Bethge, M.: Image style transfer using convolutional neural networks. In: IEEE Conference on Computer Vision and Pattern Recognition, pp. 2414–2423 (2016)
2. Horn, R.A., Johnson, C.R.: Matrix Analysis, 2nd edn. Cambridge University Press, Cambridge; New York (2013)
3. Kollo, T., von Rosen, D., Hazewinkel, M.: Advanced Multivariate Statistics with Matrices. Springer, Heidelberg (2005). https://doi.org/10.1007/1-4020-3419-9
4. Laffont, P.Y., Ren, Z., Tao, X., Qian, C., Hays, J.: Transient attributes for high-level understanding and editing of outdoor scenes. ACM Trans. Graph. (Proc. SIGGRAPH) **33**(4), 149:1–149:11 (2014)
5. Leopardi, P.: A partition of the unit sphere into regions of equal area and small diameter. Electron. Trans. Numer. Anal. **25**, 309–327 (2006)
6. Luan, F., Paris, S., Shechtman, E., Bala, K.: Deep photo style transfer. In: Proceedings of the IEEE Conference on Computer Vision and Pattern Recognition, July 2017
7. Meijer, E.: Matrix algebra for higher order moments. Linear Algebra Appl. **410**, 112–134 (2005)
8. Muirhead, R.J.: Aspects of Multivariate Statistical Theory. Wiley-Interscience, New York (2005)
9. Pei, S.C., Tseng, C.L., Wu, C.C.: Using color histogram normalization for recovering chromatic illumination-changed images. J. Opt. Soc. Am. A **18**(11), 2641–2658 (2001)

10. Pitié, F., Kokaram, A.: The linear Monge-Kantorovitch linear colour mapping for example-based colour transfer. In: 4th European Conference on Visual Media Production, pp. 1–9 (2007)
11. Pitié, F., Kokaram, A., Dahyot, R.: N-dimensional probability density function transfer and its application to color transfer. In: IEEE International Conference on Computer Vision, vol. 2, pp. 1434–1439 (2005)
12. Pitié, F.: Advances in colour transfer. IET Comput. Vis. **14**, 304–322 (2020)
13. Reinhard, E., Ashikhmin, M., Gooch, B., Shirley, P.: Color transfer between images. IEEE Comput. Graph. Appl. **21**(5), 34–41 (2001)
14. Roth, W.E.: On direct product matrices. Bull. Am. Math. Soc. **40**(6), 461–468 (1934)
15. Ruderman, D.L., Cronin, T.W., Chiao, C.C.: Statistics of cone responses to natural images: implications for visual coding. J. Opt. Soc. Am. A **15**(8), 2036–2045 (1998)
16. Santellani, E., Maset, E., Fusiello, A.: Seamless image mosaicking via synchronization. ISPRS Ann. Photogramm. Remote Sens. Spatial Inf. Sci. **4**(2), 247–254 (2018)
17. Shih, Y., Paris, S., Barnes, C., Freeman, W.T., Durand, F.: Style transfer for headshot portraits. ACM Trans. Graph. **33**(4), 1–14 (2014)
18. Shih, Y., Paris, S., Durand, F., Freeman, W.T.: Data-driven hallucination of different times of day from a single outdoor photo. ACM Trans. Graph. **32**(6), 1–11 (2013)
19. Tkačik, G., et al.: Natural images from the birthplace of the human eye. PLOS ONE **6**(6), 1–12 (2011)
20. Trussell, H., Vrhel, M.: Color correction using principle components. In: Image Processing Algorithms and Techniques II. SPIE Proceedings, vol. 1452, pp. 2–9 (1991)

Digital Line Segment Detection for Table Reconstruction in Document Images

Phuc Ngo$^{(\boxtimes)}$ (D)

Université de Lorraine, CNRS, LORIA, 54000 Nancy, France
`hoai-diem-phuc.ngo@loria.fr`

Abstract. Table detection is often involved in many applications of document analysis as tables are frequently used to present structured information. In this context, we are interested in extracting table regions in document images. More precisely, we propose a method for table detection based on a recent edge line detector which is developed in the context of digital geometry and it allows to handle noisy document images. The extracted lines are then used to reconstruct the tables contained in the image. The method has been evaluated and compared to other state-of-the-art methods and shown a very competitive result.

Keywords: Line detector · Blurred segment · Adaptive directional scan · Materialized table extraction · Digital geometry

1 Introduction

The process of table analysis involves table detection and table structure recognition. The task of table detection aims to locate regions of image that contain tables using the bounding boxes, while table structure recognition involves the identification of row and column layout information for individual table cells. It is an important task in many document analysis applications. Thank to their compact form, tables are frequently used to summarized information in a structured and relational way. They are present in a large variety of documents such as reports, scientific papers, business documents, invoice, ...

This paper addresses the table detection in document images. It is a difficult problem due to the variety of layouts and formats of tables in images (see Fig. 1). Furthermore, when tables are correctly extracted from images, it allows to perform the recognition task more efficiently, and thus improve the table information extraction such as the text recognition within each cell and the interpretation of tabular data in the document.

In the context of table detection, several approaches have been proposed in the literature. One of the earliest works on identifying tabular regions in document images is of Watanabe *et al.* [27] in which the authors proposed to identify blocks enclosed by vertical and horizontal line segments. More precisely, line segments are first detected and corner points are subsequently determined. The method then exploits the connective relationships between the extracted

S. Sclaroff et al. (Eds.): ICIAP 2022, LNCS 13232, pp. 211–224, 2022.
https://doi.org/10.1007/978-3-031-06430-2_18

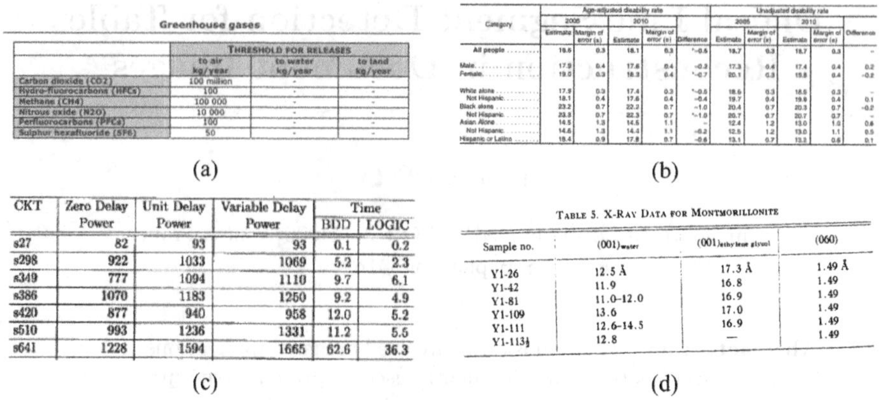

Greenhouse gases

	THRESHOLD FOR RELEASES		
	to air kg/year	to water kg/year	to land kg/year
Carbon dioxide (CO2)	100 million	-	-
Hydro-fluorocarbons (HFCs)	100	-	-
Methane (CH4)	100 000	-	-
Nitrous oxide (N2O)	10 000	-	-
Perfluorocarbons (PFCs)	100	-	-
Sulphur hexafluoride (SF6)	50	-	-

(a)

Category	Age-adjusted disability rate					Unadjusted disability rate				
	2005		2010			2005		2010		
	Estimate	Margin of error (±)	Estimate	Margin of error (±)	Difference	Estimate	Margin of error (±)	Estimate	Margin of error (±)	Difference
All people	18.6	0.3	18.1	0.3	*-0.5	18.7	0.3	18.7	0.3	-
Male	17.9	0.4	17.6	0.4	-0.2	17.3	0.4	17.4	0.4	0.2
Female	19.0	0.3	18.3	0.4	*-0.7	20.1	0.3	19.8	0.4	-0.2
White alone	17.9	0.3	17.4	0.3	*-0.5	18.6	0.3	18.6	0.3	-
Not Hispanic	18.1	0.4	17.6	0.4	-0.4	19.7	0.4	19.8	0.4	0.1
Black alone	23.2	0.7	22.2	0.7	-1.0	20.4	0.7	20.3	0.7	-0.2
Not Hispanic	23.3	0.7	22.3	0.7	*-1.0	20.7	0.7	20.7	0.7	-
Asian Alone	14.5	1.3	14.5	1.1	-	12.4	1.2	13.0	1.0	0.6
Not Hispanic	14.6	1.3	14.4	1.1	-0.2	12.5	1.2	13.0	1.1	0.5
Hispanic or Latino	18.4	0.9	17.8	0.7	-0.6	13.1	0.7	13.2	0.6	0.1

(b)

CKT	Zero Delay Power	Unit Delay Power	Variable Delay Power	Time	
				BDD	LOGIC
s27	82	93	93	0.1	0.2
s298	922	1033	1069	5.2	2.3
s349	777	1094	1110	9.7	6.1
s386	1070	1183	1250	9.2	4.9
s420	877	940	958	12.0	5.2
s510	993	1236	1331	11.2	5.5
s641	1228	1594	1665	62.6	36.3

(c)

TABLE 5. X-RAY DATA FOR MONTMORILLONITE

Sample no.	$(001)_{water}$	$(001)_{ethylene\ glycol}$	(060)
Y1-26	12.5 Å	17.3 Å	1.49 Å
Y1-42	11.9	16.8	1.49
Y1-81	11.0-12.0	16.9	1.49
Y1-109	13.6	17.0	1.49
Y1-111	12.6-14.5	16.9	1.49
Y1-113½	12.8	—	1.49

(d)

Fig. 1. Examples of materialized tables in document images. Images extracted from ICDAR 2013 table competition [13] (a–b) and from UNLV table dataset [26] (c–d).

corner points for the individual blocks using global and local tree structures. Following this idea, the methods in [6,10,12,14] use parallel lines, the horizontal and vertical space as features to extract the table regions. Other conventional techniques for table analysis have been studied based on the textual layout analysis of documents. We can mention the work of T. Kieninger [18] which relies on the detection of text block units for extracting the table cells. More precisely, the method starts with an arbitrary word as block seed, and recursively extends this block to all the words that interlace with their vertical neighbors. Then, it uses the gap information to prevent the table columns. Similarly, S. Mandal *et al.* in [21] supposed that tables are present in distinct columns and the spaces between the fields are larger than those between words in text lines. Then, the table extraction relies on the formation of word blob text lines and on finding the set of text lines that may form a table. In [25], the method is based on the layout analysis module of Tesseract OCR (Optical Character Recognition) to detect tables. More specifically, it uses the tab-stops to inform text aligns and to extract blocks composing the layout of a document. These blocks are then evaluated as table candidates using a specific strategy. Recently, the deep-learning approaches based on convolutional neural network (CNN) have been significantly investigated for table analysis. T. Kieninger and A. Dengel [28] can be considered as the pioneers to apply unsupervised learning method to the table detection using a bottom-up clustering of given word segments. After that, we have the supervised methods such as in [5,11] that used Faster R-CNN based model for table detection, or [22,23] used CNN to performed table detection and table structure recognition at the same time.

The present work aims at designing a method of table detection using image processing techniques. In particular, we are interested in *materialized tables* in documents. Such tables have rectangular form and are composed by straight line segments intersecting at right angle. They may not be closed but must contain at least two surrounded horizontal lines and inside vertical lines (see Fig. 1). For detecting these tables, the method relies on a recent line detector called *Fast Blurred Segment Detector* (FBSD) which is developed in the context of digital geometry. This detector is fast, accurate and robust to noise. It can be run in pure automatic mode, with quite few parameter settings. The extracted lines are then used as candidates to reconstruct the tables in document images. The method has been evaluated and compared to other state-of-the-art methods on different public datasets and shown a very competitive result.

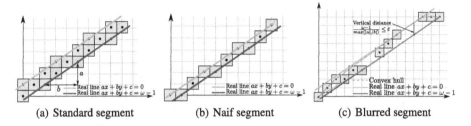

|(a) Standard segment|(b) Naif segment|(c) Blurred segment|

Fig. 2. Examples of (a) a standard segment belongs to the 4-connected digital line $\mathcal{D}(2,-3,3,5)$, (b) a naif segment belongs to the 8-connected digital line $\mathcal{D}(2,-3,0,3)$ and (c) a blurred segment of thickness $\varepsilon = 1.5$ belongs to the digital line $\mathcal{D}(3,-4,3,7)$.

The paper is organized as follows. Section 2 recalls the main theoretical notions used in this work. Section 3 describes the FBSD detector for extracting straight line segments in images. The proposed method of table extraction is presented in Sect. 4. Then, Sect. 5 shows the experimental results, conclusion and future works are provided in Sect. 6.

2 Background Notions

We recall hereafter several notions of digital geometry [20] used in this work. We refer the reader to the given references for more details.

2.1 Digital Straight Line and Blurred Segment

Definition 1 [24]**.** *A **digital straight line** of integer characteristics* $(a, b, c, \omega) \in \mathbb{Z}^4$ *with a and b relatively prime, denoted by* $\mathcal{D}(a, b, c, \omega)$, *is the set of digital points* $p = (x, y) \in \mathbb{Z}^2$ *satisfying the inequalities:*

$$0 \leq ax + by + c < \omega \tag{1}$$

Hereafter, we note $\vec{V}(\mathcal{D}) = (-b, a)$ the director vector of $\mathcal{D}$, $w(\mathcal{D}) = \omega$ its arithmetical width, $h(\mathcal{D}) = -c$ its shift to origin, and $p(\mathcal{D}) = max(|a|, |b|)$ its period. The thickness $\mu = \frac{\omega-1}{p(\mathcal{D})}$ of $\mathcal{D}$ is the minimum of the vertical and horizontal distances between lines $ax + by + c = 0$ and $ax + by + c = \omega - 1$. When $\omega = max(|a|, |b|)$ then $\mathcal{D}$ is the narrowest 8-connected line and called *naif digital line*, and $\omega = |a| + |b|$ then $\mathcal{D}$ is 4-connected line and called *standard digital line*.

A *digital straight segment* is a finite subset of $\mathcal{D}$. Figure 2 (a–b) show examples of digital straight lines and segments.

Definition 2 [16]. *A **blurred segment** $\mathcal{B}$ of assigned thickness ε is a sequence of points that are all covered by a digital straight line $\mathcal{D}$ of thickness $\mu \leq \varepsilon$. The covering digital line with minimal thickness is called the* optimal line *of $\mathcal{B}$.*

Example of blurred segment is given in Fig. 2 (c). In [16], it is shown that blurred segments can be detected in linear time by a recognition algorithm based on an incremental growth of the convex hull of added points.

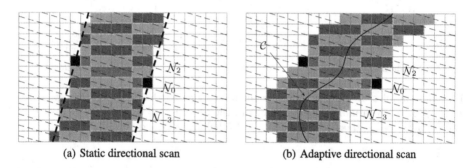

(a) Static directional scan   (b) Adaptive directional scan

Fig. 3. (a) A static directional scan with the start scan S_0 in red, odd scans in green, even scans in blue, bounds of scan lines $\mathcal{N}_i$ with dashed lines and bounds of scan strip $\mathcal{D}$ with bold dashed lines. (b) An adaptive directional scan in which the scan strip is dynamically fit to the curve $\mathcal{C}$, and the scans are accordingly shifted to cover $\mathcal{C}$. (Color figure online)

2.2 Adaptive Directional Scan

From the notion of blurred segment, two notions of directional scan are proposed in [7,17] for thick line detection in gray-level images.

Definition 3 [17]. *A **static directional scan** is an ordered partition into scans S_i restricted to the grid domain $\mathcal{G} \subset \mathbb{Z}^2$ of a thick digital straight line $\mathcal{D}$, called* scan strip. *Each scan S_i is a segment of a naive line $\mathcal{N}_i$, called* scan line, *orthogonal to $\mathcal{D}$. The directional scan is defined as:*

$$DS = \left\{ S_i = \mathcal{D} \cap \mathcal{N}_i \cap \mathcal{G} \, \middle| \, \begin{array}{l} \vec{V}(\mathcal{N}_i) \cdot \vec{V}(\mathcal{D}) = 0 \\ h(\mathcal{N}_i) = h(\mathcal{N}_{i-1}) + p(\mathcal{D}) \end{array} \right\} \tag{2}$$

In this definition, $\vec{V}(\mathcal{N}_i) \cdot \vec{V}(\mathcal{D}) = 0$ expresses the orthogonality between the scan lines $\mathcal{N}_i$ and the scan strip $\mathcal{D}$. The shift $p(\mathcal{D})$ between successive scans $\mathcal{N}_{i-1}$ and $\mathcal{N}_i$ guarantees that all points of $\mathcal{D}$ are traversed only one time. The scans S_i can be iteratively parsed from the start scan S_0 to both ends (see Fig. 3 (a)).

An **adaptive directional scan** is a dynamical version of the directional scan with an on-line registration to a moving search direction. Compared to static directional scans where the scan strip remains fixed to the initial line $\mathcal{D}_0$, here the scan strip $\mathcal{D}_i$ follows a curve $\mathcal{C}$ to track while scan lines $\mathcal{N}_i$ remain fixed (see Fig. 3 (b)).

Definition 4 [7]. *An **adaptive directional scan** is defined by:*

$$ADS = \left\{ S_i = \mathcal{D}_i \cap \mathcal{N}_i \cap \mathcal{G} \left| \begin{array}{l} \vec{V}(\mathcal{N}_i) \cdot \vec{V}(\mathcal{D}_0) = 0 \\ h(\mathcal{N}_i) = h(\mathcal{N}_{i-1}) + p(\mathcal{D}_0) \\ \mathcal{D}_i = \mathcal{D}(\widehat{C}_i, w(\mathcal{D}_0)), i > 0 \end{array} \right. \right\} \tag{3}$$

where $\widehat{C}_i$ is a triplet composed of the director vector (a_i, b_i) and the shift to origin c_i of an estimate at position i of the tangent to the curve $\mathcal{C}$ to track.

The obtained thick digital line is used to update the scan strips and lines. The last clause expresses the scan bounds update at iteration i.

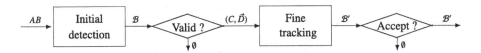

Fig. 4. Flowchart of the thick line segment detection process by FBSD [7].

3   FBSD: Fast Blurred Segment Detector

In this section, we describe briefly the efficient tool FBSD [7] to detect blurred segments from gray-level images based on discrete geometry notions presented previously. More precisely, the method examines the gradient image of the input for edge detection. Firstly, it uses gradient magnitude to select image points for the edge candidates. Then, it relies on gradient orientation to track edges and to detect the blurred segments. The obtained segments encode straight edges, and its thickness parameter characterizes the edge sharpness. In particular, a linear-time blurred segment recognition algorithm is applied to incrementally extend the segment and the adaptive directional scan allows to orientate the edge candidates in appropriate directions in gradient image.

The process for a single detection, given an input seed segment AB crossing the edges to be detected, is summarized in Fig. 4. It is composed of two steps.

The initial detection consists in building and extending a rough blurred segment $\mathcal{B}$ of assigned thickness ε_0, based on points with highest gradient magnitude found in each scan of the directional scan defined by AB with a large tolerance to gradient direction. Validity tests, for small and sparse segments, are then applied to decide the detection pursuit.

In the second step of fine tracking, the final blurred segment $\mathcal{B}'$ is built and extended with points that correspond to local maxima of the gradient, ranked by magnitude order, and with gradient direction close to start point gradient direction. At this refinement step, the adaptive directional scan, defined by the found center position C and the rough blurred segment direction $\vec{D}$, is used in order to extend the segment in the appropriate direction. As soon as the thickness of the expanding blurred segment gets stable, a procedure is applied to focalize the search direction and to avoid further insertion of outliers. Final length and sparsity thresholds of the output segment $\mathcal{B}'$ can be set accordingly. They are the only parameters of this local detector, together with the input assigned thickness ε_0.

By considering the prominent local maxima of the gradient magnitude under AB, the previous process also allows the detection of all the edges crossed by the seed segment AB, called *multi-detection*. Furthermore, in order to avoid multiple detections of the same edge, an occupancy mask is used to collect the dilated points of all the detected blurred segments, so that these points can not be used any more in the detection.

In [7], a full automatic detection of all straight lines in image is also proposed. More precisely, a line segment, that crosses the whole image, is swept in both directions, vertical then horizontal, from the center to the borders of the image. At each position, the multi-detection algorithm is run to collect all the blurred segments found under the sweeping segment. Then, small segments are rejected in order to avoid the formation of misaligned segments when the sweeping segment crosses an image edge near one of its ends. For more details of the FBSD detector, we refer the reader to the paper [7].

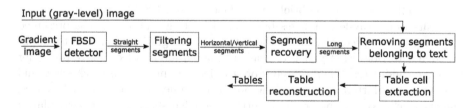

Fig. 5. Flowchart of the proposed method for table extraction from document images.

4   Main Table Extraction Framework

This section presents the proposed method for extracting tables from document images. In [14], a table is said *fully materialized* if it is composed and surrounded by straight line segments intersecting at right angle. In our work, we concern with tables of rectangular form which are also composed by straight line segments intersecting at right angle. However, they may not be closed, compared to fully materialized tables, but must contain at least two surrounded horizontal lines and inside vertical lines. We call them *materialized tables*. Such tables are quite frequent in a wide variety of documents such as invoices, scientific papers, commercial/administrative documents, question forms, ... Some examples of materialized tables are given in Fig. 1.

The proposed method relies on the edge line detector FBSD described in the previous section. The lines in document images are not all table edges, they can be part of graphics, images or separators. It should be mentioned that the acquisition noise during the sequential scanning/printing process may disrupt, damage or misalign the straight line segments (see Fig. 1 (c–d)) and it makes the task of table reconstruction and detection more difficult. This leads us to a proposal of line segment recovery before the reconstruction of tables from document images.

The automatic table extraction process is composed of 6 steps. These steps are detailed in the following. The overview of the whole process is given in Fig. 5 and illustrated in Fig. 6.

Line Segment Detection: This first step consists in extracting line segments in document images (see Fig. 6 (b)). For this, we use the FSBS detector [7]. These lines consist candidate segments of tables to be extracted and also other segments belonging to graphics, images, separators, texts, ...

Horizontal and Vertical Segment Extraction: As we address only materialized tables which are composed of horizontal and vertical segments intersecting at right angle, we thus remove all segments which are not horizontal and vertical (see Fig. 6 (c)). A tolerance of $\alpha = \pm 5°$ angle is applied for such segment selection due to impact of scanning procedure of documents.

Line Segment Recovery: In order to deal with broken lines, especially in degraded conditions and due to the acquisition noise, and to ensure a better stability of table detection, the extracted horizontal (resp. vertical) segments are prolongated and connected if they are close to each other (see Fig. 6 (d)). A distance threshold $\delta = 30$ pixels is used to test the closeness between two horizontal (resp. vertical) segments.

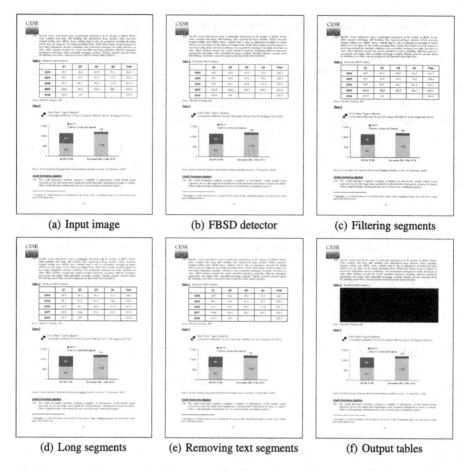

(a) Input image   (b) FBSD detector   (c) Filtering segments

(d) Long segments   (e) Removing text segments   (f) Output tables

Fig. 6. Different steps of the table extraction process: (a) input image, (b) straight segments detected by FBSD (green), (c) candidate horizontal (blue) and vertical (red) segments, (d) recovering nearby segments by extending and connecting horizontal (blue) and vertical (red) segments, (e) removing segments associated to text and (f) final result of table extraction (black box). (Color figure online)

Suppression of Segments Belonging to Text: The extracted segments from the previous step do not always belong to borders of cell tables. In particular, most of them are associated to text in document image. In order to eliminate such segments, we consider the original gray-level image and verify the line model of border cell tables. A local analysis around the extracted segment is performed. It can be observed that the pattern of the 1D gray intensity profile the segment of a cell border is typical with a bottom (of very low intensity) in the middle and high

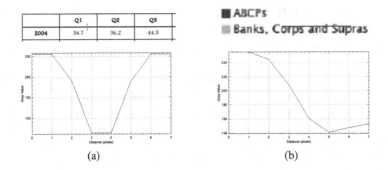

Fig. 7. Gray intensity profile (red segment) analysis along extracted segment (blue segment): (a) border cell table, and (b) text. (Color figure online)

intensity on both sides[1], and it is not always the case for segments associated to text (see Fig. 7). This profile pattern is repetitive along the segment associated to the border cell tables, but not in case of text. Therefore, by simply computing the ratio of the number of profile patterns corresponding to the border cell segment over the segment length, we can distinguish the segments associated to text. In our experiments, the ratio is set at 75% for eliminating text segments, and $w = 7$ pixels around the extracted segment is used for the local analysis of intensity profile. The result of this step is illustrated in Fig. 6 (e).

Table Cell Extraction: This step aims at detecting table cells before the final step of table reconstruction. The cell extraction is based on the vertical and horizontal segments limiting the table cells. Indeed, for the materialized tables, these cells are limited by four segments defining a rectangle. In case of open tables, $i.e.$, they are not bounded by vertical segments (see Fig. 1 (c–d)), the cells at extremities are surrounded by only three segments (two horizontal and one vertical segments), then we can close the cell by using the end points of the horizontal segments. By that way, we can eliminate, from the extracted segments, the separator lines which do not form a close rectangle.

Table Reconstruction: Finally, a table is built as unions of all its cells (see Fig. 6 (f)). For this step, we compute the connected components from the extracted cells, and the corresponding table is obtained as the bounding box of each connected component. A verification of table size is performed at the end to remove small bounding boxes related to noise or other elements which may not be the tables. For this, we keep only the bounding boxes of size bigger than 1% of image size.

---

[1] A similar line model has been used in [14] for line verification and table extraction. Their model consists in separating the 1D intensity profile, from left to right, into 3 zones: the intensity should begin to increase then (potentially) stabilize and decrease.

5   Experimental Results

This section presents results of the proposed method for table extraction from document images. In the experimental stage, the proposed approach is validated through comparisons with other recent methods in the literature: S. Faisal and S. Ray [25], H. Alhéritière et al. [14], D. Prasad et al. [23] and M. Li et al. [22]. In [25], the authors proposed a method based on components of the layout analysis module of Tesseract to locate tables in document images. In [14], the method relies on the local Radon Transform for line extraction and then the pattern recognition techniques for reconstructing tables from the extracted lines. [23] and [22] are deep-learning approaches based on convolution networks which allow to detect the regions of tables and recognize the structural body cells from the detected tables at the same time. For more details about these methods, we refer the reader to the given references.

Table 1. Evaluation results of table detection in document images on different datasets.

Dataset	Method	P	R	F_1
ICDAR 2013 dataset	Faisal et al. [25][a]	0.25	0.33	0.28
	Alhéritière et al. [14][b]	0.98	0.83	0.90
	Prasad et al. [23]	1	1	1
	M. Li et al. [22]	0.97	0.80	0.88
	Our proposal	0.91	0.85	0.88
UNLV table dataset	Faisal et al. [25]	0.86	0.79	0.82
	Alhéritière et al. [14]	0.73	0.66	0.70
	M. Li et al. [22]	**0.92**	**0.96**	**0.94**
	Our proposal	0.83	0.82	0.83
Marmot dataset	M. Li et al. [22]	0.77	**0.98**	0.86
	Our proposal	**0.93**	0.81	**0.87**

[a]Results extracted from [14].
[b]Results on a subset of ICDAR 2013 containing 88 tables in [14].

Comparison results are run on three publicly available datasets for table detection: ICDAR 2013 Table Competition [13], UNLV Table Dataset [26] and Marmot Dataset [8]. More precisely, the ICDAR 2013 [1] contains 128 documents with a total of 150 tables: 75 tables in 27 excerpts from the EU and 75 tables in 40 excerpts from the US Government. The UNLV Table Dataset [4] contains 427 examples in scanned image format from a variety of sources ranging from technical reports and business letters to newspapers and magazines. The Marmot Dataset [2] has 2000 documents mainly from research papers composed of both Chinese and English pages, in which 1000 pages containing at least one table, while the other 1000 pages do not contain tables, but have complex layout where some page components may be mistakenly recognized as tables, such as matrices and figures. All these datasets are provided with information, in XML format, of table structure ground-truth data such as bounding box coordinates, rows, columns, cells, ...

To evaluate the performance of the table detection algorithms, we use the standard evaluation metrics in the literature [15, 19]: precision (P), recall (R) and F_1-measure (F_1) which is an harmonic mean of P and R. The metrics are computed by summing up the overlapping area of the obtained result and the ground-truth. More precisely, let D be the set of pixels of the extracted tables, and G of the ground-truth. Then, the considered measure are computed as:

$$P = \frac{D \cap G}{D}, \quad R = \frac{D \cap G}{G}, \quad F_1 = \frac{2 * R * P}{R + P} \tag{4}$$

The results are reported in Table 1 on the overall average score of each dataset. We observe that our method archives well-scored and competitive results comparing to the others on the considered datasets. More precisely, for the ICDAR 2013 dataset, [23] performed best at the absolute score of 100% on the evaluation measures, and our proposal is as good as [22]. In case of the UNLV Table dataset, [22] provides the best score and our method comes in second place. This can be explained by the very low quality of scanned document images, and also this dataset contains no materialized tables (see Fig. 8). Finally, on the Marmot dataset, the proposed method has the best precision and F_1 measures, but less good than [22] on recall. This lower recall can be explained by the non detection of materialized tables and the false detection of graphics in this dataset. Though, it should be mentioned that this dataset contains high quality images.

Further comparisons with other methods in the literature are also available in the referenced papers [8, 22, 23, 25].

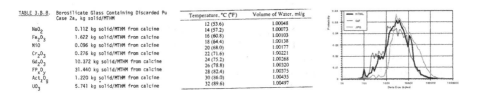

Fig. 8. Limit cases of our proposal. Images from UNLV [26] and Marmot [8] datasets.

6 Conclusion

In this paper, we presented a method for table extraction from document images based on notions of digital geometry. The method relies on the FBSD detector [7] to detect the straight line segments in image and then reconstruct the tables from the extracted lines. This detector is fast, accurate and robust to noise. In particular, it allows to handle noisy document images. Indeed, the sequential printing and scanning of a document may deteriorate its contents, including the table structures such as disrupting table lines. The proposed approach has been evaluated and compared with other methods on different table datasets. The results are encouraging and competitive to the existing methods. The source code of the proposed method, based on OpenCV library [3], is available at the *GitHub* repository: https://github.com/ngophuc/TableExtraction.

In the field of document analysis and interpretation of tabular data in document images, the table structure recognition [9,28] is also an active research topic. It involves identifying the rows, columns, individual table cells. In an advantage context, it is to retrieve the textual content of the cells. It is widely used in image captioning, video description, and many other applications. In this context, the proposed method can be helpful as a first step of table detection before extracting other information. On the other hand, we can use the method to address the need for generating data for the machine learning approaches which need many data for the training process.

As mentioned previously, we address the materialized tables in this work. However, in the considered datasets, there are tables not for this case study, and also graphics or matrices that our algorithm mistakenly detects as tables, see Fig. 8 for some examples. In future, we would like to extend the method for more general cases of tables and to conduct more comparisons with the state-of-the-art methods on more data.

References

1. ICDAR 2013 table competition dataset. https://www.tamirhassan.com/html/competition.html
2. Marmot dataset. https://www.icst.pku.edu.cn/cpdp/sjzy/index.htm
3. OpenCV: Open computer vision. https://opencv.org/
4. UNLV table dataset. https://github.com/tesseract-ocr/
5. Arif, S., Shafait, F.: Table detection in document images using foreground and background features. In: 2018 Digital Image Computing: Techniques and Applications (DICTA), pp. 1–8 (2018). https://doi.org/10.1109/DICTA.2018.8615795
6. Cesarini, F., Marinai, S., Sarti, L., Soda, G.: Trainable table location in document images. In: 2002 International Conference on Pattern Recognition, vol. 3, pp. 236–240 (2002). https://doi.org/10.1109/ICPR.2002.1047838
7. Even, P., Ngo, P., Kerautret, B.: Thick line segment detection with fast directional tracking. In: Ricci, E., Rota Bulò, S., Snoek, C., Lanz, O., Messelodi, S., Sebe, N. (eds.) ICIAP 2019. LNCS, vol. 11752, pp. 159–170. Springer, Cham (2019). https://doi.org/10.1007/978-3-030-30645-8_15
8. Fang, J., Tao, X., Tang, Z., Qiu, R., Liu, Y.: Dataset, ground-truth and performance metrics for table detection evaluation. In: 2012 10th IAPR International Workshop on Document Analysis Systems, pp. 445–449 (2012). https://doi.org/10.1109/DAS.2012.29
9. Farrukh, W., et al.: Interpreting data from scanned tables. In: 2017 14th IAPR International Conference on Document Analysis and Recognition (ICDAR), vol. 2, pp. 5–6 (2017). https://doi.org/10.1109/ICDAR.2017.250

10. Gatos, B., Danatsas, D., Pratikakis, I., Perantonis, S.: Automatic table detection in document images. vol. 3686, pp. 609–618 (08 2005). https://doi.org/10.1007/11551188_67

11. Gilani, A., Qasim, S.R., Malik, I., Shafait, F.: Table detection using deep learning. In: 2017 14th IAPR International Conference on Document Analysis and Recognition (ICDAR), vol. 1, pp. 771–776 (2017). https://doi.org/10.1109/ICDAR.2017.131

12. Green, E., Krishnamoorthy, M.: Model-based analysis of printed tables, pp. 214–217, January 1995. https://doi.org/10.1109/ICDAR.1995.598979

13. Göbel, M., Hassan, T., Oro, E., Orsi, G.: ICDAR 2013 table competition. In: 2013 12th International Conference on Document Analysis and Recognition, pp. 1449–1453 (2013). https://doi.org/10.1109/ICDAR.2013.292

14. Alhéritière, H., Amaïeur, W., Cloppet, F., Kurtz, C., Ogier, J.-M., Vincent, N.: Straight line reconstruction for fully materialized table extraction in degraded document images. In: Couprie, M., Cousty, J., Kenmochi, Y., Mustafa, N. (eds.) DGCI 2019. LNCS, vol. 11414, pp. 317–329. Springer, Cham (2019). https://doi.org/10.1007/978-3-030-14085-4_25

15. Hu, J., Kashi, R., Lopresti, D., Wilfong, G.: Medium-independent table detection, December 1999

16. Isabelle, D.R., Fabien, F., Jocelyne, R.D.: Optimal blurred segments decomposition of noisy shapes in linear time. Comput. Graph. **30**(1), 30–36 (2006)

17. Kerautret, B., Even, P.: Blurred segments in gray level images for interactive line extraction. In: Wiederhold, P., Barneva, R.P. (eds.) IWCIA 2009. LNCS, vol. 5852, pp. 176–186. Springer, Heidelberg (2009). https://doi.org/10.1007/978-3-642-10210-3_14

18. Kieninger, T.: Table structure recognition based on robust block segmentation, pp. 22–32 (1998)

19. Kieninger, T., Dengel, A.: An approach towards benchmarking of table structure recognition results, pp. 1232–1236, August 2005. https://doi.org/10.1109/ICDAR.2005.47

20. Klette, R., Rosenfeld, A.: Digital Geometry - Geometric Methods for Digital Picture Analysis. Morgan Kaufmann, San Francisco (2004)

21. Mandal, S., Chowdhury, S., Das, A., Chanda, B.: Simple and effective table detection system from document images. Int. J. Doc. Anal. Recogn. **8**, 172–182 (2006). https://doi.org/10.1007/s10032-005-0006-5

22. Minghao, L., Lei, C., Shaohan, H., Furu, W., Ming, Z., Zhoujun, L.: TableBank: a benchmark dataset for table detection and recognition (2019)

23. Prasad, D., Gadpal, A., Kapadni, K., Visave, M., Sultanpure, K.: Cascade TabNet: an approach for end to end table detection and structure recognition from image-based documents (2020)

24. Reveillès, J.P.: Géométrie discrète, calcul en nombres entiers et algorithmique. Thèse d'état, Université Strasbourg 1 (1991)

25. Shafait, F., Smith, R.: Table detection in heterogeneous documents. In: DAS 2010, pp. 65–72. Association for Computing Machinery, New York, NY, USA (2010). https://doi.org/10.1145/1815330.1815339

26. Shahab, A., Shafait, F., Kieninger, T., Dengel, A.: An open approach towards the benchmarking of table structure recognition systems. In: Proceedings of the 9th IAPR International Workshop on Document Analysis Systems, DAS 2010, pp. 113–120. Association for Computing Machinery, New York, NY, USA (2010). https://doi.org/10.1145/1815330.1815345

27. Watanabe, T., Naruse, H., Luo, Q., Sugie, N.: Structure analysis of table-form documents on the basis of the recognition of vertical and horizontal line segments. In: Proceedings of International Conference on Document Analysis and Recognition (ICDAR 1991), pp. 638–646 (1991)
28. Kieninger, T., Dengel, A.: The T-Recs table recognition and analysis system. In: Lee, S.-W., Nakano, Y. (eds.) DAS 1998. LNCS, vol. 1655, pp. 255–270. Springer, Heidelberg (1999). https://doi.org/10.1007/3-540-48172-9_21

A GAN Based Approach to Compare Identical Images with Generative Noise

Damien Simonnet$^{(\boxtimes)}$ and Ahmad-Montaser Awal

ARIADNEXT - Research Department, Rennes, France
{damien.simonnet,montaser.awal}@ariadnext.com

Abstract. Generative Adversarial Networks (GAN) have shown impressive results in the generation and translation of images, for instance to generate a painting image with a specific style from a realistic photo. This ability to change of style makes it usable to denoise an image and to the best of our knowledge, GAN has not been used for such applications. This paper presents a generic approach to compare identical images but of which one has been modified by an external process which makes it very noisy to be directly comparable to the original image. This noise results from the process of creating a document and is called *generative noise*. First, the noisy image is transformed with the generator to get a denoised image more similar to the original one. However, the denoising with the GAN is not perfect due to strong noise processes which requires to transform the original and denoised images into a comparable space, and to check that the quality is good enough to make the comparison. This approach has been applied on the Romanian identity card to compare the identity photo to the ghost image and shows significantly better results than standard comparison approaches.

Keywords: Generative Adversarial Network · Image denoising · Image comparison · Quality analysis

1 Introduction

Generative Adversarial Networks (GAN) [11,27] have been widely studied since 2014. These algorithms are composed of a generator and a discriminator. The generator generates realistic images of a specific type while the discriminator detects if an image has been generated by the generator (fake image) or if the image is real. The loss function to train the GAN is designed to generate the most realistic images and having the best discriminator. This approach has been successfully applied to the case of image translation to generate for instance realistic images from a segmented region [10], or painting of Monet from realistic photo [30].

In the comparison of identical images, the image to compare can be modified by adding an additional pattern which makes it noisy (*e.g.* in identity documents, some patterns are modified by the document generation process). This paper deals with the comparison with this type of noise which is different from the

S. Sclaroff et al. (Eds.): ICIAP 2022, LNCS 13232, pp. 225–236, 2022.
https://doi.org/10.1007/978-3-031-06430-2_19

Fig. 1. Romanian identity card with an identity photo on the left, and a ghost photo on the right. An unknown process has been applied to the identity photo to generate this 'ghost photo'. This process has notably added the writing 'Identity card', and degrading the quality by adding some wavelet pattern corresponding to the background texture of the identity card.

noise resulting from the acquisition process [4,7,9]. The noise handled in this paper is referred as *generative noise* and corresponds to any noise applied to an image during the generation process. A typical example of such *generative noise* is the noise introduced in the ghost photo (on the right) on the romanian identity card as depicted in Fig. 1. The ghost photo is identical to the identity photo but the noise introduced makes the comparison difficult using standard approaches.

This paper presents an approach based on GAN to handle difficult scenarios with *generative noise*. To the best of our knowledge it is the first time that a GAN approach is used for denoising an image and making the comparison. This approach could also be used to compare patterns when there is text inside to denoise the image before doing the comparison. It is noted that recent GAN approaches [19,21,26,28,29] have been used to solve the problem of face restoration of the image due to time degradation, or noise during acquisition but not in a case of a *generative noise* where the noise impact is more important.

The rest of this paper is organised as follows. First, related works about comparison of images and faces are presented, before describing the generic method introduced to compare images with *generative noise*. Finally, experiments and results are presented.

2   Related Work

The problem of image comparison presented in this paper is considered in the more generic way, *i.e.* checking that two images are similar without having many samples to learn a model. In the literature, it corresponds to the problem of the one shot learning.

The idea of one shot learning is that we can learn to identify if two images are similar by learning features that are applicable for any category. Thus, once the learning has been done for some categories it is easily transferable to new categories. Early work on one shot learning has been done by Fei-Fei et al. [18,20] by using a variational bayesian framework, and representing object categories with probabilistic models. Lake et al. [17] learn a generative stroke model of characters. The idea for managing new characters is that the model tries to infer latent strokes that explain pixels in the images. Koch et al. [13] presented a siamese CNN network architecture that takes as input two images and as output a probability that the images are the same. Recently, other works on deep learning, presented advances for the one shot learning by proposing a transfer learning approach to understand how to obtain a better classification boundary [16], or by introducing a memory matching network [5] that is a novel deep learning architecture to do one shot learning as inference.

The application presented in this paper is related to the face comparison, therefore related works to face recognition are also presented as a specific case of comparison methods. To resolve the problem of 2D facial recognition, approaches use a feature extraction approach combined with a classifier [1]. These methods can be divided in four categories: holistic methods, local geometrical methods, local texture based methods and deep learning based methods [1]. Holistic methods consider a subspace of features characterising the face, allowing to reconstruct the face. It uses methods such as principal component analysis [25], linear discriminative analysis [3] and independent component analysis [24]. Performance of these algorithms is sensitive to context changes and misalignment. The geometrical approach uses spatial landmarks to describe the face as reference points [8,15]. The drawback of these methods is the need to perfectly align the reference points. The local texture based approaches [2,12,22] allow to have a high efficiency in terms of time analysis and recognition rate [1], and to be invariant to scale and misalignment. Deep learning approaches, of which a review is presented in [1], outperform the holistic, geometrical and local texture approaches, by showing their robustness to pose variation, orientation, partial occlusion, misalignement and facial expression. To give an example, FaceNet [23] is a deep neural network based on a triplet loss that is using a 128 dimensional representation to characterise the face. Recent reviews about face recognition [1,14] show that there are some limitations for face recognition systems in real world application. Mainly, unconstrained environments with changes in lighting, posture or facial expression with partial occlusions or camera movement are still a challenge.

In conclusion, current one shot learning and face recognition methods focus on challenges such as posture change, facial expression change, partial occlusion and context variation. Therefore, the generalisation provided by these methods does not correspond to the problem of comparing identical images with *generative noise* where there are no posture or context variation but only a *generative noise* added.

The next section presents the novel approach to compare images with *generative noise* based on GAN. To the best of our knowledge, it is the first work that uses GAN to remove this type of noise. Other types of noises exist such as acquisition or compression artifact but they are not handled in this paper.

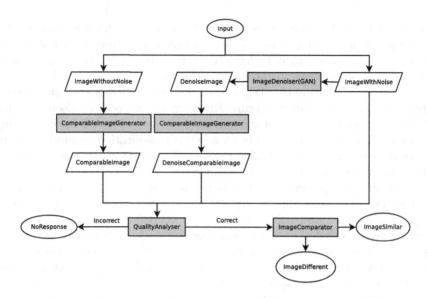

Fig. 2. Overview of the approach to compare images of which one has *generative noise*.

3   Generic Comparison of Images with *Generative Noise*

3.1   A Generic Framework

This section introduces a novel generic framework to compare images with *generative noise*. An overview of this approach is presented in Fig. 2. First, the noisy image is denoised using a GAN. Then, the denoised image and the image without noise are converted to comparable images (as GAN does not reconstruct with accuracy all the details which make images not directly comparable). Before carrying out the comparison, an analysis of the quality is done to check that the comparison is meaningful. Incorrect reconstruction of GAN, acquisition conditions or noisy processes applied on the image may change the quality on the input that is why a quality check is required. Finally, a comparison method is applied on comparable images.

The next sections present in details the use of this framework to compare images in the romanian identity card.

3.2   Application to the Comparison of Images in the Romanian Identity Card

In the context of the KYC (Know Your Customer) regulation, identity documents have to be verified to prove their authenticity and though the identity of the final user. One important security feature on identity documents is the use of ghost photo images. This latest is a duplication of the document holder's photo with additional *generative noise* (overlapping patterns, text, holograms ...). The main goal is to prevent identity photo replacement (identity theft). As

Table 1. Quality of the ghost photo with the repartition in three levels (high, medium and low) with the face and shape visibility criteria on a dataset of 651 Romanian identity cards. With 76.5% of low quality for the face, and 50.05% of low quality for shape, the Romanian identity has a very poor quality of the ghost image which makes it very difficult to use to compare to the identity photo.

	High	Medium	Low
Shape	20.05%	29.9%	50.05%
Face	5.6%	17.9%	76.5%

the ghost photo is small and hard to reproduce, fraudsters often only replace the identity photo. Measuring the similarity between these two photos allows to spot this kind of forgery. However, the comparison of the original identity photo with the ghost photo is a challenging task for the Romanian identity card (see Fig. 1) as explained in the next subsection.

3.3 Difficulty for Comparing with Ghost Images

In addition to the *generative noise* added to the ghost photo during the document creation process, capturing noise is also added to the image (as images are captured in a non-constrained manner via a smartphone or any other device) such as illumination changes, blur and compression artifacts. In addition, the quality of the produced ghost photo is related to the printing process (and the used printer) and might intrinsically produce photos with low details. To estimate the usability of the ghost photo for the comparison, a case study is realised on the Romanian identity document.

A dataset of 651 images has been annotated with two criteria: the visibility of the face and the shape of the face. For the face visibility, three levels are defined: high (the face is fully visible, the details of the ghost image are close to the identity photo), medium (the face is partially visible, in the ghost image some details of the face are missing but it is possible to do a partial identification), low (the face is not visible, it is not possible to recognise with absolute certainty the face from the ghost image). Similarly, there are three levels for the visibility of the shape of the face: high (contours of elements characterising the shape are fully visible), medium (contours of elements characterising the shape are partially visible) and low (contours of elements characterising the shape are either not visible or principaly not visible, it is very difficult to deduce the shape of the face). With 76.5% of low quality for the face, and 50.05% of low quality for shape (see Table 1), using the ghost photo to check the similarity is very challenging, and a method based on the shape of the face will probably provide better results than on the face only.

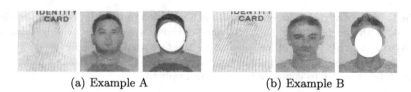

(a) Example A   (b) Example B

Fig. 3. Examples of the use of the generator to denoise the ghost image and produce the identity photo. For the A and B examples, the images shown from the left to the right are the ghost image, the denoised image and the identity photo associated to the ghost photo. Thus, the comparison is carried out between the identity photo and the denoised image. The white circle is there to mask the real face as the full data cannot be shown in this paper.

3.4   Extraction of Images in a Comparison Space

To denoise the ghost image, the GAN approach presented in [10] is used to transform the ghost image into an identity-photo-like. Some examples of this approach are presented in Fig. 3, where the shape of the person is globally kept and usable for a comparison. However, when we look at the details of the face, there are some imperfections and differences compared to the source photo. For instance, in the example A, it reconstructs a beard which is not present in the original image but effectively can be guessed from the ghost photo. In the example B, the hairstyle is different but is effectively not really visible in the ghost photo. In conclusion, the GAN approach reconstructs a face with what is visible in the ghost image but the face generated is close but different from the original image. Due to the quality of the ghost photo, the general shape of the generated face is close to the original one, but still different if we look into details (notably the eyes and the mouth, that cannot be shown in this paper for confidentiality reasons, are not reconstructed with enough details to apply a face comparison approach).

To compare identity photos and denoised images (with GAN), a transformation is applied to keep the only information shared by these types of images: the shape. This transformation is based on the k-means algorithm applied to the pixel value in the image in gray levels. The idea with this clustering is to separate the background, from the face, the hair and the clothes, therefore a k-means with four clusters is used. Some examples of this transformation are shown in Fig. 4. In the rest of this paper, these images will be called hash images as it represents a signature of the face based on the shape. The size of the hash image is 32×32. The problem with those images is that they are sensitive to illumination changes (see Fig. 4 (d)) and bad GAN reconstruction (see Fig. 4 (c)) that make them not comparable. Moreover, GAN also reconstructs a face from an empty image (see Fig. 4 (e)). Therefore, an analysis of the quality of these images is done before performing the comparison.

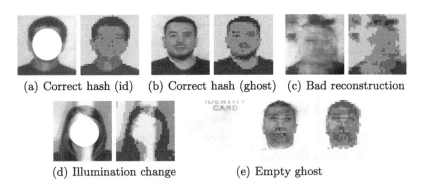

(a) Correct hash (id)   (b) Correct hash (ghost)   (c) Bad reconstruction

(d) Illumination change   (e) Empty ghost

Fig. 4. Examples of correct and incorrect (due to bad GAN reconstruction or illumination changes) hash images. Correct hash can also be reconstructed from an empty ghost image (in (e), left to right, the ghost image, the reconstruction and the hash image).

3.5 Quality Analysis

The quality analyser is composed of two steps, an analyser of the quality of hash images to detect bad GAN reconstruction and illumination changes, and an analyser of empty ghost images. These two analysers are based on SVM classifiers with dedicated features. The quality features correspond to the concatenation of HOG features [6] (window size: 32×32, block size: 8×8 and cell size 8×8) of four images associated to each level where the value is 255 when it belongs to the current level and 0 to another level, which makes in total a feature vector of size 576. This allows us to learn the expected shape of correct hash images. For detecting empty ghost images, corresponding ghost images of very bad quality, the ghost image is converted to gray level and resized at 64×64 and k-means is applied to keep only the four main gray levels. Then, HOG features [6] are computed (window size: 64×64, block size: 16×16 and cell size: 8×8) which makes a feature vector of size 576. In addition of these two criteria, the quality checks the color of identity photo by comparing pixels values in the RGB space. More specifically, the comparisons R/G, R/B and G/B are done by checking if the difference is greater than a threshold c_{diff}, and counting the proportion of the comparison meeting this condition. If it is greater than c_p, the photo is considered as color.

3.6 Comparison of Images in a Common Representation Space

In the comparable image called hash image, there are four gray levels sorted by intensity. To compare identity and ghost hash images, we consider that the level i of the identity hash corresponds to the level i of the ghost hash. The intensity value is not considered because the reconstruction process with the GAN can introduce some variations for the intensity. The more stable parts are the lighter and darker parts of the image corresponding to the background and the hair or

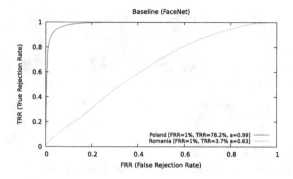

Fig. 5. The baseline FaceNet [23] works on the polish identity card but fails on the romanian identity card. A working point at 1% of FRR and the surface area under the ROC curve a are presented.

Table 2. Results for the quality in terms of TRR/FRR relatively to the empty ghost, hash quality, colour and full quality (empty ghost, hash quality and colour) analysers.

Analyser	EmptyGhost	HashQuality	Colour	FullQuality
TRR	69.5%	73.2%	99.7%	80.3%
FRR	2.3%	4.8%	0.04%	8.2%

clothes. The two other levels are less stable and correspond generally to variation of the intensity in the face. Therefore, the two first features that are computed correspond to the intersection over the union in terms of pixels of the darker and lighter level. We also modeled the change of levels between images by computing the proportion of pixels passing from the level i to j between the identity hash and the ghost hash with $(i, j) \in [\![1, 4]\!]^2$. A mean vector and a covariance matrix are learnt with these 18 features. Then, the mahalanobis distance is used to define a confidence to estimate the image similarity.

4   Experiments and Results

Table 3 presents the train and test datasets used to perform the experiments with the quality detector and image comparator. The comparison task in this paper consists in checking that images are different. Positive examples refer to different images and negative examples to similar images. For quality analysers, positive examples correspond to bad quality examples, and negative to good quality examples. The generator of the GAN used to denoise ghost images has been trained on a dataset composed of 2931 pairs of identity photo/ghost photo.

The baseline of this work is the neural network FaceNet [23] that has been retrained on a database of 45000 images (about 10000 persons) corresponding to identity photos. Figure 5 shows the performance of this baseline on the *comparisonTest* dataset (Romania) and on a Polish test dataset (8540 positives and

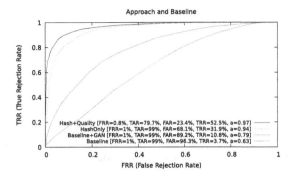

Fig. 6. Results of the proposed approach against the baseline FaceNet [23] and the baseline combined to a GAN (to denoise the ghost image). A working point at 1% of FRR and the surface area under the ROC curve a are presented. It is noted that the TAR (True Acceptance Rate) FAR (False Acceptance Rate) are also given for the working point as when the quality is not good enough no response is given.

1220 negatives), where the approach works on the Polish identity cards (78.2% of TRR - True Rejection Rate - at 1% of FRR - False Rejection Rate -) but fails on the Romanian identity cards due to the poor quality of ghost images (3.7% of TRR at 1% of FRR).

We can observe from the Table 2 that the quality analyser correctly detects 80.3% of bad quality images while only rejecting 8.3% of good quality ones. For the colour detector, the difference between B/G/R channels c_{diff} has been empirically set to 14 and the ratio c_p to 0.08.

Finally, the performance of the whole system, the image comparator (hash based) using the quality analyser, is presented in Fig. 6. For comparison, the baseline is also combined with the generator of the GAN, to denoise first the ghost image before doing the face comparison, this method is called baseline+GAN. The comparison is done with two criteria: the global comparison of the classification system (discrimination rate), and a comparison of a final user working point. Firstly, the proposed approach outperforms significantly the baselines with a discrimination rate (measured with the area under the roc curve) at 97% and 94% with and without quality analyser compared to 63% for the baseline [23] and 79% for the combination of the baseline with the GAN. This significant improvement can be explained by the strategy adopted in this work, focusing on the shape. A low rate of FRR of at maximum 1% has been chosen as a working point to use this approach in a final user system. The results are FRR $= 0.8\%$, TAR $= 79.7\%$, FAR $= 23.4\%$, TRR $= 52.5\%$ for the analysis with the quality and FRR $= 1\%$, TAR $= 99\%$, FAR $= 68.1\%$, TRR $= 31.9\%$ for the analysis without the quality. The use of the quality allows to improve the TRR by 20.6 points, the FRR by 0.2 points and the FAR by 44.7 points but with a decrease of the TAR of 19.3 points. The dataset contains 15.7% of negative pairs of bad quality, so knowing that 79.3% of bad quality samples are detected, the increase of TAR in only of 6.8%. The proposed approach with the quality detects more different images

(a) False Rejection   (b) False Acceptance

Fig. 7. Example of false rejection and false acceptance.

Table 3. Train and test datasets for the quality (emptyGhost and hashQuality) and photoswap problems with the romanian identity card. *Train datasets allow to learn the model and *TrainConf allow to learn the decision threshold.

(a) Train

Name	Positive	Negative
emptyGhostTrain	65	2845
emptyGhostTrainConf	20	2061
hashQualityTrain	539	5281
hashQualityTrainConf	265	3897
comparisonTrain	2910	NA

(b) Test

Name	Positive	Negative
emptyGhostTest	23	1741
hashQualityTest	189	2529
colourTest	23	616
fullQualityTest	204	1155
comparisonTest	9500	1359

(20.6%), with less FRR (0.2%) but with an absence of response in 24.1% and 19.5% for positive and negative pairs. Finally, Fig. 7 gives examples of errors in the system. In (a), low quality ghost image produces a false rejection due to bad reconstruction. In the meantime, some hash images of different persons which are close produces false acceptance (b). With a noisy input, it is not possible to solve all the problems as the shape is the only available information.

5   Conclusion

In conclusion, this paper has presented a novel generic approach to compare images with a *generative noise* based on a GAN algorithm to denoise the image and the use of a transformation to have comparable image. This last step is essential when the *generative noise* is high. Indeed, GAN based approaches would infer on the data without necessarily reconstruct the exact expected image. This approach has been successfully applied to compare the identity photo and the noisy ghost image of the Romanian identity card, and shows significantly better results than state of the art face comparison methods.

References

1. Adjabi, I., Ouahabi, A., Benzaoui, A., Taleb-Ahmed, A.: Past, present, and future of face recognition: a review. Electronics **9**(8), 1188 (2020)
2. Ahonen, T., Rahtu, E., Ojansivu, V., Heikkila, J.: Recognition of blurred faces using local phase quantization. In: International Conference on Pattern Recognition, pp. 1–4 (2008)

3. Belhumeur, P.N., Hespanha, J.P., Kriegman, D.J.: Eigenfaces vs. fisherfaces: recognition using class specific linear projection. IEEE Trans. Pattern Anal. Mach. Intell. **19**(7), 711–720 (1997)
4. Buades, A., Coll, B., Morel, J.M.: A review of image denoising algorithms, with a new one. Multiscale Model. Simul. **4**(2), 490–530 (2005)
5. Cai, Q., Pan, Y., Yao, T., Yan, C., Mei, T.: Memory matching networks for one-shot image recognition. In: International Conference on Computer Vision and Pattern Recognition, pp. 4080–4088 (2018)
6. Dalal, N., Triggs, B.: Histograms of oriented gradients for human detection. In: International Conference on Computer Vision and Pattern Recognition, pp. 886–893 (2005)
7. Das, A.K.: Review of image denoising techniques. Int. J. Emerg. Technol. Adv. Eng. **4**(8), 519–522 (2014)
8. Duc, B., Fischer, S., Bigun, J.: Face authentication with Gabor information on deformable graphs. IEEE Trans. Image Process. **8**(4), 504–516 (1999)
9. Gu, S., Timofte, R.: A brief review of image denoising algorithms and beyond. In: Inpainting and Denoising Challenges, pp. 1–21 (2019)
10. Isola, P., Zhu, J.Y., Zhou, T., Efros, A.A.: Image-to-image translation with conditional adversarial networks. In: International Conference on Computer Vision and Pattern Recognition, pp. 1125–1134 (2017)
11. Gui, J., Sun, Z., Wen, Y., Tao, D., Ye, J.: A review on generative adversarial networks: algorithms, theory, and applications. arXiv (2020)
12. Kannala, J., Rahtu, E.: BSIF: binarized statistical image features. In: International Conference on Pattern Recognition, pp. 1363–1366 (2012)
13. Koch, G., Zemel, R., Salakhutdinov, R.: Siamese neural networks for one-shot image recognition. In: International Conference on Machine Learning (Deep Learning Workshop) (2015)
14. Kortli, Y., Jridi, M., Al Falou, A., Atri, M.: Face recognition systems: a survey. Sensors **20**(2), 342 (2020)
15. Kotropoulos, C., Tefas, A., Pitas, I.: Frontal face authentication using morphological elastic graph matching. IEEE Trans. Image Process. **9**(4), 555–560 (2000)
16. Kozerawski, J., Turk, M.: Clear: cumulative learning for one-shot one-class image recognition. In: International Conference on Computer Vision and Pattern Recognition, pp. 3446–3455 (2018)
17. Lake, B., Salakhutdinov, R., Gross, J., Tenenbaum, J.: One shot learning of simple visual concepts. In: Annual Meeting of the Cognitive Science Society, vol. 33 (2011)
18. Li, F.F., Fergus, R., Perona, P.: One-shot learning of object categories. IEEE Trans. Pattern Anal. Mach. Intell. **28**(4), 594–611 (2006)
19. Li, X., Chen, C., Zhou, S., Lin, X., Zuo, W., Zhang, L.: Blind face restoration via deep multi-scale component dictionaries. In: European Conference on Computer Vision, pp. 399–415 (2020)
20. Li Fei-Fei, R.F., Perona, P.: A Bayesian approach to unsupervised one-shot learning of object categories. In: International Conference on Computer Vision, pp. 1134–1141 (2003)
21. Menon, S., Damian, A., Hu, S., Ravi, N., Rudin, C.: Pulse: self-supervised photo upsampling via latent space exploration of generative models. In: International Conference on Computer Vision and Pattern Recognition, pp. 2437–2445 (2020)
22. Rodriguez, Y., Marcel, S.: Face authentication using adapted local binary pattern histograms. In: European Conference on Computer Vision, pp. 321–332 (2006)

23. Schroff, F., Kalenichenko, D., Philbin, J.: FaceNet: a unified embedding for face recognition and clustering. In: International Conference on Computer Vision and Pattern Recognition, pp. 815–823 (2015)
24. Stone, J.V.: Independent component analysis: an introduction. Trends Cogn. Sci. **6**(2), 56–64 (2002)
25. Turk, M., Pentland, A.: Eigenfaces for recognition. J. Cogn. Neurosci. **3**(1), 71–86 (1991)
26. Wan, Z., et al.: Bringing old photos back to life. In: International Conference on Computer Vision and Pattern Recognition, pp. 2747–2757 (2020)
27. Wang, K., Gou, C., Duan, Y., Lin, Y., Zheng, X., Wang, F.Y.: Generative adversarial networks: introduction and outlook. CAA J. Automatica Sinica **4**(4), 588–598 (2017)
28. Wang, X., Li, Y., Zhang, H., Shan, Y.: Towards real-world blind face restoration with generative facial prior. In: International Conference on Computer Vision and Pattern Recognition, pp. 9168–9178 (2021)
29. Yang, L., et al.: HiFaceGAN: face renovation via collaborative suppression and replenishment. In: ACM International Conference on Multimedia, pp. 1551–1560 (2020)
30. Zhu, J.Y., Park, T., Isola, P., Efros, A.A.: Unpaired image-to-image translation using cycle-consistent adversarial networks. In: International Conference on Computer Vision, pp. 2223–2232 (2017)

Egocentric Human-Object Interaction Detection Exploiting Synthetic Data

Rosario Leonardi[1(✉)], Francesco Ragusa[1,2], Antonino Furnari[1,2],
and Giovanni Maria Farinella[1,2]

[1] FPV@IPLAB, DMI - University of Catania, Catania, Italy
rosario.leonardi@phd.unict.it
[2] Next Vision s.r.l. - Spinoff of the University of Catania, Catania, Italy

Abstract. We consider the problem of detecting Egocentric Human-Object Interactions (EHOIs) in industrial contexts. Since collecting and labeling large amounts of real images is challenging, we propose a pipeline and a tool to generate photo-realistic synthetic First Person Vision (FPV) images automatically labeled for EHOI detection in a specific industrial scenario. To tackle the problem of EHOI detection, we propose a method that detects the hands, the objects in the scene, and determines which objects are currently involved in an interaction. We compare the performance of our method with a set of state-of-the-art baselines. Results show that using a synthetic dataset improves the performance of an EHOI detection system, especially when few real data are available. To encourage research on this topic, we publicly release the proposed dataset at the following url: https://iplab.dmi.unict.it/EHOI_SYNTH/.

Keywords: Egocentric human-object interaction detection · Synthetic data · Active object recognition

1 Introduction

Understanding Human-Object Interactions (HOI) from the first-person perspective allows to build intelligent systems able to understand how humans interact with the world. The use of wearable cameras can be highly relevant to understand users' locations of interest [9], to assist visitors in cultural sites [5,28], to provide assistance to people with disabilities [32], or to improve the safety of workers in a factory [29]. Despite the rapid growth of wearable devices [2], the task of Egocentric Human-Object Interaction (EHOI) detection is still understudied in this domain due to the limited availability of public datasets [29]. We note that in an industrial domain, in which the set of objects of interest is known a priori (e.g., the tools and instruments the user is going to interact with), the ability to detect the user's hands, find all objects and determine which objects are involved in an interaction, can inform on the user's behavior and provide

Supplementary Information The online version contains supplementary material available at https://doi.org/10.1007/978-3-031-06430-2_20.

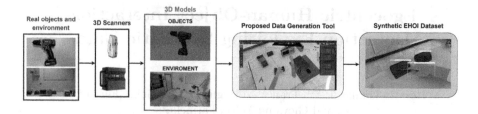

Fig. 1. Synthetic EHOIs generation pipeline. We first use 3D scanners to obtain 3D models of the set of objects and the environment. We hence use the proposed data generation tool to create the synthetic EHOI dataset.

useful information for other tasks such as object interaction anticipation [10,11]. Extending the definition proposed in [31], we hence consider the problem of detecting an EHOI as the one of predicting a quadruple <*hand, contact_state, active_object, <other_objects>>*.

To develop a system able to tackle this task in a specific industrial scenario, it is generally required to collect and label large amounts of data. To reduce the significant costs usually required for data collection and annotation, we investigated whether the use of synthetic images can help to achieve good performance when models are trained on synthetic data and tested on real one. To this end, we propose a pipeline and a tool to generate a large number of synthetic EHOIs from 3D models of a real environment and objects. Unlike previous approaches [15,26], we generate EHOIs simulating a photo-realistic industrial environment. The proposed pipeline (Fig. 1) allows to obtain 3D models of the objects and the environment using 3D scanners. Such models are then used with the proposed data generation tool to automatically produce labeled images of EHOIs. Even though some works provide datasets to study HOI in general domains [1,6,7,24,31] and in industrial contexts [29] to the best of our knowledge, this is the first attempt to define a pipeline for the generation of a large-scale photo-realistic synthetic FPV dataset to study EHOIs with rich annotations of active and non-active objects in an industrial scenario. To assess the suitability of the generated synthetic data to tackle the EHOI detection task, we acquired and labeled 8 real egocentric videos in an industrial laboratory, in which subjects perform test and repair operations on electrical boards (see Fig. 2). To address the problem of EHOI detection, we propose a method inspired by [31] that detects and recognizes all the objects in the scene, determining which of these are involved in an interaction, as well as the hands of the camera wearer (see Fig. 2). To investigate the usefulness of exploiting synthetic data for EHOIs detection, we trained the proposed method using all the synthetic images together with variable amounts of real data. In addition, we compared the results of the proposed approach with different instances of the method proposed in [31]. The results show that using synthetic data improves the performance of the EHOI method when tested on real images.

In sum, the contributions of this paper are as follows: 1) we present a new photo-realistic synthetic FPV dataset for EHOIs detection considering an industrial

Fig. 2. Example output of the proposed system. The figure on the left shows an example of a real image whereas a synthetic image is shown on the right.

scenario with rich annotations of the hands, and the active/non-active objects, including class labels and semantic segmentation masks; 2) we propose a method inspired by [31] which detects and recognizes all the objects in the scene, the hands of the camera wearer, and determines which objects are currently involved in an interaction; 3) we perform several experiments to investigate the usefulness of synthetic data for the EHOI detection task when the method is tested on real data and compare the obtained results with baseline approaches based on the state-of-the-art method described in [31].

2 Related Work

Datasets for Human Behavior Understanding. In recent years, many works focused on the Human-Object Interaction detection task considering the third-person point of view. Several datasets have been proposed to explore this problem. The authors of [14] proposed the V-COCO dataset, which adds 26 verbs to the 80 object classes of the popular COCO dataset [23]. HICO-DET [3] includes over 600 distinct interaction classes, while HOI-A [21] considers 10 action categories and 11 object classes. Previous works have also proposed datasets of videos to address the action recognition task. We can mention the ActivityNet dataset [17] which focuses on 200 different action classes as well as Kinetics [18] which contains over 700 human action classes. With the rapid growth of wearable devices, different datasets of images and videos captured from the first-person point of view have been proposed. Among these, the work of [1] provided a dataset of 48 FPV videos of people interacting with objects, including segmentation masks for 15,000 hand instances. EPIC-Kitchens [6,7] is a series of egocentric datasets focused on unscripted activities of human behavior in kitchens. EGTEA Gaze+ [19] is a dataset of 28 h of video of cooking activities. The dataset 100 d of Hands (100DOH) [31] is composed of both Third Person Vision (TPV) and FPV images and is suitable to study object-class agnostic HOI detection. The authors of [24] labeled images collected from different FPV datasets [7,12,20] providing annotations for hands, objects and their relation. The MECCANO dataset [29] contains videos acquired in an industrial-like domain also annotated with bounding boxes around active objects, together with the related classes. A massive-scale egocentric dataset named Ego4D[1] has

---

[1] Ego4D Website: https://ego4d-data.org/.

been acquired in various domains and labeled with several annotations to address different challenges. Since the annotation phase of EHOIs is expensive in terms of costs and time, the use of synthetic datasets for training purposes is desired. A few works explored the use of synthetic images generated from the first-person point of view [15,27]. The authors of such works used different strategies to customize various aspects of the scene, such as lights and backgrounds. However, these approaches tend to produce non-photorealistic images. Differently from the aforementioned works, we generate a dataset of photo-realistic synthetic images of EHOIs in an industrial environment with rich annotations of hands, including hand side (*Left/Right*), contact state (*In contact with an object/No contact*), and all the objects in the images with bounding boxes. We also provide a class label for each object and indicate whether it is an active object as well as semantic segmentation masks.

Understanding Human-Object Interactions. There has been a lot of research in computer vision focusing on understanding Human-Object Interactions. The work of [13] presented a multitask learning system to tackle HOI detection. The proposed system consists of an object detection branch, a human-centric branch, and an interaction branch. The authors of [31] tackled the HOI detection task from both TPV and FPV predicting different information about hands (i.e., bounding box, hand side and contact state) and a box around the interacted object. PPDM [21] defines an HOI as a point triplet <*human point, interaction point, object point*> where these points represent the center of the related bounding boxes. The authors of [33] proposed a new two-stage detector called Unary-Pairwise Transformer. This approach exploits unary and pairwise representations to detect Human-Object Interactions. However, all these works mainly consider third-person view scenarios. Indeed, this task is still understudied in the FPV domain. Previous FPV works focused on the detection of hands interacting with an object without recognizing it [1,25]. Other recent works focused on object-class agnostic EHOI detection [8,24]. The authors of [29] defined Egocentric Human-Object Interaction (EHOI) detection as the task of producing <*verb, objects*> pairs. The paper investigated the problem of recognizing active objects in industrial-like settings without considering hands. The authors of [4] considered the usage of synthetic data for recognizing the performed Human-Objects Interactions. In this paper, we tackle the EHOIs detection task in an industrial domain and investigate the usefulness of using synthetic data for training when the system needs to be tested on real data. In addition, our approach aims to detect both active and non-active objects as well as infer their classes.

3   Dataset

Industrial Context. We set up a laboratory to study the EHOIs detection task in a realistic industrial context. In the considered laboratory there are different objects, such as a power supply, a welding station, sockets, and a screwdriver. In addition, there is an electrical panel that allows powering on and off the

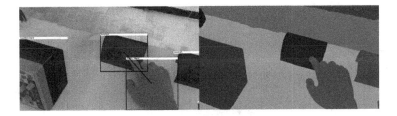

Fig. 3. Example of a synthetic EHOI generated with the developed tool. On the left, the figure shows the synthetic RGB image automatically labeled (left) as well as the semantic segmentation mask generated for the same EHOI (right).

sockets[2]. To generate synthetic data compliant to the considered real space, we acquire 3D scans of all objects and of the environment. It is worth noting that for the small objects, high-quality reconstructions are required to generate realistic EHOIs, whereas for the reconstruction of the environment, a high accuracy is not needed. Hence, to create 3D models, we used two different 3D scanners. In particular, we used an Artec Eva[3] structured-light 3D scanner, which has a 3D resolution of up to 0.2 mm, to scan the objects, and a MatterPort[4] device to scan the 3D model of the environment.

Synthetic Data. We adopted the pipeline shown in Fig. 1 to generate the synthetic data of EHOIs in the considered industrial context. We developed a tool in Blender which takes as input the 3D models of the objects and the environment and generates synthetic EHOIs along with different data, including 1) photorealistic RGB images (see Fig. 3 - left), 2) depth maps, 3) semantic segmentation masks (see Fig. 3 - right), 4) objects bounding boxes and categories indicating which of them are active, 5) hands bounding boxes and attributes, such as the hand side (*Left/Right*) and the contact state (*In contact with an object/No contact*), and 6) distance between hands and objects in the 3D space. The tool allows to customize different aspects of the virtual scene, including the camera position, the lighting, and the color of the hands for automatic acquisition. Figure 3 shows an example of synthetic EHOIs and related labels generated with our tool. The generated synthetic dataset contains a total of 20,000 images, 29,034 hands (of which 14,589 involved in an interaction), 123,827 object instances (14,589 of which are active objects), and 19 object categories including portable industrial tools (e.g., screwdriver, electrical boards) and instruments (e.g., power supply, oscilloscope, electrical panels) (See footnote 2).

Real Data. The real data consists in 8 real videos acquired using a Microsoft Hololens 2 wearable device. To this aim, we asked 7 different subjects to perform test and repair operations on electrical boards in the industrial laboratory.

---

[2] See supplementary material for more details.

[3] https://www.artec3d.com/portable-3d-scanners/artec-eva-v2.

[4] https://matterport.com/.

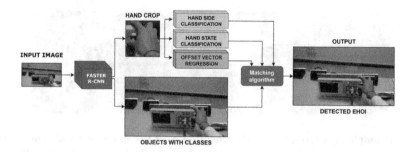

Fig. 4. The proposed system takes an egocentric RGB image as input and outputs several predictions about the status of hands and objects involved in the interactions.

To simplify the acquisition process, we defined different sequences of operations that subjects have to follow (e.g., turning on the oscilloscope, connecting the power cables to the electrical board, etc.). To make data collection consistent and more natural, we developed a Mixed-Reality application for Hololens 2 that guides the subjects through audio and images to the next operation they have to perform. The set of operations has been randomized in order to be less scripted. The average duration of the captured videos is 28.37 min. In total, we acquired 3 h and 47 min of video recordings at a resolution of 2272 × 1278 pixels and with a framerate of 30fps. An example of the captured data is shown in Fig. 2 - left. We manually annotated the real videos with all the EHOIs performed by the subjects. We used the following approach to select the image frames to be annotated: 1) we considered the first frame in which the hand touches the interacted object (i.e., contact frame), and 2) we selected the first frame that appears immediately after the hand released the object (i.e., non contact frame). For each of the considered frames we annotated: 1) hand bounding boxes and attributes, such as hand side and contact state *(In Contact with an object/No contact)*, 2) active and non-active object bounding boxes and their categories, and 3) the relationships between the hands and the active objects (e.g. *in contact with the right hand*) (See footnote 2).

4   Proposed Approach

Similarly to [31], our method extends the popular two-stage detector Faster R-CNN [30] to address the considered EHOIs detection task. However, differently than [31], the proposed method is able to detect all the objects in the image together with the active/no active object class. Figure 4 shows the architecture of the proposed approach. The proposed method detects the hands and the objects in an egocentric RGB image and infers: 1) object categories, 2) hands side, 3) hands contact state, and 4) EHOIs as <*hand, contact_state, active_object, <other_objects>>* quadruplet. Similarly to [31], we extend the object detector with four additional components: 1) the hand side classification module, 2) the hand state classification module, 3) the offset vector regression module, and 4) the matching algorithm. The modules composing our method are described in the following (See footnote 2).

Hands and Objects Detection: For objects and hands detection, we adopted a Faster R-CNN detector [30] based on a ResNet-101 backbone [16] and a Feature Pyramid Network (FPN) [22] due to their state-of-the-art performance. The network predicts a *(x,y,w,h,c)* tuple for each object/hand in the image, where the *(x,y,w,h)* tuple represents the bounding box coordinates, and *c* is the predicted object class.

Hand Side Classification Module: The hand side classification module consists of a Multi-Layer Perceptron (MLP) composed of two fully connected layers. Starting from the detected hands, it takes as input a ROI-pooled feature vector of the hand crop and predicts the side of the hand *(left/right)*.

Hand State Classification Module: We consider two contact state classes: *In Contact* and *No contact*. Other information about the contact state is embedded in the object category, which is predicted by our method, as opposed to [31] which predicts several types of contact states such as "in contact with a mobile object" or "in contact with a fixed object". The hand state classification module is composed of a MLP with two fully connected layers. We also enlarge the hand crop by 30% relative to the detected bounding box to include information of the surrounding context (e.g., nearby objects). The module takes as input the ROI-pooled feature vectors to infer the hands contact state.

Offset Vector Regression Module: Following the approach proposed in [31], we predict an offset vector that links the center of each hand bounding box to the center of the corresponding active object bounding box. The offset vector is represented by a versor v and a magnitude m. This module is composed of a MLP with two fully connected layers. It takes as input a ROI-pooled feature vector extracted from the enlarged hand crop and infers the $< v_x, v_y, m >$ triplet, where v_x and v_y represent the components of the versor v.

Matching Algorithm: The last component of the proposed system is a matching algorithm that takes as input the outputs from the previous modules to predict the *<hand, contact_state, active_object, <other_objects>>* quadruplet. The algorithm computes for each hand in contact with an object an image point ($p_{interaction}$) using the coordinates of the center of the hand bounding box and the corresponding offset vector. This point represents the predicted center of the active object bounding box. The active object is selected considering the object bounding box whose center is closest to the inferred $p_{interaction}$ point and also checking if the bounding box has a nonzero intersection with the bounding box of the considered hand.

5 Experiments and Results

We split the real dataset into training, validation, and test sets. Table 1 reports statistics about these splits. We trained our models in two stages. In the first stage, the models have been trained using only synthetic data (i.e., 0% of real data). In the second stage, we finetuned the models considering different amount of the real training data, namely, 10%, 25%, 50%, and 100%.

Table 1. Statistics of the three splits: Train, Validation and Test.

Split	Train	Val	Test
#Videos	2	2	4
#Images	992	734	1,330
%Images	32.46	24.01	43.53
#Hands	1,653	1,036	1,814
#Objects	6,483	4,337	6,778
#Active objects	1,090	662	1,120

Table 2. Object detection results using different amounts of real data.

Pretraining	Real data%	mAP%
Synthetic	0	66.44
–	10	53.27
Synthetic	10	72.69
–	25	52.34
Synthetic	25	76.19
–	50	71.17
Synthetic	50	**77.29**
–	100	70.84
Synthetic	100	<u>77.14</u>

5.1 Object Detection Performance

We evaluated the object detection performance of our method considering 19 object categories. We used the mean Average Precision metric, with an *Intersection over Union (IoU)* threshold of 0.5 (*mAP@50*)[5]. We report the results in Table 2. The *"Pretraining"* column indicates whether synthetic data were used to pretrain the models. The *"Real Data%"* column reports the percentage of real data used to finetune the models. The table shows the best results in bold, whereas the second best results are underlined. The results show that using only synthetic data to train the model (first row of Table 2) allows to achieve reasonable performance for this task (*mAP* of 66.44%). The best result (*mAP* of 77.29%) was obtained by the model pretrained on the synthetic dataset and finetuned with 50% of the real dataset, while the second best result (*mAP* of 77.14%) comes from the model pretrained on the synthetic dataset and finetuned with 100% of the real dataset. The results also highlight how combining synthetic and real data allows to increase the performance for the object detection task. Indeed, all the models which have been pretrained using synthetic data outperformed the corresponding models trained only with real data, especially when little real data is available. Furthermore, it is worth noting that the model pretrained on the synthetic dataset and finetuned with 10% of the real dataset obtained a higher performance (*mAP* of 72.69%) than all the models trained

---

[5] We used the following implementation: https://github.com/cocodataset/cocoapi.

using only the real data (*mAP* of 70.84% using 100% of real data), which supports the usefulness of synthetic data. See supplementary material for qualitative results.

5.2   Egocentric Human Object Interaction Detection

We evaluated our method considering the following metrics: 1) *AP Hand*: Average Precision of the hand detections; 2) *mAP Obj*: mean Average Precision of the active objects; 3) *AP H+Side*: Average precision of the hand detections when the correctness of the side (*Left/Right*) is required; 4) *AP H+State*: Average precision of the hand detections when the correctness of the contact state (*In contact/No contact*) is required; 5) *mAP H+Obj*: mean Average Precision of the active objects when the correctness of the associated hand is required, and 6) *mAP All*: mean Average Precision of the hand detections when the correctness of the side, contact state, and associated active object are required. Note that, while most of these metrics are based on [31], we modified the metrics influenced by active objects (i.e., *mAP Obj*, *mAP H+Obj*, and *mAP All*) to include the recognition of the object categories (switching from AP to mAP). The results summarized in Table 3 highlight that using synthetic data allows to achieve the best performance. Indeed, the model pretrained with synthetic data and finetuned with 100% of the real dataset (last row) obtained the best results considering all the evaluation measures, except for the mAP Obj measure, in which it obtains the second best result of 35.43%. In particular, considering the mAP all measure (32.61%), it outperforms the model trained using 100% of real data (22.70%) by a significant margin of 9.91%. The model trained using only synthetic data (first row) outperforms all the models using only real data with respect to the evaluation measures influenced by the active objects. Indeed, the aforementioned model obtains the best results with respect to mAP Obj (29.52%), mAP H+Obj (26.29%), and mAP All (23.78%). These performance scores are higher as compared to those achieved by the model trained with 50% of real data (i.e., 27.08%, 25.54% and 23.27%). Nevertheless, for the measures related to the hands (i.e., AP Hand, AP H+Side and AP H+State), the dis-

Table 3. Results for the EHOI detection task.

Pretraining	Real data%	AP hand	mAP Obj	AP H+Side	AP H+State	mAP H+Obj	mAP All
Synthetic	0	80.89	29.52	78.65	36.16	26.29	23.78
–	10	90.48	23.26	79.46	_50.44_	21.79	18.59
Synthetic	10	81.69	34.19	80.28	48.59	30.98	28.14
–	25	90.46	18.83	80.28	49.25	17.50	15.92
Synthetic	25	_90.61_	31.17	80.50	48.90	28.38	26.60
–	50	90.38	27.08	79.98	48.95	25.54	23.27
Synthetic	50	_90.61_	**36.23**	79.69	48.81	_31.87_	_30.50_
–	100	90.47	26.29	_89.20_	50.13	25.04	22.70
Synthetic	100	**90.67**	_35.43_	**89.37**	**50.58**	**34.09**	**32.61**

cussed method achieves limited performance, probably due to the gap between the real and synthetic domains. Please see Fig. 2-left for a qualitative example of the proposed method.

We also compare the proposed method with different baselines based on the state-of-the-art method introduced in [31], which was pretrained on the large-scale dataset 100DOH [31] which contains over 100K labeled frames of HOIs. To be able to compare the proposed method with [31], we extend the former to recognize the class of active objects following to two different approaches. The first approach consists in training a Resnet-18 CNN [16] to classify image patches extracted from the active object detections. We trained the classifier with four different sets of data: 1) BS1: 19 videos, one per object class, in which only the considered object is observed. This provides a minimal training set that can be collected with a modest labeling effort; 2) BS2: images sampled from the proposed real dataset; 3) BS3: synthetic data and 4) BS4: both real and synthetic data. Note that this set requires a significant data collection and labeling effort. The second approach (BS5) uses the YOLOv5[6] object detector to assign a label to the active objects predicted by [31]. In particular, to each active object, we assign the class of the object with the highest IoU among those predicted by YOLOv5, otherwise, if there are no box intersections, the proposal is discarded. Table 4 reports the obtained results. Considering the measures based on the active objects (i.e., mAP Obj, mAP H+Obj, and mAP all), our method trained with 50% of the real data (496 images) outperforms all the baselines based on [31] obtaining performances of 36.23%, 31.87% and 30.50% respectively. Moreover, using 100% of the real data, our approach obtains comparable performance considering the measures based on the hands (i.e., AP Hand, AP H+Side, and AP H+State). See supplementary material for qualitative comparison.

Table 4. Comparison between the proposed method and different baseline approaches based on [31].

	Pretraining	Real data%	AP hand	mAP Obj	AP H+Side	AP H+State	mAP H+Obj	mAP All
Proposed method	Synthetic	0	80.89	29.52	78.65	36.16	26.29	23.78
Proposed method	Synthetic	50	90.61	**36.23**	79.69	48.81	_31.87_	_30.50_
Proposed method	Synthetic	100	_90.67_	_35.43_	_89.37_	_50.58_	**34.09**	**32.61**
BS1	–	100	**90.76**	14.10	**89.78**	**59.23**	12.42	11.23
BS2	–	100	**90.76**	22.17	**89.78**	**59.23**	19.76	18.05
BS3	Synthetic	0	**90.76**	09.44	**89.78**	**59.23**	08.51	07.49
BS4	Synthetic	100	**90.76**	26.36	**89.78**	**59.23**	21.00	19.48
BS5	Synthetic	100	**90.76**	30.26	**89.78**	**59.23**	28.77	27.12

6 Conclusion

We considered the EHOI detection task in the industrial domain. Since labeling images is expensive in terms of time and costs, we explored how using synthetic

---

[6] YOLOv5: https://github.com/ultralytics/yolov5.

data can improve the performance of EHOIs detection systems. To this end, we generated a new dataset of automatically labeled photo-realistic synthetic EHOIs in an industrial scenario and collected 8 egocentric real videos, which have been manually labeled. We proposed a method to tackle EHOI detection and compared it with different baseline approaches based on the state-of-the-art method of [31]. Our analysis shows that exploiting synthetic data to train the proposed method greatly improves performance when tested on real data. Future work will explore how the knowledge of active/no active objects, inferred by our system, can provide useful information for other tasks such as next active object prediction.

Acknowledgements. This research has been supported by Next Vision (https://www.nextvisionlab.it/) s.r.l., by the project MISE - PON I&C 2014–2020 - Progetto ENIGMA - Prog n. F/190050/02/X44 - CUP: B61B19000520008, and by Research Program Pia.ce.ri. 2020/2022 Linea 2 - University of Catania.

References

1. Bambach, S., Lee, S., Crandall, D.J., Yu, C.: Lending a hand: detecting hands and recognizing activities in complex egocentric interactions. Int. Conf. Comput. Vis. (2015)
2. Betancourt, A., Morerio, P., Regazzoni, C.S., Rauterberg, M.: The evolution of first person vision methods: a survey. IEEE Trans. Circuits Syst. Video Technol. (2015)
3. Chao, Y.W., Liu, Y., Liu, X., Zeng, H., Deng, J.: Learning to detect human-object interactions. Winter Conf. Appl. Comput. Vis. (2018)
4. Chen, Y., Huang, S., Yuan, T., Qi, S., Zhu, Y., Zhu, S.C.: Holistic++ scene understanding: single-view 3d holistic scene parsing and human pose estimation with human-object interaction and physical commonsense. Int. Conf. Comput. Vis. (2019)
5. Cucchiara, R., Del Bimbo, A.: Visions for augmented cultural heritage experience. IEEE Multim. (2014)
6. Damen, D., et al.: Rescaling egocentric vision: collection, pipeline and challenges for epic-kitchens-100. Int. J. Comput. Vis. (2021)
7. Damen, D., et al.: Scaling egocentric vision: the epic-kitchens dataset. Eur. Conf. Comput. Vis. (2018)
8. Fu, Q., Liu, X., Kitani, K.M.: Sequential voting with relational box fields for active object detection. arXiv preprint arXiv:2110.11524 (2021)
9. Furnari, A., Farinella, G.M., Battiato, S.: Temporal segmentation of egocentric videos to highlight personal locations of interest. In: Hua, G., Jégou, H. (eds.) ECCV 2016. LNCS, vol. 9913, pp. 474–489. Springer, Cham (2016). https://doi.org/10.1007/978-3-319-46604-0_34
10. Furnari, A., Battiato, S., Grauman, K., Farinella, G.M.: Next-active-object prediction from egocentric videos. J. Vis. Commun. Image Represent. (2017)
11. Furnari, A., Farinella, G.M.: Rolling-unrolling LSTMS for action anticipation from first-person video. IEEE Trans. Pattern Anal. Mach. Intell. (2021)
12. Garcia-Hernando, G., Yuan, S., Baek, S., Kim, T.K.: First-person hand action benchmark with RGB-D videos and 3D hand pose annotations. Conf. Comput. Vis. Pattern Recogn. (2018)

13. Gkioxari, G., Girshick, R., Dollár, P., He, K.: Detecting and recognizing human-object interactions. Conf. Comput. Vis. Pattern Recogn. (2018)
14. Gupta, S., Malik, J.: Visual semantic role labeling. arXiv preprint arXiv:1505.04474 (2015)
15. Hasson, Y., et al.: Learning joint reconstruction of hands and manipulated objects. Conf. Comput. Vis. Pattern Recogn. (2019)
16. He, K., Zhang, X., Ren, S., Sun, J.: Deep residual learning for image recognition. Conf. Comput. Vis. Pattern Recogn. (2016)
17. Heilbron, F.C., Escorcia, V., Ghanem, B., Niebles, J.C.: Activitynet: a large-scale video benchmark for human activity understanding. Conf. Comput. Vis. Pattern Recogn. (2015)
18. Kay, W., et al.: The kinetics human action video dataset. arXiv preprint arXiv:1705.06950 (2017)
19. Li, Y., Liu, M., Rehg, J.M.: In the eye of the beholder: Gaze and actions in first person video. IEEE Trans. Pattern Anal. Mach. Intell.(2021)
20. Li, Y., Ye, Z., Rehg, J.M.: Delving into egocentric actions. Conf. Comput. Vis. Pattern Recogn. (2015)
21. Liao, Y., Liu, S., Wang, F., Chen, Y., Feng, J.: PPDM: parallel point detection and matching for real-time human-object interaction detection. Conf. Comput. Vis. Pattern Recogn. (2020)
22. Lin, T.Y., Dollár, P., Girshick, R., He, K., Hariharan, B., Belongie, S.: Feature pyramid networks for object detection. Conf. Comput. Vis. Pattern Recogn. (2017)
23. Lin, T.Y., et al.: Microsoft coco: common objects in context. Eur. Conf. Comput. Vis. (2014)
24. Lu, Y., Mayol-Cuevas, W.: The object at hand: Automated editing for mixed reality video guidance from hand-object interactions. Int. Symp. Mixed Augment. Real. (2021)
25. Lu, Y., Mayol-Cuevas, W.W.: Understanding egocentric hand-object interactions from hand pose estimation. arXiv preprint arXiv:2109.14657 (2021)
26. Mueller, F., et al.: Ganerated hands for real-time 3D hand tracking from monocular RGB. Conf. Comput. Vis. Pattern Recogn. (2018)
27. Mueller, F., Mehta, D., Sotnychenko, O., Sridhar, S., Casas, D., Theobalt, C.: Real-time hand tracking under occlusion from an egocentric RGB-D sensor. Int. Conf. Comput. Vis. (2017)
28. Ragusa, F., Furnari, A., Battiato, S., Signorello, G., Farinella, G.: Egocentric visitors localization in cultural sites. J. Comput. Cult. Herit. (2019)
29. Ragusa, F., Furnari, A., Livatino, S., Farinella, G.M.: The meccano dataset: understanding human-object interactions from egocentric videos in an industrial-like domain. Winter Conf. Appl. Comput. Vis. (2021)
30. Ren, S., He, K., Girshick, R., Sun, J.: Faster R-CNN: towards real-time object detection with region proposal networks. Adv. Neural Inf. Process. Syst. (2015)
31. Shan, D., Geng, J., Shu, M., Fouhey, D.F.: Understanding human hands in contact at internet scale. Conf. Comput. Vis. Pattern Recogn. (2020)
32. Wang, H., et al.: Learning a generative model for multi-step human-object interactions from videos. Comput. Graph. Forum (2019)
33. Zhang, F.Z., Campbell, D., Gould, S.: Efficient two-stage detection of human-object interactions with a novel unary-pairwise transformer. arXiv preprint arXiv:2112.01838 (2021)

Skew Angle Detection and Correction in Text Images Using RGB Gradient

Bruno Rocha[1] , Gabriel Vieira[1,2] , Helio Pedrini[3] , Afonso Fonseca[1] ,
Deborah Fernandes[1] , Júnio César de Lima[2] , Júlio César Ferreira[2] ,
and Fabrizzio Soares[1(✉)]

[1] Institute of Informatics, Universidade Federal de Goiás, Goiânia, GO, Brazil
`rocha3runo@ufj.edu.br`, {`afonsoueslei,fabrizzio`}`@ufg.br`,
`deborah@inf.ufg.br`
[2] Computer Vision Laboratory, Federal Institute Goiano, Urutaí, GO, Brazil
{`gabriel.vieira,junio.lima,julio.ferreira`}`@ifgoiano.edu.br`
[3] Institute of Computing, University of Campinas, Campinas, SP, Brazil
`helio@ic.unicamp.br`

Abstract. Detecting and correcting skew angles is critical to success
in document layout analysis and optical character recognition tasks, as
they are more susceptible to failure when on uneven skews. In automation
such as postal systems, library management, office business, and
banking data entry, skew angle estimation is crucial to improve procedures
response. Although different works have addressed this subject,
due to the variability in the input data, many solutions are restricted to
a specific language, texts whose contents are within a controlled scope,
and entries that differentiate printed from handwritten texts. This paper
introduces a new method based on RGB gradient capable of detecting
and correcting skew angles in different types of documents. We evaluate
the proposed method using two public databases and compare our results
with other techniques cited in the literature. In general, our proposal
achieved results superior to the approaches compared in all groups of
documents in the database. Furthermore, we show that our method can
work accurately in various text orientations, and it can work efficiently
against documents containing short and sparse text lines, non-textual
objects, and image noises caused by imperfect scanning.

Keywords: RGB gradient · Skew detection · Image rotation · Text
image processing

1 Introduction

Currently, computers are more than work tools, as in addition to supporting business
activities, they also contribute to the consolidation of commercial, social,
and scientific strategies [32]. Computing optimizes tasks, increases productivity,
and reduces costs. In this sense, the digitization of physical documents is essential
in the computerization process and digital migration to huge collections from
universities, libraries, bookstores, registry offices, and museums offering at least

© The Author(s), under exclusive license to Springer Nature Switzerland AG 2022
S. Sclaroff et al. (Eds.): ICIAP 2022, LNCS 13232, pp. 249–262, 2022.
https://doi.org/10.1007/978-3-031-06430-2_21

three advantages: (i) ensuring that these documents will be preserved against natural degradation, (ii) reducing physical space to safeguard them, and (iii) maintaining and preserving access to these documents efficiently.

A widely adopted digitization processes is bibliographic indexes registration, which, drastically simplifying content searching and statistical data collection [30], but it limits them to bibliographic data. To overcome this limitation, the process of transcribing physical to digital data includes manually relevant document parts for use in text-based content searches. Although this process is slow, expensive and unfeasible for huge amounts of data, an alternative adopted is photography or digital scanning to produce documents in Portable Document Format (PDF) or images. In this way, it is possible to automatically extract and interpret textual representations by computer through Optical Character Recognition (OCR) systems [9], often used to store and transmit documents.

However, OCR techniques may not present satisfactory results due to degradation resulting from the digital document acquisition process [6]. Besides, the incorrect positioning of digital documents causes misinterpretations in automatic processes, provoking errors and making document digitization projects unfeasible [13]. Hence, an important task is detecting and correcting skew angle in digital documents to ensure better analysis and accuracy of OCR systems [5].

In this work, we aim to detect and correct skew angles in different documents as shown in Fig. 1 where a document is correctly aligned after rotation operation. The novelty of our method is that it uses the RGB gradient to indicate projection profiles (PPs) and align text images to a correct position. The RGB gradient explores the gradient vector directions, which indicates the current skew level of the digital image, allowing one to align texts at different orientation levels properly. We evaluate the proposed method using two public databases whose images (color and grayscale) have a variety of content that includes figures and graphics, mathematical equations, tables and texts, and manuscripts.

Fig. 1. Input image [skew angle: 6.34°] (Left) and result after applying our method (Right).

2   Related Work

In the analysis of digital documents [6], the correct positioning of text entries plays a crucial role in extracting and recognizing the characteristics intrinsic to their content. For this reason, the detection and correction of skew angles are essential to increase the robustness in the analysis and evaluation of digital documents. Furthermore, the image alignment process effectively contributes to the accuracy gain in the human-machine interaction.

Most skew estimation and correction methods available in the literature are based on Hough transform (HT), projection profiles (PPs), and nearest-neighbor clustering [1]. Ptak et al. [22] developed a method based on a horizontal projection profile to segment text lines from cursive handwriting. As a result of applying their method, line separators can be obtained in parts of the document to split the text into horizontal fragments. Khidhir [15] applied the Radon Transform (RT) to adjust the orientation of printed text angle. In this study, RT turns text lines and line spacing into points, and then the correct angle is the one with the most occurrences of points. Ramegowda [23] proposed an approach based on morphological operation and linear regression process using a connected component analysis of binarized text lines. For each shape structure, a fit line is obtained using linear regression to determine document skew angle.

Bafjaish et al. [4] used HT for skew detection and correction of Arabic text with morphological dilation operations in binary images. Analogously, the approach proposed by Boudraa et al. [7] used the variation of HT, denominated of the Progressive Probabilistic HT, which was combined with Morphological Skeleton extraction to detect and correct the angle in historical documents.

Sun and Si [29] proposed a technique to identify text image skew and slant angle using only the gradient orientation histogram. In the work of Sun and Si was assumed that the main direction of the gradient orientation is perpendicular to the document text lines, to find the correct skew angle of the image.

Finally, Avila and Lins [3] proposed labeling black pixels to create bounding boxes, defined two lines of text by computing the least squares over the lower and upper midpoints of these boxes. Finally, the skew angle is identified by accumulating the angles of each line formed in a histogram.

3   Materials

MediaTeam Oulu Document Database [25] and Tobacco800 [17] are the databases selected to evaluate the proposed approach. MediaTeam provides file groups with challenging properties, such as figures and graphics, mathematical equations, documents filled with different colors, tables, cursive handwritten, and texture. This diversity denotes the heterogeneity and complexity of the database we use to evaluate our method. However, we did not use the following groups: *Color Segmentation Images*, *Linedrawing*, *Music*, *Streetmap*, and *Terrainmap*. These groups contain documents without text or text in small amounts, which reduces the possibility of identifying orientation in methods, similarly with our method which is based on skew angles of lines. Thus, from this database, we work with 907 images. The Tobacco800 database contains a single document type and is composed of 1,290 images collected and scanned from business documents. Other dataset characteristics and their documents types are shown in Table 1.

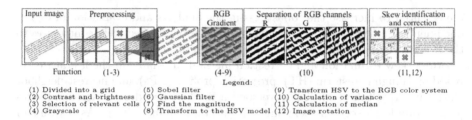

Fig. 2. Proposed method for detecting and correcting skew angle in documents.

4   Proposed Method

The proposed method takes an image (rotated or not) as input and presents an adjusted image with text lines fully aligned horizontally. Three main steps aggregate operations on images: (1) preprocessing, (2) application of RGB Gradient, and (3) skew identification and correction. The flowchart of these steps is illustrated in Fig. 2, local operations for each step are from 1 to 12.

Table 1. Information concerning the document type and their characteristics.

DB	Document type	Images	Dimension (pixels)	Figures	Equations	Colorful	Table	Handwritten	Texture
MED	Addresslist	15	$1,787 \times 2,765$	✓		✓			✓
	Advertisement	50	$2,062 \times 2,821$	✓		✓			✓
	Article	437	$1,942 \times 2,785$	✓	✓		✓		✓
	Businesscard	9	$1,077 \times 597$	✓		✓			✓
	Check	3	$1,138 \times 959$	✓				✓	✓
	Correspondence	66	$1,878 \times 2,342$	✓		✓	✓	✓	✓
	Dictionary	30	$1,491 \times 2,197$			✓			
	Faxcover	11	$2,281 \times 3,112$					✓	
	Form	38	$1,628 \times 1,964$	✓			✓		
	Mailpiece	4	$2,232 \times 1,640$						✓
	Manual	64	$1,668 \times 2,344$	✓	✓		✓		✓
	Math	25	$1,635 \times 2,485$	✓	✓		✓		
	Newsletter	73	$2,160 \times 3,138$	✓		✓			✓
	Outline	38	$1,712 \times 2,479$						✓
	Phonebook	15	$2,193 \times 3,268$	✓		✓			✓
	Programlisting	30	$1,642 \times 2,682$						
	Total	907							
T800	Business	1,290	$1,810 \times 2,306$	✓			✓	✓	✓

Databases: MED - MediaTeam Oulu Document [25]; T800 - Tobacco800 [17]

4.1   Preprocessing

In the preprocessing step, input image is divided into smaller regions, equal to 100×100 pixels. Each region (cell) is processed as an independent image. First, image regions with only pixels-white or only pixels-black will be discarded, as they do not contain relevant information for processing. For example, in Fig. 2 the image cells with only white pixels were marked (with an 'x') to illustrate image regions that are not considered in the processing. Then, we apply contrast and brightness correction in each relevant image region to remove noise. The brightness was adjusted to 1.8 followed by the contrast adjustment to 3.0. Different values of contrast and brightness were tested, and the chosen ones reflect the best values for these parameters. In the last step of the proposed method, these regions are combined with the other results to generate the final result.

4.2 RGB Gradient

The gradient RGB filter is applied to each one of the image regions. We proposed an adaptation of the RGB gradient filter in which parameters and some parts of the original algorithm were adapted to support our proposal. This filter identifies the gradient directions in the image regions and represents these directions by color. The original algorithm is available on the G'MIC website[1] in the contour category, and our adapted version is explained as follows. Let $\mathbf{I}_{RGB}$ the cell from the RGB image, where $\mathbf{I}_R$, $\mathbf{I}_G$, and $\mathbf{I}_B$ are the Red, Green, and Blue channels of $\mathbf{I}_{RGB}$, respectively. Initially, $\mathbf{I}_{RGB}$ is converted to a grayscale image $\mathbf{I}_{gray}$.

Subsequently, the Sobel filter [18,27] is applied in the x and y axes of $\mathbf{I}_{gray}$, to obtain the corresponding matrices $\mathbf{S}_x$ and $\mathbf{S}_y$, in the respective order. After that, the Gaussian filter [10] is applied to reduce the details of the matrices $\mathbf{S}_x$ and $\mathbf{S}_y$, generating new matrices $\mathbf{G}_x$ and $\mathbf{G}_y$. Given $g_{xi}^{(j)} \in \mathbf{G}_x$ and $g_{yi}^{(j)} \in \mathbf{G}_y$, they are used in Eq. (1) to find the magnitude.

$$\forall i,j \quad c_i'^{(j)} = \sqrt{(g_{xi}^{(j)})^2 + (g_{yi}^{(j)})^2} \tag{1}$$

The new RGB image with the magnitudes (refers to Fig. 2 (4–9)) is created where the $\hat{r}_i^{(j)} \in \hat{\mathbf{R}}$, $\hat{g}_i^{(j)} \in \hat{\mathbf{G}}$, and $\hat{b}_i^{(j)} \in \hat{\mathbf{B}}$ channels are calculated:

$$\forall i,j \quad \hat{r}_i^{(j)} = \frac{r_i^{(j)}}{c_i'^{(j)}}, \quad \forall i,j \quad \hat{g}_i^{(j)} = \frac{g_i^{(j)}}{c_i'^{(j)}}, \quad \forall i,j \quad \hat{b}_i^{(j)} = \frac{b_i^{(j)}}{c_i'^{(j)}}$$

The next step is to define the channels of the Hue Saturation Value (HSV) color model, using Eqs. (2), (3) and (4).

$$\forall i,j \quad \tilde{h}_i^{(j)} \in \mathbf{H} = \left\{ \frac{\text{atan2}(\hat{g}_i^{(j)}, \hat{r}_i^{(j)})180}{\pi} \right\} \quad \text{mod } 360 \tag{2}$$

where $\pi = 3.14$ and atan2$(.,.)$ returns argument coefficient's arc-tangent, as a value in radians between π to $-\pi$. The channel $\mathbf{S}$ of the HSV color space is just a matrix whose coordinates have values equal to 1, according to Eq. 3.

$$\forall i,j \quad \tilde{s}_i^{(j)} \in \mathbf{S} = 1 \tag{3}$$

---

[1] https://gmic.eu/oldtutorial/_gradient2rgb.shtml.

$$\forall i,j \quad \tilde{v}_i^{(j)} \in \mathbf{V} = \begin{cases} 1 & \text{if } \max(\hat{\mathbf{V}}) = \min(\hat{\mathbf{V}}) \\ \dfrac{\hat{v}_i^{(j)} - \min(\hat{\mathbf{V}})}{\max(\hat{\mathbf{V}}) - \min(\hat{\mathbf{V}})} & \text{otherwise} \end{cases} \quad (4)$$

Thus, $\forall i,j \; \hat{v}_i^{(j)} \in \hat{\mathbf{V}} = \sqrt{(\hat{r}_i^{(j)})^2 + (\hat{g}_i^{(j)})^2}$, whereas min(.) and max(.) are the minimum and maximum values of the set, respectively. Terms $\mathbf{H} \in [0°, 360°]$, $\mathbf{S} \in [0,1]$, and $\mathbf{V} \in [0,1]$ are the components that represent the channels H, S, and V of the HSV color system, in the respective order. Finally, the last step is to transform $\mathbf{H}$, $\mathbf{S}$, and $\mathbf{V}$ to the RGB color system as in [26], with channels $\mathbf{I}_{\dot{R}}$, $\mathbf{I}_{\dot{G}}$ and $\mathbf{I}_{\dot{B}}$ channels of the RGB color system of the RGB gradient image $\mathbf{I}_{\varphi}$.

4.3   Skew Identification and Correction

This step extracts B-channel from RGB image gradient. According to our experimentation (Table 2), channel B has a more excellent representation of the text direction in the image than channels R and G. Based on that, the channel $\mathbf{I}_{\dot{B}}$ is transformed into a binary image $\mathbf{I}_{\dot{B}}^b$ using a threshold value of 128.

Table 2. Absolute error using RGB channels.

Channel	Skew angle									
	0°	5°	10°	15°	20°	25°	30°	35°	40°	45°
R	16.48	1.61	3.56	6.70	7.98	9.16	7.28	10.30	9.41	12.03
G	45.06	44.78	43.53	42.16	39.77	39.27	37.44	34.30	32.27	26.16
B	3.27	1.09	1.16	1.47	2.28	2.72	2.48	2.27	6.23	5.17

After that, the $\mathbf{I}_{\dot{B}}^b$ image regions are rotated to all angles of $\theta \in \{0°, 1°, .., 179°\}$. For each angle, the image region is rotated and the values of the image columns are summed up to create a corresponding vector $\mathbf{v}^\theta$. The variance s_θ^2 is calculated for the vector $\mathbf{v}^\theta$ and the angle θ^v is chosen using Eq. (5).

$$\theta^v = \left(\underset{\theta \in \{0°,...,179°\}}{\arg\max} \; s_\theta^2 \right) - 90° \quad (5)$$

Thus, $\mathbf{I}_{\dot{B}}^b$ is used to create an angle θ^v, represented by θ_f. The θ_f is the best angle of a region of the $\mathbf{I}_{RGB}$ image. In this sense, we obtain a θ_f for each cell of the input image. Then, the final angle θ_F is obtained by the variance of the vector $\mathbf{v}_{\theta_f} = (\theta_f^{(1)}, \theta_f^{(2)}, ..., \theta_f^{(k)})$, where $\mathbf{v}_{\theta_f}$ is the vector of all angles of the cells of the input image. Finally, θ_F is used to rotate the image to make the text lines fully aligned horizontally.

4.4   Evaluation of the Method

The evaluation of our method was inspired by the methodology proposed by Wang et al. [31], but instead of using the angle range from $1°$ to $20°$, we increased the analysis range from $0°$ to $45°$. That is, the images were manually rotated to angles $0°$, $5°$, $10°$, $15°$, $20°$, $25°$, $30°$, $35°$, $40°$ and $45°$. The result of our proposal is the angle θ_F, which is later compared to the angle manually rotated, which will be called θ_M. Then, we calculated the absolute error $E_a = |\theta_M - \theta_F|$ for each image, which corresponds to the absolute value of the difference between the angles obtained with the automatic and manual methods. Finally, to analyze the results of the database, the mean of absolute error $\bar{E}_a$ was calculated for all images in the database for each manually rotated angle and document type.

5   Experimental Analysis and Discussion

5.1   Quantitative Evaluation

The experiments were performed in all 907 and 1,290 images from the databases MediaTeam Oulu Document and Tobacco800, respectively, as described in Sect. 3. The evaluation follows the methodology presented in Sect. 4.4, and the Mean of Absolute Error (MAE, Eq. (6)) is used as a metric to evaluate the performance of our method concerning each type of document contained in the database, as well as its ability to deal with different skew movements.

$$\bar{E}_a = \frac{\sum_{i=1}^{n} E_a^{(i)}}{n} \qquad (6)$$

where n corresponds to the number of images to be evaluated.

In this work, we also compared our method to three other approaches available in the literature, which are the Project Profile (PP), as presented by Al-Khatatneh et al. [1], Radon Transform (RT) [15] and HT [11]. PP is the baseline for comparison with the proposed method due to their similarity in rotating the image to find angles. The $\bar{E}_a$ and the standard deviation (std) were computed for all images in the document type group and the results are shown in Table 3. The values highlighted in bold correspond to the best results achieved with the methods for a given group of documents.

We observe in Table 3 that our proposal obtained better MAE than PP, RT, and HT methods in all groups of documents for the MediaTeam Oulu database. Furthermore, our method achieved the smallest standard deviation in most groups of documents, except the *Businesscard, Manual, Newsletter*, and *Phonebook* documents type. Our proposal has obtained the lowest MAE and std values (equal to 4.9 and 9.6, respectively) in the Tobacco800 database compared to PP, RT, and HT methods. In general, the results achieved evidence the capacity of our method for different types of documents and complexities.

Table 3. Absolute error - comparison of results.

DB	Doc. type	Project profile		Radon transform		Hough transform		Our proposal	
		Mean	std	Mean	std	Mean	std	Mean	std
MED	1	15.4	43.9	9.0	23.6	18.4	13.1	**1.2**	**3.0**
	2	31.9	46.7	24.7	40.2	19.6	14.3	**3.8**	**7.9**
	3	12.5	36.1	10.3	28.4	18.5	13.6	**0.4**	**2.1**
	4	89.8	6.1	90.0	**0.2**	14.4	11.6	**3.3**	10.0
	5	60.6	42.6	63.0	41.9	17.9	11.4	**5.8**	**10.5**
	6	32.4	47.0	29.9	40.7	18.8	13.2	**1.0**	**3.1**
	7	19.9	41.2	18.3	36.3	17.8	12.4	**1.3**	**4.6**
	8	19.9	43.4	14.9	32.2	6.1	4.3	**1.2**	**2.7**
	9	37.3	47.0	35.8	42.8	19.1	14.5	**0.7**	**2.7**
	10	73.4	40.3	69.7	38.0	7.0	5.8	**0.2**	**0.5**
	11	11.9	36.6	8.5	25.7	7.9	**4.5**	**1.7**	5.1
	12	12.7	40.0	7.3	25.7	24.1	20.7	**2.5**	**5.2**
	13	14.3	36.3	12.8	31.3	5.5	**3.9**	**0.7**	4.3
	14	10.9	34.2	7.7	25.0	18.1	13.0	**5.2**	**12.5**
	15	9.0	32.7	2.7	18.4	5.0	**3.7**	**2.2**	5.5
	16	9.5	33.9	5.2	22.1	6.5	4.1	**1.8**	**3.1**
T800	17	18.0	42.6	10.9	29.4	23.6	22.9	**4.9**	**9.6**

Databases: MED - MediaTeam Oulu Document [25], T800 - Tobacco800 [17]; **Doc. Type:** 1. Addresslist, 2. Advertisement, 3. Article, 4. Businesscard, 5. Check, 6. Correspondence, 7. Dictionary, 8. Faxcover, 9. Form, 10. Mailpiece, 11. Manual, 12. Math, 13. Newsletter, 14. Outline, 15. Phonebook, 16. Programlisting, 17. Business documents

The histogram provides an overview of the variation and occurrence of the E_a achieved with the methods. Figure 3 shows the absolute error distribution histories of all manually rotated images (Sect. 4.4) of the databases (MediaTeam Oulu Document and Tobacco800) using PP, RT, HT, and our methods. The bin value for the histogram was 15°.

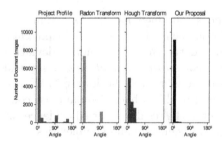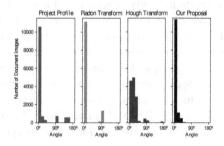

Fig. 3. Histogram of Absolute Error (E_a) from all rotated images of the databases: MediaTeam (Left) and Tobacco800 (Rigth).

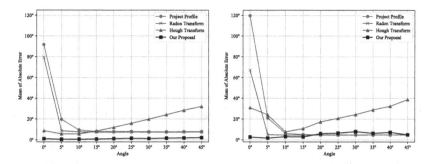

Fig. 4. Absolute error ($\bar{E}_a$) of rotated images from $0\,°$ to $45\,°$ of the databases: MediaTeam (Left) and Tobacco800 (Right).

From Fig. 3(a) (Left), it is possible to notice that our proposal was able to find a more significant number of images with an angle close to the correct one (E_a equal to $0\,°$) when compared to PP, RT, and HT. Furthermore, our method achieved a variation and occurrence of absolute error closer to $0\,°$ compared to the other three methods. This observation shows that our approach achieves promising results in identifying the skew angle of complex images in the MediaTeam Oulu Document database. By observing Fig. 3(b) (Right), our method achieved a more significant number of occurrences of E_a close to $0°$ in the Tobacco800 database. This way, our method may be a good choice for finding a skew angle in business document images.

Finally, Fig. 4 presents the graphs of $\bar{E}_a$ for each angle manually rotated from $0\,°$ to $45\,°$ (as shown in Sect. 4.4) using PP, RT, HT, and our proposal in the datasets. The graph makes it possible to visualize the behavior of the methods as the skew angle increases in the image. Figure 4(a) (Left), it is possible to note that our approach achieved lower mean errors than all other methods (PP, RT, and HT) for all angles, showing promising results for images rotated at different angles in the MediaTeam Oulu Document database. However, we also highlight that RT and PP methods presented similar behaviors, with higher errors when angle is equal to $0°$ and more minor errors for the other angles. Furthermore, while HT achieved lower results for the initial angles ($0\,°$ to $10\,°$), on the other hand, the error increases progressively from the $10°$ angle onwards. On the other hand, in Fig. 4(b) (Right), our method achieved better results for the images that are rotated from $0°$ to $15°$ when compared with the PP, RT, and HT methods.

However, from 20° to 40° angles, the results were slightly lower than the PP and RT. In general, the proposed method obtained a more stable behavior (mean errors close to 0°) for all evaluated angles, being a good option for identifying skew angles in the business documents presented in the Tobacco800 database.

The developed approach has some advantages to be highlighted. For example, the simplicity of the calculations used to identify and correct text document angles. Furthermore, the proposal can be used in color and grayscale images, and achieved adequate results with challenging digitization conditions such as blurry and various illumination conditions (both direction and intensity). In addition, it is a method that achieves good accuracy in different types, sizes, and scales of documents. Another advantage of our method is its ability to deal with multi-oriented texts in a single document, in which the method will choose the predominant orientation. We also emphasize that the proposal developed does not require prior training, which requires extensive databases and a large capacity of computational resources for training with quality, for example, approaches that use Deep Learning. However, the proposed method has higher computational cost to find the correct angle. Notwithstanding, parallelization techniques can minimize computational cost, distributing and increasing the processing speed.

Furthermore, we demonstrate our method can accurately work in a variety of orientations ranging from −45° to 45°, motivated by the method presented by Wang et al. [31]. In addition, due to RGB gradient, the proposed method can operate efficiently in documents containing diagrams, graphics, short and sparse text lines, non-textual objects and image noises caused by imperfect scanning.

Table 4 presents a comparison between the characteristics of our proposal and some related work. The main methods adopted in these studies are Randon Transform, Morphological Operations, and Hough Transform. Sun and Si [29] uses only gradient orientation histogram to find the angle skew from the image, unlike our work that used PP combined RGB gradient technique to find the correct skew angle. Differently, we investigated RGB gradient, which is a novelty in document orientation correction applications. Furthermore, some of the authors applied their methods in specific types of documents, such as handwritten texts or printed texts, while in our proposal, we deal with text documents that include figures, graphs, mathematical symbols, tables, and handwritten texts. Also, there is no standard database for evaluating skew detection methods. Therefore, some authors preferred to create their image databases, in some cases with few image samples. In our case, we opted for two public databases that together add up to more than 2,000 images. Besides that, our proposal operates on an extensive range of inclinations, similar to the related work.

Table 4. Characteristics of our proposal and other related work.

Research work	Method	Skew angle range (°)	Document characteristics	Dataset	Year
Khidhir [15]	Randon transform	[0, 50]	Text-lines	LD	2011
Ramegowda [23]	Morphological operations	[5, 25]	Text-lines, handwritten, bank forms	LD	2015
Bafjaish et al. [4]	Hough transform	[−20, 20]	Mushaf Al-Quran image	QURAN	2018
Kar et al. [14]	Core region detection, linear regression	[−25, 25]	Handwritten Devanagari, Bangla and Roman	LD	2019
Boudraa et al. [7]	Hough transform	[−15, 15] [0, 20] [−15, 15]	Text-lines, pictures, charts, diagrams and synthectic docs	DISEC2013 [20], PRIMA 2011 [2], Epshtein [12]	2020
Khuman et.al [16]	Projection profile and entropy	[−30, 30]	Meitei/Meetei image	LD, ICDAR2013 [28]	2021
Cai et al. [8]	Rule based	[−45, 45]	Text-lines, pictures, charts, diagrams	LD, DISEC2013 [20]	2021
Pramanik and Bag [21]	Salient feature	[−55, 55]	Handwritten	ICDAR 2013 [28], PHDIndic-11 [19]	2021
Salagar and Patil [24]	PCA	–	Text-lines,	60 docs from Tobacco-800 [17]	2022
Our proposal	RGB gradient	[−45, 45]	Handwritten, text-lines, tables, diagrams, equations and texture	MedianTeam [25], Tobacco-800 [17]	2022

LD: Local dataset, **QURAN:** https://www.nourelquran.com/quranforall/fahd/index.php, **docs:** documents

5.2 Visual Evaluation of the Processing Stages

Some visual results are presented in Fig. 5. We can observe the results after applying the RGB gradient and correcting the skew angles. These examples show the diversity of the database with examples of texts with figures, graphs, mathematical symbols, tables, and handwritten texts. In the first row, the original images are displayed after rotating at different inclination angles. As can be noticed,

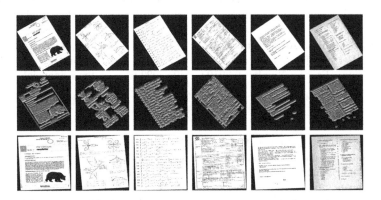

Fig. 5. Visual results: From rows 1 to 3: Original rotated images (row 1 - first), results after applying the RGB gradient (row 2), and final results (row 3).

the content of the images is not horizontally aligned, and the RGB gradient indicates the variation of intensities and color fluctuations of these images (row 2). Our method identifies regions of the most considerable significance based on this information. Then, an estimated angle is presented, allowing us to rotate the images to a position very close to the expected ones (row 3).

6   Conclusion and Future Work

This work presented a novel method for skew angle detection and correction in-text images using the RGB gradient. We prepared some experiments that evaluated the potential of our method in adjusting skew images to a suitable position. In addition, we included varied text entries in the experiments so that different scenarios were evaluated. Two public databases of text images with very diversified characteristics were used for this.

A result has shown average error is lesser than $2.3°$ in evaluations corresponding to rotations from $0°$ to $45°$. That means our method can adjust rotated images to a position closer to what was expected. Furthermore, although the results showed that the error increases according to the degree of inclination, the results are still promising and show that the RGB gradient can contribute to this task. Finally, corroborating to this statement, the results achieved in group evaluation show that the proposal satisfactorily deals with documents such as address lists, mathematical equations, newsletters, articles, and correspondences.

As guidelines for future work, we intend to improve our method by reducing the influence of the distortion caused by rotation transformations so that the process of detecting the skew angles can be more accurate. Furthermore, parallelization techniques can be used to distribute and increase the processing speed, minimizing the high cost of processing the proposal. As next step, we plan to evaluate the proposed method using other public image databases as the DISEC image dataset [20]. Also, we intend to apply our proposal exclusively on text images, printed and manuscripts, written in different languages, and keep developing this proposal to segment and classify letters encoded in characters.

Acknowledgments. Authors thank to Coordination for the Improvement of Higher Education Personnel (CAPES Finance Code #001) and Instituto Federal Goiano, câmpus Urutaí (Process Number: 23219.000404.2022-67), for their financial support.

References

1. Al-Khatatneh, A., Pitchay, S.A., Al-Qudah, M.: A review of skew detection techniques for document. In: 17th UKSim-AMSS International Conference on Modelling and Simulation, pp. 316–321. IEEE, Cambridge (2015)
2. Antonacopoulos, A., Clausner, C., Papadopoulos, C., Pletschacher, S.: Historical document layout analysis competition. In: 2011 International Conference on Document Analysis and Recognition, pp. 1516–1520 (2011). https://doi.org/10.1109/ICDAR.2011.301

3. Avila, B., Lins, R.: A fast orientation and skew detection algorithm for monochromatic document images. In: ACM Symposium on Document Engineering, pp. 118–126 (2005)
4. Bafjaish, S.S., Sanusi, M., Nasser, M., Ramzani, A., Mahdin, H.: Skew detection and correction of Mushaf Al-Quran script using Hough transform. Int. J. Adv. Comput. Sci. Appl. **9**(8), 402–409 (2018)
5. Bezmaternykh, P., Nikolaev, D.P.: A document skew detection method using fast Hough transform. In: Twelfth International Conference on Machine Vision (ICMV 2019), vol. 11433, p. 114330J. International Society for Optics and Photonics (2020)
6. Boiangiu, C.A., Dinu, O.A., Popescu, C., Constantin, N., Petrescu, C.: Voting-based document image skew detection. Appl. Sci. **10**(7), 2236 (2020)
7. Boudraa, O., Hidouci, W.K., Michelucci, D.: Using skeleton and Hough transform variant to correct skew in historical documents. Math. Comput. Simulat. **167**, 389–403 (2020)
8. Cai, C., Meng, H., Qiao, R.: Adaptive cropping and deskewing of scanned documents based on high accuracy estimation of skew angle and cropping value. Visual Comput. **37**(7), 1917–1930 (2020). https://doi.org/10.1007/s00371-020-01952-z
9. Clausner, C., Antonacopoulos, A.: Efficient and effective OCR engine training. Int. J. Doc. Anal. Recogn. **23**(1), 73–88 (2020)
10. Delibasis, K.: Efficient implementation of Gaussian and Laplacian Kernels for feature extraction from IP fisheye cameras. J. Imaging **4**(6), 1–21 (2018)
11. Dengel, A., Ahmad, R.: A novel skew detection and correction approach for scanned documents. In: International IAPR Workshop on Document Analysis Systems (2016)
12. Epshtein, B.: Determining document skew using inter-line spaces. In: 2011 International Conference on Document Analysis and Recognition, pp. 27–31 (2011). https://doi.org/10.1109/ICDAR.2011.15
13. Huang, K., Chen, Z., Yu, M., Yan, X., Yin, A.: An efficient document skew detection method using probability model and Q test. Electronics **9**(1), 55 (2020)
14. Kar, R., Saha, S., Bera, S.K., Kavallieratou, E., Bhateja, V., Sarkar, R.: Novel approaches towards slope and slant correction for tri-script handwritten word images. Imag. Sci. J. **67**(3), 159–170 (2019)
15. Khidhir, D.A.M.: Use of Radon transform in orientation estimation of printed text. In: 5th International Conference on Information Technology, pp. 1–5 (2011)
16. Khuman, Y.L.K., Devi, H.M., Singh, N.A.: Entropy-based skew detection and correction for printed meitei/meetei script ocr system. Mater. Today Proc. **37**, 2666–2669 (2021)
17. Lewis, D., Agam, G., Argamon, S., Frieder, O., Grossman, D., Heard, J.: Building a test collection for complex document information processing. In: Proceedings of the 29th Annual International ACM SIGIR Conference on Research and Development in Information Retrieval, pp. 665–666 (2006)
18. Liu, Y., Zheng, C., Zheng, Q., Yuan, H.: Removing Monte Carlo noise using a Sobel operator and a guided image filter. Visual Comput. **34**(4), 589–601 (2018)
19. Obaidullah, S.M., Halder, C., Santosh, K., Das, N., Roy, K.: Phdindic_11: page-level handwritten document image dataset of 11 official indic scripts for script identification. Multim. Tools Appl. **77**(2), 1643–1678 (2018)
20. Papandreou, A., Gatos, B., Louloudis, G., Stamatopoulos, N.: ICDAR 2013 document image skew estimation contest (DISEC 2013). In: 2013 12th International Conference on Document Analysis and Recognition, pp. 1444–1448. IEEE (2013)

21. Pramanik, R., Bag, S.: A novel skew correction methodology for handwritten words in multilingual multi-oriented documents. Multim. Tools Appl. **80**(18), 27323–27342 (2021). https://doi.org/10.1007/s11042-021-10822-2
22. Ptak, R., Żygadło, B., Unold, O.: Projection-based text line segmentation with a variable threshold. Int. J. Appl. Math. Comput. Sci. **27**(1), 195–206 (2017)
23. Ramegowda, D.: A novel method for document skew detection and correction: application to handwritten document and bank documents. Int. J. Appl. Eng. Res. **10** (2015)
24. Salagar, Rajashekhar, Patil, Pushpa: Analysis of PCA usage to detect and correct skew in document images. In: Joshi, Amit, Mahmud, Mufti, Ragel, Roshan G.., Thakur, Nileshsingh V.. (eds.) Information and Communication Technology for Competitive Strategies (ICTCS 2020). LNNS, vol. 191, pp. 687–695. Springer, Singapore (2022). https://doi.org/10.1007/978-981-16-0739-4_65
25. Sauvola, J., Kauniskangas, H.: Mediateam Document Database II, A CD-rom Collection of Document Images. University of Oulu, Finland (1999)
26. Smith, A.R.: Color gamut transform pairs. ACM Siggraph Comput. Graph. **12**(3), 12–19 (1978)
27. Sobel, I., Feldman, G.: A 3x3 Isotropic Gradient Operator for Image Processing. Stanford Artificial Intelligence Project (SAIL) (1968)
28. Stamatopoulos, N., Gatos, B., Louloudis, G., Pal, U., Alaei, A.: ICDAR 2013 handwriting segmentation contest. In: 2013 12th International Conference on Document Analysis and Recognition, pp. 1402–1406. IEEE (2013)
29. Sun, C., Si, D.: Skew and slant correction for document images using gradient direction. In: Proceedings of the Fourth International Conference on Document Analysis and Recognition, vol. 1, pp. 142–146 (1997)
30. Tzogka, C., et al.: OCR workflow: facing printed texts of ancient, medieval and modern greek literature. In: Paschke, A., Rehm, G., Qundus, J.A., Neudecker, C., Pintscher, L. (eds.) Proceedings of the CEUR Workshop, Conference on Digital Curation Technologies (Qurator 2021), Berlin, 8th–12th February 2021, vol. 2836. CEUR-WS.org (2021)
31. Wang, D., Wang, X., Liu, J.: A skew angle detection algorithm based on maximum gradient difference. In: International Conference on Transportation, Mechanical, and Electrical Engineering, pp. 1747–1750. IEEE, ChangChun (2011)
32. Zhang, D., Liu, Y., Wang, Z., Wang, D.: OCR with the deep CNN model for ligature script-based languages like Manchu. Sci. Program. **2021**, 1–9 (2021)

Weakly Supervised Attended Object Detection Using Gaze Data as Annotations

Michele Mazzamuto[1(✉)], Francesco Ragusa[1,2], Antonino Furnari[1,2], Giovanni Signorello[3], and Giovanni Maria Farinella[1,2,3]

[1] FPV@IPLAB, DMI - University of Catania, Catania, Italy
michele.mazzamuto@phd.unict.it
[2] Next Vision s.r.l. - Spinoff of the University of Catania, Catania, Italy
[3] CUTGANA - University of Catania, Catania, Italy

Abstract. We consider the problem of detecting and recognizing the objects observed by visitors (i.e., attended objects) in cultural sites from egocentric vision. A standard approach to the problem involves detecting all objects and selecting the one which best overlaps with the gaze of the visitor, measured through a gaze tracker. Since labeling large amounts of data to train a standard object detector is expensive in terms of costs and time, we propose a weakly supervised version of the task which leans only on gaze data and a frame-level label indicating the class of the attended object. To study the problem, we present a new dataset composed of egocentric videos and gaze coordinates of subjects visiting a museum. We hence compare three different baselines for weakly supervised attended object detection on the collected data. Results show that the considered approaches achieve satisfactory performance in a weakly supervised manner, which allows for significant time savings with respect to a fully supervised detector based on Faster R-CNN. To encourage research on the topic, we publicly release the code and the dataset at the following url: https://iplab.dmi.unict.it/WS_OBJ_DET/.

Keywords: Egocentric vision · Weakly supervised object detection

1 Introduction

Egocentric vision offers a convenient setting to collect visual information from the user's point of view which can be leveraged to understand human behavior [8] and intent [7]. The collected information can be in turn used to enable user-centered applications capable of assisting the camera wearer, e.g., providing information about the objects present in the scene or the potential activities which can be initiated. Wearable devices, such as smart glasses, allow to acquire

Supplementary Information The online version contains supplementary material available at https://doi.org/10.1007/978-3-031-06430-2_22.

Fig. 1. Training and inference stages of both a supervised (left) and a weakly supervised (right) approach. In the supervised approach based on object detection and gaze signal, bounding box annotated around all objects, object classes associated to each box (numbers next to the boxes in the image) and information about the attended object (gaze, represented as a green dot) are both needed. In the weakly supervised setting, only gaze and per-frame labels (numbers next to the green dots) are needed. (Color figure online)

different signals including RGB, depth and gaze, which is the pupil's fixation point of the user in the image. Understanding where the user is looking allows to gather insights into their behavior and intention. For example, in the industrial domain, an intelligent system can analyze the attention of the worker to provide information on how to use a specific tool (e.g., for continuous training) or to anticipate which objects they will interact with. In the context of healthcare, gaze data can be used to rehabilitate patients with cognitive or attention deficits. In the cultural heritage domain, wearable devices can be used to improve the fruition of artworks and the visitors experience [26]. For instance, knowing that the visitor of a museum is looking at a specific artwork gives information on the ongoing activity (observing that artwork), as well as the user's intention (e.g., the user may want to receive information on the artwork or a specific detail of it). Moreover, the collected information about the attended objects can be useful to better manage cultural sites.

A common approach to detect attended objects in the presence of gaze data consists in running a regular object detector and select the object which best overlaps with gaze data. However, this supervised approach requires a significant quantity of images labeled with bounding boxes around all objects in the scene in order to train the object detector. The involved annotation process is expensive in terms of costs and time. To streamline data collection and annotation, in this paper we propose to predict attended objects in a weakly supervised way, by relying only on gaze data and a general frame-level label about the currently attended object, which is much cheaper to obtain. Figure 1 summarizes the two approaches.

To study the problem, we propose a dataset of images collected by visitors in a cultural site, paired with gaze data and frame-level labels about the attended objects. Frames are also labeled with object bounding boxes for comparison with respect to supervised approaches. Figure 2(a) reports some images sampled from the proposed dataset. Apart from the dataset, we investigate and compare three methods which can be trained in a weakly supervised manner using only gaze data and frame-level labels. Experiments show that the fully-convolutional baseline obtains competitive results at a much lower labeling cost.

(a) Sample frames   (b) Example annotations

Fig. 2. (a) Sample frames of the proposed dataset and (b) Examples of bounding box annotations of the considered objects of interest. The green circle represents the gaze of the visitor. (Color figure online)

The proposed dataset and the code are publicly available at the following url: https://iplab.dmi.unict.it/WS_OBJ_DET/. To summarize, the main contributions of this paper are: 1) a new publicly available dataset to study weakly supervised attended object detection and 2) a weakly supervised benchmark for attended object detection.

The remainder of the paper is organized as follows. In Sect. 2, we discuss related work. Section 3 reports the proposed dataset. Section 4 presents the considered approaches, whereas Sect. 5 discusses the obtained results. Section 6 concludes the paper and summarises the main accomplishments of our study.

2   Related Works

Our work is related to different lines of research: egocentric vision in cultural sites, weakly supervised object detection and segmentation and analysis of gaze data. The following sections discuss some relevant works related to these research lines.

2.1   Datasets in Cultural Sites

There are some public datasets in the cultural heritage context. The Open-MIC [18] dataset contains photos captured in ten different exhibition spaces of several museums. The authors of [18] used the OpenMIC dataset to explore the problem of artwork identification. NoisyArt [4] is a dataset composed of artwork images collected from Google Images and Flickr containing also metadata (e.g., artwork title, comments, description and creation location) gathered from DBpedia. A recent work presented the AQUA [9] (Art QUestion Answering) dataset which contains question-answer pairs automatically generated using state-of-the-art question generation methods on the basis of paintings and comments provided by the SemArt [9] dataset. EGO-CH [27] is a dataset of egocentric videos for visitors' behaviour understanding in cultural sites. The dataset has been proposed to study four tasks related to visitors' behaviour understanding, including point of interest/object recognition task, room-based localization, object retrieval and survey prediction.

Unlike the aforementioned works, we propose a dataset composed of egocentric videos and gaze signal, which is suitable to tackle the tasks of both fully and weakly supervised attended object detection in cultural sites.

2.2   Object Detection and Its Use in Cultural Sites

Object detection is the task of detecting and recognizing all the objects present in an image. State-of-the-art methods can be grouped into two main groups: two-stage methods and one-stage methods. Two-stage detectors [10,11,29] first predict candidate bounding boxes and then extract visual features from each box to perform the classification of the object class and regress accurate bounding box coordinates. One-stage detectors [6,20,22] predict bounding boxes and classify them in a single step without the need of producing candidate bounding boxes. These detectors usually prioritize the speed of the detection at the expense of prediction accuracy.

Cultural sites generally include both three-dimensional (e.g., statues or artifacts) and two-dimensional (e.g., paintings or book pages) objects, which are often exposed in particular light conditions and protected by glass cases. The authors of [30] designed a smart audio guide that adapts to the actions and interests of museum visitors using a YOLO object detector [28] which localizes and recognizes all objects in the images collected through a smartphone. The authors of [26] investigated the problem of detecting and recognizing points of interest in cultural sites relying on a 2D CNN and a YOLO object detector. The authors of [25] propose a pipeline to detect damages in images of heritage sites, and subsequently localize the detected damage in the corresponding 3D models. The authors of [14] present a new dataset called MMSD for medieval musicological studies, which has been manually annotated by a group of musicology experts. The paper also presents a method for black box few-shot object detection.

The discussed works are mainly devoted to find all objects in the image following a supervised approach, whereas in this work, we want to detect the object the user is looking at (i.e., the attended object) in a weakly-supervised fashion. All the approaches based on object detection need data labeled with bounding boxes which is domain-specific (i.e., specific to the given cultural site) and hence hard to obtain. To reduce the annotation burden, we explore three different baselines for attended object detection which do not rely on bounding boxes in the training stage.

2.3   Weakly Supervised Object Detection and Segmentation

Weakly supervised learning (WSL) approaches are exploited when incomplete or imperfect information is available at training time. These approaches need to operate weakly-labeled training data which allow to reduce the time and cost needed for the data annotation process. To reduce the annotation time for semantic segmentation and object detection, recent works have focused on training models in a weakly- or semi-supervised setting, such as points [1,3,33] and free-form squiggles [1]. In particular, the work of [3] combines small portions of box annotations and points uniformly sampled for each bounding box.

Recent successes in semantic segmentation have been driven by methods that train CNNs originally designed for image classification to assign semantic labels to each pixel of an image [2,5,12,31]. Inspired by [31,36], we investigate three weakly supervised semantic segmentation approaches to tackle the attended object detection task. Importantly, differently from past works [1,3,24,34], we explored the attended object detection task without using bounding box annotations [24,33]. Our experiments show that it is possible to tackle the addressed task using only gaze data as supervision, thus reducing annotation time and costs.

2.4 Analysis of Gaze Data

In recent works, gaze data has been increasingly used to tackle various computer vision tasks. Gaze-tracking data has been used to perform weakly supervised training of object detectors [16,35], to infer the semantics of the scene [32], to detect salient objects in images [19] and videos [17] or to perform semantic segmentation of images [23]. Typically, training an object detector requires a large set of images annotated with bounding boxes, which is an expensive and time consuming step. The work of [24] can reduce the annotation time by defining a model that takes human eye movement data as input and infers the spatial support of the target object by labeling each pixel as either object or background.

In this work, we propose to use gaze data and a frame-based label as the only supervision. The gaze signal, which represents the user's attention, indicates the rough position of the object attended by the visitor, which we leverage for training.

3 Dataset

We asked 7 subjects (aged between 24 and 40) to capture egocentric videos while visiting a cultural site containing 15 objects of interest. Videos have been acquired using a head-mounted Microsoft HoloLens2 device[1] in the room V of the Palazzo Bellomo located in Siracusa. In order to acquire the videos and the gaze coordinates, we developed a HoloLens2 application based on Unity. The developed application allows two acquisition modalities: guided tour and free tour. The guided tour modality guides the user during the visit, generating a path to follow and providing additional audio information for each of the artworks of the site. Instead, the free tour modality leaves the visitor free to explore the environment with no particular structure or restriction about the duration of the visit[2]. The egocentric videos have been captured by the subjects who followed the two acquisition modalities. In this way, we acquired 14 videos

---

[1] https://www.microsoft.com/it-it/hololens.
[2] See supplementary material for more details.

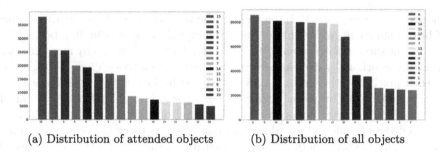

(a) Distribution of attended objects   (b) Distribution of all objects

Fig. 3. Dataset statistics.

with a resolution of 2272×1278 pixels at 30 fps. The average duration of the videos is 11.25 min for the guided tour modality, and 5.66 min for the free tour modality. We considered 15 objects of interest(See footnote 2). Note that 8 of the considered objects of interest represent details of the artwork "Annunciazione". Figure 2(a) shows some sample frames of the proposed dataset. Each frame of the dataset is associated to 2D coordinates (x,y) relative to the gaze of the visitor, as well as to a class related to the observed (attended) object or a negative class "other" if the user was not looking at any of the considered objects. The dataset contains 178,977 frames, which we manually labeled with bounding box annotations related to 15 objects of interest for evaluation purposes. Figure 2(b) shows some examples of annotations. The distribution of the considered attended objects is shown in Fig. 3(a), whereas the distribution of all objects appearing in the dataset is shown in Fig. 3(b).

4   Methods

Differently from the standard object detection task which aims to detect and recognize all the objects present in the scene, attended object detection consists in detecting and recognizing only the object observed by the camera wearer. Formally, let $O = \{o_1, o_2, ..., o_n\}$ be the set of objects in the image and let $C = \{c_1, c_2, ..., c_n\}$ be the corresponding set of object classes. The proposed task consists in detecting the attended object o_{att}, predicting its bounding box coordinates $(x_{att}, y_{att}, w_{att}, h_{att})$ and assigning them the correct class label c_{att}. We have investigated three different approaches to predict the attended objects in a weakly supervised manner, by relying only on gaze data and a frame level label of the attended object. The approaches are described in the following sections.

4.1   Sliding Window Approach

We first investigate a simple approach consisting in exploiting a ResNet18 CNN to obtain a semantic segmentation mask by classifying each image patch to infer whether it belongs to one of the objects of interest or to none of them ("other"). We first train the CNN to classify image patches of size 300×300

Fig. 4. Sliding window approach example.

pixels sampled around gaze coordinates. Each image patch is classified using the frame-level annotations associated to the fixation points (hence the attended object class or a negative "other" class if no object is observed). Note that, in order to optimize this algorithm, only fixation points and temporal annotations indicating the attended objects are needed (e.g. "the user has looked at object c from frame n to frame $n + m$"). At test time we classify all image patches (with a size of 300×300 pixels) within the image using a sliding window. The result is a segmentation mask where each element is an integer between 0 and 15 (the ID of the considered classes). Figure 4(b) shows an example of classified windows (colors indicate predicted labels) for the input image shown in Fig. 4(a), whereas Fig. 4(c) reports the related segmentation mask. While image patches are analyzed at a pre-defined scale, objects may be present at multiple scales, depending on the distance between the observer and the objects. To correctly draw a box around the attended object, we consider a patch of 100×100 pixels centered at the 2D gaze coordinates and fit a box to the connected component of the median class of the patch.

4.2 Fully Convolutional Attended Object Detection

The method described in the previous section has the main drawback of being slow. Indeed, processing an image at full resolution (e.g., 2272×1278 pixels) takes up to 168 s on a Tesla-K80 GPU. To speed up the approach, inspired by [31,36], we modify the trained ResNet by removing the Global Average Pooling operation and replacing it with a fully connected classifier with a 1×1 convolutional layer. This allows the network to predict a semantic segmentation mask of the whole image in a single step. Note that the introduced convolutional layers are initialized with the same weights of the fully connected layer in order to obtain consistent predictions. Given an input frame, the model outputs the class probability distributions for each pixel, which we then use to obtain a pixel-wise classification map. We up-sample the predicted mask to the original resolution to obtain a coarse segmentation mask of the whole image. We hence obtain the predicted bounding box of the attended object from the coarse segmentation map from gaze coordinates using the box fitting approach discussed in the previous section. Figure 5 presents an overview of the discussed approach.

Fig. 5. The proposed fully convolutional approach.

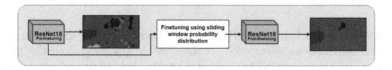

Fig. 6. The proposed finetuning strategy. Note that finetuning tends to clean the masks from artifacts related to classes that are not present in the image.

4.3   Finetuned Fully Convolutional Attended Object Detection

While the fully convolutional approach is much faster than the one based on the sliding window, we found the latter to be more accurate in our experiments. This is probably due to the fact that, during training, the CNN has observed only "decontextualized" image patches, rather than the full input image. We hence investigate whether fine-tuning the fully convolutional model with the output of the sliding window method can boost performance. To this aim, we use a sample of the coarse segmentation masks extracted from the training set using the sliding window approach (Sect. 4.1) to finetune the fully convolutional model (Fig. 6). For this finetuning, we used the Kullback-Leibler Divergence score (KL) [15] to encourage the fully convolutional model to predict at each pixel location the same probability distribution predicted by the sliding window approach.

4.4   Fully-supervised Attended Object Detector

We compare the proposed weakly supervised approaches with a fully-supervised baseline based on a Faster-RCNN [29] object detector. The detector gives as output the bounding boxes (2D coordinates) of all objects of interest present in the image. Using the 2D coordinates of the gaze, we select the bounding box which includes the gaze coordinates.

5   Experimental Settings and Results

We split the dataset into 185,457 training frames and 46,132 test frames, which is used to perform all experiments. The ResNet [13] CNN has been pre-trained on ImageNet and then finetuned on training image patches sampled around gaze locations till convergence using a batch size of 8 and a learning rate of $1e^{-3}$. This model was then used to evaluate both the sliding windows approach (Sect. 4.1) and the fully convolutional approach (Sect. 4.2). To finetune the fully

Table 1. Results obtained by the compared method. Per-column best results are reported in bold, whereas second-best results are underlined. Inference times are measured using a NVIDIA K80 GPU.

Model	Inference time (seconds)	mAP	mAP50
Sliding window approach	168	0.19	0.43
Fully convolutional approach	0.31	0.18	0.34
Fully convolutional approach + fine-tuning	0.31	**0.19**	0.41
Faster-RCNN (baseline) [29]	0.80	0.42	0.60

convolutional approach, we selected 1659 frames from the training set where the attended object was different than the "other" class. The finetuning using the KL divergence has been performed till convergence with a batch size of 1 and a learning rate of $1e^{-3}$.

The performances obtained with the four compared methods are reported in Table 1. We evaluated the methods using the mAP and mAP50 measures [21], which are commonly used for the evaluation of object detectors. We also computed inference times using a NVIDIA K80 GPU. The sliding window approach allows to achieve a decent performance with an mAP of 0.19 and an mAP50 of 0.43, relying only on gaze and attended object frame-wise annotations as supervision. Despite these results, the approach cannot be exploited in practice due to the high inference time of 168 s per image. The proposed fully convolutional approach allows to greatly reduce inference time to 0.31 s per image, however it achieves a weaker performance with an mAP 0.18 and an mAP50 of 0.34. Finetuning the fully convolutional model with the proposed optimization procedure allows to achieve a performance similar to one obtained with the sliding window approach, with an mAP of 0.19 and an mAP50 of 0.41, while retaining the reduced inference time of 0.31 s per image. The fully supervised object detector obtains better performance than the weakly supervised method (0.42 mAP vs. 0.19 mAP of the finetuned fully convolutional approach), but it requires labeled bounding boxes at training time, which would greatly affect the deployment cost of the final system. It is worth noting that, besides requiring less supervision, the finetuned fully convolutional approach also brings significant time savings, with a runtime reduced of about 62% with respect to the fully superivised model (0.31 s vs. 0.8 s). Qualitative results are shown in Fig. 7. The examples show as the masks obtained by the sliding window approach (Fig. 7 row (2)) are the most homogeneous and compact, while those of the fully convolutional approach (Fig. 7 row (3)) are more noisy. After finetuning (Fig. 7 row (4)), it is possible to observe an improvement due to the removal of spurious pixels.

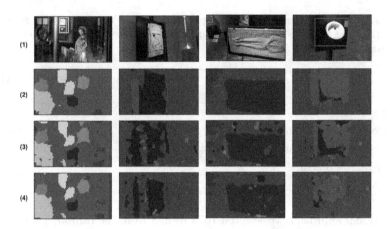

Fig. 7. Qualitative examples of the compared approaches. (1) Input image with the object detections obtained by the fully convolutional + finetuning approach. (2) Output mask of the sliding window approach. (3) Mask produced by the fully convolutional approach without finetuning. (4) Mask produced by the fully convolutional method after finetuning.

6   Conclusion

We investigated the problem of weakly supervised attended object detection in cultural sites starting from the consideration that manually labeling images with bounding boxes around all objects is a time-consuming process. To study the problem, we acquired a new dataset of egocentric videos in a cultural site including 15 objects of interest and a negative class ("other"). The dataset is composed of egocentric videos and gaze signals, and includes 178,977 frames manually labelled with bounding box annotations for evaluation purposes. We investigated three different approaches to detect attended objects in a weakly supervised way, by leaning only on gaze data and a frame-level label indicating the class of the attended object. Results show that our fully convolutional approach, after a fine-tuning stage, obtains good performance and has a lower inference time (0.31 s) with respect to a fully supervised detector. The proposed dataset can be also used to address the task of fully supervised object detection in cultural sites. Future works will focus on improving our approach to obtain comparable performance in terms of accuracy with respect to a fully-supervised method.

Acknowledgements. This research has been supported by Next Vision (https:// www.nextvisionlab.it/) s.r.l., by the project VALUE (N. 08CT6209090207 - CUP G69J18001060007) - PO FESR 2014/2020 - Azione 1.1.5., and by Research Program Pia.ce.ri. 2020/2022 Linea 2 - University of Catania.

References

1. Bearman, A.L., Russakovsky, O., Ferrari, V., Fei-Fei, L.: What's the point: semantic segmentation with point supervision. In: European Conference on Computer Vision, pp. 549–565 (2016)
2. Chen, L.C., Papandreou, G., Kokkinos, I., Murphy, K., Yuille, A.L.: Deeplab: semantic image segmentation with deep convolutional nets, atrous convolution, and fully connected CRFS. Trans. Pattern Anal. Mach. Intell. **40**(4), 834–848 (2018)
3. Cheng, B., Parkhi, O., Kirillov, A.: Pointly-supervised instance segmentation. arXiv preprint arXiv:2104.06404 (2021)
4. Chiaro, R.D., Bagdanov, A.D., Bimbo, A.: Noisyart: a dataset for webly-supervised artwork recognition. In: International Conference on Computer Vision Theory and Applications (2019)
5. Farabet, C., Couprie, C., Najman, L., LeCun, Y.: Learning hierarchical features for scene labeling. Trans. Pattern Anal. Mach. Intell. **35**(8), 1915–1929 (2013)
6. Farhadi, A., Redmon, J.: Yolov3: an incremental improvement. In: Computer Vision and Pattern Recognition, vol. 1804 (2018)
7. Furnari, A., Farinella, G.: Rolling-unrolling LSTMs for action anticipation from first-person video. Trans. Pattern Anal. Mach. Intell. **43**(11), 4021–4036 (2021)
8. Furnari, A., Farinella, G.M., Battiato, S.: Temporal segmentation of egocentric videos to highlight personal locations of interest. In: Hua, G., Jégou, H. (eds.) ECCV 2016. LNCS, vol. 9913, pp. 474–489. Springer, Cham (2016). https://doi.org/10.1007/978-3-319-46604-0_34
9. Garcia, N., et al.: A dataset and baselines for visual question answering on art. In: European Conference on Computer Vision Workshops, pp. 92–108 (2020)
10. Girshick, R.B.: Fast R-CNN. In: International Conference on Computer Vision, pp. 1440–1448 (2015)
11. Girshick, R.B., Donahue, J., Darrell, T., Malik, J.: Rich feature hierarchies for accurate object detection and semantic segmentation. Computer Vision and Pattern Recognition, pp. 580–587 (2014)
12. Hariharan, B., Arbeláez, P., Girshick, R., Malik, J.: Simultaneous detection and segmentation. In: European Conference on Computer Vision, pp. 297–312 (2014)
13. He, K., Zhang, X., Ren, S., Sun, J.: Deep residual learning for image recognition. In: Computer Vision and Pattern Recognition, pp. 770–778 (2016)
14. Ibrahim, B.I.E., Eyharabide, V., Le Page, V., Billiet, F.: Few-shot object detection: application to medieval musicological studies. J. Imaging **8**(2), 18 (2022)
15. Joyce, J.M.: Kullback-Leibler Divergence, pp. 720–722. Springer, Heidelberg (2011). https://doi.org/10.1007/978-3-642-04898-2_327
16. Karthikeyan, S., Jagadeesh, V., Shenoy, R., Ecksteinz, M., Manjunath, B.: From where and how to what we see. In: International Conference on Computer Vision, pp. 625–632 (2013)
17. Karthikeyan, S., Ngo, T., Eckstein, M., Manjunath, B.: Eye tracking assisted extraction of attentionally important objects from videos. In: Computer Vision and Pattern Recognition, pp. 3241–3250 (2015)
18. Koniusz, P., Tas, Y., Zhang, H., Harandi, M., Porikli, F., Zhang, R.: Museum exhibit identification challenge for the supervised domain adaptation and beyond. In: European Conference on Computer Vision (2018)
19. Li, Y., Hou, X., Koch, C., Rehg, J.M., Yuille, A.L.: The secrets of salient object segmentation. In: Computer Vision and Pattern Recognition), pp. 280–287 (2014)

20. Lin, T.Y., Goyal, P., Girshick, R.B., He, K., Dollár, P.: Focal loss for dense object detection. In: Transactions on Pattern Analysis and Machine Intelligence (2020)

21. Lin, T.Y., et al.: Microsoft COCO: common objects in context. In: European Conference on Computer Vision (2014)

22. Liu, W., et al.: Single shot multibox detector. In: European Conference on Computer Vision (2016)

23. Mishra, A., Aloimonos, Y., Fah, C.L.: Active segmentation with fixation. In: International Conference on Computer Vision, pp. 468–475 (2009)

24. Papadopoulos, D.P., Clarke, A.D.F., Keller, F., Ferrari, V.: Training object class detectors from eye tracking data. In: European Conference on Computer Vision (2014)

25. Pathak, R., Saini, A., Wadhwa, A., Sharma, H., Sangwan, D.: An object detection approach for detecting damages in heritage sites using 3-d point clouds and 2-d visual data. J. Cult. Herit. **48**, 74–82 (2021)

26. Ragusa, F., Furnari, A., Battiato, S., Signorello, G., Farinella, G.M.: Egocentric point of interest recognition in cultural sites. In: International Conference on Computer Vision Theory and Applications (VISAPP) (2019)

27. Ragusa, F., Furnari, A., Battiato, S., Signorello, G., Farinella, G.M.: EGO-CH: dataset and fundamental tasks for visitors behavioral understanding using egocentric vision. Pattern Recogn. Lett. **131**, 150–157 (2020)

28. Redmon, J., Divvala, S.K., Girshick, R.B., Farhadi, A.: You only look once: unified, real-time object detection. In: Computer Vision and Pattern Recognition, pp. 779–788 (2016)

29. Ren, S., He, K., Girshick, R., Sun, J.: Faster R-CNN: towards real-time object detection with region proposal networks. In: Advances in Neural Information Processing Systems (2015)

30. Seidenari, L., Baecchi, C., Uricchio, T., Ferracani, A., Bertini, M., Bimbo, A.D.: Deep artwork detection and retrieval for automatic context-aware audio guides. Trans. Multim. Comput. Commun. Appl. **13**, 1–21 (2017)

31. Shelhamer, E., Long, J., Darrell, T.: Fully convolutional networks for semantic segmentation. Trans. Pattern Anal. Mach. Intell. **39**, 640–651 (2017)

32. Subramanian, R., Yanulevskaya, V., Sebe, N.: Can computers learn from humans to see better? inferring scene semantics from viewers' eye movements. In: International Conference on Multimedia (ACM), pp. 33–42 (2011)

33. Wang, Y., Hou, J., Hou, X., Chau, L.P.: A self-training approach for point-supervised object detection and counting in crowds. Trans. Image Process. **30**, 2876–2887 (2021)

34. Yoo, I., Yoo, D., Paeng, K.: Pseudoedgenet: nuclei segmentation only with point annotations. In: Medical Image Computing and Computer Assisted Intervention (MICCAI), pp. 731–739 (2019)

35. Yun, K., Peng, Y., Samaras, D., Zelinsky, G.J., Berg, T.L.: Studying relationships between human gaze, description, and computer vision. In: Computer Vision and Pattern Recognition, pp. 739–746 (2013)

36. Zhou, B., Khosla, A., Lapedriza, À., Oliva, A., Torralba, A.: Learning deep features for discriminative localization. In: Computer Vision and Pattern Recognition, pp. 2921–2929 (2016)

Panoptic Segmentation in Industrial Environments Using Synthetic and Real Data

Camillo Quattrocchi[1], Daniele Di Mauro[1], Antonino Furnari[1,2(✉)], and Giovanni Maria Farinella[1,2]

[1] FPV@IPLAB, DMI - University of Catania, Catania, Italy
`furnari@dmi.unict.it`
[2] Next Vision s.r.l. - Spinoff of the University of Catania, Catania, Italy

Abstract. Being able to understand the relations between the user and the surrounding environment is instrumental to assist users in a worksite. For instance, understanding which objects a user is interacting with from images and video collected through a wearable device can be useful to inform the worker on the usage of specific objects in order to improve productivity and prevent accidents. Despite modern vision systems can rely on advanced algorithms for object detection, semantic and panoptic segmentation, these methods still require large quantities of domain-specific labeled data, which can be difficult to obtain in industrial scenarios. Motivated by this observation, we propose a pipeline which allows to generate synthetic images from 3D models of real environments and real objects. The generated images are automatically labeled and hence effortless to obtain. Exploiting the proposed pipeline, we generate a dataset comprising synthetic images automatically labeled for panoptic segmentation. This set is complemented by a small number of manually labeled real images for fine-tuning. Experiments show that the use of synthetic images allows to drastically reduce the number of real images needed to obtain reasonable panoptic segmentation performance.

Keywords: Panoptic segmentation · Instance segmentations · Object detection · Industrial domain · Synthetic dataset

1 Introduction

In recent years, the increasing development of wearable devices has sparked new interest on First Person Vision [2]. Among the different application contexts, First-Person Vision is a must-go approach in industrial environments, where it can be useful to improve productivity and increase safety. Possible applications include operator training, anticipation of the imminent use of a potentially dangerous object, and predictive maintenance via the estimation of machinery usage. Developing systems able to tackle the aforementioned tasks requires large amounts of data to develop, train, and evaluate computational models. Despite

S. Sclaroff et al. (Eds.): ICIAP 2022, LNCS 13232, pp. 275–286, 2022.
https://doi.org/10.1007/978-3-031-06430-2_23

datasets such as Ego4D[1], EPIC-KITCHENS [7], EGTEA-Gaze+ [18] and MEC-CANO [23] have been recently proposed, they only offer a partial answer to the need of data in industrial contexts. Indeed, these datasets only partially consider industrial scenarios [23] and even if they do, the environment-specific nature of industrial applications often requires models to be trained on domain-specific data acquired on purpose. Moreover, these datasets do not contain labels for the assessment of semantic segmentation models.

This lack of data limits the development of algorithms in industrial settings because building a dataset in such contexts is a challenging task. Indeed, designing a proper data acquisition and labeling protocol requires time and expertise. Furthermore, the data acquisition process can be long and requires the availability of a diverse set of subjects, environments and representative situations. Additionally, once the images of the dataset have been acquired, the required labeling stage (usually a manual procedure) can become a long and costly process. Data acquisition is even more challenging in the industrial domain where it is easy to run into privacy and confidentiality issues which may slow down or prevent the collection of appropriate data. These negative aspects related to the generation of real image datasets can be outrun using synthetic data [22]: once a 3D model of the target environment and objects has been acquired, the data generation and labeling can be automated. This allows to greatly increase the speed and reduce the costs of the data collection process. While the generated data can be directly used to train the model, one of the main concerns is to reach a degree of photorealism of the synthetic data to achieve good results when the trained model is tested on real world data. The amount of photorealism depends mainly on the quality of the 3D model, the software used for the generation and the computational capabilities of the adopted 3D acquisition hardware. If properly designed, synthetic data also allows to increase diversity, for instance including rare events which would be difficult to observe during the acquisition of a training set composed of real images.

The drop in performance of computer vision algorithms between synthetic and real data is known as *domain shift* [6]. When available, real data can be used to fill the gap between the real and synthetic domains through fine-tuning or more complex domain adaptation techniques [21]. Even in the presence of real data, synthetic data allows to greatly reduce the amount of real data needed to reach good results, which brings benefits in terms of the reduction of costs in the acquisition and labeling stages.

In this paper, we investigate the impact of synthetic data for the development of domain-specific applications in industrial environments. As a primary task, which can be useful in many downstream applications, we propose to study panoptic segmentation [14], which consists in identifying the main semantic elements in the scene, including both structural parts, such as walls, and object instances such as tools and equipment. Specifically, we study the suitability of training a panoptic segmentation approach using a large amount of labeled synthetic data and very small amount of labeled real data. Since both the label taxonomy and the data are inherently domain-specific in industrial settings, we

---

[1] https://ego4d-data.org/.

propose a pipeline to generate synthetic data compliant to a specific real industrial environment. We hence generate and publicly release a dataset[2] containing synthetic images generated from the 3D model of a real industrial space, as well as labeled real images collected in the same space.

In sum, the contributions of this work are twofold: (1) a pipeline for the generation of a synthetic dataset specifically designed for the panoptic segmentation task in an industrial site; (2) a labeled dataset generated using the proposed pipeline. The dataset consists of labeled real and synthetic data where each frame has four different annotations: Semantic Label, Panoptic Image, Instance Annotation, Panoptic Annotation.

To explore the usefulness of the proposed pipeline and synthetic data, we carry out a benchmark of Panoptic Segmentation [14]. The benchmark analyzes the performances of the baseline model proposed in [14] when the synthetic data is used for model training and variable amount of real data are used to fine-tune the model. The results show that the synthetic data can be useful to obtain a good Panoptic Segmentation model.

The remainder of this paper is structured as follows: in Sect. 2 we discuss related works; Sect. 3 presents the proposed dataset generation pipeline; Sect. 4 reports and discusses the results; Sect. 5 concludes the paper.

2   Related Work

The work in this paper is related to the computer vision tasks of Semantic Segmentation, Instance Segmentation, Panoptic Segmentation and to the use of synthetic datasets for learning purposes.

2.1   Virtual Dataset Generation

The use of synthetic data to train machine learning models has become increasingly popular. Ragusa et al. [22] collected both a synthetic and a real dataset related to a cultural site to study the performance of semantic segmentation models in the presence of varying amounts of synthetic and real data. Chang et al. [3] proposed to obtain synthetic images of an indoor environment from a Matterport scan to train a semantic segmentation model, highlighting the difficulty to collect large amounts of real data in order to train scene understanding algorithms. Orlando et al. [20] introduced a tool for the generation of synthetic tours in a virtual environment. The developed tool allows the navigation of an agent within a simulated environment.

Di Benedetto et al. [8] acquired a synthetic dataset using a Game Engine: the authors managed to obtain good results for object detection even with a small amount of real data to be used for fine-tuning. The works in [10–12,15,25] demonstrate the power of generating synthetic data automatically annotated for multiple tasks, such as optical flow estimation, semantic instance segmentation, object detection and tracking, object-level 3D scene layout estimation and visual odometry.

---

[2] The dataset is available at https://iplab.dmi.unict.it/ENIGMA_SEG/.

Similarly to these works, we propose to use synthetic data to tackle a supervised learning task. Differently from the previous approaches, we propose a pipeline that allows the generation of synthetic data aligned to real industrial spaces.

2.2   Semantic, Instance and Panoptic Segmentation

Semantic segmentation is the task of assigning a class to each pixel in an image [17]. Semantic segmentation aims to give more information than object detection: in addition to the position of objects, shape and geometry can also be derived. To solve this task, various approaches have been proposed in the last years such as encoder-decoder, dilated convolutional models, multiscale and pyramid networks [1,16,27].

One of the first works using deep learning to tackle semantic segmentation is known as SegNet [1], a deep network which belongs to the category of encoder-decoder models. DeepLab, presented in [5], became a seminal work with the introduction of "atrous convolution", a powerful tool in dense prediction tasks, and "atrous spatial pyramid pooling (ASPP)" to robustly segment objects at multiple scales. Chang et al. [27] proposed a semantic segmentation network that exploits global context information using a pyramid pooling layer.

The instance segmentation task consists in detecting and segmentating objects in images. The tasks includes the localization of specific objects and the inference of whether pixels belong to each object in order to discern an instance from another. Chen et al. in [4] proposed a cascade method for instance segmentation, taking full advantage of the relationship between detection and segmentation. Tian et al. [26] proposes a very powerful method capable of performing instance segmentation using only the bounding boxes at training time.

Kirillov et al. [14] introduced the panoptic segmentation task. The authors of [14] introduced the difference between "thing" and "stuff" classes and how semantic segmentation and instance segmentation, typically distinct tasks, can be unified. The paper also proposed a new metric, Panoptic Quality (PQ), to measure performance. Hwang et al. [13] proposed a variant of panoptic segmentation, called open-set panoptic segmentation which aims to recognize also the classe unknown at training time.

In this work, we consider Panoptic Segmentation as a reference task to test the value of the generated synthetic dataset. We choose this task because it enables a general understanding of the scene at a semantic level, including structural elements ("stuff") and object instances ("things"), which can be useful in industrial contexts.

3   Dataset

To study the considered problem, we have created a dataset comprised of two parts: real images with segmented masks manually annoted and synthetic images with automatically generated annotations. For each image, we have 4 annotations:

Dataset classes & number of instances

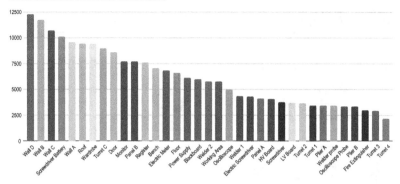

Fig. 1. The classes of the dataset with the associated number of instances. The color of the bar also indicates the color associated to the class in the semantic image. (Color figure online)

Semantic Label. Semantically annotated images, where each pixel is associated to a given semantic label (indicated by a specific color). We use a palette of 35 colors, each corresponding to a specific class (Fig. 1 reports the classes along with the number of instances and associated colors of the synthetic dataset);

Panoptic Image. Annotated images which separate "thing" classes from "stuff" classes, assigning 0 to pixels belonging to the "thing" class and a contiguous id (from 0 to the nth class) to pixels belonging to the "stuff" class. Unlabeled pixels are associated to the value 255. By "thing" we mean all those entities that can be counted (e.g. welder, piler, screwdriver, etc.), while the "stuff" class comprise those classes that represent amorphous regions of similar textures (e.g. walls, floor, etc.);

Instance Annotation. Textual annotation (a JSON in COCO format) that contains information about the segments within each image. This type of annotation also gives information about the bounding boxes and the class labels;

Panoptic Annotation. Textual annotation (a JSON in COCO format) that gives information on the area, class and bounding box of each instance. This annotation also contains an *id* for each instance, calculated as $id = R + G \times 256 + B \times 256^2$, using the color of the belonging class as an RGB triplet.

After generating the semantic annotations (as described in the following paragraphs) instance annotations and panoptic annotations were generated from the semantic annotations computing the coordinates of the bounding boxes. Finally, panoptic images were generated using Detectron2[3].

---

[3] https://github.com/facebookresearch/detectron2.

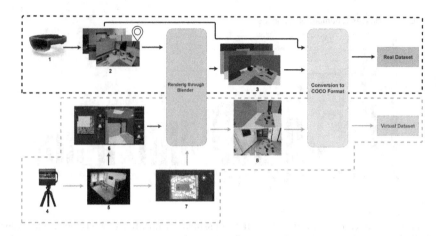

Fig. 2. Dataset generation pipeline. **Red box** generation of the real dataset: (1) acquisition of real images using HoloLens2; (2) extraction of frames and related camera poses; (3) annotation of the segmentation masks. **Blue box** generation of the synthetic dataset: (4) acquisition of the 3D model using a Matterport 3D scanner; (5) generation of the 3D model; (6) semantic labelling of the 3D model using Blender; (7) generation of a random tour inside the 3D model; (8) generation of synthetic frames and semantic labels. (Rendering through Blender) the 3D model and the positions are processed by a script for the generation of frames and semantic labels; (Conversion in COCO format) semantic labels are processed by a script for extracting JSON annotations in COCO format. (Color figure online)

3.1   Real Images Acquisition and Labeling

The real images contained in the dataset consist of 868 RGB images with a resolution of 1280×720 pixels. The images have been divided into a training set and a test set. Specifically, we used 668 images as a test and the remaining 200 images as a train. We choose an asymmetric split because we assume few real images are available for training, but we still want to have enough test images for evaluation. The acquisition of real images follows the pipeline shown in Fig. 2 (red box). The frames and corresponding camera poses, i.e. the position of the camera within the environment, were captured using a HoloLens2 device with a custom-made application which relies on HoloLens2 slam abilities to reconstruct accurate camera poses. All real images have been annotated for panoptic segmentation. To simplify the process, we use Blender in order to render semantic segmentation masks from the labeled 3D model of the scene (see next sections for details). The generated mask only contain static objects included in the 3D model. The movable objects present in the real images were annotated manually using the VGG Image Annotator (VIA) [9] enabling to easily add the masks of the mobile objects to the automatically generated segmentation masks (Fig. 3).

Fig. 3. A real image containing objects and corresponding segmentation mask.

Fig. 4. A synthetic image and the corresponding segmentation mask.

3.2   Synthetic Images Generation

The generation of the synthetic images follows the pipeline shown in Fig. 2 (blue box). This part of the dataset consists of 25,079 RGB images with a resolution of 1280 × 720 pixels. The images have been divided into a training set, a test set and a validation set following a 70:20:10 proportion (see an example in Fig. 4).

The first step was to acquire the 3D model of the environment using the Matterport 3D scanner[4]. Once the 3D model was obtained, a tool [20] developed with Unity3D[5] was used to simulate an agent that navigates within the 3D space varying the position and rotation of the camera in order to acquire images from several positions and points of view. The position and rotation of the camera are automatically obtained from the position of the the agent's head.

The 3D model obtained through the Matterport scan was fed to Blender[6] in order to obtain the semantic labels by manual labelling of the 3D model. In this stage, using the Blender brush tool, the 3D model was colored accordingly to the considered classes showed in Fig. 1 to obtain a fine-grained semantic label of the overall model. This labeling has been be used to automatically generate semantic images from given positions within the model aligned with the synthetic rgb images.

Movable Objects Scanning and Labeling. To model the presence of movable objects positioned on the workbench (e.g. power supply, oscilloscope, clamp, etc.), a 3D model of each object was first acquired using an Artec scanner[7].

---

[4] https://matterport.com/.
[5] http://unity3d.com/unity/.
[6] https://www.blender.org/.
[7] https://www.artec3d.com/.

Fig. 5. Two examples of the acquired movable items in the first two images and the corresponding monochrome textures in the last two images.

Table 1. Results of panoptic segmentation with real (R) and synthetic (S) data.

Panoptic quality [PQ]								
# of real training img.	0	25	50	75	100	125	150	175	200
Training with R img.		8.95	13.48	14.55	15.04	15.58	16.04	16.75	17.72
Training with R+S img.	**15.88**	**16.30**	**19.55**	**20.05**	**20.49**	**20.63**	**20.8**	**21.19**	**22.38**
Segmentation quality [SQ]								
# of real training img.	0	25	50	75	100	125	150	175	200
Training with R img.		25.57	40.02	44.05	40.44	41.33	41.97	42.58	42.93
Training with R+S img.	**49.15**	**43.36**	**44.95**	**46.03**	**47.10**	**48.75**	**48.91**	**49.17**	**49.47**
Recognition quality [RQ]								
# of real training img.	0	25	50	75	100	125	150	175	200
Training with R img.		12.03	18.39	20.15	20.45	21.65	22.13	23.01	24.36
Training with R+S img.	**22.58**	**23.36**	**27.46**	**28.14**	**28.6**	**28.67**	**29.16**	**29.55**	**31.69**
Average precision bounding box [AP (bbox)]								
# of real training img.	0	25	50	75	100	125	150	175	200
Training with R img.		11.99	23.10	26.33	27.46	27.68	27.79	27.81	28.82
Training with R+S img.	**22.16**	**28.86**	**34.42**	**38.59**	**39.86**	**39.91**	**41.81**	**42.57**	**43.56**
Average precision segmentation [AP (seg)]								
# of real training img.	0	25	50	75	100	125	150	175	200
Training with R img.		10.50	20.54	23.35	24.68	25.30	26.72	27.93	26.25
Training with R+S img.	**19.84**	**26.58**	**32.49**	**34.47**	**34.51**	**36.14**	**37.84**	**38.13**	**39.05**
Mean intersection over union [mIoU]								
# of real training img.	0	25	50	75	100	125	150	175	200
Training with R img.		40.25	60.88	69.47	74.47	75.11	76.38	77.10	77.93
Training with R+S img.	**55.43**	**66.63**	**71.67**	**78.90**	**80.84**	**82.95**	**83.06**	**83.29**	**84.39**

Photorealism was obtained during post processing: the real texture was preserved and a semantic monochrome texture was generated for the semantic label (Fig. 5).

The scanned objects were inserted into the 3D model on Blender, with the constraint of being over the workbench. Finally, annotated frames were extracted through two indipendent renderings in Blender: one with the original textures and one with the semantic textures (Fig. 4). During data generation, the position of the objects on the workbench has been randomly changed every 5 frames.

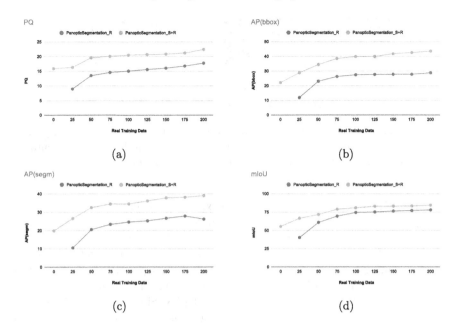

Fig. 6. Trends of four metrics for different quantities of real data.

4   Experimental Settings and Results

We perform experiments to evaluate a panoptic segmentation model trained on synthetic data and finetuned on real images considering a variable amount of real data, i.e. 200 images in our case. The experiments compare two different approaches that use synthetic and real data.

Real Data (R). In this experiment, we train the model using only real data. To assess the amount of real data needed to obtain good performances, the model was trained with different amounts of real data, and specifically 25, 50, 75, 100, 125, 150, 175, and 200 images.

Real and Synthetic Data (R+S). In this experiment, we uses both real and synthetic data to train the model. The training procedure consists of two steps: in the first one, we train the model using only synthetic data; in the second one, we fine-tune the model with real data. As in the aforementioned experiment, the amount of real data varies between 25 and 200 images. We also include the case in which no real data is used for fine-tuning, and hence the model is only trained on synthetic data.

The metrics used for model evaluation are the classic metrics used for the instance segmentation and semantic segmentation tasks, namely Intersection-Over-Union (IoU) [24] and Average Precision (AP) [19]. The results of the experiments in terms of these metrics are used to calculate the Panoptic Quality (PQ) [14] using two terms: Segmentation Quality (SQ) and Recognition Quality (RQ).

Fig. 7. An example of qualitative result obtained with the best performing model (Synthetic + 200 real images). On the right, the result of the Instance Segmentation task, whereas, on the left, the result of Semantic Segmentation.

In Table 1, we report the results of the experiments. The numbers show that the use of synthetic data is key to reduce the need of real images. For instance, in the case of the Panoptic Quality, using only synthetic images performs better than using only 125 real images (15.88% vs. 15.58%), and using synthetic images plus 50 real images is better than using only 200 real images (19.55% vs. 17.72%). This allows to obtain significant savings in the amount of human work needed to label images, which translates also to a faster deploy. Similar observations apply to all the measures except for mIoU, in which case the real images needed to overcome results obtained with 200 real images are only 75 (78.9% vs. 77.93%).

Figure 6(a)–(d) summarize the results graphically for four metrics. The curves show that PanopticSegmentation_S+R achieves much better results using the same amounts of real data than the baseline PanopticSegmentation_R. Figure 7 shows a qualitative result generated using the best performing model.

5   Conclusion

In this work, we have considered the problem of training a panoptic segmentation model in an industrial environment exploiting a large amount of synthetic data and a small set of real images. We presented a pipeline to generate semantically labeled synthetic data and proposed a dataset suitable to study the problem. We hence benchmarked a panoptic segmentation baseline on the proposed data in the presence of all synthetic images and varying amounts of real images. The results show that the use of synthetic data is key to bootstrap the training of the model, decreasing the cost of manual labeling, which speeds up the deploy.

Acknowledgements. This research is supported by Next Vision (https://www.nextvisionlab.it/) s.r.l., and the project MEGABIT - PIAno di inCEntivi per la RIcerca di Ateneo 2020/2022 (PIACERI) - linea di intervento 2, DMI - University of Catania.

References

1. Badrinarayanan, V., Kendall, A., Cipolla, R.: Segnet: a deep convolutional encoder-decoder architecture for image segmentation. IEEE Trans. Pattern Anal. Mach. Intell. **39**(12), 2481–2495 (2017)

2. Betancourt, A., Morerio, P., Regazzoni, C.S., Rauterberg, M.: The evolution of first person vision methods: a survey. IEEE Trans. Circuits Syst. Video Technol. **25**(5), 744–760 (2015)
3. Chang, A., et al.: Matterport3D: learning from RGB-D data in indoor environments. In: International Conference on 3D Vision (3DV), pp. 667–676 (2017)
4. Chen, K., et al.: Hybrid task cascade for instance segmentation. In: IEEE/CVF Conference on Computer Vision and Pattern Recognition, pp. 4974–4983 (2019)
5. Chen, L.C., Papandreou, G., Kokkinos, I., Murphy, K., Yuille, A.L.: Deeplab: semantic image segmentation with deep convolutional nets, atrous convolution, and fully connected CRFs. IEEE Trans. Pattern Anal. Mach. Intell. **40**(4), 834–848 (2017)
6. Csurka, G.: A comprehensive survey on domain adaptation for visual applications. In: Csurka, G. (ed.) Domain Adaptation in Computer Vision Applications. ACVPR, pp. 1–35. Springer, Cham (2017). https://doi.org/10.1007/978-3-319-58347-1_1
7. Damen, D., et al.: The EPIC-KITCHENS dataset: collection, challenges and baselines. IEEE Trans. Pattern Anal. Mach. Intell. **43**(11), 4125–4141 (2021)
8. Di Benedetto, M., Meloni, E., Amato, G., Falchi, F., Gennaro, C.: Learning safety equipment detection using virtual worlds. In: International Conference on Content-Based Multimedia Indexing (CBMI), pp. 1–6. IEEE (2019)
9. Dutta, A., Zisserman, A.: The VIA annotation software for images, audio and video. In: Proceedings of the 27th ACM International Conference on Multimedia, pp. 2276–2279 (2019)
10. Fabbri, M., et al.: Motsynth: how can synthetic data help pedestrian detection and tracking? In: Proceedings of the IEEE International Conference on Computer Vision, pp. 10849–10859 (2021)
11. Hu, Y.T., Chen, H.S., Hui, K., Huang, J.B., Schwing, A.G.: Sail-vos: semantic amodal instance level video object segmentation-a synthetic dataset and baselines. In: IEEE/CVF Conference on Computer Vision and Pattern Recognition, pp. 3105–3115 (2019)
12. Hu, Y.T., Wang, J., Yeh, R.A., Schwing, A.G.: Sail-vos 3d: a synthetic dataset and baselines for object detection and 3d mesh reconstruction from video data. In: IEEE/CVF Conference on Computer Vision and Pattern Recognition, pp. 1418–1428 (2021)
13. Hwang, J., Oh, S.W., Lee, J.Y., Han, B.: Exemplar-based open-set panoptic segmentation network. In: IEEE/CVF Conference on Computer Vision and Pattern Recognition, pp. 1175–1184 (2021)
14. Kirillov, A., He, K., Girshick, R., Rother, C., Dollar, P.: Panoptic segmentation. In: IEEE/CVF Conference on Computer Vision and Pattern Recognition, pp. 9396–9405 (2019)
15. Krähenbühl, P.: Free supervision from video games. In: IEEE/CVF Conference on Computer Vision and Pattern Recognition, pp. 2955–2964 (2018)
16. Lan, M., Zhang, Y., Zhang, L., Du, B.: Global context based automatic road segmentation via dilated convolutional neural network. Inf. Sci. **535**, 156–171 (2020)
17. Lateef, F., Ruichek, Y.: Survey on semantic segmentation using deep learning techniques. Neurocomputing **338**, 321–348 (2019)
18. Li, Y., Liu, M., Rehg, J.: In the eye of the beholder: Gaze and actions in first person video. IEEE Trans. Pattern Anal. Mach. Intell. (2021). https://ieeexplore.ieee.org/document/9325929

19. Lin, T.-Y., et al.: Microsoft COCO: common objects in context. In: Fleet, D., Pajdla, T., Schiele, B., Tuytelaars, T. (eds.) ECCV 2014. LNCS, vol. 8693, pp. 740–755. Springer, Cham (2014). https://doi.org/10.1007/978-3-319-10602-1_48

20. Orlando, S.A., Furnari, A., Battiato, S., Farinella, G.M.: Image based localization with simulated egocentric navigations. In: International Conference on Computer Vision Theory and Applications, pp. 305–312 (2019)

21. Pasqualino, G., Furnari, A., Signorello, G., Farinella, G.M.: An unsupervised domain adaptation scheme for single-stage artwork recognition in cultural sites. Image Vis. Comput. **107**, 104098 (2021)

22. Ragusa, F., Di Mauro, D., Palermo, A., Furnari, A., Farinella, G.M.: Semantic object segmentation in cultural sites using real and synthetic data. In: 25th International Conference on Pattern Recognition (ICPR), pp. 1964–1971 (2021)

23. Ragusa, F., Furnari, A., Livatino, S., Farinella, G.M.: The meccano dataset: understanding human-object interactions from egocentric videos in an industrial-like domain. In: IEEE Winter Conference on Applications of Computer Vision, pp. 1569–1578 (2021)

24. Rezatofighi, H., Tsoi, N., Gwak, J., Sadeghian, A., Reid, I., Savarese, S.: Generalized intersection over union: a metric and a loss for bounding box regression. In: IEEE/CVF Conference on Computer Vision and Pattern Recognition, pp. 658–666 (2019)

25. Richter, S.R., Hayder, Z., Koltun, V.: Playing for benchmarks. In: Proceedings of the IEEE International Conference on Computer Vision, pp. 2213–2222 (2017)

26. Tian, Z., Shen, C., Wang, X., Chen, H.: Boxinst: high-performance instance segmentation with box annotations. In: IEEE/CVF Conference on Computer Vision and Pattern Recognition, pp. 5443–5452 (2021)

27. Zhao, H., Shi, J., Qi, X., Wang, X., Jia, J.: Pyramid scene parsing network. In: IEEE/CVF Conference on Computer Vision and Pattern Recognition, pp. 2881–2890 (2017)

Investigating Bidimensional Downsampling in Vision Transformer Models

Paolo Bruno, Roberto Amoroso$^{(\boxtimes)}$ ⓘ, Marcella Cornia ⓘ, Silvia Cascianelli ⓘ,
Lorenzo Baraldi ⓘ, and Rita Cucchiara ⓘ

University of Modena and Reggio Emilia, Modena, Italy
225975@studenti.unimore.it, {roberto.amoroso,marcella.cornia,
silvia.cascianelli,lorenzo.baraldi,rita.cucchiara}@unimore.it

Abstract. Vision Transformers (ViT) and other Transformer-based
architectures for image classification have achieved promising performances in the last two years. However, ViT-based models require large
datasets, memory, and computational power to obtain state-of-the-art
results compared to more traditional architectures. The generic ViT
model, indeed, maintains a full-length patch sequence during inference,
which is redundant and lacks hierarchical representation. With the goal
of increasing the efficiency of Transformer-based models, we explore the
application of a 2D max-pooling operator on the outputs of Transformer
encoders. We conduct extensive experiments on the CIFAR-100 dataset
and the large ImageNet dataset and consider both accuracy and efficiency
metrics, with the final goal of reducing the token sequence length without affecting the classification performance. Experimental results show
that bidimensional downsampling can outperform previous classification
approaches while requiring relatively limited computation resources.

Keywords: Vision Transformer · ViT · Bidimensional downsampling

1 Introduction

Computer vision tasks such as image classification [14,31], object detection [28,29], or semantic segmentation [5,13] have been tackled for years by
employing convolutional neural networks (CNNs), which combine the use of a
local operator (*i.e.* the convolution) and of strategies for building hierarchical
representations through spatial downsampling, which is usually carried out either
via pooling layers or by adopting strided convolutions. Recently, the Transformer
architecture [35] has received a relevant interest from the natural language processing community [10,27] and is now being adopted to solve computer vision
tasks as well [4,11,34,44]. In the case of a Transformer, the architecture features an infinite receptive field, which is achieved through the computation of
content-based pairwise similarities and, differently from CNNs, does not feature
a hierarchical structure.

© The Author(s), under exclusive license to Springer Nature Switzerland AG 2022
S. Sclaroff et al. (Eds.): ICIAP 2022, LNCS 13232, pp. 287–299, 2022.
https://doi.org/10.1007/978-3-031-06430-2_24

While a Transformer can achieve non-locality and infinite receptive field without requiring a significant increase in the number of parameters, its overall architecture usually features an increased computational cost in terms of multiply-add operations [11,34]. This can be explained by the fact that Transformers maintain a full-length sequence across all layers so that the computational cost of running a single layer does not decrease with the layer depth. While this appears to be an issue from a computational point of view, it also hinders the fact that Transformers lacks a structure that is specifically designed for images.

In this paper, we address both issues and investigate the use of bidimensional pooling in Vision Transformers. Our proposal is in line with recent literature which has tackled either the optimization of the Vision Transformer model [6, 19,33] or the application of one-dimensional pooling [26]. In our approach, the sequence of visual tokens is re-arranged into its original spatial configuration at each architectural block, and bidimensional pooling is then applied to connect different patches together and downsample the sequence length. In this way, we both decrease the computational requirements of the architecture and create a hierarchy in the feature extraction process which resembles that of a CNN.

Experiments are carried out on both a small scale scenario, that of CIFAR-100 [21], and in a larger scale setup, that of ImageNet [30]. By comparing with both a baseline Vision Transformer and with the usage of one-dimensional pooling, we demonstrate the effectiveness of our proposal, both in terms of accuracy and reduction of the number of operations.

2   Related Work

Despite having been initially proposed for machine translation and natural language processing, Transformer-based architectures [10,27,35] have recently demonstrated their effectiveness also in many computer vision tasks [20,32], either combining self-attention with convolutional layers or using pure attention-based solutions. Approaches based on these architectures reached state-of-the-art results in several tasks, including image classification [11,34], object detection [4,46], semantic segmentation [1,2,17,44], image and video captioning [7,8,42], and image generation [18].

Although many improvements of convolutional neural networks for image classification are inspired by Transformers (*e.g.* squeeze and excitation [16], selective kernel [22], and split-attention networks [40]), the first work that transfers a pure Transformer-based architecture to the classification task has been proposed by Dosovitskiy et al. [11] with the introduction of the Vision Transformer model. This architecture takes as input a sequence of square image patches and directly applies Transformer layers over raw pixels. While it has achieved promising results on ImageNet [30] bridging the gap with state-of-the-art convolutional neural networks, a pre-training stage on a large amount of data is required to achieve these remarkable performances. To solve this issue and to manage the high computational requirements typical of Transformer-based models, many different solutions have been proposed including the use of low-bit quantization [43], network pruning [25,33], and knowledge distillation [34].

Other solutions, more specific to Transformer models, tackle the quadratic complexity of the self-attention operator. For example, Child *et al.* [6] proposed to factorize the self-attention matrix into smaller and faster attention operators, thus obtaining a $\mathcal{O}(n\sqrt{n})$ complexity. Further, linear complexity can be obtained via a kernel-based formulation of self-attention, as proposed by Katharopoulos *et al.* [19], or by performing the self-attention operator on non-overlapping local windows, as proposed in [23].

On a different line, some works go in the direction of limiting the length of the input sequence to process [12,26,37,38]. For example, the approach proposed by Yuan *et al.* [38] consists in structuring the image into tokens for capturing local patterns. Other strategies consist of downsampling the sequence length, either merging 2D patches, as done in [37], or applying 1D pooling to the intermediate tokens, as done in [26]. This work follows this direction and proposes to apply bidimensional downsampling to visual tokens at the intermediate blocks of the Vision Transformer architecture, which is closer to what is commonly done with 2D max-pooling layers in convolutional neural networks.

3   Proposed Method

We first revisit the Vision Transformer (ViT, in the following) architecture [11] and, in the following sections, introduce our proposal which applies 2D downsampling in Vision Transformers.

3.1   Preliminaries

The ViT model [11] has shown that attention and feed-forward mechanisms can be employed to solve image classification tasks. Given an input image $\mathcal{I} \in \mathbb{R}^{H \times W \times C}$, where H, W and C denote height, width, and the number of channels, this is transformed into a sequence of N square patches $[\boldsymbol{x}_p^1; \boldsymbol{x}_p^2; ..., \boldsymbol{x}_p^N]$, where $\boldsymbol{x}_p^i \in \mathbb{R}^{P^2 C}$ is the i-th patch of the input image. Being $P \times P$ the resolution of square patches, H and W usually are multiple of P and the number of patches is $N = (H \cdot W)/P^2$. A linear layer is then applied to each flattened patch to project it to the input dimensionality of the model so that patches can be employed as input tokens for a Transformer encoder.

An additional classification token $\boldsymbol{x}_{\text{class}}$ is usually added to the sequence of patch embeddings. This is implemented as a trainable vector that goes through the Transformer layers and is then projected with a linear layer to predict the final output class. Positional embeddings are then added to the patch embeddings to inject information about the position of patches inside the image. Formally, the model input can be written as:

$$z_0 = [\boldsymbol{x}_{\text{class}}, \boldsymbol{x}_p^1 \boldsymbol{E}, \boldsymbol{x}_p^2 \boldsymbol{E}, ..., \boldsymbol{x}_p^N \boldsymbol{E}] + \boldsymbol{E}_{\text{pos}}, \tag{1}$$

where $\boldsymbol{E}$ indicates the token embedding matrix and $\boldsymbol{E}_{\text{pos}}$ the positional embedding matrix.

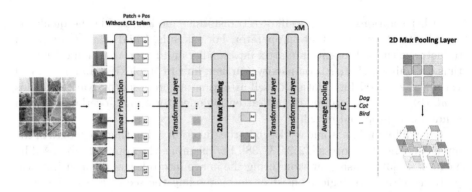

Fig. 1. Overview of the proposed architecture. To reduce redundancy and computational complexity, we progressively shrink the patches sequence length through 2D max-pooling. To this aim, we divide the ViT [11] layers into M stages. Each stage is composed of a Transformer layer, a 2D max-pooling, and a variable number of Transformer layers after the pooling operator. Instead of using the CLS token, the output of the last stage is average pooled and given as input to an FC layer to compute the final prediction. Best seen in color. (Color figure online)

The Transformer encoder [35] implemented in ViT consists of L identical layers, each being composed of a multi-head self-attention layer (MSA) and a multi-layer perceptron (MLP). Layer normalization (LN) is applied before every layer [36] and residual connections are applied after every MSA and MLP. Given the input sequence z_0, the classification output of the model can be written as:

$$z_l' = \mathsf{MSA}(\mathsf{LN}(z_{l-1})) + z_{l-1} \qquad l = 1, ..., L \tag{2}$$

$$z_l = \mathsf{MLP}(\mathsf{LN}(z_l')) + z_l' \qquad l = 1, ..., L \tag{3}$$

$$y = \mathsf{LN}(z_L^0), \tag{4}$$

where z_L^0 is the first output of the last encoder layer, corresponding to x_{class}.

Scaled Dot-Product Attention. The attention function performed in MSA layers can be seen as a mapping between queries and a set of key-value pairs with an output. Queries and keys have size d_k, the values dimension is d_v, and all are obtained as linear projections of the input sequence. Given the matrix of queries Q, and those of keys and values K and V, the output of the attention operator is:

$$\mathsf{Attention}(Q, K, V) = \mathrm{softmax}\left(\frac{QK^\mathsf{T}}{\sqrt{d_k}}\right) V. \tag{5}$$

Multi-head Attention. The above-defined attention function is computed for h different sets of keys, values, and queries, each obtained from separate and learned linear projections. The results after the h parallel operations are concatenated and projected, as follows:

$$\mathsf{MultiHead}(Q, K, V) = \mathrm{Concat}(\text{head}_1, ... \text{head}_h) W^O. \tag{6}$$

Each head is defined by the following equation:

$$\text{head}_i = \text{Attention}(\boldsymbol{QW}_i^Q, \boldsymbol{KW}_i^K, \boldsymbol{VW}_i^V), \tag{7}$$

where $\boldsymbol{W}_i^*$ indicates the parameters of each attention head.

3.2 How Pooling Layers Can Help Vision Transformers

The ViT model maintains a fixed-length sequence that passes through all the layers of the network. This choice, although simple, neglects two considerations: (i) each layer contributes differently to the accuracy and efficiency of the model, and (ii) using a sequence with a fixed length can introduce excessive redundancy, with a consequent increase in memory consumption and FLOPs without a corresponding benefit in performance. A multi-level hierarchical representation that would solve both issues is, indeed, missing. CNNs achieve this through intensive use of the pooling layer (or of the stride mechanism) to reduce the spatial dimension of the inputs [14,31] and, at the same time, significantly reduce the computational cost and increase the scalability of the model.

Bidimensional Downsampling. Inspired by Pan *et al.* [26], who investigated the usage of one-dimensional pooling in ViT-like structures, we propose to apply *bidimensional* pooling to shrink the patch embeddings sequence and create a hierarchical representation.

Without loss of generality, a max-pooling operation is considered for all the experiments. Clearly, while a 1D max-pooling can only collapse adjacent tokens, in a 2D configuration the kernel window includes all the neighboring elements with respect to the application point and considers the spatial arrangement of tokens in the input image. Our pooling strategy is thus capable of summarizing intermediate features in a spatially-aware manner. The result is a better localization of relevant features inside the feature map. While the rest of the architecture is left unchanged, we also replace the class token by average pooling the entire sequence after the last encoder layer [26].

To perform a 2D operation over a mono-dimensional sequence of intermediate activations, we firstly re-arrange the sequence of activations in matrix form, thus recovering the original spatial arrangement, and then apply the 2D max-pooling and flatten the result back into a sequence of vectors. The spatial dimensions (H_{out}, W_{out}) obtained after the application of the 2D pooling operation, and before flattening, are thus:

$$H_{out} = \left\lfloor \frac{H_{in} - K}{S} + 1 \right\rfloor \qquad W_{out} = \left\lfloor \frac{W_{in} - K}{S} + 1 \right\rfloor, \tag{8}$$

where K indicates the kernel size and S the stride. We do not apply any padding or dilation. Considering that the pooling operation alters the relative spatial positions of the activations, positional embeddings are re-computed and added after each pooling stage.

3.3 Overall Architecture

To build our architecture, we conceptually divide the encoder layers into M stages, as shown in Fig. 1. Before the first stage, the input is arranged in flattened patches, linearly projected into a sequence of tokens. A learnable positional encoding, initialized as in DeiT [11], is also added to inject information about the positions of the patches. Each stage is composed of a Transformer layer, a max-pooling 2D, and a variable number of Transformer layers. Note that the sequence length is reduced only after the pooling layer. The first Transformer layer input is described by the following equation:

$$z_0 = [x_p^1 E, x_p^2 E, ..., x_p^N E] + E_{\text{pos}}, \tag{9}$$

where E_{pos} represents the learnable position embedding. We define the output after a layer b which precedes a max-pooling layer, with also the addition of a new positional encoding, as:

$$\hat{z}_b = \text{MaxPool2D}\,(z_b) + E_b, \tag{10}$$

where z_b is defined as in Eq. 2 and E_b is the new learnable positional embeddings applied after the application of the 2D max-pooling. An average pooling is applied after the last Transformer layer of the last stage, then a fully-connected layer (FC) is used to make the final predictions. Formally, predictions can be formulated as:

$$y = \text{FC}(\text{AvgPool}(\text{LN}(z_L))). \tag{11}$$

4 Experiments

To evaluate the effectiveness of our proposal, we perform experiments on two image classification datasets and compare our approach with different variants and baselines. In the following, we first provide implementation and experimental details and then describe our experimental results.

4.1 Experimental Setting

In the following experiments, we consider different configurations of our ViT-based model equipped with 2D-pooling (which we refer to as VT2D) and compare against a ViT model with no pooling and a ViT-based model with 1D-pooling [26]. In the following, we refer to these baselines as VT and VT1D, respectively.

Datasets and Evaluation Protocol. We perform experiments on the CIFAR-100 dataset [21], which contains 100 classes and 60k images, and the ILSVRC-2012 ImageNet dataset [30], which has $1,000$ classes and 1.3M images. All trained models are compared in terms of FLOPs and number of parameters and evaluated in terms of top-1 and top-5 accuracy on the considered datasets.

Table 1. Experimental results on the CIFAR-100 dataset [21]. For each experiment, we indicate the indexes of network stages in which we perform 1D or 2D pooling and the max-pooling kernel size.

	Pooling stages	Kernel size	Params (M)	FLOPs (G)	Top-1 Acc. (%)	Top-5 Acc. (%)
VT-Ti (no pooling)	–	–	5.54	1.25	72.92	92.88
VT1D-Ti-4	0,1,2,3	3	5.58	0.38	72.76	92.67
VT2D-Ti-1	0	3 × 3	5.55	0.31	71.86	92.03
VT2D-Ti-1	1	3 × 3	5.55	0.57	73.39	92.94
VT2D-Ti-1	2	3 × 3	5.55	0.82	73.04	92.78
VT2D-Ti-1	3	3 × 3	5.55	1.08	71.49	92.53
VT2D-Ti-2	0,2	3 × 3	5.55	**0.24**	70.31	91.27
VT2D-Ti-2	0,2	2 × 2	5.55	0.29	70.92	91.82
VT2D-Ti-2	1,3	3 × 3	5.55	0.54	72.25	92.32
VT2D-Ti-2	1,3	2 × 2	5.55	0.58	72.17	92.36
VT2D-Ti-4	0,1,2,3	3 × 3	5.61	0.61	72.87	92.61
VT2D-Ti-4	0,1,2,3	2 × 2	5.65	0.88	**75.31**	**93.47**
VT-S (no pooling)	–	–	21.70	4.58	75.62	93.01
VT1D-S-4	0,1,2,3	3	21.77	1.39	76.09	93.43
VT2D-S-1	0	3 × 3	21.71	1.15	75.31	92.32
VT2D-S-1	1	3 × 3	21.71	2.08	76.59	93.16
VT2D-S-1	2	3 × 3	21.71	3.02	76.18	93.35
VT2D-S-1	3	3 × 3	21.71	3.95	75.13	93.34
VT2D-S-2	0,2	3 × 3	21.71	**0.86**	73.31	91.65
VT2D-S-2	0,2	2 × 2	21.73	1.04	74.44	92.02
VT2D-S-2	1,3	3 × 3	21.71	1.97	76.26	92.91
VT2D-S-2	1,3	2 × 2	21.73	2.13	75.51	93.13
VT2D-S-4	0,1,2,3	3 × 3	21.83	2.28	75.68	92.26
VT2D-S-4	0,1,2,3	2 × 2	21.91	3.26	**77.61**	**93.57**

Implementation Details. Following recent literature on ViT-based models [11, 34], we devise two model configurations varying the model dimensionality d and the number of attention heads h: Tiny ($d = 192$, $h = 3$) and Small ($d = 384$, $h = 6$). Regardless of the model configuration, we always employ 12 layers, divided in $M = 4$ stages with 3 Transformer layers each.

For the experiments on the CIFAR-100 dataset, we use a batch size of 128 and an initial learning rate of $1.25 \cdot 10^{-4}$, while for the ImageNet dataset, the batch size is set to 1024 and the initial learning rate is equal to $5 \cdot 10^{-4}$. The input image resolution is set to 224×224 for both datasets. For training, we use the AdamW optimizer [24], with momentum and weight decay set to 0.9

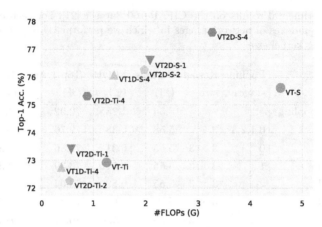

Fig. 2. Performance comparison in terms of top-1 accuracy and FLOPs on the CIFAR-100 dataset [21].

and 0.25 respectively, and train the models for 300 epochs on both the datasets. Note that, during the training phase, we use a cosine scheduler, so that the initial learning rate is reached only after 5 warm-up epochs, and a stochastic depth of 0.1 to facilitate convergence. In all experiments, model weights are initialized with a truncated normal distribution.

To obtain the required amount of data to train Transformer-based models, we follow the same data augmentation strategy used to train the DeiT model [34]. In particular, we apply rand-augment [9] and random erasing [45], together with mixup [41] and cutmix [39]. The magnitude and standard deviation of rand-augment are set to 9 and 0.5, respectively. Random erasing is applied with a probability equal to 0.25. We also employ repeated augmentation [3, 15, 34]. We run our experiments on 4 RTX 2080 GPUs.

4.2 Experimental Results

Experiments on CIFAR-100. To identify the best strategy to apply the 2D pooling for reducing the model complexity while maintaining competitive performance, we conduct an ablation study in which we vary the kernel size and stride of the pooling layers and the network configuration. The configurations considered differ in terms of the stages in which the pooling layer is applied. In particular, we vary the number of stages, from 1 to 4, and the depth of the stage in which 2D pooling is performed in the model. For all VT2D configurations, we use 2D pooling with stride 2 except when applying the 2D pooling in all four stages of the model, where we use stride equal to 1. As already mentioned, as our baselines, we also consider the VT1D approach, in which 1D pooling with stride 2 is applied at all four stages of the model, and the VT model, which has no pooling layers for downsampling. Results on the CIFAR-100 dataset are reported in Table 1, using Tiny and Small configurations. A noteworthy aspect

Table 2. Experimental results on the ImageNet dataset [30].

	Pooling stages	Kernel size	Params (M)	FLOPs (G)	Top-1 Acc. (%)	Top-5 Acc. (%)
VT1D-Ti-4	0,1,2,3	3	5.75	0.38	67.28	87.70
VT2D-Ti-1	0	3 × 3	5.72	**0.31**	65.56	86.47
VT2D-Ti-1	1	3 × 3	5.72	0.57	70.39	**89.84**
VT2D-Ti-4	0,1,2,3	2 × 2	5.82	0.88	**70.60**	89.83
VT1D-S-4	0,1,2,3	3	22.12	1.40	74.83	92.05
VT2D-S-1	0	3 × 3	22.06	**1.15**	73.92	91.17
VT2D-S-1	1	3 × 3	22.06	2.08	**78.19**	**93.87**
VT2D-S-4	0,1,2,3	2 × 2	22.26	3.26	78.02	93.66

that emerges from the presented experimental results is that applying downsampling at early stages brings the most noticeable saving in terms of computational complexity. Moreover, performing a finer-grained pooling by applying kernels of size 2 compared to 3 benefits the most the performance, at the cost of a slightly higher computational complexity.

Our approach with 2D pooling applied at all four stages obtains better performance compared to both the model without pooling and the 1D pooling version, with a significant reduction of the computational complexity, especially when compared to the VT model without pooling layers. Specifically, it can be noticed that some of the tested configurations with bidimensional pooling in one or two stages perform on par with the VT model in terms of accuracy but require on average 50% fewer FLOPs. The best-performing variant, the VT2D with four stages featuring pooling, brings to an accuracy increase of 2.39% and 1.99% for the Tiny and Small configurations, respectively, while reducing the FLOPs by one third. When comparing with the VT1D version, instead, it can be seen that the best configuration of the VT2D model is always significantly better in terms of accuracy but brings to an increase of the number of FLOPs. However, considering the variants with a single pooling stage, we can notice a computational complexity similar to the 1D pooling version, but with better performance in terms of accuracy, thus further demonstrating the effectiveness of our approach.

In Fig. 2, we show the performance comparison of our approach and the considered baselines in terms of top-1 accuracy and FLOPs, using both Tiny and Small configurations. From the graph, we can notice that our model obtains the best trade-off between overall performance and computational complexity, outperforming the VT and VT1D models in terms of accuracy while keeping FLOPs comparable or even reduced.

Experiments on ImageNet. As a further analysis, we explore the effects of applying 2D pooling in the case of a bigger and more complex dataset than CIFAR-100, and consider the ImageNet dataset. In this analysis, we include a subset of variants previously described, *i.e.* those with best accuracy/FLOPs

Table 3. Comparison with state-of-the-art models on the CIFAR-100 [21] and ImageNet [30] datasets.

	FLOPs (G)	CIFAR-100		ImageNet	
		Top-1 Acc. (%)	Top-5 Acc. (%)	Top-1 Acc. (%)	Top-5 Acc. (%)
DeiT-Ti [34]	1.25	–	–	72.20	91.10
DeiT-Ti+SCOP [33]	0.80	–	–	68.90	89.00
DeiT-Ti+PoWER [12]	0.80	–	–	69.40	89.20
HVT-Ti-1 [26]	0.64	–	–	69.64	89.40
HVT-Ti-4 [26]	0.38	69.51	91.78	–	–
VT2D-Ti-1	0.57	73.39	92.94	70.39	89.84
VT2D-Ti-4	0.88	75.31	93.47	70.60	89.83
DeiT-S [34]	4.60	–	–	79.80	95.00
DeiT-S+SCOP [33]	2.60	–	–	77.50	93.50
DeiT-S+PoWER [12]	2.70	–	–	78.30	94.00
HVT-S-1 [26]	2.40	74.27	93.07	78.00	93.83
HVT-S-4 [26]	1.39	75.43	93.56	75.23	92.30
VT2D-S-1	2.08	76.59	93.16	78.19	93.87
VT2D-S-4	3.26	77.61	93.57	78.02	93.66

trade-off, and the baseline model with 1D pooling. Again, for all VT2D configurations, we use 2D pooling with stride equal to 2, except for the model variant that applies bidimensional pooling in all four stages of the network. The results of this comparison are reported in Table 2.

Considering the Tiny configuration, our best model is VT2D-Ti-4 which applies a 2D pooling with kernel size 2 in all four stages, followed by VT2D-Ti-1 in which a single bidimensional downsampling is applied in the second stage of the network. Noticeably, VT2D-Ti-1 outperforms the one-dimensional pooling baseline with a slight increase in terms of FLOPs. Similar results can be also observed when turning to the Small configuration. Specifically, the VT2D-S-1 with two-dimensional pooling at the second stage of the network overcomes the VT1D baseline by 3.36% and 1.82% in terms of top-1 and top-5 accuracy, respectively.

Comparison with State-of-the-Art Models. As a final analysis, we report the comparison of our best variants and other state-of-the-art models based on the Vision Transformer architecture that apply different strategies to achieve efficiency (see Table 3). For the competitors, we use the same notation as for our models. In particular, we consider the knowledge distillation-based approach proposed in [34] (DeiT), two variants of DeiT that additionally perform pruning (one following the strategy proposed in [33] - SCOP, the other the strategy proposed in [12] - PoWER), and the monodimensional pooling-based approach proposed in [26] (HVT). From the table, it can be observed that bidimensional pooling is comparable to the considered state-of-the-art approaches both in terms of FLOPs saving and accuracy.

5 Conclusion

In this work, we proposed to apply the concept of bidimensional pooling, which is commonly used in CNNs, to the intermediate patch sequences processed in ViT architectures. As happens in CNNs, this downsampling strategy allows reducing the computational requirement and memory footprint of the model. Moreover, it benefits the performance by enforcing a hierarchical input representation. We evaluated our proposal on two commonly used classification datasets and demonstrated its effectiveness for reducing the computational complexity and increasing the classification accuracy. In fact, with respect to standard ViT, the required FLOPs can be almost halved for equal performance while the classification accuracy can be increased by around 2% with 30% FLOPs reduction in both the Tiny and Small configurations.

Acknowledgment. This work has been partially supported by the project "ROAD-STER: Road Sustainable Twins in Emilia Romagna", funded by the International Foundation Big Data and Artificial Intelligence for Human Development.

References

1. Amoroso, R., Baraldi, L., Cucchiara, R.: Assessing the role of boundary-level objectives in indoor semantic segmentation. In: CAIP (2021)
2. Amoroso, R., Baraldi, L., Cucchiara, R.: Improving indoor semantic segmentation with boundary-level objectives. In: IWANN (2021)
3. Berman, M., Jégou, H., Vedaldi, A., Kokkinos, I., Douze, M.: Multigrain: a unified image embedding for classes and instances. arXiv preprint arXiv:1902.05509 (2019)
4. Carion, N., Massa, F., Synnaeve, G., Usunier, N., Kirillov, A., Zagoruyko, S.: End-to-end object detection with transformers. In: ECCV (2020)
5. Chen, L.C., Papandreou, G., Kokkinos, I., Murphy, K., Yuille, A.L.: DeepLab: semantic image segmentation with deep convolutional nets, atrous convolution, and fully connected CRFs. IEEE Trans. PAMI **40**(4), 834–848 (2017)
6. Child, R., Gray, S., Radford, A., Sutskever, I.: Generating long sequences with sparse transformers. arXiv preprint arXiv:1904.10509 (2019)
7. Cornia, M., Baraldi, L., Cucchiara, R.: Explaining transformer-based image captioning models: an empirical analysis. In: AI Communications, pp. 1–19 (2021)
8. Cornia, M., Baraldi, L., Fiameni, G., Cucchiara, R.: Universal captioner: long-tail vision-and-language model training through content-style separation. arXiv preprint arXiv:2111.12727 (2021)
9. Cubuk, E.D., Zoph, B., Shlens, J., Le, Q.: RandAugment: practical automated data augmentation with a reduced search space. In: NeurIPS (2020)
10. Devlin, J., Chang, M.W., Lee, K., Toutanova, K.: BERT: pre-training of deep bidirectional transformers for language understanding. In: NAACL (2018)

11. Dosovitskiy, A., et al.: An image is worth 16x16 words: transformers for image recognition at scale. In: ICLR (2021)
12. Goyal, S., Choudhury, A.R., Raje, S., Chakaravarthy, V., Sabharwal, Y., Verma, A.: PoWER-BERT: accelerating BERT inference via progressive word-vector elimination. In: ICML (2020)
13. He, K., Gkioxari, G., Dollár, P., Girshick, R.: Mask R-CNN. In: ICCV (2017)
14. He, K., Zhang, X., Ren, S., Sun, J.: Deep residual learning for image recognition. In: CVPR (2016)
15. Hoffer, E., Ben-Nun, T., Hubara, I., Giladi, N., Hoefler, T., Soudry, D.: Augment your batch: improving generalization through instance repetition. In: CVPR (2020)
16. Hu, J., Shen, L., Sun, G.: Squeeze-and-excitation networks. In: CVPR (2018)
17. Huang, Z., Wang, X., Huang, L., Huang, C., Wei, Y., Liu, W.: CCNet: criss-cross attention for semantic segmentation. In: CVPR (2019)
18. Jiang, Y., Chang, S., Wang, Z.: TransGAN: two pure transformers can make one strong GAN, and that can scale up. In: NeurIPS (2021)
19. Katharopoulos, A., Vyas, A., Pappas, N., Fleuret, F.: Transformers are RNNs: fast autoregressive transformers with linear attention. In: ICML (2020)
20. Khan, S., Naseer, M., Hayat, M., Zamir, S.W., Khan, F.S., Shah, M.: Transformers in vision: a survey. ACM Comput. Surv. (2021)
21. Krizhevsky, A., Hinton, G.: Learning multiple layers of features from tiny images (2009)
22. Li, X., Wang, W., Hu, X., Yang, J.: Selective kernel networks. In: CVPR (2019)
23. Liu, Z., et al.: Swin transformer: hierarchical vision transformer using shifted windows. In: ICCV (2021)
24. Loshchilov, I., Hutter, F.: Decoupled weight decay regularization. In: ICLR (2019)
25. Michel, P., Levy, O., Neubig, G.: Are sixteen heads really better than one? In: NeurIPS (2019)
26. Pan, Z., Zhuang, B., Liu, J., He, H., Cai, J.: Scalable vision transformers With hierarchical pooling. In: ICCV (2021)
27. Radford, A., Narasimhan, K., Salimans, T., Sutskever, I.: Improving language understanding by generative pre-training (2018)
28. Redmon, J., Divvala, S., Girshick, R., Farhadi, A.: You only look once: unified, real-time object detection. In: CVPR (2016)
29. Ren, S., He, K., Girshick, R., Sun, J.: Faster R-CNN: towards real-time object detection with region proposal networks. In: NeurIPS (2015)
30. Russakovsky, O., et al.: ImageNet large scale visual recognition challenge. Int. J Comput. Vis. **115**(3), 211–252 (2015). https://doi.org/10.1007/s11263-015-0816-y
31. Simonyan, K., Zisserman, A.: Very deep convolutional networks for large-scale image recognition. In: ICLR (2015)
32. Stefanini, M., Cornia, M., Baraldi, L., Cascianelli, S., Fiameni, G., Cucchiara, R.: From show to tell: a survey on deep learning-based image captioning. IEEE Trans. PAMI 1–20 (2022)
33. Tang, Y., et al.: Scop: scientific control for reliable neural network pruning. In: NeurIPS (2020)
34. Touvron, H., Cord, M., Douze, M., Massa, F., Sablayrolles, A., Jégou, H.: Training data-efficient image transformers & distillation through attention. In: ICML (2021)
35. Vaswani, A., et al.: Attention is all you need. In: NeurIPS (2017)
36. Wang, Q., et al.: Learning deep transformer models for machine translation. In: ACL (2019)
37. Wang, W., et al.: Pyramid vision transformer: a versatile backbone for dense prediction without convolutions. In: ICCV (2021)

38. Yuan, L., et al.: Tokens-to-token ViT: training vision transformers from scratch on imageNet. In: ICCV (2021)
39. Yun, S., Han, D., Oh, S.J., Chun, S., Choe, J., Yoo, Y.: Cutmix: regularization strategy to train strong classifiers with localizable features. In: ICCV (2019)
40. Zhang, H., et al.: ResNeSt: split-attention networks. arXiv preprint arXiv:2004.08955 (2020)
41. Zhang, H., Cisse, M., Dauphin, Y.N., Lopez-Paz, D.: mixup: beyond empirical risk minimization. In: ICLR (2018)
42. Zhang, P., et al.: VinVL: revisiting visual representations in vision-language models. In: CVPR (2021)
43. Zhang, W., et al.: TernaryBERT: distillation-aware ultra-low bit BERT. In: EMNLP (2020)
44. Zheng, S., et al.: Rethinking semantic segmentation from a sequence-to-sequence perspective with transformers. In: CVPR (2021)
45. Zhong, Z., Zheng, L., Kang, G., Li, S., Yang, Y.: Random erasing data augmentation. In: AAAI (2020)
46. Zhu, X., Su, W., Lu, L., Li, B., Wang, X., Dai, J.: Deformable DETR: deformable transformers for end-to-end object detection. In: ICLR (2021)

Unsupervised Anomaly Localization Using Locally Adaptive Query-Dependent Scores

Naoki Kawamura[✉][iD]

Toshiba Corporation, 1, Komukai-Toshiba-cho, Saiwai-ku, Kawasaki 212-8582, Japan
naoki5.kawamura@toshiba.co.jp

Abstract. We propose a framework of reducing falsely-high anomaly scores on the unsupervised anomaly localization method. In recent years, the performance of unsupervised anomaly detection has been improved significantly by using pretrained deep features. A pretrained feature extractor can make appropriate descriptions for unsupervised anomaly detection, with extractors using embedding-based methods outperforming conventional deep-learning-based extractors. However, the intra-class variance of anomaly-free patterns has not been well considered yet by the existing methods while it is important to prevent misdetection and achieve more accurate localizations. In this paper, a framework is proposed that addresses this problem by leveraging the subtraction of deep pretrained features among anomaly-free images. The anomaly scores at the locations with higher intra-class variances are reduced through a novel and simple framework to tackle with misdetection. The experimental results using the standard industrial dataset show the outperformances of the proposed method especially for the overdetection reductions.

Keywords: Unsupervised anomaly detection and localization · Out of distribution

1 Introduction

Image anomaly detection and localization are major pattern recognition tasks. Anomaly detection is image-level classification problem of whether an image contains anomalous region or not. On the other hand, anomaly localization is pixel-level classification, or segmentation, that is a more challenging task. There are many practical applications such as industrial inspection, infrastructure maintenance, and surveillance system [1–6]. Although its usability in a variety of the applicable area, conventional supervised approaches often suffer from practical applications. The major difficulty is that there exists an unpredictably large number of anomalous patterns, and furthermore, these anomalous patterns rarely appear and are severely long-tailed samples [1,7–10]. Therefore, the sufficient annotation of anomalous data is not viable in many practical cases. For these reasons, unsupervised anomaly detection and localization have recently attracted the attention of many researchers, where only unlabeled anomaly-free (normal) images are collected for training.

S. Sclaroff et al. (Eds.): ICIAP 2022, LNCS 13232, pp. 300–311, 2022.
https://doi.org/10.1007/978-3-031-06430-2_25

Deep learning has significantly advanced research in unsupervised anomaly detection/localization, and a variety of different types of deep-learning-based approaches have been proposed [10–14]. In recent years, embedding-based approaches [7,15,16] have achieved remarkable performance, in which anomalous locations are identified based on the distance between an input image and normal images in the pretrained deep feature space.

For example, SPADE [15] is a powerful and very simple method. It uses a pyramid-level description of a pretrained model and calculates the anomaly scores based on the Euclidean distance between an input image and resembling normal images. PaDim [16] trains the Gaussian distribution at each location in the image spatial domain. Using a normal training dataset, it efficiently describe the anomaly scores as the Mahalanobis distance of an input image. These embedding-based methods outperform conventional reconstruction-based methods in an industrial dataset [1] by using only a pretrained deep feature space without the need for deep learning of the target domain.

Under the unsupervised setting, it is important to tackle with intra-class variance in the normal class. That is, the locally higher intra-class variance in the training normal class would lead to the misdetection, and not only feature description but also some out-of-distribution modeling frameworks are necessary for the accurate localization. In spite of the fact, it is not considered in SPADE while it sometimes results in severe overdetections as shown in Fig. 1. Those misdetections would be occurred by lacks of normal appearance samples for training, especially at image boundaries and edges. There are several approaches that solve this out-of-distribution problem through distribution modeling [7,16,17]. These methods estimate anomaly scores based on the unimodal distribution modeling of the training normal class, leading to the better localization performances. On the other hand, there is the distance calculation approaches of the sample-based modeling instead of the unimodal distribution modeling, in the metric learning topics [18,19]. Those approaches calculate or correct the distances based on the neighbor samples of each query in order to overcome the long-tails of out-of-distribution.

In this paper, we consider there is a room for the improvement for the intra-class problem of the unsupervised anomaly localization, and tackled with by sample-based modeling instead of unimodal distribution modeling. In the proposed method, we leverage the local relationship of normal data samples in order to perform the finer localization.

The contributions of this paper include the following:

- We propose a simple and novel framework of reducing the misdetection of unsupervised anomaly localization without any additional learning process.
- It is confirmed that the proposed sample-based modeling method outperformed the existing methods and achieved comparable performances with some unimodal distribution modeling approaches.

The remainder of this paper is organized as follows. Section 2 introduces work related to this study. Section 3 describes the proposed approach, and Sect. 4 presents our experimental results. We give our conclusions in Sect. 5.

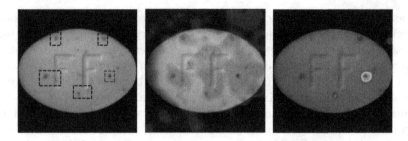

Fig. 1. Example of the proposed method. Left: an image sample from MVTec AD. The orange dotted box and the blue dotted boxes respectively indicate a true anomaly region and *misleading* normal regions. Middle: The anomaly score map calculated by SPADE [15], where many overdetected regions. Right: The anomaly score map calculated by the proposed method, which is a simple expansion of SPADE, and it could successfully localize the anomaly region. (Color figure online)

2 Related Work

There are a variety of unsupervised anomaly detection and localization methods based on deep neural networks. One-class SVDD [9,20], encoder-decoder [11,12, 21], and GAN based methods [8,13] are conventional learning-based approaches.

Robust-PCA [22] and one-class SVM [23] were the major anomaly detectors before the appearance of methods using deep learning. They train a subspace of anomaly-free features to which the input data are projected for classifying anomalous/normal data. One-class SVM models the decision surface between anomaly/normal classes as a sphere that is trained using an anomaly-free dataset. One-class SVDD [9,20] has been modeled by deep learning in the recent years, with the deep feature description used for the domain of one-class SVM.

Encoder-decoder-based and GAN-based methods, in which the normal counterpart of an input image is reconstructed to identify anomaly locations, are the most popular and widely employed. These methods reconstruct the input image inappropriately if there exist anomalous regions in the image, and the anomaly score is calculated based on the reconstruction error. The anomaly score maps are also calculated using the intermediate latent space, which represents the anomaly score of each pixel location. Despite the computational efficiency, reconstruction-based methods have limitations in terms of accuracy performance since they often reconstruct anomalous regions as normal ones because of the nature of auto-encoders and the variety of anomaly patterns.

To overcome this problem, ImageNet [24] features have come to be leveraged and known as relevant feature descriptions for anomaly detection and localization [10,25] for distinguishing between anomaly and normal images using a pretrained feature space. Those of *embedding-based* methods [7,15,16] leverage a pre-trained deep model as a feature descriptor. They calculate some kind of similarity between input images and galleries in feature space to obtain anomaly

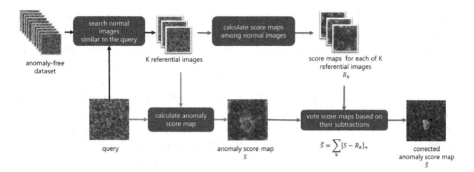

Fig. 2. Flow of the proposed framework. The anomaly score map calculated by SPADE (green arrows) is corrected using the referential normal images (red arrows). The score maps of the referential images means *fluctuation of normal patterns*, or intra-class variances of each location, and are leveraged to reduce misdetections of the input anomaly score map. (Color figure online)

scores. The embedding-based leverage a pretrained model that is trained in advance using a huge general dataset like ImageNet. SPADE [15] was the first of these to be proposed. It uses a pyramid-level description of a pretrained model and calculates the subtraction between an input image and its k-nearest normal images by the Euclidean distance. PaDim [16] trained the Gaussian distribution at each location in the spatial domain of an image using a normal training dataset to calculate the Mahalanobis distance of an input image. CFLOW-AD [17] is based on the concept of normalizing-flow, which generalizes the pixel-wise representations of z-distributions in the form of network architecture. Through CFLOW-AD, both low-resolutional and high-resolutional inputs could successfully processed and more accurate anomaly scores could be achieved. These embedding-based methods outperform conventional reconstruction-based methods in MVTec AD [1] and Shanghai Tech Campus datasets [2].

The above embedding-based methods mainly focus on the novelties of feature description [10,14,15], and distribution modeling [7,16]. In this paper, we focus on a sample-based approach for embedding-based methods, which is based on SPADE in order to achieve accurate localizations in a simple manner.

3   Proposed Method

The motivation of the proposed framework is to prevent misdetection and improve the localization accuracy. For the out-of-distribution problem in SPADE, the misdetection occurs when a normal test pattern exists in/nearby the tail of a distribution, or the decision surface, calculated among the normal training dataset. We focus on correcting the anomaly scores for the misleading normal pattern by using the relationships among the misleading normal samples in the training dataset.

Given a normal image dataset, it is assumed that the dataset contains several normal samples, each of which resemble the input image. Here, the distances among the normal images represent the intra-class variations of normal patterns at each location in the image space. It can be assumed that as the anomaly score at a location increases, the possibility of misdetection increases as well. The proposed method therefore corrects the anomaly score map calculated by a conventional method by using the intra-class anomaly scores that are calculated among normal images resembling the input image. Figure 2 shows an overview of the proposed framework. In Fig. 2, first the anomaly score map is calculated by using SPADE [15] and then it is corrected according to our motivation.

Let X be an input image and $\mathcal{G}$ be a training dataset composed of normal images. In imitation of the existing embedding-similarity-methods [15,16], each of the images is first converted to pretrained deep features, with the feature description denoted by $F(\cdot)$. The anomaly score map S is calculated from X by a conventional method as SPADE [15]. The objective of the proposed method is to improve S to output $\hat{S}$. The K-nearest samples to X in $\mathcal{G}$ are denoted by R_k $(k = 1, 2, \cdots, K)$. R_k is obtained by a traditional k-NN search in the pretrained deep feature domain using the Euclidean distance. For each R_k, the feature subtraction by other images in $\mathcal{R}$ is calculated to obtain a intra-class score map. The intra-class score map for R_k is denoted by V_k, and is calculated by a similar approach to the calculation of S. Following SPADE, the anomaly score map S is calculated as [15]:

$$S(p) = \min_k ||F(R_k, p) - F(X, p)||^2, \tag{1}$$

where p is the coordinate index.[1] Moreover, let each image in $\mathcal{Q} = \mathcal{R} \backslash \{R_k\}$ be denoted by Q_τ $(\tau = 1, 2, \cdots, K - 1)$ and the intra-class score map V_k be calculated as

$$V_k(p) = \min_\tau ||F(Q_\tau, p) - F(R_k, p)||^2. \tag{2}$$

Using V_k, the anomaly score map S is corrected by Eq. (3). Note that each score map is interpolated to the original image size by bilinear interpolation.

$$\hat{S} = \sum_k [S - \alpha V_k]_+, \tag{3}$$

where α is a weight parameter that controls the balance between the anomaly and the fluctuation, and $[\cdot]_+$ is a tensor operator that sets negative elements to zero.

4   Experimental Results

We examined the performance of the proposed method using the MVTec AD dataset [1] and the ShanghaiTech Campus (STC) dataset [2]. The MVTec AD

---

[1] Note that Eq. (1) is the feature subtraction that is the same as [15] in the case of $\kappa = 1$.

dataset contains images of 15 categories of single target objects. Each category set is composed of 50–200 query images for testing and several hundred normal images for training. The images were resized to 256×256 and center-cropped to 224×224 according to the experimental protocol [15,16]. The STC dataset contains videos of surveillance captured by static cameras at 12 scene locations. Each scene consists of several hundred test frames and tens of thousands of anomaly-free training frames. Following the protocol [15,16], all the image frames were resized to 256×256, and test frames were sub-sampled by a factor of five. Following to the standard protocol, we sub-sampled the training frames of each scene to roughly 5000 images.

We implemented and applied the proposed framework to SPADE [15]. For the computational acceleration, we employed the Faiss library [26] to obtain the approximated kNN results. The proposed method was compared with several kinds of approaches including embedding-based methods based on a sample-based modeling (SPADE [15]) or distribution-based modeling (PaDim [16], CFLOW-AD [17]). For each method, the training dataset contains normal images of the same category from the same dataset as those of input images. The parameter for the number of k-NN was set to 5, and the backbone was Wide-ResNet-50 [27] following to the SOTA comparisons [15,16].

Table 1 shows AUPRO scores (PRO) evaluated using the MVTech AD. The embedding-based methods totally outperformed the reconstruction-based methods. Compared to SPADE, which is the main baseline of the proposed method, the average PRO performance was improved by 1.8%. In comparing to the unimodal distribution modeling methods, the proposed method outperformed PaDim by 1.4% while underperforming CFLOW-AD.

We consider the sample-based modeling of the proposed method and SPADE have the following advantages compared to the unimodal distribution modeling. First, they do not need training process of the models and just a normal dataset itself are necessary. Thus they are easy to test and more *interactive* for the practical applications. For example, when users wants to add new data for the training, all they have to do is just putting the data in the dataset. Second, the sample-based modeling could be robust under using the severely noisy dataset. The noise samples would be excluded before calculating the anomaly scores because both the proposed method and SPADE search similar normal samples to the input images, The sample-based modeling success under the noisy training dataset contains many noisy samples or unsuitable data such as the other category images.

Also Table 2 show the pixel-AUROC scores on the STC dataset. The proposed method outperformed SPADE, and have the comparable performance with PaDim while CFLOW-AD outperforms the others.

4.1 Ablation Study

We also show several aspects of the proposed method compared to SPADE baseline.

Table 1. Comparison of the proposed method with SOTA anomaly localizers in PRO [%] results for each category in MVTec AD.

	OC-SVM [23]	AE (L2-norm) [11]	Student [10]	SPADE [15]	Padim [16]	CFLOW-AD [17]	Proposed
Bottle	85.0	91.0	91.8	95.5	94.8	96.8	94.8
Cable	43.1	82.5	86.5	90.9	88.8	93.53	88.5
Capsule	55.4	86.2	91.6	93.7	93.5	93.4	95.2
Hazelnut	61.6	91.7	93.7	95.4	92.6	96.7	95.1
Metal_nut	31.9	83.0	89.5	94.4	85.6	91.7	93.9
Pill	54.4	89.3	93.5	94.6	92.7	95.4	96.3
Screw	64.4	75.4	92.8	96.0	94.4	95.3	97.4
Toothbrush	53.8	82.2	86.3	93.5	93.1	95.1	93.5
Transister	49.6	72.8	70.1	87.4	84.5	81.4	84.7
Zipper	35.5	83.9	93.3	92.6	95.9	96.6	96.0
Carpet	35.5	45.6	69.5	94.7	96.2	97.7	96.4
Grid	12.5	58.2	81.9	86.7	94.6	96.7	94.5
Leather	30.6	81.9	81.9	97.2	97.8	99.4	98.4
Tile	72.2	89.7	91.2	75.9	86.0	94.3	84.6
Wood	33.6	72.7	72.5	87.4	91.1	95.8	92.5
Average	47.9	79.1	85.7	91.7	92.1	94.6	93.5

Table 2. Comparison on the STC [2] dataset in pixel-AUROC [%].

CAVGA [30]	SPADE [15]	Padim [16]	CFLOW-AD [17]	Proposed
85.5	89.9	91.2	94.5	91.7

Backbones. The performance differences compared to the baseline method were investigated with respect to the backbones of the conventional anomaly detectors and their feature extractors. We also applied the proposed framework to a third party implementation of SPADE [28], denoted by SPADE_glb in this paper. SPADE and SPADE_glb differ in terms of where the k-NN features are collected for the feature subtraction. In Eq. (1), SPADE collects corresponding k-NN features from the same location in the referential score maps while SPADE_glb collects from any location. This fact results in differences in the performance trends of different categories [15], and in the computational costs. In applying the proposed framework to SPADE_gld, Eq. (1) can be rewritten as follows:

$$S(p) = \min_{k,q} ||F(\boldsymbol{R}_k, q) - F(\boldsymbol{X}, p)||^2, \tag{4}$$

and also Eq. (2) is replaced by

$$V_k(p) = \min_{\tau,q} ||F(\boldsymbol{Q}_\tau, q) - F(\boldsymbol{R}_k, p)||^2. \tag{5}$$

For further investigation, Wide-ResNet-50 [27] and MixNet-L [29] are employed for the feature extractors. Table 3 shows the resulting evaluation indices of pixel-level-AUROC (p-AUC). The proposed method improved the overall detection accuracies, particularly for severely unaligned data for the respective categories. For example, the transistor category has severely large intra class variances

Table 3. Pixel-AUROC [%] results for each category in MVTec AD. The backbones of Wide-ResNet-50 and MixNet-L are denoted by WR50 and MixL.

	SPADE-WR50	SPADE_glb-WR50	SPADE-MixL	SPADE_glb-MixL	Prop-SPADE-WR50	Prop-SPADE_glb-WR50	Prop-SPADE-MixL	Prop-SPADE_glb-MixL
Bottle	98.4	97.0	97.7	97.5	98.2	97.7	98.1	98.0
Cable	97.2	92.3	95.1	94.2	97.2	94.8	97.2	95.3
Capsule	99.0	98.4	98.9	98.8	98.4	98.4	98.9	98.8
Carpet	97.5	98.9	98.7	98.8	97.5	99.1	98.9	98.9
Grid	93.7	98.3	98.4	97.8	94.1	94.4	98.9	98.9
Hazelnut	99.1	98.5	97.8	97.6	98.7	98.5	97.7	97.7
Leather	97.6	99.3	99.3	99.3	98.4	99.3	99.4	99.3
Metal_nut	98.1	97.1	97.0	96.7	98.1	97.1	97.3	97.2
Pill	96.5	95.0	93.7	93.2	95.9	93.5	94.4	93.9
Screw	98.9	99.1	99.4	99.3	98.9	98.9	99.5	99.5
Tile	87.4	92.8	92.0	91.7	91.3	92.2	92.6	92.4
Toothbrush	97.9	98.8	98.8	98.8	97.9	99.0	98.8	98.8
Transistor	94.1	86.6	89.2	88.1	96.3	95.9	92.3	91.7
Wood	88.5	95.3	93.9	93.8	94.3	95.4	94.7	95.3
Zipper	96.5	98.6	98.6	98.4	98.0	98.4	98.8	98.7
Average	96.0	96.4	96.5	96.3	96.9	96.8	97.2	97.0

Table 4. Pixel-AUROC [%] results with respect to the scale of the training dataset on MVTec AD.

	×0.95	×0.75	×0.50	×0.25	×0.10	×0.05
Prop	96.8	96.7	96.7	96.5	96.0	95.2
SPADE	96.4	96.4	96.3	96.0	95.8	94.5

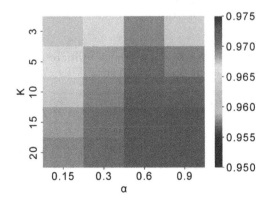

Fig. 3. Parametric behaviors of the proposed method with respect to (K, α). The heatmap value is p-AUC.

depending on SPADE_glb description, and the resultant accuracy is relatively low (86.6%). The proposed framework reduced the misdetection and improved the p-AUC score to 95.9%. In some image categories where there was less intra class variances, however, the performance was degraded by the proposed method.

Scale of Datasets. Next, Table 4 shows a preliminary experiment on MVTec AD with a reduced number of training data. The number of training data for each category was reduced by the factors in the first row in Table 4, and the performance was evaluated using the smaller training datasets. The p-AUC performance of the proposed method outperformed the baseline in all cases. It can be said that the smaller scale dataset is more acceptable with the proposed method compared to the baseline.

Feature Descriptions. We also show the behaviors with respect to the pre-trained feature extractors and the layer indices employed for the feature descriptions. Table 5 shows the changes in p-AUC with respect to the feature descriptors. There were three kinds of backbones (EfficientNet-B5, WideResNet-50, or MixNet-L), and five usage patterns of the first three layers (Layer1, Layer2, Layer3, Layer2 + Layer3, or Layer1 + Layer2 + Layer3). MixNet-L achieved the highest accuracy performance. The use of layers with larger scales led to higher computational costs while Layer1 did not affect the accuracy improvements.

Parameters. The behaviors of parameters of the proposed method are shown. Figure 3 show the robustness with respect to the parameters K and α, which respectively control the number of reference images and the weights of fluctuations. The p-AUC performance was stably improved by higher K values while there exists a trade-off between accuracy and computational cost. Note that the proposed method intuitively requires approximately $K+1$ times longer computational time than the original SPADE method because a computational bottleneck exists in the neighbor search for the feature subtraction (Eq. (1)).

Table 5. Ablation study of the p-AUC performance with respect to feature extractors (rows), and different CNN layers (columns) on the MVTec AD.

	Layer 1	Layer 2	Layer 3	Layer 2 + Layer 3	Layer 1 + Layer 2 + Layer 3
EfficientNet-B5	95.5	95.8	96.5	96.0	94.2
WideResNet-50	96.0	95.9	95.2	96.8	96.5
MixNet-L	96.8	96.8	96.7	97.2	97.1

Input   Groundtruth   SPADE   Prop

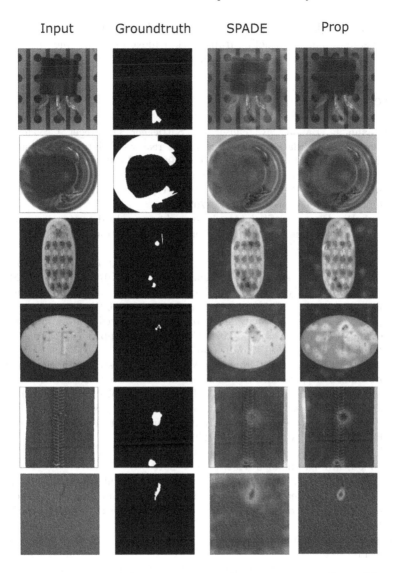

Fig. 4. Illustration of the resultant anomaly score maps in MVTec AD.

5   Conclusion

In this paper, we proposed a simple and novel anomaly localization framework under unsupervised settings. The proposed framework tackled with the out-of-distribution problem corrects the conventional anomaly scores using intra-class variance score maps calculated among normal images. The experimental results showed that the accuracy performances of SPADE, which is a type of state-of-the-art anomaly detectors based on sample-based modeling, are stably improved independent of feature extractors, backbones, and data categories. The proposed sample-based modeling framework achieved comparable performances with unimodal distribution modeling methods.

References

1. Bergmann, P., Batzner, K., Fauser, M., Sattlegger, D., Steger, C.: The MVTec anomaly detection dataset: a comprehensive real-world dataset for unsupervised anomaly detection. Int. J. Comput. Vis. **129**(4), 1038–1059 (2021)
2. Liu, W., Luo, W., Lian, D., Gao, S.: Future frame prediction for anomaly detection-a new baseline. In: Proceedings of the IEEE Conference on Computer Vision and Pattern Recognition, pp. 6536–6545 (2018)
3. Sultani, W., Chen, C., Shah, M.: Real-world anomaly detection in surveillance videos. In: Proceedings of the IEEE Conference on Computer Vision and Pattern Recognition, pp. 6479–6488 (2018)
4. Santhosh, K.K., Dogra, D.P., Roy, P.P.: Anomaly detection in road traffic using visual surveillance: a survey. ACM Comput. Surv. (CSUR) **53**(6), 1–26 (2020)
5. Lee, T., Chun, C., Ryu, S.K.: Detection of road-surface anomalies using a smartphone camera and accelerometer. Sensors **21**(2), 561 (2021)
6. Gasparini, R., et al.: Anomaly detection for vision-based railway inspection. In: Bernardi, S., et al. (eds.) EDCC 2020. CCIS, vol. 1279, pp. 56–67. Springer, Cham (2020). https://doi.org/10.1007/978-3-030-58462-7_5
7. Rippel, O., Mertens, P., Merhof, D.: Modeling the distribution of normal data in pre-trained deep features for anomaly detection. In: 2020 25th International Conference on Pattern Recognition (ICPR), pp. 6726–6733. IEEE (2021)
8. Schlegl, T., Seeböck, P., Waldstein, S.M., Langs, G., Schmidt-Erfurth, U.: f-AnoGAN: fast unsupervised anomaly detection with generative adversarial networks. Med. Image Anal. **54**, 30–44 (2019)
9. Ruff, L., et al.: Deep one-class classification. In: International Conference on Machine Learning, pp. 4393–4402. PMLR (2018)
10. Bergmann, P., Fauser, M., Sattlegger, D., Steger, C.: Uninformed students: student-teacher anomaly detection with discriminative latent embeddings. In: Proceedings of the IEEE/CVF Conference on Computer Vision and Pattern Recognition, pp. 4183–4192 (2020)
11. Kingma, D.P., Welling, M.: Auto-encoding variational bayes. arXiv preprint arXiv:1312.6114 (2013)
12. Gong, D., et al.: Memorizing normality to detect anomaly: memory-augmented deep autoencoder for unsupervised anomaly detection. In: Proceedings of the IEEE/CVF International Conference on Computer Vision, pp. 1705–1714 (2019)
13. Sabokrou, M., Khalooei, M., Fathy, M., Adeli, E.: Adversarially learned one-class classifier for novelty detection. In: Proceedings of the IEEE Conference on Computer Vision and Pattern Recognition, pp. 3379–3388 (2018)
14. Kim, J.H., Kim, D.H., Yi, S., Lee, T.: Semi-orthogonal embedding for efficient unsupervised anomaly segmentation. arXiv preprint arXiv:2105.14737 (2021)
15. Cohen, N., Hoshen, Y.: Sub-image anomaly detection with deep pyramid correspondences. arXiv preprint arXiv:2005.02357 (2020)
16. Defard, T., Setkov, A., Loesch, A., Audigier, R.: PaDiM: a patch distribution modeling framework for anomaly detection and localization. arXiv preprint arXiv:2011.08785 (2020)

17. Gudovskiy, D., Ishizaka, S., Kozuka, K.: CFLOW-AD: real-time unsupervised anomaly detection with localization via conditional normalizing flows. In: Proceedings of the IEEE/CVF Winter Conference on Applications of Computer Vision, pp. 98–107 (2022)

18. Wen, Y., Zhang, K., Li, Z., Qiao, Y.: A discriminative feature learning approach for deep face recognition, pp. 499–515 (2016)

19. Tian, Y., Sun, C., Poole, B., Krishnan, D., Schmid, C., Isola, P.: What makes for good views for contrastive learning? Adv. Neural Inf. Process. Syst. **33**, 6827–6839 (2020)

20. Chalapathy, R., Menon, A.K., Chawla, S.: Anomaly detection using one-class neural networks. arXiv preprint arXiv:1802.06360 (2018)

21. Daniel, T., Kurutach, T., Tamar, A.: Deep variational semi-supervised novelty detection. arXiv preprint arXiv:1911.04971 (2019)

22. Kwitt, R., Hofmann, U.: Robust methods for unsupervised PCA-based anomaly detection. In: Proceedings of IEEE/IST Workshop on Monitoring, Attack Detection and Mitigation, pp. 1–3 (2006)

23. Li, K.L., Huang, H.K., Tian, S.F., Xu, W.: Improving one-class SVM for anomaly detection. In: Proceedings of the 2003 International Conference on Machine Learning and Cybernetics (IEEE Cat. No. 03EX693), vol. 5, pp. 3077–3081. IEEE (2003)

24. Deng, J., et al.: ImageNet: a large-scale hierarchical image database. In: 2009 IEEE Conference on Computer Vision and Pattern Recognition, pp. 248–255. IEEE (2009)

25. Yi, J., Yoon, S.: Patch SVDD: patch-level SVDD for anomaly detection and segmentation. In: Proceedings of the Asian Conference on Computer Vision (2020)

26. Johnson, J., Douze, M., Jégou, H.: Billion-scale similarity search with GPUs. IEEE Trans. Big Data **7**, 535–547 (2019)

27. Zagoruyko, S., Komodakis, N.: Wide residual networks. arXiv preprint arXiv:1605.07146 (2016)

28. https://github.com/byungjae89/spade-pytorch

29. Tan, M., Le, Q.V.: MixConv: mixed depthwise convolutional kernels. arXiv preprint arXiv:1907.09595 (2019)

30. Venkataramanan, S., Peng, K.-C., Singh, R.V., Mahalanobis, A.: Attention guided anomaly localization in images. In: Vedaldi, A., Bischof, H., Brox, T., Frahm, J.-M. (eds.) ECCV 2020. LNCS, vol. 12362, pp. 485–503. Springer, Cham (2020). https://doi.org/10.1007/978-3-030-58520-4_29

Toward a System for Post-Earthquake Safety Evaluation of Masonry Buildings

Giovanni Giacco[1]([✉]) [iD], Giulio Mariniello[2] [iD], Stefano Marrone[1] [iD],
Domenico Asprone[2] [iD], and Carlo Sansone[1] [iD]

[1] Department of Electrical Engineering and Information Technology (DIETI),
University of Naples Federico II, Via Claudio 21, 80125 Naples, Italy
giovanni.giacco@unina.it
[2] Department of Structures for Engineering and Architecture (DIST),
University of Naples Federico II, Via Claudio 21, 80125 Naples, Italy

Abstract. A quick and accurate post-earthquake safety assessment is critical for emergency management and reconstruction. Accurate knowledge of the scenario enables optimal use of human and economic resources. In Earth-quake prone countries, National Emergency Management Agency defines standard forms to collect information during inspections (e.g., Italian AeDES form, New Zealand Earthquake rapid assessment form, American ATC-20 Rapid evaluation safety assessment form). Assisting the technicians in the compilation of the cards and assessing their correctness guarantees a faithful reconstruction of the reality. We propose a Deep Learning-based tool that can recognize, localize, and quantify damages starting from a set of photos of the building to be assessed. The analysis results are expressed in terms of a Damage Assessment Matrix, which allows a quick association to the safety form.

Keywords: Crack detection · Deep-learning · Masonry buildings

1 Introduction

A quick and reliable assessment of structure and infrastructure is crucial for emergency management. In seismic areas, after an event occurs, rescue procedures are initiated by public protection authorities. The organization of rescues requires an expeditious assessment of the areas most subjected to damage and the identification of the safest access routes. Numerous studies about seismic damages simulation in inhabited areas are based on building characteristics. Despite the high simulation reliability, detecting actual damage conditions is essential to better characterize the scenario. In this context, systems for rapid damage assessment using Unmanned Aerial Vehicles (UAVs) could better characterize maps and define the details of rescue missions. When the rescue operations are over, rapid assessments of damaged structures are mandatory before allowing access to residents. The structures are evaluated by groups of volunteer engineers, who compile synthetic reports on the state and define the practicability of the building. Technicians quickly evaluate hundreds of buildings in very

S. Sclaroff et al. (Eds.): ICIAP 2022, LNCS 13232, pp. 312–323, 2022.
https://doi.org/10.1007/978-3-031-06430-2_26

complex contexts; this determines a high risk of error in verifying and compiling reports.

Since 1997, in Italy, the assessment report document used is the Aedes form, whose acronym in English can be translated as "Practicability and damage in the seismic emergency" [1]. It is a form for the fast detection of damages, the definition of prompt intervention measures, and evaluating post-seismic practicability of buildings with ordinary structural typology (in masonry, reinforced concrete, steel, or wood) of the building for housing and services [2]. Similar forms have been approved by North American and New Zealand authorities.

Machine Learning and Deep Learning techniques are highly used in the context of structural engineering, especially for Structural Health Monitoring and damage detection applications [3–5]. These applications are based on the analysis of the different types of data; the most common are vibration data [6–8], highly sensitive to mass and stiffness, and static data [9,10], correlated to change of stress state. These more advanced techniques are challenging to use in the post-earthquake fast assessment phase because they require an ad-hoc sensors system during all operational phases of the structure or an accurate study to estimate the pre-seismic structural condition. Therefore, for ordinary structures, there is a need for a system that can provide results quickly, albeit with a lower level of accuracy. The diffusion and development of Deep Learning (DL) algorithms for image analysis are enabling the development of several solutions based on the use of camera images or surveys for assessing the state of structures [11,12]. Such methodologies can directly detect damages from a photo or video frames or process video and derive its dynamic properties.

Several approaches have been defined for the support of engineers during a visual inspection, using a Deep Learning-based system [13,14]. Most of them are focused on bridges since codes require periodic inspections for this kind of structure. On the other hand, for masonry buildings that are very susceptible to earthquakes, there are no approaches in the literature to support or replace the on-site inspection in order to detect, localize and quantify the structural damage. This study is the first attempt to give a synthetic report of the structural damage for a masonry building through a DL-based approach to the authors' best knowledge.

In this context, we propose the use of a vision-based system to support post-disaster Building Safety Evaluations for masonry buildings. An automatic approach can speed up a process mainly based on manual inspections. In addition, the introduction of a support system based on machine learning in error-prone activities, such as the safety report filling, permits more precise correspondence between the actual state and the detected state.

The main contributions of this work can be summarised as follows:

1. A DL-based approach for crack detection, localization, and quantification on masonry buildings utilizing photos of buildings facades;
2. The preparation from scratch of a dataset of concrete cracks for model training;

3. A baseline methodology to produce a synthetic assessment of the structural damage to support the compilation of inspection forms or highlight compilation errors by technicians.

The rest of the paper is structured as follows: Sect. 2 describes the proposed approach for crack detection, localization, and quantification in masonry buildings; Sect. 3 reports and analyses obtained results, while Sect. 4 provides some conclusions and future perspectives.

2   Materials and Methods

This section, starting from the overall description of the methodology, illustrates the components that contributed to the whole approach.

2.1   The Proposed Approach

We propose an image analysis approach, using a set of pictures of the building to be assessed, that provides a synthetic tool to support the compilation and validation of the safety assessment forms filled on site. Figure 1 shows a diagram for the overall approach we proposed.

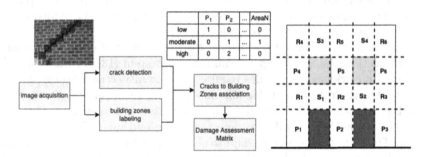

Fig. 1. The workflow of the proposed approach. On the right there is an example of the building zones subdivision: P_i represents a pier, S_i represents a spandrel, R_i a rigid zone. On the right there is an example of the matrix of damages.

This approach is based on the interaction of two different vision-based predictive models that have the objective to detect, localize and quantify the damage. The images are acquired during the inspections of the technicians, using high-resolution cameras or by UAVs that detect the different facades of the building following a predetermined path. Both two methodologies can achieve good performance. However, the first allows us to emphasize the details and then highlight the minor damage in the most accessible areas by the technician; vice versa, the drone allows a constant degree of accuracy regardless of the presence of damages. The acquired images are then processed by the two components of the Damage

Assessment system. The first component deals with the identification of the cracks and their intensity. In contrast, the second deals with the localization of the cracks concerning the schematization adopted in the macro-modeling approach [15]. Starting from detecting the openings, piers, spandrels, and rigid zones will be identified. In particular, as depicted in Fig. 1, we adopted the following labeling scheme: wall piers are areas on left and right of openings, wall spandrels are areas on top and bottom of openings, rigid zones are areas of intersection between piers and spandrels.

The result of the two components is synthesized in the matrix of damages (matrix in Fig. 1). This matrix provides information on the amount, intensity, and location of the damage. The generic element of this matrix has the number of cracks detected belonging to the i-th class of damage intensity and located in the j-th macro area. This matrix allows a rapid correlation with the state of the structure. Knowing the extent and quantity of damage, it is possible to estimate the reduction of stiffness of the different macro-areas and, through simplified approaches, evaluate the overall extent of loss of strength of the structure.

2.2   Data Preparation

Deep Learning (DL) algorithms are data-driven techniques; thus, they profoundly rely on data quality and the amount of data. Primarily, it needs to pay attention to the model's ability to adapt appropriately to new, previously unseen data, i.e., avoid overfitting training data. This aspect is of particular importance in this work. The images to be analyzed will be taken by instrumentation on drones rather than by technicians during the inspection. So, photos will be highly heterogeneous in brightness, shot angles, and backgrounds. These features make the task analyzed in this work more challenging because, in past works, special consideration was frequently paid when collecting data so that photos are taken in a homogeneous way keeping stable conditions, such as distance, angle, etc. [16–18]. These rules cannot apply strictly to photos taken by UAVs rather than by technical personnel. In addition, already existing datasets do not provide data about the severity of the cracks.

Therefore, a generic dataset, able to include several scenarios, is critical for increasing the chance of developing a tool that can generalize well and perform adequately in real cases. Considering all these requirements, a new dataset was prepared for this study. The training dataset was populated by labeling photos of post-earthquake surveys in central Italy. Additional photos were obtained from the Internet. In order to emulate the scenario where different users and instrumentations will work to the data collection, various people photographed cracks with their phones or DSLR cameras. We provided them with simple guidelines, e.g., photographing a whole single facade of a building at a time, being in the front of the facade when possible. The dataset also contained various photos collected by camera drones. In total 115 photos containing cracks, positive patches, and 222 without any crack, negative patches, were gathered. They include several strong light, distortion, and darkness types and have different resolutions. These photos were divided into 224 × 224 pixels patches, which led to 1652 patches containing cracks. Moreover, 11490 non-crack patches were randomly selected from the gathered photos. For the negative patches, two patches could not overlap. In order to increase the number of crack samples, because crack patches only consist of a small proportion of the collected images, we generated new patches with the following steps: (i) for a patch containing crack, we got all adjacent patches with an overlapping both horizontal and vertical of 50%; (ii) filtering out all generated patches which not contain cracks; (iii) rotation of each candidate patch by a random angle $\alpha \in [0°, 360°]$. Out of the generated samples from the above steps, the final dataset contained 5423 positive patches.

Multiple annotators labeled the dataset, and they were asked to annotate several pieces of information beyond the crack segmentation. In particular, they annotated the severity of the crack, i.e., low, moderate, high. Low severity was assigned to surface damage visible in close-up photos. Medium severity was assigned to small cracks (<5 mm) for which the space is not such that shadows are evident between the crack edges. High severity was associated with all damage effects with significant crack opening with shadows or other major disruption. Of the 5423 positive patches, 813 had a low severity, 1861 moderate, and 2749 high.

In addition, all building openings, i.e., doors and windows, were labeled too. From the 337 photos taken, 2250 building openings were labeled. Windows and doors were considered with their frames. We had to take care of occlusion during the labeling phase, an inevitable feature in the captured data. An occlusion occurs when objects appear in front of objects we label. In that case, we employed a simple strategy: if an object is occluded, the area was still labeled with its corresponding category. This choice arises from the fact that, in this work, we are mainly interested in identifying the whole area occupied by the object.

2.3   Crack Detection with Convolutional Neural Networks

In this research, different state-of-the-art CNNs pre-trained on ImageNet [19] were examined for their effectiveness in classifying images from building surfaces on patch level. The addressed problem was a multi-class classification where the model must discriminate between (i) non-crack, (ii) low severity crack, (iii) moderate severity crack (iv) high severity crack. As the reference dataset described in Sect. 2.2 is relatively small to enable a robust training of a complete deep learning model, transfer-learning via fine-tuning was applied. Fine-tuning was implemented by training only the fully connected (FC) layers at the top of a pre-trained network, using the new data with a low learning rate. In details, an FC layer with 512 features and rectified linear unit (ReLU) activation was added, followed by batch normalization and a dropout layer with a probability of 0.5. Batch normalization is an approach that improves the speed, performance, and stability of artificial neural networks and it is used to normalize the input layer by adjusting and scaling the activations. At the same time, dropout temporarily disconnects the neural connections between connected layers during training. Finally, an FC layer with softmax activation was placed to classify the images into one of the four classes. CNNs taken into account were: VGG16 [20], DenseNet121 [21], ResNet34 [22], ResNet50 [22].

We trained all our models using Adam optimizer with $\beta_1 = 0.9$, $\beta_2 = 0.99$, $\epsilon = 1^{-5}$, a weight decay of 1^{-2} and a batch size of 64. The considered loss function was the cross entropy (CE) loss function (L_{ce}) and is given as:

$$L_{ce} = -(ylog(p) + (1 - y)log(1 - p))$$

where y is the ground-truth and p is the probability for that class.

In order to find an optimal learning rate (LR) value, we made use of the *learning rate range test* [23]. LR range test is a technique to estimate the reasonable minimum and maximum boundary values for LR. It runs the model for several epochs while letting the LR increase linearly between low and high LR values after each mini-batch, until the loss value explodes. Plotting the accuracy versus the LR allows choosing the LR one order lower than the point where the loss is minimum.

The performance of the networks was evaluated based on the values of accuracy and F1-score, which are defined as:

$$Accuracy = \frac{TP + TN}{TP + TN + FP + FN}$$

$$F1 = 2 \times \frac{precision \times recall}{precision + recall} = \frac{TP}{TP + \frac{1}{2}(FP + FN)}$$

where TP, TN, FP and FN correspond to true positive, true negative, false positive and false negative, respectively.

Since the problem is a multi-class classification task, the single per-class scores were combined, averaging them. The dataset was split into training, validation, and test sets with ratios 70%, 20%, and 10%, respectively.

2.4   Building Zones Labeling

Section 2.1 described how, knowing building openings position, a building facade was subdivided into piers, spandrels, and rigid zones. In this work, building facade openings detection is implemented by means of image segmentation via Convolutional Neural Networks. We categorized the facade image data into two classes: windows and doors. A pseudo-class representing the background categorizes all features that do not belong to other classes. The first deep-learning approach for the semantic segmentation task was based on a fully convolutional neural network (FCN) [24]. In an FCN, the fully connected layers are replaced by convolution layers that act as deconvolution operators. The deconvolution operations restore the output feature maps to the original input resolution, resulting in a class label corresponding to each pixel, i.e., a pixel-wise mask. The spatial resolution of the feature maps, i.e., the outputs of each convolution layer, decreases throughout the feature extraction process, allowing the learned feature maps to be more invariant to small translations of the inputs. However, this downsampling process becomes a considerable concern because the process can potentially erase much information. Herein, we used a U-Net-based architecture, an improvement of FCN. U-Net [25] introduced operations called skip connections, outperforming the FCN approach. The architecture has an efficient symmetric encoder-decoder structure, with a downsampling part, i.e., the encoder, and an upsampling part, i.e., the decoder. Skip connections concatenate feature maps from the contracting path to the symmetric feature maps in the expansive path. The design allows for features representing small object information to be transmitted to higher levels of the network, better preserving small object information. Because windows and doors are small objects to detect in a building facade dataset, the benefits of the symmetric U-Net architecture are highly relevant to our problem.

Instead of using the encoder as published in the original paper, U-Net was employed as a backbone architecture, but several models replaced the encoder part. We experimented with the same models adopted in the classification task: VGG16, DenseNet121, ResNet34, ResNet50. As the reference dataset is relatively small to enable a robust training of a complete deep learning model, transfer learning via fine-tuning was applied in this task too. In particular, weights computed with the crack classification task described before were adopted as a starting point.

Models were trained with image patches of 512×512 and using Adam optimizer with $\beta_1 = 0.9$, $\beta_2 = 0.99$, $\epsilon = 1^{-5}$, a weight decay of 1^{-2} and a batch size of 10. The considered loss function was the dice loss function which is given as:

$$Dice = \frac{2\,|A \cap B|}{|A| + |B|} \qquad L_{dice} = 1 - Dice$$

where A is the predicted segmentation mask, B the ground-truth one and $|.|$ represents the number of elements in the set.

The learning rate was found with the LR range test technique discussed before. One cycle policy [26] variated the LR during the training. The technique requires an initial interval of values: we choose the maximum value using the range test and the lower as 1/5th or 1/10th of the maximum LR. Starting from this interval, the algorithm went from the lower to the higher value in step one and from the higher back to the lower in step two. We used this approach for each epoch, considering a lower value of 1/10th or 1/100th w.r.t. the maximum one in the few last epochs. An early stopping criterion was used to stop training once the model performance stopped improving on the validation dataset. The dataset was split into training, validation, and test sets with ratios 70%, 20%, and 10%, respectively.

A series of post-processing operations were applied to the segmentation masks in order to reduce imperfections by accounting for the standard structure of doors and windows, i.e., rectangular shape. Primarily, noisy data were filtered out using *remove_small_holes* and *remove_small_objects* operations from the Scikit-Image library. Later, using the OpenCV library, all contours in the binary mask were located, and for each one, the bounding rectangle was found.

3 Experimental Results

This section presents the results obtained by the trained networks for crack detection and building opening segmentation.

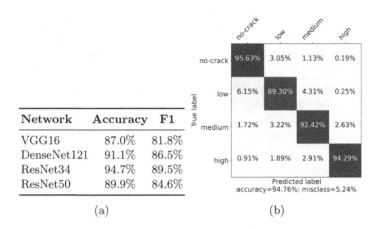

Network	Accuracy	F1
VGG16	87.0%	81.8%
DenseNet121	91.1%	86.5%
ResNet34	94.7%	89.5%
ResNet50	89.9%	84.6%

(a)   (b)

Fig. 2. Results for the patch classification task. (a) Confusion Matrix for ResNet34 model; (b) Metrics of the networks

Figure 2a enlists the obtained metrics from the trained models on the test set for the crack detection task of Sect. 2.2. While all the considered networks got high accuracy, 87% or more, ResNet34 surpassed the rest by scoring 94.7%, with an F1 score of 89.5%. In order to examine the benefit of transfer learning, ResNet34 was also evaluated without pretraining with randomly initialized weights. The accuracy and F1 dropped to 86.2% and 81.8%, revealing that transfer learning helped boost performance. Based on the collected metrics, it can be concluded that the network learns rich features that allow for correct classifications on the produced dataset.

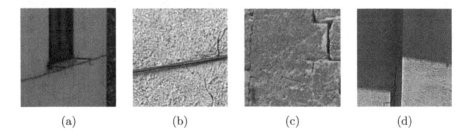

(a)   (b)   (c)   (d)

Fig. 3. (a) FP example; (b) Wrong FP: patch labeled in the dataset as "no-crack", but it contains a crack that the model detects; (c) (d) FN examples

Different cases of FP and FN predicted with ResNet34 from the validation set are displayed in Fig. 3. In Fig. 3a, part of an electrical wire is wrongly classified as crack. Evidently, a further expansion of the masonry dataset should better represent these cases. On the contrary, the situation shown in Fig. 3b needs attention: the model detects a crack, which is not correctly reported in the dataset. In this case, the model shows a superior generalization capacity detecting cracks not annotated during the dataset's creation. The confusion matrix in Fig. 2b highlights the performance of the best model. It is deduced that the model excels in predicting the non-crack case correctly. Instead, some difficulty emerges classifying patches with low severity cracks. These issues may be partially explained by the lower number of low-class examples and possibly by some mislabeling introduced by the annotators.

Table 1. Metrics of the networks used for building openings segmentation

Network	Pretrained	Parameters	Dice
U-Net-VGG16	Yes	46.1M	70.6%
U-Net-DenseNet121	Yes	41.6M	70.2%
U-Net-Resnet34	Yes	48.0M	71.8%
U-Net-Resnet50	Yes	73.7M	67.9%

Table 1 shows the segmentation results from the trained networks. Dice values were very similar among the models, but U-Net-ResNet34 obtained the best, with a value of 71.8%. Although dice value is not very high and other solutions could be investigated, e.g., different segmentation models trained with different loss functions, it is important to note that in this work, the building openings segmentation was of interest only for identifying the areas of the building. This work's goal was not to find the best segmentation approach for a building facade.

Figure 4 shows how the interaction between the zone labeling and the crack detection provides the final output of the proposed approach. The damage assessment matrix summarizes the presence of cracks in the building structure, highlighting their localization in the piers, spandrels, or rigid zone of the whole structure. Technicians can easily use this information to compile the safety form or assess the correctness of an already compiled form.

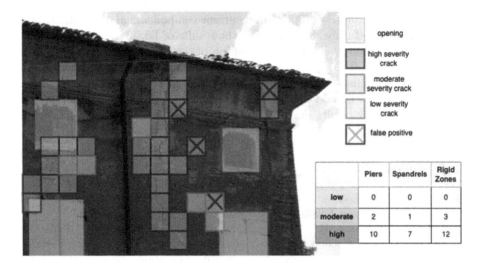

	opening
	high severity crack
	moderate severity crack
	low severity crack
	false positive

	Piers	Spandrels	Rigid Zones
low	0	0	0
moderate	2	1	3
high	10	7	12

Fig. 4. Final result example of the overall proposed approach.

4 Conclusions

In this study, a methodology to define a support system for the compilation of forms for the safety assessment of masonry buildings after a seismic event was proposed. In an emergency context, this tool is fundamental to guarantee the precision and accuracy of the data collected by field engineers, and therefore, the quality of emergency operations and reconstruction plans. The system is founded on an integrated approach based on two DL-based models that work together, evaluating cracks and localizing them in the macro-areas of the facades of a masonry building. For the training of the predictive models, a dataset with photos from masonry structures was assembled containing complex backgrounds and various crack types and sizes. Different Deep Learning (DL) networks were evaluated, and by leveraging the effect of transfer learning, crack detection on masonry surfaces was performed on a patch level. ResNet34 obtained the highest accuracy, which was 94.7%. In addition, a DL model for building openings segmentation was assessed in order to subdivide the building facade into piers, spandrels, and rigid zones. U-Net-ResNet34 obtained the best result with a dice score of 71.8%. Mixing these pieces of information, the damage assessment matrix was proposed as a synthetic perspective of the structural damage of the building.

Currently, the system has some constraints related to the characterization of the cracks discovered on the building surface. The proposed approach was founded on patch-based crack detection. Although it permitted promising results in terms of accuracy with a small dataset, it allowed quantifying the damage just in terms of the number of the detected areas interested by a crack. This number depends on the patch's size and cannot describe proper measurements of crack characteristics, including the area, perimeter, width, length, and orientation. So, the system can be improved both in its current components and by adding new components in the future. In particular, the results of DL methods heavily rely on the dataset's quality. Thus, the expansion of the current masonry dataset is highly recommended. A semantic segmentation approach could be analyzed to detect cracks and characterize them in terms of area, width and length. Finally, new modules could be added to allow the automatic compilation of forms, the digitization of their results in a Building Information Modeling (BIM) environment, and the optimized design of rehabilitation interventions.

References

1. Italian Civil Protection Department, AeDES form. www.protezionecivile.gov.it/en/normativa/dpcm-dell8-luglio-2014. Accessed 20 Mar 2022
2. Dolce, M., Speranza, E., Dalla Negra, R., Zuppiroli, M., Bocchi, F.: Constructive features and seismic vulnerability of historic centres through the rapid assessment of historic building stocks. The experience of Ferrara, Italy. In: Toniolo, L., Boriani, M., Guidi, G. (eds.) Built Heritage: Monitoring Conservation Management. RD, pp. 165–175. Springer, Cham (2015). https://doi.org/10.1007/978-3-319-08533-3_14
3. Farrar, C.R., Worden, K.: Structural Health Monitoring: A Machine Learning Perspective. Wiley, Hoboken (2012)
4. Mondal, T.G., et al.: Deep learning-based multi-class damage detection for autonomous post-disaster reconnaissance. Struct. Control Health Monit. **27**(4), e2507 (2020)
5. Mariniello, G., et al.: Layout-aware extreme learning machine to detect tendon malfunctions in prestressed concrete bridges using stress data. Autom. Constr. **132**, 103976 (2021)
6. Mariniello, G., et al.: Structural damage detection and localization using decision tree ensemble and vibration data. Comput. Aided Civ. Infrast. Eng. **36**, 1129–1149 (2020)
7. Abdeljaber, O., et al.: Real-time vibration-based structural damage detection using one-dimensional convolutional neural networks. J. Sound Vib. **388**, 154–170 (2017)
8. Huseynov, F., et al.: Bridge damage detection using rotation measurements-experimental validation. Mech. Syst. Signal Process. **135**, 106380 (2020)
9. Glisic, B., et al.: Damage detection and characterization using long-gauge and distributed fiber optic sensors. Opt. Eng. **52**(8), 087101 (2013)
10. Bocherens, E., et al.: Damage detection in a radome sandwich material with embedded fiber optic sensors. Smart Mater. Struct. **9**(3), 310 (2000)
11. Dworakowski, Z., et al.: Vision-based algorithms for damage detection and localization in structural health monitoring. Struct. Control Health Monit. **23**(1), 35–50 (2016)
12. Patterson, B., et al.: Deep learning for automated image classification of seismic damage to built infrastructure. In: Eleventh US National Conference on Earthquake Engineering (2018)
13. Koch, C., et al.: Achievements and challenges in machine vision based inspection of large concrete structures. Adv. Struct. Eng. **17**(3), pp. 303–318 (2014)
14. Yuan, C., et al.: A novel intelligent inspection robot with deep stereo vision for three-dimensional concrete damage detection and quantification. Struct. Health Monit., 14759217211010238 (2021)

15. Brencich, A., Gambarotta, L ., Lagomarsino, S.: A macroelement approach to the three-dimensional seismic analysis of masonry buildings. In: 11th European Conference on Earthquake Engineering, vol. 90, pp. 1–10 (1998)
16. Li, S., Zhao, X., Zhou, G.: Automatic pixel-level multiple damage detection of concrete structure using fully convolutional network. Comput. Aided Civ. Infrastruct. Eng. **34**(7), 616–634 (2019)
17. Dorafshan, S., Thomas, R.J., Maguire, M.: Comparison of deep convolutional neural networks and edge detectors for image-based crack detection in concrete. Constr. Build. Mater. **186**, 1031–1045 (2018)
18. Özgenel, Ç.F., Gönenç Sorguç, A.: Performance comparison of pretrained convolutional neural networks on crack detection in buildings. In: ISARC. Proceedings of the International Symposium on Automation and Robotics in Construction, vol. 35, pp. 1–8. IAARC Publications (2018)
19. Deng, J.: ImageNet: a large-scale hierarchical image database. In: IEEE Conference on Computer Vision and Pattern Recognition, pp. 248–255. IEEE (2009)
20. Liu, S., Deng, W.: Very deep convolutional neural network based image classification using small training sample size. In: 3rd IAPR Asian Conference on Pattern Recognition (ACPR), pp. 730–734 (2015)
21. Huang, G., et al.: Densely connected convolutional networks. In: Proceedings of the IEEE Conference on Computer Vision and Pattern Recognition, pp. 4700–4708 (2017)
22. He, K., et al.: Deep residual learning for image recognition. In: Proceedings of the IEEE conference on Computer Vision and Pattern Recognition, pp. 770–778 (2016)
23. Smith, L.N.: Cyclical learning rates for training neural networks. In: 2017 IEEE Winter Conference on Applications of Computer Vision (WACV), pp. 464–472. IEEE (2017)
24. Long, J., Shelhamer, E., Darrell, T.: Fully convolutional networks for semantic segmentation. In: Proceedings of the IEEE Conference on Computer Vision and Pattern Recognition, pp. 3431–3440 (2015)
25. Zhou, Z., Rahman Siddiquee, M.M., Tajbakhsh, N., Liang, J.: UNet++: a nested U-net architecture for medical image segmentation. In: Stoyanov, D., et al. (eds.) DLMIA/ML-CDS -2018. LNCS, vol. 11045, pp. 3–11. Springer, Cham (2018). https://doi.org/10.1007/978-3-030-00889-5_1
26. Smith, L.N.: A disciplined approach to neural network hyper-parameters: Part 1-learning rate, batch size, momentum, and weight decay. arXiv preprint arXiv:1803.09820 (2018)

Multi-varied Cumulative Alignment for Domain Adaptation

Marc Oliu[1]([✉])[iD], Sarah Adel Bargal[3], Stan Sclaroff[3][iD], Xavier Baró[1][iD], and Sergio Escalera[1,2][iD]

[1] Universitat Oberta de Catalunya, 08018 Barcelona, Spain
moliusimon@uoc.edu
[2] Universitat de Barcelona, 08007 Barcelona, Spain
[3] Boston University, Boston, MA 02215, USA

Abstract. Domain Adaptation methods can be classified into two basic families of approaches: non-parametric and parametric. Non-parametric approaches depend on statistical indicators such as feature covariances to minimize the domain shift. Non-parametric approaches tend to be fast to compute and require no additional parameters, but they are unable to leverage probability density functions with complex internal structures. Parametric approaches, on the other hand, use models of the probability distributions as surrogates in minimizing the domain shift, but they require additional trainable parameters to model these distributions. In this work, we propose a new statistical approach to minimizing the domain shift based on stochastically projecting and evaluating the cumulative density function in both domains. As with non-parametric approaches, there are no additional trainable parameters. As with parametric approaches, the internal structure of both domains' probability distributions is considered, thus leveraging a higher amount of information when reducing the domain shift. Evaluation on standard datasets used for Domain Adaptation shows better performance of the proposed model compared to non-parametric approaches while being competitive with parametric ones. (Code available at: https://github.com/moliusimon/mca).

Keywords: Domain Adaptation · Computer vision · Neural networks

1 Introduction

Domain Adaptation (DA) is a sub-domain of transfer learning focused on applying a model trained in one or more source domains to a related target domain. Both domains share the same feature space, but the sample distribution in that space is different. The distribution discrepancy, which commonly entails some kind of transform of the feature space, is known as the domain shift. The goal of DA is to reduce said domain shift such that the method originally trained on the labeled source domain improves its accuracy on the target domain. One example

S. Sclaroff et al. (Eds.): ICIAP 2022, LNCS 13232, pp. 324–334, 2022.
https://doi.org/10.1007/978-3-031-06430-2_27

may be the difference between hand-written and computer-generated characters. While both of them share the same basic feature space, it will inevitably shift due to the different appearance between images in both domains.

DA approaches can be classified into non-parametric and parametric. While both families model the probability distribution of the samples, they do so in two different ways. Non-parametric methods compute statistical indicators such as feature-space covariances in order to approximate the Probability Density Function (PDF) of each domain, while parametric methods use learnable parameters, usually in the form of neural networks, to create a more complex model of the distribution. These two families and their methods are explained in Sect. 2.

We propose a new non-parametric DA approach. Differently from the others, it takes into account the complex internal structure of the distribution, while still retaining the main advantages of such approaches: no learnable parameters and a minimal memory footprint. This is done by minimizing, through an auxiliary loss, the discrepancy between random Cumulative Distribution Function (CDF) estimates as measured in both domains' feature spaces. In order to do away with problems caused by the high dimensionality of the feature space, a series of 1-dimensional random projections of the same are used instead. An overall alignment of the non-collapsed PDFs is obtained by using this metric on a high number of both random sample points and directions of projection.

2   Related Work

There are two main ways of tackling DA. The first uses statistical metrics to model either the probability distributions of both domains or their discrepancy. This then allows us to either find a transform between both domain distributions, or to define an auxiliary loss to minimize the domain shift during training. This family of methods can be considered more classical, many predating the popularization of deep learning. Following are some of the main approaches.

Sun et al. propose CORAL [8], a simple technique that brings the target domain feature representations of a trained model to those of the source. The target features are whitened and then re-colored using the source covariance through the transform $S = C_S^{-\frac{1}{2}} C_T^{\frac{1}{2}}$, which must be applied to each target sample at test time. Here, $C_{S/T}$ are the source and target covariances. Deep CORAL [9] is an extension of this method also proposed by Sun et al.. It bypasses this problem by jointly minimizing the classification loss alongside the squared Frobenius norm of the difference in covariance matrices. This results in a single-step training process that does not need to consider the domain shifts at test time.

MK-MMD minimization [6] introduced by Long et al., similarly to Deep CORAL, minimizes the discrepancy between target hidden representations through the use of an auxiliary loss. In this case, multiple kernel maximum mean discrepancy (MK-MMD) [3] is used instead of the domain covariances.

The above methods suffer from precision problems due to approximating the domain distributions to simplified models such as multi-variate Gaussians, which disregard any kind of internal structure the distributions may exhibit. To

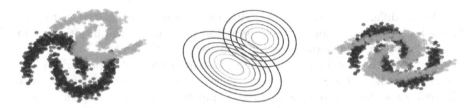

Fig. 1. Toy example of the source and target domain distributions of a two-class classification problem (moons dataset). Left: Samples from both domains, with a domain shift that includes rotation, translation and skewing. Center: PDFs of both domains. Right: Corrected domain shift given the PDFs.

more accurately represent the domain distributions, more modern approaches use auxiliary parametric models such as deep neural networks instead.

Deep Adaptation Networks (DAN) [2], put forward by Ganin *et al.*, was the first to use adversarial learning. The method consists of a feature extractor and two classifiers predicting the sample class and domain. Instead of two-stage adversarial training, the authors introduce the Gradient Reversal Layer (GRL), which inverts the gradients during back-propagation. Placing it in-between the feature extractor and domain classifier results on the domain classifier being trained to distinguish between both domains, with the feature extractor reducing the domain shift in order to fool the classifier.

Bousmalis *et al.* proposed Deep Separation Networks (DSN) [1], an extension of DANs where the feature representation is stratified by using three feature extractors: one extracts features common to both domains, while the other two are domain-specific. The class and domain classifiers work solely on the common features, while an auxiliary convolutional network uses both the common and domain-specific features to reconstruct the input images, ensuring that the full representation contains all of the information in the image.

Saito *et al.* [7] proposed an adversarial approach that considers the internal structure of the probability distributions by aligning the classification decision boundaries. The architecture consists of a feature extractor and twin class classifiers. The training consists of two stages. First, the feature extractor and both classifiers are minimized to fit the source domain samples. Second, the discrepancy of both classifiers on the target domain is maximized while the feature extractor adversarially minimizes that same discrepancy. This brings the feature space of the target domain closer to that of the source domain, the samples of the latter serving as supports defining the valid regions of the feature space. Lee *et al.* [4] use the same approach as in [7], but using the Sliced Wasserstein Distance (SWD) as the similarity metric between target domain samples.

Other works, such as ADA-DM by Xu *et al.* [10], propose alternative variants of adversarial training. Here, a neural network generates a mean and standard deviation for each sample feature. This allows both to directly evaluate the source samples with a classification network, as well as combine both distributions into a mixed domain generator. A generator-discriminator pair of modules is then

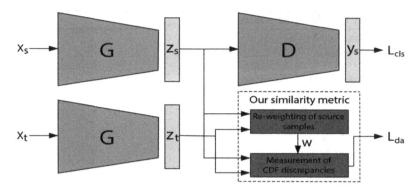

Fig. 2. General training pipeline. Feature extractor G generates features for all samples. Source samples are fed to discriminator D and evaluated with classification loss L_{cls}. DA is done in two steps. First a re-weighting of the source samples is found to correct for potential cross-domain imbalances. The CDF discrepancies between both domains are then estimated at various points and minimized through the DA loss L_{da}.

used to generate and evaluate samples of both the source, target, and mixed domains, providing an additional classification loss as well as both adversarial domain classification and custom triplet learning losses.

A more recent approach by Li *et al.* [5] combines metric and adversarial learning. In said work, a new statistical metric is used to minimize both the inter-domain divergence and maximize the intra-class density of both domains feature spaces. Additionally, a domain classifier over the label probability distributions is used as an adversarial loss.

3 Proposed Method

Non-parametric DA approaches typically use simple statistical metrics, e.g. mean and covariance, to align both domains through a simplified model of their PDFs. While this works to a degree, either most or all information regarding the internal structure of the distributions is lost. This causes two main problems:

1. The main covariance directions may be different across domains, resulting in misalignment. This is more common for low covariance components, which tend to encode less information but can still have an impact on the classifier.
2. One might obtain a generally correct alignment in terms of the major variance directions, but with the internal structure being distorted either through stretching or twisting of certain regions.

These problems are illustrated in Fig. 1 for a simple two-class classification task. If we were to remove the domain shift in the example by matching the maximum variance axes across domains, we would end up with a misalignment in the internal structure of both domains, resulting in a high classification error.

Algorithm 1. Weighting of source samples

Input: X_S, X_T, N_S, N_T

1: $\widetilde{X}_T \leftarrow X_T - \frac{1}{N_T}\mathbb{1}^{<1 \times N_T>} X_T$

2: $w \leftarrow \mathbb{1}^{<N_S \times 1>}$

3: **for** $iter$ in $[0 \ldots k]$ **do**

4:      $\widetilde{X}_S \leftarrow X_S - \frac{1}{N_T}w^T X_S$

5:      $A \leftarrow \frac{1}{N_S}(\widetilde{X}_S \widetilde{X}_S^T)^{\circ 2}$

6:      $B \leftarrow \frac{1}{N_T}(\widetilde{X}_S \widetilde{X}_T^T)^{\circ 2}$

7:      $w \leftarrow A^{-1}B\mathbb{1}^{N_T \times 1}$

8: **end for**

9: $w \leftarrow \frac{N_S}{|w|_1}w$

The proposed DA approach consists of two steps, as illustrated in Fig. 2. First, we introduce an algorithm to re-balance the source samples such that the source covariance matches that of the target (see Sect. 3.1). This mitigates problems related to relative imbalances between domains such as class imbalance. Second, we propose a novel approach based on CDF alignment that avoids the challenges associated with high dimensionality spaces. We achieve this by working on random 1-d projections of the feature space (see Sect. 3.2).

3.1   Balancing Sample Distributions

When aligning the PDFs $P_s(x_i)$ and $P_t(x_i)$, it is important to have similar sample distributions. In the case of a classification task, for instance, class imbalances between domains would hinder the alignment process. Yet, not knowing the target labels during training prevents us from balancing the datasets. To address this, we propose estimating a weighting of the source samples that increases the contribution of samples in under-represented areas of the PDF. This effectively re-balances the source domain such that it more closely matches the target PDF. We obtain these weights by minimizing the difference between the target and re-weighted source covariance matrices. Given $\widetilde{X}_S = \{x_i^{(s)} - \overline{x}^{(s)}\}$ and $\widetilde{X}_T = \{x_i^{(t)} - \overline{x}^{(t)}\}$, our goal is:

$$\arg\min_w ||\frac{1}{N_S}\widetilde{X}_S^T D_w \widetilde{X}_S - \frac{1}{N_T}\widetilde{X}_T^T \widetilde{X}_T||_F^2$$
$$= \frac{1}{N_S^2}w^T(\widetilde{X}_S\widetilde{X}_S^T)^{\circ 2}w - \frac{2}{N_S N_T}\mathbb{1}^T(\widetilde{X}_T\widetilde{X}_S^T)^{\circ 2}w + \frac{1}{N_T^2}\mathbb{1}^T(\widetilde{X}_T\widetilde{X}_T^T)^{\circ 2}\mathbb{1} \tag{1}$$

Here, D_w is the diagonal matrix whose entries are the source sample weights, N_S and N_T correspond to the number of source and target samples, and $\mathbb{1}$ is a column vector of ones. Thus, this equation aims to find the weights w minimizing the squared Frobenius norm between both covariances. This problem can be solved with the next closed form solution, where $\circ$ denotes the Hadamard power:

$$w = \left(\frac{1}{N_S}\left(\widetilde{X}_S\widetilde{X}_S^T\right)^{\circ 2}\right)^{-1}\frac{1}{N_T}\left(\widetilde{X}_S\widetilde{X}_T^T\right)^{\circ 2}\mathbb{1}^{<N_T \times 1>} \tag{2}$$

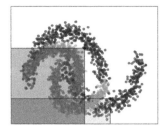

Fig. 3. Left: The CDF estimate at a given point corresponds to the fraction of samples falling on the negative octant relative to the point. Right: Proposed projection method, where a random hyper-plane containing the point cuts the space in two, effectively estimating the CDF on a 1-dimensional projection of the space.

Note that we defined the centered source samples as $\widetilde{x}_i^{(s)} = x_i^{(s)} - \overline{x}^{(s)}$. This is not accurate, since the mean changes with the weighting w such that $\widetilde{x}_i^{(s)} = x_i^s - \frac{1}{N_S} w^T X_S$. Plugging this into Eq. 1 would lead to a quadratic form for w, forcing us to resort to quadratic optimization techniques. Instead of that, we initialize $w_i = \frac{1}{N_S}$ for each sample, then iteratively compute the centered sample features and re-estimate the weights using Eq. 2. This is shown in Algorithm 1.

3.2 Aligning the Cumulative Distribution Functions

In order to align the source and target CDFs $C_s(x)$ and $C_t(x)$, given a batch of samples, we will evaluate both functions at random points of the distribution and minimize the difference across domains. Unfortunately, due to the number of samples falling into the negative octant becoming statistically insignificant, this cannot be done for high-dimensional feature spaces. To side-step this issue we evaluate the CDFs at random cut-off points and 1-d projections of both functions. Each projection then has a meaningful number of samples on each side of the cut-off point and, through many such random projections, we are able to better map the feature space. A comparison between regular CDF estimation and the proposed approach is shown in Fig. 3.

Lets define two functions $C_s(x, v)$ and $C_t(x, v)$ which take as input a point $x \in \mathbf{R}^d$ for which the CDF is evaluated, and a vector $v \in \mathbf{R}^d$ over which function $C_{\{s,t\}}(x)$ is projected. Then $C_{\{s,t\}}(x, v)$ corresponds to the probability of a sample falling to the negative side of the hyper-plane with normal v and containing the point x. Such function is represented by the following equation:

$$\widetilde{C}_s(x, v) = \frac{1}{N_s} \delta \left(X_s v - x^T v \right) \mathbb{1}^{<N_s \times 1>} \tag{3}$$

Here, $\delta(x)$ is a step function taking the value 1 for negative values and 0 otherwise. Optimizing this function through back-propagation is not possible due to $\delta(x)$ not being differentiable. Instead, we propose using a sigmoid function to approximate it. We will also want to use the sample weighting defined in Sect. 3.1

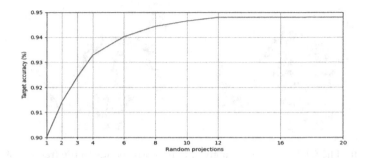

Fig. 4. Calibration of parameter n_p on the SVHN to MNIST DA problem. Each value corresponds to the average accuracy over 32 runs, relative to the number of random 1-dimensional projections of the feature space per sample.

when computing $\hat{C}_s(x, v)$ in order to find a CDF estimate over the re-balanced source dataset, which more closely matches the CDF of the target. This gives the following pair of equations:

$$\hat{C}_s(x, v) = \frac{1}{N_S}\sigma\left(X_S v - x^T v\right) w$$
$$\hat{C}_t(x, v) = \frac{1}{N_T}\sigma\left(X_T v - x^T v\right) \mathbb{1}^{<N_T \times 1>}$$

(4)

Note that these functions approximate the probability of a sample falling on the *positive* side of the hyper-plane. Since the projection vector v is picked at random and is as likely to point in one direction as the other, this does not affect the method. In order to match the projected CDF estimates of both domains, we use entropy maximization over the normalized estimates. Given a set of random hyper-planes $P = \{(x_1, v_1), .., (x_k, v_k)\}$, the DA loss L_{da} is:

$$L_{da} = \frac{1}{k}\sum_{(x,v)\in P}\overline{C}_s(x, v)log(\overline{C}_s(x, v)) + \overline{C}_t(x, v)log(\overline{C}_t(x, v))$$

$$\overline{C}_s(x, v) = \frac{\hat{C}_s(x, v)}{\hat{C}_s(x, v) + \hat{C}_t(x, v)}, \quad \overline{C}_t(x, v) = \frac{\hat{C}_t(x, v)}{\hat{C}_s(x, v) + \hat{C}_t(x, v)}$$

(5)

Multiple random projection vectors v can be used for each sample when evaluating $\overline{C}_s(x, v)$, obtaining multiple evaluations of the DA loss functions per sample. This allows us to map the probability space without needing to increase the batch size. We use this during training, with hyper-parameter n_p defining the number of random 1-dimensional space projections per sample.

Calibration is conducted for n_p in Fig. 4 using the SVHN to MNIST task (see Sect. 4 for more information). As shown, the accuracy gradually increases with the number of random feature space projections per sample, until it plateaus at around $n_p = 12$. We obtain similar results for the other datasets. As such the parameter is set to 12 for all experiments.

MNIST
USPS
SVHN

Fig. 5. Samples from the digits datasets. They consist of ten classes for the digits from 0 to 9. MNIST is the simplest, each pixel being either completely black or white. USPS introduces variations in contrast, blurriness and changes in intensity across the digit's stroke. SVHN consists of home numbers captured from the street, with varying typographic styles and containing distractors in the form of neighboring numbers.

3.3 Training Loss

The final training loss $L = L_{cls} + \lambda L_{da}$ is obtained by linearly combining the classification loss L_{cls} and DA loss L_{da}. The first is a regular cross-entropy loss, commonly used in classification problems, while the later is obtained as shown in Eq. 5. The weighting parameter λ is obtained through annealing, its value growing from 0 through the training process to asymptotically converge to Λ. This gives the model time to converge to a useful feature representation using the source domain samples, before gradually enforcing that representation to match that of the target domain. The value of λ is obtained as below, where t is the current mini-batch training iteration.

$$\lambda = \left(\frac{2}{1 + e^{-\frac{t}{1000}}} - 1 \right) \Lambda \tag{6}$$

Please note that this function follows a sigmoid curve centered at zero. Its value steadily increases, eventually converging at 1, at which point $\lambda = \Lambda$. In all of our experiments, we used $\Lambda = 100$.

4 Experiments

We focus on four common DA classification tasks on standard digit and traffic sign benchmark datasets. Figure 5 displays visual samples of the three digit classification datasets, namely MNIST, USPS and SVHN. For the traffic sign datasets, samples are shown in Fig. 6. We used the same proposed topologies for the classification models and training/testing data partitions, changing only the DA loss. We note that our method does not work properly when the DA loss is applied to the output of a convolutional layer. As such, while the network topology has remained intact in every case, the DA loss has been moved to an appropriate layer when necessary, resulting in slightly different feature extraction/classification partitions. The datasets, data partitions, and topology used for each task, are explained below.

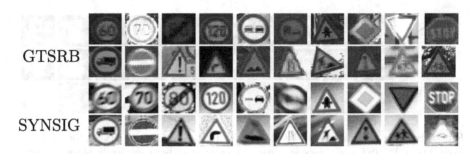

Fig. 6. Samples from both traffic sign datasets. GTSRB: Real world images under different illumination conditions and in some cases minor obstructions. SYNSIG: Synthetic dataset with the signs drawn over random backgrounds. Blur, lightning, saturation and geometric distortions are introduced to the samples.

SVHN → MNIST. SVHN consists on color images of house numbers. We use 73257 images for training and 26032 for testing. The target dataset is MNIST, which consists of hand-written monochrome digits. We use 55000 images for training and 10000 for testing. The images are resized to 32 × 32 pixels for both datasets. The feature extractor consists of three convolutional layers with convolutions of size 5 × 5, respectively with 64, 64 and 128 kernels, and a fully-connected layer with 3072 units. All layers are followed by batch normalization and, in the case of convolutional layers, a 2 × 2 max. pooling. The feature extractor consists of a dropout layer with $p = 0.5$ followed by two fully-connected layers with 2048 and 10 neurons respectively.

MNIST ↔ USPS. Both datasets display hand-written digits from 0 to 9. In the case of MNIST, we use 55000 images for training and 10000 for testing. For USPS, we have 7438 and 1860 train and test samples, respectively. In both cases the images are resized to 28 × 28 pixels. The feature extractor consists of two 5 × 5 conv. layers with 32 and 48 kernels respectively, followed by a fully-connected layer with 100 hidden units. The convolutional layers are followed by 2 × 2 max. pooling layers, and all layers implement batch normalization. The classifier consists of two fully-connected layers with 100 and 10 hidden units respectively, each trailed by a $p = 0.5$.

SYNSIG → GTSRB. Both datasets display 43 common traffic signs. SYNSIG consists of synthetic images. We use 100000 images for training and 2000 for testing. GTSRB consists of real-world images taken in Germany. We use 31367 images for training and 32171 for testing. In both cases the images are resized to 40 × 40 pixels. The feature extractor consists of three convolutional layers with kernels of size 5 × 5 × 96, 3 × 3 × 144 and 5 × 5 × 256, and a fully-connected layer with 512 hidden units. All layers are followed by batch normalization and, in the case of the convolutional layers, a 2 × 2 max. pooling. The classifier consists of a single fully-connected layer with 43 neurons, trailed by a $p = 0.5$ dropout layer.

Figure 1 shows a comparison of state-of the-art methods against our proposed approach. Methods are divided into non-parmetric distribution matching, and

Table 1. Classification accuracy of various DA models on the four most common DA tasks. The proposed approach outperforms all non-parametric approaches, obtaining results comparable to state-of-the-art parametric methods.

	SVHN *to* MNIST	MNIST *to* USPS	USPS *to* MNIST	SYNSIG *to* GTSRB
Source only	67.1%	79.4%	63.4%	85.1%
Non-parametric distribution matching methods				
MMD [6]	71.1%	81.1%	–	91.1%
DeepCORAL [9]	63.1%	80.7%	–	–
MCA (Ours)	**94.8%**	**94.6%**	**94.7%**	**95.1%**
Parametric distribution matching methods				
DAN [2]	71.1%	85.1%	73.0%	88.7%
DSN [1]	82.7%	91.3%	–	93.1%
MCD [7]	96.2%	96.5%	94.1%	94.4%
SWD [4]	**98.9%**	**98.1%**	**97.1%**	**98.6%**
DM-ADA [10]	95.5%	96.7%	94.2%	–

parametric approaches, based on the usage of additional neural modules plus adversarial training. As it can be seen, our approach far surpasses the accuracy of non-parametric methods on all datasets, despite belonging to that same category. It also outperforms many of the parametric methods, such as DAN [2] and DSN [1], surpasses MCD [7] for both SYNSIG to GTSRB and USPS to MNIST, and surpasses DM-ADA [10] for the latter. The only approach that consistently surpasses our method is SWD [4]. This is despite our approach being non-parametric, meaning that it does not require additional neural modules nor a multi-stage training approach. Overall, our method achieves accuracies comparable to the state-of-the-art without any of its disadvantages (Table 1).

5   Conclusions

We propose a non-parametric DA approach based on a novel CDF alignment metric, achieving results comparable to the state-of-the-art without the need for additional neural modules to model the PDFs nor multi-stage training approaches. This results in lower memory requirements, a simpler training pipeline and the ability to consider the internal structure of the probability distributions. The overall structure of the method is in line with classical non-parametric distribution matching methods, compared to which our approach greatly surpasses their accuracy. Furthermore, our model could be directly applied to any DA problem, regardless of whether it is a regression or classification task. This is demonstrated by the fact that DA is performed on the latent feature space instead of the label space, and it does not depend on the classification labels whatsoever, contrary to most other state-of-the-art approaches. We consider evaluating this method on other regression-based tasks as interesting future work.

Acknowledgments. This work has been partially supported by the Spanish project PID2019-105093GB-I00 and by ICREA under the ICREA Academia programme.

References

1. Bousmalis, K., Trigeorgis, G., Silberman, N., Krishnan, D., Erhan, D.: Domain separation networks. In: Advances in Neural Information Processing Systems, pp. 343–351 (2016)
2. Ganin, Y., Lempitsky, V.: Unsupervised domain adaptation by backpropagation. arXiv preprint arXiv:1409.7495 (2014)
3. Gretton, A., et al.: Optimal kernel choice for large-scale two-sample tests. In: Advances in Neural Information Processing Systems, pp. 1205–1213. Citeseer (2012)
4. Lee, C.Y., Batra, T., Baig, M.H., Ulbricht, D.: Sliced Wasserstein discrepancy for unsupervised domain adaptation. In: Proceedings of the IEEE Conference on Computer Vision and Pattern Recognition, pp. 10285–10295 (2019)
5. Li, J., et al.: Maximum density divergence for domain adaptation. IEEE Trans. Pattern Anal. Mach. Intell. **43**, 3918–3930 (2020)
6. Long, M., Cao, Y., Wang, J., Jordan, M.I.: Learning transferable features with deep adaptation networks. arXiv preprint arXiv:1502.02791 (2015)
7. Saito, K., Watanabe, K., Ushiku, Y., Harada, T.: Maximum classifier discrepancy for unsupervised domain adaptation. In: Proceedings of the IEEE Conference on Computer Vision and Pattern Recognition, pp. 3723–3732 (2018)
8. Sun, B., Feng, J., Saenko, K.: Return of frustratingly easy domain adaptation. In: Proceedings of the AAAI Conference on Artificial Intelligence, vol. 30 (2016)
9. Sun, B., Saenko, K.: Deep CORAL: correlation alignment for deep domain adaptation. In: Hua, G., Jégou, H. (eds.) ECCV 2016. LNCS, vol. 9915, pp. 443–450. Springer, Cham (2016). https://doi.org/10.1007/978-3-319-49409-8_35
10. Xu, M., et al.: Adversarial domain adaptation with domain mixup. In: Proceedings of the AAAI Conference on Artificial Intelligence, vol. 34, pp. 6502–6509 (2020)

First Steps Towards 3D Pedestrian Detection and Tracking from Single Image

Gianluca Mancusi[1,2(✉)], Matteo Fabbri[1,3], Sara Egidi[2], Mattia Verasani[2], Paolo Scarabelli[2], Simone Calderara[1], and Rita Cucchiara[1]

[1] University of Modena and Reggio Emilia, Modena, Italy
{gianluca.mancusi,matteo.fabbri,simone.calderara,
rita.cucchiara}@unimore.it
[2] Tetra Pak Packaging Solutions S.P.A., Modena, Italy
{gianluca.mancusi,sara.egidi,mattia.verasani,
paolo.scarabelli}@tetrapak.com
[3] GoatAI S.r.l., Modena, Italy
matteo.fabbri@goatai.it

Abstract. Since decades, the problem of multiple people tracking has been tackled leveraging 2D data only. However, people moves and interact in a three-dimensional space. For this reason, using only 2D data might be limiting and overly challenging, especially due to occlusions and multiple overlapping people. In this paper, we take advantage of 3D synthetic data from the novel MOTSynth dataset, to train our proposed 3D people detector, whose observations are fed to a tracker that works in the corresponding 3D space. Compared to conventional 2D trackers, we show an overall improvement in performance with a reduction of identity switches on both real and synthetic data. Additionally, we propose a tracker that jointly exploits 3D and 2D data, showing an improvement over the proposed baselines. Our experiments demonstrate that 3D data can be beneficial, and we believe this paper will pave the road for future efforts in leveraging 3D data for tackling multiple people tracking. The code is available at (https://github.com/GianlucaMancusi/LoCO-Det).

Keywords: Multiple-object tracking · Synthetic data · 3D people detection

1 Introduction

People detection and tracking in crowded scenarios are highly challenging tasks with a mature literature, with applications ranging from surveillance to autonomous driving. To effectively advance the field, the community has been adopting neural networks since the advent of deep learning [1,3,11,12,14,26].

The research history of people detection and tracking in computer vision revolve around two-dimensional approaches, where the subjects are tracked on the image plane using 2D bounding box detections. However, humans move and interact in a three-dimensional world. Not considering this important point might

S. Sclaroff et al. (Eds.): ICIAP 2022, LNCS 13232, pp. 335–346, 2022.
https://doi.org/10.1007/978-3-031-06430-2_28

lead to overcomplicated solutions to problems that could have simpler answers in a three-dimensional space. For example, disambiguating people that occlude each other might be easier in 3D than in 2D: when two people are overlapped in the 2D image plane, their spatial location is typically separable in 3D [22].

However, in crowded scenarios, people detection and tracking is still addressed using 2D data only. In fact, common trackers completely ignore the depth of the targets and only reason in terms of image coordinates. The absence of methods that leverage 3D information is mainly due to the absence of available datasets that can provide adequate annotations. Only recently, promising works on 3D people tracking have been proposed [21,22]. However, those methods are not suitable for crowded scenarios as they are trained on non-crowded datasets and do not handle strong occlusions effectively.

Here, we take advantage of the recently proposed MOTSynth [8] dataset for designing a first attempt to exploit 3D annotations for the task of people detection and tracking in crowded scenarios. MOTSynth is a large, synthetic dataset specifically designed for the tasks of pedestrian detection and tracking, that already showed outstanding generalization capabilities on real world scenarios.

Following LoCO [6] and CenterNet [36], we propose a novel 3D people detector that, given a single RGB image, is able to predict the image plane bounding boxes, as long as the camera distances of every pedestrian in the scene. The training scheme has been carried out leveraging synthetic data only (no fine-tuning on real data) while performances are evaluated on real datasets.

The newly provided detections are then used to perform an extensive study involving multiple trackers over real and synthetic test sets. Results show that 3D data is mostly beneficial, especially for sequences with strong occlusions. To the best of our knowledge, until now, no previous attempts to exploit 3D data to tackle people detection and tracking in crowded scenarios have been proven successful. We strongly believe that leveraging the third dimension could be a key factor to solve the most important challenges of multiple people tracking.

2   Related Works

Multiple Object Tracking. The problem of multiple object tracking has challenged computer vision researchers for many years. The techniques proposed are wide-ranging, and the most significant are summarized in [15]. It is possible to track any type of object, and there are trackers suitable for different targets. This is because tracked entities can have different types of motion and behavior. For example, ants follow a Brownian motion [10], while people walk in a more linear fashion. In this paper, we study tracking related to people in crowded scenarios, where people freely move and interact in indoor or outdoor locations.

One of the most popular trackers is SORT [2]. SORT is a barebones implementation of a visual multiple object tracking framework based on rudimentary data association and state estimation techniques. Its purpose is to serve as a baseline for more recent and sophisticated trackers. SORT is based on the well-known linear Kalman filter [24] to predict the state of the targets in the next time step. A famous extension of SORT is DeepSORT [29], which uses a deep neural network to compute re-ID features for the association step. In this work, we are not interested in exploiting visual cues, as we are more concerned of evaluating how the spatial location impact the performance of the tracker.

More recently, multiple successful trackers have been proposed in literature in the past few years [1,3,11,18,21,27,30–33]. Despite the ever-growing literature in the tracking community, none of these recently proposed methods utilize 3D cues to improve the performance of their trackers.

On the other hand, some attempts have been made to exploit three-dimensional information for designing tracking methodologies. However, they assume multiple camera setups [13,19,20,34], or dedicated sensors [25,28].

Only recently, promising works that target 3D people detection and tracking have been proposed [21,22]. However, those methods do not perform well on crowded scenarios, as they have been modeled on non-crowded datasets. For this reason, they do not handle strong occlusions effectively.

In this work, instead, we advocate for a novel approach that fully exploit 3D information for the task of multiple people tracking in particularly crowded scenarios from single RGB cameras.

Prediction of People's Distance. Predicting the camera distance of a person from a monocular image has always been challenging in computer vision, but with the advent of deep learning, it has become possible to tackle the problem more effectively. In particular, many works perform a multi-person 3D human pose estimation by predicting not only the distance but also the pose of every person [9,17,35]. However, those methods are not suitable for crowded scenarios peculiar of surveillance applications.

Among the recently proposed 3D methods for assessing the people's camera distance, the paper of Fabbri *et al.* [6] is indeed the most relevant to our work. In fact, the proposed LoCO architecture is able to handle 3D multi-person human pose estimation in crowded scenarios from a single image. In this work, we accordingly modified the LoCO architecture to deliver multiple 3D people detection instead of multiple pose estimation.

MOT Datasets. Visual surveillance centric datasets aim at crowded scenarios where pedestrians are interacting and occluding each other. MOTChallenge [5] is the reference benchmark for assessing the performance of multi-object tracking methods, as it provides consistently labeled crowded tracking sequences. In particular, MOT17 [16] and MOT20 [4] provide challenging sequences of crowded urban scenes, capturing severe occlusions and scale variations. None of the dataset provided in the MOTChallenge suite, however, provide 3D annotations.

Among the synthetic datasets, JTA [7], and its improved version MOTSynth [8], are the most relevant ones for surveillance applications. Both JTA and MOT-Synth have been collected utilizing the Grand Theft Auto V video-game, which simulates a city and its inhabitants in a three-dimensional world. The provided sequences are highly photorealistic while providing temporally consistent bounding boxes and instance segmentation labels, 3D poses with occlusion information, and depth maps.

3 Method

The following subsections summarize the key elements of our approach. Following the tracking-by-detection paradigm, we propose to split the tracking problem into two distinct tasks: detection and tracking. Section 3.1 describes our proposed 3D detector, while Sect. 3.2 illustrates our tracking methodology.

3.1 3D People Detection

At the core of our proposal lies our 3D people detector: LoCO-Det. We modified the original architecture in Fabbri *et al.* [6] in order to perform 3D people detection rather than 3D pose estimation.

The original architecture of LoCO consists of two separate networks, a Volumetric Heatmap Autoencoder (VHA), which compresses three-dimensional volumetric tensors $\mathfrak{H}$, representing the 3D space into a compressed 2D code $e(\mathfrak{H})$, and a Code Predictor, which takes the RGB image I of shape $3 \times H \times W$ as input and predicts the compressed 2D codes $f(I)$.

VHA is trained with the ground truth three-dimensional volumetric tensors $\mathfrak{H}$ with the objective of minimizing the reconstruction loss. Once the VHA training converges, the 2D codes $e(\mathfrak{H})$ obtained from the VHA encoder are used as supervision signal for the Code Predictor, which is trained using RGB images. At testing time, the 2D codes $f(I)$ computed by the Code Predictor are fed to the VHA decoder to obtain the original three-dimensional volumetric representation $\tilde{\mathfrak{H}} = d(f(I))$.

In the original version of LoCO, 14 volumetric tensors are predicted, one for each human joint. Our idea, inspired by the CenterNet [36] approach and based on LoCO, is to predict 3 volumetric tensors instead of 14 volumetric heatmaps. More specifically, our representation consists of one tensor for the centers, one tensor for the widths and one tensor for the heights of the bounding boxes. The new input of the modified VHA architecture takes the following structure:

$$\mathfrak{M} = [\boldsymbol{C}, \boldsymbol{S_w}, \boldsymbol{S_h}]^T$$

where $\boldsymbol{C}$ is the tensor containing the centers of the bounding boxes, $\boldsymbol{S_w}$ and $\boldsymbol{S_h}$ are tensors containing, in the same positions of the corresponding centers, the value of width and height of the bounding box.

Given a person k, we define the 2D bounding box $\boldsymbol{b}_k = (u_k, v_k, w_k, h_k)$, where u_k and v_k are the image plane coordinates of the center, w_k and h_k are the width and height respectively. For each person, we also define the camera distance $d_k = \sqrt{x_k^2 + y_k^2 + z_k^2}$ where x_k, y_k, and z_k are the 3D coordinates of the person position.

We further define a 2.5D location $\boldsymbol{p}_k = (u_k, v_k, d_k)$ where $u_k \in \{1, ..., H'\}$ and $v_k \in \{1, ..., W'\}$ are respectively the row and column indexes of the center pixel on the image plane, and $d_k \in \{1, ..., D\}$ is the quantized distance from the camera. In our experiments, we used $D = 316$, $H' = 136$ and $W' = 240$.

For $\boldsymbol{C}$, the value $\mathfrak{c}_k$ at a generic location $\boldsymbol{p}$ is obtained by centering a Gaussian with sigma $\sigma = 2$ in $\boldsymbol{p}_k$:

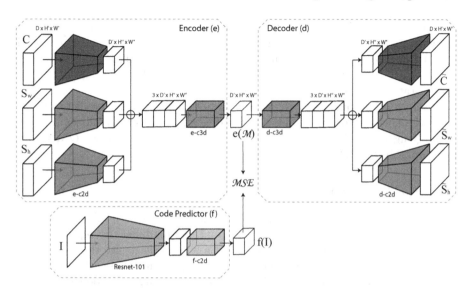

Fig. 1. Schematization of the proposed LoCO-Det pipeline. At training time, the Encoder e produces the compressed representation $e(\mathfrak{M})$ which is used as ground truth from the Code Predictor f. At test time, the intermediate representation $f(I)$ computed by the Code Predictor is fed to the Decoder d for the final output. In our case, $H' = H/8$ and $W' = W/8$.

$$\mathfrak{c}_k(\boldsymbol{p}) = e^{-\frac{\|\boldsymbol{p} - \boldsymbol{p}_k\|^2}{\sigma^2}}$$

It is possible to re-obtain the coordinates Q_C of the centers as follows:

$$Q_C = \bigcup_{n=1,\dots,N} \{\boldsymbol{p} : \mathfrak{c}_n(\boldsymbol{p}) > \boldsymbol{p}' \;\; \forall \boldsymbol{p}' \in \mathfrak{N}_{\bar{p}}\} \tag{1}$$

where $\mathfrak{N}_{\bar{p}}$ is the 6-connected neighborhood of $\bar{\boldsymbol{p}}$, i.e. the set of coordinates $\mathfrak{N}_{\bar{p}} = \{\boldsymbol{p} : \|\boldsymbol{p} - \bar{\boldsymbol{p}}\| = 1\}$ at unit distance from $\bar{\boldsymbol{p}}$.

For $\boldsymbol{S}_w$ and $\boldsymbol{S}_h$, the values $\mathfrak{s}_{w_k}$ and $\mathfrak{s}_{h_k}$ at a generic location $\boldsymbol{p}$ are obtained by centering a sphere in $\boldsymbol{p}_k$. In our experiment, we used a diameter of 5. The values inside the spheres are w_k and h_k respectively.

We modified the VHA architecture in LoCO by reducing the e-c2d and d-c2d blocks from 14 to 3. The blocks associated to width and height have their weight shared. We further modified both e-c3d and d-c3d by replacing the two 3D convolutions in each block with one 3D convolutional layer. Figure 1 top depicts our modified version of the VHA, called VTA (Volumetric Tensor Autoencoder). Finally, the VTA is trained by minimizing the loss function defined as follows:

$$L = \lambda_c L_{\mathrm{MSE}}(\boldsymbol{C}, \tilde{\boldsymbol{C}}) + \lambda_w \, L_{\mathrm{MASK}}(\boldsymbol{S}_w, \tilde{\boldsymbol{S}}_w) + \lambda_h \, L_{\mathrm{MASK}}(\boldsymbol{S}_h, \tilde{\boldsymbol{S}}_h)$$

where L_{MSE} is the MSE loss function, L_{MASK} is the masked MSE loss function and $\lambda_c \in \mathbb{R}$, $\lambda_w \in \mathbb{R}$, $\lambda_h \in \mathbb{R}$ are scaling weights.

The Code Predictor, Fig. 1 bottom, is composed by a pre-trained feature extractor (convolutional part of Resnet-101), and a fully convolutional block (f-c2d) composed of four convolutions. We trained the Code Predictor by minimizing the MSE loss between $f(I)$ and $e(\mathfrak{M})$, where $\mathfrak{M}$ is the ground truth representation associated with the image I.

During evaluation, the $Q_{\tilde{C}}$ coordinates obtained from the predicted $\tilde{C}$ are used to look up the corresponding width and height values in S_w and S_h. Finally, by applying the pinhole camera model to the computed locations $Q_{\tilde{C}}$, we are able to obtain the 3D coordinates in the standard camera system.

3.2   Tracking in 3D

Our first attempt in addressing multiple people tracking with 3D data consists in extending the SORT baseline. To this end, we modified the representation and the motion model used to propagate a target's identity into the next frame. As for SORT, we approximate the inter-frame displacements of each object with a linear constant velocity model. The state s of each target is modelled as:

$$s = (x, y, z, h, w, \dot{x}, \dot{y}, \dot{z})$$

where x, y, and z are the 3D standard camera system coordinates, h and w represent the height and width of the 2D bounding box. When a detection is associated to a target, the detected 3D coordinates and bounding box sizes are used to update the target state and the velocity components are solved optimally via a Kalman filter. If no detection is associated to the target, its state is simply predicted without correction using the linear velocity model.

In assigning detections to existing targets, each target's bounding box geometry is estimated by projecting the 3D coordinates into the image plane and using width and height to create the predicted 2D bounding box. Thus, predicting its new location in the current frame. Following SORT, the assignment cost matrix is then computed as the intersection-over-union (IoU) distance between each detection and all predicted bounding boxes from the existing targets. The assignment is solved optimally using the Hungarian algorithm.

4   Experiments

The following subsections describe the experiments and results obtained in this work. In Sect. 4.1 the implementation details of the LoCO-Det architecture and the related detection results are presented, while in Sect. 4.2 the experiments carried out with different trackers and the results obtained in the comparison between 2D and 3D are discussed.

4.1   LoCO-Det Results and Implementation Details

As explained in Sect. 3.1, the VTA is trained by taking as input 3 volumetric tensors, one containing the centers of the bounding boxes, one containing the width

Table 1. Metrics computed on MOT17 and MOTSynth datasets to evaluate the LoCO-Det detector after projecting the 3D detections into the 2D image plane. Comparison are made against Yolov3, that directly provides 2D detection.

Dataset	Detector	MODA	MODP	AP	Rcll	Prcn	TP	FP	FN	F1	FAR
MOT17 (train-set)	LoCO-Det (3D)	59.00	77.85	0.57	**68.36**	87.96	**45387**	6215	**21006**	**76.93**	1.17
	Yolov3 (2D)	**60.14**	**82.82**	**0.61**	64.58	**93.57**	42875	**2946**	23518	76.42	**0.55**
MOTSynth (test-set)	LoCO-Det (3D)	50.71	79.27	0.50	57.26	89.72	3110471	356203	2321461	69.91	1.03
	Yolov3 (2D)	**58.35**	**87.66**	**0.53**	**59.87**	**97.52**	**3266731**	**83170**	**2189196**	**74.19**	**0.24**

values and one containing the height values. During the VTA training, affine transformations, i.e. scaling and translation, are applied to the input tensors. During the Code Predictor training, the affine transformations of the augmentation applied to the volumetric tensors $\mathfrak{M}$ are also applied to the corresponding image I.

MOTSynth pedestrian with less than 20% of visible joints are ignored. If the joint of the head is visible, we consider the annotation even if it does not have 20% of visible joints. To train the Code Predictor, we utilized an ADAM optimizer with a learning rate of 0.0001, and a batch size of 12. The Code Predictor takes as input the MOTSynth images halved to 960×540.

Some results of LoCO-Det compared to Yolov3 [23] are shown in Table 1. Both architectures are trained exclusively on MOTSynth and evaluated on MOTSynth and MOT17.

Performance is measured using 2D detection metrics, as MOT17 only provides 2D annotations. From the table, it can be seen that both Yolov3 and LoCO-Det perform comparably on MOT17. However, on MOTSynth, Yolov3 outperforms LoCO-Det. The reason is due to the fact that Yolov3 is a much more robust 2D detector. Additionally, LoCO-Det has a more complex task to solve, which is to estimate the camera distances of every pedestrian in the scene. Some qualitative results on MOT17 and MOTSynth are depicted in Fig. 2.

4.2 Tracking 2D vs 3D

To study how 3D trackers can be better than 2D trackers, we decided to utilize one of the simplest trackers, namely the Nearest Neighbor tracker (NN). The reason behind this adoption is that we want to firstly study the association performance of the tracker in the 3D space, neglecting the next state prediction performance.

Nearest Neighbor Tracker. The NN tracker considers the current frame bounding box centers and computes the distance to the previous frame's bounding box centers. Bounding boxes at minimum distance will be part of the same tracklet. We implemented one version that utilize the bounding box centers in 2D and another version that utilize them in 3D.

The results in Table 2 show results obtained on MOTSynth. In the ideal case, i.e. utilizing ground truth detections, the best performing NN tracker is

Fig. 2. Qualitative results of LoCO-Det. The bounding box distance from the camera increases as the color of the bounding box changes from green to red. Images (a, b) are taken from MOT17. While (c, d) are taken from MOTSynth. (Color figure online)

definitely the one that exploit 3D information. This indicates that, by improving the detector, the 3D tracker can greatly outperform its 2D counterpart. By replacing the ground truth detection with the ones obtained with LoCO-Det, the 3D tracker still outperforms the 2D one. Using the observations provided by LoCO-Det and evaluating on MOT17, we see in Table 3 that the 3D tracker, even in the real case, can outperform the 2D tracker.

Table 3 similarly shows the performances on MOT17, divided by sequence. The only sequence of MOT17 where the 3D tracker is significantly outperformed by the 2D tracker is the sequence 4, which is captured with a high angle of view. In this particular case, the 3D tracker does not bring significant advantage, as the people rarely occlude each other.

SORT 3D. We defined the SORT tracker as our baseline, which we compare against our first 3D adaptation called SORT-3D. Here, the Kalman filter perform the prediction in the 3D space, neglecting w and h, that are only updated with the new observations. The Hungarian algorithm is applied to a cost matrix created using 3D Euclidean distances between the centers of the targets and the new observations. The maximum distance between two centers is set to 2 m. Tracklets are deleted after being lost for 11 frames. For a tracklet to be valid, it must be matched to an observation at least twice.

Table 2. The Nearest Neighbor (NN) tracker on MOTSynth test set evaluated with the detections provided by LoCO-Det and ground truth detections.

Observations	Tracker	MOTA ↑	HOTA ↑	IDF1 ↑	IDSW ↓	FP ↓	FN ↓
Ground-Truth	NN-2D	88.9	58.5	47.9	205592	732713	90095
	NN-3D	**90.6**	**76.7**	**67.7**	**65850**	**717358**	**90412**
LoCO-Det	NN-2D	28.6	16.1	11.7	267344	265950	6103199
	NN-3D	**28.9**	**16.7**	**12.5**	**258167**	**260773**	**6091272**

Table 3. Results of the NN tracker on MOT17, divided by sequence.

	Tracker	MOTA ↑	HOTA ↑	IDF1 ↑	IDSW ↓	FP ↓	FN ↓
MOT17-02	NN-2D	18.7	22.4	17.3	**342**	1077	**13672**
	NN-3D	**18.8**	**23.9**	**18.9**	345	**1042**	13699
MOT17-04	NN-2D	**51.5**	**32.6**	**26.9**	**1087**	**1767**	20211
	NN-3D	50.7	28.3	21.6	1388	1931	**20109**
MOT17-05	NN-2D	39.9	**37.4**	**34.5**	341	336	**3478**
	NN-3D	**40.9**	35.1	33.8	**295**	**288**	3504
MOT17-09	NN-2D	59.6	34.1	30.2	185	**50**	1911
	NN-3D	**61.5**	**41.1**	**36.7**	**144**	54	**1851**
MOT17-11	NN-2D	**33.5**	23.5	19.1	358	451	7725
	NN-3D	34.2	**27.7**	**23.3**	**299**	**428**	**7721**
MOT17-13	NN-2D	60.4	43.8	37.5	186	250	3298
	NN-3D	**61.0**	**46.0**	**40.4**	**156**	**249**	**3278**

The experiments in Table 4 show that with an ideal detector, i.e. using ground truth observations, SORT-3D is much better than SORT, especially in terms of IDSW. However, using the observations from LoCO-Det, SORT-3D is outperformed by the traditional implementation of SORT (2D version). This is because SORT associates targets and new observation utilizing the IoU. In fact, the IoU allows the association step to be aware not only of the location of people, but also of their shape. Instead, during the associations, SORT-3D only considers the distance between 3D centers, that might have noise.

SORT 2D + 3D. To take the most out of both 2D and 3D information, we designed a tracker that predicts the new state in the 3D space, then moves into 2D space for the association phase, as explained in Sect. 3.2. The IoU threshold used in our experiments is 0.15.

Our newly proposed SORT 2D+3D tracker gets better results than the classic version of SORT, as shown in Table 5. The improvement, although not pronounced, demonstrates that 3D data can be a viable solution to tackle multiple people tracking with another perspective.

Table 4. Comparison on MOTSynth between SORT 2D based on IoU association, SORT 3D based on Euclidean Distance association, and SORT 2D+3D which performs the Kalman filter predictions in 3D and project the bounding boxes in 2D, performing an IoU association.

Detector	Tracker on MOTSynth	MOTA ↑	HOTA ↑	IDF1 ↑	IDSW ↓	FP ↓	FN ↓
Ground-Truth	SORT 2D (IoU-based)	86.7	68.6	59.8	134033	705092	396998
	SORT 3D (distance-based)	85.7	**76.2**	**67.4**	**56263**	910784	**356740**
	SORT 2D+3D (IoU)	**86.8**	69.7	60.9	131011	**701964**	397139
LoCO-Det	SORT 2D (IoU-based)	**29.0**	21.8	18.1	160415	181044	**6266115**
	SORT 3D (distance-based)	27.2	19.9	16.6	198500	234060	6344827
	SORT 2D+3D (IoU)	**29.0**	**21.9**	**18.3**	**159389**	**180777**	6267953

Table 5. Comparison on MOT17 using the same trackers of Table 4

Tracker on MOT17-train	MOTA ↑	HOTA ↑	IDF1 ↑	IDSW ↓	FP ↓	FN ↓
SORT 2D (IoU-based)	**41.6**	40.7	37.8	1512	**3189**	**60889**
SORT 3D (distance-based)	39.3	35.8	32.5	2337	3766	62008
SORT 2D+3D (IoU)	**41.6**	**41.0**	**38.1**	**1497**	3210	60928

5   Discussion

In this paper, we proposed an effective way of addressing 3D people detection from single images. We trained our novel LoCO-Det on synthetic data only without fine-tuning on real data, and provide state-of-the-art results on real data.

We further show that 3D information can be beneficial, especially when people are occluding each other, as ID switches are consistently higher when relying on 2D data only.

Finally, we showed a first attempt in designing a competitive tracker able to perform a 3D reasoning, with the hope to inspire future researcher in exploiting 3D information for solving the most important challenges of multiple people tracking.

Acknowledgements. Partially supported by the PREVUE "PRediction of activities and Events by Vision in an Urban Environment" project (CUP E94I19000650001), PRIN National Research Program, Italian Ministry for Education, University and Research (MIUR), by ROADSTER "Road Sustainable Twins in Emilia Romagna" project, International Foundation Big Data and Artificial Intelligence for Human Development, and by Tetra Pak Packaging Solutions S.P.A.

References

1. Bergmann, P., Meinhardt, T., Leal-Taixé, L.: Tracking without bells and whistles. In: ICCV (2019)

2. Bewley, A., et al.: Simple online and realtime tracking. In: 2016 IEEE International Conference on Image Processing (ICIP), September 2016
3. Brasó, G., Leal-Taixé, L.: Learning a neural solver for multiple object tracking. In: CVPR (2020)
4. Dendorfer, P., et al.: Mot20: a benchmark for multi object tracking in crowded scenes. arXiv preprint arXiv:2003.09003 (2020)
5. Dendorfer, P., et al.: Motchallenge: a benchmark for single-camera multiple target tracking. Int. J. Comput. Vis. **129**(4), 845–881 (2021)
6. Fabbri, M., et al.: Compressed volumetric heatmaps for multi-person 3D pose estimation. In: Proceedings of the IEEE/CVF Conference on Computer Vision and Pattern Recognition, pp. 7204–7213 (2020)
7. Fabbri, M., et al.: Learning to detect and track visible and occluded body joints in a virtual world. In: Ferrari, V., Hebert, M., Sminchisescu, C., Weiss, Y. (eds.) ECCV 2018. LNCS, vol. 11208, pp. 450–466. Springer, Cham (2018). https://doi.org/10.1007/978-3-030-01225-0_27
8. Fabbri, M., et al.: MOTSynth: how can synthetic data help pedestrian detection and tracking? In: International Conference on Computer Vision (ICCV) (2021)
9. Fan, T., et al.: Revitalizing optimization for 3D human pose and shape estimation: a sparse constrained formulation. In: Proceedings of the IEEE/CVF International Conference on Computer Vision (ICCV) (2021)
10. Gordon, D.M., Paul, R.E., Thorpe, K.: What is the function of encounter patterns in ant colonies? Anim. Behav. **45**(6), 1083–1100 (1993). ISSN: 0003-3472
11. Huang, Y., et al.: SQE: a self quality evaluation metric for parameters optimization in multi-object tracking. In: CVPR (2020)
12. Kim, C., Li, F., Rehg, J.M.: Multi-object tracking with neural gating using bilinear LSTM. In: Ferrari, V., Hebert, M., Sminchisescu, C., Weiss, Y. (eds.) ECCV 2018. LNCS, vol. 11212, pp. 208–224. Springer, Cham (2018). https://doi.org/10.1007/978-3-030-01237-3_13
13. Kwon, O.-H., Tanke, J., Gall, J.: Recursive Bayesian filtering for multiple human pose tracking from multiple cameras. In: Ishikawa, H., Liu, C.-L., Pajdla, T., Shi, J. (eds.) ACCV 2020. LNCS, vol. 12623, pp. 438–453. Springer, Cham (2021). https://doi.org/10.1007/978-3-030-69532-3_27
14. Leal-Taixé, L., Canton-Ferrer, C., Schindler, K.: Learning by tracking: Siamese CNN for robust target association. In: CVPR Workshops (2016)
15. Luo, W., et al.: Multiple object tracking: a literature review. Artif. Intell. **293**, 103448 (2021)
16. Milan, A., et al.: MOT16: a benchmark for multi-object tracking. arXiv preprint arXiv:1603.00831 (2016)
17. Moon, G., Chang, J.Y., Lee, K.M.: Camera distance-aware top-down approach for 3D multi-person pose estimation from a single RGB image. In: ICCV, pp. 10133–10142 (2019)
18. Pang, J., et al.: Quasi-dense similarity learning for multiple object tracking, June 2021
19. Pham, N.T., Huang, W., Ong, S.H.: Probability hypothesis density approach for multi-camera multi-object tracking. In: Yagi, Y., Kang, S.B., Kweon, I.S., Zha, H. (eds.) ACCV 2007. LNCS, vol. 4843, pp. 875–884. Springer, Heidelberg (2007). https://doi.org/10.1007/978-3-540-76386-4_83
20. Quach, K.G., et al.: DyGLIP: a dynamic graph model with link prediction for accurate multi-camera multiple object tracking. In: CVPR, pp. 13784–13793, June 2021

21. Rajasegaran, J., et al.: Tracking people by predicting 3D appearance, location & pose. ArXiv abs/2112.04477 (2021)
22. Rajasegaran, J., et al.: Tracking people with 3D representations. In: NeurIPS (2021)
23. Redmon, J., Farhadi, A.: YOLOv3: an incremental improvement. arXiv preprint arXiv:1804.02767 (2018)
24. Reid, D.B.: An algorithm for tracking multiple targets. IEEE Trans. Autom. Control **24**, 843–854 (1979)
25. Sato, S.: Multilayer lidar-based pedestrian tracking in urban environments. In: IEEE Intelligent Vehicles Symposium, pp. 849–854. IEEE (2010)
26. Son, J., et al.: Multi-object tracking with quadruplet convolutional neural networks. In: CVPR (2017)
27. Tokmakov, P., et al.: Learning to track with object permanence (2021)
28. Weng, X., et al.: GNN3DMOT: graph neural network for 3D multi-object tracking with 2D-3D multi-feature learning. In: CVPR, pp. 6499–6508 (2020)
29. Wojke, N., Bewley, A., Paulus, D.: Simple online and realtime tracking with a deep association metric. In: ICIP, pp. 3645–3649. IEEE (2017)
30. Xu, Y., et al.: How to train your deep multi-object tracker. In: CVPR (2020)
31. Yin, J., et al.: A unified object motion and affinity model for online multi-object tracking. In: CVPR (2020)
32. Zeng, F., et al.: MOTR: end-to-end multiple-object tracking with transformer. arXiv preprint arXiv:2105.03247 (2021)
33. Zhang, Y., et al.: ByteTrack: multi-object tracking by associating every detection box. arXiv preprint arXiv:2110.06864 (2021)
34. Zhang, Y., et al.: 4D association graph for realtime multi-person motion capture using multiple video cameras. In: CVPR, pp. 1324–1333 (2020)
35. Zheng, C., et al.: 3D human pose estimation with spatial and temporal transformers. In: Proceedings of the IEEE/CVF International Conference on Computer Vision (ICCV), pp. 11656–11665, October 2021
36. Zhou, X., Wang, D., Krähenbühl, P.: Objects as points arXiv preprint arXiv:1904.07850 (2019)

Connected Components Labeling on Bitonal Images

Federico Bolelli[(✉)], Stefano Allegretti, and Costantino Grana

Dipartimento di Ingegneria "Enzo Ferrari", Università degli Studi di Modena e Reggio Emilia, Modena, Italy
{federico.bolelli,stefano.allegretti,costantino.grana}@unimore.it

Abstract. Several algorithmic solutions for the optimization of Connected Components Labeling have been proposed in literature. Among them, one of the most effective is a block-based mask to drastically reduce the number of memory accesses during the labeling procedure. This paper proposes a systematic approach for labeling multiple pixels at once, automatically generating the actions to be performed on the current pixel/block given the mask values. The proposed strategy allows to extend existing techniques for the generation of optimal decision trees to much more complex masks, where the connectivity between pixels inside a block is not guaranteed. A showcase application, consisting in the design of an efficient CCL algorithm for bitonal images, demonstrates the effectiveness of our proposal in terms of speed and memory footprint.

Keywords: Connected components labeling · Bitonal images · Optimal decision trees

1 Introduction

The task of labeling connected components, also known as Connected Components Labeling or CCL in short, aims at producing a description of the objects inside binary images, by generating a symbolic image where each pixel of a single connected component (object) is assigned a unique identifier. Objects inside binary images are usually defined according to the pixel neighborhood, which can be either *4-* or *8-neighborhood* for 2D-images. The rest of the paper will focus on the 8-neighborhood.

Connected components labeling represents a fundamental pre- and post-processing step for many Computer Vision and Image Processing pipelines [3, 6,8,11,12,14,18,25,26,28,29,31]. CCL has an exact output, and therefore different algorithmic solutions are only compared in term of speed and memory footprint. After the introduction of the task in the Sixties, several proposals were made in the course of decades to optimized its computational load, both for sequential [5,20,24,32] and parallel architectures [1,23,27,33]. Among the different algorithmic solutions, block-based scan approaches (i.e. label a block of 2×2 pixels at once) [9,10,17], decision trees [19,32] and state prediction [15,22] (i.e. reuse the information gathered during the previous step when labeling the

S. Sclaroff et al. (Eds.): ICIAP 2022, LNCS 13232, pp. 347–357, 2022.
https://doi.org/10.1007/978-3-031-06430-2_29

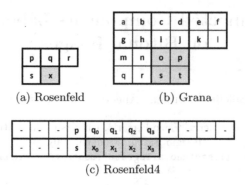

(a) Rosenfeld   (b) Grana

(c) Rosenfeld4

Fig. 1. Example of scan masks. Gray squares identify current pixels to be labeled using information extracted from white pixels. (a) and (b) are very common masks employed in CCL (c) is an extended version of the Rosenfeld mask that is proposed and analyzed in this paper. In this specific case, the current pixels, i.e. x_0, x_1, x_2, x_3, do not necessarily share the same label. Dashes identify meaningless pixels.

current pixel/block) revealed to be some of the most valuable strategies, especially when combined together [4].

Binary images can be efficiently stored with only 1 bit per pixel ("1-bit graphics" format, or bitonal images). This representation is especially useful in embedded systems with limited resources, where memory usage must be reduced to a minimum. In banking, as an example, the bitonal image is the legally recognized standard for electronic check clearing in the United States and many other countries. Working with 1-bit per pixel images (also denoted as 1-bpp or bitonal images) allows to considerably reduce the amount of memory accesses; on the other hand, it also requires additional bitwise operations for retrieving single pixel values.

In the context of 1-bpp images, being able of labeling an entire byte as a single block would guarantee a significant performance improvement, without requiring to convert the input into a 1-byte per pixel image. Unfortunately, the assumption that foreground pixels are always connected inside a block does not hold in such a case, and algorithms proposed in literature for the automatic generation of binary decision trees are not feasible. This paper introduces a systematic approach for generating all the possible actions associated to a scanning mask, which is employed to design an extremely fast CCL algorithm for bitonal images capable of labeling four consecutive pixels at once.

The rest of this paper is organized as follows. Section 2 resumes the latest contributions on connected components labeling; the proposed strategy is described in Sect. 3, and the result is evaluated in Sect. 4. Finally, in Sect. 5 conclusions are drawn.

2   Related Work

Originally introduced by Rosenfeld and Pfaltz [30], connected components labeling has a very long story, full of different strategies and proposals. Since its first appearance in 1966, many papers showed algorithms to improve the efficiency of the task. Traditionally, the fastest CCL algorithms employ a two scan strategy. In the first scan, each pixel is assigned a provisional label determined using a mask of already visited pixels, such as the one in Fig. 1a, and possible equivalences between labels are recorded. Then, a representative label is established for each connected component, and the second scan replaces provisional labels with final ones.

Several strategies have been proposed for the resolution of label equivalence, and the most commonly seen in literature employ some variation of the union-find. The union-find data structure, first applied to CCL by Dillencourt *et al.* [13], provides two convenient procedures to deal with equivalence classes of labels: *Find*, which retrieves the representative label of an equivalence class, and *Union*, which merges two equivalence classes into one, ensuring that they share the same representative label.

After the introduction of union-find, a significant improvement was provided by Wu *et al.* in [32]. The authors proved an optimal strategy, based on a manually identified decision tree, to reduce the average number of load/store operations during the first scan of the input image, driven by the Rosenfeld mask (Fig. 1a). The resulting algorithm have been christened *Scan Array-based Union Find*, or SAUF in short.

In 2010, Grana *et al.* [17] introduced a major breakthrough, consisting in a 2 × 2 block-based approach (Fig. 1b). The problem was modeled as a *command execution metaphor*: values of pixels in the scanning mask constitute a *rule* (binary string), which is associated to a set of equivalent actions in an *OR*-decision table (Fig. 2). Given this decision table, an algorithm can simply read all the pixels inside the mask, identify the rule, and find the action to be performed in the corresponding column. In [19], a dynamic programming approach to convert *OR*-decision tables into optimal binary decision trees was proposed, in order to minimize the average number of conditions to be checked when choosing the correct action to be performed. The resulting algorithm is denoted as *Block-Based Decision Tree*, or BBDT. Many improvements were published since then [21].

In 2014, He *et al.* [22] demonstrated the possibility to use a finite state machine to summarize the value of already inspected pixels during the horizontal mask movement.

In [15], decision trees and configuration transitions are combined in a decision forest, where each previous pattern allows to "predict" some of the current configuration pixel values, thus allowing for automatic code generation. The first scan phase of the algorithm is ruled by a forest of decision trees connected into a single graph, where each tree derives from a reduction of the complete optimal decision tree.

Conditions																				condition outcomes
	x	0	1	1	1	1	1	1	1	1	1	1	1	1	1	1	1	1	
	p	-	0	1	0	0	0	1	1	1	0	0	0	1	1	1	0	1	
	q	-	0	0	1	0	0	1	0	0	1	1	0	1	1	0	1	1	
	r	-	0	0	0	1	0	0	1	0	1	0	1	1	0	1	1	1	
	s	-	0	0	0	0	1	0	0	1	0	1	1	0	1	1	1	1	
Actions	no op.	1																		action entries
	new		1																	
	p			1				1		1				1	1			1	
	q				1			1			1	1		1	1		1	1	
	r					1			1				1			1	1		
	s						1			1	1			1			1	1	
	p + r							1							1				
	r + s												1			1			

Fig. 2. *OR*-decision table for the Rosenfeld mask.

Additionally, in [7] Bolelli *et al.* demonstrated that switching from decision trees to Directed Rooted Acyclic Graphs (DRAGs) allows for a reduction of machine code footprint, thus lessening the impact on instruction cache.

Finally, in [4] authors managed to combine the block-based mask with state prediction and code compression: the resulting algorithm, known as *Spaghetti Labeling*, was modeled as a Directed Rooted Acyclic Graph with multiple entry points, automatically generated without manual intervention.

3   Method

In this section, the proposed method for labeling multiple pixels at once is presented. As usual, CCL is performed with two raster scans of the input, here briefly summarized.

The first scan employs a mask moving in discrete steps, which highlights the current pixel(s) to be labeled and its neighborhood, composed of already analyzed and labeled pixels; at each step, the current pixel is assigned a provisional label, and if it connects two or more connected components, their labels are recorded as equivalent by means of some variation of Union-Find [7]. This set of operations carried out for a certain mask position is known as *action*, and depends on the values of pixels inside the mask, which form a binary word known as *command* [17].

Then, the second scan simply replaces each provisional label with the chosen representative for the equivalence class, thus completing the task. While the second scan is usually fixed and nearly identical for most algorithms, the first scan is where algorithmic proposals differ the most: here several optimizations can take place, such as the aforementioned decision tree (decide the action without reading the whole command word), block-based strategy (label multiple connected pixels at once), prediction (avoid to re-read neighbor pixels known from the previous step), and compression (reduce machine code size by merging equivalent subtrees of a larger decision tree).

The new technique proposed with this work extends the block-based approach, by overcoming the limitation that all pixels to be labeled at once must be connected. In fact, with respect to previous proposals [9,10,17], limited to a block size of at most 2×2 pixels, the devised algorithm can be applied to blocks of any shape.

In the following, the term *macroblock* identifies the pixels of the mask to be labeled during the current step. A macroblock can be divided into disjoint *blocks*, each of which always contains pixels connected to each other. This ensures that a block can always be assigned only one label.

On the other hand, it is not possible to assume that only one single action is performed on the current macroblock. Taking the mask of Fig. 1c as an example, the macroblock is composed by pixels x_0, x_1, x_2, and x_3 which can be divided into two different blocks: $x_0, x_1 \in X_0$ and $x_2, x_3 \in X_1$.

In this context, it may happen that, e.g., block X_0 requires a *new label*, while X_1 must be assigned the result of a *merge* —the union procedure of the union-find data structure— of two different existing label classes.

In literature, no attempt has been made to systematically generate the set of actions associated to a macroblock-based scan mask. The block-based mask proposed in [17] (Fig. 1b), for example, shares the same actions of the Rosenfeld mask, with the only addition of some merge operations that have no effect in a pixel-based context.

Let us start with some formal definitions. Be $I : \mathcal{L} \rightarrow \{B, F\}$ a binary image, *i.e.*, a function defined over a 2-dimensional square lattice $\mathcal{L}$, where pixels only assume two possible values, background (B) and foreground (F), usually represented by integers 0 and 1 respectively.

The 8-neighborhood of pixel $p = (p_r, p_c) \in \mathcal{L}$, denoted as $\mathcal{N}_8(p)$, is the set of pixels sharing an edge or vertex with p:

$$\mathcal{N}_8(p) = \{q = (q_r, q_c) \in \mathcal{L} \mid max(|p_r - q_r|, |p_c - q_c|) \leq 1\} \tag{1}$$

Given $S \subset \mathcal{L}$, pixels $p, q \in S$ are *connected in* S, denoted as $p \diamond_S q$, if a path of neighbor foreground pixels exists, all belonging to S and leading from p to q:

$$p \diamond_S q \Leftrightarrow \exists \{s_i \in S \mid I(s_i) = F \wedge s_1 = p, s_{n+1} = q, s_{i+1} \in \mathcal{N}_8(s_i), \forall i = 1, ..., n\} \tag{2}$$

Connectivity in S is an equivalence relation, since the properties of *reflexivity*, *symmetry* and *transitivity* hold. Equivalence classes of this relation are called *Connected Components (CCs)* of S. When $S = \mathcal{L}$, we omit the subscript in the notation $p \diamond q$, and CCs of $\mathcal{L}$ are referred to as just CCs.

To better detail the proposed algorithmic solution, we divide the pixels in the mask in two subsets:

- *Outer Pixels (P_O)*, pixels inside the mask but outside the macroblock. Outer pixels already have a provisional label, since they have already been analyzed by the algorithm.
- *Inner Pixel (P_I)*, pixels inside the macroblock. Inner pixels must be assigned a provisional label in the current step.

As an example, in Fig. 1c, $P_O = \{p, q_0, q_1, q_2, q_3, r, s\}$, while $P_I = \{x_0, x_1, x_2, x_3\}$. In order to proceed with the generation of the action set, the following operations are required for each configuration of the mask:

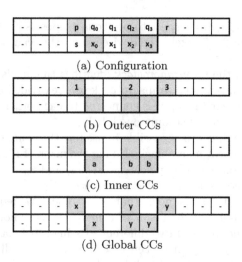

(a) Configuration

(b) Outer CCs

(c) Inner CCs

(d) Global CCs

Fig. 3. Example of the proposed action(s)-generation algorithms when applied to the mask configuration depicted in (a). Gray squares identify foreground pixels inside the mask. The pixels to be labeled using the information extracted from all the others are x_0, x_1, x_2, and x_3. Together they constitute the inner part of the mask. Remaining pixels are the outer part of the mask. Dashes identify meaningless values.

- Identify the CCs of P_O; this set of outer connected components is denoted as C_O;
- Identify connected components of P_I; these inner connected components are denoted as C_I;
- Identify the connected components for the whole mask, i.e., CCs for $P_O \cup P_I$; these are denoted as C_T;
- For each $t \in C_T$, consider all inner pixels belonging to t (i.e. the set $t \cap P_I$); all these pixels must be assigned the same label l. This label is determined analyzing the set of outer connected components contained in t, denoted as $C_O^t = \{c \in C_O \mid c \subset t\}$. If $C_O^t = \varnothing$, then a new label is created for l. If $\#C_O^t = 1$, then l can be assigned the label of any pixel of $c \in C_O^t$. Finally, if $\#C_O^t > 1$, all CCs in C_O^t must be merged, meaning that their labels are marked as equivalent and l is set to any of them (typically the smallest).

It is important to stress that $C_T \neq C_O \cup C_I$, and that the external components are defined without considering connections through pixels currently under examination (the pixels of the macroblock). The same goes for internal components, where we do not consider connections due to external pixels. Those connections are considered only for C_T.

An example of action generation is reported in Fig. 3. In Fig. 3a, a mask configuration is shown: the gray squares represent foreground pixels, x_0, x_1, x_2, x_3 are the pixels in the "current" macroblock, and dashes identify meaningless pixels. The process starts with the detection of the connected components in the outer part of the mask, i.e. ignoring the current macroblock (Fig. 3b). In this specific example, three different objects are in C_O, respectively named 1, 2, 3. Then, inner connected components are identified inside the macroblock: a and b as in Fig. 3c. Finally global connected components (C_T), x and y, are depicted in Fig. 3d. In particular, CC x contains both the inner component a and the outer component 1, so from this configuration we can derive the first operation to be performed, $a = 1$, that easily translates into action $x_0 = p$. This action means: *assign to x_0 the same label previously assigned to p*. Moreover, CC y contains inner component b and outer components 2 and 3. In this case the operation is $b = 2 + 3$: *assign to all the pixels of connected component b the result of the merge between components 2 and 3*. Translating this operation into an action, we obtain $x_2\ x_3 = q_2 + r$, that is *assign to pixels x_2 and x_3 the merge of labels previously assigned to pixels q_2 and r*.

In order to give the reader an additional example, we can consider the same configuration of Fig. 3a and add q_1 as foreground. In this case, outer and inner components are the same of Fig. 3b and Fig. 3c, but component 2 is made of two pixels (q_1 and q_2) instead of just one. On the other hand, we obtain just one single global component. This causes the algorithm to generate the action $x_0\ x_2\ x_3 = p + q_1\ q_2 + r$: x_0, x_2, and x_3 *must be assigned the result of the merge between p, one between q_1 and q_2, and r*. Actually, the *or* between q_1 and q_2 is responsible for the generation of two equivalent actions, $x_0\ x_2\ x_3 = p + q_1 + r$ and $x_0\ x_2\ x_3 = p + q_2 + r$. Most equivalence cases can be resolved with the block-based approach described in the following; the others are treated during the generation of the optimal decision tree, as explained in [19]. Basically, each global connected component generates one or multiple equivalent actions, responsible for the labeling of all pixels belonging to one or more internal components.

3.1   Reducing Actions with Blocks

The number of actions generated with the proposed approach grows very quickly as the mask size increases, making the generation of the optimal decision tree extremely hard or even impossible. In order to reduce the number of actions and simplify the problem, we introduce a block-based approach. As described above, macroblocks are divided into blocks, and pixel-based actions are replaced with block-based ones, eliminating possible duplicates. This way, many of the previous actions translate into the same one, and can be removed.

As an example, let us consider the following pixel-based actions: $x_0 = q_0$ and $x_1 = q_0\ q_1$, the latter actually representing the two equivalent actions $x_1 = q_0$ and $x_1 = q_1$. Since x_0 and x_1 are always connected, they can be viewed as part of the single block X_0, and the same applies to q_0 and q_1, which are part of the block Q_0. Thus all the three aforementioned pixel-based actions can be fused into $X_0 = Q_0$, producing the same outcome.

As previously described in literature [4,17], when working with block-based actions the second scan of an algorithm requires a slight overhead to correctly handle blocks, i.e. assigning the block label to all foreground pixels belonging to that block. Obviously, the reduction in the number of actions can be more or less significant depending on the mask features. For what concerns the mask of

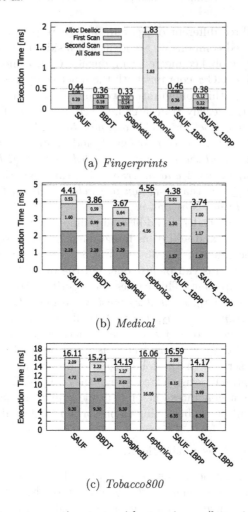

(a) *Fingerprints*

(b) *Medical*

(c) *Tobacco800*

Fig. 4. Average run-time tests with steps in ms (lower is better).

Fig. 1c, actions reduce of about 80% (from 413 to 85 actions) when moving to the block-based approach. After generating all the possible actions associated to a scan mask and the corresponding OR-decision table, the algorithm described in [19] can be employed for the generation of an optimal decision tree, which maps the mask configuration to an action, minimizing the average number of load/store operations required.

The described approach has been employed to generate an optimal decision tree for the mask of Fig. 1c, which constitutes the core of a new CCL algorithm, specifically designed for 1-bpp images. Since it shares the general structure of SAUF, but operates on 4-pixel macroblocks, it is referred to as SAUF4_1BPP.

4   Experimental Results

The performance evaluation of the proposed algorithm has been carried out with an extended version of the YACCLAB benchmark [7,16], an open source C++ framework specifically designed to test CCL algorithms. In order to incorporate a standard well-known implementation of bitonal images, the benchmark has been integrated with Leptonica, an open-source image processing library employed in several projects (e.g. Tesseract OCR by Google). The extended version of the YACCLAB benchmark, including the proposed algorithm implementation, is available at https://github.com/prittt/YACCLAB.

Experimental results discussed in the following were obtained on an Intel(R) Core(TM) i7-4790 CPU @ 3.60 GHz with Windows 10.0.17134 (64 bit) OS and MSVC 19.15.26730 compiler. Our proposal is evaluated on three datasets: *Fingerprints*, *Medical* and *Tobacco800*, which cover the most common CCL application fields, a full description can be found in [2]. Figure 4 highlights how the performance of algorithms is influenced by the different phases they are composed of: memory management, first scan and second scan.

The selected algorithms for comparison are SAUF, BBDT, Spaghetti, and the CCL implementation available in Leptonica. The first three algorithms, mentioned in Sect. 2, represent the state of the art regarding 1-byte per pixel images; for a fair comparison, their first scan times also include a conversion of the input to their preferred format. SAUF_1BPP and SAUF4_1BPP, finally, are the 1-bpp algorithms introduced in this paper. The former is a simple adaptation of SAUF, which iterates over the eight pixels stored in each byte; the latter employs a decision tree generated starting from the mask of Fig. 1c, employing the action generation algorithm introduced with this paper. All the algorithms employ the classic union-find label solver [32].

The Leptonica algorithm is based on a seedfill approach which, as can be observed, is extremely inefficient when connected components extend vertically (e.g. *Fingerprints*), causing a series of non cache-friendly memory accesses. On the other hand, when small size CCs constitute the images, Leptonica has comparable performance with SAUF_1BPP.

The main proposal of this work, SAUF4_1BPP, considerably exceeds the performance of Leptonica, with a speedup ranging from 1.13 to 4.81 depending on the dataset, and thus represents the currently most efficient CCL algorithms designed to work on bitonal images. Moreover, SAUF4_1BPP has comparable performance to Spaghetti (current state of the art for CCL on binary images), when the latter needs a prior conversion of the input. However, SAUF_1BPP only requires about $1/9\times$ memory for the input data, making it an excellent choice for use cases where memory size is constrained.

5   Conclusion

An effective solution to automatically map a connected components labeling scan mask configuration with the actions to be performed has been presented, which allows to label multiple pixels at each mask shift. A CCL algorithm generated using this systematic approach is presented, which outperforms competitors on bitonal images, confirming the effectiveness of the method.

References

1. Allegretti, S., Bolelli, F., Cancilla, M., Pollastri, F., Canalini, L., Grana, C.: How does Connected Components Labeling with Decision Trees perform on GPUs? In: Computer Analysis of Images and Patterns. vol. 11678, pp. 39–51 (2019)
2. Allegretti, S., Bolelli, F., Grana, C.: Optimized block-based algorithms to label connected components on GPUs. IEEE Trans. Parallel Distrib. Syst., 423–438 (2019). https://doi.org/10.1109/TPDS.2019.2934683
3. Allegretti, S., Bolelli, F., Pollastri, F., Longhitano, S., Pellacani, G., Grana, C.: Supporting skin lesion diagnosis with content-based image retrieval. In: 2020 25th International Conference on Pattern Recognition (ICPR), January 2021. IEEE (2021)
4. Bolelli, F., Allegretti, S., Baraldi, L., Grana, C.: Spaghetti Labeling: Directed Acyclic Graphs for Block-Based Connected Components Labeling. IEEE Trans. Image Process. **29**(1), 1999–2012 (2019)
5. Bolelli, F., Baraldi, L., Cancilla, M., Grana, C.: Connected components labeling on DRAGs. In: 2018 24th International Conference on Pattern Recognition (ICPR), pp. 121–126 (2018)
6. Bolelli, F., Borghi, G., Grana, C.: XDOCS: An Application to Index Historical Documents. In: Digital Libraries and Multimedia Archives. pp. 151–162. Springer (2018)
7. Bolelli, F., Cancilla, M., Baraldi, L., Grana, C.: Towards reliable experiments on the performance of Connected Components Labeling algorithms. J. Real Time Image Proc. **17**(2), 229–244 (2018)
8. Canalini, L., Pollastri, F., Bolelli, F., Cancilla, M., Allegretti, S., Grana, C.: Skin lesion segmentation ensemble with diverse training strategies. In: Computer Analysis of Images and Patterns, pp. 89–101 (2019)
9. Chang, W.Y., Chiu, C.C.: An efficient scan algorithm for block-based connected component labeling. In: 22nd Mediterranean Conference on Control and Automation, pp. 1008–1013 (2014)
10. Chang, W.Y., Chiu, C.C., Yang, J.H.: Block-Based Connected-Component Labeling Algorithm Using Binary Decision Trees. Sensors **15**(9), 23763–23787 (2015)
11. Cipriano, Marco: Deep Segmentation of the Mandibular Canal: A New 3D Annotated Dataset of CBCT Volumes. IEEE Access **10**, 11500–11510 (2022). https://doi.org/10.1109/ACCESS.2022.3144840
12. Cipriano, M., Allegretti, S., Bolelli, F., Pollastri, F., Grana, C.: Improving segmentation of the inferior alveolar nerve through deep label propagation. In: Proceedings of the IEEE/CVF Conference on Computer Vision and Pattern Recognition (CVPR), pp. 1–10. IEEE (2022)
13. Dillencourt, M.B., Samet, H., Tamminen, M.: A General Approach to Connected-Component Labeling for Arbitrary Image Representations. J. ACM **39**(2), 253–280 (1992)
14. Fabbri, M., et al.: MOTSynth: how can synthetic data help pedestrian detection and tracking? In: Proceedings of the IEEE/CVF International Conference on Computer Vision, pp. 10849–10859 (2021)

15. Grana, Costantino, Baraldi, Lorenzo, Bolelli, Federico: Optimized connected components labeling with pixel prediction. In: Blanc-Talon, Jacques, Distante, Cosimo, Philips, Wilfried, Popescu, Dan, Scheunders, Paul (eds.) ACIVS 2016. LNCS, vol. 10016, pp. 431–440. Springer, Cham (2016). https://doi.org/10.1007/978-3-319-48680-2_38

16. Grana, C., Bolelli, F., Baraldi, L., Vezzani, R.: YACCLAB - yet another connected components labeling benchmark. In: 2016 23rd International Conference on Pattern Recognition (ICPR), pp. 3109–3114 (2016)

17. Grana, C., Borghesani, D., Cucchiara, R.: Optimized Block-based Connected Components Labeling with Decision Trees. IEEE Trans. Image Process. **19**(6), 1596–1609 (2010)

18. Grana, C., Borghesani, D., Cucchiara, R.: Automatic segmentation of digitalized historical manuscripts. Multimedia Tools and Applications **55**(3), 483–506 (2011)

19. Grana, C., Montangero, M., Borghesani, D.: Optimal decision trees for local image processing algorithms. Pattern Recognition Letters 33(16), 2302–2310 (2012)

20. He, L., Chao, Y., Suzuki, K.: A linear-time two-scan labeling algorithm. In: 2007 IEEE International Conference on Image Processing, pp. 241–244 (2007)

21. He, L., Ren, X., Gao, Q., Zhao, X., Yao, B., Chao, Y.: The connected-component labeling problem: A review of state-of-the-art algorithms. Pattern Recogn. **70**, 25–43 (2017)

22. He, L., Zhao, X., Chao, Y., Suzuki, K.: Configuration-transition-based connected-component labeling. IEEE Trans. Image Process. **23**(2), 943–951 (2014)

23. Hennequin, A., Lacassagne, L., Cabaret, L., Meunier, Q.: A new direct connected component labeling and analysis algorithms for GPUs. In: 2018 Conference on Design and Architectures for Signal and Image Processing (DASIP), pp. 76–81. IEEE (2018)

24. Lacassagne, L., Zavidovique, B.: Light speed labeling: efficient connected component labeling on risc architectures. J. Real-Time Image Proc. **6**(2), 117–135 (2011)

25. Laradji, I.H., Rostamzadeh, N., Pinheiro, P.O., Vazquez, D., Schmidt, M.: Where are the Blobs: Counting by Localization with Point Supervision. In: Computer Vision – ECCV 2018. pp. 547–562 (2018)

26. Pham, H.V., Bhaduri, B., Tangella, K., Best-Popescu, C., Popescu, G.: Real time blood testing using quantitative phase imaging. PLOS ONE **8**(2), e55676 (2013)

27. Playne, D., Hawick, K.: A new algorithm for parallel connected-component labelling on GPUs. IEEE Trans. Parallel Distrib. Syst. **29**(6), 1217–1230 (2018). https://doi.org/10.1109/TPDS.2018.2799216

28. Pollastri, F., Bolelli, F., Paredes, R., Grana, C.: Augmenting data with GANs to segment melanoma skin lesions. Multimedia Tools Appl. **79**(21–22), 15575–15592 (2019)

29. Porrello, A., Abati, D., Calderara, S., Cucchiara, R.: Classifying signals on irregular domains via convolutional cluster pooling. In: The 22nd International Conference on Artificial Intelligence and Statistics. pp. 1388–1397. PMLR (2019)

30. Rosenfeld, A., Pfaltz, J.L.: Sequential operations in digital picture processing. J. ACM **13**(4), 471–494 (1966)

31. Uslu, F., Bharath, A.A.: A recursive Bayesian approach to describe retinal vasculature geometry. Pattern Recognition 87, 157–169 (2019)

32. Wu, K., Otoo, E., Suzuki, K.: Two strategies to speed up connected component labeling algorithms. Pattern Anal. Appl. 0(LBNL-59102) (2005)

33. Zavalishin, S., Safonov, I., Bekhtin, Y., Kurilin, I.: Block equivalence algorithm for labeling 2D and 3D images on GPU. Electron. Imaging **2016**(2), 1–7 (2016). Society for Imaging Science and Technology

A Deep Learning Based Framework for Malaria Diagnosis on High Variation Data Set

Luca Zedda, Andrea Loddo$^{(\boxtimes)}$, and Cecilia Di Ruberto

Department of Mathematics and Computer Science,
University of Cagliari, Cagliari, Italy
l.zedda12@studenti.unica.it, {andrea.loddo,dirubert}@unica.it

Abstract. Malaria is a globally widespread disease caused by parasitic protozoa transmitted by infected female Anopheles mosquitoes. It is caused in humans only by the parasite Plasmodium, further classified into four different species. Identifying malaria parasites is possible by analyzing digital microscopic blood smears, which is tedious, time-consuming, and error-prone. Therefore, automation of the process has assumed great importance as it helps the laborious manual process of review and diagnosis. This work proposes a real-time malaria parasite detector and classification system after studying the YOLOv5 detector and comparing different off-the-shelf convolutional neural network architectures for four-class classification on Plasmodium Falciparum life stages. The results show that the use of the networks YOLOv5 and DarkNet-53 reaches great accuracy in detecting and classifying the life stages of Plasmodium Falciparum, achieving an accuracy of 95.2% and 96.02%, respectively, and outperforming the state-of-art. The obtained results enable broad improvements geared explicitly towards recognizing types and life stages of less common species of malaria parasites, even in mobile environments.

Keywords: Computer vision · Deep learning · Image processing · Malaria parasites detection · Malaria parasites classification

1 Introduction

Malaria is a globally widespread disease caused by parasitic protozoa transmitted to humans by infected female mosquitoes of Anopheles. In 2019, there were an estimated 229 million malaria cases worldwide, with an estimated 409,000 deaths due to malaria [18]. Parasites of the genus Plasmodium cause malaria in humans by attacking red blood cells (RBCs) and are grouped into five different species: P. falciparum, P. vivax, P. ovale, P. malariae, and P. knowlesi. Malaria plasmids within the human host have the following life stages: ring, trophozoite, schizont, and gametocyte. The World Health Organisation (WHO) defines human malaria as preventable and treatable if diagnosed promptly.

Blood cell analysis using peripheral blood slides under a light microscope is considered the gold standard for the detection of leukaemia [30], blood cell counting [5,6,31], or the diagnosis of malaria [12]. Manual microscopic examination of peripheral blood smears (PBS) for malaria diagnosis has advantages such as high sensitivity and specificity compared to other methods. However, it requires about 15 min for microscopic examination of a single blood sample [16], and the quality of the diagnosis depends solely on the experience and knowledge of the microscopist. In addition, the images analyzed may be subject to variations in illumination and staining that can affect the results.

This work contributes to developing an algorithm for the automatic identification of malaria parasites. The main contributions of this research are summarized as follows: (a) development of a pipeline to localize and classify malaria parasites in real-time and (b) study on the most prevalent parasite, P. falciparum, to build a baseline. The identification algorithm is divided into parasite detection and stage of life classification. We opted for You Only Look Once (YOLO) architecture in the initial detection phase because it is smaller than regional-based detection models. Due to its simple architecture, YOLOv5 is faster and more robust than the previous versions. It converges reasonably quickly, providing a generalized model for the issue, while in the classification phase, we used the DarkNet-53 model, as it obtained the best performance.

The rest of the manuscript is organized as follows. Section 2 presents the literature on computer-aided diagnostic (CAD) systems for malaria. In Sect. 3 we illustrate the data set and our approach. The results are given in Sect. 4. Finally, in Sect. 5 we draw the findings and directions for future works.

2 Related Work

Several CAD-based solutions for automatic detection of malaria parasites have been developed in recent years to reduce the problems of manual analysis and provide more robust and standardized interpretation of blood samples, reducing the costs of diagnosis [19]. They are built by combining image processing and machine learning techniques [3], or deep learning approaches [10,21], especially after the proposal of AlexNet's Convolutional Neural Network (CNN) [9].

Since malaria parasites always affect the RBCs, any automatic malaria detection needs to analyze the erythrocytes to discover if they are infected or not by the parasite and, further, categorize the life stages. Most of the related works face only the classification issue without considering the detection problem.

Regarding more recent CNN-based approaches, Liang et al. [10] proposed a new CNN-based model for classifying single cells as infected or uninfected. In contrast, Rajaraman et al. [21] found that some pre-existing networks, through transfer learning (TL), can be more efficient than ad-hoc designed networks. In particular, ResNet50 achieved the best performance. Subsequently, they improved the performance through an ensemble of CNNs. Rahman et al. [20] also exploited TL using both natural and medical images for binary classification.

Some other techniques rely on a combination of features extracted from CNNs and handcrafted features [17] or direct use of object detectors [1].

Recently, much effort is on mobile devices, which allow for cheaper and faster malaria diagnosis in underdeveloped areas of the world, where more expensive

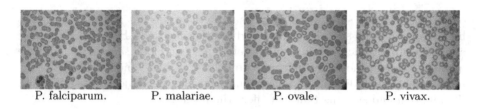

| P. falciparum. | P. malariae. | P. ovale. | P. vivax. |

Fig. 1. Example of the four types of malaria parasites of MP-IDB.

laboratories do not exist. As an example, Bias et al. [3] created a histogram-based edge detection technique coupled with easily accessible hardware.

[14,20] adopted the same data set of our study. The former adopted a semantic segmentation followed by a CapsNet to classify P. falciparum rings, while the latter compared different off-the-shelf networks for binary classification. The main differences between our work and the stat-of-art are that we exploited a detector and an extended set of off-the-shelf CNNs with dual purposes: to detect unhealthy RBCs and classify the different life stages of parasites in a real-time framework. Finally, it is the first baseline provided for life stage classification on the Malaria Parasite Image Database (MP-IDB) [13].

3 Materials and Methods

Here we describe the data set, the proposed approach, and the experimental setup employed in our study.

3.1 Data Set

MP-IDB [13] consists of four malaria parasite species: *P. falciparum, P. malariae, P. ovale, P. vivax*, represented by 104, 37, 29, and 40 images, respectively, for a total amount of 210. Each one has its ground-truth. The images have been acquired with 2592×1944 resolution and 24-bit color depth (see Fig. 1). Every species contains four distinct life stages: ring, trophozoite, schizont, and gametocyte. We used the P. falciparum portion for this work, with the following distribution per stage: 1,230 rings, 18 schizonts, 42 trophozoites, and 7 gametocytes.

3.2 Our Approach

Here, we describe the pipeline of our proposed approach, represented in Fig. 2. This work focuses on the P. falciparum class to build a baseline for any other species.

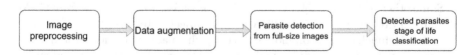

Fig. 2. Pipeline of our approach.

Fig. 3. Examples of MP-IDB-FC images before and after preprocessing step.

Step #1: Image Preprocessing. Figure 1 shows that MP-IDB images have many background color variations due to the different acquisition conditions and depending on the status of the blood smear, which degrades even if well preserved [13]. For this reason, we designed a preprocessing step. In particular, we adjusted the three RGB color components using a color-balancing technique through:

$$C_{out} = C_{in} \frac{m_{im}}{m_c}, \tag{1}$$

where C_{out} is the processed component, C_{in} is the component to be processed, m_{im} is the average of the three mean intensities, and finally m_c is the average intensity of the component to be processed.

Moreover, we manually extracted single-parasite images from the full-size ones. From now on, we refer to this data set as MP-IDB-Falciparum-Crops (MP-IDB-FC). Examples are shown in Fig. 3.

Step #2: Data Augmentation. We applied different data augmentation techniques to overcome the limited data availability issue and to give our approach more generalization capability, reducing the error rate [9,15]. We produced: a) 100 new images for each of the original sets with the parameters given in Table 1 for the detection stage (from now on, MP-IDB-Falciparum-Full-Augmented, MP-IDB-FF-A), and b) an augmented version of

Table 1. Image augmentation parameters for YOLOv5m6 training.

Augmentation	Parameters	Probability
Rotation	Range: [−90, 90] degrees	0.5
Scale	Range: [0.9, 1.1]	0.5
Resized crop	Range: [0.6, 1], final size: [1,280, 1,280]	0.5
Vertical flip	–	0.5
Horizontal flip	–	0.5
Shear	Range: [−20, 20]	0.5
Gaussian noise	Variance range: [50, 100]	0.3
HSV - Hue	Shift limit: 20	0.3
HSV - Saturation	Shift limit: 30	0.3
HSV - Value	Shift limit: 20	0.3

MP-IDB-FC (from now on, MP-IDB-Falciparum-Crops-Augmented, MP-IDB-FC-A) to train the classifier, leaving MP-IDB-FC as the validation set. In the latter case, we dealt with single crops, and we did not need to create a significant variance since we intended to classify by exploiting the fine-grained details. So, we applied only random rotations between $-90°$ and $90°$, random translations between -10 and 10 pixels on the X- and Y-axes, and a 50% reflection around the X- and Y-axes. In addition, given the reduced availability of different parasite stages, we also oversampled all classes except the ring class.

Step #3: Parasite Detection. Recently, YOLO [22] has been used for different object detection tasks [33]. Instead of using the region selection method [6], it merges bounding box estimation and object identification into an end-to-end differentiable network. Thus, YOLO uses a CNN to predict bounding boxes and their classes. It divides the image into an $S \times S$ grid where S is constant. For each grid, YOLO generates a serial number of bounding boxes. If a bounding box has confidence greater than a certain threshold, the bounding box is selected to locate the object within the image.

In this work, we faced parasite detection from full-size images with YOLOv5, a recent upgrade of the YOLO family of detectors. It needs only one step of forwarding propagation through the CNN to make predictions, so it needs to analyze the image only once. It then outputs the known objects along with the bounding boxes after non-maximum suppression, ensuring that it recognizes each object only once. It is a well-known fact that the YOLO architecture struggles to detect small objects that appear in groups [22]. However, the scenario addressed does not represent such cases since smaller objects, i.e., rings, are large enough to be considered by the detector. Moreover, they do not produce clumps since they occupy only one RBC at a time.

More specifically, we employed the medium version, i.e., *YOLOv5m6*, which contains 35.7 million trainable parameters, less than the 88.5 million of the standard version. It was trained using the Common Object in Context (COCO) [11] data set to detect 80 classes. We changed the final layer to address our task, i.e., to detect the four classes corresponding to the four phases of malaria life. We picked this model to limit the number of parameters and let the detector be employable with low-end machines, primarily considering the possible limitations of the mobile devices used for real-time malaria parasite detection [3,12].

Step #4: Parasites Classification. We evaluated fourteen different CNN architectures to classify the detected parasites to choose the most appropriate classification method. They were all pre-trained on the well-known natural image data set ImageNet [4] and adapted to our task with a fine-tuning procedure, as proposed in [26]. We retained all the layers except the last fully connected, replaced with a layer set up to accommodate the four parasite classes.

AlexNet [9], VGG-16, and VGG-19 [27] are architectures composed of 8, 16, and 19 layers, respectively. GoogLeNet [28] and Inception-v3 [29] are based on the inception layer and composed of 100 and 140 layers, while Inception-ResNet-v2 combines the inception structure with residual connections and

Table 2. Training hyperparameters.

Parameters	Detector values	Classifier values
Solver	Adam	Adam
Max epochs	300	100
Mini-batch size	4	32
Initial learning rate	3×10^{-3}	1×10^{-4}
Momentum	0.8	–
Optimizer weight decay	0.00036	–
Box loss gain	0.0296	–
Obj loss gain	0.301	–
IoU training threshold	0.2	–
Anchor-multiple threshold	2.91	
Learn rate drop period	–	10
Learn rate drop factor	–	0.1
L_2 regularisation	–	0.1

includes 164 layers. The three ResNet architectures have 18, 50, 101 layers for ResNet-18, ResNet-50, and ResNet-101, respectively, based on residual learning. They are easier to optimize even when the depth increases considerably [7]. DenseNet-201 is comprised of multiple densely connected layers, and their outputs are connected to all the successors in a dense block [8]. ShuffleNet [32], and MobileNetV2 [25] are lighter networks, suited for real-time executions and mobile devices usage. Finally, DarkNet-19 and DarkNet-53 [23] are the foundation for YOLO object detectors. They are 19 and 53 layers deep, and the latter improves the former by using residual connections and more layers.

3.3   Experimental Setup

We used the PyTorch implementation of YOLOv5[1], developed by the Ultralytics LLC team [2] for the detection of P. falciparum parasites. We conducted the experiments on a machine with the following configuration: Intel (R) Core (TM) i9-8950 HK @ 2.90 GHz CPU, 32 GB RAM and NVIDIA GTX1050 Ti 4GB GPU.

We first divided the data set into 70, 20, and 10% for the training, validation, and testing set, respectively. We obtained 73, 21, and 10 images. As explained in Sect. 3.2, we trained the detector on a training set taken from MP-IDB-FC-A, composed only of 7,300 augmented images. Then, considering that YOLO chooses the bounding box based on the size of the input image, we resized the images to a resolution of 1280 × 1280 pixels to be compatible with the chosen model. Although it increases the training time, the advantage of this method

---

[1] Available at: https://github.com/ultralytics/yolov5.

Table 3. Performance and inference time of the YOLOv5m6 detection module.

Test set	P	R	$mAP@0.5$	Inference time (s)
TS1	87.4	90.1	84.6	0.1
TS2	92.8	95.2	95.2	0.1

is that our model can detect and classify both large and small objects on an image. It took almost 45 min per epoch to train the model. Table 2 presents the training parameters adopted.

This algorithm proposes a set of default parameters tuned for generic data sets. For MP-IDB, after a preliminary study on the parasites, we had to refine some of them to improve detection quality. In particular, we devoted a significant effort to choosing the most appropriate Intersection over Union (IoU) value for neighboring training candidate boxes. Moreover, during the validation, we tuned the Non-Maximal Suppression (NMS) threshold of an object proposal, i.e., if a neighboring box is with an IoU greater than the set threshold, the lower-ranked box is removed. We found that they were the most influential at changing the model proposals. So, we set either IoU and NMS values to 0.2.

Afterward, we designed a multiclass classification to determine the life stage of the parasites found by the detection module by fine-tuning the CNN architectures with the hyperparameters in Table 2. We set the L_2 regularization to avoid overfitting during training.

4   Experimental Results and Discussion

We evaluated the detection performance on the ground-truth provided by the authors with three standard metrics: the PASCAL VOC's mean Average Precision (mAP@0.5) (computed for IoU threshold equal to 0.5) [24], precision (P), recall (R). We also used the last two to evaluate the classification performance, along with accuracy (A), specificity (S), and F1-measure (F1). We evaluated each metric for classification as a weighted average on the number of classes and the only parasite class for detection.

4.1   Detection Results

We computed the final inference of the detection module on two different test sets: the first one formed of 1,000 augmented images representing the test set of MP-IDB-FC-A (TS1) and the entire original data set, composed of 104 images (TS2). On these tests, our module achieved $mAP@0.5$ value of 0.846 and 0.952, respectively. The performance results in detection are shown in Table 3. Numerically, this module correctly detected all the 42 trophozoites, 18 schizonts, and 7 gametocytes. It correctly detected 1,159 rings out of 1,230.

We performed further analysis on the detection results to understand why the detector prevented from uncovering 71 rings. In particular, in Fig. 4 we report

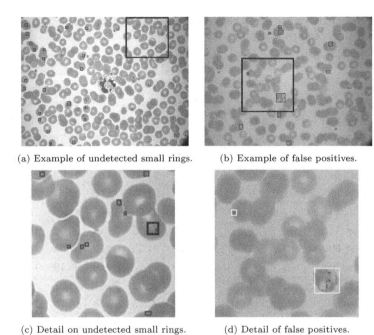

(a) Example of undetected small rings.   (b) Example of false positives.

(c) Detail on undetected small rings.   (d) Detail of false positives.

Fig. 4. Detection issues revealed by detection results. At the top, green boxes indicate ground-truth, while blue boxes indicate the detected parasites. At the bottom left, undetected rings are shown in red, while at the bottom right, two false positives are indicated with yellow boxes. (Color figure online)

the two worst-case scenarios. Figure 4c emphasizes the low performance of the detector when the crops are tiny, as envisaged in Sect. 3.2. The main reasons for this error are the image contrast and, above all, the ambiguity of the false positives' cell structures, very similar to a true parasite. This condition affects most of the errors made during this phase due to some images' particular challenging ring sizes. Instead, Fig. 4d shows how the detector was not able to detect all the parasites. In addition, it produced three false positive predictions.

4.2   Multiclass Classification Performance on the Detection Results

Table 4 shows the classification performance of all compared networks while Fig. 5a represents the confusion matrix of DarkNet-53. Some misclassification between rings and trophozoites occurred because of their slight shape, structure, and color differences. Under certain circumstances, a mature ring can easily be misclassified with a trophozoite even by humans, the latter being the natural evolution of the former. Therefore, the misclassification of 30 trophozoites as rings and of 19 rings as trophozoites can be easily motivated by their strong similarities and does not affect the significance of the results obtained. Also, three gametocytes were misclassified as rings, which needs further insights. As

Table 4. Weighted average performance computed using the 14 tested CNNs in classifying the parasites provided by the detection module.

Network	Time (min)	A	P	S	R	F1
AlexNet	25	95.85	**95.71**	67.61	95.53	95.46
DarkNet-19	88	82.11	92.57	74.96	79.98	84.98
DarkNet-53	348	**96.05**	95.48	62.03	**95.83**	**95.58**
DenseNet-201	202	94.30	94.69	69.76	93.78	94.15
GoogLeNet	50	69.47	93.29	**85.19**	67.93	76.63
Inception-v3	420	91.98	94.50	69.61	91.66	92.73
Inception-ResNet-v2	160	93.29	94.90	73.00	92.87	93.67
MobileNetV2	188	83.92	93.96	73.65	83.47	87.53
ResNet-18	121	91.00	93.85	68.46	90.52	91.87
ResNet-50	1,375	93.37	94.35	68.57	92.95	93.53
ResNet-101	106	92.72	94.11	65.29	92.19	93.02
ShuffleNet	1,903	92.69	94.92	76.31	92.27	93.33
VGG-16	299	58.27	93.13	83.48	56.79	68.03
VGG-19	62	68.04	93.86	87.34	66.72	76.06

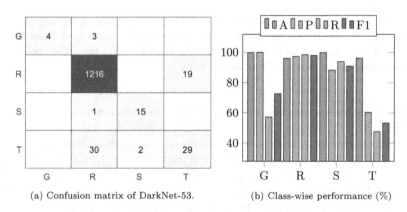

(a) Confusion matrix of DarkNet-53.   (b) Class-wise performance (%)

Fig. 5. Overall system performance.

a general rule, our classification module achieved 96.05% accuracy and 95.58% F1. These results confirmed that addressing the ring-trophozoite problem can be crucial to improving performance. Although we only trained the classification module with augmented images, infected cells are detected and classified correctly with minor loss.

4.3 System Comparison

We assess that our proposal is the first attempt to realize a real-time malaria parasites detection and classification on the MP-IDB. Indeed, Table 5 shows that the other methods did not use any detection strategy. In particular, Rahman et al. [20] directly performed a binary classification of single cells, segmented from MP-IDB employing the watershed transform. Unlike them, we did not focus on healthy RBCs to not affect the focus on the parasites. For this reason, a direct comparison with our work is not possible. In any case, on the entire MP-IDB, they reported 85.18% accuracy with a fine-tuned VGG-19.

On the other hand, Maity et al. [14] built a complete system capable of segmenting infected RBCs with a multilayer feedforward ANN from full-size images and, then, applied a CapsNet to classify the obtained crops. In their work, the authors reported a correct classification of 885 P. falciparum rings, out of 927 considered, leading to an accuracy of 95.46%. The authors did not report any classification of gametocytes, trophozoites, or schizonts.

The MP-IDB contains 1,255 Falciparum rings, all of which were addressed in our approach. Notably, our method is missing only 42 small rings; therefore, our approach outperformed the state of the art on the ring stage with an accuracy of 97.30% (see Fig. 5b). Finally, to better show the class performance, Sect. 4.2 reports the final confusion matrix and the weighted class accuracy for our approach, indicating that it detects all stages with an overall accuracy of 96.05%.

Table 5. Comparison of performance with the state of the art.

Work	Task	Method	A (%)
[20]	Single cells stages classification	VGG-19	85.18
[14]	Segmentation+classification of ring stage	ANN + CapsNet	95.46
Our	Detection+classification of ring stage	YOLOv5+DarkNet-53	97.30
Our	Detection+classification of all stages	YOLOv5+DarkNet-53	96.05

5 Conclusions

In this study, we have investigated the performance of deep learning models for automated malaria diagnosis. In particular, we have proposed a framework for malaria parasite detection and further stage of life classification. We consider our proposed two-step approach a novel, rapid, and robust malaria screening method, supported by the excellent results obtained. The parasite detection is performed by the YOLOv5m6 detector, adequately tuned to detect the parasite stages. We produced more than 10,000 augmented images to train the detector and classification modules. The first one considered only the parasite regions, comprehensive of all the stages of life, achieving 95.2% $mAP@0.5$ for parasite detection. Further, the classification process was performed using the DarkNet-53, considering the four stages of life as classes. It reached a weighted accuracy of 96.05%, only missing 42 tiny ring-stage parasites. The performance of the proposed approach shows significant and promising results with respect to the state of the art, even though we planned several further investigations. First of all, we aim to address the smallest parasites, adopting a multi-scale detection. Secondly, to get rid of the false positives, we aim to include an attention-based module that can discriminate meaningful features in such a way as to give the modules more specific information belonging to parasites. Third, we aim to introduce an intermediate module to give DarkNet-53 the ability to share features found during detection with the classification module. Finally, we aim to further study our framework to test its robustness across multiple data sets and assess its feasibility in clinical practice.

References

1. Abdurahman, F., Fante, K.A., Aliy, M.: Malaria parasite detection in thick blood smear microscopic images using modified YOLOV3 and YOLOV4 models. BMC Bioinform. **22**(1), 112 (2021)
2. Jocher, G., et al.: ultralytics/yolov5: v5.0 - YOLOv5-P6 1280 models, April 2021. https://doi.org/10.5281/zenodo.4679653
3. Bias, S., Reni, S., Kale, I.: Mobile hardware based implementation of a novel, efficient, fuzzy logic inspired edge detection technique for analysis of malaria infected microscopic thin blood images. Proc. Comput. Sci. **141**, 374–381 (2018)
4. Deng, J., Dong, W., Socher, R., Li, L.J., Li, K., Fei-Fei, L.: ImageNet: a large-scale hierarchical image database In: CVPR, pp. 248–255 (2009)
5. Di Ruberto, C., Loddo, A., Puglisi, G.: Blob detection and deep learning for leukemic blood image analysis. Appli. Sci. **10**(3), 1176 (2020)
6. Di Ruberto, C., Loddo, A., Putzu, L.: Detection of red and white blood cells from microscopic blood images using a region proposal approach. Comput. Biol. Med. **116**, 103530 (2020)
7. He, K., Zhang, X., Ren, S., Sun, J.: Deep residual learning for image recognition. In: 2016 IEEE Conference on Computer Vision and Pattern Recognition, CVPR, pp. 770–778 (2016)
8. Huang, G., Liu, Z., Weinberger, K.Q.: Densely connected convolutional networks. CoRR abs/1608.06993 (2016)
9. Krizhevsky, A., Sutskever, I., Hinton, G.E.: ImageNet classification with deep convolutional neural networks. In: Proceeding of the 25th International Conference on Neural Information Processing Systems, NIPS 2012, vol. 1, pp. 1097–1105 (2012)
10. Liang, Z., et al.: CNN-based image analysis for malaria diagnosis. In: IEEE International Conference on Bioinformatics and Biomedicine, BIBM 2016, Shenzhen, China, 15–18 December, pp. 493–496. IEEE Computer Society (2016)
11. Lin, T.Y., et al.: Microsoft coco: common objects in context. In: Fleet, D., Pajdla, T., Schiele, B., Tuytelaars, T. (eds.) Computer Vision - ECCV 2014, pp. 740–755. Springer, Cham (2014). https://doi.org/10.1007/978-3-319-10602-1_48
12. Loddo, A., Di Ruberto, C., Kocher, M.: Recent advances of malaria parasites detection systems based on mathematical morphology. Sensors **18**(2), 513 (2018)
13. Loddo, A., Di Ruberto, C., Kocher, M., Prod'Hom, G.: MP-IDB: the malaria parasite image database for image processing and analysis. In: Lepore, N., Brieva, J., Romero, E., Racoceanu, D., Joskowicz, L. (eds.) SaMBa 2018. LNCS, vol. 11379, pp. 57–65. Springer, Cham (2019). https://doi.org/10.1007/978-3-030-13835-6_7

14. Maity, M., Jaiswal, A., Gantait, K., Chatterjee, J., Mukherjee, A.: Quantification of malaria parasitaemia using trainable semantic segmentation and capsnet. Pattern Recognit. Lett. **138**, 88–94 (2020)
15. Mikołajczyk, A., Grochowski, M.: Data augmentation for improving deep learning in image classification problem. In: 2018 International Interdisciplinary PhD Workshop, IIPhDW, pp. 117–122 (2018)
16. Moody, A.: Rapid diagnostic tests for malaria parasites. Clin. Microbiol. Rev. **15**(1), 66–78 (2002)
17. Nanni, L., Ghidoni, S., Brahnam, S.: Handcrafted vs. non-handcrafted features for computer vision classification. Pattern Recogn. **71**, 158–172 (2017)
18. World Health Organization. https://www.who.int/news-room/fact-sheets/detail/malaria (2021). Accessed 13 Sept 2021
19. Poostchi, M., Silamut, K., Maude, R.J., Jaeger, S., Thoma, G.: Image analysis and machine learning for detecting malaria. Transl. Res. **194**, 36–55 (2018); in-Depth Review: Diagnostic Medical Imaging
20. Rahman, A., Zunair, H., Reme, T.R., Rahman, M.S., Mahdy, M.: A comparative analysis of deep learning architectures on high variation malaria parasite classification dataset. Tissue Cell **69**, 101473 (2021)
21. Rajaraman, S., Jaeger, S., Antani, S.K.: Perf. eval. of deep neural ensembles toward malaria parasite detection in thin-blood smear images. PeerJ **7**, e6977 (2019)
22. Redmon, J., Divvala, S., Girshick, R., Farhadi, A.: You only look once: unified, real-time object detection. In: 2016 IEEE Conference on Computer Vision and Pattern Recognition, CVPR, pp. 779–788 (2016)
23. Redmon, J., Farhadi, A.: Yolov3: an incremental improvement. CoRR abs/1804.02767 (2018)
24. Ren, S., He, K., Girshick, R., Sun, J.: Faster R-CNN: towards real-time object detection with region proposal networks. Adv. Neural. Inf. Process. Syst. **28**, 91–99 (2015)
25. Sandler, M., Howard, A.G., Zhu, M., Zhmoginov, A., Chen, L.: Mobilenetv 2: inverted residuals and linear bottlenecks. In: 2018 IEEE Conference on Computer Vision and Pattern Recognition, CVPR 2018, Salt Lake City, UT, USA, 18–22 June, pp. 4510–4520. IEEE Computer Society (2018)
26. Shin, H., et al.: Deep convolutional neural networks for computer-aided detection: CNN architectures, dataset characteristics and transfer learning. IEEE Trans. Med. Imaging **35**(5), 1285–1298 (2016)
27. Simonyan, K., Zisserman, A.: Very deep convolutional networks for large-scale image recognition. In: Bengio, Y., LeCun, Y. (eds.) 3rd International Conference on Learning Representations, ICLR 2015, San Diego, CA, USA, 7–9 May, Conference Track Proceedings (2015)
28. Szegedy, C., et al.: Going deeper with convolutions. In: 2015 IEEE Conference on Computer Vision and Pattern Recognition, CVPR, pp. 1–9 (2015)
29. Szegedy, C., Vanhoucke, V., Ioffe, S., Shlens, J., Wojna, Z.: Rethinking the inception architecture for computer vision. In: 2016 IEEE Conference on Computer Vision and Pattern Recognition, CVPR, pp. 2818–2826 (2016)
30. Vogado, L.H., Veras, R.M., Araujo, F.H., Silva, R.R., Aires, K.R.: Leukemia diagnosis in blood slides using transfer learning in CNNS and SVM for classification. Eng. Appl. Artif. Intell. **72**, 415–422 (2018)
31. Xie, W., Noble, J.A., Zisserman, A.: Microscopy cell counting and detection with fully convolutional regression networks. Comput. Methods Biomech. Biomed. Eng. Imaging Vis. **6**(3), 283–292 (2018)

32. Zhang, X., Zhou, X., Lin, M., Sun, J.: ShuffleNet: an extremely efficient convolutional neural network for mobile devices. In: 2018 IEEE Conference on Computer Vision and Pattern Recognition, CVPR 2018, Salt Lake City, UT, USA, 18–22 June, pp. 6848–6856. IEEE Computer Society (2018)

33. Zhu, X., Lyu, S., Wang, X., Zhao, Q.: Tph-yolov5: improved yolov5 based on transformer prediction head for object detection on drone-captured scenarios. In: Proceedings of the IEEE/CVF International Conference on Computer Vision, ICCV Workshops, pp. 2778–2788, October 2021

A Deep Learning-Based System for Product Recognition in Intelligent Retail Environment

Rocco Pietrini[1]([✉])[ID], Luca Rossi[1][ID], Adriano Mancini[1][ID], Primo Zingaretti[1][ID], Emanuele Frontoni[1,2][ID], and Marina Paolanti[1,2][ID]

[1] VRAI Lab, Dipartimento di Ingegneria dell'Informazione (DII), Università Politecnica delle Marche, 63100 Ancona, Italy
{r.pietrini,l.rossi}@pm.univpm.it,
{a.mancini,p.zingaretti}@staff.univpm.it
[2] Department of Political Sciences, Communication and International Relations, University of Macerata, Via Don Minzoni 22/A, 62100 Macerata, Italy
{emanuele.frontoni,marina.paolanti}@unimc.it
http://www.univpm.it

Abstract. This work proposes a pipeline that aims to recognize the products in a shelf, at the level of the single SKU (Stock Keeping Unit), starting from a photo of that shelf. It is composed of a first neural network that detects the individual products on the shelf and has been trained with the SKU110K dataset and a second network, designed and built within this work that associates to the single image created by the first network, an embedding vector, which describes its distinctive features. By obtaining this vector of the input image, it is possible to measure the similarity, by means of the cosine similarity, between this vector and all the embedding vectors in the comparison dataset. The vector with the highest cosine similarity is associated to an image labeled with the EAN (European Article Number) code and, therefore, this EAN will be that of the input image. Given the particular task, there are not currently any dataset able to meet our requirements as they have not such a granular level of detail (EAN labeled), so a new properly designed dataset is created to solve this task.

Keywords: Deep learning · Retail · Planogram compliance · Dataset

1 Introduction

In the retail environment, the automatic recognition of products has an important role in increasing their management by the sellers and the buyers, with the following improvements and satisfaction of the customers' shopping experience [12]. Computer vision has proven fruitful for human behavior understanding in Intelligent Retail Systems with the use of different technologies such as RGB-D and RGB cameras [3], the latter now available on commercial devices and basically ubiquitous. In this work, the recognition of the products has not

S. Sclaroff et al. (Eds.): ICIAP 2022, LNCS 13232, pp. 371–382, 2022.
https://doi.org/10.1007/978-3-031-06430-2_31

the aim to automate the purchases but to optimize the management of the planogram (or its automatic detection). The planogram represents the positioning of products on a shelf, and its management has the purpose of increasing performance and, consequently, customer purchases. For example, a planogram can be designed to minimize the out-of-stock situation of a specific product and therefore its profit loss [23]. Monitor the planogram compliance represents a challenge because the planogram is not always respected, for example, when local consumers have different needs or merely there are mistakes during the deployment of the planogram on the shelf (planogram is generally set up by the retailer headquarter and sent to the points of sale) or when customers put back items on the shelf. The incorrect positioning of the products determines a low conformity of the planogram and inconsistency of the data collected about a shelf since the products location is no longer known [22]. The state of the art of the shoppers behaviour understanding propose technologies to analyze for example the time spent by customers in front of a certain shelf, or even the interactions with the products and the failure to comply with the planogram prevents these data from being used [10]. To evaluate the planogram compliance, the shelf image at a given instant of time and a "master" image are compared. Currently, this activity is almost entirely done manually, and therefore the automation of the whole process would lead to a reduction in time, costs and errors. In fact, the manual verification of each product on a shelf is an extremely time consuming operation since the barcodes (European Article Number, EAN) are printed in different positions and therefore the time required is not negligible for large numbers, for their localization and acquisition. Some retailers puts barcodes on the price tag but they can often refer to internal codes (not uniquely identifying a product worldwide) instead of the EAN and sometimes it happens that the products are not positioned correctly and the price tags may refer to a different product. It is hence required to take every product out of the shelf, locate the bar barcode, scan it and place the product back on the shelf. Finally, RFID (Radio Frequency IDentification) is sometimes used to identify the products, but they proved to be expensive and radio waves can interfere with other sources [14] resulting inadequate for the planogram check. During the years, research has been conducted for product recognition starting from images of the retail environment using convolutional neural networks. For example, in [20], the authors use the Inception V3 [19] network to classify eight different product classes on real store shelves. The accuracy in real shops is 87.5% and, considering the limited number of classes, this result can certainly be improved. The most recent work closely related to our problem is proposed in [21] which illustrates an approach able to recognize a large number of classes. In the article, the authors present a pipeline composed of a "Detector" that extracts the individual products and an "Embedder" which is the extractor network of feature consisted of a VGG16 [17], a Convolutional Neural Network (CNN) trained with Triplet Loss. The dataset used is *GroZi-3.2k* [4], created specifically for this research [4] which contains 3,288 different classes of food products.

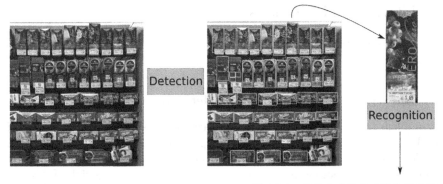

EAN: 8006380315699

Fig. 1. Functional diagram of the pipeline.

In a similar retail context, our work proposes an approach able to recognize the products present on a shelf, at the level of a single EAN, starting from the image of the shelf, and overcoming the current limits imposed by Computer Vision algorithms such as SIFT [9] and becoming independent from the sensors. SIFT has numerous critical issues such as poor performance from both a computational point of view and quality of the result [11]. In particular, we propose a pipeline that uses as a first stage the method presented by Goldman et al. [5], that is able to detect products in a such densely packed scene of a shelf using a RetinaNet together with a Soft-IoU layer to estimate the Jaccard index between the detected box and the (unknown) ground truth box and EM-Merger unit, which converts detections and Soft-IoU scores into a MoG (Mixture of Gaussians), and resolves overlapping detections in packed scenes. Then we introduce a second network, specifically designed in this work, that assigns an embedding vector to each detection generated by the first network. The embedding vector is used to compare the product image (generated cropping the detections of the first network), using the cosine similarity, with all the embedding vectors of a new dataset, collected in this work.

Figure 1 shows the proposed pipeline. During the development of the architecture we have faced and solved several problems such as: the excessive similarity of products that have different EAN, or problems related to lighting, the point of view of the photographer, the occlusions, the quality of the images. The approach used has set itself the goal of being as robust as possible to these issues. The remainder of the paper is organized as follows. Section 2 presents the reasons that led to the realization of a new dataset, how it was built and what is the architecture of the neural network presented. Section 3 describes the results obtained and related comments are reported and in Sect. 4 we identify possible future developments aimed at improving the achieved results.

2   Materials and Methods

This section aims to describe how the dataset was collected, the issues encountered during the process and how the neural architecture was designed.

2.1   Dataset Collection: Definition, Method, and Design

The problem treated in this paper is not particularly studied in the context of Deep Learning; for this reason, there are no pre-existing and accessible datasets, suitable for our aim, containing a large number of images of products present on the shelves of different points of sale and labeled with the right EAN. To this end, the idea is to create a new dataset composed of images obtained in real conditions of products on the shelves, and labeled with the corresponding EAN. We obtained real images from all the shelves in the stores of "Acqua & Sapone", "Oasi (Gabrielli Group)" and "Si con Te (CE.DI.MARCHE)" as well as the chocolate department of an "Iper" (all of them are Italian Mass Market Retailers). The EAN of the products present in the corresponding shelves were also collected using a barcode scanner. In particular, the following number of images of bay (a bay represent a single module of a shelf) were taken: 176 of "Acqua & Sapone", 333 of "Oasi", 160 for "Si con Te" and 4 for "Iper".

Then, the neural network of Goldman et al. [5][1] was used to detect the products present in the images, obtaining several images of individual products contained in the image of a shelf. This network was used because it solves a particularly complex task, that is to accurately detect the individual products in densely populated shelves and, at the current state of the art, it is one of the most efficient and effective. Moreover, this network was trained with SKU110K dataset, consisting of 11762 images of dense shelves of products from thousands of stores around the world, from the United States to East Asia, for a total of 1.74 million bounding boxes[2]. The bounding boxes produced by this network do not always meet our requirements (some false positives such as price tags, or single products splitted into several products) and, for this reason, it was necessary to create an annotator software, created in Python and equipped with a graphical interface created with tkinter[3], which allowed to manually create customized bounding box to create the dataset with the best images and that, finally, can associate the right EAN to each image. Furthermore, each image has been associated with a category deriving from a category tree developed by GS1 in 1999 as a result of the work of a group of companies that wanted to respond to the need to have a common goods classification for products[4]. In particular, this categorization consists of five hierarchical levels, where the fifth level has a degree of maximum granularity. Giving an overview of these categories, first level

---

[1] https://github.com/eg4000/SKU110K_CVPR19.
[2] https://retailvisionworkshop.github.io/detection_challenge_2020/.
[3] https://docs.python.org/library/tkinter.html.
[4] https://gs1it.org/migliorare-processi/relazione-industria-distribuzione-best-practice-ecr/albero-categorie-classificazione-condivisa-prodotti/.

can be grocery, beverage, second level can be bakery, cereals, water, third level can be wafer, sparkling water, fourth level vanilla, 0–50 centilitre, and the last level portioned, plastic etc. We decided to use up to the second level as it was considered a sufficient level of granularity. During the creation of the dataset and the labeling of the images, the categories of the second level were converted into numbers. The categorization is in the form $xxyy$ where xx represents the first level of the category tree and yy the second level. This categorization is useful in the inference phase as it is possible to make comparisons only on a sub-set of the dataset, thus increasing the time performance due to fewer comparisons to be made and reducing errors. The final result was to obtain a dataset consisting of $35,802$ images of $14,426$ EAN distinct, divided into 57 categories. On average, 2.5 images of each EAN were collected.

(a) EAN: 8002910057701

(b) EAN: 7640110704721

(c) EAN: 8002190002644

(d) EAN: 4005900694843

Fig. 2. Dataset sample

A minimum requirement about the shelf photo resolution has been set to 12 megapixels, in order to have a sufficient quality to visibly distinguish similar products. In addition, in order to improve the overall accuracy pictures are perspective-corrected and manually cropped to the shelf area before feeding the detection network, this is possible thanks to a mobile application specifically developed as part of the pipeline. During the design of the system, several factors had to taken into account, mostly related the packaging of the products, which can be summarized as follows:

– Even with the same EAN, there may be different packaging due to physiological changes due to the need to update the graphics or brands present therein, also to adapt the packaging to new promotions or external company collaborations. In these cases, we have kept the new packaging by discarding images of the old packaging and images of packaging that do not contain parts dedicated to promotions.

- The products can be grouped in transparent envelopes and the network of Goldman et al. [5] can perform the detection of the single products in the package (even if can not be sold separately) and not the entire package. In particular, the problem was identified with the soft drink crates.
- There are numerous non-rigid packaging (such as pasta and biscuits) and the images of product differ by how the product is distributed inside them.
- Some packaging is highly reflective; reflections greatly reduce the visibility of the product, making impossible to adopt OCR techniques for these types of products as a hypothetical future development.
- Problems of non-frontal positioning and/or incorrect alignment. Recognition is practically impossible since the distinctive features are generally present in the frontal face.
- Products with the same EAN can have different packaging (in terms of graphics).

Problems encountered in creating the dataset will also occur in the inference phase. To obtain the best possible result, it is necessary to acquire the image of the shelf in the best possible way, taking care of the position of the products and, if possible, the lighting, going to reduce reflections to facilitate the work of the network of Goldman et al. [5].

2.2 Neural Network

Since the products in the different stores continuously vary both in number and in type, it has not been possible to use a classification network with a fixed number of classes otherwise, for each new article or variation of a pre-existing packaging, the classification network would have to be trained again. For these reasons, it was necessary to use a network with an independent number of classes.

Therefore, our approach associates the right EAN to the image of the product considering the similarity between the images, and making it independent of the number of classes and changes to the packaging. However, the recognition of a new product requires that within the dataset used for the comparison of the embeddings there is the new image with the new packaging, and with the corresponding EAN. In fact, the network will generate, for each image, an embedding vector which will then be compared, by calculating the cosine similarity, with the embeddings of the images used as a comparison database and the corresponding EAN. The approach used to solve this problem is strongly inspired by the work of Schroff et al. [15], moving the application domain, from the recognition of faces to the identification of product EAN.

Since our dataset is not excessively large, we used a convolutional neural network with fewer parameters than FaceNet: from 140 million parameters of the network used in FaceNet to just over 18 million of our network. This assumption was also supported by the experimental results. In particular, the neural network MobileNetV2 [13] (without final classification layer) was used, followed by a layer of dense with a dimensionality of 256. Therefore, the output of the network will be an embedding of size 256 on which the L2 normalization has also been applied, which has the task of making the norm of each embedding equal to 1, making the space of the embeddings of radius 1, as in FaceNet. Once the network has been trained, the cosine similarity between the embedding of the input image and the embedding of each image used for similarity verification is calculated to determine the image most similar to the one taken in input.

The cosine similarity between two generic numerical vectors A and B is represented by the following equation and a score between -1 e $+1$, where values close to $+1$ represent a greater similarity ($+1$ corresponds to equal vectors):

$$cosine\ similarity = \frac{A \cdot B}{\|A\| \, \|B\|} \tag{1}$$

The EAN of the image with the greatest cosine similarity will ideally be the EAN of the image in input to the network. The neural network was implemented via the Tensorflow framework using the Python programming language. During the training, the following parameters were used: the network used for the extraction of features is the MobileNetV2, the size of the embedding is equal to 256, the size of each image is 224×224 with 3 channels, the loss is the Triplet Hard Loss with a margin of 1.0 (soft margin), the loss optimizer is Adam [8] with a Learning Rate of 0.001, the training took place in 80 epochs and, finally, the batch size is equal to 128. Furthermore, the ImageNet [2] weights were loaded, applying the transfer learning technique to speed up and improve learning and early stopping was also used to avoid overfitting by evaluating, for each epoch, the trend of the validation loss and saving the weights with the lowest validation loss. The choice of calculating the triplets online is also derived from the conclusions of Hermans et al. [6]. In fact, they show how the use of triplets calculated online can greatly increase the accuracy of the model and reduce training times. In addition to this, the training with all possible triplets would not have been a viable option anyway as the complexity would become $\mathcal{O}(\frac{N^3}{C})$, where N is the number of samples and C is the number of classes, making the training nearly impossible for large datasets. The authors of the paper [6] introduce a variant of the Triplet Loss called Batch Hard (Triplet Hard Loss) where, for each image in the batch, the most difficult positive and negative images are selected within the batch.

3 Experimental Results

The dataset was divided into training and validation sets with percentages of 80% and 20%. Furthermore, before this subdivision, 196 EANs (357 images of 17 categories) present in the dataset of the "Acqua & Sapone" store and present in the other stores were removed from the total dataset: we used these isolated EAN as a test set, since these are images that the neural network has never seen, thus simulating a real situation. In the inference phase the whole new dataset is used for the comparison, in other words each image produced by the detection network of [5] is given as input to our network which returns an embedding vector as output. For each of these vectors the cosine similarity is calculated between all the embeddings of the dataset used for comparison in order to identify the vector (which is associated with an EAN) with the highest cosine similarity, and the most similar image. In this way the EAN can be obtained.

The accuracy is determined as follows:

$$Accuracy(set) = \frac{\#images\,of\,correct\,EAN}{\#\,set} \qquad (2)$$

3.1 Comparison Between Backbone

MobileNetV2 network was chosen as the backbone (the network that extracts the features), after the evaluation of some state-of-the-art CNN models for feature extraction in terms of accuracy. The networks tested were MobileNetV2, MobileNetV3 Large and Small [7], VGG16 [18] and Xception [1]. The VGG16 showed a constant loss without any learning, probably due to the large number of parameters compared to the size of the dataset. Table 1 shows the accuracy in terms of Top-1, Top-5 and Top-10 of each backbone tested. To have a consistent comparison, the same data and parameters were used during the different training sessions.

Table 1. Comparison between the networks used as backbone

Accuracy	Top-1(%)	Top-5(%)	Top-10(%)
MobileNetV2	43.4	69.2	75.6
MobileNetV3Small	41.1	60.5	70.5
MobileNetV3Large	40.6	55.7	64.4
Xception	18.7	36.13	42.5

As the Table shows, the MobileNetV2 network has the best accuracy.

Analyzing only on two exemplary categories (of the 17 categories of the test dataset) using the network with MobileNetV2 as backbone, we obtain these accuracy values:

- Category "Personal care/hygiene": Top-1: 50.0%, Top-5: 68.42%, Top-10: 76.31%
- Category "Personal care/Oral hygiene": Top-1: 35.41%, Top-5: 68.75%, Top-10: 79.16%

Observing Fig. 3 we can see that there are products extremely similar to each other. In this example there are very small changes in the writings on the labels and in the color of the bottles. However, lamps of different colors that illuminate the shelf or shadows positioned in a different way make it difficult for our network to recognize the right product. Moreover, the category "Personal care/Oral hygiene" has many items hanging on hooks and not resting directly on the shelves, further complicating recognition due to variations in the angle of the product.

3.2   Special Cases

In Figs. 3 and 4 we can see two different problems. In the first figure there is the problem of recognizing extremely similar products (fine-grained Classification). We note how some images are particularly noisy due to the use of smartphones with poor quality lenses, making recognition work difficult even for a human. In particular, the network associates, to a noisy input, an EAN of a noisy image, so the noise has been interpreted as a feature. The second figure shows a nice mistake: our network confuses chickpeas with hazelnuts (being visually similar patterns) and we believe that this problem can be solved by increasing the training dataset.

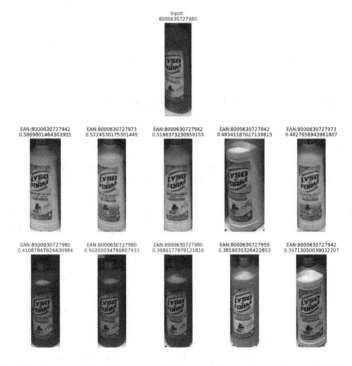

Fig. 3. Top-10 of a specific detergent. Left to right, top to down, EAN and cosine similarity are reported. Very similar products are not correctly detected (detected in 6th position).

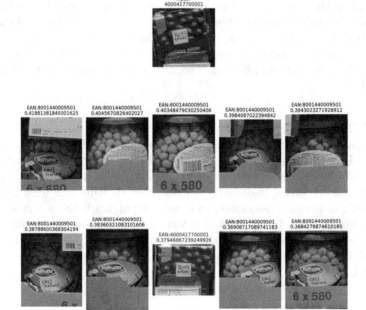

Fig. 4. Top-10 of a chocolate bar. Left to right, top to down, EAN and cosine similarity are reported. Hazelnuts are confused with chickpeas.

4   Conclusions and Future Researches

This paper aims to present a novel pipeline to recognize the EAN of the products in the shelves. The results are very interesting and future developments can increase these results by reducing false positives deriving from price signs and metal parts of shelves or the divisions of products into sub-products. The results can be optimized by increasing the dataset size, with the following network re-training and the increase of the embedding in the comparison dataset. A following optimization of the optical quality of smartphones and the resulting improvement in the image quality will allow to apply OCR methods to increase the accuracy of the pipeline. Such investments in the retail field mean that, when the planogram is completely correct, there will be a sales increase of 7.8% and an improvement in profit of 8.1% in just two weeks [16]. Furthermore, it may be interesting to proceed with the creation or purchase from specialized companies of an EAN database containing, for each of these, the product information such as the name, the brand, the weight of the package, etc. This would make it possible to obtain new and interesting information such as, for example, statistics on the largest brand on a shelf (share of shelf). Furthermore, currently the network does not detect "holes" in the shelves resulting from the out-of-stocks; therefore, the idea is to implement an algorithm that evaluates the distance between each contiguous bounding box and, if greater than a certain threshold, infer a missing product. A further future development for dataset augmentation is the use of generative networks (such as GANs) to increase the number of samples and improve the entire pipeline.

References

1. Chollet, F.: Xception: deep learning with depthwise separable convolutions (2017)
2. Deng, J., Dong, W., Socher, R., Li, L.J., Li, K., Fei-Fei, L.: ImageNet: a large-scale hierarchical image database. In: 2009 IEEE Conference on Computer Vision and Pattern Recognition, pp. 248–255 (2009). https://doi.org/10.1109/CVPR.2009.5206848
3. Frontoni, E., Paolanti, M., Pietrini, R.: People counting in crowded environment and re-identification. In: Rosin, P.L., Lai, Y.-K., Shao, L., Liu, Y. (eds.) RGB-D Image Analysis and Processing. ACVPR, pp. 397–425. Springer, Cham (2019). https://doi.org/10.1007/978-3-030-28603-3_18
4. George, M., Floerkemeier, C.: Recognizing products: a per-exemplar multi-label image classification approach. In: Fleet, D., Pajdla, T., Schiele, B., Tuytelaars, T. (eds.) ECCV 2014. LNCS, vol. 8690, pp. 440–455. Springer, Cham (2014). https://doi.org/10.1007/978-3-319-10605-2_29
5. Goldman, E., Herzig, R., Eisenschtat, A., Goldberger, J., Hassner, T.: Precise detection in densely packed scenes. In: Proceeding Conference Computer Vision Pattern Recognition, CVPR (2019)
6. Hermans, A., Beyer, L., Leibe, B.: In defense of the triplet loss for person re-identification (2017)
7. Howard, A., et al.: Searching for mobilenetv3 (2019)
8. Kingma, D.P., Ba, J.: Adam: a method for stochastic optimization (2017)
9. Lowe, D.: Object recognition from local scale-invariant features. In: Proceedings of the Seventh IEEE International Conference on Computer Vision, vol. 2, pp. 1150–1157 (1999). https://doi.org/10.1109/ICCV.1999.790410
10. Paolanti, M., Pietrini, R., Mancini, A., Frontoni, E., Zingaretti, P.: Deep understanding of shopper behaviours and interactions using RGB-D vision. Mach. Vis. Appl. **31**(7), 1–21 (2020)
11. Pietrini, R., Manco, D., Paolanti, M., Placidi, V., Frontoni, E., Zingaretti, P.: An IOT edge-fog-cloud architecture for vision based planogram integrity. In: Proceedings of the 13th International Conference on Distributed Smart Cameras, pp. 1–5 (2019)
12. Puccinelli, N.M., Goodstein, R.C., Grewal, D., Price, R., Raghubir, P., Stewart, D.: Customer experience management in retailing: understanding the buying process. J. Retail. **85**(1), 15–30 (2009)
13. Sandler, M., Howard, A., Zhu, M., Zhmoginov, A., Chen, L.C.: Mobilenetv 2: inverted residuals and linear bottlenecks (2018)
14. Santra, B., Mukherjee, D.P.: A comprehensive survey on computer vision based approaches for automatic identification of products in retail store. Image Vis. Comput. **86**, 45–63 (2019)
15. Schroff, F., Kalenichenko, D., Philbin, J.: Facenet: a unified embedding for face recognition and clustering. In: 2015 IEEE Conference on Computer Vision and Pattern Recognition, CVPR. IEEE (2015). https://doi.org/10.1109/cvpr.2015.7298682
16. Shapiro, M.: Executing the Best Planogram, Norwalk, CT (2009)

17. Simonyan, K., Zisserman, A.: Very deep convolutional networks for large-scale image recognition. arXiv preprint arXiv:1409.1556 (2014)

18. Simonyan, K., Zisserman, A.: Very deep convolutional networks for large-scale image recognition (2015)

19. Szegedy, C., et al.: Going deeper with convolutions. In: Proceedings of the IEEE Conference on Computer Vision and Pattern Recognition, pp. 1–9 (2015)

20. Timothy Chong, I.B., Wee, M.: Deep learning approach to planogram compliance in retail stores, Stanford, CA, USA (2016)

21. Tonioni, A., Serra, E., Di Stefano, L.: A deep learning pipeline for product recognition on store shelves. In: 2018 IEEE International Conference on Image Processing, Applications and Systems, IPAS, pp. 25–31. IEEE (2018)

22. Topps, J., Taylor, G.: Managing the Retail Supply Chain: Merchandising Strategies that Increase Sales and Improve Profitability. Kogan Page Publishers (2018)

23. Vaira, R., Pietrini, R., Pierdicca, R., Zingaretti, P., Mancini, A., Frontoni, E.: An IOT edge-fog-cloud architecture for vision based pallet integrity. In: Cristani, M., Prati, A., Lanz, O., Messelodi, S., Sebe, N. (eds.) ICIAP 2019. LNCS, vol. 11808, pp. 296–306. Springer, Cham (2019). https://doi.org/10.1007/978-3-030-30754-7_30

LDD: A Grape Diseases Dataset Detection and Instance Segmentation

Leonardo Rossi[1]([✉])(iD), Marco Valenti[1], Sara Elisabetta Legler[2],
and Andrea Prati[1](iD)

[1] IMP Lab - D.I.A., University of Parma, Parma, Italy
{leonardo.rossi,andrea.prati}@unipr.it, marco.valenti@studenti.unipr.it
[2] Horta s.r.l., Università Cattolica del Sacro Cuore spin off in Piacenza,
Piacenza, Italy
s.legler@horta-srl.com
https://implab.ce.unipr.it/, https://www.horta-srl.it/

Abstract. The Instance Segmentation task, an extension of the well-known Object Detection task, is of great help in many areas, such as precision agriculture: being able to automatically identify plant organs and the possible diseases associated with them, allows to effectively scale and automate crop monitoring and its diseases control. To address the problem related to early disease detection and diagnosis on vines plants, a new dataset has been created with the goal of advancing the state-of-the-art of diseases recognition via instance segmentation approaches. This was achieved by gathering images of leaves and clusters of grapes affected by diseases in their natural context. The dataset contains photos of 10 object types which include leaves and grapes with and without symptoms of the eight more common grape diseases, with a total of 17,706 labeled instances in 1,092 images. Multiple statistical measures are proposed in order to offer a complete view on the characteristics of the dataset. Preliminary results for the object detection and instance segmentation tasks reached by the models Mask R-CNN [6] and R^3-CNN [10] are provided as baseline, demonstrating that the procedure is able to reach promising results about the objective of automatic diseases' symptoms recognition.

Keywords: Object detection · Instance segmentation · Supervised learning · Grape diseases

1 Introduction

Recognizing plant diseases is a deeply-felt problem in the agricultural sector. Early disease detection and diagnosis is a key aspect for a correct and sustainable disease control. Traditionally, the diagnosis of leaf and cluster diseases of grapes relies on expert judgment based on visual inspection of the disease symptoms and signs, which usually leads to high cost and potential errors. With the rapid development of artificial intelligence, machine learning methods have been

---

L. Rossi and M. Valenti—Contributed equally.

S. Sclaroff et al. (Eds.): ICIAP 2022, LNCS 13232, pp. 383–393, 2022.
https://doi.org/10.1007/978-3-031-06430-2_32

applied to plant disease detection to make it smarter. In recent years, deep learning approaches led to a huge improvement of image analysis with the tasks of object detection [2] and its evolution, instance segmentation [5]. A main problem related with these deep learning tasks is the need of a labeled dataset with a large amount of images in order to obtain an acceptable performance. The purpose behind the creation of this dataset is the implementation of an automatic system for the analysis of the leaves and bunches of vines through images, in order to identify the diseases that affect them. The project was born in collaboration with Horta s.r.l, a spin off company of the Università Cattolica del Sacro Cuore in Piacenza (Italy), that develops and provides web based services to agricultural and agro-industrial chains, with the aim of increasing competitiveness and sustainability of crop management and ensuring greater food safety. The final purpose is to give the possibility to an inexperienced user to analyze the general status of the plants and identify any diseases through a photo taken from a mobile device. Our goal is to be able to identify each disease in a very precise way, and provide the user with information about the disease severity. Plant diseases cause significant losses to farmers and the possibility of an early and efficient disease symptoms detection is of fundamental importance for a correct disease management. The information provided could be integrated automatically in broader management tools, such as Decision Support Systems (DSSs), and increase the support given to farmers for the sustainable management of vineyards. Instance segmentation represents a good tool since it allows to examine the leaves and grapes accurately at segmentation level.

2   Related Works

Within the literature, there are several attempts to automatically analyze plants and their diseases by means of images. In [1], authors created a dataset of grapevine leaves in order to detect the Esca disease through an image classification algorithm. With the same scope, authors of [9] addressed object detection for Esca disease, by providing 6,000 labelled images of Bordeaux vineyards. In [16], the authors proposed a dataset for three common grape leaf diseases (Black rot, Black measles, also called Esca, and Leaf blight) and for the Mites of grape; the dataset contains 4,449 original images of grape leaf diseases labeled with bounding boxes. Similarly, [11] gathered 300 images for five grape varieties: Chardonnay, Cabernet Franc, Cabernet Sauvignon, Sauvignon Blanc and Syrah; the corresponding dataset is publicly available and contains label for grape detection and instance segmentation. Another example of dataset on disease image recognition is the one proposed in [15], where labels to classify four diseases of wheat (stripe rust, leaf rust, grape downy mildew, grape powdery mildew) are also provided. Finally, in [4], authors labeled with bounding boxes 5,000 images of tomato diseases. Recently, a dataset for Apple Orchards diseases has been proposed in [12], which contains handmade annotated segmentation of a image subset (142) from Plant Pathology Challenge 2020 database [13]. The [14] is an example of a dataset for instance segmentation for cotton leaf disease detection, with 2,000 photos made with a smartphone camera in a real-world scenario.

All the datasets that take grape diseases into consideration are limited to image classification or, at most, object detection. On the other hand, the datasets that offer segmentation consider grape varieties or diseases of other plant. As far as we know, no one has ever composed a dataset containing bounding boxes and segmentations for grape diseases before.

3 Leaf Diseases Dataset

(a) bunch of grapes affected by powdery mildew (oidio).

(b) leaves affected by downy mildew (peronospora).

Fig. 1. Example of segmentation with the help of Label Studio.

Our dataset contains 1,092 RGB images of grapes and 17,706 annotations in COCO-like [8] format for the tasks of Object Detection and Instance Segmentation. The images were collected from Horta's internal databases, the competition *Grapevine Disease Images*[1], and from search engine scraping; all segmentations were manually made by the authors. Figure 1 shows some examples of labeled images. The annotations are based on 10 categories of objects: leaf, grape and eight different diseases. Table 1 shows the class id, label key and name, the type of disease (together with leaf and grape), and the total available annotations (column "Count").

Annotations are split in training (80%) and validation (20%) dataset, with the help of the *cocosplit*[2] utility.

Figure 2 reports some examples for each class.

Annotation Procedure. To insert the annotations, the online tool *Label Studio v1.5*[3] was used with the following configuration:

---

[1] https://www.kaggle.com/piyushmishra1999/plantvillage-grape.
[2] https://github.com/akarazniewicz/cocosplit.
[3] https://labelstud.io/.

Table 1. LDD annotations description.

	Key	Count	Label	Name	Type
1	LBR	5478	vl_black_rot	Black Rot	Leaf disease
2	GBR	1443	vg_black_rot	Black Rot	Grape disease
3	LGM	88	vl_grey_mould	Grey mould (Botrite)	Leaf disease
4	GGM	612	vg_grey_mould	Grey mould (Botrite)	Grape disease
6	VL	1647	vines_leaf	Vine leaf	Leaf
7	VG	993	vines_grape	Bunch of grapevines	Grape
9	LPM	2086	vl_powdery_mildew	Powdery mildew (Oidio)	Leaf disease
10	GPM	1740	vg_powdery_mildew	Powdery mildew (Oidio)	Grape disease
12	LDM	2565	vl_downy_mildew	Downy mildew (Peronospora)	Leaf disease
13	GDM	1054	vg_downy_mildew	Downy mildew (Peronospora)	Grape disease
		17706	Total		

Listing 1.1. "Label Studio Instance Segmentation configuration"

```
<View>
    <Image name="image" value="$image" zoom="true" zoomControl="true"/>
    <PolygonLabels name="label" toName="image" strokeWidth="3"
            pointSize="small" opacity="0.9">
        <Label value="vl_black_rot" background="#D4380D"/>
        <Label value="vg_black_rot" background="#FFC069"/>
        <Label value="vl_grey_mould" background="#AD8B00"/>
        <Label value="vg_grey_mould" background="#D3F261"/>
        <Label value="vl_powdery_mildew" background="#096DD9"/>
        <Label value="vg_powdery_mildew" background="#ADC6FF"/>
        <Label value="vl_downy_mildew" background="#F759AB"/>
        <Label value="vg_downy_mildew" background="#FFA39E"/>
        <Label value="vines_leaf" background="#AD8B00"/>
        <Label value="vines_grape" background="#D3F261"/>
    </PolygonLabels>
</View>
```

Example of web interface can be seen in Fig. 3 where, for each image, the user can select the proper category and then the polygon which identifies the instance.

Since the annotation of this type of images also requires expertise in agronomy, we first had a student creating coarse-grain annotations, and then have an agronomist expert in this field to review and fix them. At the end, the annotations have been exported in COCO-like format, using the built-in export function of label studio that for each polygonal mask auto-generate the corresponding bounding box.

Statistics. In Fig. 4a, the percentage of instances for each category are reported. As it can be noticed, the dataset is strongly unbalanced, but the same distribution can be appreciated in both training and validation dataset. As shown

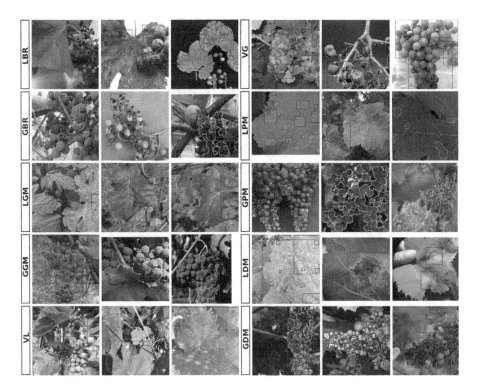

Fig. 2. Samples of annotated images in LDD.

in Fig. 4b, the maximum number of unique categories for each image can vary from one to five, with more than 80% of the images containing two classes. This could be explained by the fact that, usually, the image contains a leaf or a grape affected by a particular disease. Contrary to the number of classes per image, the number of annotations within a single image is much more varied, as shown in Fig. 5, with more than one hundred different annotations for complex images. Figures 6 offer a different point of view, showing that the mean percentage of the instance size with respect to the image size is very low, with more than 80% of the instances being in the range [0%, 0.05%] (see, for instance Fig. 6b and 6c). Figure 6a shows the distributions for leaves and grapes, which are much more diverse with respect to diseases.

Fig. 3. Label Studio web interface.

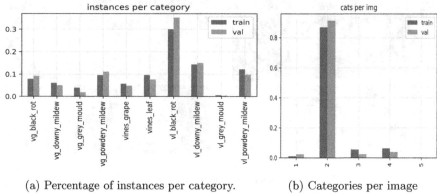

(a) Percentage of instances per category.   (b) Categories per image

Fig. 4. Dataset LDD: (a) Percentage of instances for each category. (b) Percentage of annotated categories per image. Better seen in color.

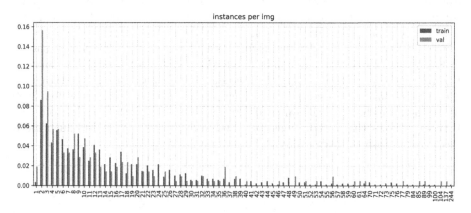

Fig. 5. Annotations per image: percentage of images in LDD.

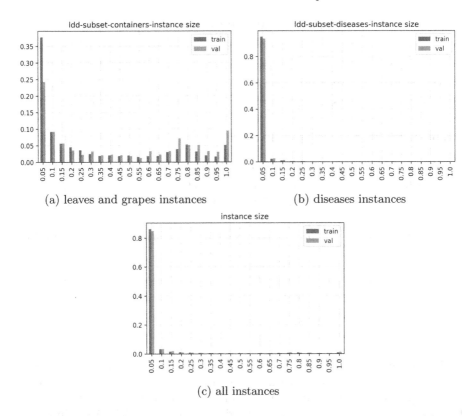

(a) leaves and grapes instances

(b) diseases instances

(c) all instances

Fig. 6. Dataset LDD instance sizes distribution: (a) only leaves and Grapes. (b) only diseases. (c) all dataset. Better seen in color.

4 Experiments

Evaluation Metrics. All the tests are done on LDD validation dataset, which contains 212 images and 3,234 annotations. We report mean Average Precision (mAP) and AP_{50} and AP_{75} with 0.5 and 0.75 minimum IoU thresholds, and AP_s, AP_m and AP_l for small, medium and large objects, respectively.

Implementation Details. All the values are obtained running the training with the same hardware and hyper-parameters. When available, the original code released by the authors is used. Our code is developed on top of the MMDe-tection [3] source code. We perform the training on a single machine with 1 NVIDIA GeForce RTX 2070 with 8GB of memory. The train lasts 12 epochs with a batch size of 2 images. We use the Stochastic Gradient Descent (SGD) optimization algorithm with a learning rate of 0.00125, a weight decay of 0.0001, and a momentum of 0.9. The learning rate decays at epoch 8 and 11. We use the ResNet 50 [7] backbone for the models, initialized by pre-trained weights on ImageNet.

Table 2. Performance of Mask R-CNN and R^3-CNN models.

#	Method	Bounding box						Mask					
		AP	AP_{50}	AP_{75}	AP_s	AP_m	AP_l	AP	AP_{50}	AP_{75}	AP_s	AP_m	AP_l
1	Mask R-CNN	21.0	36.8	21.9	9.1	18.8	21.2	20.2	35.6	21.6	8.8	17.2	20.1
2	R^3-CNN	22.7	38.4	22.8	9.5	19.9	30.9	22.2	36.2	24.3	9.4	19.3	29.8

Table 3. BBox AP per category. See Table 1 for class key. AP: Average Precision.

#	Method	LBR	GBR	LGM	GGM	VL	VG	LPM	GPM	LDM	GDM	AP
1	Mask R-CNN	45.8	14.3	41.7	6.0	51.3	14.0	4.8	16.9	14.2	0.9	21.0
2	R^3-CNN	49.1	15.2	40.4	7.0	55.3	16.7	6.0	18.2	17.4	1.4	22.7

Table 4. Segmentation AP per category. See Table 1 for class key. AP: Average Precision.

#	Method	LBR	GBR	LGM	GGM	VL	VG	LPM	GPM	LDM	GDM	AP
1	Mask R-CNN	48.6	15.0	40.4	5.4	50.3	10.3	4.4	14.3	12.1	0.7	20.2
2	R^3-CNN	52.0	16.2	44.6	4.6	54.8	11.6	5.0	15.8	15.7	1.3	22.2

For the experiment, basic Mask R-CNN [6] and R^3-CNN [10] models were used, with the intention to test these baseline models on the proposed dataset.

Table 2 shows the training results for the task of Object Detection and Instance Segmentation. Best values are given by the R^3-CNN. More in detail, Tables 3 and 4 show AP values for each category for object detection and instance segmentation, respectively. Figure 7 shows some examples of R^3-CNN predictions

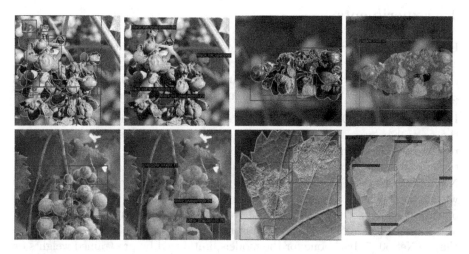

Fig. 7. R^3-CNN prediction examples (right) compared with ground-truth (left).

Normalized Confusion Matrix

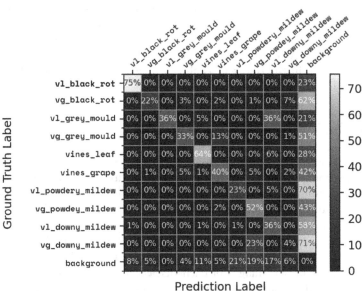

Fig. 8. R^3-CNN confusion matrix.

compared with ground-truth. For black rot on leaves, one of most represented disease in our dataset, we reached promising results. The same for leaves, even though there were much fewer instances. Conversely, a grape of wine is more difficult to detect achieving a third of performances compared to leaves with two thirds of its instances. Despite a fair number of instances, most difficult disease to detect is *peronospora_grappolo* (Downy mildew on bunches). Figure 8 shows the confusion matrix for the R^3-CNN model.

5   Discussion on the Dataset

We introduced a new dataset for segmenting the most common leaf and grapes diseases in their natural environment. The main idea behind the creation of this dataset was to use images with few leaves or bunches affected by a few diseases in foreground in order to facilitate and speed-up the manual process of labeling, but obtaining an accurate annotation. This choice explains the number of categories per image shown in Fig. 4b and can possibly led to a degradation of performances in case of the segmentation of images with an high numbers of leaves or grapes, that are very common into a natural environment.

Emphasis was placed on generating ground truth labels where the disease's polygon was entirely contained inside the leaf or cluster polygon, in order to encourage the strict connection between infected organ and disease, trying to avoid cases where an isolated disease is shown.

Another peculiarity of this dataset is related to the choice of decomposing large ground-truth polygons into smaller ones in order to realize the most accurate labeling. In particular, this approach was used for the labeling of clusters and their diseases in order to exclude the small gaps that are present between berries in a bunch. A consequence of this label partitioning is the predominance of small instances in the dataset, as shown in Fig. 6c, which can led to a more challenging segmentation task.

6   Conclusions

In this paper, we have described the newly-proposed LDD dataset, which contains images coming mostly from the natural environment in which the vines are cultivated. Images are cataloged in terms of objects found in the images, by means of bounding boxes and segmentations in the COCO-like format.

As future work, we plan to increase the amount of images, in particular for those categories that under-represented like *Powdery (Oidio)* and *Downy (Peronospora)* mildews on both grapes and leaves. The final goal is to collect at least the amount of 500 images of each disease in order to provide a common and public ground of comparison for instance segmentation architectures in this kind of applications.

Acknowledgments. We acknowledge Horta s.r.l for the collaboration and help to build the dataset. This study was performed within the project "AGREED - Agriculture, Green & Digital" (ARS01 00254 on PON "Ricerca e Innovazione" 2014–2020 and FSC funds).

References

1. Alessandrini, M., Rivera, R.C.F., Falaschetti, L., Pau, D., Tomaselli, V., Turchetti, C.: A grapevine leaves dataset for early detection and classification of ESCA disease in vineyards through machine learning. Data Brief **35**, 106809 (2021)
2. Amit, Y., Felzenszwalb, P., Girshick, R.: Object detection. In: Computer Vision: A Reference Guide, pp. 1–9 (2020)
3. Chen, K., et al.: Mmdetection: open MMLab detection toolbox and benchmark. arXiv preprint arXiv:1906.07155 (2019)
4. Fuentes, A., Yoon, S., Kim, S.C., Park, D.S.: A robust deep-learning-based detector for real-time tomato plant diseases and pests recognition. Sensors **17**(9), 2022 (2017)
5. Hafiz, A.M., Bhat, G.M.: A survey on instance segmentation: state of the art. Int. J. Multi. Inf. Retrieval **9**, 171–181 (2020)
6. He, K., Gkioxari, G., Dollár, P., Girshick, R.: Mask R-CNN. In: Proceedings of the IEEE International Conference on Computer Vision, pp. 2961–2969 (2017)
7. He, K., Zhang, X., Ren, S., Sun, J.: Deep residual learning for image recognition. In: Proceedings of the IEEE Conference on Computer Vision and Pattern Recognition, pp. 770–778 (2016)

8. Lin, T.Y., et al.: Microsoft coco: common objects in context. In: Fleet, D., Pajdla, T., Schiele, B., Tuytelaars, T. (eds.) Computer Vision-ECCV 2014, pp. 740–757. Springer, Cham (2014). https://doi.org/10.1007/978-3-319-10602-1_48

9. Rançon, F., Bombrun, L., Keresztes, B., Germain, C.: Comparison of sift encoded and deep learning features for the classification and detection of ESCA disease in bordeaux vineyards. Remote Sens. **11**(1), 1 (2019)

10. Rossi, L., Karimi, A., Prati, A.: Recursively refined R-CNN: instance segmentation with self-ROI rebalancing. In: Tsapatsoulis, N., Panayides, A., Theocharides, T., Lanitis, A., Pattichis, C., Vento, M. (eds.) Computer Analysis of Images and Patterns, pp. 476–486. Springer, Cham (2021). https://doi.org/10.1007/978-3-030-89128-2_46

11. Santos, T.T., de Souza, L.L., dos Santos, A.A., Avila, S.: Grape detection, segmentation, and tracking using deep neural networks and three-dimensional association. Comput. Electron. Agric. **170**, 105247 (2020)

12. Storey, G., Meng, Q., Li, B.: Leaf disease segmentation and detection in apple orchards for precise smart spraying in sustainable agriculture. Sustainability **14**(3), 1458 (2022)

13. Thapa, R., Zhang, K., Snavely, N., Belongie, S., Khan, A.: The plant pathology challenge 2020 data set to classify foliar disease of apples. Appli. Plant Sci. **8**(9), e11390 (2020)

14. Udawant, P., Srinath, P.: Cotton leaf disease detection using instance segmentation. J. Cases Inf. Technol. (JCIT) **24**(4), 1–10 (2022)

15. Wang, H., Li, G., Ma, Z., Li, X.: Application of neural networks to image recognition of plant diseases. In: 2012 International Conference on Systems and Informatics, ICSAI2012, pp. 2159–2164. IEEE (2012)

16. Xie, X., Ma, Y., Liu, B., He, J., Li, S., Wang, H.: A deep-learning-based real-time detector for grape leaf diseases using improved convolutional neural networks. Front. Plant Sci. **11**, 751 (2020)

Inpainting Transformer for Anomaly Detection

Jonathan Pirnay[(⊠)] and Keng Chai

Digital Incubation, Fujitsu Technology Solutions GmbH, Munich, Germany
`digital.incubation@ts.fujitsu.com`

Abstract. Anomaly detection in computer vision is the task of identifying images which deviate from a set of normal images. A common approach is to train deep convolutional autoencoders to inpaint covered parts of an image and compare the output with the original image. By training on anomaly-free samples only, the model is assumed to not being able to reconstruct anomalous regions properly. For anomaly detection by inpainting we suggest it to be beneficial to incorporate information from potentially distant regions. In particular we pose anomaly detection as a patch-inpainting problem and propose to solve it with a purely self-attention based approach discarding convolutions. The proposed Inpainting Transformer (InTra) is trained to inpaint covered patches in a large sequence of image patches, thereby integrating information across large regions of the input image. When training from scratch, in comparison to other methods not using extra training data, InTra achieves results on par with the current state-of-the-art on the MVTec AD dataset for detection and segmentation.

Keywords: Anomaly detection · Self-attention · Transformer

1 Introduction

Anomaly detection and localization in vision describe the problem of deciding whether a given image is atypical with respect to a set of normal samples, and to identify the respective anomalous subregions within the image. Both problems have strong implications for industrial inspection [3] and medical applications [8]. In practical industrial applications, anomalies occur rarely. Due to the lack of sufficient anomalous samples, and as anomalies can be of unexpected shape and texture, it is hard to deal with this problem with supervised methods. Current approaches follow unsupervised methods and try to model the distribution of normal data only. At test time an anomaly score is given to each image to indicate how much it deviates from normal samples. For anomaly localization a similar score is assigned to subregions or individual pixels of the image.

A common approach following this paradigm is to use deep convolutional autoencoders or generative models such adversarial networks in order to model the manifold of normal training data. The difference between the input and reconstructed image is then used to compute the anomaly scores. In practice

S. Sclaroff et al. (Eds.): ICIAP 2022, LNCS 13232, pp. 394–406, 2022.
https://doi.org/10.1007/978-3-031-06430-2_33

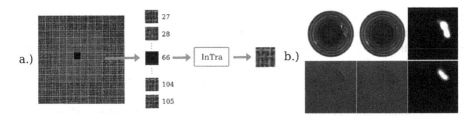

Fig. 1. Schematic overview of the proposed method. a.) The image is split into square patches. An inpainting transformer model (InTra) is trained to reconstruct a covered patch (black) from a long sequence of surrounding patches (red). Positional embeddings are added to the patches to include spatial context. b.) Examples: By reconstruction of all patches of an input image (left), a full reconstruction is obtained (middle). Comparison of original and reconstruction yields a pixel-wise anomaly score (right). (Color figure online)

this approach often suffers from the drawback that convolutional autoencoders generalize strongly and anomalies are reconstructed well, leading to misdetection. Recent methods propose to mitigate this effect by posing the generative part as an inpainting problem: Parts of the input image are covered and the model is trained to reconstruct the covered parts in a self-supervised way [4,10,13,22]. By conditioning on the neighborhood of the excluded part only, small anomalies get effectively retouched. Due to their limited receptive field, fully convolutional neural networks (CNNs) are partially ineffective in modeling distant contextual information, which makes the removal of larger anomalous regions difficult. For inpainting in general settings, this can be effectively addressed by introducing contextual attention in the model [21]. For inpainting in the context of anomaly detection we suggest it to be beneficial to learn the relevant patterns alone by combining information from large regions around the covered image part via attention.

Inspired by the recent success of self-attention based models such as Transformers [17] in image recognition [7], we pose anomaly detection as a patch-inpainting problem and propose to solve it without convolutions: images are split into square patches, and a Transformer model is trained to reconstruct covered patches on the basis of a long sequence of neighboring patches. By recovering the whole image in this way, a full reconstructed image is obtained where the reconstruction of an individual patch incorporates a large context and not only the appearance of its immediate neighborhood. Thus patches are not reconstructed by simply mimicking the local neighborhood, leading to high anomaly scores even for spacious anomalous regions.

Our contributions enfold the modeling of anomaly detection as a patch-sequence inpainting problem which we solve using a deep Transformer network consisting of a simple stack of multiheaded self-attention blocks. Within this network convolutional operations are removed entirely. Furthermore we propose to employ long residual connections between the Transformer blocks and to perform a nonlinear dimension reduction for keys and queries when computing self-attention in order to improve the network's reconstruction capabilities for difficult surfaces. By adding embeddings of the position of individual patches within an image to the sequence of patches, it is possible to perform the inpainting in a global context even if the sequence of patches does not cover the full image.

We evaluate our method on the challenging MVTec AD dataset [3] for both detection and segmentation. Although Transformer networks are usually trained on huge amounts of data, we effectively train our networks with ∼55M parameters from scratch only on the 60–400 images available for each category in MVTec AD. We compare our results to the current state-of-the-art not using any extra training data. Our proposed method InTra achieves on par results on the detection task and slightly better results on the segmentation task.

2   Related Work

In anomaly detection, reconstruction-based methods try to model only normal, defect-free samples. For this, deep CNN autoencoders are widely used to learn the manifold of defect-free images in a latent bottleneck. Given defective test data, these models should not be able to properly reconstruct the anomalous image since they only model normal data [2,5,16]. An anomaly map for segmentation is usually generated via pixel-wise difference or similarity measures between the input image and its model reconstruction, leading to noticeable anomalies.

Even though in reconstruction-based methods the models are trained on defect-free samples only, they often generalize well to anomalies in practice [9]. An inpainting scheme can be used to effectively hide anomalous regions to further restrict a model's capability to reconstruct anomalies [4,10,13,22]. By covering parts of the original image, the reconstruction method needs to have semantic understanding of the image to be able to generate a coherent and realistic image. Zavrtanik et al. propose to use a U-Net architecture [15] taking advantage of long residual connections. Their reconstruction-based method randomly selects multiple parts of the image to inpaint, yielding the current state-of-the-art results for anomaly detection via inpainting for different benchmarks [22].

Anomalies which span over a large area may still cause problems as these will not be covered up sufficiently enough. As such we propose to add global context by replacing CNNs with a Transformer-based framework applied in vision.

Transformer models were originally introduced in natural language processing (NLP) and have since evolved to be the modern design for various sequence tasks like text translation, generation and document classification [6,17,20].

In a Transformer model, self-attention is used to relate elements of a sequence to each other. Based on the relative weighted importance a shared representation is calculated taking into account the relative dependencies between sequence elements. This is able to replace recurrent neural networks in sequence-to-sequence modeling because long-range dependencies are processed globally. The general architecture can be found in the original work [17].

While Transformer architectures have been widely studied in NLP and sequence modeling, convolutional architectures have been essentially the standard tool in recent years due to weight sharing, translation equivariance and locality. Due to the induced bias in fully convolutional autoencoders, the restricted receptive field limits global context [21]. Even though in theory the self-attention framework may mitigate this problem, running self-attention on the whole image without further simplifications is not feasible [11,14].

Recently Dosovitskiy et al. have proposed Vision Transformer [7], where the image data is split up into square non-overlapping uniform patches. Each patch and position gets embedded into a latent space and every image is treated as a sequence of these embedded patches. A Transformer architecture is applied on the restructured data achieving comparable results to state of the art CNNs and even surpassing them on some tasks while reducing model bias.

3   Inpainting Transformer for Anomaly Detection

Our approach is based on a simple stack of Transformer blocks which are trained to inpaint covered image patches based on neighboring patches. An overview of the method is shown in Fig. 1.

3.1   Patch Embeddings and Multihead Feature Self-attention

We use a similar notation as in [7]. Let $\mathbf{x} \in \mathbb{R}^{H \times W \times C}$ be an input image, where (H, W) denotes the (height, width)-size and C the number of channels of the image. Let K be the desired side length of a square patch and $N := \frac{H}{K}, M := \frac{W}{K}$ (the image is resized such that K divides H and W). We split the image $\mathbf{x}$ into a $N \times M$ grid of flattened square patches

$$\mathbf{x}_p \in \mathbb{R}^{(N \times M) \times (K^2 \cdot C)},$$

where $\mathbf{x}_p^{(i,j)} \in \mathbb{R}^{K^2 \cdot C}$ is the patch in the i-th row and j-th column. Our aim is to choose square subgrids of some side length L in this patch grid and train a network to reconstruct any covered patch in the subgrid based on the rest of the subgrid's patches. Formally, this inpainting problem is as follows:

Let $\left(\mathbf{x}_p^{(i,j)}\right)_{(i,j) \in S}$ be such a square subgrid ("window") of patches defined by some index set $S = \{r, \ldots, r+L-1\} \times \{s, \ldots, s+L-1\}$. Here L is the side length of the window, and (r, s) is the grid position of the window's upper left patch. If $(t, u) \in S$ is the position of some patch, the formal task to inpaint (t, u) given S is to approximate the patch $\mathbf{x}_p^{(t,u)}$ using only the content and positions of all other patches $\left(\mathbf{x}_p^{(i,j)}\right)_{(i,j) \in S \backslash \{(t,u)\}}$ in the window.

As by definition Transformers are invariant with respect to reorderings of the input, the one-dimensional positional information $f(i, j) := (i - 1) \cdot N + j$ of a patch $\mathbf{x}_p^{(i,j)}$ is used. To use as a sequence input to the Transformer model, we map the window of patches and their positional information into some latent space of dimension D, i.e. for each patch $\mathbf{x}_p^{(i,j)}$ with $(i,j) \in S \setminus \{(t,u)\}$ we set

$$\mathbf{y}^{(i,j)} := \mathbf{x}_p^{(i,j)} \mathbf{E} + \text{posemb}(f(i,j)) \in \mathbb{R}^D \tag{1}$$

with learnable weight matrix $\mathbf{E} \in \mathbb{R}^{(K^2 \cdot C) \times D}$, and where posemb denotes a standard learnable one-dimensional position embeddings.

To account for the patch at position (t,u) to inpaint, we add a *single* learnable embedding $\mathbf{x}_{\text{inpaint}} \in \mathbb{R}^D$ to the position embedding via

$$\mathbf{z} := \mathbf{x}_{\text{inpaint}} + \text{posemb}(f(t,u)) \in \mathbb{R}^D. \tag{2}$$

The embedding $\mathbf{x}_{\text{inpaint}}$ is comparable to the class token in [6].

The vectors $\mathbf{z}$ and $\mathbf{y}^{(i,j)}$ for all $(i,j) \in S \setminus \{(t,u)\}$ build the final sequence of embedded patches which serves as an input sequence of length L^2 to the Inpainting Transformer model.

Applying multihead self-attention (MSA) to an input sequence forms the heart of a standard Transformer block as in [17]. For this, queries $\mathbf{q}$, keys $\mathbf{k}$ and values $\mathbf{v}$ are obtained by mapping the input sequence with learnable weight matrices $\mathbf{W}_q, \mathbf{W}_k, \mathbf{W}_v \in \mathbb{R}^{D \times D}$. Self-attention is then computed over slices of $\mathbf{q}, \mathbf{k}, \mathbf{v}$. In cases where the patches of the training images are very similar but indistinct the dot product of queries and keys are very close to each other, leading to an almost uniform softmax-weighted sum in the calculation of MSA. To mitigate this, we propose to perform a nonlinear dimension reduction when computing $\mathbf{q}$ and $\mathbf{k}$. For this, $\mathbf{W}_q$ and $\mathbf{W}_k$ are swapped with multilayer perceptrons (MLP) with a single hidden layer. In all our models we used an output dimension of $\frac{D}{2}$ and a hidden layer dimension of $2 \cdot D$ with GELU non-linearity. We refer to this modified MSA as multihead feature self-attention (MFSA). We experienced improved detection results with MFSA (see Sect. 4.3). However, depending on the output and hidden dimension of the MLPs, the number of learnable parameters increases strongly with MFSA.

3.2   Network Architecture and Training

Our network architecture for inpainting is composed of a simple stack of n Transformer blocks. Figure 2 illustrates the architecture. The structure of each Transformer block mainly follows [7] and consists of MFSA followed by a multilayer perceptron (MLP). Layer normalization is applied before ("pre-norm" [18]), and residual connections after MFSA and MLP. Each MLP has a single hidden layer with GELU nonlinearity and maps $\mathbb{R}^D \to \mathbb{R}^{4 \cdot D} \to \mathbb{R}^D$. In particular the input and output of each Transformer block is a sequence in $\mathbb{R}^{L^2 \times D}$ (see Fig. 2).

To obtain the inpainted patch, we average over the output sequence of the last Transformer block to get a single vector in $\mathbb{R}^D$ which is mapped back to the pixel space of the flattened patches $\mathbb{R}^{K^2 \cdot C}$ via a learnable affine transformation followed by a sigmoidal.

In early experiments an inspection of the attention weights showed that a large spatial context is present in earlier layers. In addition to that, Attention

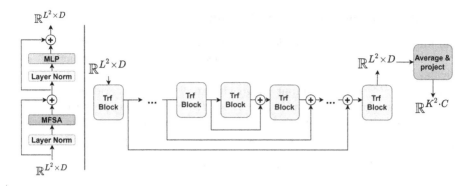

Fig. 2. Overview of the proposed architecture. Left: Parts of an individual transformer block. Right: A stack of Transformer blocks builds the full architecture. Long residual connections are used to add information from earlier blocks to later ones.

Rollout [1] has been used in [7] to illustrate that information across the entire input image is integrated already in the lowest layers. In order to carry this early information to deeper blocks of the network, we put additional long residual connections between early and late layers in a U-Net fashion [15]. We found that the use of long residual connections leads to more structural detail in the overall reconstruction, slightly improving both detection and segmentation (see Sect. 4.3).

The network is trained by randomly sampling batches of patch windows with a fixed side length L from normal image data. In each window a random patch position (t, u) is chosen, which is inpainted by the network as described in the previous sections.

For the loss function, we compare the original and reconstructed patch with pixel-wise L_2 loss. To account for perceptual differences, we also include structural similarity [23] and gradient magnitude similarity [19].

Given an original and reconstructed patch $\mathbf{x}_p, \hat{\mathbf{x}}_p \in \mathbb{R}^{K \times K \times C}$, the full loss function $\mathcal{L}$ is given by

$$\mathcal{L}(\mathbf{x}_p, \hat{\mathbf{x}}_p) = L_2(\mathbf{x}_p, \hat{\mathbf{x}}_p) + \frac{\alpha}{K^2} \sum_{(i,j) \in K \times K} (1 - \mathrm{GMS}_{\mathrm{avg}}(\mathbf{x}_p, \hat{\mathbf{x}}_p)^{(i,j)})$$
$$+ \frac{\beta}{K^2} \sum_{(i,j) \in K \times K} (1 - \mathrm{SSIM}_{\mathrm{avg}}(\mathbf{x}_p, \hat{\mathbf{x}}_p)^{(i,j)}) \tag{3}$$

where α, β are individual scaling parameters, $\mathbf{1}$ is a matrix of ones and $\mathrm{GMS}_{\mathrm{avg}}$ (resp. $\mathrm{SSIM}_{\mathrm{avg}}$) denotes the gradient magnitude similarity maps (resp. structural similarity maps) averaged over the color channels.

3.3   Inference and Anomaly Detection

The inferencing process is divided into two steps: First a complete inpainted image is generated, afterwards the difference between the reconstruction and original is used to compute a pixel-wise anomaly map.

Let $\mathbf{x} \in \mathbb{R}^{H \times W \times C}$ be an input image with an $N \times M$ patch grid as introduced above. For each patch position $(t, u) \in N \times M$, we choose an appropriate patch window of side length L which is used as a basis to inpaint the patch at position $\mathbf{x}_p^{(t,u)}$. In particular we define the window by its upper left patch $\mathbf{x}_p^{(r,s)}$ via

$$r = g(t) - \max(0, g(t) + L - N - 1), \tag{4}$$
$$s = g(u) - \max(0, g(u) + L - M - 1), \tag{5}$$

where the map g is given by $g(c) := \max\left(1, c - \lfloor \frac{L}{2} \rfloor\right)$.

The above equations choose (r, s) such that (t, u) is as much centered in the $L \times L$ patch-window as possible. Using this window, the patch $\mathbf{x}_p^{(t,u)}$ is reconstructed by the network as described. By reconstructing all patches in the $N \times M$ grid, we obtain a full reconstruction $\hat{\mathbf{x}}$ of the whole image.

For the generation of an expressive anomaly map from $\mathbf{x}$ and $\hat{\mathbf{x}}$ we use a simplified variant of the GMS-based scheme proposed in [22]. Denote by $\mathbf{x}_l$ an image $\mathbf{x}$ resized to scale l. Now for original and reconstructed images $\mathbf{x}, \hat{\mathbf{x}} \in \mathbb{R}^{H \times W \times C}$ and scale $l \in \{\frac{1}{2}, \frac{1}{4}\}$, we set

$$m_l(\mathbf{x}, \hat{\mathbf{x}}) := \text{bluravg}_l(1 - \text{GMS}_{\text{avg}}(\mathbf{x}, \hat{\mathbf{x}})) \in \mathbb{R}^{l \cdot H \times l \cdot W} \tag{6}$$

for a scaled and smoothed version of the gradient difference. To ease notation, we denote by bluravg_l the application of an averaging filter followed by some Gaussian blur operation, both with a predefined kernel size and variance. As in [22], smoothing improves robustness with respect to small, poorly reconstructed anomalous regions. We resize the two-dimensional maps $m_{\frac{1}{2}}$ and $m_{\frac{1}{4}}$ back to the image's original size and take the pixel-wise mean which yields a difference map $\text{diff}(\mathbf{x}, \hat{\mathbf{x}}) \in \mathbb{R}^{H \times W}$.

To finally obtain an anomaly map for $\mathbf{x}$ during inference, we take the squared deviation of the difference map to the normal training data, i.e.

$$\text{anomap}(\mathbf{x}) := \left(\text{diff}(\mathbf{x}, \hat{\mathbf{x}}) - \frac{1}{|T|} \sum_{\mathbf{z} \in T} \text{diff}(\mathbf{z}, \hat{\mathbf{z}})\right)^2 \in \mathbb{R}_{\geq 0}^{H \times W}, \tag{7}$$

where T is the set of normal training samples. The pixel-wise maximum of $\text{anomap}(\mathbf{x})$ is taken as a scalar anomaly score for detection on the image level. An example of an anomaly map can be seen in Fig. 1b.

4   Experiments

We evaluate our method on the MVTec AD dataset which contains high resolution samples of 5 texture and 10 object categories stemming from manufacturing

Table 1. Detection/Segmentation results for MVTec AD. Results are presented in ROC AUC % on image level for detection, and on pixel level for segmentation.

Category	RIAD [22]	CutPaste [12]	InTra (Ours)	InTra Image Size
	Det./Seg.	Det./Seg.	Det./Seg.	
Carpet	84.2/96.3	93.1/98.3	**98.8/99.2**	512×512
Grid	99.6/**98.8**	99.9/97.5	**100.0**/98.8	256×256
Leather	**100.0**/99.4	**100.0/99.5**	**100.0/99.5**	512×512
Tile	93.4/89.1	93.4/90.5	**98.2/94.4**	512×512
Wood	93.0/85.8	**98.6/95.5**	97.5/88.7	512×512
Avg. textures	95.1/93.9	97.0/**96.3**	**98.9**/96.1	
Bottle	99.9/**98.4**	98.3/97.6	**100.0**/97.1	256×256
Cable	**81.9/94.2**	80.6/90.0	70.3/91.0	256×256
Capsule	88.4/92.8	**96.2**/97.4	86.5/**97.7**	320×320
Hazelnut	83.3/96.1	**97.3**/97.3	95.7/**98.3**	256×256
Metal Nut	88.5/92.5	**99.3**/93.1	96.9/**93.3**	256×256
Pill	83.8/95.7	**92.4**/95.7	90.2/**98.3**	512×512
Screw	84.5/98.8	86.3/96.7	**95.7/99.5**	320×320
Toothbrush	**100.0/98.9**	98.3/98.1	**100.0/98.9**	256×256
Transistor	90.9/87.7	95.5/93.0	**95.8/96.1**	256×256
Zipper	98.1/97.8	99.4/**99.3**	99.4/99.2	512×512
Avg. objects	89.9/94.3	**94.3**/95.8	93.0/**96.9**	
Avg. all categories	91.7/94.2	**95.2**/96.0	95.0/**96.6**	

[3]. The dataset has been a widely used benchmark for anomaly detection and localization in the manufacturing domain. Each category consists of around 60 to 400 normal, defect-free samples for training. For each test image there is a ground-truth binary image labeled on pixel-level for segmentation of anomalous test images.

Based on an image's anomaly score we report standard ROC AUC as a detection metric. For localisation, the image's anomaly map (7) is used for an evaluation of pixel-wise ROC AUC.

4.1   Implementation Details

We train our model on each product category from scratch. We randomly choose 10% of images from the normal training data (however a maximum of 20) and use them as a validation set to control the quality of reconstructions. In each epoch 600 patch windows are sampled randomly per image. To augment the dataset, random rotation and flipping is used.

The choice of three parameters has an obvious significant impact on the performance: Side length K of square patches, side length L of a patch window and the choice of height H and width W (with $H = W$, as all images are square) to which the original image is resized during training and inference. The patch size determines how much of the image is covered, the size of the patch window determines the dilation of context we include during inpainting, the image size implicitly influences both. For all models we choose $K = 16$, $L = 7$. For the choice of image size in our pipeline, a balance needs to be struck between enlarging the image context of the 7×7 window, quality of patch reconstructions and computation time, as Transformer models usually take a long time to train. The heuristics is to choose the image as small as possible while keeping patch reconstructions at a high level of detail. Hence we train the model with image dimensions $256 \times 256, 320 \times 320, 512 \times 512$ for 200 epochs and compare the best (epoch-wise) validation losses (averaged over ± 5 epochs). If there is no significant improvement in the validation loss of at least 10^{-4} for an image dimension with the next in size, the smaller dimension is chosen. The rightmost column of Table 1 shows the resulting image sizes for each category. We note that in practice K, L, H, W could be tuned for the detection task at hand if prior knowledge about possible defects is present.

The Inpainting Transformer model trained consists of 13 blocks with 8 attention heads each and a latent dimension of $D = 512$, using MFSA. In total this amounts to $\sim$55M learnable parameters.

Given an image size of 512×512, a kernel size of 21 (resp. 11) is used for averaging and Gaussian blur (with $\sigma = 2$) for $\text{bluravg}_{\frac{1}{2}}$ (resp. $\text{bluravg}_{\frac{1}{4}}$) in (6). The kernel sizes are scaled linearly for smaller image sizes. Anomaly maps are resized back to their original high image resolution for proper segmentation comparison.

For the loss function $\mathcal{L}$ in (3) we set $\alpha = \beta = 0.01$. The network is trained using the Adam optimizer with a learning rate of 0.0001 and a batch size of 256 until no improvement on the validation loss is observed for 50 consecutive epochs. The model weights at the epoch with the best validation loss are chosen for evaluation. Although common for training Transformers, we don't apply dropout at any point.

4.2 Results and Discussion

The results for detection and segmentation are reported in Table 1. We compare our method to RIAD [22], which also uses an inpainting reconstruction-based method and computes anomaly maps based on GMS. Furthermore we compare the results to CutPaste [12] which uses a special data augmentation strategy to train a one-class classifier in a self-supervised way. CutPaste also offers results using pretrained representations, however we focus on the results without extra training data in accordance to our training procedure.

To our knowledge RIAD offers the current best performing model based on an inpainting scheme, whereas CutPaste is the best performing model on the MVTec AD benchmark not using extra training data. Our method outperforms

Table 2. Detection/Segmentation results for the ablation studies. 'Regular' refers to the architecture as described in the previous sections, and 'NLR' to no long residual connections'.

	Regular	NLR	MSA	$L = 5$	$L = 9$
	Det./Seg.	Det./Seg.	Det./Seg.	Det./Seg.	Det./Seg.
Avg. text.	98.2/95.3	98.6/95	98.4/95.5	98.4/95.8	98.3/96.2
Avg. obj.	90.7/95.5	90.0/95.3	89.6/95.5	90.0/95.2	90.8/95.8
Avg. all	93.2/95.4	92.9/95.2	92.5/95.5	92.8/95.4	93.3/95.9
Diff. to 'Regular'		**−0.3/−0.2**	**−0.7/+0.1**	**−0.5/0.0**	**+0.1/+0.5**

RIAD on both detection and segmentation. On the detection task, CutPaste is superior to our method by 0.2%, however on the segmentation task, we can improve the result by 0.6%.

It is worth noting that there are two strongly underperforming categories: *Cable* contains many anomalous images where the defect lies in the overall constitution of the product (such as missing pieces). Combined with noise in large areas this makes these anomalies hard to detect via inpainting. Although the defects in *capsule* are per-se easily visible on the generated anomaly maps, our method does not learn the reconstruction of the typography sufficiently well, leading to high anomaly scores also on normal samples.

4.3 Ablation Studies

We examine the influence of certain building blocks in the architecture. All categories except *Leather* are trained for 200 epochs with settings as described in Sect. 4.1, if not stated otherwise. *Leather* takes the longest until details start to show in the inpainted patches, so for comparability the network is trained for 700 epochs. It should be noted that training for only 200 epochs for most categories does not lead to results comparable to Table 1. The average results of the ablation studies are reported in Table 2. We observe a decline in both detection and segmentation when omitting long residual connections. We furthermore examine the effect of using normal multihead self-attention (MSA) instead of MFSA as described in Sect. 3.1. Segmentation results do not improve using MFSA over MSA, however the average detection results improve by 0.7% when using MFSA. Lastly we test how the side length L of a patch window influences the performance by training the network also with $L \in \{5, 9\}$. Detection and segmentation improve with growing patch windows, as more information from distant pixels can be used for the inpainting task. This comes with high computational cost however, as the computation of the dot product in self-attention is quadratic in the sequence length.

5   Conclusion

Inspired by the success of self-attention in vision tasks, we have used a Transformer model for visual anomaly detection by using an inpainting reconstruction approach. We argued that by discarding convolutions and using only self-attention to incorporate global context into reconstructions, anomalies can be successfully detected and localized. Hyperparameters such as the input image size, patch sequence length and patch dimension have a strong impact on the overall performance, and including the detection of good values for them in the training pipeline is paramount. With a simple pipeline as proposed, we have shown that InTra can reach state-of-the-art results on the popular MVTec AD dataset not using extra training data. Further extensions may include the usage of hidden self-attention output embeddings in the anomaly map generation leading to a hybrid reconstruction-embedding based approach.

References

1. Abnar, S., Zuidema, W.H.: Quantifying attention flow in transformers. In: Proceedings of the 58th Annual Meeting of the Association for Computational Linguistics, ACL 2020, Online, 5–10 July 2020, pp. 4190–4197. Association for Computational Linguistics (2020). https://doi.org/10.18653/v1/2020.acl-main.385
2. Baur, C., Wiestler, B., Albarqouni, S., Navab, N.: Deep autoencoding models for unsupervised anomaly segmentation in brain MR images. In: Crimi, A., Bakas, S., Kuijf, H., Keyvan, F., Reyes, M., van Walsum, T. (eds.) BrainLes 2018. LNCS, vol. 11383, pp. 161–169. Springer, Cham (2019). https://doi.org/10.1007/978-3-030-11723-8_16
3. Bergmann, P., Fauser, M., Sattleger, D., Steger, C.: MVTec AD - a comprehensive real-world dataset for unsupervised anomaly detection. In: Proceedings of the IEEE Conference on Computer Vision and Pattern Recognition, pp. 9592–9600 (2019)
4. Bhattad, A., Rock, J., Forsyth, D.A.: Detecting anomalous faces with 'no peeking' autoencoders. CoRR abs/1802.05798 (2018). http://arxiv.org/abs/1802.05798
5. Chow, J., Su, Z., Wu, J., Tan, P., Mao, X., Wang, Y.: Anomaly detection of defects on concrete structures with the convolutional autoencoder. Adv. Eng. Inform. **45**, 101105 (2020). https://doi.org/10.1016/j.aei.2020.101105
6. Devlin, J., Chang, M., Lee, K., Toutanova, K.: BERT: pre-training of deep bidirectional transformers for language understanding. In: Proceedings of the 2019 Conference of the North American Chapter of the Association for Computational Linguistics: Human Language Technologies, NAACL-HLT 2019, Minneapolis, MN, USA, 2–7 June 2019, vol. 1 (Long and Short Papers), pp. 4171–4186. Association for Computational Linguistics (2019). https://doi.org/10.18653/v1/n19-1423
7. Dosovitskiy, A., et al.: An image is worth 16×16 words: transformers for image recognition at scale. In: International Conference on Learning Representations (2021)
8. Fernando, T., Gammulle, H., Denman, S., Sridharan, S., Fookes, C.: Deep learning for medical anomaly detection - a survey (2021). https://doi.org/10.1145/3464423

9. Gong, D., et al.: Memorizing normality to detect anomaly: memory-augmented deep autoencoder for unsupervised anomaly detection. In: IEEE International Conference on Computer Vision (ICCV) (2019)

10. Haselmann, M., Gruber, D.P., Tabatabai, P.: Anomaly detection using deep learning based image completion. In: 2018 17th IEEE International Conference on Machine Learning and Applications (ICMLA), pp. 1237–1242 (2018). https://doi.org/10.1109/ICMLA.2018.00201

11. Ho, J., Kalchbrenner, N., Weissenborn, D., Salimans, T.: Axial attention in multi-dimensional transformers (2019). http://arxiv.org/abs/1912.12180

12. Li, C.L., Sohn, K., Yoon, J., Pfister, T.: CutPaste: self-supervised learning for anomaly detection and localization. In: Proceedings of the IEEE/CVF Conference on Computer Vision and Pattern Recognition (CVPR), pp. 9664–9674, June 2021

13. Nguyen, B., Feldman, A., Bethapudi, S., Jennings, A., Willcocks, C.G.: Unsupervised region-based anomaly detection in brain MRI with adversarial image inpainting. In: 2021 IEEE 18th International Symposium on Biomedical Imaging (ISBI), pp. 1127–1131 (2021). https://doi.org/10.1109/ISBI48211.2021.9434115

14. Parmar, N., et al.: Image transformer. In: Proceedings of the 35th International Conference on Machine Learning. Proceedings of Machine Learning Research, vol. 80, pp. 4055–4064. PMLR, 10–15 July 2018

15. Ronneberger, O., Fischer, P., Brox, T.: U-Net: convolutional networks for biomedical image segmentation. In: Navab, N., Hornegger, J., Wells, W.M., Frangi, A.F. (eds.) MICCAI 2015. LNCS, vol. 9351, pp. 234–241. Springer, Cham (2015). https://doi.org/10.1007/978-3-319-24574-4_28

16. Sakurada, M., Yairi, T.: Anomaly detection using autoencoders with nonlinear dimensionality reduction. In: Proceedings of the MLSDA 2014 2nd Workshop on Machine Learning for Sensory Data Analysis, MLSDA 2014, pp. 4–11. Association for Computing Machinery, New York, NY, USA (2014). https://doi.org/10.1145/2689746.2689747

17. Vaswani, A., et al.: Attention is all you need. In: Proceedings of the 31st International Conference on Neural Information Processing Systems, pp. 6000–6010, NIPS 2017. Curran Associates Inc., Red Hook, NY, USA (2017)

18. Wang, Q., et al.: Learning deep transformer models for machine translation. In: Proceedings of the 57th Annual Meeting of the Association for Computational Linguistics, pp. 1810–1822. Association for Computational Linguistics, Florence, Italy, July 2019. https://doi.org/10.18653/v1/P19-1176

19. Xue, W., Zhang, L., Mou, X., Bovik, A.C.: Gradient magnitude similarity deviation: a highly efficient perceptual image quality index. IEEE Trans. Image Process. **23**(2), 684–695 (2014). https://doi.org/10.1109/TIP.2013.2293423

20. Yang, Z., Dai, Z., Yang, Y., Carbonell, J., Salakhutdinov, R.R., Le, Q.V.: XLNet: generalized autoregressive pretraining for language understanding. In: Advances in Neural Information Processing Systems: Annual Conference on Neural Information Processing Systems 2019, vol. 32, NeurIPS 2019, 8–14 December 2019, Vancouver, BC, Canada, pp. 5754–5764 (2019)

21. Yu, J., Lin, Z., Yang, J., Shen, X., Lu, X., Huang, T.S.: Generative image inpainting with contextual attention. In: 2018 IEEE Conference on Computer Vision and Pattern Recognition, CVPR 2018, Salt Lake City, UT, USA, 18–22 June 2018, pp. 5505–5514. Computer Vision Foundation/IEEE Computer Society (2018). https://doi.org/10.1109/CVPR.2018.00577

22. Zavrtanik, V., Kristan, M., Skočaj, D.: Reconstruction by inpainting for visual anomaly detection. Pattern Recogn. **112**, 107706 (2021). https://doi.org/10.1016/j.patcog.2020.107706

23. Wang, Z., Bovik, A.C., Sheikh, H.R., Simoncelli, E.P.: Image quality assessment: from error visibility to structural similarity. IEEE Trans. Image Process. **13**(4), 600–612 (2004). https://doi.org/10.1109/TIP.2003.819861

FrameNet: Tabular Image Preprocessing Based on UNet and Adaptive Correction

Yufei Wang[1,2], Chen Du[1,2], and Baihua Xiao[1(✉)]

[1] The State Key Laboratory of Management and Control for Complex Systems, Institute of Automation, Chinese Academy of Sciences, Beijing 100190, China
{wangyufei2020,duchen2016,baihua.xiao}@ia.ac.cn
[2] University of Chinese Academy of Sciences, Beijing, China

Abstract. Detecting and recognizing objects in images with complex backgrounds and deformations is a challenging task. In this work, we propose FrameNet, while a deep table lines segmentation network based on our Res18UNet with an adaptive deformation correction algorithm for correcting the table lines. We use Itinerary/Receipt of E-ticket for Air Transport to evaluate our methods. The experiment results show that our Res18UNet can reduce the number of parameters and improve the speed of image segmentation without significantly reducing the segmentation accuracy, and our correction method can better correct the perspective deformation and some distorted tabular images with no dependence on pixel-level ground truth image. In addition, we also apply our model and method to VAT invoice dataset and prove that they also have better transfer ability.

Keywords: Computer vision · Deep learning · Image rectification

1 Introduction

The complex background and the deformation of the object of the image is usually the difficulties in detection and recognition tasks. In the real world, such problems exist on scanning QR code, extracting information from the picture of invoices. Once a good preprocessing method can be designed, the subsequent steps such as detection and recognition will be more efficient.

Invoices usually have texture, shade and red seal, and may be distorted since the improper operation of photographers (shown in Fig. 1), which leads to difficulties to obtain the information on this kind of invoice image effectively. However, most invoices are structured, the horizontal and vertical lines of the invoices will divide the information into specific areas. Therefore, the performance of detection and correction of table blocks becomes a significant step in preprocessing. Traditional line detection methods are usually based on edge detection and Hough transform [1,2], but it is not robust due to the interference from complex background and deformation of the image. In the recent years, researchers tackled this problem using deep neural networks, which resulted in huge performance leaps in computer vision tasks [3–5] and was extended to other

© The Author(s), under exclusive license to Springer Nature Switzerland AG 2022
S. Sclaroff et al. (Eds.): ICIAP 2022, LNCS 13232, pp. 407–417, 2022.
https://doi.org/10.1007/978-3-031-06430-2_34

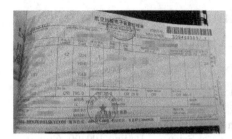

Fig. 1. Examples of Itinerary/Receipt of E-ticket for Air Transport, it can be seen that since the invoices were bound into a book, the image has some deformation.

fields [6,7]. Thus, detecting table lines using deep neural network from complex background and deformable image become realizable.

In this work, we propose FrameNet, which contains a deep neural network for table line segmentation and an adaptive deformation correction for table line rectification, can be used as a preprocessing step for distorted tabular image. The contributions of this work are two-fold.

1. We propose a table line segmentation network based on Res18UNet to predict the table line on pixel level. The network can effectively reduce the amount of parameters and computation without a significant drop of localization accuracy.
2. An adaptive deformation correction method is proposed to correct the table line without pixel-level corrected ground truth images. Results show that our method can better deal with the obstacles of perspective deformation and slight distortion in images and is suitable for tabular image than deep neural network method DewarpNet [8].

2   Methods

2.1   Res18UNet for Table Line Segmentation

UNet [9] is an end-to-end semantic segmentation network that consists of a contracting path and an expansive path. The contracting path extracts feature and downsample the feature map, while the expansive path upsamples the feature map and combines it to the origin feature map from contracting path at the same stage.

However, UNet has a huge number of parameters and mass computation. To this end, following ResNet [5], we propose Res18UNet where use ResNet18 as the downsampling backbone of UNet (shown in Fig. 2) for our table lines segmentation task. Unlike ResNet, we remove max pooling layer in our Res18UNet, and build a convolutional block by stacking two BasicBlocks. Our contracting path consists of four convolutional blocks, the output of each convolutional block is concatenated with the output of the upsampling layer at the same stage. Then

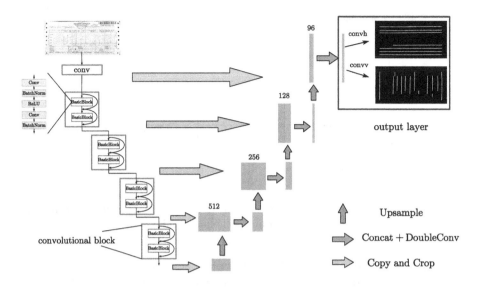

Fig. 2. The architecture of Res18UNet.

the image scale will be expanded through a double convolutional layers and an upsampling layer. In the output layer, we build two convolutional layers followed by a Sigmoid function to predict the probability of horizontal and vertical lines for each pixel respectively.

We use binary cross-entropy loss function (Eq. 1) to calculate the horizontal and vertical loss for each pixel respectively, and add them together to obtain the total loss (Eq. 2).

$$L\left(x_i, y_i\right) = -\left(y_i \cdot \log x_i + (1 - y_i) \cdot \log\left(1 - x_i\right)\right) \tag{1}$$

$$L = \sum_i L_h\left(x_i, y_i\right) + \sum_i L_v\left(x_i, y_i\right) \tag{2}$$

where x_i is the output probability of horizontal line or vertical line of one pixel, and y_i is the ground-truth value of the pixel.

2.2 Adaptive Correction Based on TPS

After obtaining the prediction of the probability of horizontal and vertical lines for each pixel, we use binarization to get binary images where white pixels represent table lines and black pixels represent the background (shown in the upper right part of Fig. 2). Then we apply *bitwise-and* operation to obtain the intersection points so that we can find the moment for each intersection contour by Eq. 3:

$$M_{ij} = \sum_x \sum_y x^i y^j I(x, y) \tag{3}$$

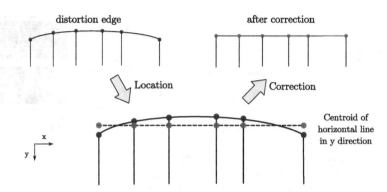

Fig. 3. Main steps of our adaptive correction method, we only show how to correct the horizontal line, correcting the vertical line is similar to this.

where $I(x, y)$ denotes to the value of the pixel at (x, y), and M_{00} is the area (for binary image) or sum of grey level. Therefore, we can calculate the centroid $(\bar{x}, \bar{y})$ for each intersection contour using the following equation:

$$\{\bar{x}, \bar{y}\} = \left\{ \frac{M_{10}}{M_{00}}, \frac{M_{01}}{M_{00}} \right\} \tag{4}$$

We use centroid above to represent the corresponding intersection contour, and the subsequent processes are Adaptive Localization and Correction.

1. **Adaptive Localization:** For the centroid $(\bar{x}_i, \bar{y}_i)$ of the i-th contour, first traverse all horizontal lines and vertical lines to find which one it belongs to (we assume h for the horizontal line and v for the vertical line), then use Eq. 4 to calculate the centroid in the x direction of v (called $\bar{x}_i^{(v)}$) and the centroid in the y direction of h (called $\bar{y}_i^{(h)}$), and $(\bar{x}_i^{(v)}, \bar{y}_i^{(h)})$ is the corrected coordinates for $(\bar{x}_i, \bar{y}_i)$.
2. **Correction:** Use thin plane spline (TPS [10]) to transfer image from $(\bar{x}_i, \bar{y}_i)$ to $(\bar{x}_i^{(v)}, \bar{y}_i^{(h)})$.

Our adaptive correction method is illustrated in Fig. 3 and Algorithm 1.

3 Experiment

3.1 Dataset

We use Itinerary/Receipt of E-ticket for Air Transport dataset in our experiment. In order to show the correction effect, we also made some synthetic data by distorting and deforming the invoice template image. The training set contains 51 images, 47 of which are real invoice images and 4 of which are synthetic invoice images. Test sets are divided into real test set and synthetic test set, where real test set contains 46 real images and synthetic test set contains 13 synthetic images.

Algorithm 1: Adaptive Correction based on TPS.

Input: I denotes the original image, I_h, I_v denotes the prediction feature map
of horizontal and vertical lines respectively.

1   $I_p = $ `bitwise_and`(I_h, I_v) ;   `// Extract intersection points.`

2   $C = $ `findContours`(I_p) ;   `// Find set of points for each contour.`

   `// Use Eq4 to get the centroid of each contour.`

3   **for** $i = 1$ **to** `length`(C) **do** $(\bar{x}_i, \bar{y}_i)=$`moment`$(C[i])$;

4   **for** *each* $(\bar{x}_i, \bar{y}_i)$ **do**   `// Adaptive Localization.`

     | `// Find the contour mask of horizontal and vertical line where`

     | $(\bar{x}_i, \bar{y}_i)$ `belongs to and calculate its centroid` (h, v).

5 | $h=$`findHContour`$((\bar{x}_i, \bar{y}_i), I_h)$;

6 | $v=$`findVContour`$((\bar{x}_i, \bar{y}_i), I_v)$;

7 | $(_, \bar{y}_i^{(h)}) = $ `moment`(h);

8 | $(\bar{x}_i^{(v)}, _) = $ `moment`(v);

9   **end**

10   $I' = $ `tps`$(I, (\bar{x}, \bar{y}), (\bar{x}^v, \bar{y}^h))$;   `// Use TPS to correct the image.`

11   **return** I' ;

Table 1. Comparison of FLOPS (Floating-point operations per second) and parameters of different networks. Res18UNet has far fewer parameters and computations than the other two networks.

Network	GFLOPS	Parameters $(\times 10^6)$
UNet	1371.8	17.27
MobileUNet	1056.92	10.04
Res18UNet	655.4	7.25

3.2   Experiment Settings

Applying data augmentation is essential to help network learn the desired invariance and robustness properties when few training samples are available. In our experiment, we randomly flipped horizontally and vertically with a probability of 50%, and randomly added motion blur. At last, we resized the training images uniformly to 256×512.

We trained by RMSprop with a minibatch size of 5 images, learning rate of 0.0001 with a decay of $1/2$ at 50 and 100 epoch, weight_decay of 0.5 and momentum of 0.9. Our model was trained and tested on a single NVIDIA GTX 1080Ti. We trained for 1000 epochs and saved the checkpoint for testing.

3.3   Results

Backbone. We compared our Res18UNet to UNet and MobileUNet (use MobileNetv1 [11] as the backbone of the subsampling section in UNet). The comparison of parameters and computation operations of different backbone is shown in Table 1.

Table 2. Comparison of segmentation accuracy of different networks, where f.w. IU denotes the frequency weighted intersection over union.

Network	Pixel acc.	Mean acc.	Mean IU	f.w. IU
UNet	97.1	97.4	81.8	95.1
MobileUNet	96.4	96.8	81.0	94.5
Res18UNet	96.8	97.2	81.0	94.8

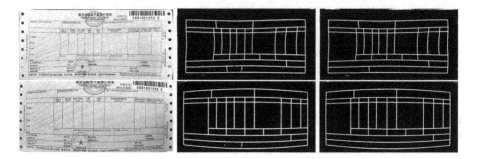

Fig. 4. Comparison of UNet and Res18UNet on table line segmentation task. The first column shows the original invoice image, the second and third columns show the segmentation result from UNet and Res18UNet respectively. We can find from middle column that UNet may predict the red seal at the bottom of invoice to be lines.

Table 3. Comparison of different loss function.

Loss	Pixel acc.	Mean acc.	Mean IU	f.w. IU
Dice loss	95.6	97.1	79.4	93.4
bce loss	96.8	97.2	81.0	94.8

We use four metrics proposed in FCN [12] for our table lines segmentation: pixel accuracy, mean accuracy, mean IU and frequency weighted IU. The comparison of segmentation accuracy is shown in Table 2.

Combined with the above results, we note that in the case of little difference in accuracy, the number of parameters and computation of UNet is at least twice than that of Res18UNet, which make it better to distinguish the edge area. However, since the lack of training data and residual module, UNet tends to overfit and predict the edge of red seal at the bottom of invoice to be lines in some images (shown in Fig. 4). While compared with MobileUNet, our Res18UNet is superior in computation, storage and accuracy.

Loss Function. We also compared binary cross-entropy (bce) loss with another loss function, dice loss. Results in Table 3 show that bce loss is more suitable for the task.

Table 4. Change in variance before and after correction in test set.

Test set		σ_y	σ_x
Real images	Before correction	77.06	73.31
	After correction	44.99	55.16
Synthetic images	Before correction	62.26	69.47
	After correction	42.77	62.8

Correction. After performing the adaptive correction steps, we send the corrected image into Res18UNet again to detect table lines, then calculate the variance in the y direction of the obtained horizontal lines as well as the variance in the x direction of the obtained vertical lines, and finally compare them with lines variance before correction.

The details of calculating variance for horizontal lines are as follows: Once we obtain the horizontal segmentation map (the horizontal output of Res18UNet), we first find the contour for all horizontal line segmentation results, then for each contour, we calculate the variance of all pixels in the y direction using Eq. 5, finally we sum up variance (Eq. 6) from different contours so that we obtain the variance of horizontal lines (σ_y).

$$\sigma_y^{(i)} = \frac{1}{N_i} \sum_{j=1}^{N_i} \left(y_j - \mathrm{E}\left[\boldsymbol{y}^{(i)}\right]\right)^2, (x_j, y_j) \in \Omega_i \tag{5}$$

$$\sigma_y = \sum_{i=1}^{N} \sigma_y^{(i)} \tag{6}$$

where Ω_i denotes to the i-th contour (N contours in total), $\boldsymbol{y}^{(i)}$ denotes to the y coordinate values of all points in Ω_i, N_i and $\sigma_y^{(i)}$ denotes to the number of pixels and the variance of the y coordinate in Ω_i respectively. Analogous to this, we can also calculate the variance of vertical lines (σ_x).

We test our correction algorithm on both real dataset and synthetic dataset, and use variance to evaluate our algorithm. On the test set of real images, the variance is reduced by 41% and 30% in the y and x direction respectively, where on the test set of synthetic images, the variance is reduced by 31% and 10% in the y and x direction respectively. The reason why the variance is smaller on the synthetic test set than on the real test set is that the perspective deformation on synthetic images are less obvious than that of real images. Although the edges of the image are affected during the correction process and will result in the information lost of some pixel (shown as black edge area in the middle column of Fig. 5 and 6), the table line areas can be well corrected and the image looks flat. So the result shows that our correction method can handle the perspective deformation and slight distortion well, and the performance is better reflected in the synthetic test set than real test set. The variance of real dataset and

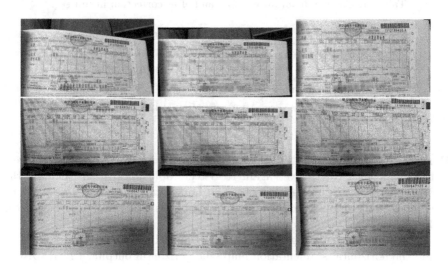

Fig. 5. Comparison of different correction methods for real test images. First column: original image. Second column: image processed by our FrameNet. Third column: image processed by DewarpNet [8].

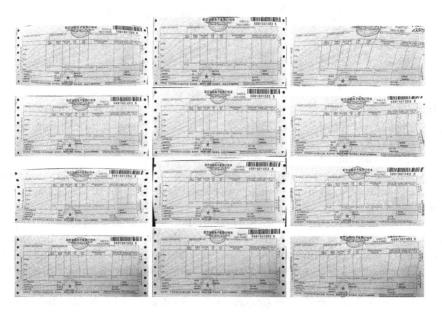

Fig. 6. Comparison of different correction methods for synthetic test images. First column: original image. Second column: image processed by our FrameNet. Third column: image processed by DewarpNet [8].

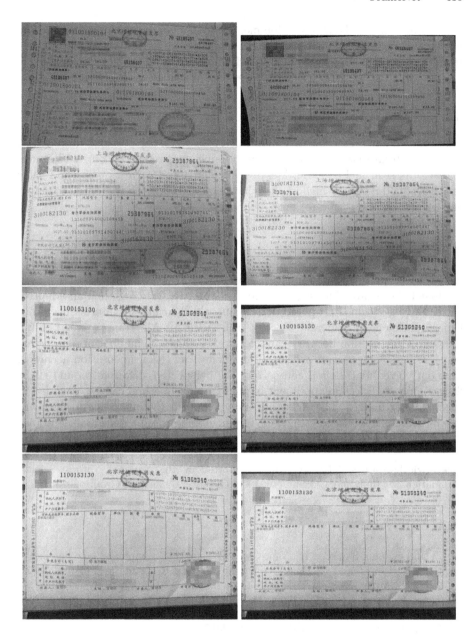

Fig. 7. Results of FrameNet applied to VAT invoices. The left column shows the origin VAT image while the right column shows the corrected results.

synthetic dataset are shown in Table 4, while the performance of correcting on real and synthetic dataset is shown in Fig. 5 and 6.

We also compared our correction method with a famous document unwarping network, DewarpNet [8]. We use the pre-trained proposed in [8] on our test set. Results show that DewarpNet may fail to solve the correction for table line images while our method can better deal with it (shown in Fig. 5 and 6).

3.4 Extended Experiment

We extended our FrameNet to VAT invoices. Based on the previous training set, we added 10 labeled VAT samples for training, and tested on our VAT image test set. As seen in Fig. 7, our Res18UNet model and correction method show relatively good results, which implies that our method has the potential to detect and correct different types of tabular images.

4   Conclusion

In this work, we propose FrameNet, a preprocessing method for tabular image. We first build a table line segmentation network, Res18UNet, then an adaptive correction algorithm is proposed to overcome the deformation interference problem. Experiment on Itinerary/Receipt of E-ticket for Air Transport first shows that our segmentation network can effectively reduce the number of parameters and have less computation without highly affecting the segmentation accuracy, then our correction method can correct the perspective deformation and slight distortion case. Extend experiment implies that our method can be extended to other tabular images. Therefore, our model could be helpful to perform downstream tasks such as detecting and recognizing words for deformable tabular image.

Acknowledgements. This work was supported by the National Natural Science Foundation of China under Grant 71621002, Grant 62071469 and Grant 62073326.

References

1. Zheng, F., Luo, S., Song, K., Yan, C.-W., Wang, M.-C.: Improved lane line detection algorithm based on Hough transform. Pattern Recognit Image Anal. **28**(2), 254–260 (2018)
2. Tong, J., Shi, H., Wu, C., Jiang, H., Yang, T.: Skewness correction and quality evaluation of plug seedling images based on Canny operator and Hough transform. Comput. Electron. Agric. **155**, 461–472 (2018)
3. Krizhevsky, A., Sutskever, I., Hinton, G.E.: ImageNet classification with deep convolutional neural networks. In: Advances in Neural Information Processing Systems, pp. 1097–1105 (2012)
4. Goodfellow, I., et al.: Generative adversarial nets. In: Advances in Neural Information Processing Systems, pp. 2672–2680 (2014)
5. He, K., Zhang, X., Ren, S., Sun, J.: Deep residual learning for image recognition. In: Proceedings of the IEEE Conference on Computer Vision and Pattern Recognition, pp. 770–778 (2016)

6. Vaswani, A., et al.: Attention is all you need. In: Advances in Neural Information Processing Systems, pp. 5998–6008 (2017)

7. Devlin, J., Chang, M.-W., Lee, K., Toutanova, K.: BERT: pre-training of deep bidirectional transformers for language understanding, arXiv preprint arXiv:1810.04805 (2018)

8. Das, S., Ma, K., Shu, Z., Samaras, D., Shilkrot, R.: DewarpNet: single-image document unwarping with stacked 3D and 2D regression networks. In: Proceedings of the IEEE/CVF International Conference on Computer Vision, pp. 131–140 (2019)

9. Ronneberger, O., Fischer, P., Brox, T.: U-Net: convolutional networks for biomedical image segmentation. In: Navab, N., Hornegger, J., Wells, W.M., Frangi, A.F. (eds.) MICCAI 2015. LNCS, vol. 9351, pp. 234–241. Springer, Cham (2015). https://doi.org/10.1007/978-3-319-24574-4_28

10. Bookstein, F.L.: Principal warps: thin-plate splines and the decomposition of deformations. IEEE Trans. Pattern Anal. Mach. Intell. **11**(6), 567–585 (1989)

11. Howard, A.G., et al.: MobileNets: efficient convolutional neural networks for mobile vision applications, arXiv preprint arXiv:1704.04861 (2017)

12. Long, J., Shelhamer, E., Darrell, T.: Fully convolutional networks for semantic segmentation. In: Proceedings of the IEEE Conference on Computer Vision and Pattern Recognition, pp. 3431–3440 (2015)

Coarse-to-Fine Visual Question Answering by Iterative, Conditional Refinement

Gertjan J. Burghouts$^{(\boxtimes)}$ ⓘ and Wyke Huizinga

TNO, The Hague, The Netherlands
gertjan.burghouts@tno.nl
https://gertjanburghouts.github.io

Abstract. Visual Question Answering (VQA) is a very interesting technique to answer natural language questions about an image. Recent methods have focused on incorporating knowledge into an improved VQA model, by augmenting the training set, representing scene graphs, or including reasoning. We also leverage knowledge to make VQA more robust. Yet we take a different route: we take the VQA model as-is and extend it with a novel algorithm called Guided-VQA that guides the questioning by leveraging knowledge to obtain better answers. This enables knowledge-extended VQA while not having to retrain the VQA model. This is beneficial when computing resources and/or time to adapt to new knowledge are limited. We start with the observation that VQA has difficulties with answering compositional and finegrained questions. We propose to solve this by a coarse-to-fine scheme of posing questions. The proposed Guided-VQA algorithm is an iterative, conditional refinement that decomposes a compositional, finegrained question into a sequence of coarse-to-fine questions by leveraging taxonomic knowledge about the involved objects. On Visual Genome, we show that it improves the answers significantly over standard VQA. This is relevant for robust deployment of VQA where resources or adaptation time are limited.

Keywords: Visual Question Answering · External knowledge · Iterative refinement · Image analysis

1 Introduction

Visual Question Answering (VQA) [2] is a technique that answers natural language questions about an image. In this way, VQA extends the capabilities of image interpretation beyond the classification labels of images and the objects and events therein [6], which makes it a very interesting technique. The learning of joint representations of image content and natural language has led to powerful multimodal VQA models such as ViLBERT (Vision-and-Language BERT) [8], e.g., by co-attentional transformer layers. The availability of the 12-in-1 language and vision multi-task dataset [9] has improved VQA models further, enabling better visually-grounded language understanding, for visual question answering, caption-based image retrieval, grounding referring expressions and multi-modal verification. An adaptation of ViLBERT that learns all 12 tasks altogether, is one of the top performers in VQA [9]. Although powerful, state-of-the-art VQA models may struggle when asked a logical composition including negation, conjunction, disjunction, and antonyms [4]. This is further supported by examples that we encountered when experimenting with a state-of-the-art VQA model (ViLBERT [8] trained on the 12-in-1 multi-task dataset [9]) on the Visual Genome dataset [7]. Figures 1a and 1b illustrate that inconsistent answers are produced for respectively a compound question and a question that requires an understanding of the underlying taxonomy. Indeed, explicit knowledge about the world may be needed to resolve such issues [4].

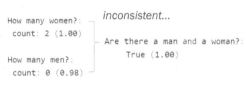

(a) VQA has difficulty with compound questions.

(b) VQA has difficulty with questions that require an understanding of the underlying taxonomy: Boy and Man are indistinguishable for VQA while it distinguishes Child from Adult.

Fig. 1. VQA and two illustrations of failure cases. Explicit knowledge about the world, such as composition and taxonomy, may be needed to resolve such issues.

Recent methods have focused on incorporating knowledge into a VQA model by various approaches. One approach is to curate or extend the training set using external knowledge sources, which is shown to lead to an improved VQA model [4]. Another approach is to include intermediate representations such as scene graphs [13], potentially backed by a knowledge base [15]. In [11] a reasoning module was included, to relate knowledge about the involved objects and relations. These advancements have led to better VQA models, such as ERNIE-ViL, which is now leading the VQA benchmark [13]. The described approaches require retraining the VQA model, which is not always possible. For instance, computing resources may be limited, while VQA models have many parameters hence require heavy compute. Moreover, the time to adapt to new knowledge may be limited, while retraining a large VQA model is time-consuming.

Our goal is to leverage external knowledge to improve VQA, while avoiding retraining the model, and at the same time enabling quick adaptation to new knowledge. We consider a VQA model as-is. We explore a route to incorporate knowledge by the way that the VQA is deployed. To that end, we consider the two flaws from Fig. 1 as a basis and propose solutions to both of them in a single model. The first improvement is to decompose a compound questions into its constituent questions. We propose a scheme to decompose a compound question and to aggregate the confidences of the decomposed sub-questions into a single confidence for the compound question. The second improvement is to include a taxonomy to guide questions that may require knowledge about parent-child relationships. For instance, Boy and Man are highly similar to VQA, while it is able to distinguish Child and Adult. Breaking down the question about 'Is there a Boy?' into asking first 'Is there a Child?' is very helpful here (see Fig. 2).

Fig. 2. VQA, 'Are there one man and one boy?' Answer: 'No' (wrong). Reason: VQA struggles with compound questions, and Boy and Man are indistinguishable for VQA, although it is able to distinguish Adult vs. Child. This paper: decompose questions and use knowledge of taxonomy (e.g., a Boy is a Child) to guide VQA to get better answers.

We propose a coarse-to-fine scheme to answer finegrained questions, that invokes a taxonomy for resolving the finegrained question. We hypothesize that coarser questions are easier to answer, because they are less specific and more common, hence VQA may have learned a better answer. The coarse-to-fine scheme is an algorithm that performs iterative, conditional refinement of the question. Iterative refinement refers to starting with a coarse question and moving to increasingly more specific questions, of which the confidences are combined. The answer to the coarser question is conditional to the next, more specific, question. For instance, if there is no Child, there cannot be a Boy. Our proposed method, called Guided-VQA, combines the decomposition of compound questions and coarse-to-fine answering of finegrained questions. This enables knowledge-extended VQA, while not having to re-train the VQA model. This is beneficial when computing resources and/or time to adapt to new knowledge are limited. On Visual Genome, we show that it improves the answers significantly over VQA as-is.

2   Related Work

Recent methods have focused on incorporating knowledge into a VQA model. One approach is to leverage knowledge to curate or extend the training set, by augmentations that include logical compositions and linguistic transformations (negation, disjunction, conjunction and antonyms) [4]. In [10] image descriptions were enriched by more expressive descriptions from external knowledge sources. Both approaches improved the resulting VQA model. Another approach is to include intermediate representations such as scene graphs. In [3] an image is represented by multiple knowledge graphs from the visual, semantic and factual views. The visual and semantic graphs are regarded as image-conditioned instantiation of the factual graph. A recurrent process collects complementary evidence from the graphs to refine the answer. In [5], a reinforcement agent was proposed that learns to navigate over the extracted scene graph to generate paths, which are then the basis for deriving answers. Other works have focused on enriching the graph. In [15] the scene graph's entities were linked to a knowledge base, to add facts about the entities. A combination of the scene graph with a concept graph was proposed in [15] to learn a question-adaptive graph representation of related instances. The current state-of-the-art on VQA benchmarks is ERNIE-ViL [13], which also uses semantic connections. These connections model objects, attributes of objects and relationships between objects across vision and language. ERNIE-ViL utilizes scene graphs of visual scenes to construct prediction tasks that improve the modelling. These tasks are about predicting objects, attributes and relations, implemented by predicting nodes of different types in the scene graph parsed from the sentence.

Another area of VQA research is focused on reasoning. In [11] a chain of reasoning (CoR) was proposed. This enables multi-step and dynamic reasoning on the possible relations and objects. In an iterative way, its relational reasoning forms new, improved relations between objects thereby refining the answer. In a similar vein, a tiered reasoning method was proposed by [12] to select dynamically object candidates based on language representations to generate pairwise relations between selected candidates. In [1], Probabilistic Soft Logic (PSL) was used as part of an engine that reasons over a set of inputs: visual relations, the semantically parsed question, and background ontological knowledge from word2vec and ConceptNet.

The advances described above have enables better learning of VQA models. Our paper differs in that focuses on leveraging existing VQA models as-is, by a method on top of a VQA model that adds (updated) knowledge to improve its answers.

3   Method

The proposed method is called Guided-VQA. The rationale is that it guides the collection and order of questions that are asked to the given VQA model, thereby improving its answer. The guidance is achieved in multiple, sequential steps. (1) Decomposition of a compound question into its constituent questions (Sect. 3.1). Hypothesis: VQA produces more often correct answers to such partial questions than to a compound question. (2) A coarse-to-fine refinement of each partial question (Sect. 3.2). Hypothesis: VQA produces more often correct answers to coarser questions than finegrained questions. (3) Contradiction removal (Sect. 3.3.). Possibly the refinement of multiple partial questions may lead to inconsistencies which are removed. (4) Conditioning (Sect. 3.4). For a more specific question, the confidence should not exceed the coarser question, for which a conditioning is proposed. (5) The given VQA model is invoked to answer all expanded partial, refined questions. After obtaining the confidences, the conditional rule is applied and an overall single confidence is obtained after aggregation (Sect. 3.5).

3.1   Decomposition

A question Q may be compound and is defined by this grammar rule:

$$Q \rightarrow q(o, n) \, [\, c \, Q \,] \tag{1}$$

where $q(o, n)$ is a partial question, $[\cdot]$ an optional expansion, and $c \in \{and, or\}$. The partial question $q(o, n)$ refers to whether there are n instances of object o in the image. The possible values of n are:

$$n \begin{cases} > 0, & \text{exactly } n \text{ instances of } o \\ = 0, & \text{there is an instance of } o \\ = -1, & \text{there is no instance of } o \end{cases} \tag{2}$$

For example, let $Q = $ 'Are there one boy and one girl?', then $o \in \{boy, girl\}$, $n \in \{one\}$ and $c = and$. The question Q will be mapped to partial questions $q(o, n) \in \{$'Is there one boy?', 'Is there one girl?'$\}$.

3.2   Iterative Refinement

Each (partial) question $q(o, n)$ in the (compound) question Q, is refined in a coarse-to-fine manner by expansion:

$$q(o, n) \rightarrow \begin{cases} \{\, q(o,0),\ q(o,n)\,\}, & \text{if } n > 0 \\ \{\, q(p(o,T),n),\ q(o,n)\,\}, & \text{if } n \in \{0,-1\} \text{ and } p(o,T) \neq \emptyset \quad (3) \\ <\!end\!>, & \text{if } n \in \{0,-1\} \text{ and } p(o,T) = \emptyset \end{cases}$$

where $p(o, T)$ is the parent of o according to the taxonomy T describing parent-child relations that are guiding the coarse-to-fine questioning.

For example, let T be the following taxonomy:

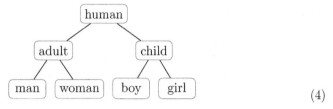

$$(4)$$

The partial question $q(o, n) = q(boy, 1) =$ 'Is there one boy?' is iteratively expanded to the set of coarser questions according to Eq. 3:

$$q(boy, 1) = \text{'Is there one boy?'}$$
$$\rightarrow \{\, q(human, 0),\ q(child, 0),\ q(boy, 0),\ q(boy, 1)\,\}$$
$$= \{\text{'Is there a human?', 'Is there a child?',}$$
$$\text{'Is there a boy?', 'Is there one boy?'}\} \qquad (5)$$

3.3 Contradiction Removal

The refinement from Subsect. 3.2 leads to an expansion of partial questions q, resulting in coarse-to-fine sequences. That expansion can be flattened, yielding:

$$\hat{q} = \{\, q_i(o_i, n_i)\,\}_{i \in [0\,..\,m]} \qquad (6)$$

with m the total amount of sub-questions. Contradictions have the form:

$$q_j(o_j, -1), \text{ where } q_k(o_j, n) \in \hat{q} \text{ with } n \in \{0,1\} \qquad (7)$$

which are removed from $\hat{q}$ to obtain $\hat{q}'$ without contradictions.

For example, let $\hat{q} = \{$'Is there a human?', 'Is there a child?', 'Is there a boy?', 'Is there no human?', 'Is there no adult?', 'Is there no woman?'$\}$. For o_j = human, there is a contradiction $q_j(o_j, -1) =$ 'Is there no human?' and $q_k(o_j, 1)$ = 'Is there a human?'. The contradiction 'Is there no human?' is removed from $\hat{q}$ to obtain a consistent expansion: $\hat{q}' = \{$'Is there a human?', 'Is there a child?', 'Is there a boy?', 'Is there no adult?', 'Is there no woman?'$\}$.

3.4   Conditioning

More specific questions cannot be answered with more confidence than the coarser preceding question in the coarse-to-fine sequence. To enforce this, a monotonic decrease is applied across the coarse-to-fine sequence, by this iterative adjustment:

$$s'(q_i \mid s(q_{i-1})) = min(s(q_{i-1}), \ s(q_i)) \tag{8}$$

where q_{i-1} precedes q_i in the coarse-to-fine sequence, and $s(q_i)$ is the confidence obtained for q_i.

For example, let q_{i-1} = 'Is there a human?' and its confidence $s(q_{i-1})$ = 0.75, and let q_i = 'Is there a child?' and its confidence $s(q_i)$ = 0.80, then the confidence of $s(q_i)$ is corrected into $s'(q_i)$ = min(0.75, 0.80) = 0.75.

3.5   Prediction

The predicted confidence $s(Q)$ for the answer to (compound) question Q (Eq. 1) is obtained by aggregating the conditioned confidences $s(q_i)$ for all $q_i \in \hat{q}'$ as follows. First, each coarse-to-fine expanded sequence is collapsed by taking the average of all $s(q_j)$ in that sequence. Next, the confidences for all sequences are combined, according to Q's definition and its logical connectors $c_k \in \{and, or\}$. Following fuzzy logic's Zadeh operators [14], each $and(\cdot)$ is computed by $min(\cdot)$, and each $or(\cdot)$ by $max(\cdot)$.

For example, let Q = 'Is there a boy and no woman?', then $\hat{q}'_1$ = {'Is there a human?', 'Is there a child?', 'Is there a boy?'} and $\hat{q}'_2$ = {'Is there no adult?', 'Is there no woman?'}, with corresponding confidence sequences s_1 and s_2, where $s_1(q_1) > s_1(q_2) > s_1(q_3)$ and $s_2(q_1) > s_2(q_2)$, e.g., $s_1(q_i)$ = {0.85, 0.80, 0.75} and $s_2(q_i)$ = {0.70, 0.70}. The collapsed confidences are $\overline{s_1}$ = 0.80 and $\overline{s_2}$ = 0.70. The corresponding logical connector is and, yielding the final predicted confidence $s(Q)$ = min(0.80, 0.70) = 0.70.

4   Experiments

4.1   Dataset

For experiments, we consider Visual Genome [7]. It is a very large dataset of a wide variety of everyday images. Each image is accompanied by descriptions of the depicted objects and the relations between them. From this metadata, we extracted 25 queries concerning compositions of four objects: {man, woman, boy, girl}, where each object occurs {0, 1, 2} times. These objects are especially interesting, because we found (see Introduction) that VQA has problems to distinguish man vs. boy and woman vs. girl, where Guided-VQA can help to resolve the issue because it includes taxonomic knowledge about adult vs. child and that man and boy are both male and woman and girl are female. An example query is {1 man, 2 women, 0 boys, 1 girl}. Note that not every composition is available in the dataset. For each query, the images were manually checked, because the dataset is noisy. For instance, some For each query, ~10 images were collected, resulting in a total of 25 queries x ~10 images. For parent-child relationships we use taxonomy T from Eq. 4.

4.2 Setup

The experimental setup is one query vs. all other queries, repeated for all queries. The objective is that the target query scores high confidences for the images of that query, whereas for the remaining images (relating to other queries) the confidences are low. This is evaluated for the set of queries by the mean average precision. We evaluate the results for the set of queries $\{Q_1..Q_n\}$ by the mean average precision (mAP) score: $mAP = \sum_{i=1}^{n} AP(Q_i) \, / \, n$, with $AP(Q_i) = \sum_{j=1}^{m} P_{Q_i}(k) \, \Delta r_{Q_i}(k)$, where k is the rank in the sorted sequence of confidences for Q_i, n is the number of images in the dataset, $P(k)$ is the precision at cut-off k in the sequence, and $\Delta r(k)$ is the change in recall from items $k-1$ to k. Precision and recall are determined from the positive (target) images (i.e., depicting the query Q_i) and the negative (non-target) images (i.e., not depicting the query Q_i).

We compare the three VQA variants: VQA as-is, Sub-VQA (decomposition from Subsect. 3.1 but no refinement) and Guided-VQA (decomposition and refinement, Subsects. 3.1 to 3.5) in order to study the merit of the proposed decomposition and refinement (note that refinement is only possible after decomposition).

4.3 Implementation

For the VQA model, we consider ViLBERT[1] (Vision-and-Language BERT) [8] trained on the 12-in-1 language and vision multi-task dataset [9], containing many types of questions covering visual question answering, caption-based image retrieval, grounding referring expressions, and multi-modal verification. It is one of the top performers in VQA [9]. It outputs logits to the set of 3129 possible answers. For yes/no questions, we extract the respective logits for yes and no. For counting questions, we extract the respective logits for numbers 0–9. The logits are passed through softmax to obtain confidences. In the experiments, ViLBERT is considered as-is (i.e., VQA as-is), and also inside the proposed Guided-VQA to answer the decomposed questions (Sub-VQA) and decomposed plus expanded questions (Guided-VQA).

4.4 Results

Table 1 shows the performance (mean average precision, mAP) for the 25 compound queries from the selected Visual Genome images. Clearly, Guided-VQA outperforms VQA as-is. Sub-VQA performs better than VQA as-is, which

---

[1] https://github.com/facebookresearch/vilbert-multi-task..

demonstrates that applying a decomposition is useful. Guided-VQA performs better than Sub-VQA, which demonstrates that it is useful to incorporate the taxonomy for coarse-to-fine refinement of a finegrained question. Both proposed components of Guided-VQA (i.e., decomposition and refinement) improve VQA's answers.

Table 1. Visual Genome results (mAP) for VQA variants.

VQA variant	Decomposition	Refinement	mAP (↑)
VQA as-is [9]			0.826
Sub-VQA (ours)	✓		0.863
Guided-VQA (ours)	✓	✓	**0.934**

4.5   Analysis

Errors by Guided-VQA happen when VQA makes mistakes on the taxonomic dependencies. For instance, when a girl is mistaken for an adult. We inspect the performance for the hard queries: the queries composed of at least 2 objects. Best hard queries: 'Are there one man and one boy?' scores mAP=0.989, and 'Are there one man and one woman?' scores mAP = 0.979. Worst hard queries: 'Are there one man and one girl?' scores mAP = 0.760, and 'Are there two boys and one girl?' scores mAP = 0.852. From the best and worst hard queries, we conclude that it is easier to assess a woman as a adult and a boy as a child, and harder to assess a girl as a child. An example is shown in Fig. 3 (left), where the girl is mistaken for an adult. This pushes down the confidence for the girl (i.e., child) hence the prediction for the compound query.

Illustrations of improvements by Guided-VQA (compared to VQA as-is), are shown in Fig. 3 (right). We inspect the rankings of both VQA variants. The shown target images are ranked first by Guided-VQA (correct), while VQA ranked it between the non-targets (erroneous).

4.6   Meaning of Confidences

We inspect the confidences of VQA as-is, Sub-VQA and Guided-VQA, for target queries and non-target (other) queries. Figure 4 shows the confidences for the three variants, respectively. The confidences of Guided-VQA are well separable for target and non-target queries. Moreover, the confidences are typically larger than 0.5 for target queries and smaller than 0.5 for non-target queries. This makes interpretation of the confidences by Guided-VQA easier. This is a useful property when Guided-VQA is integrated into larger systems, such as robotics and (semi-)autonomous platforms, as in such systems autonomous decisions can be made based on the confidences.

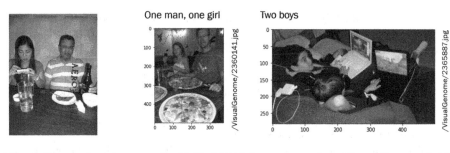

Fig. 3. Illustration of an error by Guided-VQA for an image from Visual Genome (left): the girl is mistaken by the VQA model for an adult. Illustrations of improvements by Guided-VQA over VQA as-is (right).

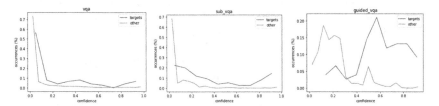

Fig. 4. Confidences on target queries ('targets') and non-targets ('other') for Visual Genome for resp. VQA as-is (left), Sub-VQA (middle) and Guided-VQA (right). Confidences by Guided-VQA are better separated and more interpretable.

5   Conclusions

In this paper, we presented an algorithm called Guided-VQA to incorporate knowledge into a given VQA model to improve its answers to questions that are compound or require understanding of the underlying taxonomy. The approach is to reason on top of the given VQA model, using external knowledge, instead of retraining it. This is relevant for robust deployment of VQA, where resources or adaptation time are limited. Guided-VQA decomposes the question and follows a coarse-to-fine scheme including taxonomic knowledge. The algorithm is an iterative, conditional refinement. We demonstrated the added value of Guided-VQA over VQA as-is, by experiments on Visual Genome. In an ablation study, we showed that both the decomposition and the refinement are helpful for obtaining better answers. Moreover, the confidences of to positives and negatives are better separated and calibrated. This is useful for further reasoning in larger systems such as robots and (semi-)autonomous platforms.

References

1. Aditya, S., Yang, Y., Baral, C.: Explicit reasoning over end-to-end neural architectures for visual question answering. In: Thirty-Second AAAI Conference on Artificial Intelligence (2018)

2. Antol, S., et al.: VQA: visual question answering. In: Proceedings of the IEEE International Conference on Computer Vision, pp. 2425–2433 (2015)
3. Garcia, N., Otani, M., Chu, C., Nakashima, Y.: KnowIT VQA: answering knowledge-based questions about videos. In: Proceedings of the AAAI Conference on Artificial Intelligence, vol. 34, pp. 10826–10834 (2020)
4. Gokhale, T., Banerjee, P., Baral, C., Yang, Y.: VQA-LOL: visual question answering under the lens of logic. In: Vedaldi, A., Bischof, H., Brox, T., Frahm, J.-M. (eds.) ECCV 2020. LNCS, vol. 12366, pp. 379–396. Springer, Cham (2020). https://doi.org/10.1007/978-3-030-58589-1_23
5. Hildebrandt, M., Li, H., Koner, R., Tresp, V., Günnemann, S.: Scene graph reasoning for visual question answering. arXiv preprint arXiv:2007.01072 (2020)
6. Hudson, D.A., Manning, C.D.: GQA: a new dataset for real-world visual reasoning and compositional question answering. In: Proceedings of the IEEE/CVF Conference on Computer Vision and Pattern Recognition, pp. 6700–6709 (2019)
7. Krishna, R., et al.: Visual genome: connecting language and vision using crowd-sourced dense image annotations (2016). https://arxiv.org/abs/1602.07332
8. Lu, J., Batra, D., Parikh, D., Lee, S.: ViLBERT: pretraining task-agnostic visiolinguistic representations for vision-and-language tasks. In: Proceedings of the 33rd International Conference on Neural Information Processing Systems, pp. 13–23 (2019)
9. Lu, J., Goswami, V., Rohrbach, M., Parikh, D., Lee, S.: 12-in-1: multi-task vision and language representation learning. In: Proceedings of the IEEE/CVF Conference on Computer Vision and Pattern Recognition, pp. 10437–10446 (2020)
10. Voruganti, S.E.A.: Visual question answering with external knowledge. In: Proceedings of the International Conference on Smart Data Intelligence (ICSMDI 2021) (2021)
11. Wu, C., Liu, J., Wang, X., Dong, X.: Chain of reasoning for visual question answering. Adv. Neural Inf. Process. Syst. **31**, 275–285 (2018)
12. Yang, X., Lin, G., Lv, F., Liu, F.: TRRNet: tiered relation reasoning for compositional visual question answering. In: Vedaldi, A., Bischof, H., Brox, T., Frahm, J.-M. (eds.) ECCV 2020. LNCS, vol. 12366, pp. 414–430. Springer, Cham (2020). https://doi.org/10.1007/978-3-030-58589-1_25
13. Yu, F., et al.: ERNIE-ViL: knowledge enhanced vision-language representations through scene graphs. In: Proceedings of the AAAI Conference on Artificial Intelligence, vol. 35, pp. 3208–3216 (2021)
14. Zadeh, L.A., Klir, G.J., Yuan, B.: Fuzzy Sets, Fuzzy Logic, and Fuzzy Systems: Selected Papers, vol. 6. World Scientific, Singapore (1996)
15. Ziaeefard, M., Lécué, F.: Towards knowledge-augmented visual question answering. In: Proceedings of the 28th International Conference on Computational Linguistics, pp. 1863–1873 (2020)

DASP: Dual-autoencoder Architecture for Skin Prediction

Igor L. O. Bastos[1]([✉]) [iD], Victor H. C. Melo[1] [iD], Raphael F. Prates[2] [iD], and William R. Schwartz[1] [iD]

[1] Smart Sense Laboratory, UFMG, Belo Horizonte, Brazil
igorcrexito@gmail.com
[2] Department of Computer Science, UniCamp, Campinas, Brazil

Abstract. This paper introduces a novel human skin detection approach based on the application of a dual autoencoder architecture, composed of models to detect background and skin zones concomitantly. This method, named Dual-Autoencoder Skin Predictor (DASP), associates the outputs of two autoencoders through a composite loss that minimizes the error between predicted skin/background areas and the groundtruth. More importantly, the composite loss penalizes overlapping zones between autoencoders predictions, leading our approach to better capture fine-grained and complementary information between skin and background. To combine semantic information with the skin color distribution, heavily tackled by handcrafted skin detection methods, our architecture relies on a main input that considers multiple colorspaces. Besides, a secondary input provides a standard skin/background patch vector to the model, granting information regarding their color distribution. Our experiments support the accurate performance of the proposed architecture and highlights the contributions of the composite loss and multiple inputs. For instance, DASP achieves the best and the second best results on Pratheepan and Mutual Guidance datasets, respectively.

Keywords: Human skin detection · Autoencoders · Composite loss

1 Introduction

Skin detection is a semantic segmentation task that assigns skin or background labels to every pixel in an image [13,21]. It is an important problem due to its applicability as a pre-processing step for different tasks, as face detection [24], human body parts segmentation [18], skin lesion detection [17,31,33], biometric identification [29], gesture recognition [3] and human activity recognition [1,4,5]. Despite its relevance, skin detection is still an open problem in the presence of shadows, backgrounds that resemble the skin and challenging scenarios (i.e., large variations of skin appearance and illumination conditions) [2,19,21].

The application of skin detection on realistic scenarios motivated a shift from methods based uniquely on color distribution of individual pixels [22,25] to modern skin detection approaches based on deep image segmentation networks

S. Sclaroff et al. (Eds.): ICIAP 2022, LNCS 13232, pp. 429–441, 2022.
https://doi.org/10.1007/978-3-031-06430-2_36

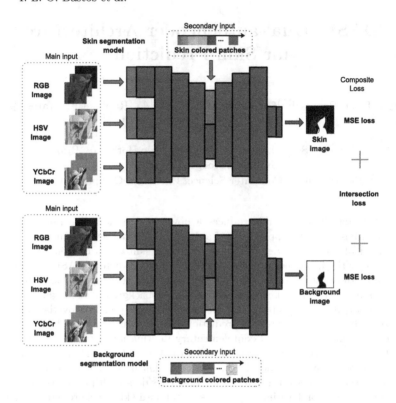

Fig. 1. Dual-autoencoder architecture proposed in the present study (DASP). The final loss function penalizes the MSE of each autoencoder and the intersection between their predictions.

[8,23,28], which weigh the local appearance of the images and incorporate semantic information [13,35]. For semantic information, one can understand as the spatial information that characterizes an image or parts of it, relating their appearance to the context where it is inserted into. This information might help to characterize a region with skin-like coloring but with a dissimilar appearance, or a region located next to objects that do not correspond to parts of the human body as a non-skin zone.

In the literature, many methods do not explore the skin color distribution along with the semantic information [3,13,35]. In addition, they do not tackle the modeling of the background as a complementary task to the skin detection and do not explore any correlation between skin and background, which is a crucial information to better tune deep skin segmentation models.

Focusing on the aforementioned issues, the present work introduces the Dual-Autoencoder Skin Predictor (DASP), illustrated in Fig. 1. This approach relies on an architecture composed of two autoencoder models that are responsible to predict skin and background zones by mutually learning their tasks and exploring correlations between them. To contemplate the color distribution of skin and background zones, both models present two inputs: (i) a main input that corresponds to an image converted to RGB, HSV and YCbCr colorspaces; and (ii) a secondary input composed of patches from skin and background zones that are provided to their respective autoencoder. As a result, our approach combines appearance/semantic information, gathered through their multiple layers, and skin color distribution information, producing a wealthier representation of the input data, used to output the skin/background correspondent output.

The present work also introduces a composite loss function that minimizes the error between skin/background predictions and their groundtruth images while penalizing any intersection between the skin and background models responses. Although applied to the skin segmentation problem, the proposed loss function and architecture could be applied to other image segmentation problems.

The main contributions of this work are the dual autoencoder architecture, aimed at modeling complementary segmentation tasks, in addition to the composite loss function, responsible to enhance the contribution between the models. Furthermore, we evaluate the performance improvement achieved with multiple input modalities, represented by multiple colorspaces and a skin/background standard patch vector. DASP architecture is validated through experiments conducted on the Pratheepan [27] and Mutual Guidance [13] datasets, for which the approach achieves 84.73% and 73.72% of Intersection over Union (IoU), respectively, being comparable to the-state-of-art methods.

2 Related Work

Over the past years, skin detection methods have been proposed and applied as a preliminary step for different studies, obtaining promising results. These methods range from (i) approaches focused on the skin color distribution [21], mostly based on the employment of threshold values [9,15,30] and statistical modelings of skin regions considering multiple colorspaces [10,14], to (ii) approaches based on traditional and deep machine learning techniques to learn a skin color model [11,13,19,32], incorporating information about spatial locality of image pixels and, as a consequence, semantic information about the images.

The methods proposed by Peer and Solina [20] and Kovac et al. [15], for instance, exploit the discriminability of skin pixels based on the color distribution, argumenting for the creation of skin clusters by defining margins at different colorspaces. These margins were obtained by an extensive empirical evaluation of skin images. In the method of Peer and Solina [20], applied to face detection, boundaries are defined on the RGB colorspace to group skin pixels. In the study of Kovac et al. [15], margins are defined on the YCbCr colorspace.

Although establishing accurate margins for skin clusters, methods that purely rely on skin color distribution fail due to the significant overlap between skin and non-skin pixels in colorspaces [13], making room for approaches that employ machine learning strategies to detect skin pixels by gathering semantic information regarding skin spatial distribution in images [13,17,35]. Zuo et al. [35] proposed an architecture composed of convolutional and recurrent layers to emphasize spatial semantic contextual dependencies on the images. Their architecture combines image representations at different semantic and spatial scales, producing a final prediction map based on the sum of these representations. In turn, Arsalan et al. [1] proposes a similar approach, assembling an encoder-decoder architecture composed of convolutional layers and skip-connections.

In the method proposed by He et al. [13], a dual task architecture is designed to perform skin and body detection, composed of layers to encode the input image into a higher semantic representation followed by two decoder branches, responsible to output image and body predictions. The method proposed by He et al. [13], named Mutual Guidance, invests on a multi-task architecture, exploring the correlation between skin and body coordinates. Besides the skin masks, Mutual Guidance requires body mask annotations for training, which are used as a prior for higher probability of skin location. As a result, the method achieves high IoU values on skin detection datasets, such as Pratheepan [27] and on the own assembled Mutual Guidance dataset [13].

On a recent trend, deep networks for image segmentation, such as U-NET [23] and Deeplab-v3 [8] architectures, have been applied to skin detection, leading to accurate results. Despite similarities between the proposed DASP and previously proposed image segmentation networks, such as the presence of skip-connections and multi-scale spatial convolutions, the proposed method differs from them due to the application of a unique architecture, composed of two autoencoder branches that are associated by a composite loss function, which penalizes errors on the mapping of background and skin pixels. In addition, DASP takes into account inputs considering different colorspaces and counts on a secondary input, representing a standard vector of skin/background patch colors. As a result, our model is able to gather information from skin color distribution and also incorporates semantic information through its deep autoencoder models.

3   Proposed Approach

This paper proposes a novel architecture to perform skin segmentation: the Dual-Autoencoder Skin Predictor (DASP)[1], which is composed of a *skin segmentation* and a *background segmentation* autoencoders, as depicted in Fig. 1. While the former predicts skin regions, the latter detects background regions, i.e., any pixel that does not belong to skin.

Despite the apparent redundancy between the tasks performed by these two networks, the design of the present architecture aims at exploring the complementary appearance between skin and background distributions. This point is emphasized by the loss function of DASP, which penalizes the mean squared error obtained from each model, in addition to any intersecting prediction between the skin and background models. Equation 1 details the loss function employed in DASP for a set of n images.

---

[1] Source code and trained models are available at github.com/igorcrexito/skin detector.

$$loss = \frac{1}{n}\sum_{i=1}(gtSKIN_i - pSKIN_i)^2 +$$

$$\frac{1}{n}\sum_{i=1}(gtBACK_i - pBACK_i)^2 + \tag{1}$$

$$H(Inter(pSKIN_i, pBACK_i), BFrame),$$

where H represents the Huber loss computed over the intersection between skin and background predictions and a black frame ($BFrame$). As a result, any pixel highlighted by both models is penalized according to the Huber loss formulation. The first two terms of the sum represent the MSE between each model output (pSKIN and pBACK) and their corresponding groundtruth images (gtSKIN and gtBACK). The reason for the employment of a Huber loss derives from the superior results obtained on experiments with different losses, in which Huber outperformed MAE and MSE, as porteriorly described in Sect. 4.

The architecture of each autoencoder model is similar to image segmentation networks, such as U-NET [23], composed of an encoding and a decoding part, as illustrated in Fig. 2. Hyperparameters, such as the number of convolutional filters on layers, depth of the network, optimizer and batch size, were determined by the search space, showed in Table 1, associated with the Hyperband technique [16]. The details about DASP are outlined in the following subsection.

3.1   Autoencoder Architecture

Inputs and Outputs. DASP autoencoders expect two inputs. The main input corresponds to an image at RGB, HSV and YCbCr colorspaces. Different colorspace information could contribute to skin/background separation, since each of them presents desirable features that can characterize skin, in accordance with approaches based on skin color distribution [15,20,21]. In Fig. 2, a secondary

Table 1. Search space for DASP. *Enc. Factor* is a coefficient to be multiplied by the number of filters on encoding layers of the models, considering an initial architecture. In turn, Dec. Factor is multiplied by the decoding layers. Encoding and Decoding layers represent the number of convolutional layers on these parts of the network. Best hyperparameters are presented in bold.

	v1	v2	v3
Weight initializer	Normal	Uniform	**Glorot normal**
Optimizer	**Adam**	Adagrad	RMSProp
Enc. Factor	0.75	**1.00**	1.50
Dec. Factor	0.75	**1.00**	1.50
Batch size	6	9	**12**
Encoding layers	7	**9**	12
Decoding layers	7	**9**	12

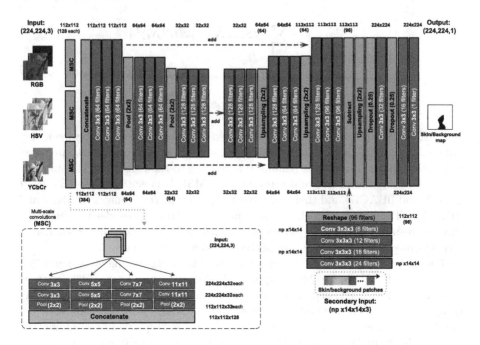

Fig. 2. DASP architecture. The dimensionality and disposition of each layer is presented. On the secondary input, *np* corresponds to the number of skin/background patches used to compose an input vector. ReLU and sigmoid activation layers are depicted in blue and purple, respectively. Gray layers present no activation function. (Color figure online)

input is shown. This input represents a vector composed of skin or background image patches, obtained from the SFA dataset [6], with dimensionality (number of patches) determined by the hyperparameter *np*. This patch vector acts as a standard skin/background distribution. As a result, the main input representation, assembled over the model layers, ponders this standard color/appearance of skin and background patches before outputting their predictions. In terms of outputs, DASP employs two autoencoders: one to predict skin and other to predict background. As a result, our approach counts on two outputs, both capable of detecting skin or background by a simple inversion of their pixel values.

Multi-scale Convolutions. The main input of the model, represented by images at different colorspaces, is provided to convolutional layers with different filter sizes, similarly to Inception modules [26]. With that, spatial information at different scales can be captured, producing a wealthier representation of the input and incorporating more semantics on the segmentation process.

Encoding. The encoding part of the DASP autoencoders gathers spatial information from input images and weighs the skin color distribution from different color spaces. To accomplish that, multiple convolutional blocks are designed to encode the data into a higher semantic representation. Each convolutional block

Table 2. DASP results on Mutual Guidance (left) and Pratheepan (right) datasets.

	IoU (%)	Precision (%)	Recall (%)	IoU (%)	Precision (%)	Recall (%)
Chen et al. [7]	51.44	76.11	63.18	62.43	76.36	77.89
Zuo et al. [35]	63.98	81.21	74.99	73.91	85.19	82.88
U-NET [23]	69.62	83.96	80.61	79.62	89.55	87.87
ResNet50 [12]	66.03	84.73	74.82	77.97	88.30	86.87
Deeplab-v3-MobileNET [8]	67.95	81.92	79.90	77.63	86.93	87.59
Or-SkipNet [19]	n.a	n.a	n.a	81.60	**93.60**	87.73
He et al. [13]	**75.29**	**87.34**	84.64	81.89	92.58	87.51
Ours (DASP - skin)	73.72	85.79	84.16	**84.73**	90.22	92.50
Ours (DASP - background)	73.58	83.24	**87.14**	84.39	87.40	**95.30**

is followed by pooling layers, which are responsible to reduce the spatial dimensionality of the data besides enhancing scale and positional invariance.

Decoding. The decoding part of the autoencoders gathers spatial information from the representation produced by the encoder part and produces information with a higher spatial dimensionality, being next to the initial resolution of the input images. For the decoding part of the model, the secondary input is provided, being merged with the main input representation by a subtraction layer.

Skip-connections. Similarly to image segmentation networks, DASP autoencoders rely on skip-connections, important for gradient maintenance. In addition, these connections allow the decoding layers of the model to handle information at different semantic scales, since these layers receive information from their precedent layers and from layers at the beginning of the model.

Dropout and Output Layers. The decoded representation of the information is provided to the final layers of the model, being submitted to dropout and upsampling operations. These operations lead the representation to assume the same spatial dimensionality of input images. A final convolutional layer outputs the skin/background map according to a logistic sigmoid function. Therefore, the pixels of output images have their values associated with the probability of them to belong to skin or background. Since each autoencoder employs an error-based loss function (MSE, for example), the skin/background detection is treated in DASP as a regression problem.

4   Experimental Results

To evaluate the proposed approach, tests are conducted on the Pratheepan [27] and Mutual Guidance [13] datasets. The entire DASP architecture was trained from scratch and the evaluation followed each dataset protocols.

Evaluation. The results of DASP regarding IoU, Precision and Recall are showed in Table 2, along with literature methods for Pratheepan and Mutual Guidance datasets. The experiments evidence an accurate performance of DASP, obtaining results comparable to state-of-the-art methods for all evaluated metrics. The results of the two autoencoders are showed in Table 2, since both models could predict skin and background by an inversion of their outcomes.

DASP Strategy and Loss on a Different Model. As a final experiment, we evaluated the application of DASP segmentation strategy on U-NET [23], a widely employed image segmentation network. To accomplish that, we composed an architecture similar to DASP, employing two U-NET models, responsible to segment skin and background. We also evaluated the employment of our composite loss considering U-NET. Table 3 shows the IoU (%) values achieved with this experiment on Pratheepan dataset. The baseline U-NET result is compared to a dual architecture that weighs the sum of MSE losses, without penalizing the intersection between predictions (NI). In addition, results regarding different losses to penalize intersection (i.e., Huber, MSE and MAE) are presented.

Table 3. U-NET results with DASP strategy. *NI* stands for a final loss using only the error of the skin and background models, without penalizing intersecting predictions, while MSE, MAE and Huber consider the intersection error to compute the final loss.

	IoU (%)
U-NET (baseline)	79.62
U-NET (NI)	79.87
Huber	80.83
MSE	80.58
MAE	**81.06**

5   Discussion and Concluding Remarks

The accurate performance of DASP, showed in Table 2, led our approach to be compared with state-of-the-art methods for both datasets. Despite being surpassed by the approach of He et al. [13] for some metrics, our method achieved results that are comparable to theirs, with advantages regarding the no-requirement for body mask annotations, besides presenting a less complex architecture and an easier to converge/train loss. Figure 3 shows outputs for state-of-the-art methods along with DASP predictions. The performance of one of the DASP models can strongly surpass the other in some cases, even with their similar results showed in Table 2. This fact can be associated to the difficult to segment the background or skin for some images, as illustrated in Fig. 3.

An ablation study was performed and it is presented in Table 4, evaluating different features of DASP approach. The employment of a single autoencoder to model skin, for instance, promoted a relevant drop on IoU, Precision and Recall metrics. With that, no correlation could be extracted from the modeling of background and skin, with no gathering of complementary information from them. The consideration of a dual autoencoder architecture, even without the

Table 4. Ablation study regarding variations on the architecture depicted in Fig. 2. *No intersection loss* stands for the sum of MSE losses, for both models, with no penalization for intersecting predictions to compute the final loss of the model. *Inc. Single autoencoder model* represents a single autoencoder with the number of filters increased, obtaining a number of trainable parameters similar to a dual autoencoder architecture.

	IoU (%)	Precision (%)	Recall (%)
No secondary input	77.12	85.21	86.08
Single autoencoder model (SM)	70.97	79.55	83.48
Inc. Single autoencoder model	70.71	80.09	82.87
No intersection loss (NI)	80.43	86.17	84.39
MAE instead of Huber	82.58	88.96	90.56
MSE instead of Huber	83.84	**90.22**	91.48
No skip-connections	76.64	84.36	88.40
No HSV and YCbCr (only RGB)	74.61	82.20	85.49
Full Architecture (DASP)	**84.73**	90.20	**92.50**

penalization of intersection zones, led to more accurate results. Without this penalization, there is no direct association between skin and background models. Therefore, the performance gain could be associated to factors such as the mutual learning of both tasks, in accordance to the research of Zhang et al. [34], besides a smoother convergence on the training of the models and an implicit correlation of their outputs through the sum of the losses. Figure 4 shows skin and background predictions for single independent autoencoder (SM), a dual-autoencoder architecture with no penalization on intersections (NI) and full DASP autoencoders. Intersecting predictions are clearly noticed in SM predictions, while less perceived in the remaining ones. In addition, NI models are not accurate to predict fine-grained details and still present overlapping predictions between skin and background models. Lastly, DASP autoencoders present more accurate results, with almost no intersecting predictions.

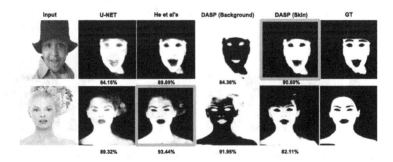

Fig. 3. Predictions of DASP and state-of-the-art methods. Background predictions must be inverted. IoU (%) rates are presented along with images. Even not outperforming the approach of He et al. [13] for some images, our performance is very comparable.

Fig. 4. Predictions for single independent models (SM), dual-architecture models with no penalization on intersection (NI) and DASP models (DASP). Red boxes highlight a zone predicted as skin and background by both SM models, while green boxes highlight the lack of accuracy on the prediction of fine-grained details by NI models. (Color figure online)

The ablation study also supported that information from different colorspaces enhanced DASP results, in consonance with approaches based on skin-color distribution. In turn, the improvement from the correlated information between skin and background could also be noticed on tests evaluated on the U-NET architecture. Despite not outperforming our results, the application of a strategy based on two networks to model skin and background promoted an IoU improvement. With our composite loss, U-NET also had its performed enhanced.

DASP achieved accurate results for most images on both evaluated datasets. Figure 5(a) shows an accurate result achieved despite color similarities between the person skin and the background. This can be associated with the gathering of semantic information, which ponders that background appearance is not similar to human body parts. In Fig. 5(b), even for a grayscale image, DASP explores the appearance (semantic) information to produce a reasonable response.

Although the accurate results for most images, DASP also presented defective outcomes. In Fig. 6(a), the model was not effective to detect skin in the shadowed zones. In Fig. 6(b), the cloth in the arm region presents appearance and skin color similar to a human arm, being wrongly predicted as skin.

The high IoU, Precision and Recall suggest that DASP properly models skin and background zones in images. Our results are comparable to the most accurate methods on literature, with DASP emerging as the best reported approach on the Pratheepan database. Further, we intend to evaluate variations on our architecture, such as the application of atrous convolutions [8] instead of pooling layers. In addition, we intend to employ DASP as a pre-processing step for different approaches, such as skin lesion detection and gesture recognition.

Fig. 5. DASP background and skin results due to semantic information. a) Background color similar to skin colors. b) Grayscale image.

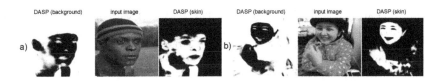

Fig. 6. Inaccurate predictions of DASP. a) Lighting and shadowing conditions. b) Appearance and color similarity with human body parts.

Acknowledgement. The authors would like to thank the National Council for Scientific and Technological Development – CNPq (Grant 309953/2019-7) and the Minas Gerais Research Foundation – FAPEMIG (Grant PPM-00540-17).

References

1. Arsalan, M., Kim, D., Owais, M., Park, K.: Or-skip-net: outer residual skip network for skin segmentation in non-ideal situations. Expert Syst. Appl. **141** (2020)
2. Baldissera, D., Nanni, L., Brahnam, S., Lumini, A.: Postprocessing for skin detection. J. Imaging **7** (2021)
3. Bastos, I.L.O., Angelo, M.F., Loula, A.C.: Recognition of static gestures applied to Brazilian sign language (libras). In: 2015 28th SIBGRAPI Conference on Graphics, Patterns and Images, pp. 305–312 (2015)
4. Bastos, I.L.O., Melo, V.H.C., Schwartz, W.R.: Bubblenet: a disperse recurrent structure to recognize activities. In: 2020 IEEE International Conference on Image Processing (ICIP), pp. 2216–2220 (2020)
5. Bastos, I.L.O., Soares, L.R., Schwartz, W.R.: Pyramidal Zernike over time: a spatiotemporal feature descriptor based on Zernike moments. In: Mendoza, M., Velastín, S. (eds.) CIARP 2017. LNCS, vol. 10657, pp. 77–85. Springer, Cham (2018). https://doi.org/10.1007/978-3-319-75193-1_10
6. Casati, J., Moraes, D., Rodrigues, E.: SFA: a human skin image database based on feret and AR facial images. In: IX Workshop de Visão Computacional (2013)
7. Chen, L., Zhou, J., Liu, Z., Chen, W., Xiong, G.: A skin detector based on neural network. In: IEEE 2002 International Conference on Communications, Circuits and Systems and West Sino Expositions, vol. 1, pp. 615–619 (2002)
8. Chen, L.-C., Zhu, Y., Papandreou, G., Schroff, F., Adam, H.: Encoder-decoder with atrous separable convolution for semantic image segmentation. In: Ferrari, V., Hebert, M., Sminchisescu, C., Weiss, Y. (eds.) ECCV 2018. LNCS, vol. 11211, pp. 833–851. Springer, Cham (2018). https://doi.org/10.1007/978-3-030-01234-2_49
9. Dawod, A., Abdullah, J., Alam, M.: Adaptive skin color model for hand segmentation. In: 2010 International Conference on Computer Applications and Industrial Electronics, pp. 486–489 (12 2010)
10. Dhantre, P., Prasad, R., Saurabh, P., Verma, B.: A hybrid approach for human skin detection. In: 2017 7th International Conference on Communication Systems and Network Technologies (CSNT), pp. 142–146 (2017)
11. Hajiarbabi, M., Agah, A.: Human skin color detection using neural networks. J. Intell. Syst. **24**(4), 425–436 (2015)
12. He, K., Zhang, X., Ren, S., Sun, J.: Deep residual learning for image recognition. In: 2016 IEEE Conference on Computer Vision and Pattern Recognition (CVPR), pp. 770–778 (June 2016)

13. He, Y., et al.: Semi-supervised skin detection by network with mutual guidance. In: The IEEE International Conference on Computer Vision (ICCV) (October 2019)
14. Jones, M., Rehg, J.: Statistical color models with application to skin detection. In: Proceedings. 1999 IEEE Computer Society Conference on Computer Vision and Pattern Recognition (Cat. No PR00149), vol. 1, pp. 274–280 (1999)
15. Kovac, J., Peer, P., Solina, F.: Human skin color clustering for face detection. In: IEEE EUROCON 2003. Computer as a Tool, vol. 2, pp. 144–148 (2003)
16. Li, L., Jamieson, K.G., DeSalvo, G., Rostamizadeh, A., Talwalkar, A.: Efficient hyperparameter optimization and infinitely many armed bandits. CoRR abs/1603.06560 (2016)
17. Liu, L., Mou, L., Zhu, X., Mandal, M.: Skin lesion segmentation based on improved u-net. In: 2019 IEEE Canadian Conference of Electrical and Computer Engineering (CCECE), pp. 1–4 (2019)
18. Micilotta, A., Bowden, R.: View-based location and tracking of body parts for visual interaction. In: BMVC (2004)
19. Minhas, K., et al.: Accurate pixel-wise skin segmentation using shallow fully convolutional neural network. IEEE Access 8, 156314–156327 (2020)
20. Peer, P., Solina, F.: An automatic human face detection method. In: Proceedings of the 4th Computer Vision Winter Workshop (CVWW 1999) (December 1999)
21. Phung, S.L., Bouzerdoum, A., Chai, D.: Skin segmentation using color pixel classification: analysis and comparison. IEEE Trans. Pattern Anal. Mach. Intell. 27(1), 148–154 (2005)
22. Rahmat, R., Chairunnisa, T., Gunawan, D., Sitompul, O.: Skin color segmentation using multi-color space threshold. In: 2016 3rd International Conference on Computer and Information Sciences (ICCOINS) (August 2016)
23. Ronneberger, O., Fischer, P., Brox, T.: U-Net: convolutional networks for biomedical image segmentation. In: Navab, N., Hornegger, J., Wells, W.M., Frangi, A.F. (eds.) MICCAI 2015. LNCS, vol. 9351, pp. 234–241. Springer, Cham (2015). https://doi.org/10.1007/978-3-319-24574-4_28
24. Senior, A., Hsu, R.L., Mottaleb, M.A., Jain, A.K.: Face detection in color images. IEEE Trans. Pattern Anal. Mach. Intell. 24(5), 696–706 (2002)
25. Sigal, L., Sclaroff, S., Athitsos, V.: Skin color-based video segmentation under time-varying illumination. Trans. Patt. Anal. Mach. Intell. 26(7), 862–877 (2004)
26. Szegedy, C., et al.: Going deeper with convolutions. In: IEEE Conference on Computer Vision and Pattern Recognition (CVPR), pp. 1–9 (June 2015)
27. Tan, W., Chan, C., Yogarajah, P., Condell, J.: A fusion approach for efficient human skin detection. IEEE Trans. Ind. Inform. 8(1), 138–147 (2012)
28. Tarasiewicz, T., Nalepa, J., Kawulok, M.: Skinny: a lightweight u-net for skin detection and segmentation. In: 2020 IEEE International Conference on Image Processing (ICIP), pp. 2386–2390 (2020)
29. Uemori, T., Ito, A., Moriuchi, Y., Gatto, A., Murayama, J.: Skin-based identification from multispectral image data using CNNS. In: Proceedings of the IEEE/CVF Conference on Computer Vision and Pattern Recognition (CVPR) (June 2019)
30. Wenjun, T., Gaoyang, D., Han, S., Ziyi, F.: Gesture segmentation based on YCb'Cr' color space ellipse fitting skin color modeling. In: 2012 24th Chinese Control and Decision Conference (CCDC), pp. 1905–1908 (2012)
31. Wu, H., Pan, J., Li, Z., Wen, Z., Qin, J.: Automated skin lesion segmentation via an adaptive dual attention module. Trans. Med. Imaging 40(1), 357–370 (2021)
32. Wu, Q., Cai, R., Fan, L., Ruan, C., Leng, G.: Skin detection using color processing mechanism inspired by the visual system. In: IET Conference on Image Processing (IPR 2012), pp. 1–5 (January 2012)

33. Yang, J., Sun, X., Liang, J., Rosin, P.L.: Clinical skin lesion diagnosis using representations inspired by dermatologist criteria. In: Proceedings of the IEEE Conference on Computer Vision and Pattern Recognition (CVPR) (June 2018)
34. Zhang, Y., Xiang, T., Hospedales, T.M., Lu, H.: Deep mutual learning. In: 2018 IEEE Conference on Computer Vision and Pattern Recognition, pp. 4320–4328 (2018)
35. Zuo, H., Fah, H., Blasch, E., Ling, H.: Combining convolutional and recurrent neural networks for human skin detection. Signal Process. Lett. **24**(3), 289–293 (2017)

Learning Advisor Networks for Noisy Image Classification

Simone Ricci , Tiberio Uricchio$^{(\boxtimes)}$, and Alberto Del Bimbo

University of Florence, Florence, Italy
{simone.ricci,tiberio.uricchio,alberto.bimbo}@unifi.it

Abstract. In this paper, we introduced the novel concept of advisor network to address the problem of noisy labels in image classification. Deep neural networks (DNN) are prone to performance reduction and overfitting problems on training data with noisy annotations. Weighting loss methods aim to mitigate the influence of noisy labels during the training, completely removing their contribution. This discarding process prevents DNNs from learning wrong associations between images and their correct labels but reduces the amount of data used, especially when most of the samples have noisy labels. Differently, our method weighs the feature extracted directly from the classifier without altering the loss value of each data. The advisor helps to focus only on some part of the information present in mislabeled examples, allowing the classifier to leverage that data as well. We trained it with a meta-learning strategy so that it can adapt throughout the training of the main model. We tested our method on CIFAR10 and CIFAR100 with synthetic noise, and on Clothing1M that contains real-world noise, reporting state-of-the-art results.

Keywords: Meta learning · Advisor network · Noisy labels

1 Introduction

Modern image classification systems are based on using deep neural networks models that are trained on a huge number of labeled images [11]. Due to the extreme cost of labeling such an amount of images and difficulty in covering many concepts, researchers recently have looked into methods that generate labels automatically. One significant line of research exploits available labeled images from non-experts (e.g. from social networks, online stores) that can be easily retrieved in large quantities but may have mislabeled [1].

Deep neural networks typically consist of a large number of parameters that are highly shared among feature dimensions and states, enabling flexibility in learning different tasks and classes. This flexibility has the advantage to lead to strong discriminative models unless data annotations are corrupted by noise, leading to performance reduction and overfitting problems [9]. Recent methods tried to address the problem by using curriculum learning [4], directly estimating

S. Sclaroff et al. (Eds.): ICIAP 2022, LNCS 13232, pp. 442–453, 2022.
https://doi.org/10.1007/978-3-031-06430-2_37

the labels noise in the set [8], or measuring the confidence of the network during training [12], also using another co-trained network [7]. The idea was usually to understand mislabeled samples out of distribution and reduce their influence on the learning by dampening their loss or decreasing their impact directly from the training set.

In this paper, we proposed a meta-learning approach to address the problem of noisy labels in image classification based on an advisor network, developed to help the classifier. While a standard image classification model is trained, the advisor network observes the main network activations and adjusts features at training time when noisy label images are identified as input. This allows the classifier model to get information even from mislabeled samples where some noise structure is present. We only retained the main model as the final classifier, while the advisor was discarded. Unlike the teacher-student paradigm, the advisor network was not trained to solve the image classification task, but only to help the learning process of the classifier model by its altering activations.

In summary, our contribution is:

- We propose the use of an advisor network, i.e. the use of an additional network at training time, learned by meta-learning, that can adjust activations and gradient of the main network that is being trained.
- We develop such concept for the task of image classification, allowing the training of an image classification network in presence of artificial label noise.
- We test our approach in presence of artificial label noise and on a popular noisy dataset, obtaining state-of-the-art performance.

2 Related Works

2.1 Noisy Training Labels

Numerous works deal with the problem of noisy labels in training data. It has been shown that the performance of machine learning systems degrades in the presence of label noise [18,21]. A first solution involves a loss correction to mitigate the effect of mislabeled samples on the classifier network. For example GLC [8], Reed [22], M-correction [2], F-correction [6] and S-adaptation [20] estimated the matrix of corruption probabilities used to change the wrong labels to the correct ones. Instead, [17,25,32] modeled the annotations noise distribution linearly combining the output of the network and the noisy label to estimate true labels. Another different approach was assigning a weight to each sample. A lower weight value avoids the contribution of that sample to the training of the network. In this way, it is possible to assign low values to mislabeled examples and high values to correct ones. MentorNet [10] and MentorMix [9] found the latent weights with data-driven curriculum learning and the student-teacher paradigm. Other contributions include data augmentation strategies like Mixup [33], Advaug [5] and DevideMix [13]. Differently from these methods, we modified the network activation using an advisor instead of the loss value.

2.2   Meta Learning

There are methods [3,15,26,27] that needs supplemental clean label to handle the noise. This assumption of clean data is also true for a solution that exploits the Meta-learning paradigm. It consists of the use of machine learning algorithms to assist the training and optimization of other machine learning models. Meta-learning [14,23,24,28] had used to address the noisy labels problem. With small clean validation data, the meta-model learns how to correct the biased training labels. For example, L2R [23] weighed each example giving less importance to the mislabels samples. MLNT [14] simulated regular training with synthetic noisy labels. MW-Net [24] learned an explicit weighting function that can be easily adapted to different types of annotations noise. MLC [28] estimated the label noise transition matrix. Contrary to all aforementioned meta-learning solutions, our method does not act by directly modifying the loss of the neural network. We applied a meta-attention layer inside a neural network. The weights of the attention are learned by the advisor network. In this way, the mislabeled data can be leveraged to improve the overall performance of the main model.

3   Method

3.1   Task

In this paper, we developed a method that can handle images with noisy labels when training networks for image classification. We started from the idea that also a mislabeled example contain information that can contribute to a greater generalization of the network. The model should concentrate only on some convenient parts of these data. Our idea was to exploit the attention mechanism to enhance the useful parts of the information and lower the rest. We made use of an auxiliary advisor network that learns automatically a function that weighs the features extracted from a DNN during its training. This advisor network should be aware of the state of the main model and the meta-learning training solves this constraint. Our method Meta Feature Re-Weighting (MFRW) acts like a meta-attention layer. Different from weighting loss methods that tend to completely remove the influence of mislabeled examples during the training our MFRW can take advantage of them.

We first introduce meta-learning basics and formulation typical of methods that learn robust deep neural networks from noisy labels. Then in Sect. 3.3, we explain our method showing the architecture of the whole process.

3.2   Meta Learning for Noisy Image Classification

In general meta-learning (ML) is referred to the process of improving a learning algorithm over multiple learning episodes, also called commonly learning to learn. Usually, ML is divided into two learning algorithms: an inner (or base) algorithm that solves a task, such as images classification, defined by a training dataset and objective function; an outer (or upper/meta) algorithm that updates the inner one, such that the main model it learns improves an outer objective function. ML was applied to solve the problem of noisy labels in training data [23,24]. We introduce the symbols useful for understanding ML in this particular setting and the three basic steps into which the entire learning process is divided.

Let $D^{train} = \{x_i^{tra}, y_i^{tra}\}_{i=1}^N$ be the noisy annotated training set, where N is the total number of examples, composed of an image x_i and the correspondent one-hot label y_i over c classes. In general if we have a deep neural network (DNN) model $\Phi(\cdot; w)$, where w are its parameters and $\hat{y} = \Phi(x; w)$ is its prediction on the input image x, we can obtain the optimal parameters w^* by minimizing the softmax cross-entropy loss $\ell(\hat{y}, y)$ on the training set D^{train}. ML, applied to the Noisy Image Classification task, requires the presence of an additional verified dataset. This validation set $D^{val} = \{x_j^{val}, y_j^{val}\}_{j=1}^M$ is much smaller than the training set, $M \ll N$.

In [24] a meta-model was used to implement the ML process. A multilayer perceptron network with only one hidden layer learns how to weigh each training example. Let $\Psi(\cdot; \theta)$, parameterized by θ, be the meta-model that maps a loss value to a scalar weight. In this case, the optimal parameters w^* are derived using the following weighted loss:

$$w^*(\theta) = \underset{w}{\mathrm{argmin}} \frac{1}{N} \sum_{i=1}^N \mathcal{V}_i^{tra}(\theta) \mathcal{L}_i^{tra}(w) \tag{1}$$

with $\mathcal{V}_i^{tra}(\theta) = \Psi(\mathcal{L}_i^{tra}(w); \theta)$ as the weight predicted by the meta model for the i-th training example. Instead the meta model is trained minimizing the validation loss:

$$\theta^* = \underset{\theta}{\mathrm{argmin}} \frac{1}{M} \sum_{j=1}^M \mathcal{L}_j^{val}(w^*(\theta)) \tag{2}$$

where $\mathcal{L}_j^{val}(w^*(\theta)) = \ell(\Phi(x_j^{val}; w^*(\theta), y_j^{val}))$ is the loss for the j-th validation example.

Equations (1) and (2) can be minimized alternating optimization via gradient descent. One solution that ensures the efficiency of the algorithm and its convergence [24] adopt an online strategy to update θ and w through a single optimization loop, which is divided into three main steps.

The first step is called Virtual-Train because the original DNN will not be updated and the optimization is carried out on a virtual model that is the copy of the original one. Consider the t-th iteration and associated mini batches $\mathcal{B}^{train} = \{(x_i^{tra}, y_i^{tra})\}_{i=1}^n$ and $\mathcal{B}^{val} = \{(x_j^{val}, y_j^{val})\}_{j=1}^m$, where n and m are the size of mini-batch respectively. The virtual update can be derived by:

$$\hat{w}(\theta) = w - \alpha \frac{1}{n} \sum_{i=1}^n \mathcal{V}_i^{tra}(\theta) \nabla_w \mathcal{L}_i^{tra}(w) \tag{3}$$

where α is the learning rate for the DNN and w is its parameter at the current iteration. Then there is the Meta-Train step, where given the optimized virtual model the meta model is updated by:

$$\theta' = \theta - \beta \frac{1}{m} \sum_{j=1}^{m} \nabla_\theta \mathcal{L}_i^{val}(\hat{w}(\theta)) \tag{4}$$

with β the learning rate for the meta model. In last step, Actual-Train, the base DNN model is optimized taking into account the previously updated meta model.

$$w' = w - \alpha \frac{1}{n} \sum_{i=1}^{n} \mathcal{V}_i^{tra}(\theta') \nabla_w \mathcal{L}_i^{tra}(w) \tag{5}$$

w' becomes the w in Eq. (3) for the $(t+1)$-th iteration.

3.3   Meta Feature Re-Weighting (MFRW)

Attention for a DNN is a mechanism that tries to mimics the cognitive attention of the human brain. It intensifies the important parts of the input and reduces the rest. In Meta Feature Re-Weighting (MFRW) the attention is applied with a Hadamard product between the feature extracted from a DNN and a vector of weights automatically learned from a meta-model. In order to get this, we separated the main model $\Phi(\cdot; w)$ in two-part: the backbone $\Phi_b(\cdot; w_b)$, that has an image x as input and gives out a feature vector f, and the classifier $\Phi_c(\cdot; w_c)$, that has f as input and a probability score vector s as output. In this way, it was possible to manipulate the feature f directly with our meta-model Ψ.

The meta-model takes two different inputs $\Psi(f, \mathcal{L})$ and gives back a vector of weights W_f. The first input f is the feature extracted from the backbone Φ_b relative to the example x. This is important for the meta-model because it makes the W_f strictly connected to the feature that needs to be modified. The other input is the loss $\mathcal{L}$ of the example x calculated from the prediction obtained by the main full model Φ. This gives the meta-model information about how much x is a "hard" or an "easy" example for the main model. The two inputs together let the meta-model differentiate a feature belonging to a noisy x from the one related to a correct x. In dot-product attention the multiplication is done element-wise, so the W_f has to be of the same size of f, and its values must be in the range $\in (0, 1)$.

MFRW is divided into 4 main phases for each iteration. Figure 1 shows the overall process of our method divided by step. We put our method at the t-th iteration and we will describe each step to reach the $(t+1)$-th.

Our method needs an additional initial phase Loss Pre-Calculation respect to [24] and what is described in Sect. 3.2. We must calculate in advance the value of loss $\mathcal{L}^{pre}$ related to the training batch x^{train}. This is done at the beginning to obtain a loss value dependent on the original feature and not on the weighted one. It is not an expensive step because it is a direct loss inference, without gradient calculation.

The second step is the Virtual-Train. Here Φ_b^t and Φ_c^t are temporary clone of the original ones. The train batch x^{train} is passed in Φ_b^t to obtain the features f^{train}. Then f^{train} goes inside Ψ^t with the relative loss values $\mathcal{L}^{pre}$ to get the vector of weights W_f. We multiplied element-wise f^{train} with W_f to get a new

1 - Loss Pre-Calculation

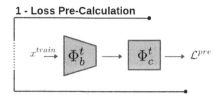

2 - Virtual-Train

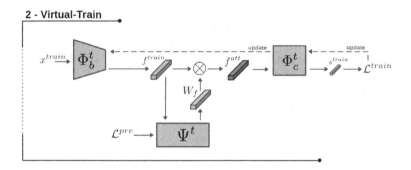

3 - Meta-Train

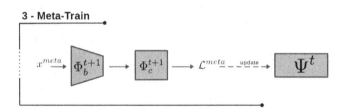

4 - Actual-Train

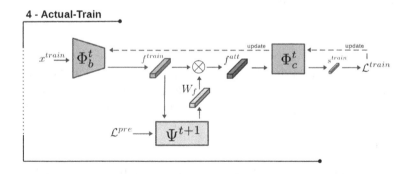

Fig. 1. Illustration of an iteration of the proposed Meta Feature Re-Weighting (MFRW) method. Each iteration is divided into four steps. First, a Loss Pre-Calculation is performed to calculate in advance the loss $\mathcal{L}^{pre}$ value of the training batch x^{train}. The second step is the Virtual-Train, where a clone of the main model is virtually updated. Here the meta-model modifies the feature of the main model multiplying it with a vector of weights. The purple color indicates the weighted features. The third step shows the Meta-Train process. With a meta batch of clean example x^{meta} the meta-model is updated minimizing the loss $\mathcal{L}^{meta}$ given by the previous virtually updated network. In the last phase Actual-Train, the real main model is trained with the meta-model optimized (yellow color). (Color figure online)

feature vector with attention f^{att}. This is given to Φ_c^t to obtain the score s^{train} and then the correspondent loss $\mathcal{L}^{train}$. We now virtually update Φ_b^t and Φ_c^t parameters, but not the one of Ψ^t.

Like [24] we have a clean and balanced meta dataset that will be used to train the meta-model Ψ in the third steps Meta-Train. Here we have Φ_b^{t+1} and Φ_c^{t+1} virtually updated from the step before. Now we pass a meta batch x^{meta} through them in order to get a validation loss $\mathcal{L}^{meta}$. Then Ψ^t is updated minimizing $\mathcal{L}^{meta}$. In this way, the meta-model is optimized to help the main model minimize its error on clean data. Here the optimization takes into consideration also the previous Virtual-Train, thus this is the most expensive part of the method.

The last phase is the Actual-Train where the original Φ_b^t and Φ_c^t are optimized taking into account the updated meta-model Ψ^{t+1}.

The meta-model is used only during the training time of the main network. It will be discarded at test time when only the main network is retained as the final model.

3.4   Meta Model Architecture

Our meta-model Ψ has a really simple architecture. The inputs of the network are a feature f and a loss value $\mathcal{L}_x$. Each input is projected in a fixed size embedding space through a separate fully connected layer. Then the embeddings are concatenated and passed to another fully connected layer that projects them into a larger common space. Its size is the sum of the dimension of each previous embedding. Finally, a linear layer is used to pass the data from the common space to a vector with a size equal to the one of the feature f, that is given as input. Because the output must be an attention weight in the range $\in (0, 1)$ we put a sigmoid activation after the last layer.

4   Experiments

To demonstrate the effectiveness of our method, we conducted experiments on synthetically generated datasets with controlled noise structure and level. Then we tested its ability to generalize with experiments on a real-world dataset.

4.1   Datasets

Following previous work [10,23,24], we used CIFAR-10 and CIFAR-100 which are the typical choice to generate synthetic datasets containing different types of noise structure. They are composed of 50K training images and 10K test images of size 32 × 32. Of the training set, 1000 images with clean labels are randomly selected to create the validation set for meta-training.

In addition to synthetic datasets, there is a collection of data containing real-world noise. Clothing1M [30] is a dataset that is composed of 1 million images of clothing taken from online shopping websites. There are 14 categories like T-shirt, Shirt, Knitwear, etc. The labels are obtained from the text of the images provided by the sellers and not from expert annotators, that's why there are errors. The validation set of $7k$ clean data is as the meta dataset. This dataset allowed our strategy to be evaluated as a concrete application for fine-grained classification with noisy training annotations.

4.2   Implementation Details

We used the same settings for the experiments on CIFAR-10 and CIFAR-100. The backbone was a Resnet-32 trained through SGD with a momentum of 0.9, weight decay of 5e−4, batch size of 128, and a starting learning rate of 0.1. Learning rate decreased to its $\frac{1}{10}$ at the 50 epoch and 70 epoch, stopped at the 100 epochs.

With Clothing1M we used as backbone a ResNet-50 pre-trained on ImageNet. It was trained through SGD with a momentum of 0.9, weight decay of 1e−3, and a starting learning rate of 0.01. The batch size was 32 and it was preprocessed resizing the image to 256×256, crop the center 224×224, and performing normalization. The learning rate was divided by $\frac{1}{10}$ after 5 epochs and stops at 10 epochs.

In every experiment, the meta-model was optimized with Adam and a learning rate of 1e−4. The embedding space size was set always to 100.

4.3   Results

Flip (or asymmetric) is a noise that is designed to mimic the structure where labels are only replaced by similar classes, e.g. dog $\leftrightarrow$ cat. We choose to test our method on that type of noise because it usually appends that the label error could depend on the ambiguity between classes and similar visual patterns [30]. We created a synthetic version of CIFAR-10 and CIFAR-100. The noise ratio was controlled by a parameter p, which represents the probability that a clean example is contaminated by noise. In this way we could test our method on different level of noise, from $p = 0.0$ (no noise), to $p = 0.8$ (heavy noise).

Table 1 shows the accuracy results on the test set of CIFAR-10 and CIFAR-100 datasets with flip label noises. The compared methods are directly cited with the result on their paper. For MW-Net [24] and the direct training (CrossEntropy) we report also our reproduced results. The accuracy gained over the other methods were significant. We can see that at a higher noise rate our result outperforms MW-Net and CrossEntropy by a large margin, indicating the effectiveness of our method on the synthetic Flip noise. From the results of Table 1 is possible to notice a limitation of our strategy that occurs when there is no noise ($p = 0.0$) in the training annotations. We obtained worse accuracy values than the training with the classic softmax cross-entropy loss on both CIFAR-10 and CIFAR-100. The advisor network introduces a bias from the distribution of the meta set to the training data. Because the training annotations are completely correct the introduction of this meta bias makes the accuracy a little worse than without.

Table 1. Top-1 accuracy on CIFAR10 and CIFAR100 dataset with Flip noise. The backbone used was a ResNet-32. p denotes the different levels of noise. The results for the cited method are reported directly from their original papers. $\dagger$ indicates the results obtained by our implementation. The first and the second best results are respectively marked with bold and underline.

Dataset	Flip CIFAR-10					Flip CIFAR-100				
Noise p	0.0	0.2	0.4	0.6	0.8	0.0	0.2	0.4	0.6	0.8
CrossEntropy [24]	92.89	76.83	70.77	–	–	70.50	50.86	43.01	–	–
Reed-Hard [22]	92.31	88.28	81.06	–	–	69.02	60.27	50.40	–	–
S-Model [6]	83.61	79.25	75.73	–	–	51.46	45.45	43.8	–	–
Self-paced [12]	88.52	87.03	81.63	–	–	67.55	63.63	53.51	–	–
Focal Loss [16]	<u>93.03</u>	86.45	80.45	–	–	70.02	61.87	54.13	–	–
Co-teaching [7]	89.87	82.83	75.41	–	–	63.31	54.13	44.85	–	–
D2L [17]	92.02	87.66	83.89	–	–	68.11	63.48	51.83	–	–
Fine-tuning [24]	**93.23**	82.47	74.07	–	–	**70.72**	56.98	46.37	–	–
MentorNet [10]	92.13	86.3	81.76	–	–	70.24	61.97	52.66	–	–
L2RW [23]	89.25	87.86	85.66	–	–	64.11	57.47	50.98	–	–
GLC [8]	91.02	89.68	<u>88.92</u>	–	–	65.42	63.07	**62.22**	–	–
MW-net [24]	92.04	90.33	87.54	–	–	70.11	<u>64.22</u>	58.64	–	–
CrossEntropy†	92.33	90.56	86.25	26.67	13.58	70.18	**65.02**	50.25	18.67	4.32
MW-net† [24]	92.19	<u>90.74</u>	87.63	<u>42.41</u>	<u>27.19</u>	<u>70.57</u>	64.13	51.23	<u>19.89</u>	<u>7.42</u>
Ours	91.87	**91.09**	**90.26**	**89.34**	**82.47**	68.93	63.54	<u>59.07</u>	**56.13**	**20.29**

We introduced also two new noise settings, namely Flip2 and Flip3. The difference from Flip is that the noise is equally distributed over multiple similar classes, two and three respectively. Table 2 and 3 show respectively the result for noise of type Flip2 and Flip3. We can see how our method performs better than the others, especially in very noisy situations.

Table 2. Accuracy result on CIFAR10 and CIFAR100 dataset with Flip2 noise. p denotes the different level of noise. $\dagger$ indicates the results obtained by our implementation. The first and the second best results are respectively marked with bold and underline.

Dataset	Flip2 CIFAR-10				Flip2 CIFAR-100			
Noise p	0.2	0.4	0.6	0.8	0.2	0.4	0.6	0.8
CrossEntropy†	<u>90.71</u>	87.83	75.83	11.86	<u>64.91</u>	57.7	36.55	7
MW-net† [24]	**90.93**	<u>88.83</u>	<u>86.85</u>	<u>27.49</u>	**65.37**	59	<u>36.97</u>	<u>7.99</u>
Ours	90.66	**89.72**	**87.75**	**73.83**	63.07	<u>57.96</u>	**45.35**	**22.41**

Table 4 shows the results on Clothing1M. As we can see our method outperforms the current state-of-the-art result.

Table 3. Result for Flip3 noise on CIFAR10 and CIFAR100 dataset. p denotes the different level of noise. † indicates the results obtained by our implementation. The first and the second best results are respectively marked with bold and underline.

Dataset	Flip3 CIFAR-10				Flip3 CIFAR-100			
Noise p	0.2	0.4	0.6	0.8	0.2	0.4	0.6	0.8
CrossEntropy†	90.13	88.44	82.31	20.34	<u>65.29</u>	<u>59.35</u>	44	<u>11.07</u>
MW-net† [24]	**90.56**	<u>88.49</u>	<u>85.65</u>	<u>22.69</u>	**65.33**	**62.74**	<u>45.77</u>	10.33
Ours	<u>90.31</u>	**88.96**	**87.73**	**75.53**	62.98	59.08	**52.28**	**25.72**

Table 4. Comparison with state-of-the-art methods in test accuracy (%) on Clothing1M dataset. Results for baselines are copied from original papers.

Method	Accuracy (%)
CrossEntropy [24]	68.94
F-correction [20]	69.84
JoCoR [29]	70.30
S-Model [6]	70.36
M-correction [2]	71.00
MLC [28]	71.06
Joint-Optim [25]	72.16
MLNT [14]	73.47
P-correction [32]	73.49
MW-Net [24]	73.72
MentorMix [9]	74.30
FaMUS [31]	74.43
DivideMix [13]	74.76
AugDesc [19]	75.11
Ours	**75.35**

5   Conclusions

In this paper, we introduced Meta Feature Re-Weighting (MFRW), which makes use of a novel concept of advisor network to mitigate the problem of training DNNs on corrupted labels. We empirically show the effectiveness of our method on a synthetic and real-world noisy dataset for the classification task. The experimental results demonstrate that advisor strategy can leverage information present in noisy data helping the main network to achieve a better generalization performance. Our method yields state-of-the-art performance on the Clothing1M dataset. Future research in this area may include adapting the advisor network to different problems than noise, like class imbalance.

References

1. Algan, G., Ulusoy, I.: Image classification with deep learning in the presence of noisy labels: a survey. Knowl.-Based Syst. **215**, 106771 (2021)
2. Arazo, E., Ortego, D., Albert, P., O'Connor, N., McGuinness, K.: Unsupervised label noise modeling and loss correction. In: International Conference on Machine Learning, pp. 312–321. PMLR (2019)
3. Azadi, S., Feng, J., Jegelka, S., Darrell, T.: Auxiliary image regularization for deep CNNs with noisy labels. arXiv preprint arXiv:1511.07069 (2015)
4. Bengio, Y., Louradour, J., Collobert, R., Weston, J.: Curriculum learning. In: Proceedings of the 26th Annual International Conference on Machine Learning, pp. 41–48 (2009)
5. Cheng, Y., Jiang, L., Macherey, W., Eisenstein, J.: AdvAug: robust adversarial augmentation for neural machine translation. In: Proceedings of the 58th Annual Meeting of the Association for Computational Linguistics, pp. 5961–5970 (2020)
6. Goldberger, J., Ben-Reuven, E.: Training deep neural-networks using a noise adaptation layer (2016)
7. Han, B., et al.: Co-teaching: robust training of deep neural networks with extremely noisy labels. In: Advances in Neural Information Processing Systems (2018)
8. Hendrycks, D., Mazeika, M., Wilson, D., Gimpel, K.: Using trusted data to train deep networks on labels corrupted by severe noise. In: Advances in Neural Information, vol. 31, pp. 10456–10465 (2018)
9. Jiang, L., Huang, D., Liu, M., Yang, W.: Beyond synthetic noise: deep learning on controlled noisy labels. In: International Conference on Machine Learning, pp. 4804–4815. PMLR (2020)
10. Jiang, L., Zhou, Z., Leung, T., Li, L.J., Fei-Fei, L.: MentorNet: learning data-driven curriculum for very deep neural networks on corrupted labels. In: International Conference on Machine Learning, pp. 2304–2313. PMLR (2018)
11. Krizhevsky, A., Sutskever, I., Hinton, G.E.: ImageNet classification with deep convolutional neural networks. In: Advances in Neural Information Processing Systems, vol. 25 (2012)
12. Kumar, M., Packer, B., Koller, D.: Self-paced learning for latent variable models. In: Advances in Neural Information Processing Systems, vol. 23, pp. 1189–1197 (2010)
13. Li, J., Socher, R., Hoi, S.C.: DivideMix: learning with noisy labels as semi-supervised learning. In: International Conference on Learning Representations (2019)
14. Li, J., Wong, Y., Zhao, Q., Kankanhalli, M.S.: Learning to learn from noisy labeled data. In: Proceedings of the IEEE/CVF Conference on Computer Vision and Pattern Recognition, pp. 5051–5059 (2019)
15. Li, Y., Yang, J., Song, Y., Cao, L., Luo, J., Li, L.J.: Learning from noisy labels with distillation. In: Proceedings of the IEEE International Conference on Computer Vision, pp. 1910–1918 (2017)
16. Lin, T.Y., Goyal, P., Girshick, R., He, K., Dollár, P.: Focal loss for dense object detection. In: Proceedings of the IEEE International Conference on Computer Vision, pp. 2980–2988 (2017)
17. Ma, X., et al.: Dimensionality-driven learning with noisy labels. In: International Conference on Machine Learning, pp. 3355–3364. PMLR (2018)
18. Nettleton, D.F., Orriols-Puig, A., Fornells, A.: A study of the effect of different types of noise on the precision of supervised learning techniques. Artif. Intell. Rev. **33**(4), 275–306 (2010)

19. Nishi, K., Ding, Y., Rich, A., Hollerer, T.: Augmentation strategies for learning with noisy labels. In: Proceedings of the IEEE/CVF Conference on Computer Vision and Pattern Recognition, pp. 8022–8031 (2021)
20. Patrini, G., Rozza, A., Krishna Menon, A., Nock, R., Qu, L.: Making deep neural networks robust to label noise: a loss correction approach. In: Proceedings of the IEEE Conference on Computer Vision and Pattern Recognition, pp. 1944–1952 (2017)
21. Pechenizkiy, M., Tsymbal, A., Puuronen, S., Pechenizkiy, O.: Class noise and supervised learning in medical domains: the effect of feature extraction. In: 19th IEEE Symposium on Computer-Based Medical Systems (CBMS 2006), pp. 708–713. IEEE (2006)
22. Reed, S., Lee, H., Anguelov, D., Szegedy, C., Erhan, D., Rabinovich, A.: Training deep neural networks on noisy labels with bootstrapping. arXiv preprint arXiv:1412.6596 (2014)
23. Ren, M., Zeng, W., Yang, B., Urtasun, R.: Learning to reweight examples for robust deep learning. In: International Conference on Machine Learning, pp. 4334–4343. PMLR (2018)
24. Shu, J., et al.: Meta-Weight-Net: learning an explicit mapping for sample weighting. In: Proceedings of the 33rd International Conference on Neural Information Processing Systems, pp. 1919–1930 (2019)
25. Tanaka, D., Ikami, D., Yamasaki, T., Aizawa, K.: Joint optimization framework for learning with noisy labels. In: Proceedings of the IEEE Conference on Computer Vision and Pattern Recognition, pp. 5552–5560 (2018)
26. Vahdat, A.: Toward robustness against label noise in training deep discriminative neural networks. In: Advances in Neural Information Processing Systems, vol. 30, pp. 5596–5605 (2017)
27. Veit, A., Alldrin, N., Chechik, G., Krasin, I., Gupta, A., Belongie, S.: Learning from noisy large-scale datasets with minimal supervision. In: Proceedings of the IEEE Conference on Computer Vision and Pattern Recognition, pp. 839–847 (2017)
28. Wang, Z., Hu, G., Hu, Q.: Training noise-robust deep neural networks via meta-learning. In: Proceedings of the IEEE/CVF Conference on Computer Vision and Pattern Recognition, pp. 4524–4533 (2020)
29. Wei, H., Feng, L., Chen, X., An, B.: Combating noisy labels by agreement: a joint training method with co-regularization. In: Proceedings of the IEEE/CVF Conference on Computer Vision and Pattern Recognition, pp. 13726–13735 (2020)
30. Xiao, T., Xia, T., Yang, Y., Huang, C., Wang, X.: Learning from massive noisy labeled data for image classification. In: Proceedings of the IEEE Conference on Computer Vision and Pattern Recognition, pp. 2691–2699 (2015)
31. Xu, Y., Zhu, L., Jiang, L., Yang, Y.: Faster meta update strategy for noise-robust deep learning. In: Proceedings of the IEEE/CVF Conference on Computer Vision and Pattern Recognition, pp. 144–153 (2021)
32. Yi, K., Wu, J.: Probabilistic end-to-end noise correction for learning with noisy labels. In: Proceedings of the IEEE/CVF Conference on Computer Vision and Pattern Recognition, pp. 7017–7025 (2019)
33. Zhang, H., Cisse, M., Dauphin, Y.N., Lopez-Paz, D.: mixup: beyond empirical risk minimization. In: International Conference on Learning Representations (2018)

Learning Semantics for Visual Place Recognition Through Multi-scale Attention

Valerio Paolicelli[1] , Antonio Tavera[2](✉) , Carlo Masone[3] ,
Gabriele Berton[2] , and Barbara Caputo[2]

[1] IIT, Genova, Italy
valerio.paolicelli@iit.it
[2] Politecnico di Torino, Torino, Italy
{antonio.tavera,gabriele.berton,barbara.caputo}@polito.it
[3] CINI, Roma, Italy

Abstract. In this paper we address the task of visual place recognition (VPR), where the goal is to retrieve the correct GPS coordinates of a given query image against a huge geotagged gallery. While recent works have shown that building descriptors incorporating semantic and appearance information is beneficial, current state-of-the-art methods opt for a top down definition of the significant semantic content. Here we present the first VPR algorithm (Code and dataset are available at: https://github.com/valeriopaolicelli/SegVPR) that learns robust global embeddings from both visual appearance and semantic content of the data, with the segmentation process being dynamically guided by the recognition of places through a multi-scale attention module. Experiments on various scenarios validate this new approach and demonstrate its performance against state-of-the-art methods. Finally, we propose the first synthetic-world dataset suited for both place recognition and segmentation tasks.

Keywords: Visual place recognition · Segmentation · Attention

1 Introduction

Visual place recognition (VPR) [17], *i.e.*, the task of recognizing the location where a photo was taken, is usually cast as an image retrieval problem: the location of a photo (*query*) is estimated by comparing it to a huge database of geo-tagged images (*gallery*). Much of the recent research on this subject has focused on finding better image representations to perform the retrieval. Despite the advances in this direction made possible by the use of deep convolutional neural networks (DCNN) [2,21,25], current VPR solutions still fail to achieve

Supplementary Information The online version contains supplementary material available at https://doi.org/10.1007/978-3-031-06430-2_38.

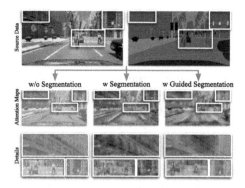

Fig. 1. Prior works have shown that the VPR task can be improved by imbuing the features with semantic information. However, when the semantic task is not controlled by the place recognition, it draws information from all the scene indistinctly (middle). By conditioning the semantic segmentation on the place recognition task (right), the model learns to draw information only from those semantic categories most discriminative for a place, like buildings or traffic lights.

the degree of generality and flexibility required to work across different environments and conditions [31]. Recent studies have found that these problems can be mitigated by building image descriptors based not only on visual appearance, but also on the semantic content in the scene [11,15,19]. Intuitively, dynamic objects or elements that are both common across all places and that lack distinctive features (*e.g.*, roads and sky) are not very informative for VPR. On the other hand, content that is stable across different conditions and that has a wide range of intra-class variations, such as buildings [19], can more robustly describe places (Fig. 1).

While this intuition holds promise, it currently faces two intertwined open challenges, *i.e.*, the need to define a priori the semantic content to use and the lack of an appropriate annotated database from where to learn it. Indeed, previous works on combining appearance and semantic information to generate global descriptors for VPR have either empirically defined what semantic classes should be used to describe places [18,19] or used all the semantic content available [11]. While this can be seen as a legitimate choice, it remains that it does not exist a public VPR dataset containing images annotated with pixel-wise semantic maps. What is currently available to the community are large-scale, multi-scenario VPR datasets without semantic annotations [3,16,27,28], or autonomous driving datasets that provide pixel-wise labels but either lack GPS annotation [6,23,24] or are too small in scale to effectively build query and gallery sets [8,9,12,30].

Here we argue that it is best to let the model figure out what semantic information is more relevant to describe and recognize a place. Hence, we present a large scale synthetic database, annotated with both GPS and pixel-wise semantic maps, jointly with a new architecture that builds global descriptors for VPR in a data driven manner, informed by both visual appearance and semantic content of the training data. To do so, we introduce an attention-based mechanism to dynamically condition, during training, the segmentation process on the recognition of places. The attention mechanism operates by leveraging features located at multiple spatial scales to capture the discriminative urban objects with different sizes in the scene. In summary, the contributions of this paper are:

- a new data-driven method to generate highly informative global descriptors for VPR, leveraging both visual appearance and semantic content at different scales;
- a new synthetic dataset for large-scale visual place recognition that also contains pixel-wise semantic labels;
- extensive validation on various real-world scenarios, demonstrating both the effectiveness of each component in our architecture and a consistent improvement over previous state-of-the-art methods.

2   Related Work

Semantically Informed Visual Place Recognition. Most modern approaches for visual place recognition rely on an image retrieval formulation [17], using DCNNs to extract appearance features which then generate global descriptors by means of aggregation [2] or pooling [21,25]. Recent works have proposed to enhance this paradigm by introducing attention mechanisms [3,14,29,33] and domain adaptation techniques to align features of different scenarios [3,13]. Few studies suggest building the global descriptors not only using the visual information in the images, but also their semantic content. Along these lines, the method presented in [19] requires segmenting an images also at inference time while [15] requires a 3D point cloud of the scene. Closely related to our work is DASGIL [11], an architecture that uses a single encoder shared by three tasks (VPR, depth mask reconstruction and semantic mask reconstruction) to create embeddings that fuse visual, geometric and semantic information. Similar to our solution, DASGIL is trained on a synthetic dataset and it uses domain adaptation to align the features extracted from the synthetic and real-world domains. Besides that, in [11] the segmentation task focuses indiscriminately on all the semantic classes. On the contrary, our solution is built on the intuition that not all the semantic content is useful for VPR and we let the place recognition task guide the segmentation one via an attention mechanism. Moreover, DASGIL builds a global descriptor by flattening and concatenating the features extracted at multi-scale, without an embedding step. This produces extremely large descriptors which are not well suited for large scale problems. Instead, we use a novel multi-scale aggregation layer which produces more compact descriptors.

3   Method

We consider having at training a collection $X = \{(x, y, z)\}$ of N triplets, where x is an RGB image composed by $\mathcal{I}$ pixels, y is the semantic map that associates

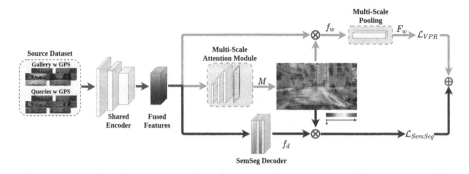

Fig. 2. Illustration of the proposed architecture. The Multi-Scale attention module (middle) is trained by the place recognition task (top) and guides the segmentation module (bottom) towards relevant semantic classes. Blue lines refer to the visual place recognition task (VPR) that implements the novel multi-scale pooling layer (top). **Purple lines** refer to the segmentation task (bottom). (Color figure online)

to each pixel i a class from a set of semantic classes C, and z is the GPS coordinate where the image was taken. We propose a novel framework for visual place recognition that leverages both the appearance and pixel-wise semantic information available during training to learn image representations that are more effective for the place recognition problem. Our architecture, depicted in Fig. 2, consists of a single encoder shared by two tasks:

- a visual place recognition task (*VPR*) (Sect. 3.1), that implements a novel multi-scale pooling layer to generate the global embeddings used for the retrieval process;
- a semantic segmentation task (*SemSeg*) (Sect. 3.2), that implements a decoder for parsing the scene according to the set of classes C.

By sharing the same encoder, the two tasks induce it to learn features that combine both the visual information used for VPR and the semantic information present in the scene. However, without a proper mechanism to control this fusion, the model would equally focus on all semantic classes, regardless of their actual relevance for the VPR task. We introduce such a mechanism in the form of a multi-scale attention module that is trained only on the VPR task but modulates also the features extracted by the SemSeg decoder. In this way, the VPR guides the segmentation, informing it where to focus on the scene. Note that at inference time the decoder is not used, thus the deployed model is quite lightweight.

3.1 Multi-scale and Attentive VPR Task

The shared encoder in our architecture is a ResNet [10] truncated after the last convolutional block. We indicate as f_4 and f_5 the outputs of the last two convolutional blocks, *conv4* and *conv5*, with shapes $C_4 \times H_4 \times W_4$ and $C_5 \times H_5 \times W_5$, respectively. These features are used as input to both the multi-scale attention module (Fig. 2, middle) and to the novel multi-scale pooling layer (Fig. 2, top).

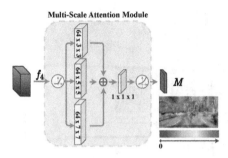

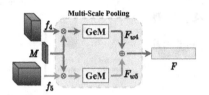

Fig. 3. The attention module leverages features at multiple spatial scales to capture objects with different sizes, and produces a map M that marks the retrieval salient regions. $\oplus$ indicates up-sampling and concatenation.

Fig. 4. Illustration of the multi-scale pooling module. Features extracted at different levels of the shared encoder are exploited to compute the global descriptors. $\otimes$ indicates up-sampling and dot product. $\oplus$ indicates concatenation.

Multi-scale Attention. Recent works have demonstrated the use of multi-scale attention mechanisms in place recognition as a way to focus on the salient regions in the image [14,29,33]. In our architecture, the attention module becomes instrumental to make place recognition guide the semantic segmentation during training. The module, depicted in Fig. 3, takes the output f_4 from the encoder and passes it through a bank of $O \times k_s \times k_s$ filters with O number of output channels and different k_s kernel sizes ($64 \times 3 \times 3$, $64 \times 5 \times 5$ and $64 \times 7 \times 7$). The outputs of these filters are upsampled and concatenated channel-wise, before passing through a $1 \times 1 \times 1$ filter and a softplus function that produces a $1 \times H \times W$ attention map M. The scores in the attention map M indicate where the retrieval is focusing.

Multi-scale Pooling. In VPR it is widely common to use pooling layers after the convolutional backbone to extract compact global descriptors for the retrieval, the state-of-the-art being GeM [21]. In order to exploit semantic and appearance information at different abstraction levels, we introduce a multi-scale GeM layer ($ms\text{-}GeM$) that uses both f_4 and f_5, as illustrated in Fig. 4. These features are first weighted by the attention scores M via dot-product and L2-normalized. Then, they are pooled using vanilla GeM layers to produce the global descriptors $F_{w4} \in \mathbb{R}^{C_4}$ and $F_{w5} \in \mathbb{R}^{C_5}$. Finally, these descriptors are concatenated to form a representation $F \in \mathbb{R}^{C_4+C_5}$.

VPR Loss. We use the weakly supervised triplet margin loss and training protocol from [2] to train the model to extract descriptors for the VPR task. For each training query we consider a positive and a negative example drawn from the gallery. The positive example is an image of the same place as the one depicted in the query, whereas the negative example is an image of a different location. Both the query and its corresponding positive/negative examples are from the source domain, therefore we use the available GPS information to select the examples. In particular, we consider as negative example the most similar

Fig. 5. Attention heatmaps from our framework. We see that attention is focused only on relevant semantic classes according to the VPR task, such as buildings.

image in the features space far from the query GPS coordinates. Finally, for each query descriptor F_q the loss is

$$\mathcal{L}_{VPR} = h(d(F_q, F_p) + m - d(F_q, F_n)) \tag{1}$$

where h is the hinge loss, $d(\cdot, \cdot)$ is the Euclidean distance, m is a fixed margin, F_p and F_n are the descriptors of the positive and negative examples, respectively. The goal pursued by $\mathcal{L}_{VPR}$ is to learn descriptors so that the distance between a training query and its positive example is smaller than the distance between the query and its negative example by at least the margin m. Additionally, it is also used to optimize the parameters of the multi-scale attention module and thus focus the segmentation task on the salient regions for place recognition.

3.2 Guided Semantic Segmentation Task

The SemSeg task informs the features extraction process with semantic information. For this purpose, we use a semantic segmentation decoder D_{seg} (see Fig. 2). However, to force the model to focus on the semantic information that is most discriminative for places, the output of the decoder D_{seg} is weighted by the attention map M, which is trained by the VPR task alone. To train the shared encoder and the decoder, the SemSeg branch uses a cross-entropy loss computed for each class y_i at pixel $\mathcal{I}$, i.e.,

$$\mathcal{L}_{SemSeg} = -\frac{1}{|\mathcal{I}|} \sum_{i \in \mathcal{I}} y_i \cdot \log p_i^{y_i}((M^i \cdot f_d^i)) \tag{2}$$

where M^i is the attention map related to the feature f_d^i extracted from the decoder D_{seg}, while $p_i^{y_i}$ denotes the probability for class y_i at pixel $\mathcal{I}$. Figure 5 shows some examples of heatmaps resulting from the attention module on the source domain images. From these heatmaps it emerges that the network learns to focus on distinctive man-made structures such as buildings or shop signs.

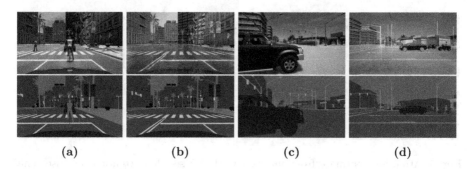

(a)   (b)   (c)   (d)

Fig. 6. a-b) show examples of the Town10 gallery and query sets from the front view, while c-d) are from the Town03 left view. Gallery and query sets are collected under the Clear Noon and Hard Rain Sunset weather conditions.

3.3   Training Loss

Summarizing, the VPR-SemSeg loss function is:

$$\mathcal{L}_{VPR-SemSeg} = \mathcal{L}_{VPR} + \alpha \cdot \mathcal{L}_{SemSeg} \tag{3}$$

where α is a scalar weight. Both $\mathcal{L}_{VPR}$ and $\mathcal{L}_{SemSeg}$ affect the encoder weights to produce features that are informative for place recognition and combine visual and semantic information. However, only $\mathcal{L}_{VPR}$ impacts the attention module weights, thus making the segmentation dependent on the place recognition.

4   A VPR and Semantic Segmentation Dataset

The proposition of using semantic information in VPR in a data-driven approach is limited by the lack of a dataset that is both built for place recognition, *i.e.*, containing multiple views of the same places tagged with GPS coordinates, and that provides fine-grained semantic labels. There is only a couple of synthetic datasets that come close to this requirement, but eventually fall short. One is Virtual KITTI 2 [4], which was used by [11], but it only guarantees 447 images per scenario. Moreover, images are not associated to GPS coordinates, so place matching is only done by the name of the images. The other is SYNTHIA [24], which contains very dense sequences of images that are suitable for the visual localization task [20] ($\leq$5 m from one gps coordinate to the other) but not for the coarser place recognition task usually considered in literature ($\leq$25 m). To develop our data-driven approach for VPR that combines both visual and semantic knowledge we created a new synthetic dataset.

This new dataset was inspired by IDDA [1], which was built from the CARLA virtual simulator [7] specifically for semantic segmentation but without GPS annotations. Following the methodology from [1] we used CARLA 0.9.10 to build a new dataset that includes both GPS/IMU information and pixel-wise semantic annotations with 25 semantic classes (with 17 of them in common to the Cityscapes [6] standard). This new synthetic dataset contains more than 40000 images (10091 per scenario) captured across two different urban maps (Town03 and Town10 from CARLA notation) and in two weather conditions, Clear Noon (CN) and Hard Rain Sunset (HRS). To collect the data we equipped the ego-vehicle with four cameras (front, rear, left and right view). We split the front and rear view frames captured in the Town10 in a gallery and a query set; the first including the CN images and the latter the HRS ones, see Fig. 6a-b). The left and right view images captured in the Town03 are used as a validation set, following the same criterion used to split the training set, see Fig. 6c-d).

4.1   Synthetic-to-Real Domain Adaptation

The new synthetic dataset provides the training data for the method presented in Sect. 3. However, there is a significant gap between the images from this simulator (*source domain* X_s) and from the real-world (*target domain* X_t). To reduce this gap we use adversarial training that aims at aligning the features extracted from the synthetic and target domains. We assume having available at training time a set of unlabeled target domain images, besides the labeled synthetic data. While the VPR and SemSeg tasks illustrated in Sects. 3.1 and 3.2 are trained using only the synthetic data X_s, we introduce a domain discriminator D similar to [22] to distinguish the source features produced by D_{seg} from the target ones. The discriminator is trained with the binary cross-entropy loss:

$$\mathcal{L}_{discr} = -\sum_{i \in \mathcal{I}}(1-z)logD(p_i(x_s)) + zlogD(p_i(x_t)) \tag{4}$$

with $x_s \in X_s$ and $x_t \in X_t$, $p_i(x)$ the features from D_{seg} and z a parameter which is 0 if the features are from the source and 1 if the features are from the target domain. Overall, the adversarial training tries to fool the encoder and the decoder, computing a binary cross-entropy loss on $\mathcal{X}_t$ labeled like source:

$$\mathcal{L}_{adv} = -\frac{1}{|\mathcal{I}|}\sum_{i \in \mathcal{I}} logD(p_i(x_t)) \tag{5}$$

In this way, the model learns to align the features for every domain and the encoder becomes able to extract discriminative local embeddings for the real world VPR task. Summarizing, the overall training loss function becomes:

$$\mathcal{L}_{tot} = \mathcal{L}_{VPR-SemSeg} + \beta \cdot \mathcal{L}_{adv} + \gamma \cdot \mathcal{L}_{discr} \tag{6}$$

where β and γ are scalar weights and $\mathcal{L}_{adv}$ affects the encoder weights.

5   Experiments

Comparisons with Other Methods. We assess the effectiveness of our approach comparing it to GeM [21] and RMAC [25], two state-of-the-art VPR methods that use global descriptors. We also compare to DASGIL [11] which, similarly to our method, uses semantic information albeit selecting it in a top-down

Table 1. Results on the Oxford RobotCar [16], using Overcast as gallery. $\star$ means ResNet101 as encoder, all the others are with ResNet50. DA stays for Domain Adaptation, DL is for DeepLabv2 [5] segmentation decoder, while PSP is for PSPNet [32]. • indicates the experiments of [11] with the pretrained models.

Method	Overcast/Rain 1/5/10	Overcast/Snow 1/5/10	Overcast/Sun 1/5/10	Overcast/Night 1/5/10	Avg 1/5/10	Avg
GeM	71.4/83.8/87.9	37.6/55.1/63.6	30.9/47.9/55.7	4.4/12.2/18.3	36.1/49.7/56.4	47.4
GeM + DA	73.6/87.4/91.2	41.2/59.9/68.6	32.6/52.0/60.4	7.3/20.5/29.4	38.7/54.9/62.4	52.0
GeM $\star$	69.6/83.8/88.3	40.3/58.4/66.3	32.7/51.2/61.1	6.0/16.5/23.6	37.2/52.5/59.8	49.8
GeM + DA $\star$	78.7/90.8/93.7	43.8/64.7/72.8	33.8/54.3/64.1	10.2/25.2/34.2	41.6/58.7/66.2	55.5
RMAC	73.9/86.9/90.9	35.1/53.5/61.6	22.7/39.9/48.0	2.5/8.2/13.0	33.6/47.1/53.4	44.7
RMAC + DA	69.8/83.2/88.0	39.8/58.4/66.5	24.9/42.6/51.7	3.9/10.3/16.3	34.6/48.6/55.6	46.3
RMAC $\star$	68.6/82.5/87.0	35.6/52.8/60.3	27.0/43.4/52.6	2.9/10.0/15.8	33.5/47.2/53.9	44.9
RMAC + DA $\star$	65.3/80.0/84.6	32.0/52.1/61.6	25.6/44.0/54.2	3.3/9.5/14.9	31.5/46.4/53.8	43.9
DASGIL_nodepth	87.2/92.9/94.5	39.3/55.3/63.1	20.8/33.3/40.6	6.5/14.2/19.5	38.4/48.9/54.4	47.3
DASGIL_FD •	76.8/83.7/86.4	53.4/65.0/70.2	46.5/59.3/65.0	2.0/6.3/11.4	44.7/53.6/58.3	52.2
DASGIL_CD •	75.3/84.7/87.9	34.0/45.7/51.9	12.0/21.7/28.0	1.0/4.1/6.4	30.6/39.1/43.6	37.7
Ours w DL	**92.2/96.8/97.8**	**68.0/83.4/89.0**	**62.6/79.6/85.0**	11.3/24.5/34.1	**58.5/71.1/76.5**	**68.7**
Ours w DL $\star$	90.5/96.9/97.7	62.5/79.6/85.7	55.1/73.7/81.0	10.8/25.3/34.7	54.7/68.9/74.8	66.1
Ours w PSP	90.7/96.5/97.3	59.5/75.5/82.4	47.0/65.2/71.9	12.0/26.4/34.7	52.3/65.9/71.6	63.3
Ours w PSP $\star$	91.2/96.4/97.7	61.4/83.1/88.4	55.4/74.0/81.0	**14.2/32.6/42.3**	55.6/**71.5/77.4**	68.2

manner. Besides semantics, DASGIL can also leverage depth information. We compare to three different versions of DASGIL: the pre-trained models released by the authors using their proposed Flatten Discriminator (DASGIL_FD •) and Cascade Discriminator (DASGIL_CD •), as well as the model with Flatten Discriminator trained on our dataset but without depth information (DASGIL_nodepth). DASGIL_FD • and DASGIL_CD • are pretrained on KITTI and VIRTUAL KITTI 2, while just DASGIL_nodepth is pretrained on our dataset. In the comparison with DASGIL we must also note that it extracts and concatenates the local features from the i-th and j-th *conv* layer of the shared encoder, producing final embeddings of variable dimension $1 \times 64 \times (H_i \times W_i + H_j \times W_j)$. On RobotCar [16], this results in 15360-D (dimensional) descriptors w.r.t the more compact 3072-D descriptors produced by our method.

Datasets and Protocol. For the evaluation we used Oxford RobotCar [16] as the inference dataset. Oxford RobotCar is a collection of images from the city of Oxford taken in different environmental conditions from a car-mounted camera. We use the Overcast scenario as the gallery, while the queries are divided into four scenarios: Rain, Snow, Sun, and Night, with one image sampled every 5 m in order to decrease data redundancy. In order to have a fair comparison and truly assess the effectiveness of our solution, which cannot be trained directly on the target domain because of the lack of semantic labels, we trained all models on our new synthetic dataset, adding to GeM and RMAC the same unsupervised domain adaptation (DA) module used in our network. On other hand, DASGIL, which also uses semantic labels, already includes a DA module. Despite the usage of DA techniques we can expect the result to be considerably lower than if the models would be trained directly on the target domain, due to

Table 2. Results using Pitts30k [27] and RTokyo [28] validation and test sets, and MapillarySLS [28] validation cities. * means ResNet101 as encoder while all the others are with ResNet50. DA stays for Domain Adaptation, DL means DeepLabv2 [5] segmentation decoder and PSP is for PSPNet [32]. • indicates the experiments of [11] with the trained models provided by the authors.

Method	Pitts30k - Val 1/5/10	Pitts30k - Test 1/5/10	RTokyo - Val 1/5/10	RTokyo - Test 1/5/10	MapillarySLS 1/5/10
GeM	37.6/56.5/64.9	39.5/59.1/67.3	28.7/43.5/51.2	8.1/17.4/24.8	26.5/38.6/44.5
GeM + DA	41.1/61.0/69.0	41.7/61.6/69.7	33.7/49.4/57.0	13.4/25.7/31.6	27.2/40.0/46.1
GeM ⋆	38.0/57.9/67.4	38.8/58.5/67.3	26.4/41.5/49.2	10.1/20.4/27.4	25.7/39.5/45.0
GeM + DA ⋆	47.1/68.9/77.3	47.7/68.4/75.9	36.0/52.6/60.3	17.6/34.6/43.5	28.9/41.5/48.4
RMAC	39.4/58.5/67.3	43.3/62.9/71.1	40.1/55.7/62.6	13.8/28.3/36.9	33.6/44.3/50.6
RMAC + DA	37.8/57.2/66.1	41.9/62.1/71.3	36.3/51.7/59.0	10.4/23.8/31.5	28.8/40.4/46.7
RMAC ⋆	33.8/53.5/63.0	37.7/57.4/66.9	31.7/46.9/54.6	10.0/20.9/27.7	31.5/44.0/48.5
RMAC + DA ⋆	32.1/51.3/60.6	36.4/56.3/65.3	30.5/45.8/53.4	14.8/25.7/33.0	30.2/41.4/46.2
DASGIL_nodepth	11.1/17.6/21.6	12.8/20.6/26.0	17.7/28.7/35.0	2.0/3.8/6.4	11.1/16.6/19.7
DASGIL_FD •	8.3/12.6/15.3	8.7/13.8/16.6	5.9/9.1/11.9	0.0/0.7/1.0	6.4/8.6/10.1
DASGIL_CD •	6.8/10.7/13.9	8.5/12.3/15.8	7.2/11.9/15.3	0.3/1.0/1.0	7.7/11.4/14.1
Ours w DL	56.3/73.8/80.1	58.9/75.2/80.4	49.6/64.1/70.3	20.7/37.1/45.5	**34.6**/45.8/52.3
Ours w DL ⋆	**57.9**/76.3/82.8	**59.4**/76.2/81.5	49.0/64.0/70.3	21.4/38.0/45.0	34.5/46.8/**53.0**
Ours w PSP	52.9/70.9/77.9	56.3/73.0/79.0	47.3/62.1/68.6	17.4/30.9/38.6	33.0/**47.2/53.0**
Ours w PSP ⋆	57.7/**76.4/83.3**	59.1/**76.6/82.2**	**51.3/66.2/72.5**	**26.9/44.6/51.9**	32.9/46.1/51.9

the strong domain shift. Finally, we also compare the generalization capability of these methods, trained on the synthetic dataset with Oxford RobotCar as target domain, and testing them on three other datasets: Pitts30k [27], the revisited version of Tokyo24/7 [26] (RTokyo) proposed in the supplementary material of [28], and the recent MapillarySLS [28] validation set (since the test set labels have not yet been released). In all experiments we used the standard VPR metric Recall@N [2], considering a retrieved gallery image as positive if it is within 25 m from the query.

Additional Info. Implementation details, real-world domain experiments, qualitative results and ablation studies are provided in the *Supplementary Material*.

5.1 Results

The results of the experiments are reported in Table 1. We observe that all methods perform worse when tested on queries from the Night and Sun scenarios, due to a stronger visual dissimilarity with respect to the gallery images that are taken from the Overcast scenario.

Overall, we see that our architecture largely outperforms all other methods, with and without DA, from a minimum of 8% to a maximum of 24%. Curiously, with RMAC [25] results seem to get worse when adding DA. We do not have a definitive explanation for this negative effect, but we confirmed it with the generalization experiments presented in Table 2. Even DASGIL, which is the most similar method to our work, Table 1 struggles to generalize to all scenarios. The model trained on our datasets and without the depth information (DAS-GIL_nodepth) performs very well on the Rain scenario, but it still results on average worse than the GeM. We suspect that this method suffers considerably from the large domain gap between our synthetic dataset and the RobotCar [16] target images. For this reason we extend the experiments by using also the pretrained models provided by the authors (DASGIL_FD • and DASGIL_CD •), that are trained on synthetic and real datasets Virtual KITTI 2 [4] and KITTI [8]. Nevertheless, these implementations still remain lower than our method by at least 11%. We further confirm our assumptions in Table 2 testing the generalization capability on very large city datasets. Finally, the generalization results shown in Table 2 demonstrate that our solution generalizes to unseen domains better than all the other methods by usually a large margin.

6 Conclusions

We have presented a new method for generating global descriptors for VPR, exploiting both visual appearance and semantic features at different scales. Our solution, unlike previous works, let the model determine what semantic information to use, in a data driven way. The key for this, is an attention mechanism that lets the VPR guide the semantic segmentation. Experiments on well-known VPR benchmarks, where we surpass the current state-of-the-art methods, validate our intuition and architecture. We also show that our model generalizes well to unseen target domains. Finally, we contribute a new dataset, rich of RGB images under different conditions, pixel-wise semantic masks and GPS coordinate, that is instrumental to explore the connection between semantics and appearance in VPR task, and that we believe will be useful to the research community.

References

1. Alberti, E., Tavera, A., Masone, C., Caputo, B.: IDDA: a large-scale multi-domain dataset for autonomous driving. IEEE Robot. Autom. Lett. **5**(4), 5526–5533 (2020). https://doi.org/10.1109/LRA.2020.3009075
2. Arandjelović, R., Gronat, P., Torii, A., Pajdla, T., Sivic, J.: NetVLAD: CNN architecture for weakly supervised place recognition. IEEE Trans. Pattern Anal. Mach. Intell. **40**(6), 1437–1451 (2018). https://doi.org/10.1109/TPAMI.2017.2711011
3. Berton, G.M., Paolicelli, V., Masone, C., Caputo, B.: Adaptive-attentive geolocalization from few queries: a hybrid approach. In: IEEE Winter Conference on Applications of Computer Vision, pp. 2918–2927, January 2021
4. Cabon, Y., Murray, N., Humenberger, M.: Virtual KITTI 2 (2020)
5. Chen, L.C., Papandreou, G., Kokkinos, I., Murphy, K., Yuille, A.: DeepLab: semantic image segmentation with deep convolutional nets, atrous convolution, and fully connected CRFs. IEEE Trans. Pattern Anal. Mach. Intell. **40**, 834–848 (2018)
6. Cordts, M., et al.: The cityscapes dataset for semantic urban scene understanding. In: IEEE Conference on Computer Vision and Pattern Recognition (2016)

7. Dosovitskiy, A., Ros, G., Codevilla, F., Lopez, A., Koltun, V.: CARLA: an open urban driving simulator. In: Levine, S., Vanhoucke, V., Goldberg, K. (eds.) Proceedings of the 1st Annual Conference on Robot Learning. Proceedings of Machine Learning Research, vol. 78, pp. 1–16. PMLR, 13–15 November 2017. http://proceedings.mlr.press/v78/dosovitskiy17a.html

8. Geiger, A., Lenz, P., Urtasun, R.: Are we ready for autonomous driving? The KITTI vision benchmark suite. In: IEEE Conference on Computer Vision and Pattern Recognition (2012)

9. Geyer, J., et al.: A2D2: audi autonomous driving dataset (2020). https://www.a2d2.audi

10. He, K., Zhang, X., Ren, S., Sun, J.: Deep residual learning for image recognition. In: IEEE Conference on Computer Vision and Pattern Recognition, pp. 770–778 (2016)

11. Hu, H., Qiao, Z., Cheng, M., Liu, Z., Wang, H.: DASGIL: domain adaptation for semantic and geometric-aware image-based localization. IEEE Trans. Image Process. 30, 1342–1353 (2021). https://doi.org/10.1109/TIP.2020.3043875

12. Huang, X., et al.: The apolloscape dataset for autonomous driving. In: IEEE Conference on Computer Vision and Pattern Recognition, June 2018

13. Jenicek, T., Chum, O.: No fear of the dark: image retrieval under varying illumination conditions. In: International Conference on Computer Vision (2019)

14. Kim, H.J., Dunn, E., Frahm, J.: Learned contextual feature reweighting for image geo-localization. In: IEEE Conference on Computer Vision and Pattern Recognition, pp. 3251–3260 (2017). https://doi.org/10.1109/CVPR.2017.346

15. Larsson, M., Stenborg, E., Toft, C., Hammarstrand, L., Sattler, T., Kahl, F.: Fine-grained segmentation networks: self-supervised segmentation for improved long-term visual localization. In: International Conference on Computer Vision, October 2019

16. Maddern, W., Pascoe, G., Linegar, C., Newman, P.: 1 year, 1000km: the Oxford RobotCar dataset. Int. J. Robot. Res. 36(1), 3–15 (2017). https://doi.org/10.1177/0278364916679498

17. Masone, C., Caputo, B.: A survey on deep visual place recognition. IEEE Access 9, 19516–19547 (2021). https://doi.org/10.1109/ACCESS.2021.3054937

18. Mousavian, A., Košecká, J., Lien, J.-M.: Semantically guided location recognition for outdoors scenes. In: IEEE International Conference on Robotics and Automation, pp. 4882–4889 (2015). https://doi.org/10.1109/ICRA.2015.7139877

19. Naseer, T., Oliveira, G., Brox, T., Burgard, W.: Semantics-aware visual localization under challenging perceptual conditions. In: IEEE International Conference on Robotics and Automation. IEEE (2017). http://lmb.informatik.uni-freiburg.de/Publications/2017/OB17

20. Pion, N., Humenberger, M., Csurka, G., Cabon, Y., Sattler, T.: Benchmarking image retrieval for visual localization. In: 2020 International Conference on 3D Vision (3DV), pp. 483–494 (2020)

21. Radenović, F., Tolias, G., Chum, O.: Fine-tuning CNN image retrieval with no human annotation. IEEE Trans. Pattern Anal. Mach. Intell. 41(7), 1655–1668 (2019)

22. Radford, A., Metz, L., Chintala, S.: Unsupervised representation learning with deep convolutional generative adversarial networks. In: Bengio, Y., LeCun, Y. (eds.) The International Conference on Learning Representations (2016). http://arxiv.org/abs/1511.06434

23. Richter, S.R., Vineet, V., Roth, S., Koltun, V.: Playing for data: ground truth from computer games. In: Leibe, B., Matas, J., Sebe, N., Welling, M. (eds.) ECCV 2016. LNCS, vol. 9906, pp. 102–118. Springer, Cham (2016). https://doi.org/10.1007/978-3-319-46475-6_7

24. Ros, G., Sellart, L., Materzynska, J., Vazquez, D., Lopez, A.M.: The SYNTHIA dataset: a large collection of synthetic images for semantic segmentation of urban scenes. In: IEEE Conference on Computer Vision and Pattern Recognition, pp. 3234–3243, June 2016

25. Tolias, G., Sicre, R., Jégou, H.: Particular object retrieval with integral max-pooling of CNN activations. In: The International Conference on Learning Representations, San Juan, Puerto Rico, pp. 1–12, May 2016. https://hal.inria.fr/hal-01842218

26. Torii, A., Arandjelović, R., Sivic, J., Okutomi, M., Pajdla, T.: 24/7 place recognition by view synthesis. In: IEEE Conference on Computer Vision and Pattern Recognition (2015)

27. Torii, A., Sivic, J., Okutomi, M., Pajdla, T.: Visual place recognition with repetitive structures. IEEE Trans. Pattern Anal. Mach. Intell. **37**(11), 2346–2359 (2015)

28. Warburg, F., Hauberg, S., Lopez-Antequera, M., Gargallo, P., Kuang, Y., Civera, J.: Mapillary street-level sequences: a dataset for lifelong place recognition. In: IEEE Conference on Computer Vision and Pattern Recognition, June 2020

29. Xin, Z., et al.: Localizing discriminative visual landmarks for place recognition. In: IEEE International Conference on Robotics and Automation, pp. 5979–5985 (2019)

30. Yu, F., et al.: BDD100K: a diverse driving dataset for heterogeneous multitask learning. In: IEEE Conference on Computer Vision and Pattern Recognition, June 2020

31. Zaffar, M., Khaliq, A., Ehsan, S., Milford, M., McDonald-Maier, K.: Levelling the playing field: a comprehensive comparison of visual place recognition approaches under changing condition. In: IEEE International Conference on Robotics and Automation Workshop, pp. 1–8 (2019)

32. Zhao, H., Shi, J., Qi, X., Wang, X., Jia, J.: Pyramid scene parsing network. In: IEEE Conference on Computer Vision and Pattern Recognition, pp. 6230–6239 (2017). https://doi.org/10.1109/CVPR.2017.660

33. Zhu, Y., Wang, J., Xie, L., Zheng, L.: Attention-based pyramid aggregation network for visual place recognition. In: ACM International Conference on Multimedia, MM 2018, pp. 99–107. Association for Computing Machinery, New York (2018)

One-Shot HDR Imaging via Stereo PFA Cameras

Tehreem Fatima[ID], Mara Pistellato[✉][ID], Andrea Torsello[ID],
and Filippo Bergamasco[ID]

DAIS, Università Ca'Foscari Venezia, 155, via Torino, Venice, Italy
{tehreem.fatima,mara.pistellato,andrea.torsello,
filippo.bergamasco}@unive.it

Abstract. High Dynamic Range (HDR) imaging techniques aim to
increase the range of luminance values captured from a scene. The literature counts many approaches to get HDR images out of low-range
camera sensors, however most of them rely on multiple acquisitions producing ghosting effects when moving objects are present.

In this paper we propose a novel HDR reconstruction method exploiting stereo Polarimetric Filter Array (PFA) cameras to simultaneously
capture the scene with different polarized filters, producing intensity
attenuations that can be related to the light polarization state. An additional linear polarizer is mounted in front of one of the two cameras, raising the degree of polarization of rays captured by the sensor. This leads
to a larger attenuation range between channels regardless the scene lighting condition. By merging the data acquired by the two cameras, we can
compute the actual light attenuation observed by a pixel at each channel
and derive an equivalent exposure time, producing a HDR picture from
a single polarimetric shot. The proposed technique results comparable
to classic HDR approaches using multiple exposures, with the advantage
of being a one-shot method.

Keywords: PFA camera · HDR · Polarimetric imaging

1 Introduction

The majority of modern digital cameras limit their capability to record the
captured scene irradiance to 8 bit per channel. This translates in a significant
weakness when acquiring high-contrast pictures, as some areas are going to be
over or under saturated regardless the exposure time. Computational photography counts several techniques to recover High Dynamic Range (HDR) images,
allowing for a better image visualisation and synthesis, as well as post-processing
operations [18]. Classical approaches recover HDR scenes with standard camera
sensors by merging multiple pictures taken with different exposure times [5,19].
This limits their applicability to static scenes because moving subjects likely
exhibit ghosting effects that must be properly accounted [11,23]. Such limitation
is not present in single-image approaches, but in this case obtaining an extended

S. Sclaroff et al. (Eds.): ICIAP 2022, LNCS 13232, pp. 467–478, 2022.
https://doi.org/10.1007/978-3-031-06430-2_39

dynamic range is more challenging [2,3]. The recent advancement of learning-based methods allowed data-driven formulations of HDR reconstruction, usually exploiting the capabilities of Convolutional Neural Networks (CNNs). For example, Kalantari et al. [10] propose to reconstruct the HDR scene from a set of three exposures with moving subjects, while [7,15] synthesise a set of images with different exposures starting from a single Low Dynamic Range (LDR) picture, recovering the HDR image in a standard way. Other alternative methods are based on encoder-decoder network architecture to recover the HDR image from saturated image regions [6,30]. Finally, recent works as [1,12,16,17,25] propose alternative pipelines or architectures to extract missing information from LDR images and recover the original scene features.

In this work we propose a novel method to recover HDR images with a single shot with two Polarimetric Filter Array (PFA) cameras. A PFA camera mounts micro-polarizers at predefined orientations (0°, 45°, 90° and 135°) per macro-pixel, and thus allow the recovering of the polarization properties of incoming light. Since the polarization state is related to the features of the acquired scene, several applications are specifically designed to work with PFA sensors [22,31,32]. When placed in front of a camera, a linear polarizer attenuates the incoming light intensity according to the ray polarization state. Such effect is exploited to recover both angle and degree of polarization for each macro-pixel (see Sect. 2), but visually the intensity observed at each channel appears like an exposure time reduction. Our method is built upon this observation and is inspired by the work presented by Wu et al. [29]. Indeed, it is the first one proposing to use polarization properties of light to increase the dynamic range of their imaging system. Their major contribution is connecting light polarization state with virtual (or equivalent) exposure times, with an additional advantage of performing HDR reconstruction in a pixel-dependent way. Unfortunately, the resulting dynamic range depends on the Degree of Linear Polarization (DoLP) of the scene. Since the majority of real-world scenes are scarcely polarized, their technique is not as effective as taking multiple shots with an arbitrarily broad range of exposure times. To overcome such drawback, we propose a model that employs two PFA cameras capturing the scene at the same time as in a typical stereo setup, one of the two mounting an additional polarizer in front of the lens. Such additional filter allows the camera to receive exclusively highly polarized light, translating in a high dynamic range for the different channels, independently from the actual scene polarization. The actual angle and degree of polarization of the scene are recorded by the secondary camera (with no additional filter) and mapped into the image plane of the main camera. We propose a model to connect the values observed by the two cameras, so that we can recover the equivalent exposure times coherently with the original scene and the intensities observed by the main camera. Moreover, we propose a robust Stokes parameter estimation aimed at recovering missing values for over or under-saturated channels exploiting the redundancy of the camera filter angles. Experimental results show that our method offers a better HDR reconstruction with respect to the single camera setup, providing an image quality comparable with state-of-the-art techniques.

2   Polarimetric HDR

Together with amplitude and frequency, polarization is a property that is common to all type of vector waves. Being electromagnetic in nature, light makes no exception on this. Interestingly, such vector field consists of only two transverse components E_x, E_y, perpendicular to each other, that could be chosen for convenience to propagate in the z direction. This allows the complete description of E in terms of 4 scalar quantities called the Stokes polarization parameters $\left(S_0\ S_1\ S_2\ S_3\right)$ [9]. In this model, S_0 is the total intensity of the light, as we commonly capture with digital grayscale cameras. S_1 and S_2 express the *linear polarization* with respect to two reference frames mutually tilted by 45°. Finally, S_3 quantifies the amount of left-right *circular polarization* of the beam.

As a consequence of the Schwarz's inequality, Stokes parameter are related by the following formula:

$$S_0^2 \geq S_1^2 + S_2^2 + S_3^2 \tag{1}$$

showing that the total amount of light radiation S_0 is generally composed by a mixture of *polarized* (either linear and circular) and *unpolarized* light. In particular, when light is fully polarized Eq. 1 becomes an equality, and when is fully unpolarized, the right-hand side is zero [9].

Stokes parameters are ubiquitous in their usage because they are directly measurable using two optical elements: *retarders* and *polarizers*. The former produces a phase-shift between E_x and E_y whereas the latter filters everything except the light radiation aligned with a certain transmission axis. This is exploited in PFA cameras to extract the first 3 Stokes parameters similar to how colour cameras use Bayer filter to recover light frequency. Pixels are masked with a repeated pattern of *linear polarizers* oriented at 0°, 45°, 90°, and 135° wrt. the horizontal direction of the image. When properly demosaiced [20], we obtain four images $I_0, I_{45}, I_{90}, I_{135}$ grouping all the pixels with a certain polarizer orientation. From that, the first three Stokes parameters are simply computed as follows:

$$S_0 = I_0 + I_{90} = I_{45} + I_{135} \tag{2}$$
$$S_1 = I_0 - I_{90} \tag{3}$$
$$S_2 = I_{45} - I_{135}. \tag{4}$$

Unfortunately, S_3 cannot be obtained without a retarder, but circular polarization is relatively rare in nature [4]. Therefore, many applications simply assume that light is only composed by either unpolarized or linearly polarized light with a certain angle. To physically describe this behaviour, the Degree of Linear Polarization (DoLP) and the Angle of Linear Polarization (AoLP) are computed from Stokes parameters as:

$$\mathrm{DoLP} = \frac{\sqrt{S_1^2 + S_2^2}}{S_0}, \quad \mathrm{AoLP} = \frac{1}{2}\arctan\frac{S_2}{S_1}. \tag{5}$$

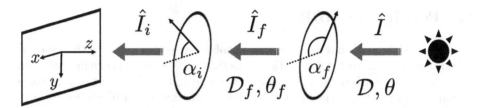

Fig. 1. Our proposed HDR imaging device: two linear polarizers are placed in the optical path from scene to sensor. The innermost is part of the PFA camera and oriented at a fixed angle $\alpha_i \in \{0, \frac{\pi}{4}, \frac{\pi}{2}, \frac{3\pi}{4}\}$, following the array pattern arrangement. The other one is placed outside the lenses and can be freely oriented with an angle α_f with respect to the camera horizontal axis x. The relative rotation of the two allows us to adjust the dynamic range of the system regardless the scene DoLP $\mathcal{D}$ and AoLP θ.

2.1   The Proposed Acquisition Model

Let the image *irradiance* $\hat{I}$ be the amount of energy flowing from the scene to the sensor. When an (ideal) linear polarizer is placed in front of the sensor, as in a PFA camera, the irradiance is attenuated as follows:

$$\hat{I}' = \frac{1}{2}\hat{I}\big(1 + \mathcal{D}\cos(2\theta - 2\alpha)\big) \tag{6}$$

where $\mathcal{D}$ and θ are the DoLP and AoLP of the light entering the filter, and α is the orientation of the filter with respect to the camera horizontal axis. As we will see later, this attenuation has the same effect of reducing the camera shutter speed (darkening the image) so it can be used as in classical HDR imaging.

From Eq. 6 it is easy to note that the attenuation varies with the relative difference between the incoming AoLP and filter orientation α. $\hat{I}'$ is maximum when $\theta \cong \alpha \pmod{\pi}$ and minimum when $\theta \cong \alpha + \frac{\pi}{2} \pmod{\pi}$. The ratio between $\hat{I}'_{\max}$ and $\hat{I}'_{\min}$ (i.e. the dynamic range of the system) depends on the amount of linear polarization $\mathcal{D}$ of the light. Unfortunately, in a typical scene without a massive amount of specular reflections, the DoLP is low and inversely proportional to the surface's albedo due to the so-called *Umov's* effect [14,27]. So, regardless the filters orientation, an HDR system using a built-in PFA is not particularly effective, because bright objects (the ones requiring a greater extent in filter attenuation) are characterized by a DoLP $\ll 1$.

Our idea is to use an additional linear polarizer, as sketched in Fig. 1. In this way, light rays with intensity $\hat{I}$, DoLP $\mathcal{D}$, and AoLP θ are attenuated by two filters before being measured by the sensor. The first filter is placed in front of the camera lens and can be freely oriented with an angle α_f. The resulting intensity $\hat{I}_f$ is modelled by Eq. 6 as:

$$\hat{I}_f = \frac{1}{2}\hat{I}\big(1 + \mathcal{D}\cos(2\theta - 2\alpha_f)\big) \tag{7}$$

The DoLP and AoLP after the filter should ideally be $\mathcal{D}_f = 1$ and $\theta_f = \alpha_f$, respectively. However, since the front polarizer is not perfect, we assume a certain variability on that and model them explicitly. Then, light rays reach a second filter embedded in the camera's PFA. Depending on the pixel, such filter can assume only 4 possible angles α_i ($0 \leq i \leq 3$), where $\alpha_0 = 0$, $\alpha_1 = \frac{\pi}{4}$, $\alpha_2 = \frac{\pi}{2}$, $\alpha_3 = \frac{3\pi}{4}$. By substituting Eq. 7 into Eq. 6 we can model the irradiance $\hat{I}_i$ after the combination of both the filters:

$$\hat{I}_i = \frac{1}{4}\hat{I}\big(1 + \mathcal{D}\cos(2\theta - 2\alpha_f)\big)\big(1 + \mathcal{D}_f\cos(2\theta_f - 2\alpha_i)\big). \tag{8}$$

For the so called *reciprocity property* [5], the actual intensity values captured by the camera I_i are related to irradiance $\hat{I}_i$ and exposure time t_c through $g(I_i) = \hat{I}_i t_c$, where $g()$ is the *Inverse Camera Response Function* (ICRF). Algebraically, we can substitute the product $\hat{I}_i t_c$ with $\hat{I} t_i$ where:

$$t_i = \frac{1}{4}t_c\big(1 + \mathcal{D}\cos(2\theta - 2\alpha_f)\big)\big(1 + \mathcal{D}_f\cos(2\theta_f - 2\alpha_i)\big). \tag{9}$$

In other words, the four images $\hat{I}_0, \ldots, \hat{I}_3$ obtained by rotating the internal polarizer are de-facto identical to the four images that would be obtained without the filter but by changing the exposure time according to Eq. 9. We call $t_0 \ldots t_3$ *equivalent exposure times*.

2.2 HDR Imaging Procedure

The equivalent exposure times for each pixel depends by the interplay between α_f, α_i and the scene polarization state $\mathcal{D}, \theta$. However, $\mathcal{D}$ and θ cannot be observed with our two-filter configuration because the external filter masks the full scene polarization to the internal one used for measuring. Therefore, we propose to use a second PFA camera (without any additional filter) to map the scene DoLP and AoLP to the former image.

The whole procedure can be summarized as follows:

1. Two PFA cameras are mounted side-by-side like in a typical stereo configuration, with a minimal baseline and hardware trigger to ensure that images are captured at the same time. Let *Cam0* be the one with an additional linear polarizer attached in front of the lenses and *Cam1* the other.
2. A picture is taken with both cameras at the same time. Let t_c be the *Cam0* shutter speed. *Cam1* shutter speed is not important as long as the image is exposed as good as possible[1].
3. Both the images are demosaiced to obtain $I_0^0, I_{45}^0, I_{90}^0, I_{135}^0$ and $I_0^1, I_{45}^1, I_{90}^1, I_{135}^1$ (subscript denote CCD filter orientation, superscript the camera number, and images are assumed to be normalized in $0 \ldots 1$ range after applying the ICRF estimated a-priori). The first three Stokes parameters are computed using Eq. 2, 3, 4.
4. Optical flow f mapping pixels from *Cam1* to *Cam0* is computed from S_0^0 and S_0^1 as described in [8].

---

[1] Considering that we are acquiring a scene with a high dynamic range, it will be unavoidable to over- or under-expose some areas.

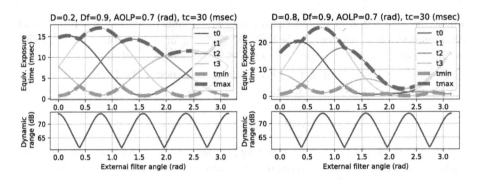

Fig. 2. Numerical examples of the equivalent exposure times (top) and dynamic range (bottom) for ideal scenes with a constant DoLP $D = 0.2$ (first column) and $D = 0.8$ (right column) varying the external filter angle α_f.

5. Scene DoLP and AoLP are computed from the Stokes parameters of *Cam1* (S_0^1, S_1^1, S_2^1) using Eq. 5. The resulting images are warped with f to obtain $\mathcal{D}$ and θ as it would have been observed from *Cam0*.
6. $\mathcal{D}_f$ and θ_f are computed from the *Cam0* Stokes parameters using Eq. 5.
7. The equivalent exposure times t_0, t_1, t_2, t_3 of $I_0^1, I_{45}^1, I_{90}^1, I_{135}^1$ are computed as in Eq. 9. Note that, unlike typical HDR techniques, exposure times are computed per-pixel as they depend on light polarization state that vary along the image.

The result is a set of 4 images $I_0^1, I_{45}^1, I_{90}^1, I_{135}^1$ as it would be obtained by bracketing the real shutter speed at times $t_0 \ldots t_3$. Albeit any method can be chosen at this point to obtain the final HDR image, we use [5] for its simplicity and overall quality. It is essentially based on averaging each image with a Gaussian weight $w(l) = \exp\left(-\frac{(l-0.5)^2}{2\sigma^2}\right)$ giving more importance to irradiance values closer to the middle of the response function[2]. The final HDR image is thus:

$$I_{\text{HDR}} = \frac{w(I_0^1)\frac{I_0^1}{t_0} + w(I_{45}^1)\frac{I_{45}^1}{t_1} + w(I_{90}^1)\frac{I_{90}^1}{t_2} + w(I_{135}^1)\frac{I_{135}^1}{t_3}}{w(I_0^1) + w(I_{45}^1) + w(I_{90}^1) + w(I_{135}^1)}. \tag{10}$$

By manually rotating the external filter (i.e. varying the angle α_f) the dynamic range of the acquisition system can be adjusted to match the imaged scene. In Fig. 2 we show how the equivalent exposure times, computed as in Eq. 9, vary depending on α_f. If the scene is mostly unpolarized (first column), $t_{\min}$ and $t_{\max}$ will be loosely affected by the external filter angle. On the contrary, if the scene is polarized, the absolute range of equivalent times will be smaller when the filter is orthogonal to the scene AoLP. In both the cases, the dynamic range $DR = 20\log\frac{255\,t_{\max}}{t_{\min}}$ vary from a minimum of ≈ 61 dB to a maximum of ≈ 74 dB (considering 8 bits per pixel images) regardless the actual scene polarization. In particular, when α_f is aligned with any α_i, we observe the maximum DR

---
[2] We empirically observed that $\sigma = 0.2$ usually gives satisfactory results.

and two out of the four images will have an equivalent exposure time equal to $0.5(t_{\max} + t_{\min})$. When α_f is in between two α_i (for example $\alpha_f = \frac{\pi}{8}$) we get the minimum dynamic range and all equivalent exposure times are equal to either $t_{\min}$ or $t_{\max}$. The important thing to notice is that such behaviour is independent by the actual scene polarization state. Indeed, the DR is only affected by: (i) the external filter angle α_f and (ii) the ability of the external filter to block all the unpolarized light (i.e. the DoLP $\mathcal{D}_f$ of light exiting the filter). This flexibility is the main advantage of our proposed method.

2.3   Robust Estimation of Stokes Parameters

The accuracy of $\mathcal{D}$ and θ is crucial to compute the equivalent exposure times especially when α_i and α_f are almost aligned. Certainly, camera thermal noise and quantization affect the computation of Stokes parameters but many works in the literature, dealing with polarization imaging, simply ignore this aspect. We propose here a simple procedure to improve S_0, S_1, S_2 by assuming that each pixel of $I_0, I_{45}, I_{90}, I_{135}$ is perturbed by zero-mean additive Gaussian noise before being clipped to range $0 \ldots 1$.

By observing Eq. 2 we see that our four channel (noisy) polarized image $\bar{I}_{\mathcal{P}} = (\bar{I}_0, \bar{I}_{45}, \bar{I}_{90}, \bar{I}_{135})^T$ is an over-parametrization over the Stokes parameters. Indeed, since:

$$I_0 + I_{90} = I_{45} + I_{135} \tag{11}$$

we can recover any of the four channels by knowing the other three. So, we start by cleaning the acquired images to recover pixels in $\bar{I}_{\mathcal{P}}$ where exactly one of the four channels are either under- or over-exposed.

If all the channels are in the proper range, we consider each pixel p as a vector of four Gaussian distributed random variables $\bar{I}_k(p) = I_k(p) + \epsilon$, $\epsilon \sim N(0, \sigma_{k,p})$ and use a Maximum Likelihood approach to find the best value of $I_k(p)$ such that Eq. 11 holds. In practice this boils down in solving, for each pixel p, the constrained least squares:

$$\operatorname*{argmin}_{x_p = (I_0(p)\ I_{45}(p)\ I_{90}(p)\ I_{135}(p))^T} \|x(p) - \bar{I}_{\mathcal{P}}(p)\|^2 \tag{12}$$

$$\text{s.t.} \quad \mathbf{C}x = 0$$

where $\mathbf{C} = (1 \ -1 \ 1 \ -1)$. Using the KKT conditions, the analytical optimal solution is given by:

$$\begin{pmatrix} I_0(p) \\ I_{45}(p) \\ I_{90}(p) \\ I_{135}(p) \end{pmatrix} = \begin{pmatrix} k_1 & k_2 & -k_2 & k_2 \\ k_2 & k_1 & k_2 & -k_2 \\ -k_2 & k_2 & k_1 & k_2 \\ k_2 & -k_2 & k_2 & k_1 \end{pmatrix} \begin{pmatrix} \bar{I}_0(p) \\ \bar{I}_{45}(p) \\ \bar{I}_{90}(p) \\ \bar{I}_{135}(p) \end{pmatrix} \tag{13}$$

with coefficients $k_1 = 0.75$ and $k_2 = 0.25$.

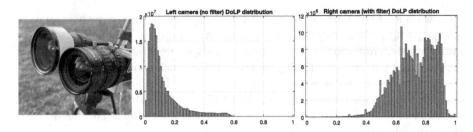

Fig. 3. Left: polarimetric stereo setup (right camera with additional filter on). Center and right: histograms of DoLP distribution for both cameras in our dataset.

3   Experimental Section

In this section we first describe the acquisition process we adopted to compute ground truth images with the proposed setup. Then, we compare the proposed HDR reconstruction technique with state-of-the-art algorithmic and learning-based methods using standard and polarimetric cameras.

An HDR polarimetric dataset was acquired using two FLIR Blackfly mono-chrome Polarization cameras, mounting a Sony IMX250MZR sensor providing PFA mosaiced images of size 2448 × 2048 pixels. ICRF was estimated as in [21] and applied to every captured image. The two cameras were mounted on a solid frame with a minimal baseline (see Fig. 3, left), and connected via an hardware trigger so that pictures are acquired simultaneously. As required by our method, an additional external linear polarizer was added to the right camera only, oriented with an angle of $\approx 15°$ wrt. the camera horizontal axis. For each scene, we first took 30 stereo shots with increasing exposure time, then we removed the additional filter and took another set of 30 pictures.

Note that (i) exposure values were set according to the scene and accounting for the external filter attenuation, i.e. the ith intensity image of each set exhibits the same average intensity with and without the filter; (ii) one set of 30 acquisitions takes around half a second and the filter removal is fast ($\sim 1s$). Therefore, the captured scene was essentially unchanged with and without the filter. For each scene, the HDR Ground Truth (GT) was computed by applying [5] to the intensities obtained from the camera with no additional filter. We acquired a total of 20 challenging HDR scenes, both outdoor and indoor.

Table 1. Comparison between HDR methods with Reinhard tone mapping [24] applied.

Method	Best PSNR (dB)	PSNR (dB)	Best MSSIM ×10⁻¹	MSSIM ×10⁻¹
Our	**26.1946 ± 3.5419**	**20.8537 ± 4.8464**	**9.8187 ± 0.1881**	**9.4217 ± 1.0714**
Wu et al. [29]	24.5501 ± 3.1613	19.5490 ± 4.7051	9.7739 ± 0.2428	9.3361 ± 1.1289
HDRCNN	14.3141 ± 1.1944	11.9532 ± 2.3152	7.7141 ± 1.3194	5.9161 ± 2.1367
Deep-HDR	14.8922 ± 1.3342	12.3500 ± 2.5288	7.8611 ± 1.3488	5.9814 ± 2.1720
Two-stage-HDR	19.1025 ± 1.6915	14.4062 ± 3.1722	9.5515 ± 0.2924	8.6506 ± 1.3224
KO	14.1248 ± 1.6047	12.2216 ± 1.8131	5.5661 ± 1.3923	4.4925 ± 1.6370

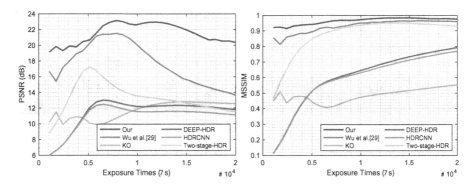

Fig. 4. PSNR (left) and MSSIM (right) for same scene varying exposure times.

3.1 Comparisons

We compared our method with state-of-the-art techniques performing HDR imaging from a single shot. The most similar work proposed in the literature is Wu et al. [29], where HDR is performed using a single PFA camera. We also compared with an inverse tone mapping technique described in [13] (KO from now on) which estimates HDR from a single image. Moreover, we considered three recent learning-based methods namely: HDRCNN [6], Two-stage-HDR [1] and Deep-HDR [26]. These methods are not designed for polarimetric cameras but perform HDR from a single intensity image.

For our method we applied the procedure described in Sect. 2.2 (i.e. considering data from both the cameras), while for Wu et al. we only used the polarimetric images captured from the camera with no external filter. Finally, for all the other methods we gave as input demosaiced intensity-only images, simulating a regular grayscale acquisition.

Output HDR images are evaluated against the GT according to their Peak Signal-to-Noise Ratio (PSNR) and Multi-Scale Structural SIMilarity (MSSIM) [28]. Results are listed in Table 1. Best PSNR and best MSSIM columns correspond to average of the best value achieved for each scene among all exposures, while PSNR and MSSIM are average of values computed on all exposures. Our proposed technique produced significantly improved PSNR and MSSIM as compared to other techniques. As expected, Wu et al. exhibits the second best result since it is the only other method taking full advantage of polarimetric information. However, the majority of pixels in the acquired scenes exhibit a DoLP ranging from 0 to 0.2 (See Fig. 3, center) so Wu et al. still suffers a limited extent of equivalent exposure times. On the other hand, adding the external linear polarizer significantly increase the DoLP (Fig. 3, right) thus increasing the dynamic range of the acquisition device.

Since all the approaches use a single-shot to produce the output HDR, it is important to assess how much a correct exposure of the acquired image is critical for the final result. To evaluate this, we plotted in Fig. 4 the resulting PSNR (left) and MSSIM (right) for the same scene varying the exposure time.

| GT | Our | Wu et al. [29] | Two-stage-HDR |

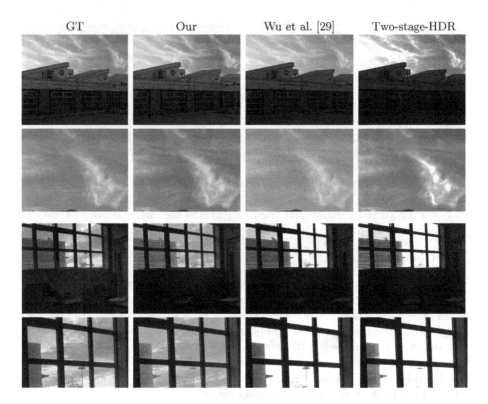

Fig. 5. Qualitative analysis of HDR reconstruction methods. Rows 2 and 4 show a zoomed in patch of scenes shown in 1 and 3 respectively.

All the curves are mostly bell-shaped, showing how there exists a single optimal exposure time maximizing the quality of the resulting HDR image. However, our method is far less sensitive to that (see for example the PSNR plot on the left) where a broad range of exposure times ranging from 6 to 15 ms equally result in a PSNR above 22. In any case, our solution consistently outperforms other techniques demonstrating its versatility in situations where it is not trivial to choose a good exposure time for a certain scene. Finally, Fig. 5 shows some selected examples to qualitatively compare various methods.

4   Conclusions

In this paper we proposed a novel HDR reconstruction technique based on a pair of PFA cameras in which one of the two mounts an additional linear polarizer in front of the lenses. We derived the imaging model for the whole system and showed that the range of equivalent exposure times is much wider than using a single camera as previously proposed by Wu et al. Experiments demonstrate how the new method outperforms state-of-the-art single shot HDR techniques, including recent Deep Learning approaches.

References

1. Sharif, A.S., Naqvi, R.A., Biswas, M., Kim, S.: A two-stage deep network for high dynamic range image reconstruction. In: Proceedings of the IEEE/CVF Conference on Computer Vision and Pattern Recognition, pp. 550–559 (2021)
2. Banterle, F., et al.: High dynamic range imaging and low dynamic range expansion for generating HDR content. In: Computer Graphics Forum, vol. 28, pp. 2343–2367. Wiley Online Library (2009)
3. Banterle, F., Ledda, P., Debattista, K., Chalmers, A.: Inverse tone mapping. In: Proceedings of the 4th International Conference on Computer Graphics and Interactive Techniques in Australasia and Southeast Asia, pp. 349–356 (2006)
4. Cronin, T.W., Marshall, J.: Patterns and properties of polarized light in air and water. Philos. Trans. Royal Soc. B Biol. Sci. **366**(1565), 619–626 (2011)
5. Debevec, P.E., Malik, J.: Recovering high dynamic range radiance maps from photographs. In: ACM SIGGRAPH 2008 Classes, pp. 1–10 (2008)
6. Eilertsen, G., Kronander, J., Denes, G., Mantiuk, R.K., Unger, J.: HDR image reconstruction from a single exposure using deep CNNs. ACM Trans. Graph. (TOG) **36**(6), 1–15 (2017)
7. Endo, Y., Kanamori, Y., Mitani, J.: Deep reverse tone mapping. ACM Trans. Graph. **36**(6), 1–10 (2017). (Proc. of SIGGRAPH ASIA 2017)
8. Farnebäck, G.: Two-frame motion estimation based on polynomial expansion. In: Bigun, J., Gustavsson, T. (eds.) SCIA 2003. LNCS, vol. 2749, pp. 363–370. Springer, Heidelberg (2003). https://doi.org/10.1007/3-540-45103-X_50
9. Goldstein, D.H.: Polarized Light. CRC Press, Boca Raton (2017)
10. Kalantari, N.K., Ramamoorthi, R., et al.: Deep high dynamic range imaging of dynamic scenes. ACM Trans. Graph. **36**(4), 144–1 (2017)
11. Khan, E.A., Akyuz, A.O., Reinhard, E.: Ghost removal in high dynamic range images. In: 2006 International Conference on Image Processing, pp. 2005–2008. IEEE (2006)
12. Khan, Z., Khanna, M., Raman, S.: FHDR: HDR image reconstruction from a single LDR image using feedback network. In: 2019 IEEE Global Conference on Signal and Information Processing (GlobalSIP), pp. 1–5. IEEE (2019)
13. Kovaleski, R.P., Oliveira, M.M.: High-quality reverse tone mapping for a wide range of exposures. In: 2014 27th SIBGRAPI Conference on Graphics, Patterns and Images, pp. 49–56. IEEE (2014)
14. Kupinski, M.K., Bradley, C.L., Diner, D.J., Xu, F., Chipman, R.A.: Angle of linear polarization images of outdoor scenes. Opt. Eng. **58**(8), 082419 (2019)
15. Lee, S., An, G.H., Kang, S.-J.: Deep recursive HDRI: inverse tone mapping using generative adversarial networks. In: Ferrari, V., Hebert, M., Sminchisescu, C., Weiss, Y. (eds.) ECCV 2018. LNCS, vol. 11206, pp. 613–628. Springer, Cham (2018). https://doi.org/10.1007/978-3-030-01216-8_37
16. Liu, Y.L., et al.: Single-image HDR reconstruction by learning to reverse the camera pipeline. In: Proceedings of the IEEE/CVF Conference on Computer Vision and Pattern Recognition, pp. 1651–1660 (2020)
17. Marnerides, D., Bashford-Rogers, T., Hatchett, J., Debattista, K.: ExpandNet: a deep convolutional neural network for high dynamic range expansion from low dynamic range content. In: Computer Graphics Forum, vol. 37, pp. 37–49. Wiley Online Library (2018)
18. McCann, J.J., Rizzi, A.: The Art and Science of HDR Imaging, vol. 26. Wiley, New York (2011)

19. Mertens, T., Kautz, J., Van Reeth, F.: Exposure fusion: a simple and practical alternative to high dynamic range photography. In: Computer Graphics Forum, vol. 28, pp. 161–171. Wiley Online Library (2009)

20. Mihoubi, S., Lapray, P.J., Bigué, L.: Survey of demosaicking methods for polarization filter array images. Sensors **18**(11), 3688 (2018)

21. Mitsunaga, T., Nayar, S.K.: Radiometric self calibration. In: Proceedings. 1999 IEEE Computer Society Conference on Computer Vision and Pattern Recognition (Cat. No PR00149), vol. 1, pp. 374–380. IEEE (1999)

22. Pistellato, M., Bergamasco, F., Fatima, T., Torsello, A.: Deep demosaicing for polarimetric filter array cameras. IEEE Trans. Image Process. **31**, 2017–2026 (2022). https://doi.org/10.1109/TIP.2022.3150296

23. Prabhakar, K.R., Babu, R.V.: Ghosting-free multi-exposure image fusion in gradient domain. In: 2016 IEEE International Conference on Acoustics, Speech and Signal Processing (ICASSP), pp. 1766–1770. IEEE (2016)

24. Reinhard, E., Heidrich, W., Debevec, P., Pattanaik, S., Ward, G., Myszkowski, K.: High Dynamic Range Imaging: Acquisition, Display, and Image-Based Lighting. Morgan Kaufmann, San Francisco (2010)

25. Santos, M.S., Ren, T.I., Kalantari, N.K.: Single image HDR reconstruction using a CNN with masked features and perceptual loss. arXiv preprint arXiv:2005.07335 (2020)

26. Santos, M.S., Tsang, R., Khademi Kalantari, N.: Single image HDR reconstruction using a CNN with masked features and perceptual loss. ACM Trans. Graph. **39**(4) (2020). https://doi.org/10.1145/3386569.3392403

27. Umow, N.: Chromatische depolarisation durch lichtzerstreuung. Phys. Z **6**, 674–676 (1905)

28. Wang, Z., Simoncelli, E.P., Bovik, A.C.: Multiscale structural similarity for image quality assessment. In: The Thrity-Seventh Asilomar Conference on Signals, Systems & Computers, vol. 2, pp. 1398–1402. IEEE (2003)

29. Wu, X., Zhang, H., Hu, X., Shakeri, M., Fan, C., Ting, J.: HDR reconstruction based on the polarization camera. IEEE Robot. Autom. Lett. **5**(4), 5113–5119 (2020)

30. Yang, X., Xu, K., Song, Y., Zhang, Q., Wei, X., Lau, R.W.: Image correction via deep reciprocating HDR transformation. In: Proceedings of the IEEE Conference on Computer Vision and Pattern Recognition, pp. 1798–1807 (2018)

31. Yu, Y., Zhu, D., Smith, W.A.: Shape-from-polarisation: a nonlinear least squares approach. In: Proceedings of the IEEE International Conference on Computer Vision Workshops, pp. 2969–2976 (2017)

32. Zappa, C.J., Banner, M.L., Schultz, H., Corrada-Emmanuel, A., Wolff, L.B., Yalcin, J.: Retrieval of short ocean wave slope using polarimetric imaging. Meas. Sci. Technol. **19**(5), 055503 (2008)

PolygloNet: Multilingual Approach for Scene Text Recognition Without Language Constraints

Àlex Solé Gómez[✉], Jorge García Castaño, Peter Leškovský, and Oihana Otaegui Madurga

Intelligent Transport Systems and Engineering, Vicomtech, Basque Research and Technology Alliance (BRTA), Donostia-San Sebastián, Spain
{asole,jgarciac,pleskovsky,ootaegui}@vicomtech.org
https://www.vicomtech.org

Abstract. The detection and recognition of text instances in camera-captured images or videos generate rich and precise semantic information for interpreting and describing the scene. However, recognizing text in the wild remains a challenging problem despite its value. Apart from the inherent problems in text detection and recognition tasks, knowing the language to be recognized is required in unsupervised forensic applications where multilingual information is frequently employed. This work proposes an unconstrained multilingual text recognition pipeline for scene text detection and recognition. The pipeline consists of multiple text recognition experts contributing to determining the output text sentence. Each expert translates the visual information into a candidate text sentence. Finally, our post-processing model, named PolygloNet, encodes and aggregates all text sentences to generate the optimal text sequence. This model can select the most likely text sequence and correct spelling errors produced in the recognition stage. Experimental results show that our model can produce accurate recognition results on relevant datasets (MLT2017 [2] and MLT2019 [1]).

Keywords: Scene text detection (STD) · Multilingual text recognition · Text analysis · Long short-term memory (LSTM)

1 Introduction

Scene text recognition is the process of detecting and recognizing text in natural images. Current approaches decompose the problems into several parts: (i) text detection [5,16], (ii) text recognition [17,23], and, in some cases, (iii) an additional post-processing step to improve and/or correct the recognized text [20,23].

One of the most widespread uses of scene text recognition is forensic analysis, where detailed investigation for detecting and documenting the course, reasons, culprits, and consequences of security incidents or violation of rules are carried out. Scene text recognition is a crucial technology in order to extract clues from suspicious multimedia sources such are cybercrime, terrorism, or pedophilia, among others [18,24]. In these cases, what is required is to extract the maximum amount of information from an image or video to find a broad set of meaningful semantic entities. In this scenario, real-time or low-cost computation methods are not explicitly required. Instead, accuracy is favored over efficient processing capabilities. Another essential consideration for forensics is that knowing the language before extraction will require human interaction in most circumstances. For the cases of a large amount of data to be analyzed, having this human interaction can considerably slow the analysis. Most of the current approaches for these tools are only in english or the language must set before the recognition step (e.g. [3,9,15,19]). Creating a single model for multi-lingual text recognition is also not feasible because of the significant language disparity. For example, most of the Latin alphabets have around 27 characters. Meanwhile, Asian languages can have thousands of different characters in their alphabets. Including all the alphabets in one neural network will also decrease the accuracy of the results.

Huang et al. [14] proposed a methodology for multilingual scene text recognition where they use a script identification model to select the best recognition model from a multiplexer of different recognition heads trained on different languages. Chet et al. [7] presented a methodology that infers the script identification and the text recognition simultaneously, sharing part of the feature extractor. Even though it is a more efficient approach, this error is propagated to the rest of the modules if the language detector fails at the language detection stage.

In recent years, new methods for text processing using deep learning have been proposed. These new techniques can be used to process the extracted text to correct the possible spell errors and improve the overall pipeline result. Zhou et al. [30] proposed a translation model for natural language processing that uses LSTM layers to translate sentences at word level but to have the context of the rest of the words. Sakaguchi et al. [22] presented a language processing mechanism based on LSTM for spell checking. Their approach uses the embeddings at a character level instead of a word level. The results show good performance correcting words when the misspelling is on a few characters or disorders.

This paper addresses the challenges of multi-language scene text recognition by proposing a novel architecture for high computing environments where the priority is accuracy. The unconstrained multilingual text recognition pipeline proposed consists of three parts. Initially, a detection module that extracts the text regions from the images. It is followed by a multi-expert recognition module that converts text regions to text format in different languages. Finally, our PolygloNet model aggregates all the text predictions candidates to one unique output.

In particular, our PolygloNet model uses multiple LSTM encoders an a single LSTM decoder inspired by the LSTM models for language translation [28,30]. We use all the predicted text candidates from a multi-expert recognition module to faithfully aggregate the independent predictions and generate the optimal text sequence. As can be noted, this approach avoids propagation errors in case the script identification fails. Our contributions can be summarized as follows:

- A novel unconstrained multilingual text recognition pipeline for scene text recognition.
- A novel PolygloNet model that allows automated text recognition for multiple languages.
- The ability to correct spelling errors produced in the recognition stage.

2   Methodology

Figure 1 shows the proposed unconstrained multilingual text recognition pipeline consisting of three stages: (i) text detection stage extracts those areas of the image where text appears. Those areas are represented by polygons. (ii) the text recognition stage executes a group of recognized *experts*. Each recognition *expert* converts the extracted text area to text format, independently from each other, and returns a text sentence with the corresponding recognition score. (iii) The last stage is represented by our PolygloNet model, which aggregates all the predicted text sentences and scores for computing the optimal text output with a global score.

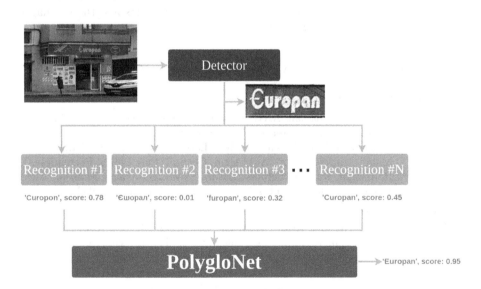

Fig. 1. The unconstrained multilingual text recognition pipeline.

2.1   Detection Module

For the scene text detection, we used the Character Region Awareness for Text Detection (CRAFT) detector [5]. Craft exploits each character's area and the affinity relations between characters. It also provides a training methodology with synthetic data and an estimation of the character-level segmentation from natural images.

The detector extracts the text region as poly and applies a four-point perspective transform to rectify the image. Also, the image is converted to grayscale, and when the contrast of the extracted patch is low, we increase the contrast.

2.2   Multi-expert Recognition Module

We propose to follow a structure of a group of *experts*. An *expert* is a model that is trained focusing on a specific alphabet. Each *expert* returns a string in text format and a score. In particular, we used *experts* based on *CRNN* [23] and *Tesseract*[1].

CRNN is an end-to-end trainable architecture that integrates image feature extraction, sequence modeling, and transcription designed for scene text recognition. The feature extractor extracts relevant information sequences from the image using convolutional neural networks. The sequence modeling uses LSTM [13] to process the extracted sequences. The transcription converts the sequence to the output string Connectionist Temporal Classification (CTC) [11]. An *expert* represents a model trained for each of the available alphabets using the *CRNN* architecture. Each model contains some of the most representative languages from the given alphabet. Table 1 shows the alphabets and the languages supported for each of the *experts*.

Table 1. Alphabets and Languages supported for the CRNN's experts

Experts	Supported Languages
Latin	Afrikaans, Albanian, Azerbaijani, Bosnian, Croatian, Danish, Dutch, English, Estonian, French, German, Hungarian, Icelandic, Indonesian, Irish, Italian, Kurdish, Latin, Latvian, Lithuanian, Malay, Maltese, Maori, Norwegian, Occitan, Pali, Polish, Portuguese, Romanian, Russian, Serbian (Latin), Slovak, Slovenian, Spanish, Swahili, Swedish, Tagalog, Turkish, Uzbek, Vietnamese, Welsh
Arabic	Arabic, Persian (Farsi), Uyghur, Urdu
Cyrillic	Abaza, Adyghe, Avar, Belarusian, Bulgarian, Chechen, Dargwa, Ingush, Kabardian, Lak, Lezghian, Mongolian, Russian, Serbian (Cyrillic), Tabassaran, Ukranian
Bengali	Assamese, Bengali
Devangari	Hindi, Marathi, Nepali, Bihari, Maithili, Angika, Bhojpuri, Magahi, Nagpuri, Newari

[1] https://github.com/tesseract-ocr/tesseract.

Tesseract OCR is an open-source OCR engine developed by Google. It is currently supporting 116 languages [4]. Nowadays is one of the states of the art for OCR [8,10,12,29], specially for printed documents [26]. Tesseract OCR version 4.00 supports LSTM. We used the Tesseract engine to create our last *expert*. Tesseract offers the option to infer using all the available language according to a priority list that has to be set. For our case, we have set the priority to focus mainly on the European, Cyrillic, and Arabic languages. Table 2 show the most important languages for Tesseract's inference.

Table 2. Priority order for Tesseract's *expert*

Priority order
English, Deutsch, French, Spanish, Dutch, Italian, Portuguese, Russian, Arabic, ...

2.3 PolygloNet

Figure 2 shows the implemented architecture of PolygloNet. The idea behind this model is to decide which is the optimal text sentence from all the *experts* and try to correct it, if needed, based on the input from other *experts*. In other words, aggregate all the sentence candidates proposed by the *experts* into an optimal text sentence. It was inspired by the translation tools that exist for natural language processing but using multiple encoders [28,30]. It uses all the text candidates and the scores from all of the *experts* to generate the optimal text sentence.

The module uses an LSTM encoder-decoder architecture at a character level. For each of the *experts*, we have an encoder that translates the sentence to a meaning vector. Each encoder has an embedding layer that converts the character into numerical features and, secondly, a set of bidirectional LSTM layers that processes the sequences. All the encoders from the different experts share weights. The output states from the encoder LSTM layers are aggregated together through a dense perceptron layer and passed to the decoder. The outputs of the encoders are multiplied by the respective scores and summed all together. The decoder part consists of bidirectional LSTM layers with the states from the encoder and a final dense layer with a Softmax activation. The model has 256 embedding dimensions, 1024 LSTM units, and 2 layers of LSTM in the Encoder and the Decoder. The parameters specifications of the model are not fixed can be dynamically set based on the complexity of the problem or the number of languages or experts.

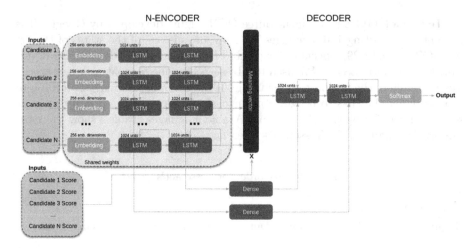

Fig. 2. PolygloNet architecture.

3 Discussion and Results

Our experiments focus only on the proposed PolygloNet model. We collect the required inputs (text sequences and scores) using the detector module and our multi-expert recognition module.

3.1 Datasets

To evaluate the performance of the proposed method, we focus on two relevant datasets: **MLT2017** [2] and **MLT2019** [1] containing multilingual data for Scene Text Detection challenges. MLT datasets offer a variety of images with text in different languages and even different alphabets. We ignored Asian languages (Chinese, Japanese, and Korean) for our task.

   Synthetic Text Images is a common used technique for generating datasets for text recognition [6,8,25]. It allows the generation of a large amount of data without any annotation cost. In our case, we only wanted to generate training and testing data for the Cyrillic alphabet cause of the lack of any known datasets for this alphabet.

3.2 Implementation Details

The experiments were realized using an Intel Xeon Gold 6230 CPU, 756 GB of RAM, and an NVIDIA Tesla V100 32 GB. With our hardware specifications and a batch size of 64 text sequences, our PolygloNet takes approximately 220 ms to infer. The Craft detector implementation and pretrained weights are from the official repository[2]. The multi-expert recognition module was implemented

---

[2] https://github.com/clovaai/CRAFT-pytorch.

using the pretrained models from EasyOCR library[3] and Tesseract OCR. For the Tesseract, we set the page segmentation mode (PSM) parameter to treat the image as a single text line. For the Synthetic Text Images in Cyrillic, we used the Text Recognition Data Generator repository[4]. We split the 20% of the MLT and Synthetic training datasets for testing, and we generated 20%.

We realized that all our training data were images cropped to a single word, but, in real life, the detector sometimes detects more than one word. When more than one word was inferred through our pipeline, the quality was decreased. To solve this problem, we created a text dataset concatenating words from the predictions of the multi-expert recognition module with some of the most typical connectors that we can find in real life (space, comma, dot, etc.). Figure 3 shows a few examples of the concatenated words. The algorithm chooses one random word from the dataset and concatenates it with between another 2 to 4 randomly selected words using the available connectors randomly. This concatenation also helps cases where the detected words are not in the same language (e.g., multilingual street sign). Figure 4 shows the output from our module when this augmentation is not used during training and when this augmentation is used.

Ground Truth	Latin	Arabic	Cyrillic	Bengali	Devangar	Tesseract
di.OMBONI	di.OMBONI	–80]0.ل	&i.OMoll	৭.0ʀ80	রা.0ঢ80	di.OMBONI
антракт (011) المحرق	Ŝl_aurpakr OID	0 0 87ل_المحرق	@->-|_антракт (1	9১ >|_১৮-ভ৫ম (07 1	৮ >_3#3ঽ- (011)	антракт (04 13 المحرق
গাঁইপুর Hamada	TpTr Hamada	3"٠٤ 9	1MK Нагада	গাঁইপুব ঈ১শ্ৰবহ	পফলনা ঊথ়ঃ	7 Hamada
Резінськии / M'BAREK	Pe3IHcbKu" / MBAREK	8ৎ° / =3i#↓8৯%	Резінськии / M'BA?EK	?৭3-৫৮ঃ৸৴	['8ঽ৳৪ঃ	৭31ঢ়+⟨̃⟩ / ঢ়843াঁঢ	резинськии / M'BAREK

Fig. 3. Examples of word augmentation algorithm

(a)   (b)

Fig. 4. a) Results training without word augmentation b) Results training with the word augmentation.

3.3 Hyper Parameter Tuning

We carried out hyperparameter tuning experiments to select the bests parameters for our model. We followed the Hyperband methodology proposed by Li et al. [21] where the best model is selected using a championship style bracket. The algorithm randomly selects the model's parameters and trains a few epochs. Only the bests models are passed to the next round to train for a few epochs more. The algorithm continues until only one model is left. The training computation time is reduced significantly using this approach.

---

[3] https://github.com/JaidedAI/EasyOCR.
[4] https://github.com/Belval/TextRecognitionDataGenerator.

Based on this approach, we obtained the best parameters for our model: 256 embedding dimensions, 1024 LSTM units, and 2 layers of LSTM in the Encoder and the Decoder. Table 3 shows the 5-top results for Hyperband tunning in our model.

Table 3. Best five models from the hyper band hypertuning.

5-top models	Number layers	Embedding dimensions	Number LSTM	Number params	Accuracy
1	2	256	1024	171.69 M	0.81
2	2	512	1024	173.99 M	0.80
3	1	128	1024	78.25 M	0.78
4	1	16	1024	77.25 M	0.77
5	3	256	256	17.39 M	0.75

It can be noted that all the architectures achieve similar accuracy scores. We selected the first top model because we do not have hardware limitations, although a simpler model can be feasible for less-resourced environments.

3.4 PolygloNet Analysis

Since our approach is based on natural language processing and post-processing of the extracted text, comparing with other recognition or detection modules is not a fair comparison because they are independent of our model, and we only reuse their predictions as our input. To make a fair comparison, we propose three scenarios:

– **Score:** We select which of the candidate sentence is the optimal one based on the highest score obtained from our multi-expert recognition module.
– **Lang:** We select which candidate sentence is the optimal based on the ground truth information. Since we do not have any pretrained model for script detection, we will suppose the ideal case where the script detection model continuously detects the correct language.
– **Ours:** Using our PolygloNet model.

The metrics used for the experiments are the mean Word Accuracy and the 1-NED (Normalized Edit Distance). The Word Accuracy is when the predicted sentence is correct, otherwise zero. It provides the percentage of the dataset we can predict without any error. The 1-NED (Normalized Edit Distance) or the Levenshtein distance measures how similar the prediction is to the ground truth. This metric accounts for the number of simple operations (adding, deleting, and changing a character) needed to achieve the correct word.

Table 4 shows the word accuracy and the 1-NED (Normalized Edit Distance) metric between our approach and the case of knowing the language and the case of choosing the language based on the highest predicted score. For the case of word augmentation, the case of knowing the language is not computed because the languages are mixed and are not a single one.

Table 4. Results comparing the different multi-lingual approaches. Word Accuracy and 1-NED for each of the datasets and available alphabets.

	MLT2019				MLT2017			Cyrillic synth	Word augmentation
	Latin	Arabic	Bangla	Hindi	Latin	Arabic	Bangla	Cyrillic	Mixed
Word accuracy									
Score	0.55	0.27	0.22	0.40	0.53	0.25	0.20	0.67	0.15
Lang	0.54	0.33	0.37	**0.57**	0.57	0.30	0.30	0.73	–
Ours	**0.92**	**0.86**	**0.86**	0.48	**0.91**	**0.87**	**0.85**	**0.78**	**0.27**
1-NED									
Score	0.70	0.38	0.34	0.50	0.74	0.36	0.30	0.77	0.69
Lang	0.73	0.70	0.68	**0.82**	0.79	0.67	0.61	0.91	–
Ours	**0.96**	**0.93**	**0.92**	0.74	**0.96**	**0.93**	**0.92**	**0.92**	**0.79**

We can observe a considerable improvement using our model, especially for the Latin alphabets. For the Hindi alphabet, a lower accuracy can be observed than selecting the language model a priori. Although the reason is not clarified, we suppose it is because of some bias in the training data or the alphabet complexity. The results obtained for the Word Accuracy in the Word Augmentation dataset were drastically low. Since we are concatenating text strings, we have larger strings, and it is more frequent to fail at least one character.

We also have to consider that the script detection modules have minimal errors in real life. In other words, in real life, the results for selecting the language approach will be lower than the ones presented by this paper.

3.5   Multisource Analysis

In the last experiment, we wanted to test if Tesseract contributes to the optimal prediction and if our model helps in multisource recognition tasks (e.g., handwritten, text in the wild, scanned documents, ...). Table 5 compares the model in the case of using the Tesseract with the case of not using it.

Based on these results, we can assume that our model is good in multisource tasks since Tesseract is more designed for documents and the rest of the *experts* for scene text detection. Having different kinds of recognition *experts* as text candidates can be helpful in the cases of more robust recognition or for multisource tasks. For example, in the case that the forensics wanted to scan a suspect device with a large number of diverse images (screenshots, text in the wild, handwritten text, ...). We can train specific *experts* for each type of text and join them using the proposed PolygloNet approach. The results would not be magnificent without using Tesseract and may be comparable with the Lang setting.

Table 5. Results comparing using separately or joint the recognition modules. Word Accuracy and 1-NED for each of the datasets and available alphabets.

	MLT2019				MLT2017			Cyrillic synth	Word augmentation
	Latin	Arabic	Bangla	Hindi	Latin	Arabic	Bangla	Cyrillic	Mixed
Word accuracy									
Ours/w Tesseract	**0.92**	**0.86**	**0.86**	**0.48**	**0.91**	**0.87**	**0.85**	**0.78**	**0.27**
Ours/wout Tesseract	0.59	0.41	0.32	0.37	0.57	0.41	0.27	0.71	0.13
Tesseract	0.60	0.19	0.10	0.21	0.58	0.18	0.10	0.40	0.15
1-NED									
Ours/w Tesseract	**0.96**	**0.93**	**0.92**	**0.74**	**0.96**	**0.93**	**0.92**	**0.92**	**0.79**
Ours/wout Tesseract	0.76	0.70	0.63	0.70	0.79	0.67	0.57	0.88	0.66
Tesseract	0.74	0.25	0.16	0.27	0.75	0.23	0.16	0.69	0.69

4   Conclusions

We have presented an unconstrained multilingual text recognition pipeline that includes the PolygloNet model to deal with scene text recognition tasks in the case of multi-language text in high-performance environments. Our PolygloNet model allows us to carry out the inference without the need to know the language, with the advantage that the *experts* from our multi-expert recognition module help each other to correct the text. We have tested some of the relevant multi-language datasets where we have obtained a notable improvement, around 30/40% of the word accuracy, compared to the ideal case of knowing the language beforehand. As such, experiments show that our PolygloNet can be used as a multisource text recognizer from different sources (handwritten text, printed text, text in the wild, ...). Given that our proposed PolygloNet model can lead to too many parameters, i.e., adding Asiatic languages with thousands of different characters, future work can explore other architectures to handle these situations better, such as transformers [27]. Also, explore more the multisource analysis by adding new *experts* that can handle new text sources, e.g., handwritten text. Finally, study the possibility to design lighter models to cope with real-time scenarios.

Acknowledgement. This work was supported by the European Union's Horizon 2020 Research and Innovation Program (AIDA Project) under Grant Agreement 883596.

References

1. Overview - ICDAR 2019 robust reading challenge on multi-lingual scene text detection and recognition. https://rrc.cvc.uab.es/?ch=15. Accessed 24 May 2021
2. Overview - ICDAR2017 competition on multi-lingual scene text detection and script identification. https://rrc.cvc.uab.es/?ch=8. Accessed 24 May 2021
3. Tesseract OCR. https://github.com/tesseract-ocr/tesseract. Accessed 17 May 2021
4. Traineddata files for version 4.00+. https://tesseract-ocr.github.io/tessdoc/Data-Files.html. Accessed 24 May 2021

5. Baek, Y., Lee, B., Han, D., Yun, S., Lee, H.: Character region awareness for text detection, pp. 9365–9374 (2019)
6. Chen, X., Jin, L., Zhu, Y., Luo, C., Wang, T.: Text recognition in the wild: a survey. ACM Comput. Surv. **54**(2) (2021). https://doi.org/10.1145/3440756
7. Chen, Z., Yin, F., Zhang, X.Y., Yang, Q., Liu, C.L.: MuLTReNets: multilingual text recognition networks for simultaneous script identification and handwriting recognition. Pattern Recognit. **108**, 107555 (2020). https://doi.org/10.1016/j.patcog.2020.107555. https://www.sciencedirect.com/science/article/pii/S0031320320303587
8. Chernyshova, Y.S., Sheshkus, A.V., Arlazarov, V.V.: Two-step CNN framework for text line recognition in camera-captured images. IEEE Access **8**, 32587–32600 (2020). https://doi.org/10.1109/ACCESS.2020.2974051
9. Du, Y., et al.: PP-OCR: a practical ultra lightweight OCR system. CoRR abs/2009.09941 (2020). https://arxiv.org/abs/2009.09941
10. Etter, D., Rawls, S., Carpenter, C., Sell, G.: A synthetic recipe for OCR. In: 2019 International Conference on Document Analysis and Recognition (ICDAR), pp. 864–869 (2019). https://doi.org/10.1109/ICDAR.2019.00143
11. Graves, A., Fernández, S., Gomez, F., Schmidhuber, J.: Connectionist temporal classification: labelling unsegmented sequence data with recurrent neural networks. In: Proceedings of the 23rd International Conference on Machine Learning, ICML 2006, pp. 369–376. Association for Computing Machinery, New York (2006). https://doi.org/10.1145/1143844.1143891
12. Hasnat, M.A., Chowdhury, M.R., Khan, M.: An open source tesseract based optical character recognizer for Bangla script. In: 2009 10th International Conference on Document Analysis and Recognition, pp. 671–675 (2009). https://doi.org/10.1109/ICDAR.2009.62
13. Hochreiter, S., Schmidhuber, J.: Long short-term memory. Neural Comput. **9**(8), 1735–1780 (1997)
14. Huang, J., et al.: A multiplexed network for end-to-end, multilingual OCR, pp. 4545–4555 (2021). https://doi.org/10.1109/CVPR46437.2021.00452
15. Kuang, Z., et al.: MMOCR: a comprehensive toolbox for text detection, recognition and understanding, pp. 3791–3794 (2021). https://doi.org/10.1145/3474085.3478328
16. Liao, M., Zou, Z., Wan, Z., Yao, C., Bai, X.: Real-time scene text detection with differentiable binarization and adaptive scale fusion. IEEE Trans. Pattern Anal. Mach. Intell., 1 (2022). https://doi.org/10.1109/TPAMI.2022.3155612
17. Liu, W., Chen, C., Wong, K.Y.K., Su, Z., Han, J.: Star-Net: a spatial attention residue network for scene text recognition. In: Richard C. Wilson, E.R.H., Smith, W.A.P. (eds.) Proceedings of the British Machine Vision Conference (BMVC), pp. 43.1–43.13. BMVA Press, September 2016. https://doi.org/10.5244/C.30.43. https://dx.doi.org/10.5244/C.30.43
18. Medina, P., Fidalgo, E., Alegre, E., Alaiz, R., Jáñez-Martino, F., Bonnici, A.: Rectification and super-resolution enhancements for forensic text recognition. Sensors, 32–37 (2020). https://doi.org/10.3390/s20205850
19. Mindee: docTR: document text recognition (2021). https://github.com/mindee/doctr
20. Pal, A., Mustafi, A.: Vartani spellcheck - automatic context-sensitive spelling correction of OCR-generated Hindi text using BERT and Levenshtein distance. CoRR abs/2012.07652 (2020). https://arxiv.org/abs/2012.07652
21. Rostamizadeh, A., Talwalkar, A., DeSalvo, G., Jamieson, K., Li, L.: Efficient hyperparameter optimization and infinitely many armed bandits (2017)

22. Sakaguchi, K., Duh, K., Post, M., Durme, B.V.: Robsut Wrod reocginiton via semi-character recurrent neural network, pp. 3281–3287 (2017)

23. Shi, B., Bai, X., Yao, C.: An end-to-end trainable neural network for image-based sequence recognition and its application to scene text recognition. IEEE Trans. Pattern Anal. Mach. Intell. **39**(11), 2298–2304 (2017). https://doi.org/10.1109/TPAMI.2016.2646371

24. Silva, S.M., Jung, C.R.: License plate detection and recognition in unconstrained scenarios. In: Ferrari, V., Hebert, M., Sminchisescu, C., Weiss, Y. (eds.) ECCV 2018. LNCS, vol. 11216, pp. 593–609. Springer, Cham (2018). https://doi.org/10.1007/978-3-030-01258-8_36

25. Smith, R.: Tesseract blends old and new OCR technology. https://github.com/tesseract-ocr/docs/tree/master/. Accessed 24 May 2021

26. Tkachenko, I., Gomez-Krämer, P.: Robustness of character recognition techniques to double print-and-scan process. In: 2017 14th IAPR International Conference on Document Analysis and Recognition (ICDAR), vol. 09, pp. 27–32 (2017). https://doi.org/10.1109/ICDAR.2017.392

27. Vaswani, A., et al.: Attention is all you need 30 (2017). https://proceedings.neurips.cc/paper/2017/file/3f5ee243547dee91fbd053c1c4a845aa-Paper.pdf

28. Yang, S., Wang, Y., Chu, X.: A survey of deep learning techniques for neural machine translation. CoRR abs/2002.07526 (2020). https://arxiv.org/abs/2002.07526

29. Zacharias, E., Teuchler, M., Bernier, B.: Image processing based scene-text detection and recognition with tesseract. CoRR abs/2004.08079 (2020). https://arxiv.org/abs/2004.08079

30. Zhou, J., Cao, Y., Wang, X., Li, P., Xu, W.: Deep recurrent models with fast-forward connections for neural machine translation. CoRR abs/1606.04199 (2016). http://arxiv.org/abs/1606.04199

Real-Time Hand Gesture Identification in Thermal Images

James M. Ballow⬭ and Soumyabrata Dey(⊠)⬭

Clarkson University, 8 Clarkson Avenue, Potsdam, NY 13699, USA
{ballowjm,sdey}@clarkson.edu

Abstract. Hand gesture-based human-computer interaction is an important problem that is well explored using color camera data. In this work we proposed a hand gesture detection system using thermal images. Our system is capable of handling multiple hand regions in a frame and process it fast for real-time applications. Our system performs a series of steps including background subtraction-based hand mask generation, k-means based hand region identification, hand segmentation to remove the forearm region, and a Convolutional Neural Network (CNN) based gesture classification. Our work introduces two novel algorithms, *bubble growth* and *bubble search*, for faster hand segmentation. We collected a new thermal image data set with 10 gestures and reported an end-to-end hand gesture recognition accuracy of ∼97%.

Keywords: Hand detection · Hand gesture classification · Human computer interaction · Center of palm · Wrist points

1 Introduction

Communication between human and computer via hand gestures has been studied extensively and continues to be a fascination in the computer vision community. This is not a surprise given the proliferation in use of artificial intelligence over the past decade to improve the lives of many through the development of *smart* systems. For humans to convey intention to computers, studies began with the use of external devices [1,6,17,20] to enhance focus on particular regions of hands so as to limit the number of features required to interpret hand movement sequences into pre-determined messages. Over time, more advanced equipment was introduced to eliminate the need for external devices to be attached to a user's hands to isolate hand movements; equipment like the use of a depth camera [14,18,19], high resolution RGB cameras [2,3,15], and (less-frequently) thermal cameras [13].

Even though hand gesture detection is a well-explored problem using some modalities of data such as RGB camera images, there have been a very limited number of works on thermal data [8,12,13]. These studies have typically required multiple sensors, a fixed location of the hand in frame to estimate wrist points, or have a user wear clothing to separate the hand from the forearm.

S. Sclaroff et al. (Eds.): ICIAP 2022, LNCS 13232, pp. 491–502, 2022.
https://doi.org/10.1007/978-3-031-06430-2_41

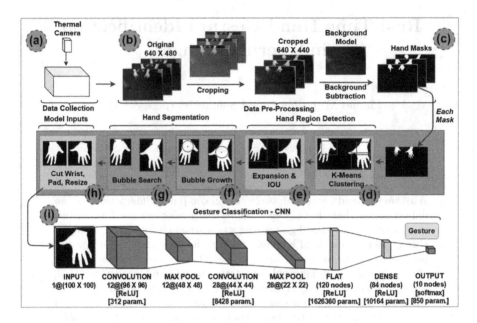

Fig. 1. End-to-end flowchart for (a) data capture, (b) frame cropping and background subtraction, (c) hand mask generation, (d-e) hand region detection, (f-g) hand segmentation using key-points, (e) final model input formatting, (i) and gesture classification.

Thermal modality can complement RGB data modality because it is not affected by different lighting conditions and skin color variance. Moreover, thermal data based analyses can be extended to swipe detection techniques by temperature tracing on natural surfaces [4]. Therefore, future research can benefit by using multi-modal (thermal and RGB) data for a robust hand gesture detection system. To this end, through research on hand gesture detection techniques using thermal data is essential.

Efficient hand segmentation is vital to the success of thermal camera based hand detection. This is because thermal images lack many distinguishable features such as color and textures, and including regions from other heated objects such as a forearm can reduce the classification accuracy. Our algorithmic pipeline uses background subtraction in the data pre-processing stage for generating a hand mask, k-means clustering and overlapping cluster grouping for hand region isolation of each potential hand region, center of palm and wrist point detection for hand segmentation (removing forearm), and a CNN-based model gesture classification (Fig. 1). In this process we introduced two novel algorithms, *bubble growth* and *bubble search*, for hand segmentation. The main contributions of this paper are as follows: 1) Collection of a new thermal hand gesture data set from 23 users performing the same 10 gestures. 2) *Bubble Growth* method, which uses a distance transform and hand-forearm contour to expand a circle (bubble) to the maximum extent possible inside the hand. 3) *Bubble Search* method, which was inspired by the use of an expansion of the maximum inscribed circle and

Fig. 2. Gestures to test the proposed algorithms and train our CNN model.

Table 1. Data set composition by gesture number (G_x) and left/right hands.

	Users	G_1	G_2	G_3	G_4	G_5	G_6	G_7	G_8	G_9	G_{10}	Left	Right
Training data	20	1536	1576	1438	1544	1183	1459	1358	1391	1524	1285	6657	7637
Test data	3	103	57	34	92	115	30	17	92	138	214	517	355

a threshold distance of two consecutive contour points as detailed in [2], but includes additional constraints, a reference point, and an evolving (in lieu of fixed) bubble expansion. 4) Developing an end-to-end real-time hand gesture detection system that can process multiple hands at a frame-rate of 8 to 10 frames per second (fps) with high accuracy ($\sim$97%.). Our *bubble growth* and *bubble search* methods are superior to other methods because they neither use nor require knowledge of any projection [5,15], angle of hand rotation [15], finger location or fixed values for bubble radii [2,21], degree of palm roundness [19]. The combined average speed of our methods (*bubble growth* = 0.012 s/hand, *bubble search* = 0.007 s/hand; total = 0.019 s/hand).

2   Methods

This paper proposes a process (Fig. 1) that can perform real-time hand detection and gesture classification from a thermal camera video feed. The process is divided into 5 major parts: (1) data collection; (2) data pre-processing; (3) hand region detection; (4) hand segmentation, and (5) gesture classification.

2.1   Data Collection

All thermal hand gesture video data is collected with a Sierra Olympic Viento-G thermal camera with a 9 mm lens. The video frames are recorded in indoor conditions (temperature between 65° to 70 °F) at 30 fps and stored as 16-bit TIFF images with a 640 × 480 pixel resolution. The camera is fixed to a wooden stand and oriented downwards towards a tabletop.

Data was collected from 23 users demonstrating 10 pre-defined gestures with left and right hands to develop training and test data sets. Separate sets of users contributed to the training and test data sets to demonstrate that our method is user-agnostic. Figure 2 illustrates all 10 gestures, and Table 1 lists the data we have collected and divided into a training data set, testing data set.

For this paper, we also used an external data called *Finger Digits 0–5* [10] set to assess how our process generalizes. From this set we tested gesture 1, 2, 4, 5, and 9 (2000 images per gesture). This data set was not used in training or in any other experimentation in this project.

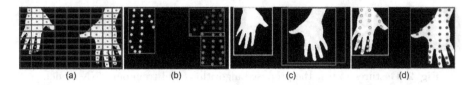

Fig. 3. Hand region detection: (a-b) centers of white pixel in all grids are clustered using k-means. (c-d) optimal number of hand regions are detected.

2.2   Data Pre-processing

Without loss of generality to image size, we reduced the size of our images from 640×480 to 640×440 to reduce camera scope to the boundaries of the table top to make background subtraction simpler. Images were also converted to 8-bit JPG images for ease of viewing and using certain python packages (e.g., *OpenCV*).

A MOG2 background subtraction model from *OpenCV* was used to generate hand masks from each thermal image. We initialized the model with frames that, at the beginning of data capture, contain no hands or heated objects and have a table at constant room temperature. The model is updated over the video sequence when there are no pixels found in the hand mask, essentially allowing only pixels associated with slight change in room temperature to be updated. Each mask is binarized (black and white pixels) using Otsu's method to select an appropriate threshold value.

2.3   Hand Region Detection

Hand regions are identified in hand masks using tightly bounding boxes around each hand object (which may or may not contain any length of forearm) using a two-step process detailed below.

First, a k-means clustering algorithm (Fig. 3) is used to identify contiguous objects in the hand mask using a number of centroids estimated with a silhouette analysis optimal cluster-finding technique. To achieve real-time processing speed, we expedite this step by reducing the number of points to consider when clustering. To do this, we encapsulate all white pixels in the hand mask with a grid that is subdivided into equally-sized partitions. For each coupon containing at least one white pixel, the center of mass is calculated to reduce a coupon's points down to a single point. The silhouette analysis [11] identifies the optimal cluster number by performing k-means clustering for different k values (k in the range of 2 to 3) and selecting the optimal k for which the highest silhouette score is calculated.

Second, a bounding box is placed initially around each cluster but is expanded in all dimensions equally until the box entirely contains a set of contiguous white pixels in the hand mask. If the number of centroids selected is larger than the number of hands in the hand mask (e.g., poor hand mask generation), then one or more regions will be bounded by multiple boxes. Removal of these duplicate boxes is performed using intersection-over-union (IOU) and a threshold of 0.7—the boxes that remain after IOU are the hand regions.

Algorithm 1. Center of Palm Detection

Output: C_{COP}, R_{COP}

```
 1: function BUBBLEGROWTH(H_part, C_est, R_est)
 2:     C_cand, R_cand ← C_est, R_est
 3:     C_COP, R_COP ← C_est, R_est
 4:     C_visited ← {}
 5:     repeat
 6:         for h ← H_part do
 7:             C_cand ← SHORTADVANCEMENT(C_COP, h, F_pace)
 8:             if C_cand is in C_visited then
 9:                 continue
10:             end if
11:             R_cand ← SHORTESTDISTANCE(C_cand, H_part)
12:             if R_cand > R_COP then
13:                 C_COP, R_COP ← C_cand, R_cand
14:                 SIGNALBUBBLEMOVED()
15:                 break
16:             end if
17:             C_visited ← C_cand
18:         end for
19:     until no SIGNALBUBBLEMOVED
20:     return C_COP, R_COP
21: end function
```

2.4 Hand Segmentation

A hand region may include variable forearm length which can hinder high-accuracy gesture classification due to lack of forearm agnosticism. This can be avoided by including gesture samples with variable lengths of forearms in the training set, but this would be a costly endeavor. Instead, as illustrated in Fig. 4 we algorithmically sever the hand region from the forearm at the wrist and thus removed all forearm-related data variation. The process uses two novel algorithms: Bubble Growth (to find COP) and Bubble Search (to find WP).

Reference Point Determination: Bubble Growth and Bubble Search require a reference point (C_{ref}) located on the edge through which the hand enters the frame, or *hand penetration edge*. To get this point, we estimate the largest contiguous array of white pixels on each of the four edges of the hand region to be the hand penetration edge. The extreme ends of the array of white pixels on the hand penetration edges are set as the reference point edges ($C_{ref,edges}$) and the midpoint of these points yields C_{ref}.

Bubble Growth: Given a subset of contours (H_{part}) from the entire set of contours for the hand region (H_{all}) and an initial estimate for COP (C_{est}), Algorithm 1 moves around the space in the palm to find the COP (C_{COP}). We form H_{part} by taking a sparse subset of H_{all} in a fashion that preserves the overall shape of the hand region; doing this optimized the cost of Algorithm 1. We compute C_{est} by first calculating the distance transform (DT) [16] of the

---

Algorithm 2. Wrist Point Detection

Output: W_1, W_2
Requires: $R_{exp}, D_{min}, D_{max}, i_{max}$

1: **function** BUBBLESEARCH($H_{all}, C_{COP}, R_{COP}, C_{Ref}, C_{Ref,edges}$)
2:     $D_{Ref} \leftarrow$ DISTANCE(C_{COP}, C_{Ref})
3:     $D_{max}, D_{min}, R_{exp} \leftarrow$ APPLYSCALARS(R_{COP})
4:     $L_{meetsCriteria} \leftarrow \{\}$
5:     **if** $R_{COP} > D_{Ref}$ **then**
6:         $W_1, W_2 \leftarrow C_{Ref,edges}$
7:     **else**
8:         **repeat**
9:             $H_{inside} \leftarrow$ POINTSINSIDECIRCLE($H_{all}, C_{COP}, R_{exp}$)
10:            **for** $i \leftarrow$ INDICIES(H_{inside}) **do**
11:                $P_1, P_2 \leftarrow$ GETELEMENTS($H_{inside}, i, i+1$)
12:                $M_{1,2} \leftarrow$ MIDPOINT(P_1, P_2)
13:                $D_1, D_2 \leftarrow$ (DISTANCE(P_1, P_2), DISTANCE($M_{1,2}, C_{Ref}$))
14:                **if** $D_{min} < D_1 < D_{max}$ **and** $D_2 < D_{Ref}$ **then**
15:                    $L_{meetsCriteria} \leftarrow ([D_1, D_2, P_1, P_2])$
16:                **end if**
17:            **end for**
18:            **if** $|L_{meetsCriteria}| > 1$ **then**
19:                $W_1, W_2 \leftarrow$ GETPOINTSWITHSMALLESTD$_2$($L_{meetsCriteria}$)
20:                **break**
21:            **else if** i_{max} reached **then**
22:                $W_1, W_2 \leftarrow C_{Ref,edges}$
23:            **else**
24:                $R_{exp} \leftarrow F_2 \times R_{exp}$
25:            **end if**
26:        **until** i_{max} reached
27:    **end if**
28:    **return** W_1, W_2
29: **end function**

---

entire hand region, obtaining the maximum DT value (η_{max}), and selecting all points in the hand region which have a DT value (η) such that $\eta \geq 0.80 * \eta_{max}$. We select the point in this set furthest from C_{ref} as the C_{est}.

Starting with C_{est} as the first C_{COP}, the algorithm tries to find the next center candidate (C_{cand}). This is performed by SHORTADVANCEMENT, which attempts to move C_{COP} by a fraction of a distance (F_{pac}) along the connecting vector between C_{COP} and each contour point $h \in H_{part}$ using the following equation: $\overrightarrow{C_{cand}} = \overrightarrow{C_{COP}} + F_{pace} * (\overrightarrow{h} - \overrightarrow{C_{COP}})$. The first h resulting in a C_{cand} where $R_{cand} > R_{COP}$, forces C_{COP} to update to C_{cand} and the remaining $h \in H_{part}$ are skipped to make way for the next iteration. A growing list of centers that have been visited ($C_{visited}$) is used to ensure no Euclidean distance (used to calculate R_{cand}) is unnecessarily performed because the resulting C_{cand} will have already been visited and assessed from a past iteration. When $\forall h \in H_{part}$ the bubble does not move, Algorithm 1 terminates.

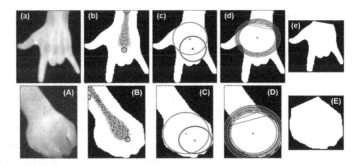

Fig. 4. Hand segmentation with C_{Ref} on a single edge of frame (a–e) and C_{Ref} split between two edges (A–E). (a, A) original image; (b, B) DT and C_{Ref} (purple) to find C_{est} (red); (c, C) C_{est} (red) and C_{COP} (green); (d, D) R_{exp} expanding to find appropriate W_1, W_2 (orange); (e, E) arm removed. (Color figure online)

Bubble Search: Given H_{all} and the C_{COP}, Algorithm 2 searches for WP (W_1, W_2). First, R_{COP} is expanded to R_{exp} (i.e., $R_{exp} = F_1 \times R_{COP}$) but continues expanding incrementally (i.e., $R_{exp} = F_2 \times R_{exp}$). R_{exp} will expand, searching for a pair of contour points that exist inside of the expanded bubble (i.e., $h \in H_{inside}$ where $H_{inside} \subseteq H_{all}$). We say that W_1 and W_2 have been found if two contiguous $h \in H_{inside}$ meet the following criteria:

1. $|\overline{W_1 W_2}| > D_{min}$, where $D_{min} = F_3 \times R_{COP}$.
2. $|\overline{W_1 W_2}| < D_{max}$, where $D_{max} = F_4 \times R_{COP}$.
3. $|\overline{M_{1,2} C_{ref}}| < |\overline{C_{COP} C_{ref}}|$, where $M_{1,2}$ is MIDPOINT(W_1, W_2).

D_{max} and D_{min} are empirically determined limits of the distance between two WP in the scale of R_{COP}.

Algorithm 2 terminates if: (1) W_1, W_2 are found; or (2) $i_{max} = 10$ is reached; or (3) if before searching for W_1 and W_2 it is determined that $R_{COP} < |\overline{C_{COP} C_{ref}}|$. In case 2 and 3, $C_{Ref,edges}$ are selected as W_1 and W_2. Once WP are identified, all points defined by an inequality whose boundary passed through W_1 and W_2 and is opposite the side containing the COP is erased (pixels set to 0) to eliminate the forearm from the hand region. Finally, the border around the region is squeezed to the tightest bounding box, and the images is padded by 5 pixels and resized to 100×100. At this point, the hand region has been satisfactorily standardized as a CNN model input.

2.5   Gesture Classification

A CNN model was trained to identify 10 pre-defined hand gestures. The architecture and the training parameters are summarized at the bottom of Fig. 1. The model was trained using a *categorical cross-entropy* loss function and an *adam* optimizer with a variable learning rate.

3   Experiments and Results

All experiments in this paper are performed on a desktop with 32 GB of RAM, AMD Ryzen 7, 3700X, 8-core processor at 3.59 GHz on a 64-bit Windows 10 platform with build number 19043.1165. All CNN model training was performed through Google Co-laboratory using a GPU farm stationed at Google.

For hand region detection, The number of centroids and the granularity of the grid in the silhouette analysis affects the cost of clustering. We observed the cost of finding an optimal number of centroids using silhouette analysis for different grid sizes on a single hand gesture over 7 frames. To keep cost low while also ensuring multiple hands are correctly isolated, the grid size should be fixed to a size of 10×10 and limited to at most 3 centroids. This limits the number of hands that can be detected in a single image to at most 3.

$|H_{part}|$ in *bubble growth* is selected to strike a balance between the cost and accuracy. Table 3 justifies the selection of $|H_{part}| = 30$. Experimentation has shown that F_{pace} affects both the convergence speed and the accuracy of the center of palm. Larger values for F_{pace} cause Algorithm 1 to terminate prematurely, while lower value increases the computation cost. While $F_{pace} = 0.10$ worked best for us, an optimal value selected is left as a future work.

A few experiments were performed to find optimize values for $F_1 - F_4$ used in Algorithm 2 using 1000 images from 3 different users. A value of 1.2 has been proposed for factor F_1 in [2], but we propose $F_1 = 1.4$ after performing a study to assess the average accuracy and cost of Algorithm 2 keeping $F_2 = 1.01$ as constant. F_1 is varied between $(1.10, 1.45)$ in increments of 0.05 (Fig. 5, (a)). Similarly, F_2 is varied between $(1.001, 1.015)$ in increments of 0.002 keeping $F_1 = 1.4$ as constant (Fig. 5, (b)). Values of $F_3 = 1.1$ and $F_4 = 1.9$ are determined by plotting values of $\|\overline{W_1 W_2}\|$ divided by R_{COP} and selecting factors that bound 99% of all plotted values (Fig. 5, (c-d)). A value of $i_{max} = 10$ is determined such that it is small enough to minimize calculation time and large enough to allow searching for a significant number of hand samples.

Bubble Growth and Bubble Search were tested using 1532 hand samples obtained using all 1,217 thermal images. The overall hand detection success rate of 95.64% is on par with the 93.1% to 98.7% reported in [7] and is detailed in Table 4. The term *success* is defined for Algorithm 1 as: (a) C_{COP} exists in palm region; (b) $R_{COP} \approx R_{max-inscribed}$; (c) bubble contains ≤ 20 black pixels. The

Table 2. Average cost (sec) to detect hand region for different grid sizes and number of centroids; average computed over 7 contiguous frames.

Centroids	5×5	10×10	15×15	20×20
2 only	0.0128	0.0161	0.0176	0.0235
2 to 3	0.0236	0.0298	0.0365	0.0397
2 to 4	0.0357	0.0451	0.0544	0.0576
2 to 5	0.0486	0.0624	0.0749	0.0822

Table 3. Number of contour points in H_{part} vs accuracy of Algorithm 1 and cost (sec); number of bad bubbles reported as percentage of 291 images tested.

Points	Bad bubbles	Avg cost
10	95.9%	0.0041
20	32.3%	0.0058
30	6.9%	0.0085
40	5.2%	0.0112
50	1.7%	0.0141

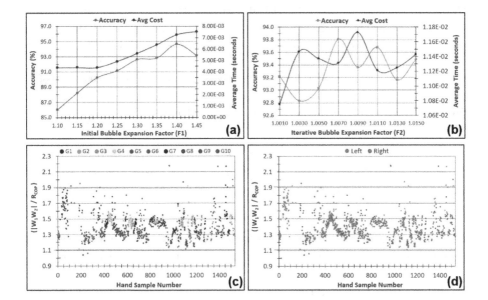

Fig. 5. Plots to determine (a) F_1, (b) F_2, (c-d) F_3, and F_4.

term *success* is defined for Algorithm 2 as: W_1 and W_2 are between C_{COP} and C_{Ref}; (b) $\overline{W_1 W_2}$ segments the hand and forearm regions as expected (Table 2).

The gesture classification model architecture was based on an LeNet-1 architecture. It was trained solely with the training data collected for this study and achieved a training accuracy of 99.9%. As seen in Table 5, this model exhibits an overall testing accuracy of 96.9% with our testing set. Our model was tested with *finger-digit-05*; results are limited to 5 gestures because these were the only gestures matching those with which our model was trained. This is slightly better than 96.7% reported in [2], which also used a model to classify 10 different gestures. This is also better than the 90.5% recognition accuracy reported in [15].

We used data augmentation to randomly rotate hand samples between 0° and 360° to fit the model to hands at any orientation. Our process standardizes the size of hands when resizing the image, making our model zoom-agnostic.

Table 4. Success and cost (sec/image) of Algorithm 1 (BG) and Algorithm 2 (BS). Values listed overall (All) and by gesture number (G_x).

	Samples	$BG_{success}$	$BS_{success}$	BG_{avg}	BG_{min}	BG_{max}	BS_{avg}	BS_{min}	BS_{max}
All	1532	95.64%	96.22%	0.012	0.001	0.120	0.007	2E−05	0.090
G_1	325	96.92%	99.69%	0.009	0.002	0.030	0.004	0.003	0.060
G_2	64	98.48%	98.48%	0.014	0.001	0.030	0.006	0.001	0.040
G_3	67	98.51%	94.03%	0.010	0.003	0.030	0.008	0.005	0.070
G_4	153	100.0%	100.0%	0.012	0.005	0.040	0.004	2E−05	0.050
G_5	182	87.36%	92.86%	0.012	0.003	0.030	0.018	0.005	0.090
G_6	71	97.22%	98.61%	0.010	0.004	0.027	0.005	0.004	0.065
G_7	55	98.18%	94.55%	0.014	0.004	0.030	0.005	0.003	0.065
G_8	127	96.06%	99.21%	0.010	0.003	0.030	0.004	0.005	0.013
G_9	239	92.86%	99.58%	0.016	0.004	0.115	0.005	0.001	0.090
G_{10}	249	97.21%	86.85%	0.010	0.004	0.120	0.011	0.005	0.080

Table 5. Accuracy of CNN gesture classification model overall (G_{model}) and by gesture number (G_x).

	G_{model}	G_1	G_2	G_3	G_4	G_5	G_6	G_7	G_8	G_9	G_{10}
Our testing set accuracy (%)	96.9	97.1	98.3	97.1	96.7	99.1	90.0	94.1	92.4	97.8	97.7
finger-digit-05 accuracy (%)	97.3	96.8	90.4	−	99.6	99.9	−	−	−	100	−

We experimented with batch sizes between 16 and 32 [9] to obtain the best training results and a variable learning rate between 1E−06 and 1E−R04.

4   Discussion and Conclusion

This paper proposes a real-time end-to-end system that can detect hand gestures from the video feed of a thermal camera. In that process the work introduces two novel methods for center of palm detection (bubble growth) and wrist point detection (bubble search) which are fast, accurate, and invariant to hand shape, hand orientation, arm length, and sizes (closeness to camera).

To maintain the real-time processing speed, our method can simultaneously detect hand gestures of 3 regions. However, this can be easily relaxed by introducing more processing power and distributed computing. This can also be achieved by using a heuristic to approximate the optimal number of centroids to use in clustering in lieu of testing a set of per-defined centroids using silhouette analysis.

Finally, we experimentally validated that our algorithm is user-agnostic (i.e. the algorithm can identify hand gestures of users that are not included in the training samples). Our system is highly accurate in detecting center of palm, wrist points, and hand gestures from hand masks produced from thermal images.

Even though there are a lot of hand gesture detection algorithms available using color video data, only a few techniques solve the problem with thermal data. Our methods show that even though limited in features, thermal video is a viable medium to capture hand gestures for accurate gesture recognition. Moreover, future research can combine thermal with other data modalities (e.g., RGB, depth) for an even more robust hand gesture detection system.

References

1. Bellarbi, A., Benbelkacem, S., Zenati, N., Belhocine, M.: Hand gesture interaction using color-based method for tabletop interfaces. In: 2011 IEEE 7th International Symposium on Intelligent Signal Processing, pp. 1–6, September 2011. https://doi.org/10.1109/WISP.2011.6051717
2. Chen, Z., Kim, J., Liang, J., Zhang, J., Yuan, Y.: Real-time hand gesture recognition using finger segmentation. Sci. World J., 9 pages (2014). https://doi.org/10.1155/2014/267872
3. Dardas, N.H., Georganas, N.D.: Realtime hand gesture detection and recognition using bag-of-features and support vector machine techniques. IEEE Trans. Instrum. Meas. 60, 11 (2011). https://doi.org/10.1109/TIM.2011.2161140
4. Gately, J., Liang, Y., Wright, M.K., Banerjee, N.K., Banerjee, S., Dey, S.: Automatic material classification using thermal finger impression. In: Ro, Y.M., et al. (eds.) MMM 2020. LNCS, vol. 11961, pp. 239–250. Springer, Cham (2020). https://doi.org/10.1007/978-3-030-37731-1_20
5. Grzejszczak, T., Kawulok, M., Galuszka, A.: Hand landmarks detection and localization in color images. Multimedia Tools Appl. 75(23), 16363–16387 (2015). https://doi.org/10.1007/s11042-015-2934-5
6. Ibarguren, A., Maurtua, I., Sierra, B.: Layered architecture for real time sign recognition: hand gesture and movement. Eng. Appl. Artif. Intell. 23(7), 1216–1228 (2010). https://doi.org/10.1016/j.engappai.2010.06.001
7. Islam, M.M., Siddiqua, S., Afnan, J.: Real time hand gesture recognition using different algorithms based on American sign language. In: 2017 IEEE International Conference on Imaging, Vision Pattern Recognition (icIVPR), pp. 1–6 (2017). https://doi.org/10.1109/ICIVPR.2017.7890854
8. Kim, S., Ban, Y., Lee, S.: Tracking and classification of in-air hand gesture based on thermal guided joint filter. Sensors 17(12), 166 (2017). https://doi.org/10.3390/s17010166
9. Meng, L., Li, R.: An attention-enhanced multi-scale and dual sign language recognition network based on a graph convolution network. Sensors 21(4) (2021). https://doi.org/10.3390/s21041120. https://www.mdpi.com/1424-8220/21/4/1120
10. O'Shea, R.: Finger digits 0–5, November 2019. https://www.kaggle.com/roshea6/finger-digits-05
11. Rousseeuw, P.J.: Silhouettes: a graphical aid to the interpretation and validation of cluster analysis. J. Comput. Appl. Math. 20, 53–65 (1987). https://doi.org/10.1016/0377-0427(87)90125-7. https://www.sciencedirect.com/science/article/pii/0377042787901257
12. Sato, Y., Kobayashi, Y., Koike, H.: Fast tracking of hands and fingertips in infrared images for augmented desk interface. In: Proceedings Fourth IEEE International Conference on Automatic Face and Gesture Recognition (CatNoPR00580), pp. 462–467 (2000). https://doi.org/10.1109/AFGR.2000.840675
13. Song, E., Lee, H., Choi, J., Lee, S.: AHD: thermal image-based adaptive hand detection for enhanced tracking system. IEEE Access 6, 12156–12166 (2018). https://doi.org/10.1109/ACCESS.2018.2810951

14. Sridhar, S., Mueller, F., Oulasvirta, A., Theobalt, C.: Fast and robust hand track-ing using detection-guided optimization. In: 2015 IEEE Conference on Computer Vision and Pattern Recognition (CVPR), pp. 3213–3221 (2015). https://doi.org/10.1109/CVPR.2015.7298941

15. Stergiopoulou, E., Papamarkos, N.: Hand gesture recognition using a neural net-work shape fitting technique. Eng. Appl. Artif. Intell. **22**, 1141–1158 (2009). https://doi.org/10.1016/j.engappai.2009.03.008

16. Strutz, T.: The distance transform and its computation (2021). https://arxiv.org/abs/2106.03503

17. Tompson, J., Stein, M., Lecun, Y., Perlin, K.: Real-time continuous pose recovery of human hands using convolutional networks, vol. 33, September 2014. https://doi.org/10.1145/2629500

18. Wu, D., et al.: Deep dynamic neural networks for multimodal gesture segmentation and recognition. IEEE Trans. Pattern Anal. Mach. Intell. **38**, 1 (2016). https://doi.org/10.1109/TPAMI.2016.2537340

19. Yao, Z., Pan, Z., Xu, S.: Wrist recognition and the center of the palm estima-tion based on depth camera. In: 2013 International Conference on Virtual Reality and Visualization, pp. 100–105, September 2013. https://doi.org/10.1109/ICVRV.2013.24

20. Yeo, H.-S., Lee, B.-G., Lim, H.: Hand tracking and gesture recognition system for human-computer interaction using low-cost hardware. Multimedia Tools Appl. **74**(8), 2687–2715 (2013). https://doi.org/10.1007/s11042-013-1501-1

21. Zhou, Y., Jiang, G., Lin, Y.: A novel finger and hand pose estimation technique for real-time hand gesture recognition. Pattern Recogn. **49**, 102–114 (2016). https://doi.org/10.1016/j.patcog.2015.07.014

Graph-Based Generative Face Anonymisation with Pose Preservation

Nicola Dall'Asen[1,2]([✉]), Yiming Wang[3][iD], Hao Tang[4], Luca Zanella[3], and Elisa Ricci[2,3][iD]

[1] University of Pisa, Pisa, Italy
nicola.dallasen@phd.unipi.it
[2] University of Trento, Trento, Italy
[3] Fondazione Bruno Kessler, Trento, Italy
[4] ETH Zürich, Zürich, Switzerland

Abstract. We propose AnonyGAN, a GAN-based solution for face anonymisation which replaces the visual information corresponding to a source identity with a condition identity provided as any single image. With the goal to maintain the geometric attributes of the source face, i.e., the facial pose and expression, and to promote more natural face generation, we propose to exploit a Bipartite Graph to explicitly model the relations between the facial landmarks of the source identity and the ones of the condition identity through a deep model. We further propose a landmark attention model to relax the manual selection of facial landmarks, allowing the network to weight the landmarks for the best visual naturalness and pose preservation. Finally, to facilitate the appearance learning, we propose a hybrid training strategy to address the challenge caused by the lack of direct pixel-level supervision. We evaluate our method and its variants on two public datasets, CelebA and LFW, in terms of visual naturalness, facial pose preservation and of its impacts on face detection and re-identification. We prove that AnonyGAN significantly outperforms the state-of-the-art methods in terms of visual naturalness, face detection and pose preservation. Code and pretrained model are available at https://github.com/Fodark/anonygan.

1 Introduction

In the era of deep learning, the availability of large scale data has undoubtedly brought technological advances. However, the very same fact has also fostered the growing concern regarding privacy issues. Visual privacy preservation is mostly achieved via video redaction methods by obfuscating the personally identifiable information (PII) of a data subject, whose face is often the most identity-informative part. Classic face anonymisation techniques, e.g., blurring [5] or pixelation [8], can effectively remove PII. However, this comes at a high

Supplementary Information The online version contains supplementary material available at https://doi.org/10.1007/978-3-031-06430-2_42.

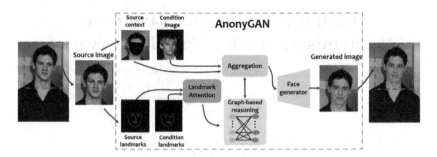

Fig. 1. AnonyGAN anonymises faces by generating similar faces to *any* condition image while preserving the facial pose of the source image, with a novel *landmark attention* model and a face generator with *graph-based landmark-landmark reasoning*.

cost of degrading other vision-related tasks, in particular for the action/emotion recognition where the poses play an essential role.

Thanks to recent advances in generative adversarial networks (GANs) [9], several anonymisation solutions have been proposed to generate *natural-looking* faces that correspond to different identities [4,7,18,26], while preserving the original facial poses using the landmarks as the guidance [13,22,27]. Yet, it is challenging to produce realistic images in this context due to the lack of ground-truth images in real-world, i.e., images of different persons with the same facial pose. This also draws a fundamental distinction to the pose-guided image generation task [6,10,28], whose ground-truth images, i.e. images of the same person with varying poses, are available to provide direct pixel-level supervision. While for the pose preservation, the main challenge lies in the relation reasoning between the condition pose and the source pose, where graph-based modelling has demonstrated its strengths in such geometric reasoning [28]. Finally, the landmarks choice impacts the visual naturalness, pose preservation and anonymisation, which is often heuristically handled with no optimality guaranteed [13,22].

To address the above-mentioned challenges, we propose a novel graph-based GAN architecture, **AnonyGAN**, to perform landmark-guided face anonymisation (see Fig. 1). Our network takes as input a *context* image and a condition image, as well as the facial landmarks extracted from the source and target images. The *context* image is the source image with the face (excluding the forehead) masked out, and it provides the necessary contextual information, e.g. the skin tone and the background, for naturally blending the generated faces to the source image. In order to improve the pose preservation, we propose to first disentangle the geometrical reasoning from the appearance, where the source landmarks and the condition landmarks are modelled as a bipartite graph with Graph Convolution Networks (GCNs). The appearance of the condition image is then aggregated to the pose reasoning module to generate more natural faces with the source facial pose preserved. Moreover, we introduce a novel landmark attention model to allow the network to automatically learn the importance of the facial landmarks, avoiding sub-optimal manual decisions. Finally, we propose a novel hybrid training strategy to address the training difficulty caused by the lack of ground-truth images. We form the source and condition pairs using both the same image and different images to facilitate the appearance learning. The former provides direct pixel-level supervision, while the latter applies a weak context-level supervision by exploiting the high-level features extracted from the appearance discriminator. We validate **AnonyGAN** on two public datasets, i.e., CelebA [21] and LFW [12], and demonstrate that **AnonyGAN** can greatly improve the generated faces in terms of visual naturalness and pose preservation compared to baselines, i.e., blurring and pixelation, and the state-of-the-art method [22]. To summarise, our main contributions are listed below:

- We propose a novel GAN-based architecture, **AnonyGAN**, to perform landmark-guided face anonymisation, achieving the *highest visual quality* with the *best pose preservation* on two benchmark dataset.
- We propose to exploit a graph formulation on the landmarks of the source and condition face images to perform geometric reasoning using GCNs, and prove its effectiveness in improving the pose preservation.
- We propose a landmark attention model to automatically weight the facial landmarks, achieving a higher visual quality and perception performance.
- We propose a hybrid training strategy with a strong pixel-level and a weak context-level supervision to address the training without ground-truth images, achieving the best visual quality.

2   Related Work

We discuss recent face obfuscation techniques for anonymising visual data and briefly cover related works addressing pose-guided image generation.

Visual Anonymisation often refers to irreversible obfuscation techniques for removing PII of the data subject in visual content, a.k.a. de-identification in some works [7]. Many works anonymise visual data by obfuscating the faces, the most privacy-concerning content, using classic techniques, such as blurring via filters [5] or pixelation by enlarging the pixels [8]. Such methods remove the most identifiable visual information, but greatly compromise other perception tasks, such as object detection and action recognition. Recently, with the progresses in GANs, face anonymisation techniques have advanced by generating realistic faces of a different identity, leaving intact most of the non-identifiable visual and geometrical attributes [7,13,22,26,27]. Sun *et al.* [26] proposed a two-step face inpainting technique for anonymisation by first generating the 68 facial landmarks, and then synthesising the faces guided by the landmarks. With a blurred face as condition, the generated faces have a rather high visual quality, yet resemble to the original face. DeepPrivacy [13] exploits the generator of StyleGAN [14] to generate a face of a fake identity, conditioned both on the context image with the face masked out and on 7 facial landmarks. The generated faces are limited in anonymisation and pose preservation. Conditional GANs are also proposed to explicitly control the identity on the generated faces [22, 27]. Recently, CIAGAN [22] has introduced an identity discriminator to enforce the generated faces to be different from the source image, achieving a better anonymisation performance, while the visual naturalness and pose preservation are not yet satisfactory although the subset of landmarks are carefully chosen. Moreover, CIAGAN cannot be easily applied to unknown condition identity, as the condition identities are encoded within the network during the training.

AnonyGAN takes any condition image as a reference for appearance, and improves the visual naturalness and facial pose preservation by first reasoning on landmark to landmark relations, and then generating the image aggregating the appearance features. Moreover, the choice of facial landmarks impacts both the image quality and the perception task with a certain trade-off. This aspect has not been properly addressed in the literature yet [22]. Instead, in this work we propose a landmark attention model to allow the network to learn the relative importance among the landmarks for face anonymisation.

Pose-Guided Image Generation is a related task to visual anonymisation where the main challenge is due to the pose deformation between the source and the target image. Modelling the relations between the source pose and the target pose is the key to solve this challenging task. Existing methods such as [1–3,19,20,25,29,31] are based on the stacking of several convolutional layers, which can only leverage the relations between the source pose and the target pose locally. For instance, Zhu *et al.* [31] proposed a Pose-Attentional Transfer Block, in which the source and target poses are simply concatenated and then fed into an encoder to capture their dependencies. Different from existing methods for modelling the relations between the source and target poses in a localised manner, BiGraphGAN [28] reasons and models the crossing long-range relations between different features of the source pose and target pose in a bipartite graph, which are then used for the person image generation process. In general, for pose-guided image generation, the ground-truth images of the same person with various poses are available. In contrast, there is no set of images of different identities with the exact same pose, thus making the face anonymisation a more challenging problem due to the lack of direct pixel-level supervision.

3   Proposed Method

The proposed **AnonyGAN**, as illustrated in Fig. 2, aims to anonymise a source image of any identity I_s by replacing the face with any condition identity provided as a condition image I_c while preserving the facial pose of source image Lm_s. Both the source image I_s and the condition image I_c are preprocessed to extract their facial landmarks Lm_s and Lm_c, respectively, for the graph-based landmark-landmark reasoning. Both facial poses are defined by 68 landmarks and encoded as 68-channel images as in [23] with a channel per landmark. In order to generate the face with a consistent context of the source image, in particular, for the skin tone, we also input to the network the context of the source image I_{sc}, i.e., the background image with the face masked out by the contour defined by the facial landmarks.

The main components of our network are: a *Landmark Attention Model* that takes as input Lm_s and Lm_c defined by the full 68 landmarks, and learns to

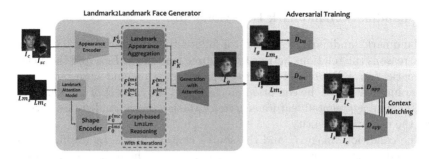

Fig. 2. Network architecture of **AnonyGAN**. The condition image I_c and source context image I_{sc} are encoded with an appearance encoder, while the source and condition landmarks Lm_s and Lm_c are encoded by a shape encoder. The shape codes F_0^{lmc} and F_0^{lms} are passed to the graph-based landmark reasoning model, and iteratively aggregated with the appearance code for K iterations. The final appearance code F_K^i is used for the face generation. The adversarial training is performed with two discriminators: The Appearance Discriminator D_{app} optimises for naturally blending the facial attributes of the target into the source image, and the Landmark Discriminator D_{lm} optimises for preserving the source pose.

weight the landmarks to optimally trade-off between the face naturalness and re-identification performance; a *Landmark2Landmark Face Generator* that follows a similar architecture of the generator of BiGraphGAN [28]. We use GCNs to learn the spatial relations between Lm_s and Lm_c, aggregated with the visual features extracted from the condition image I_c. With several iterations of the landmark appearance aggregation, the final image I_g is generated with an attention model. For *adversarial training*, we adopt a *Landmark Discriminator* $\mathbf{D}_{lm}$ and an *Appearance Discriminator* $\mathbf{D}_{app}$. $\mathbf{D}_{lm}$ is designed to preserve the facial pose of the source image, while $\mathbf{D}_{app}$ is designed to preserve the appearance of the condition face with the skin tone matched to the source face. We will provide more details on the *Landmark Attention Model, Landmark2Landmark Face Generator* and *Adversarial Training* in the following sections.

3.1 Landmark Attention Model

The landmark attention model is designed to learn the optimal weighting strategy on the landmarks to achieve jointly the best visual naturalness and pose preservation. We first concatenate the 68-channel landmark maps of both the source facial pose and the condition facial pose, and feed it to an Efficient Channel Attention module [30] formulated as:

$$\omega = \sigma(Conv1D_J(GAP(Concat(Lm_s, Lm_c)))), \qquad (1)$$

where $\sigma(\cdot)$ is the Sigmoid function, $Conv1D_J(\cdot)$ is 1-D convolution with kernel size J, $GAP(\cdot)$ represents the operation of the channel-wise Global Average Pooling, and $Concat(\cdot)$ is the concatenation operation.

3.2   Landmark2Landmark Face Generator

The landmark2landmark face generator follows the architecture of [28] that itera-tively reasons the relations between the source landmarks Lm_s and the condition landmarks Lm_c following a bipartite graph formulation, and aggregates with the appearance feature of the condition image I_c and the context of the source image I_{sc}. The final aggregated feature is used for the landmark-guided face generation with the condition identity.

The condition image I_c and the source context image I_{sc} are first concate-nated and fed to an appearance encoder to generate the initial appearance code F_0^i, while the Lm_s and Lm_c after the Landmark Attention Model are passed to a shape encoder to obtain the shape codes F_0^{lms} and F_0^{lmc}, respectively. The shape codes are then fed to a graph-based landmark-to-landmark reasoning model in a bipartite graph via GCNs to update the shape codes, which are then fed to the Landmark Appearance Aggregation model to synchronise the updates in both appearance and shape codes. Such operation of landmark-to-landmark reasoning and appearance aggregation is performed *iteratively* to form a thor-ough reasoning from low to high level. At the K-th iteration of the graph-based landmark-landmark reasoning and appearance aggregation, the final appearance code F_K^i is passed to both an image decoder to generate the intermediate result $\tilde{I}_g$, and an attention decoder to produce the attention mask A_i, a one-channel attention mask with the pixel value between 0 to 1. The final generated image is obtained by $I_g = I_c \otimes A_i + \tilde{I}_g \otimes (1 - A_i)$, where $\otimes$ denotes element-wise product.

3.3   Adversarial Training

Two discriminators are designed for the adversarial training. Specifically, the *Landmark Discriminator* $\mathbf{D}_{lm}$ is fed with the landmark-image pairs of the source $\{Lm_s, I_s\}$ and the generated $\{Lm_s, I_g\}$ to encourage the generation of similar facial pose of the source image. The *Appearance Discriminator* $\mathbf{D}_{app}$ takes the source-condition image pair $\{I_s, I_c\}$ and the generated-condition pair $\{I_g, I_c\}$ to guide the face generation with the facial attributes of the condition face within the same context of the source image by coupling I_s and I_g with I_c. In order to facilitate the appearance learning, we train the Appearance Discriminator in a hybrid manner with the source-condition image pairs composed by either the same image or a couple of images with different poses and contexts. When the source and condition images are the same, i.e. $I_s = I_c$, we apply a L1 loss on I_g and I_s to provide the strong pixel-level supervision. When the source and condition images are different, we apply a weak supervision for *Context Matching* by enforcing similar high-level features of I_g and I_s that correspond to skin tone and background.

We employ several losses to drive the network learning. We train D_{app} with the adversarial loss L_{app}:

$$L_{app} = \min_{G} \max_{D_{app}} \mathbb{E}[log(\mathbf{D}_{app}(I_s, I_c))] + \mathbb{E}[1 - log(\mathbf{D}_{app}(I_g, I_c))]. \tag{2}$$

The landmark discriminator D_{lm} is trained with L_{lm}, driving the generator towards the correct pose:

$$L_{lm} = \min_{G} \max_{D_{lm}} \mathbb{E}[log(\mathbf{D}_{lm}(I_s, Lm_s))] + \mathbb{E}[1 - log(\mathbf{D}_{lm}(I_g, Lm_s))]. \qquad (3)$$

Moreover, when the source and condition pair is of different images, the *Context Matching* supervision is realised through the weak Feature Matching (wFM) loss [4] defined as:

$$L_{wFM}(\mathbf{D}_{app}) = \sum_{i=m}^{M} \frac{1}{N_i} ||\mathbf{D}_{app}^{(i)}(I_g, I_c) - \mathbf{D}_{app}^{(i)}(I_s, I_c)||_1, \qquad (4)$$

where $D_{app}^{(i)}$ denotes the feature map produced by the i-th layer of the discriminator D_{app}; N_i is the number of elements in the feature map produced by the i-th layer; m is the first layer from which the weak feature matching loss computation starts; and M is the total number of layers of the discriminator D_{app}. When the condition and source pair is composed of the same image, we provide the generator with the direct pixel-level supervision using the image reconstruction loss $L_{Recon} = ||I_g - I_s||_1$.

The *final loss* function is expressed as a weighted sum of the above-mentioned loss terms, with the weighting parameters empirically set.

4 Experiments

We evaluate our proposed method **AnonyGAN** and its variants in comparison with state-of-the-art methods using two public face datasets. We validate the effectiveness of our design choices by evaluating the generated faces in terms of the visual naturalness, pose preservation, face detection and anonymisation.

Datasets. Our model is evaluated on two benchmark datasets created for face-related computer vision tasks, i.e., CelebA [21] and Labeled Faces in the Wild (LFW) [12,17]. **CelebA** is a large-scale dataset of 202,599 images with different poses and backgrounds of 10,177 celebrities. **LFW** is composed of 13,233 face images collected from the web covering 5,749 identities with 1,680 people having two or more images. We use CelebA for both training and testing, while LFW is used for testing only. We follow the same train/test split of CelebA as in [22], where in total 1,563 identities with more than 30 images per identity are selected. The training set is composed of 1,200 identities, where 24,000 pairs are formed for the training. We use the images of the remaining 363 identities for testing, where each source image is paired with a condition image that is randomly selected from the images of the next identity as in [22]. For the test set of LFW, we follow the protocol defined in [7], where we form 6,000 pairs of images organised in 10 different folds with each folder containing half of the pairs corresponding to the same person, and the other half corresponding to different persons. We use Dlib [16] to preprocess the 68 facial landmarks, which are then used to define the mask area of each face.

Evaluation Metrics. We evaluate the performance of our model in terms of visual quality, pose preservation, face detection, and face re-identification. The **visual quality** of generated faces is measured by the Fréchet inception distance (FID) [11], which calculates the distance between real and synthetic images in a feature space given by a specific layer of Inception Net. The lower the FID score is, the higher the quality of generated images[1]. The **pose preservation** is measured by the $L1$ distance between the detected 68 landmarks and the ground truth landmarks, normalised by the inter-ocular distance [15]. The model should generate faces that minimally impact the visual detection task. We evaluate the **face detection** performance on the generated images using two face detection algorithms, i.e., Dlib [16] and FaceNet [24]. A higher detection rate is more desired. Meanwhile, the generated faces should maximally prohibit the **face re-identification** for the best anonymisation. We report the rate of the correct matches of the generated faces and source faces using FaceNet without fine-tuning. In particular, for the test of LFW, we compute for each fold the re-identification rate, and report the mean and the standard deviation among all folds. A lower re-identification rate indicates a better face anonymisation.

Implementation Details. We set the learning rate for the generator during training to $2 * 10^{-4}$ and the learning rate for discriminators to $2 * 10^{-6}$. We perform hybrid training using the condition and source image pairs with the same images and different images to facilitate the appearance learning. During training, we we set 75% of pairs with $I_c = I_s$ for each batch.

Method Comparisons. We compare **AnonyGAN** against a set of baselines and state-of-the-art GAN-based methods that are closely related to the face obfuscation task: 1) **Blurring** applies a 101×101 Gaussian kernel to blur the face region. 2) **Pixelation** uses a 10×10 pixelation mask on the face region. 3) **CIAGAN** [22] is a state-of-the-art face anonymisation method based on conditional GAN, which accepts as input the context image and 29 facial landmarks of the source image and a conditioning ID, and generates a face matching both the context and the conditioning ID. We use the inference model provided by the authors for the evaluation.

Moreover, in order to demonstrate the validity of our design choices and, in particular, the importance of our Landmark Attention (LA) and Context Matching (CM), we ablate several variants of **AnonyGAN**: 1) **AnonyGAN-(CM, LA)$^-$** (68 lm) is trained without Landmark Attention and Context Matching, with the full 68 facial landmarks. 2) **AnonyGAN-(CM, LA)$^-$** (29 lm) is trained without Landmark Attention (LA) and Context Matching (CM), but with 29 facial landmarks that are manually selected as in CIAGAN [22], in order to show the impact of the landmark choice. 3) **AnonyGAN-(CM)$^-$** (68 lm) is trained with Landmark Attention but without the Context Matching, with 68 facial landmarks, to prove the capability of the Landmark Attention module on automatically weighting the landmarks. Finally, **AnonyGAN**(68 lm) is trained with both Landmark Attention and Context Matching, with all 68 landmarks, to justify the capability of Context Matching for improving visual fidelity.

---

[1] FID Implementation is taken from: https://github.com/mseitzer/pytorch-fid

Result Discussion. Table 1 presents the results of all methods evaluated on the test set of CelebA. Classic face obfuscation techniques, i.e. blurring and pixelation, reduce greatly the visual quality, thus hampering face detection. On the other hand, they achieve the lowest re-identification rate, thus the best anonymisation performance. The pose preservation metric is not available as the landmarks are not detectable from the anonymised faces. The visual quality of the images generated by CIAGAN [22] is in general inferior to our **AnonyGAN** models as indicated by the FID metric. The unnaturalness of the generated faces leads to a lower detection rate with the standard face detector Dlib, but also confuses the face re-identification module, leading to a better anonymisation performance. Moreover, our **AnonyGAN** models significantly improve the pose preservation performance thanks to the graph-based geometric reasoning.

Table 1. Performance of proposed **AnonyGAN** and its variants evaluated on the test set of CelebA, compared to baselines and state-of-the-art GAN-based method.

	$FID \downarrow$	Face detection $\uparrow$		Face re-identification $\downarrow$		Pose$\downarrow$
		dlib	Facenet	Casia	VGG	
Blurring	95.13	4%	4%	**0.07%**	**0.02%**	–
Pixelation	59.82	1%	28%	0.28%	0.12%	–
CIAGAN [22] (recomp.)	37.94	96%	**100%**	1.61%	0.51%	1.44
AnonyGAN-(CM, LA)$^-$ (68 lm)	43.99	**100%**	**100%**	2.63%	0.58%	**0.16**
AnonyGAN-(CM, LA)$^-$ (29 lm)	30.24	**100%**	**100%**	2.84%	0.66%	**0.16**
AnonyGAN-CM$^-$ (68 lm)	26.12	**100%**	**100%**	2.70%	0.91%	**0.16**
AnonyGAN (68 lm)	**22.53**	**100%**	**100%**	3.52%	1.60%	**0.16**

Interestingly, by choosing a subset of the landmarks (29 out of 68 landmarks) carefully as in CIAGAN, we observe that **AnonyGAN-(CM, LA)**$^-$ (29 lm) improves the visual quality without impacting the pose preservation, compared to **AnonyGAN-(CM, LA)**$^-$ (68 lm). Moreover, the result of **AnonyGAN-CM**$^-$ (68 lm) in comparison to **AnonyGAN-(CM, LA)**$^-$ (29 lm) shows that the introduction of Landmark Attention enables the network to learn the importance of landmarks automatically, achieving a better visual quality without impacting the pose preservation. Finally, **AnonyGAN** with both Landmark Attention and Context Matching achieves the best results in terms of face visual quality, face detection and pose preservation, however with a slight compromise in the anonymisation performance compared to the variants and CIAGAN. These above-mentioned observations are also in line with the results evaluated on the LFW dataset as reported in Table 2.

Table 2. Performance of the proposed **AnonyGAN** and its variants evaluated on the test set of LFW in comparison to the baselines and state-of-the-art GAN-based method.

	$FID \downarrow$	Face detection ↑		Face re-identification ↓		Pose↓
		dlib	Facenet	Casia	VGG	
Blurring	85.03	7%	17%	**0.03% ± 0.07**	**0.02% ± 0.05**	–
Pixelation	47.97	6%	22%	**0.03% ± 0.07**	**0.02% ± 0.05**	–
CIAGAN [22] (recomp.)	18.37	99%	100%	1.28% ± 0.32	0.08% ± 0.12	0.45
AnonyGAN-(CM, LA)⁻ (68 lm)	44.41	99%	100%	3.93% ± 0.67	0.38% ± 0.19	0.11
AnonyGAN-(CM, LA)⁻ (29 lm)	19.42	97%	100%	4.18% ± 0.45	0.35% ± 0.31	0.11
AnonyGAN-(CM)⁻ (68 lm)	14.66	100%	100%	3.35% ± 0.78	0.32% ± 0.27	0.06
AnonyGAN (68 lm)	**8.93**	100%	100%	3.55% ± 0.94	0.48% ± 0.29	**0.05**

Figure 3 shows the faces generated by **AnonyGAN** and the compared method CIAGAN [22], with the condition (the 1st column) and source (the 2nd column) images. For the condition identities, we use those that CIAGAN has been trained with for a fair comparison. We can observe that **AnonyGAN** generates more natural looking images with the facial attributes of the condition face better transferred to the context of the source image. The pose of the source face is better preserved, confirming the effectiveness of the proposed Landmark Attention module that allows for automatic weighting on the full 68 landmarks.

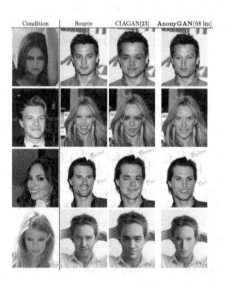

Fig. 3. Qualitative results with images from CelebA.

5 Conclusions

In this paper, we proposed **AnonyGAN**, a GAN-based solution for face anonymisation, generating faces with the appearance of any condition image and the facial pose of the source image. Our approach leverages landmark-to-landmark geometric reasoning via GCNs to model the relations between the condition and source facial landmarks. We also introduced a landmark attention model to automatically learn the importance of the facial landmarks. We compared our approach with the state-of-the-art approaches both quantitatively and qualitatively, demonstrating a better performance both in term of visual quality and pose preservation. As future work, we will explore alternatives to further improve the face anonymisation performance and adapt the model to operate in real world.

Acknowledgement. This work has been supported by the European Union's Horizon 2020 research and innovation programme under grant agreement No. 957337, and by the European Commission Internal Security Fund for Police under grant agreement No. ISFP-2020-AG-PROTECT-101034216-PROTECTOR.

References

1. AlBahar, B., Huang, J.B.: Guided image-to-image translation with bi-directional feature transformation. In: Proceedings of the IEEE/CVF International Conference on Computer Vision, pp. 9016–9025 (2019)
2. Balakrishnan, G., Zhao, A., Dalca, A.V., Durand, F., Guttag, J.: Synthesizing images of humans in unseen poses. In: Proceedings of the IEEE Conference on Computer Vision and Pattern Recognition, pp. 8340–8348 (2018)
3. Chan, C., Ginosar, S., Zhou, T., Efros, A.A.: Everybody dance now. In: Proceedings of the IEEE/CVF International Conference on Computer Vision, pp. 5933–5942 (2019)
4. Chen, R., Chen, X., Ni, B., Ge, Y.: SIM swap. In: Proceedings of the ACM International Conference on Multimedia, October 2020
5. Du, L., Zhang, W., Fu, H., Ren, W., Zhang, X.: An efficient privacy protection scheme for data security in video surveillance. J. Vis. Commun. Image Represent. **59**, 347–362 (2019)
6. Esser, P., Sutter, E., Ommer, B.: A variational U-Net for conditional appearance and shape generation. In: Proceedings of the IEEE Conference on Computer Vision and Pattern Recognition, pp. 8857–8866 (2018)
7. Gafni, O., Wolf, L., Taigman, Y.: Live face de-identification in video. In: Proceedings in IEEE/CVF International Conference on Computer Vision, pp. 9377–9386 (2019)
8. Gerstner, T., DeCarlo, D., Alexa, M., Finkelstein, A., Gingold, Y., Nealen, A.: Pixelated image abstraction with integrated user constraints. Comput. Graph. **37**(5), 333–347 (2013)
9. Goodfellow, I., et al.: Generative adversarial nets. In: Proceedings of Conference on Neural Information Processing Systems (2014)

10. Grigorev, A., Sevastopolsky, A., Vakhitov, A., Lempitsky, V.: Coordinate-based texture inpainting for pose-guided human image generation. In: Proceedings of the IEEE/CVF Conference on Computer Vision and Pattern Recognition, pp. 12135–12144 (2019)
11. Heusel, M., Ramsauer, H., Unterthiner, T., Nessler, B., Hochreiter, S.: GANs trained by a two time-scale update rule converge to a local nash equilibrium. arXiv preprint arXiv:1706.08500 (2017)
12. Huang, G.B., Mattar, M., Berg, T., Learned-Miller, E.: Labeled faces in the wild: a database for studying face recognition in unconstrained environments. In: Workshop on Faces in 'Real-Life' Images: Detection, Alignment, and Recognition. Erik Learned-Miller and Andras Ferencz and Frédéric Jurie, Marseille, France, October 2008
13. Hukkelås, H., Mester, R., Lindseth, F.: DeepPrivacy: a generative adversarial network for face anonymization. In: Bebis, G., et al. (eds.) ISVC 2019. LNCS, vol. 11844, pp. 565–578. Springer, Cham (2019). https://doi.org/10.1007/978-3-030-33720-9_44
14. Karras, T., Laine, S., Aila, T.: A style-based generator architecture for generative adversarial networks. In: Proceedings in IEEE/CVF Conference on Computer Vision and Pattern Recognition, pp. 4396–4405 (2019)
15. Kazemi, V., Sullivan, J.: One millisecond face alignment with an ensemble of regression trees. In: Proceedings of IEEE Conference on Computer Vision and Pattern Recognition, pp. 1867–1874 (2014)
16. King, D.E.: Dlib-ml: a machine learning toolkit. J. Mach. Learn. Res. 10, 1755–1758 (2009)
17. Learned-Miller, G.B.H.E.: Labeled faces in the wild: Updates and new reporting procedures. Technical report. UM-CS-2014-003, University of Massachusetts, Amherst, May 2014
18. Li, L., Bao, J., Yang, H., Chen, D., Wen, F.: FaceShifter: towards high fidelity and occlusion aware face swapping. arXiv preprint arXiv:1912.13457 (2019)
19. Liang, D., Wang, R., Tian, X., Zou, C.: PCGAN: partition-controlled human image generation. In: AAAI (2019)
20. Liu, W., Piao, Z., Min, J., Luo, W., Ma, L., Gao, S.: Liquid warping GAN: a unified framework for human motion imitation, appearance transfer and novel view synthesis. In: ICCV (2019)
21. Liu, Z., Luo, P., Wang, X., Tang, X.: Deep learning face attributes in the wild. In: Proceedings of International Conference on Computer Vision (ICCV), December 2015
22. Maximov, M., Elezi, I., Leal-Taixé, L.: CIAGAN: conditional identity anonymization generative adversarial networks. In: Proceedings in IEEE/CVF Conference on Computer Vision and Pattern Recognition, pp. 5446–5455 (2020)
23. Sagonas, C., Antonakos, E., Tzimiropoulos, G., Zafeiriou, S., Pantic, M.: 300 faces in-the-wild challenge: database and results. Image Vis. Comput. 47, 3–18 (2016). 300-W, the First Automatic Facial Landmark Detection in-the-Wild Challenge
24. Schroff, F., Kalenichenko, D., Philbin, J.: FaceNet: a unified embedding for face recognition and clustering. In: Proceedings of the IEEE Conference on Computer Vision and Pattern Recognition, pp. 815–823 (2015)
25. Siarohin, A., Sangineto, E., Lathuilière, S., Sebe, N.: Deformable gans for pose-based human image generation. In: Proceedings of the IEEE/CVF Conference on Computer Vision and Pattern Recognition (2018)

26. Sun, Q., Ma, L., Oh, S.J., Van Gool, L., Schiele, B., Fritz, M.: Natural and effective obfuscation by head inpainting. In: Proceedings of the IEEE Conference on Computer Vision and Pattern Recognition, pp. 5050–5059 (2018)
27. Sun, Q., et al.: A hybrid model for identity obfuscation by face replacement. In: Ferrari, V., Hebert, M., Sminchisescu, C., Weiss, Y. (eds.) ECCV 2018. LNCS, vol. 11205, pp. 570–586. Springer, Cham (2018). https://doi.org/10.1007/978-3-030-01246-5_34
28. Tang, H., Bai, S., Torr, P.H., Sebe, N.: Bipartite graph reasoning GANs for person image generation. In: Proceedings of the British Machine Vision Conference (2020)
29. Tang, H., Xu, D., Liu, G., Wang, W., Sebe, N., Yan, Y.: Cycle in cycle generative adversarial networks for keypoint-guided image generation. In: Proceedings of ACM Multimedia (2019)
30. Wang, Q., Wu, B., Zhu, P., Li, P., Zuo, W., Hu, Q.: ECA-Net: efficient channel attention for deep convolutional neural networks. In: Proceedings of the IEEE/CVF Conference on Computer Vision and Pattern Recognition (2020)
31. Zhu, Z., Huang, T., Shi, B., Yu, M., Wang, B., Bai, X.: Progressive pose attention transfer for person image generation. In: Proceedings of the IEEE Conference on Computer Vision and Pattern Recognition (2019)

Improving Localization for Semi-Supervised Object Detection

Leonardo Rossi$^{(\boxtimes)}$ iD, Akbar Karimi iD, and Andrea Prati iD

IMP Lab - D.I.A. - University of Parma, Parma, Italy
{leonardo.rossi,akbar.karimi,andrea.prati}@unipr.it
http://implab.ce.unipr.it/

Abstract. Nowadays, Semi-Supervised Object Detection (SSOD) is a hot topic, since, while it is rather easy to collect images for creating a new dataset, labeling them is still an expensive and time-consuming task. One of the successful methods to take advantage of raw images on a Semi-Supervised Learning (SSL) setting is the Mean Teacher technique [17], where the operations of pseudo-labeling by the Teacher and the Knowledge Transfer from the Student to the Teacher take place simultaneously. However, the pseudo-labeling by thresholding is not the best solution since the confidence value is not strictly related to the prediction uncertainty, not permitting to safely filter predictions. In this paper, we introduce an additional classification task for bounding box localization to improve the filtering of the predicted bounding boxes and obtain higher quality on Student training. Furthermore, we empirically prove that bounding box regression on the unsupervised part can equally contribute to the training as much as category classification. Our experiments show that our IL-net (Improving Localization net) increases SSOD performance by 1.14% AP on COCO dataset in limited-annotation regime. The code is available at https://github.com/IMPLabUniPr/unbiased-teacher/tree/ilnet.

Keywords: Object detection · Multi-task learning · Teacher-student technique · Semi-supervised learning · Semi-supervised object detection

1 Introduction

Supervised learning usually requires a large amount of annotated training data which can be expensive and time-consuming to produce. Semi-Supervised Learning (SSL), on the other hand, addresses this issue by taking advantage of large unlabeled data accompanied by a small labeled dataset. Usually, in an SSL setting, we have two models that act as Teacher and Student. The latter is trained in a supervised way, drawing the needed ground truths from the dataset, if exist, and from the Teacher's predictions, otherwise.

This approach can be beneficial in many machine learning tasks including object detection, whose final result is a list of bounding boxes (bboxes) and the

S. Sclaroff et al. (Eds.): ICIAP 2022, LNCS 13232, pp. 516–527, 2022.
https://doi.org/10.1007/978-3-031-06430-2_43

corresponding classes. In this task, the network is devoted to finding the position of the localized objects in an image, as well as identifying which class they belong to. Hereinafter, we will refer to this specific application as Semi-Supervised Object Detection (SSOD). In [10], authors have collected many interesting ideas applied to SSOD. Among them, the one that seems more promising is the Mean Teacher [17], which uses the Exponential Moving Average (EMA) as a knowledge transfer technique from the Student to the Teacher. In particular, while the Teacher produces pseudo-labels on unlabeled and weakly augmented images to obtain more reliable labels, the Student is trained using the pseudo-labels as ground-truth on the same images, but strongly augmented to differentiate the training. In weak augmentation, only a random horizontal flip is applied, while in strong augmentation, other transformations such as randomly adding color jittering, grayscale, Gaussian blur, and cutout patches are also used.

Although they obtain good performances with this pseudo-labeling technique, the Student model is still influenced by the Teacher's erroneous predictions. The quality of these predictions is difficult to control by applying only a simple confidence thresholding. For this reason, there is a need for an additional way to strengthen the region proposals and reduce the number of erroneous predictions by the Teacher. In the proposed architecture, we introduce a new task used for the classification of the bboxes, with the aim of distinguishing good quality ones from the others. This new score exploits the information complementary to the class score already used in these networks, allowing a different level of filtering. In addition, we show how to take advantage of the regression tasks on the unsupervised learning part. Usually, they are excluded in the unsupervised training phase. The justification is that the classification score is not able to filter the potentially incorrect bboxes [5,10]. A well-known problem during training is the Objective Imbalance [12], which is characterized by the difficulty in balancing the different loss contributions. In our hypothesis, this is the case for the unsupervised training part. In order to obtain a positive effect from regression losses, an adequate balance of the contribution of these two losses (regression and category classification) is a possible solution to the above-mentioned problem. In this way, we prevent the regression losses on unsupervised dataset from dominating the training, a phenomenon that greatly amplifies the error introduced by inaccurate Teacher predictions.

The main contributions of this paper are the following:

- a new bounding box IoU (Intersection over Union) classification task to filter out errors on pseudo-labels produced by the Teacher;
- the introduction of a regression task on the unlabeled dataset which can help the network to learn better;
- an exhaustive ablation study on all the components of the architecture.

2 Related Work

Semi-Supervised Learning for Object Detection. In [14], the authors proposed the Π-Model which is an ensemble of the predictions of the model at different epochs under multiple regularizations and input augmentation conditions. The predictions for the unlabeled images are merged together to form a better predictor. In [4], the authors proposed Consistency-based Semi-supervised Learning for Object Detection (CSD), which uses consistency constraints as a self-training task to obtain a better training of the network. In [15], an SSL framework with Self-Training (via pseudo label) and the Augmentation driven Consistency regularization (STAC) is introduced, exploiting weak data augmentation for model training and strong data augmentation for pseudo-labeling. Several works [10,16,17] proposed to use the Mean Teacher model, applying EMA to the weights instead of the predictions, facilitating knowledge transfer from Student to the Teacher model at each iteration. For one-stage object detection models, authors of [11] use an Expectation-Maximization approach, generating pseudo-labels in the expectation step and training the model on them in the maximization step, optimizing for classification in each iteration and for localization in each epoch. In [19], the authors propose a Soft Teacher mechanism where the classification loss of each unlabeled box is weighted by the Teacher classification score, also using a box jittering approach to select the most reliable pseudo-bboxes. In [6], authors utilize SelectiveNet to properly filter pseudo-bboxes trained after the Teacher. In [20], the authors propose two models, one of which generates a proposal list of bounding boxes and the second one refines these proposals, using the average class probability and the weighted average of the bounding box coordinates.

Bounding Box Intersection Over Union (BBox IoU). In [3], the authors added a new branch on top of the Fast R-CNN model to estimate standard deviation of each bounding box and use it at Non-maximum Suppression (NMS) level to give more weight to the less uncertain ones. In [5,21], the authors added a new branch on top of the Faster R-CNN to regress bounding box IoU and to multiply this value with the classification score to compute final score of the suppression criterion of the NMS. In the same direction, Fitness NMS was proposed in [18] to correct the detection score for better selecting bounding boxes which maximize their estimated IoU with the ground-truth.

3   Teacher-Student Learning Architecture

In SSOD, we have two datasets. The first set $\mathbf{D_s} = \{x_i^s, y_i^s\}_{i=1}^{N_s}$, typically smaller, contains N_s images $\mathbf{x^s}$ with the corresponding labels $\mathbf{y^s}$, and the second set $\mathbf{D_u} = \{x_i^u\}_{i=1}^{N_u}$ contains N_u images $\mathbf{x^u}$ without labels. Similarly to [10], our architecture is composed of two identical models where one behaves as Teacher and the other as the Student. At each iteration of the semi-supervised training, the Student is trained with one batch of images coming from $\mathbf{D_s}$ (supervised training) and one from $\mathbf{D_u}$ (unsupervised training), using the pseudo-labels generated by the Teacher as ground-truth and filtered by a threshold. Then, the Student transfers the learned knowledge to the Teacher using the EMA applied to the weights.

In our model, the total loss L for the Student is composed of two terms, which come from the supervised L_{sup} and unsupervised L_{unsup} training:

$$L = L_{sup} + L_{unsup}$$

$$L_{sup} = \sum_i L_{cls} + L_{reg} + \boldsymbol{L^{IoU}}$$

$$L_{unsup} = \sum_i \alpha L_{cls} + \beta \boldsymbol{L_{reg}} + \gamma \boldsymbol{L^{IoU}} \tag{1}$$

$$L_{cls} = L_{cls}^{rpn}(x_i^s, y_i^s) + L_{cls}^{roi}(x_i^s, y_i^s)$$

$$L_{reg} = L_{reg}^{rpn}(x_i^s, y_i^s) + L_{reg}^{roi}(x_i^s, y_i^s)$$

where L_{cls} contains the sum of Region Proposal Network (RPN) and Region of Interest (RoI) classification losses, while L_{reg} contains the sum of RPN and RoI regression losses. The L^{IoU} loss will be defined in Subsect. 3.2. α, β and γ represent how much weight each component of the unsupervised training has. The new terms introduced in this paper, w.r.t. the ones defined in [10], are shown in bold in the above equations and will be described in the following.

3.1 Bounding Box Regression on Unsupervised Training

The training consists of two phases, the *burn up* stage and *Teacher-Student Mutual Learning* stage. In the first stage, the Student is trained only on the labeled dataset, following the standard supervised training procedure. Then, for the second stage, at each iteration, the Student is trained on two batches at the same time, one coming from the labeled dataset and the other coming from the unlabeled dataset. For the latter, the baseline [10] trains only the RPN and RoI head classifier and disregards regression. The authors justify this choice noticing that the classification confidence thresholds are not able to filter the incorrect bounding box regression. Our hypothesis is that training also on pseudo bounding box regression could help the Student as long as the task is correctly weighted with respect to the others. Category classification and regression tasks are learned in parallel. Given that the classification performance improves during the training, we can also expect the bounding box regression to behave the same way. In other words, while the training proceeds, the average quality of the pseudo-labels increases and so does the classification confidence value.

During the Student training, although pseudo bounding boxes are filtered with a threshold on classification confidence score, we still have bounding boxes of any IoU quality. This problem is related to the uncertainty on prediction. Figure 1a visualizes the IoU distribution quality with respect to the ground-truth of the Teacher filtered predictions. We can notice that the number of pseudo bounding boxes with IoU less than the threshold (0.6 in our experiments) remains almost constant during the entire training, unlike the others which slowly increase. In Fig. 1b, we show only the filtered pseudo bounding boxes that the Teacher has wrongly classified and split by their IoU. In Fig. 1c, the same data are shown using a different graph. We can notice that the number

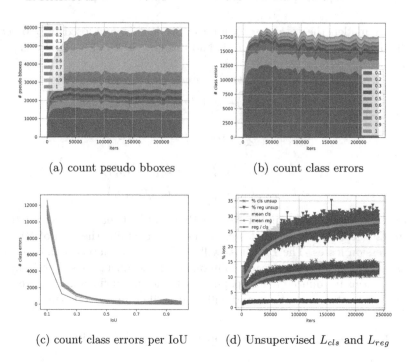

(a) count pseudo bboxes

(b) count class errors

(c) count class errors per IoU

(d) Unsupervised L_{cls} and L_{reg}

Fig. 1. The pseudo bounding boxes generated during training: (1a) IoU distribution of pseudo-bboxes, (1b) distribution of pseudo-bboxes when the predicted class is wrong. (1c) number of pseudo-bboxes per IoU collected every 5000 iterations. (1d) Unsupervised classification and regression losses comparison during the training. Better seen in color. (Color figure online)

of wrongly classified pseudo-bboxes decreases exponentially with the increase of the quality, and this trend remains the same during the training. The bounding boxes with low quality ($IoU < 0.6$) represent 45% of the total pseudo-bboxes and more than 90% of classification errors. This means that the low-quality IoU bboxes contain almost all the classification errors.

By looking at Fig. 1d, we can see that the unsupervised regression loss (from RPN and RoI heads) represents between 20% and 30% of the total loss. Conversely, the unsupervised classification loss (from RPN and RoI heads) accounts only for 15% (at most) of the total loss. This means that an error in pseudo-labels has almost three times (see blue line in Fig. 1d) more weight on regression branch than on classification branch. To avoid the amplification of the errors, an appropriate value for β is chosen, by under-weighting the regression contribution and making it comparable with the classification one.

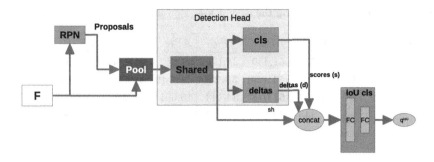

Fig. 2. Faster R-CNN architecture with our branch in red. (Color figure online)

3.2 Bounding Box IoU Classification Task

As noted by [5,21], adding a new task related to the IoU prediction could help the network to train better. Our model learns to predict the bounding box quality, using this value to further filter pseudo-labels in conjunction with the classification score. Differently from them, we train the model to make a binary classification instead of a regression since it is sufficient (and easier to learn) for the purpose of filtering.

Figure 2 illustrates the architecture of a two-stage object detection model such as Faster R-CNN [13] with our additional branch highlighted in red. For each positive bounding box (*i.e.*, the ones recognized as foreground), it concatenates the output of the shared layers (x_i^{sh}, with the size of 1024), all the classification scores (s_i, with the size of the number of classes plus the background) and all bounding box regression deltas (d_i, with the size of 4). All these features are passed to the *IoU cls* branch (also called $f(\cdot)$ in the following), which outputs a vector with the same size as s_i (*i.e.*, one for each class). This will return the ith IoU classification, called *IoU score* q^{IoU}, corresponding to the predicted bbox. The $f(\cdot)$ branch consists of two fully-connected layers with an ELU [1] activation function between them and a *sigmoid* activation function at the end.

The loss L_i^{IoU} for the new branch of IoU classification, conditioned on the class, is defined as:

$$q_i^{IoU} = f(concat(x_i^{sh}, s_i, d_i))$$
$$L_i^{IoU} = FL(q_i^{IoU}, t_i) \tag{2}$$

where the FL function is the Focal Loss [8] with its own γ value equal to 1.5 and t_i represents the binary target label. For the ith bounding box, the branch output $q_i^{IoU} \in [0,1]$ is a single value which predicts if it is a high- or low-quality bounding box. The target t_i and its IoU are defined as follows:

Table 1. Performance varying the weight loss β on unsupervised regression losses.

#	β	AP	AP_{50}	AP_{75}	AP_s	AP_m	AP_l
0	0.5	31.775	51.450	34.190	16.952	34.384	41.549
1	**1.0**	**31.947**	**51.530**	**34.270**	**16.949**	**34.900**	**41.306**
2	2.0	31.754	51.078	34.143	16.691	35.008	41.670
3	4.0	30.445	49.387	32.727	15.044	33.200	40.209

$$IoU_i = \max IoU\left(b_i^t, g\right) \quad \forall b_i \in \mathbf{B} | IoU\left(b_i, g\right) \geqslant u$$

$$t_i = \begin{cases} 1 & IoU_i > \mu \\ 0 & \text{otherwise} \end{cases} \tag{3}$$

where $\mathbf{B}$ is the list of proposals coming from the RPN, b_i are the ith predicted bounding boxes, g is the ground-truth bounding box, u is the minimum IoU threshold to consider the bounding b_i as positive example (typically set to 0.5) and μ (set to 0.75) is the minimum IoU threshold to be classified as high-quality.

In the evaluation phase, the Teacher's filtering of the predicted bounding boxes with low confidence is preceded by the IoU classification filtering, which uses a new IoU inference threshold θ (set to 0.4).

4   Experiments

Dataset. We perform our tests on the MS COCO 2017 dataset [9]. The training dataset consists of more than 117,000 images and 80 different classes of objects, where 10% of the labeled images are used.

Evaluation Metrics. All the tests are done on COCO minival 2017 validation dataset, which contains 5000 images. We report mean Average Precision (mAP) and AP_{50} and AP_{75} with 0.5 and 0.75 minimum IoU thresholds, and AP_s, AP_m and AP_l for small, medium and large objects, respectively.

Implementation Details. All the values are obtained running the training with the same hardware and hyper-parameters. When available, the original code released by the authors is used. Our code is developed on top of the Unbiased Teacher [10] source code. We perform the training on a single machine with 6 T P100 GPUs with 12 GB of memory. The train lasts 180,000 iterations with a batch size of 2 images per GPU for the supervised part and 2 images for the unsupervised part, with α set to 4. We use the Stochastic Gradient Descent (SGD) optimization algorithm with a learning rate of 0.0075, a weight decay of 0.0001, and a momentum of 0.9. The learning rate decays at iteration 179990 and 179995. We use the Faster R-CNN with FPN [7] and the ResNet 50 [2] backbone for the teacher and student models, initialized by pre-trained on ImageNet, and the same augmentation of Unbiased Teacher [10].

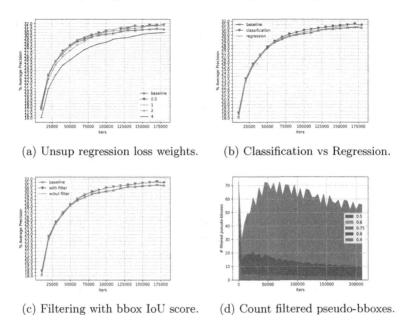

(a) Unsup regression loss weights.   (b) Classification vs Regression.

(c) Filtering with bbox IoU score.   (d) Count filtered pseudo-bboxes.

Fig. 3. Ablation studies: (3a) weights for unsupervised regression loss on RPN and RoI. (3b) classification vs regression loss on bbox IoU branch. (3c) filtering bbox on inference with bbox IoU classification score. (3d) count bboxes filtered by inference threshold μ during training.

4.1 Ablation Study

Unsupervised Regression Loss. In this experiment, we empirically show that we can also use regression losses on RPN and RoI head for the unsupervised part. We test different weights for the constant β in the loss formula (see Eq. 1). In Fig. 3a and Table 1, we can see that greatly amplifying the contribution can be deleterious, becoming counterproductive in the case of β equal to 4.

Bounding Box IoU Branch Loss Type. Our proposal involves a new IoU classification task, trained with a binary cross-entropy function. In this experiment, we test how performance changes in case our new branch learns a regression task instead of a classification, using a *smooth L1* loss. In this case, the ground-truth is represented by the real IoU value between the bbox and the ground-truth. In Fig. 3b, we can see that classification branch is a bit more stable and reaches a slightly higher performance.

With and Without Filtering Bboxes. The bbox IoU branch learns to recognize high quality bounding boxes and, as the default behavior, also to pre-filter Teacher's pseudo-bboxes depending on our new threshold score. In Fig. 3c and in Table 2 (rows #2 and #3), we see that our new branch contributes to the increase of the general performance (+0.48% mAP). Then, another small improvement is given by the filtering phase, increasing the performance by +0.1% mAP.

Table 2. Performance comparison with original Unbiased Teacher (UT) model: (2) Training with BBox IoU branch with and (3) w/out pseudo-labels filtering.

#	Method	AP	AP_{50}	AP_{75}	AP_s	AP_m	AP_l
1	UT	31.027	50.757	33.056	17.014	33.684	40.322
2	Ours (with filter)	31.604	51.181	33.962	16.816	34.283	40.809
3	Ours (w/out filter)	31.509	51.118	33.564	16.848	34.684	40.582

Table 3. Performance using BBox IoU classification branch with inference threshold θ fixed to 0.5 and varying training threshold μ.

#	μ	AP	AP_{50}	AP_{75}	AP_s	AP_m	AP_l
1	0.5	31.199	51.009	33.047	16.187	34.000	40.180
2	0.6	31.128	50.785	33.268	17.102	33.805	39.932
3	0.7	31.461	51.319	33.637	16.714	34.217	40.100
4	**0.75**	**31.604**	**51.181**	**33.962**	**16.816**	**34.283**	**40.809**
5	0.8	31.336	50.601	33.707	16.327	34.180	40.476
6	0.9	27.125	43.515	28.800	12.815	29.486	36.034

Bounding Box Training Threshold μ. In this experiment, we test the bbox IoU classification branch, setting inference threshold θ to 0.5 and varying the training threshold μ. From Table 3, it is clear that the choice of a correct threshold greatly influences the performance. On the one hand, if the threshold is too low, it does not help the network to learn more descriptive feature maps. On the other hand, if it is too high, the risk to wrongly filter out the bounding boxes will increase. As we can see in Fig. 3d, with the increase of IoU threshold μ, the number of teacher pseudo-bboxes filtered during the training increases exponentially. This is likely due to an imbalance in training, where the higher the threshold, the fewer high-quality examples are available. The best value is in the middle between the threshold u (0.5) and the IoU maximum value 1.0.

Table 4. Performance using BBox IoU classification branch with training threshold μ fixed to 0.75 and varying inference threshold θ.

#	θ	AP	AP_{50}	AP_{75}	AP_s	AP_m	AP_l
0	0.3	31.404	51.205	33.792	16.273	34.542	40.851
1	**0.4**	**31.630**	**51.185**	**34.044**	**17.387**	**34.494**	**40.784**
2	0.5	31.604	51.181	33.962	16.816	34.283	40.809
3	0.6	31.158	50.227	33.418	16.203	33.687	40.323
4	0.7	30.649	49.216	33.138	16.532	33.409	39.542

Table 5. Study on unsupervised regression losses and IoU classification loss.

#	L_{reg}^{unsup}	x^{sh}	Scores	Deltas	AP	AP_{50}	AP_{75}	AP_s	AP_m	AP_l
1					31.027	50.757	33.056	17.014	33.684	40.322
2	✓				31.947	51.530	34.270	16.949	34.900	41.306
3	✓	✓			31.754	51.189	34.032	16.850	34.657	41.320
4	✓	✓	✓		**32.166**	**51.772**	**34.765**	**16.647**	**34.999**	**41.870**
5	✓	✓	✓	✓	31.923	51.464	34.070	16.202	35.197	41.368
6		✓	✓	✓	31.630	51.185	34.044	17.387	34.494	40.784

Bounding Box Inference Filter Threshold θ. In this experiment, we test our bbox IoU branch, setting the training threshold μ to 0.75 (best value previously found) and varying the inference threshold θ. In Table 4, we see that the best value for this threshold is in the middle as expected because the branch is trained to reply 1 if the bbox is good enough and 0 otherwise.

IL-net: Improving Localization Net. Finally, we test the full architecture IL-net, composed of the unsupervised regression losses and the new IoU classification branch. Since, using both of them, the contribution of the new branch is absorbed by the loss of unsupervised regression (see rows 2 and 5 in Table 5), we performed an ablation study reducing the values in input to the new branch (see Eq. 2). This analysis has allowed us to highlight that by removing the contribution of the deltas, we can increase the general performance. This behavior could be explained by the fact that the deltas are optimized from both losses (L_{usup}^{reg} and L^{IoU}), causing the conflict as a result.

5   Conclusions

In this paper, we proposed two new architectural enhancements with respect to the network proposed in [10]: a new bounding box IoU classification task to filter out errors on pseudo-labels produced by the Teacher and the introduction of the unsupervised regression losses. For the former, we introduced a lightweight branch to predict the bounding box IoU quality. For the latter, we demonstrated how to successfully integrate it in the training, balancing the training tasks. Our new model called IL-net, which contains both, increases the general SSOD performance by a 1.14% AP on COCO dataset in a limited annotation regime.

Acknowledgments. This research benefits from the HPC (High Performance Computing) facility of the University of Parma, Italy.

References

1. Clevert, D.A., Unterthiner, T., Hochreiter, S.: Fast and accurate deep network learning by exponential linear units (elus). arXiv preprint arXiv:1511.07289 (2015)
2. He, K., Zhang, X., Ren, S., Sun, J.: Deep residual learning for image recognition. In: Proceedings of the IEEE Conference on Computer Vision and Pattern Recognition, pp. 770–778 (2016)
3. He, Y., Zhu, C., Wang, J., Savvides, M., Zhang, X.: Bounding box regression with uncertainty for accurate object detection. In: Proceedings of the IEEE/CVF Conference on Computer Vision and Pattern Recognition, pp. 2888–2897 (2019)
4. Jeong, J., Lee, S., Kim, J., Kwak, N.: Consistency-based semi-supervised learning for object detection. Adv. Neural Inf. Process. Syst. **32**, 10759–10768 (2019)
5. Jiang, B., Luo, R., Mao, J., Xiao, T., Jiang, Y.: Acquisition of localization confidence for accurate object detection. In: Proceedings of the European Conference on Computer Vision (ECCV), pp. 784–799 (2018)
6. Li, Y., Huang, D., Qin, D., Wang, L., Gong, B.: Improving object detection with selective self-supervised self-training. In: Vedaldi, A., Bischof, H., Brox, T., Frahm, J.-M. (eds.) ECCV 2020. LNCS, vol. 12374, pp. 589–607. Springer, Cham (2020). https://doi.org/10.1007/978-3-030-58526-6_35
7. Lin, T.Y., Dollár, P., Girshick, R., He, K., Hariharan, B., Belongie, S.: Feature pyramid networks for object detection. In: Proceedings of the IEEE Conference on Computer Vision and Pattern Recognition, pp. 2117–2125 (2017)
8. Lin, T.Y., Goyal, P., Girshick, R., He, K., Dollár, P.: Focal loss for dense object detection. In: Proceedings of the IEEE International Conference on Computer Vision, pp. 2980–2988 (2017)
9. Lin, T.-Y., et al.: Microsoft COCO: common objects in context. In: Fleet, D., Pajdla, T., Schiele, B., Tuytelaars, T. (eds.) ECCV 2014. LNCS, vol. 8693, pp. 740–755. Springer, Cham (2014). https://doi.org/10.1007/978-3-319-10602-1_48
10. Liu, Y.C., et al.: Unbiased teacher for semi-supervised object detection. arXiv preprint arXiv:2102.09480 (2021)
11. Nguyen, N.V., Rigaud, C., Burie, J.C.: Semi-supervised object detection with unlabeled data. In: VISIGRAPP (5: VISAPP), pp. 289–296 (2019)
12. Oksuz, K., Cam, B.C., Kalkan, S., Akbas, E.: Imbalance problems in object detection: a review. IEEE Trans. Pattern Anal. Mach. Intell. **43**(10), 3388–3415 (2020). IEEE
13. Ren, S., He, K., Girshick, R., Sun, J.: Faster R-CNN: towards real-time object detection with region proposal networks. Adv. Neural Inf. Process. Syst. **28**, 91–99 (2015)
14. Samuli, L., Timo, A.: Temporal ensembling for semi-supervised learning. In: International Conference on Learning Representations (ICLR), vol. 4, p. 6 (2017)
15. Sohn, K., Zhang, Z., Li, C.L., Zhang, H., Lee, C.Y., Pfister, T.: A simple semi-supervised learning framework for object detection. arXiv preprint arXiv:2005.04757 (2020)
16. Tang, Y., Chen, W., Luo, Y., Zhang, Y.: Humble teachers teach better students for semi-supervised object detection. In: Proceedings of the IEEE/CVF Conference on Computer Vision and Pattern Recognition, pp. 3132–3141 (2021)
17. Tarvainen, A., Valpola, H.: Mean teachers are better role models: weight-averaged consistency targets improve semi-supervised deep learning results. In: NIPS (2017)
18. Tychsen-Smith, L., Petersson, L.: Improving object localization with fitness NMS and bounded IoU loss. In: Proceedings of the IEEE Conference on Computer Vision and Pattern Recognition, pp. 6877–6885 (2018)

19. Xu, M., et al.: End-to-end semi-supervised object detection with soft teacher. arXiv preprint arXiv:2106.09018 (2021)
20. Zhou, Q., Yu, C., Wang, Z., Qian, Q., Li, H.: Instant-teaching: an end-to-end semi-supervised object detection framework. In: Proceedings of the IEEE/CVF Conference on Computer Vision and Pattern Recognition, pp. 4081–4090 (2021)
21. Zhu, L., Xie, Z., Liu, L., Tao, B., Tao, W.: IoU-uniform R-CNN: breaking through the limitations of RPN. Pattern Recognit. **112**, 107816 (2021)

Interactive Deep Annotation as DARos: Object Detection Supervision for Efficient Instance Segmentation

Lihao Wang$^{(\boxtimes)}$, Rachid Benmokhtar, and Xavier Perrotton

Valeo Mobility Tech Center - Driving Assistance Research (DAR),
6 Rue Daniel Costantini, 94000 Créteil, France
{lihao.wang,rachid.benmokhtar,xavier.perrotton}@valeo.com

Abstract. We propose DARos (Deep Annotation Region-based by object supervision), a high-quality and generic annotation solution minimizing the cost of human-computer interaction. The two-stage instance segmentation approach is designed combining full-image-based object detection, RoI-based Inside Outside Guidance approach and salient object detection. It automatically identifies the object category. Only two clicks are required for easy cases and three clicks for hard cases to obtain a first segmentation prediction. Furthermore, interactive correction is supported via additional clicks. Extensive experiments demonstrate the accuracy and robustness of DARos on diverse popular benchmarks: PASCAL VOC, GrabCut, Berkeley, MSCOCO and Cityscapes. It outperforms state-of-the-art with a reduced annotation effort by 12%.

Keywords: Deep learning · Interactive annotation · Instance segmentation · Data augmentation

1 Introduction

Deep learning-based techniques have gathered wide attraction in both scientific and industrial communities. They are actively applied to various tasks, such as object detection, image classification and semantic/instance segmentation. Image annotation plays an important role in machine learning, it is defined as the task of labeling an image with relevant metadata information depending on the project domain experts and the predefined use cases. The most used annotation techniques are oriented to object bounding box and segmentation.

However, one of the difficulties to apply deep learning is the need for a large amount of labeled corpus where the order of annotated dataset is generally difficult to estimate. Compared to bounding boxes, segmentation-based annotation is much more time-consuming because pixel-level labels are required. To decrease the annotation cost, numerous interactive instance segmentation methods like [20,23,34] have been proposed in recent years. These methods can be regarded as semi-automatic annotation techniques which require the

S. Sclaroff et al. (Eds.): ICIAP 2022, LNCS 13232, pp. 528–540, 2022.
https://doi.org/10.1007/978-3-031-06430-2_44

annotator to provide some guidance inputs such as bounding boxes [21,31,34], clicks [15,16,20] or scribbles [2,9,13] and output a segmentation mask. This process is usually interactive because the annotator can correct the predicted mask by adding new guidance inputs until an acceptable accuracy is obtained.

A comprehensive review on interactive instance segmentation approaches can be found in [27]. Interactive instance segmentation methods can be divided into two categories (Fig. 1): Full Image-based (FI) and Region of Interest-based (RoI).

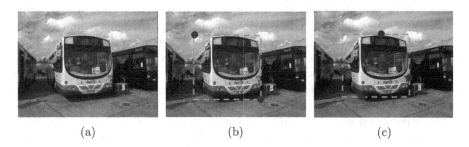

(a)   (b)   (c)

Fig. 1. Category examples: (a) FI-based (1 click), (b-c) RoI-based with bounding box (2 clicks) and extreme points (4 clicks) respectively.

FI category considers the full image as its input, where only one click is needed to generate the first mask. Depending on the object complexity, several interactive mask corrections could be necessary to reach the accuracy target.

The first interactive CNN-based segmentation is proposed by Xu *et al.* [33], where user clicking interactions are represented via Euclidean distance maps and concatenated with RGB image channels. They achieved high-quality segmentation with fewer user effort compared to computer vision approaches based on hand-crafted features [3,4,9,10]. On the other hand, a generalization limitation is also observed subject to the image properties (e.g. image size).

To tackle the ambiguity of all possible segmentation, a dual network architecture is designed by Li *et al.* [15]: one for plausible segmentations that conform to the user's input and the second to select among these. An iterative training strategy simulating annotator's behavior to improve model performance is proposed by Mahadevan *et al.* [22]. Jang *et al.* [13] followed a back-propagating refinement for interactive correction which enforces user-specified locations to have correct labels. Lin *et al.* [20] demonstrated the critical role of the first click about providing the location information of the target object. Based on this fact, FCA (First Click Attention) is introduced to make better use of the first click and achieves state-of-the-art performance among all existing FI methods.

The RoI-based category [1,5,21,23] treats the region cropped by the annotator as its input. It has three advantages: i) the output mask will not surpass the pre-defined region, and the maximum error is limited. ii) the cropped region is often small, therefore it is possible to increase the input image resolution, which can help to learn object details. iii) it's not sensitive to the image properties since the networks are trained on cropped and resized regions.

To capture the dense environment in the autonomous driving field, high resolution with large image size is employed (e.g. (1242×375) in KITTI [8] and (2048×1024) for Cityscapes [6]). Using FI methods, the image down-sampling will lose detailed information of small objects and decrease the mask quality. Therefore RoI-based methods are preferred.

A bounding box-based approach called Polygon-RNN which predicts the polygon outlining the object instead of pixel-wise result is proposed by Castrejon et al. [5]. It leverages Recurrent Neural Network (RNN) for polygon generation. Improved Polygon-RNN via Reinforcement Learning (RL) and Graph Neural Network (GCN) is introduced by Acuna et al. [1]. A regression problem is constructed for multi-layer GCN to predict simultaneously all vertices on the object polygon [21] to support splines format, facilitating curved objects annotation.

Although only two clicks are necessary to define a bounding box, Papadopoulos et al. [25] showed that the two corners are often outside the object of interest and it can be time-consuming to obtain a tight box. Then, an efficient way of extreme points clicking (left-most, right-most, top, bottom pixels) for RoI selection is modeled. Based on this approach, DEXTR (Deep Extreme) [23] integrates extreme clicking information with the CNN model input and obtains high-quality object segmentation, where Wang et al. [31] proposed an end-to-end approach combining CNN models with level set optimization based on four extreme points as input. Finally, the current state-of-the-art performance is achieved by Inside-Outside Guidance (IOG) of Zhang et al. [34]. IOG requires only three clicks to generate a first segmentation mask: two outside points at bounding box corners and one inside point near the object center.

Inspired by IOG network, we propose improvements with the aim to reduce annotation time and cost. We demonstrate that the inside guidance point is not always necessary especially for a single object in the selected bounding box. To be more specific, objects to be annotated are divided into easy (i.e. non-overlapping objects) and hard categories (i.e. overlapping objects): only two outside clicks are needed for the first one, nevertheless the third click is necessary to remove ambiguity for the hard samples.

The contribution of this work is twofold. Firstly, we formulate the limitation of today's interactive object segmentation approaches. Secondly, we propose an efficient framework which helps the model focus on hard cases and avoid unnecessary annotation effort. We also propose a novel data augmentation method to provide more hard samples during training, which can significantly improve the model's performance. Our framework is an end-to-end automatic annotation approach based on instance segmentation, coupled with 2D object detection. The annotator can correct interactively the erroneous areas.

The rest of the paper is organized as follows. Section 2 presents the architecture of the proposed framework as well as data augmentation method. Section 3 demonstrates experimental results on several public datasets as well as real-world road driving scenarios, and validates the effectiveness of the proposed smart annotation system. Finally, Sect. 4 presents conclusions and future works.

2   Proposed Annotation Framework

2.1   DARos Architecture and Workflow

The overall workflow of the proposed DARos system is presented in Fig. 2 and can be decomposed as three main blocks:

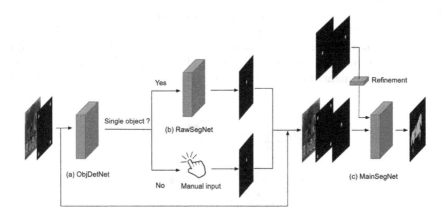

Fig. 2. Proposed DARos framework

Object Detection Network (ObjDetNet): A classic Faster R-CNN [28] with ResNet-50 [12] and (Feature Pyramid Networks) FPN [18] architecture is employed. In addition, this block doesn't need to be run in real time. Object detection results can be generated offline before starting the annotation. It hence will not slow down the inference speed during annotation.

Raw Segmentation Network (RawSegNet): It extracts an inside point for objects without ambiguity. Bounding boxes containing only one object (*i.e.* non-overlapping objects) are considered as easy ones because for them two manual clicks can already guarantee a high-quality segmentation. Therefore to save the annotator's effort, RawSegNet provides automatically the third input point for these easy samples by applying a light network. More details will be shown in the Sect. 3.2.

The light U^2-Net [26] network is employed for RawSegNet to guarantee a real-time inference speed. It is designed initially for Salient Object Detection (SOD) to segment the most visually attractive objects in the image.

Main Segmentation Network (MainSegNet): It predicts the final mask of the object. IOG [34] backbone is composed of three parts: two cascaded networks CoarseNet and FineNet as the main body and a side refinement branch for interactive correction. The latter is replaced by a ResNet-50 [12] to augment the model capacity. Similar to IOG, it takes three points as guidance: the annotator selects two outside points to define the RoI and one inside point near the object center. For easy samples, it is automatically computed by RawSegNet block, while it should be selected manually for hard samples.

Before the annotator begins to annotate, the ObjDetNet block performs object detection on the full image and outputs the bounding box of each object. The detected objects are divided into two categories: easy samples (*i.e.* without overlapping boxes) and hard samples (*i.e.* with overlapping boxes).

To comply with the model input format, the center point is provided automatically by RawSegNet for easy samples. However for hard samples, an object of this category can be confused with other objects in its bounding boxes, and therefore a third manual click on its center is necessary, as presented in Fig. 3.

Fig. 3. Easy and hard objects example: Hard objects "1" and "2" (*i.e.* their bounding boxes are overlapped). DARos requires the inside guidance point (e.g. the red point for object "1") as a third click to make the prediction. For easy object "3" (*i.e.* only one object in its bounding box) only two clicks are required. (Color figure online)

When the annotator provides a bounding box of an Object of Interest (OoI), it is compared with all detected objects by ObjDetNet on the same image. A matched detection is the one which has the highest IoU (Intersection over Union) with the OoI. An IoU threshold $\tau = 0.5$ is applied considering object detection error. If no detected object has $IoU > 0.5$, the detection is considered as failed and the object is categorized as hard. Overlapping rate is calculated with a tolerance of 0.1. These values are defined in the validation step.

2.2   Human-in-Loop Simulation

Additional information can be introduced by the annotator to interactively correct any mistakes in the predicted segmentation. The annotator iteratively clicks in the erroneous areas in order to guide the model to a better prediction. Like IOG [34], DARos supports interactive adding of new foreground and background clicks for further refinement if the user is not satisfied with the current segmentation result. To achieve this feature, an iterative training strategy is applied to simulate the interactive correction behavior. This training process is actually a multi-task learning (MTL) paradigm, minimizing the binary cross-entropy loss function.

2.3   Data Augmentation (DA)

It is a common technique to apply data augmentation techniques during the training of deep learning approaches to avoid overfitting and improve the overall performance [30]. In the interactive segmentation domain, simple techniques are used such as rotation, cropping and flipping [23,26,34]. Advanced techniques such as data mixing are mostly employed to create new datasets [17,32]. Liew *et al.* [17] constructed *ThinObject-5K* dataset of 5743 foreground thin elongated objects from the Internet and composite them on existing backgrounds.

Inspired by this approach, a novel DA technique named *Foreground Background Shuffling* is proposed to increase the number of hard samples. It extracts randomly foreground objects from training images and overlays them onto other images (the background). A minimum overlapping rate is applied to validate the created image (*i.e.* $ov_{min} = 10\%$ between the foreground and background objects). Figure 4 illustrates an example of valid and failed cases. Table 1 presents the number of synthetic images for different datasets used in our training.

Table 1. Data augmentation summary

Dataset	Number of original objects	Number of synthetic objects
PASCAL-1k train	3,507	9,308
PASCAL-10k train	25,832	43,486
Cityscapes train 10%	5,359	5,655

3   Experiments

3.1   Experimental Setup

Datasets: Following [20,23,34], two public datasets are adopted for training (PASCAL VOC [7] and SBD [11]). PASCAL train set (also called PASCAL-1k which contains 1464 images) is augmented with additional SBD labels to generate the well-known PASCAL-10k dataset (with 10,582 images).

To demonstrate the DARos effectiveness and its generalization capacity, five state-of-the-art validation datasets are employed: PASCAL VOC test, Grab-Cut [29], Berkeley [24], Cityscapes [6] and MSCOCO [19].

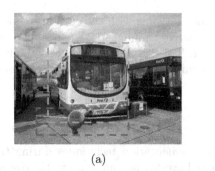

(a)   (b)

Fig. 4. Examples of the proposed DA *Foreground Background Shuffling* for an extracted airplane object. (a) Valid synthetic image (minimum overlapping rate between the airplane and the background bus object is respected). (b) Invalid synthetic image (no overlapping with background objects).

Implementation Details: To achieve a general framework as well as to reduce training effort, we adopt directly a publicly available weight pre-trained on MSCOCO dataset [19] for ObjDetNet. As mentioned in the Sect. 2.1, the detection can be performed prior to the annotation. Once the annotation begins, the stored result will be retrieved when the annotator clicks on each object.

On the other hand, RawSegNet is trained on PASCAL-1k coupled by Adam optimizer [14] using default hyper parameters (*i.e.* initial learning rate $lr = 10^{-3}$, $\beta_1 = 0.9$, $\beta_2 = 0.999$, $\epsilon = 10^{-8}$, weight decay $w_d = 0$). The training is conducted on 2 GTX 2080 Ti GPUs (batch size of 32), which takes about 36 h. The inference time requires 38 ms on a single GTX.

For *MainSegNet*, we follow [23,34] approaches. SGD optimizer is applied with $lr = 10^{-8}$, $\beta = 0.9$, $w_d = 5 \times 10^{-4}$. One training step in interactive mode is equivalent to three steps in non-interactive mode. For the latter, based on PASCAL-1k, 200 training epochs are needed ($\approx$33 h), and 100 epochs ($\approx$110 h) on PASCAL-10k. For interactive mode, 50 epochs ($\approx$54 h) with PASCAL-1k, and 30 epochs ($\approx$87 h) on PASCAL-10k. The inference time requires 77 ms on a single GTX.

3.2   Non-interactive Instance Segmentation Vs. Baseline

We compare DARos with IOG [34] as it represents the current state-of-the-art method for instance annotation. The results are illustrated in Table 2.

With the same number of clicks (*i.e.* ObjDetNet is not activated and for each object the third click is provided manually), DARos has achieved better mIoU (mean Intersection over Union) than IOG. And with reduced number of clicks (*i.e.* ObjDetNet is activated), its performance is comparable to IOG. We decrease in average 0.36 clicks processed on the 5 test datasets. This reduction accelerates annotation by 12%.

Table 2. Comparison with IOG & DEXTR in non-interactive mode

Train set	PASCAL-1k				PASCAL-10k			
Method	DEXTR	IOG	DARos	DARos	DEXTR	IOG	DARos	DARos
PASCAL	90.5	92.0	**92.5**	92.1	91.5	93.2	**93.6**	93.0
GrabCut	91.2	92.8	**93.8**	93.5	91.6	**94.5**	93.5	92.8
Berkeley	88.5	**91.7**	90.8	89.6	89.9	91.5	**91.7**	90.4
Cityscapes	72.2	72.8	**73.9**	73.3	75.9	**77.6**	76.9	76.0
MSCOCO	78.9	78.5	**79.8**	79.6	82.1	83.2	**84.3**	84.0
#Clicks	4	3	3	2.64	4	3	3	2.64

Figure 5 shows the DARos behavior on easy (blue curve) and hard samples (black curve) considering the average number of clicks $\in [2, 3]$. The mIoU of easy samples can be considered as stable when the number of clicks decreases from three to two, however mIoU of hard samples drops significantly. This drop demonstrates that if the RoI contains multiple objects, the network could be easily confused by other objects. Therefore, the third click is necessary to indicate the object of interest whereas only two clicks are sufficient to obtain a high-quality segmentation for easy samples. In practice, an overlapping tolerance of 0.1 is chosen to obtain a good trade-off between model performance and clicking effort, resulting in 2.56 as the average number of clicks on PASCAL test set.

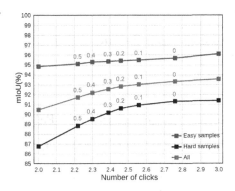

Fig. 5. mIoU results at different overlapping IoU tolerances. Overlapping IoU tolerance values are marked in green. Tested on PASCAL. (Color figure online)

3.3 Interactive Instance Segmentation Vs. RoI Methods

Extensive interactive segmentation tests are conducted to highlight the DARos performance. To align with other methods, clicking reduction with ObjDetNet and RawSegNet is not applied. Therefore only *MainSegNet* is used and the number of clicks for each object begins always from three.

To compare with state-of-the-art baselines, PASCAL VOC and Cityscapes datasets are employed. DARos outperforms IOG (Fig. 6a) from fourth clicks on PASCAL VOC. When comparing with other state-of-the-art RoI methods [21,31] focusing on automobile application, DARos is trained on PASCAL-10k and fine-tuned on 10% of Cityscapes (Table 1). As shown in Fig. 6b, fine-tuned DARos performs already significantly better than other methods trained on the full Cityscapes, which demonstrates its superior generalization capability.

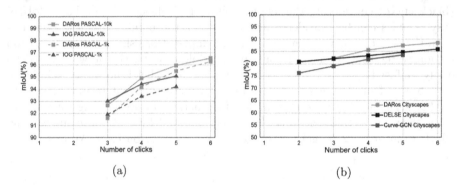

(a)   (b)

Fig. 6. Interactive instance segmentation vs. RoI methods. (a) Comparison to IOG [34] on PASCAL VOC. (b) Comparison to DELSE [31] and Curve-GCN [21] on Cityscapes.

3.4 Interactive Instance Segmentation Vs. Full-Image Methods

DARos is also compared with FI methods on PASCAL VOC, GrabCut, Berkeley, Cityscapes and MSCOCO as presented in Fig. 7. To align with other methods, clicking reduction with ObjDetNet and RawSegNet is not applied. DARos outperforms state-of-the-art FI methods particularly with FCA [20] on the different datasets except for MSCOCO (seen), where close results are obtained. Especially, the huge performance gap between FCA and DARos on Cityscapes demonstrates the limitation of FI methods when testing on high resolution images, as explained in Sect. 1.

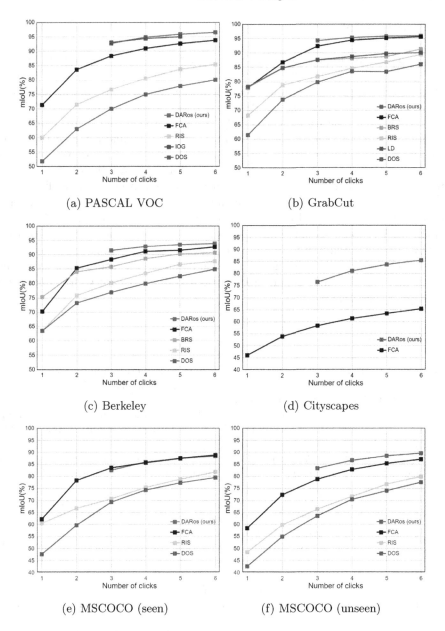

Fig. 7. Interactive instance segmentation vs. Full-image methods on PASCAL VOC, GrabCut, Cityscapes and MSCOCO (seen, unseen).

4 Conclusions

In this paper, we propose DARos, a high-quality annotation solution minimizing the cost of human-computer interaction. The two-stage instance segmentation approach is designed combining classic object detection, Inside Outside Guidance (IOG) [34] instance segmentation as well as U^2-Net [26] salient object detection. DARos automatically separates the object into easy and hard categories, where only two clicks are required for the easy category and three clicks on the hard samples to obtain a first segmentation prediction. Interactive correction is supported via additional clicks. We also propose a novel data augmentation technique to increase the number of hard samples during training. Extensive experiments demonstrate the accuracy and robustness of our DARos on diverse popular benchmarks. DARos achieves comparable performance with the state-of-the-art method IOG [34] while reducing the average number of clicks with a gain of 12% of acceleration. In addition, our proposed solutions are generic and can be applied to any other methods in the domain. Our future work will focus on integrating the three separate networks into one to reduce the annotation system complexity.

References

1. Acuna, D., Ling, H., Kar, A., Fidler, S.: Efficient interactive annotation of segmentation datasets with Polygon-RNN++. In: IEEE Conference on Computer Vision and Pattern Recognition, pp. 859–868 (2018)
2. Bai, J., Wu, X.: Error-tolerant scribbles based interactive image segmentation. In: IEEE Conference on Computer Vision and Pattern Recognition, pp. 392–399 (2014)
3. Bai, X., Sapiro, G.: A geodesic framework for fast interactive image and video segmentation and matting. In: IEEE 11th International Conference on Computer Vision, pp. 1–8 (2007)
4. Boykov, Y.Y., Jolly, M.: Interactive graph cuts for optimal boundary region segmentation of objects in N-D images. In: Proceedings Eighth IEEE International Conference on Computer Vision, vol. 1, pp. 105–112 (2001)
5. Castrejón, L., Kundu, K., Urtasun, R., Fidler, S.: Annotating object instances with a Polygon-RNN. In: IEEE Conference on Computer Vision and Pattern Recognition, pp. 4485–4493 (2017)
6. Cordts, M., et al.: The cityscapes dataset for semantic urban scene understanding. In: IEEE Conference on Computer Vision and Pattern Recognition, pp. 3213–3223 (2016)
7. Everingham, M., Gool, L.V., Williams, C.K.I., Winn, J.M., Zisserman, A.: The pascal visual object classes (VOC) challenge. Int. J. Comput. Vis. **88**(2), 303–338 (2010)
8. Geiger, A., Lenz, P., Urtasun, R.: Are we ready for autonomous driving? The KITTI vision benchmark suite. In: IEEE Conference on Computer Vision and Pattern Recognition, pp. 3354–3361 (2012)
9. Grady, L.: Random walks for image segmentation. IEEE Trans. Pattern Anal. Mach. Intell. **28**(11), 1768–1783 (2006)

10. Gulshan, V., Rother, C., Criminisi, A., Blake, A., Zisserman, A.: Geodesic star convexity for interactive image segmentation. In: IEEE Computer Society Conference on Computer Vision and Pattern Recognition, pp. 3129–3136 (2010)
11. Hariharan, B., Arbeláez, P., Bourdev, L., Maji, S., Malik, J.: Semantic contours from inverse detectors. In: International Conference on Computer Vision, pp. 991–998 (2011)
12. He, K., Zhang, X., Ren, S., Sun, J.: Deep residual learning for image recognition. In: IEEE Conference on Computer Vision and Pattern Recognition, pp. 770–778 (2016)
13. Jang, W., Kim, C.: Interactive image segmentation via backpropagating refinement scheme. In: IEEE Conference on Computer Vision and Pattern Recognition, pp. 5292–5301 (2019)
14. Kingma, D.P., Ba, J.: Adam: a method for stochastic optimization (2014)
15. Li, Z., Chen, Q., Koltun, V.: Interactive image segmentation with latent diversity. In: IEEE Conference on Computer Vision and Pattern Recognition, pp. 577–585 (2018)
16. Liew, J., Wei, Y., Xiong, W., Ong, S., Feng, J.: Regional interactive image segmentation networks. In: IEEE International Conference on Computer Vision, pp. 2746–2754 (2017)
17. Liew, J.H., Cohen, S., Price, B., Mai, L., Feng, J.: Deep interactive thin object selection. In: Winter Conference on Applications of Computer Vision (2021)
18. Lin, T., Dollár, P., Girshick, R., He, K., Hariharan, B., Belongie, S.: Feature pyramid networks for object detection. In: IEEE Conference on Computer Vision and Pattern Recognition, pp. 936–944 (2017)
19. Lin, T.-Y., et al.: Microsoft COCO: common objects in context. In: Fleet, D., Pajdla, T., Schiele, B., Tuytelaars, T. (eds.) ECCV 2014. LNCS, vol. 8693, pp. 740–755. Springer, Cham (2014). https://doi.org/10.1007/978-3-319-10602-1_48
20. Lin, Z., Zhang, Z., Chen, L.Z., Cheng, M.M., Lu, S.P.: Interactive image segmentation with first click attention. In: IEEE Conference on Computer Vision and Pattern Recognition, pp. 13336–13345 (2020)
21. Ling, H., Gao, J., Kar, A., Chen, W., Fidler, S.: Fast interactive object annotation with Curve-GCN. In: IEEE Conference on Computer Vision and Pattern Recognition, pp. 5252–5261 (2019)
22. Mahadevan, S., Voigtlaender, P., Leibe, B.: Iteratively trained interactive segmentation. In: British Machine Vision Conference, BMVC, Newcastle, UK, 3–6 September 2018, p. 212. BMVA Press (2018)
23. Maninis, K., Caelles, S., Pont-Tuset, J., Van Gool, L.: Deep extreme cut: from extreme points to object segmentation. In: IEEE Conference on Computer Vision and Pattern Recognition, pp. 616–625 (2018)
24. McGuinness, K., O'Connor, N.E.: A comparative evaluation of interactive segmentation algorithms. Pattern Recognit. **43**(2), 434–444 (2010)
25. Papadopoulos, D.P., Uijlings, J.R.R., Keller, F., Ferrari, V.: Extreme clicking for efficient object annotation. In: IEEE International Conference on Computer Vision, pp. 4940–4949 (2017)
26. Qin, X., Zhang, Z., Huang, C., Dehghan, M., Zaiane, O.R., Jagersand, M.: U2-net: going deeper with nested u-structure for salient object detection. Pattern Recognit. **106**, 107404 (2020)
27. Ramadan, H., Lachqar, C., Tairi, H.: A survey of recent interactive image segmentation methods. Comput. Vis. Media **6**(4), 355–384 (2020). https://doi.org/10.1007/s41095-020-0177-5

28. Ren, S., He, K., Girshick, R., Sun, J.: Faster R-CNN: towards real-time object detection with region proposal networks. IEEE Trans. Pattern Anal. Mach. Intell. **39**(6), 1137–1149 (2017)

29. Rother, C., Kolmogorov, V., Blake, A.: GrabCut: interactive foreground extraction using iterated graph cuts. ACM Trans. Graph. **23**(3), 309–314 (2004)

30. Shorten, C., Khoshgoftaar, T.: A survey on image data augmentation for deep learning. J. Big Data **6**, 1–48 (2019)

31. Wang, Z., Acuna, D., Ling, H., Kar, A., Fidler, S.: Object instance annotation with deep extreme level set evolution. In: IEEE Conference on Computer Vision and Pattern Recognition, pp. 7492–7500 (2019)

32. Xu, N., Price, B., Cohen, S., Huang, T.: Deep image matting. In: IEEE Conference on Computer Vision and Pattern Recognition, pp. 311–320 (2017)

33. Xu, N., Price, B., Cohen, S., Yang, J., Huang, T.: Deep interactive object selection. In: IEEE Conference on Computer Vision and Pattern Recognition, pp. 373–381 (2016)

34. Zhang, S., Liew, J.H., Wei, Y., Wei, S., Zhao, Y.: Interactive object segmentation with inside-outside guidance. In: IEEE Conference on Computer Vision and Pattern Recognition, pp. 12231–12241 (2020)

Cluster-based Convolutional Baseline for Multi-Camera Vehicle Re-identification

Rajat Mehta[1]($\boxtimes$) iD, Markus Kaechele[2] iD, Didier Stricker[3] iD,
and Naim Bajcinca[1] iD

[1] Department of Mechatronics, TU Kaiserslautern, Kaiserslautern, Germany
{rajat.mehta,naim.bajcinca}@mv.uni-kl.de
[2] Ikara Vision Systems UG, Kaiserslautern, Germany
markus.kaechele@ikara.ai
[3] DFKI, Kaiserslautern, Germany
didier.stricker@dfki.de

Abstract. Increased traffic-related problems like traffic rule violations, traffic congestions, road accidents, etc., have resulted in huge demands for smart traffic management systems worldwide. Vehicle re-identification is a key component of such systems where the goal is to identify vehicles across images that have been captured by different cameras. In this work, we propose a Cluster-based Convolutional Baseline (CCB) model that extracts part-level features from the vehicle mages using a partitioning method based on K-means clustering. The extracted part-level features provide fine-grained information present in the images that is necessary to differentiate between two very similar looking vehicle images and helps to overcome some of the biggest challenges of vehicle re-identification like intra-class differences i.e. when two images of the same vehicle (captured from different viewpoints) look very different and inter-class similarities in which two different vehicles look very similar to each other. We evaluated our model using mean average precision (mAP) and cumulative characteristics curve (CMC) evaluation metrics and achieved an mAP of 84.5%, Rank@1 and Rank@5 accuracy of 96.7% and 98.7% respectively, beating the existing state-of-the-art models on VeRi-776 dataset. Code is available at https://github.com/Rajat-Mehta/Vehicle_Reidentification.

Keywords: Re-identification · Convolutional neural networks · Object tracking

1 Introduction

A lot of work has been done in the past decade focusing on smart traffic management systems and problems like road accidents, traffic congestions and traffic rule violations, etc. Some of these works include person/pedestrian re-identification, traffic anomaly detection, vehicle re-identification, and vehicle speed detection, etc. Vehicle re-identification (re-ID) is one of the least studied topics among these but has huge research significance and real-life applications. These models can be used in traffic surveillance systems to track vehicles in real-time, to analyze the

S. Sclaroff et al. (Eds.): ICIAP 2022, LNCS 13232, pp. 541–552, 2022.
https://doi.org/10.1007/978-3-031-06430-2_45

traffic flow patterns within a city and to study the behavior of individual drivers and vehicles, etc. The goal of a vehicle re-identification model is to re-identify a query vehicle image from a set of images called gallery image set. The gallery set might consist of thousands of images of different vehicles captured by multiple cameras from different viewpoints. It faces many challenges, for example, inter-class similarities and intra-class differences between vehicles, bad lighting conditions, low-resolution images, etc. That means the minute details present in target images need to be analysed to distinguish between vehicles in such challenging environments.

In this work, we propose a convolutional neural network called cluster-based convolutional baseline (CCB) that is specifically designed for the re-identification task and uses part-level features present in the vehicle images. The proposed model deals with the challenges that we mentioned above and helps in capturing the minute details present in the vehicle images that are useful in distinguishing between two very similar looking vehicles and in learning discriminative feature representation from different vehicles. The main idea behind CCB is to divide the vehicle image into p different parts in a way such that the similar parts of the vehicle (for example, parts of the vehicle with similar texture like roof, headlights, and sides of the vehicle) are grouped together. Then, we extract the part-level features from each part and combine them to form a global representation of the vehicle images. It creates a unique representation for each vehicle that is capable of capturing viewpoint invariant features which are useful when a query image and the gallery images are taken from different viewpoints. The detailed explanation of CCB is given in Sect. 4.

The remainder of this paper consists of the following sections: Sect. 2 covers the background and related work. In Sect. 3 we explain different variants of our baseline models along with their experimental results. Then, in Sects. 4 and 5 we explain our cluster-based convolutional baseline (CCB) model followed by results and a comparison with other approaches. Finally, in Sect. 6 we conclude our work and also potential avenues for future improvements are investigated before the paper is concluded.

2   Related Work

In this section, we will review the related work and research on re-identification with a focus on three main aspects: feature representation, metric learning, and vehicle re-identification.

Feature Representation: The goal of representation learning is to train a machine learning algorithm to learn useful representation from the given data. It is very important for the re-identification model to extract fine-grained information present in the vehicle images and learn a discriminative feature representation that takes into consideration viewpoint variations, illumination changes, inter-class similarities, scale and color invariance, etc. Prior to deep learning, most feature learning approaches were based on handcrafted features like HSV color histogram [2], LBP histogram [3], SIFT [4], etc. An extensive use of these features can be seen in [2,5]. More recent approaches are based on deep learning models that use large amounts of training data and learn to map high dimensional images to low dimensional image embeddings without manual intervention. In [23], authors utilize different directional pooling layers, i.e. horizontal average pooling (HAP), vertical average pooling (VAP), diagonal average pooling (DAP) and anti-diagonal average pooling (AAP) pooling layers to extract quadruple directional features from vehicle images to reduce the adverse effect of viewpoint variations which is one of the biggest challenges in re-identification problems. In [24], a two branch neural network is proposed that consists of a stripe-based branch that extracts part-level features using a horizontal average pooling and dimension-reduced convolutional layers. The attribute-aware branch, on the other hand, is used to extract the global features. Both local and global features are then combined to form the final image descriptor for vehicle re-ID.

Metric Learning: Metric learning is closely related to re-ID and its goal is to learn a feature space and transform vehicle images to that space in a way such that samples belonging to the same class are closer and samples of the different classes are far away. In [7], the authors proposed a siamese network that consists of two identical networks. These two networks share weights and learn to map input images to the new feature space. Contrastive loss [8] and triplet loss [9] are the two widely used loss functions for metric learning. The idea behind these loss functions is to assist the re-ID models in transforming data samples to the new feature space by utilizing the relationships between image pairs or triplets respectively. These methods often suffer from slow convergence because they use only one negative example and do not interact with other negative classes. In [25], this problem is addressed by a new objective function that generalizes triplet loss by allowing joint comparisons among multiple negative examples.

Vehicle Re-identification: In the past, many approaches using handcrafted feature representations have been proposed for vehicle re-ID. For example, Arth et al. [11] use PCA-SIFT to extract information from vehicle images and represent it as a vocabulary tree. Many approaches have been proposed that rely on license plate verification, for example, [13] proposed a method which uses a combination of vehicle's appearance-based features, license plate and spatio-temporal information to build a progressive retrieval pipeline. Wu et al. [12] extract license plate features and fuse them with global features of the vehicle using a weight learned by the network. A problem with such models is that the license plates are not always visible, for example, in bad illumination conditions and in case of criminal activities license plates might be intentionally occluded or forged. Therefore, methods based on the complete appearance of vehicles have more research significance.

Most of the methods that we have mentioned above are based on global features of vehicles and focus less on local discriminative regions. These local regions, however, can be useful to discriminate between two very similar-looking vehicles. In [14], Zhang et al. introduced a part-guided attention network (PGAN) that detects prominent parts in the vehicle image and then finds their importance using a separate part attention module (PAM). PAM provides the model with abilities to calculate attention in important regions and at the level of different car parts. Authors of [26] introduce a part detector to recognize different vehicle parts, such as front, back, left, and right which provides geometric information about the structure of the vehicles. With this information, different part feature extractors are used to filter the hard negative candidates improving the overall accuracy. A multi-scale feature network with an attention module has been proposed in [27] to integrate local and global features. It helps in reducing the loss of information and preserving the discriminative features of the vehicle using an attention module. Inspired by the advantages of deep metric learning and the part-level feature extraction models [1], we propose a new vehicle re-identification method which is explained in the next chapters.

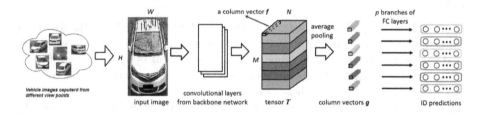

Fig. 1. PCB model adapted to vehicle re-ID task. Image adapted from [1].

3   Experiments with PCB

Part-based convolutional baseline (PCB) model was introduced in [1] as an advancement of the part-level feature extraction models. It was first used in person re-identification to extract part-level features from the target images of persons for facilitating the re-ID and image retrieval tasks. The idea behind PCB is to divide the vehicle image into coarsely partitioned parts that capture the part-level features corresponding to the actual parts of the vehicle image. As shown in Fig. 1, the input to the PCB model is an ROI extracted from the input image using classical object detection approaches, for example, YOLO [21] and Faster R-CNN [22]. The extracted ROI is then fed to the PCB network and processed by a stack of convolutional layers to generate a tensor T. This tensor (T) is then divided into p parts and each of those parts is passed on to an average pooling layer to obtain column vectors g. Finally, each of the column vectors g is fed to the corresponding classifiers which are implemented with a fully-connected layer followed by a softmax layer producing the probability distributions of the output classes. Each of these p classifiers is being specialised in classifying the individual parts into one of the output classes (vehicle ids). The model is optimized by minimizing the Cross-Entropy loss for each part individually. Once the model has been trained, the column vectors g are concatenated to form a single vector G which acts as the final descriptor for the input vehicle image.

3.1   Limitations of PCB

PCB turns out to be a good baseline for the task of vehicle re-identification but it suffers from the within-part content inconsistency problem as discussed in [1]. This problem occurs because of the uniform partitioning that is performed on tensor T which generates a large number of outliers within each part. These outliers turn out to be closer to the contents of other parts leading to the content inconsistencies among the generated parts. The uniform partitioning scheme works well for person images because of the fixed structure of human bodies (which always consists of the head at the top, then shoulders and chest, thighs, legs and feet at the bottom). Unlike pedestrian images, the content present in a vehicle's image can vary significantly depending on the viewpoint from which the vehicle is captured.

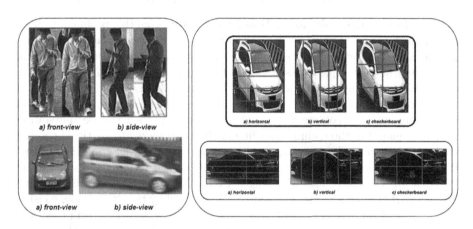

Fig. 2. (Left) Two images of a person and a car captured from two different viewpoints. (Right) Comparison of different partitioning schemes.

As shown in Fig. 2 (left), two images of the same person taken from two different viewpoints have a lot of overlap but if we compare this with two images of the same vehicle taken from front and side views, the vehicle looks completely different in both images. That means, the content present in the vehicle images can vary significantly depending on the viewpoint from which it is captured leading to a bad feature representation created by uniform partitioning. We performed experiments with different partitioning schemes like vertical, horizontal and checkerboard partitioning and found that horizontal partitioning works well for vehicle images captured from a front view and the vertical partitioning works better for images captured from a side view as shown in Fig. 2 (right). However, to solve the within-part content inconsistency problem we propose a new partitioning scheme that generates partitions by clustering the visually similar parts together irrespective of where they are present in the image.

4   Cluster-Based Convolutional Baseline

The main idea behind partitioning the feature maps is to generate p parts that are semantically similar to each other, and train the corresponding classifiers specialised on these parts. As discussed in the last section and shown in Fig. 2 (left), this can not be achieved if we apply uniform partitioning to vehicle images because it generates these parts based on fixed partitioning and thus ignores the semantic information present in them. An intuitive way to understand this problem is as follows: the column vectors f present in the same part of tensor T (see Fig. 1) should be similar to each other and dissimilar to vectors present in the other parts. If this is not the case, we can say that our model is not generating semantically consistent partitions which could otherwise help us in training p classifiers each focused on the corresponding part-level features (for example, headlight, tires, license plates, and vehicle roof, etc.) of the vehicles.

Cluster-Based Convolutional Baseline: To overcome this problem, we propose a cluster-based partitioning method that uses K-means clustering [15] for generating p partitions of the feature maps (tensor T). The new method aims at reinforcing the within-part consistency in the newly generated parts. The model is trained in two stages: first, we train a clustering model which learns to group a given set of column vectors (f) into p groups (clusters) in a way such that each group contains column vectors that are similar, then we use the trained clustering model in the training pipeline of the CCB model to partition the feature maps. Figure 3 shows a visual representation of the clustering mechanism. We randomly sample a set of vehicle images from the training set and extract their feature vectors (tensor T) using a PCB model that is pre-trained with a uniform partitioning scheme. The tensor T generated by the PCB model consists of multiple column vectors, all of which are collected together to form a dataset for our clustering method. Once we have trained the K-means algorithm on the collected data and generated p cluster centroids, we take our CCB model and train it with the new clustering-based partitioning scheme. As shown in Fig. 4, we feed the training images to our model that generates the feature vector or tensor T as output which consists of multiple column vectors (f). To assign each of those column vectors to their respective groups, we calculate the cosine similarity between vector f and the p clusters generated in the previous step.

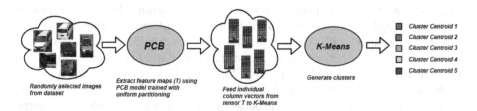

Fig. 3. Using a PCB model trained with uniform partitioning to extract feature maps which are then used to generate p clusters.

All vectors which have the highest cosine similarity with cluster p_i are assigned to the ith cluster group respectively. Once we have group assignments, we use an average pooling layer to generate the column vectors g which are then fed to the classifiers for predicting the class Ids.

4.1   Training Process of CCB

As a prerequisite to training a cluster-based convolutional baseline model, we need a PCB model trained with the uniform partitioning scheme. Then we take a subset of training images and feed them to this pre-trained model which generates as many feature vectors (tensor T) as many images are fed to it. All of these feature vectors, in turn, consist of multiple column vectors (f) which need to be grouped based on their similarities. We collect all such column vectors and create a new dataset where each column vector f with n number of dimensions is a single data item. With this data, we train a K-Means clustering model which groups the given set of column vectors into p clusters.

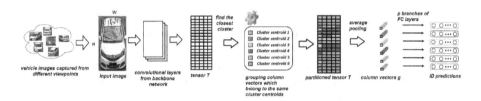

Fig. 4. Architecture of cluster-based convolutional baseline model.

The cluster centroids act as the representatives of the p parts that we are going to create next. The column vectors present in tensor T are then grouped based on their cosine similarities with these centroids. The vectors which have the highest similarity with the same centroid are assigned to the same part and this gives us p unique parts. Given these newly created p parts, we perform average pooling on them which gives us the column vectors g which are then fed to the corresponding classifiers to predict class Ids. The introduced clustering module in the CCB pipeline helps to overcome the content inconsistency problem without needing to train the separate p-part classifiers as proposed in the original paper.

5   Experiments

5.1   Dataset

We used VeRi-776 [10] dataset for our experimental purposes. It consists of images captured in real-time and unconstrained traffic settings. It is divided into two parts: a test set consisting of 11579 images of 200 different vehicles and a training set consisting of 37781 images of 576 vehicles. In addition to this, a subset of the gallery/test set consisting of 1678 images is kept as a query image set that will be used for testing the model. In total, it consists of 51,035 images of vehicles belonging to 776 classes captured by 20 different cameras. Each vehicle is captured by 2–18 separate cameras during different times of the day.

5.2 Evaluation Metrics

We evaluated our vehicle re-identification system on widely used metrics named cumulative matching characteristics (CMC) and mean average precision (mAP). The CMC score gives a probability value that tells if our model is able to fetch correct images in top-1 or top-5 result lists. The advantage of using CMC is that it not only returns Rank@1 accuracy but also tells us if our model is able to fetch correct images in top-5 and top-10 lists. Mean average precision (mAP) uses precision and recall of the model to quantify how good our model is at performing the re-identification for the given query images.

5.3 Implementation Details

We trained the CCB model on VeRi-776 and compared its performance with the two variants of PCB (one with uniform partitioning and the other with checkerboard partitioning). It was trained in two steps: first, we trained a K-means clustering model to generate six group representatives, then we plugged it in our CCB model and used it to partition tensor T into six parts.

Training K-means Clustering: We trained a K-means clustering model that would be used to cluster the individual vectors present in the feature maps into six clusters. To collect data for training K-means, we fed 3,136 randomly selected vehicle images to a pre-trained PCB model and extracted their feature maps. It returned a $(24 \times 12 \times 2048)$ dimensional tensor for each input image i.e. 288 vectors of dimension 2048 were extracted from a single image which in total gave us 903,168 vectors. These vectors were fed to the clustering model and after training, it returned six cluster centroids c which represented the corresponding cluster groups.

Cluster-Based Convolutional Baseline: Given the six cluster centroids, we trained our CCB model on VeRi-776 for 40 epochs by plugging in our custom cluster-based partitioning block in the previous architecture shown in Fig. 4. As shown in the figure, the input images go through the convolutional layers of the trained baseline model which generates a tensor T ($24 \times 12 \times 2048$). For each vector in this tensor, we find the nearest cluster centroid among the six centroids (c). All of these vectors are assigned with a label of their nearest cluster centroid which results in six different groups of vectors (see Fig. 5 for visualization of cluster plots).

These groups are then fed to the average pooling layer which performs their spatial downsampling and returns six column vectors g which are fed to the classifiers implemented with fully connected layers. The layers before our partitioning block remain unchanged and have already been fine-tuned on VeRi-776, which induces the training for our CCB model. So, after adding the new clustered partitioning block we perform another induced training for all layers of the CCB model which converges very fast.

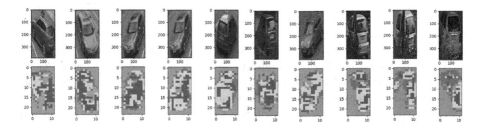

Fig. 5. Visualization of cluster plots/partitions generated by K-means clustering for tensor T (best viewed in color). (Color figure online)

Unlike PCB, the cluster-based convolutional baseline allows for non-uniform partitions. It doesn't put any restrictions on how partitions should be created. Instead, it creates six cluster representatives and assigns vectors present in tensor T to each of them based on a distance-based similarity measure. That means, irrespective of whether a vector is at the top of tensor T or bottom or wherever, it will be assigned to the part to which it is most similar. That means, the vectors in each part will be similar to each other and dissimilar to vectors in other parts and that solves the within-part inconsistency problem. On evaluation, we achieved Rank@1 accuracy of 96.7%, Rank@5 accuracy of 98.7% and 84.5% mAP which is 3.9% and 6.6% higher than PCB-checkerboard (80.3%) and PCB (77.9%) respectively. We validated the choice of $c = 6$ i.e. creating six clustered partitions by training the CCB model with three values of c: 4, 6 and 8 and $c = 6$ performed best followed by 4 and then 8, as shown in Table 1.

Table 1. Experiments with different number of cluster partitions on VeRi-776 dataset.

# of clusters	Rank@1	Rank@5	Rank@10	mAP
4	96.9%	**98.8%**	99%	83.3%
6	**96.7%**	98.7%	**99.1%**	**84.5%**
8	96.7%	98.6%	99%	82.1%

Cluster Plots: As discussed above, the CCB model does not restrict the partitioning block to divide feature maps into fixed patterns and is focused to create partitions such that similar vectors are grouped together. Figure 5 shows the cluster plots generated by the CCB model and it can be seen that the patterns present in the cluster plots generated by the model resemble the patterns seen in the vehicle image. It is also evident from these plots that the model can recognize the background and occlusions very well and group them in the same cluster (labeled with green color).

Table 2 shows mAP and CMC results achieved in our experiments in comparison to the state-of-the-art vehicle re-identification models on VeRi-776 dataset. We use cross-camera-search evaluation protocol suggested in [6]. It can be seen that our CCB model performs best in terms of mAP value followed by the Multi granularity net and VehicleNet. The top three results other than the CCB and PCB-checkerboard i.e. Multi-granularity net [20], VehicleNet [18] and PGAN [16] models show that learning part and cluster-based features gives the model the ability to differentiate between vehicle images irrespective of view-point, illumination conditions, inter-class similarities, and natural occlusions, etc., which are some of the biggest challenges of vehicle re-identification systems.

Table 2. Comparing results of different methods on VeRi-776.

Methods	Rank@1	Rank@5	Rank@10	mAP
ResNet50-baseline	94.5%	97.8%	98.7%	70.3%
PCB	96.8%	98.5%	99%	77.9%
PCB-checkerboard	95.8%	98.6%	98.9%	80.3%
PGAN [16]	96.5%	–	–	79.3%
Part-regularized [17]	94.3%	**98.7%**	–	74.3%
VehicleNet [18]	96.8%	–	–	83.41%
PAMTRI [19]	92.86%	96.97%	–	71.88%
Multi-granularity [20]	**96.9%**	–	–	83.9%
CCB [Ours]	96.7%	**98.7%**	**99.1%**	**84.5%**

6   Concluding Remarks

In this work, we proposed a cluster-based convolutional baseline (CCB) model to learn part-level discriminative features from the vehicle images which can be used to find similarities between vehicle images. CCB uses a new partitioning scheme that generates semantically similar parts and overcomes the within-part content inconsistency problem discussed in Sect. 3.1. We used a K-means clustering model (trained on more than 900,000 vectors) to generate six parts. These part-level features are then fed to the corresponding classifiers that are specialised in classifying them into one of the output classes. Unlike the previous partitioning methods in which the model is restricted to generate partitions in fixed patterns, the clustered partitioning generates semantically consistent partitions leading to an improvement in the mAP scores.

Currently, the re-identification system consists of two parts: a vehicle detection model and a re-ID model, both of which are being trained separately. In the future, we can extend our approach by training both of these models simultaneously and improve them together. We also showed in our work that part-based features play an important role in re-identification and image retrieval tasks. This approach can be further improved by using attention-based models for proposing important regions in the image, that can then be combined with part-based features to capture fine-grained information and details of different parts of the vehicles.

References

1. Sun, Y., Zheng, L., Yang, Y., Tian, Q., Wang, S.: Beyond part models: person retrieval with refined part pooling (and a strong convolutional baseline). In: Proceedings of the European Conference on Computer Vision (ECCV), pp. 480–496 (2018)
2. Farenzena, M., Bazzani, L., Perina, A., Murino, V., Cristani, M.: Person re-identification by symmetry-driven accumulation of local features. In: IEEE Computer Society Conference on Computer Vision and Pattern Recognition, pp. 2360–2367 (June 2010)
3. Li, W., Wang, X.: Locally aligned feature transforms across views. In: Proceedings of the IEEE Conference on Computer Vision and Pattern Recognition, pp. 3594–3601 (2013)
4. Zhao, R., Ouyang, W., Wang, X.: Unsupervised salience learning for person re-identification. In: Proceedings of the IEEE Conference on Computer Vision and Pattern Recognition, pp. 3586–3593 (2013)
5. Ma, B., Su, Y., Jurie, F.: Local descriptors encoded by fisher vectors for person re-identification. In: Fusiello, A., Murino, V., Cucchiara, R. (eds.) ECCV 2012. LNCS, vol. 7583, pp. 413–422. Springer, Heidelberg (2012). https://doi.org/10.1007/978-3-642-33863-2_41
6. Liu, X., Liu, W., Mei, T., Ma, H.: A deep learning-based approach to progressive vehicle re-identification for urban surveillance. In: Leibe, B., Matas, J., Sebe, N., Welling, M. (eds.) ECCV 2016. LNCS, vol. 9906, pp. 869–884. Springer, Cham (2016). https://doi.org/10.1007/978-3-319-46475-6_53
7. Bromley, J., Guyon, I., LeCun, Y., Säckinger, E., Shah, R.: Signature verification using a siamese time delay neural network. In: Advances in Neural Information Processing Systems, pp. 737–744 (1994)
8. Hadsell, R., Chopra, S., LeCun, Y.: Dimensionality reduction by learning an invariant mapping. In: IEEE Computer Society Conference on Computer Vision and Pattern Recognition (CVPR 2006), vol. 2, pp. 1735–1742 (June 2006)
9. Schroff, F., Kalenichenko, D., Philbin, J.: Facenet: a unified embedding for face recognition and clustering. In: Proceedings of the IEEE Conference on Computer Vision and Pattern Recognition, pp. 815–823 (2015)
10. Liu, X., Liu, W., Mei, T., Ma, H.: A deep learning-based approach to progressive vehicle re-identification for urban surveillance. In: Leibe, B., Matas, J., Sebe, N., Welling, M. (eds.) ECCV 2016. LNCS, vol. 9906, pp. 869–884. Springer, Cham (2016). https://doi.org/10.1007/978-3-319-46475-6_53
11. Arth, C., Leistner, C., Bischof, H.: Object reacquisition and tracking in large-scale smart camera networks. In: 2007 First ACM/IEEE International Conference on Distributed Smart Cameras, pp. 156–163 (September 2007)

12. Wu, H., Li, D., Zhou, Y., Hu, Q.: Multi-view feature fusion network for vehicle re-identification. In: International Conference on Advanced Information Technologies and Applications, pp. 39–47. Taipei, China (October 2017)

13. Li, D., Chen, X., Zhang, Z., Huang, K.: Learning deep context-aware features over body and latent parts for person re-identification. In: Proceedings of the IEEE Conference on Computer Vision and Pattern Recognition, pp. 384–393 (2017)

14. Zhang, X., Zhang, R., Cao, J., Gong, D., You, M., Shen, C.: Part-Guided Attention Learning for Vehicle Re-Identification. arXiv preprint, arXiv:1909 (2019)

15. Lloyd, S.: Least squares quantization in PCM. IEEE Trans. Inf. Theory **28**(2), 129–137 (1982)

16. Zhang, X., Zhang, R., Cao, J., Gong, D., You, M., Shen, C.: Part-guided attention learning for vehicle re-identification. arXiv preprint arXiv:1909.06023 (2019)

17. He, B., Li, J., Zhao, Y., Tian, Y.: Part-regularized near-duplicate vehicle re-identification. In: Proceedings of the IEEE Conference on Computer Vision and Pattern Recognition, pp. 3997–4005 (2019)

18. Zheng, Z., Ruan, T., Wei, Y., Yang, Y.: VehicleNet: learning robust feature representation for vehicle re-identification. In: CVPR Workshops, vol. 2, p. 3 (June 2019)

19. Tang, Z., et al.: Pamtri: pose-aware multi-task learning for vehicle re-identification using highly randomized synthetic data. In: Proceedings of the IEEE International Conference on Computer Vision, pp. 211–220 (2019)

20. Chen, Y., Jing, L., Vahdani, E., Zhang, L., He, M., Tian, Y.: Multi-camera vehicle tracking and re-identification on AI city challenge 2019. In: Proceedings of the IEEE Conference on Computer Vision and Pattern Recognition Workshops, pp. 324–332 (2019)

21. Redmon, J., Farhadi, A.: YOLOv3: an incremental improvement. arXiv preprint arXiv:1804.02767 (2018)

22. Ren, S., He, K., Girshick, R., Sun, J.: Faster R-CNN: towards real-time object detection with region proposal networks. Adv. Neural Inf. Process. Syst. **28**, 91–99 (2015)

23. Zhu, J., et al.: Vehicle re-identification using quadruple directional deep learning features. IEEE Trans. Intell. Transp. Syst. **21**(1), 410–420 (2019)

24. Qian, J., Jiang, W., Luo, H., Yu, H.: Stripe-based and attribute-aware network: a two-branch deep model for vehicle re-identification. Meas. Sci. Technol. **31**(9), 095401 (2020)

25. Sohn, K.: Improved deep metric learning with multi-class n-pair loss objective. In: Advances in Neural Information Processing Systems, pp. 1857–1865 (2016)

26. Jiang, M., et al.: Robust vehicle re-identification via rigid structure prior. In: Proceedings of the IEEE/CVF Conference on Computer Vision and Pattern Recognition, pp. 4026–4033 (2021)

27. Li, L., Xu, Y., Zhang, X.: Weighted local feature vehicle re-identification network. In: Proceedings of the 4th International Conference on Computer Science and Application Engineering, pp. 1–5 (October 2020)

Road Quality Classification

Martin Lank[ID] and Magda Friedjungová[(⊠)][ID]

Faculty of Information Technology, Czech Technical University in Prague,
Prague, Czech Republic
{lankmart,magda.friedjungova}@fit.cvut.cz

Abstract. Road quality significantly influences safety and comfort
while driving. Especially for most kinds of two-wheelers, road damage
is a real threat, where vehicle components and enjoyment are heavily
impacted by road quality. This can be avoided by planning a route considering the surface quality. We propose a new publicly available and
manually annotated dataset collected from Google Street View photos.
This dataset is devoted to a road quality classification task considering
six levels of damage. We evaluated some preprocessing methods such as
shadow removal, CLAHE, and data augmentation. We adapted several
pre-trained networks to classify road quality. The best performance was
reached by MobileNet using augmented dataset (75.55%).

Keywords: Road quality · Image classification · Data preprocessing ·
Google Street View

1 Introduction

Every time we travel, we tend to optimise the route to get the best directions
based on our parameters - typically time, fluency, absolute distance, or low traffic.
Considering these parameters, we try to choose the most suitable route using
our knowledge or modern navigation software.

However, this navigation software is not suitable for everyone. If we consider
two-wheelers or vintage cars, the typical main optimisation parameter is road
quality. In this case the least damaged roads are most desirable. Motorcyclists
typically want the road to be as smooth as possible. It should be pointed out
that even one pothole or crack on an overall smooth road can make it dangerous.
Although we can find such quality roads quite easily on highways and main roads,
it goes against the second main parameter - playfulness. That includes low traffic
and winding roads, which is quite the opposite of what the highway and main
roads look like.

Nowadays, services like Google Street View can be used to visually check the
road and get an idea of what it looks like. Unfortunately, it can easily take an hour
to plan a hundred kilometre long journey. Using machine learning approaches to
automate this visual check could streamline the process even more.

The main contribution of this work is a new publicly available dataset devoted
to road quality classification. The dataset was collected from Google Street

S. Sclaroff et al. (Eds.): ICIAP 2022, LNCS 13232, pp. 553–563, 2022.
https://doi.org/10.1007/978-3-031-06430-2_46

View photos and manually annotated into six road quality-descriptive categories. Proposed dataset brings several challenges in data preprocesing as is described below. We evaluated some preprocessing methods and also present a comparison of several pre-trained networks to classify road quality.

1.1 Related Work

Automated evaluation of road quality can be done by classifying acceleration data [2,12,30], or images [18,21]. However, using acceleration data lacks scalability as there is no public provider of such data on a similar scale like e.g. Google Street View for images. Also, the vehicle needs to go through the defects, which is something riders typically want to avoid. Due to this we focus on images only.

Many recent works [8,15,19] do not primarily solve classification problems, but focus on damage detection, where every corruption such as a pothole or a crack is detected (as object segmentation). Many recent solutions were developed thanks to the Global Road Damage Detection Challenge 2020[1].

Let us mention the following works as examples focusing on classification. In [18] the objective was to use images and Convolutional Neural Networks (CNN) to classify road surface by type. They defined six classes: asphalt, dirt, grass, wet asphalt, cobblestone, snow. Furthermore, they combined several datasets (KITTI [9], RobotCar [17], NREC [20], New College [25], Stadtpilot (private)). The best model was based on ResNet50 with an average accuracy of 92%.

In [21] road type and quality classification were done using two existing datasets KITTI and CaRINA [24] and a new RTK dataset created by taking frames from a video and assigning them categories: asphalt, paved and unpaved, where each was further subcategorised by their surface quality into good, regular and bad. The proposed ensemble model reached an accuracy of 95.73%.

In [16] the authors created their own large-scale dataset with about 700000 samples by matching government road inspection data and images from Google Street View. The images were classified into three categories: good, fair, and poor surface. The authors utilised Fisher-Vector CNN [6] and reached an accuracy of 58%.

To conclude, the typical image classification approach is to use CNN [10]. Neural networks in general can be trained using supervised learning to determine the predefined categories. However, they usually require large datasets with thousands of samples. If the dataset is relatively small and does not cover enough diversity in a given domain, data augmentation can sometimes be helpful.

1.2 Publicly Available Datasets

In Table 1 we summarize publicly available image datasets that could potentially be useful for road quality classification.

As can be seen, majority of the datasets are not designed for classification purposes. The annotated datasets discretise quality into only three classes. Also,

---

[1] https://rdd2020.sekilab.global/, https://github.com/sekilab/RoadDamageDetector.

Table 1. Publicly available image datasets devoted to road quality detection.

Dataset	Task	Num. of Classes	Samples
RTK [21]	Type + quality classification	3 (good, regular, bad)	6297
Paris-Saclay[16]	Quality classification	3 (good, fair, poor)	∼700 000
KITTI [9]	Autonomous driving	–	–
CaRINA [24]	Surface detection	3 (easy, medium, difficult)	900
RobotCar [17]	Autonomous driving	–	–
GRDDC 2020 [3]	Damage segmentation	–	26620
DIPLODOC [31]	Road segmentation	–	865
RQ Dataset (ours)	Quality classification	6 (described in 2)	7247

the Paris-Saclay dataset might have misleading labels (due to a 1.2 year time gap in data records) and only focuses on urban roads, which motorcycle riders typically want to avoid due to higher traffic. Class imbalance is also a problem for RTK and Paris-Saclay datasets as the poor class contains only 0.7% of all data. Furthermore, due to the wide field of view, the images contain unwanted objects such as buildings, cars, trees or traffic signs. In [21] the authors showed that combining the RTK dataset with CaRINA and KITTI did not always lead to increased performance on RTK test set, but did lead to an increase in performance on the CaRINA and KITTI test set. To sum up, all available datasets would have to be re-annotated and combined or extended with more data. However, that might not necessarily increase the performance.

2   Road Quality Dataset

In regards to the perception and riding experience of those that ride two-wheeled vehicles, we propose the following six categories:

1. Top-quality roads - smooth surface without any damage. The rider can fully enjoy the ride.
2. Overall smooth surface with tiny longitudinal cracks and/or repairs not affecting the safety and riding pleasure.
3. Roads with patches, moderate linear cracks in any directions. Riders must pay some attention to the road.
4. Bumpy roads with potholes or multiple (layered) repairs. Riders must pay very close attention to the road, and the fun factor is gone.
5. Stone paved or unpaved roads. Extremely dangerous for motorcyclists, especially in the rain due to low adhesion. Only plausible in case of emergency with extra caution.
6. Generally pictures that do not contain roads, or only a small fraction of it not allowing to determine the quality. E.g. images with cars, trees, fields or buildings.

Based on these categories, we collected and annotated a new dataset called Road Quality (RQ) Dataset. The dataset is available in the Github repository[2]. We collected road images from the Czech Republic with different surfaces using Google Street View and its APIs. One of our ideal dataset requirements was for images to contain minimum unwanted objects. The field of view and the pitch were empirically set to 40° and –30°, respectively. With these values, the images depict the road right ahead of a car and do not contain unwanted objects along the road. In addition, the dataset covers various lighting conditions (shadow, brightness, dark, light), various weather conditions and various road types. Next, it covers roads with and without surface markings. Several dataset examples of each category can be seen in Fig. 1.

To wrap up, we collected 7247 road images along 1115 km. The pictures have a resolution of 640 × 480. All the routes were chosen based on the authors' riding experience and then manually annotated into the six categories mentioned above. The number of images in each category is following: 2064 images per first category, 1791 per second, 1665 per third, 1353 per fourth, 214 per fifth, and 160 images per sixth category.

3   Experiments

Here we present several experiments using the original and preprocessed RQ Dataset. We also present the results of several pre-trained classification models. All the implementations were done using Anaconda Data Science Platform [1] and Python as programming language, on a machine running Windows 10 20H2 equipped with Core i7 CPU, NVIDIA GeForce GTX 1070 GPU and 32 GB RAM.

3.1   Dataset Preparation

During dataset preparation, we focused on shadow removal and contrast improvement. We also experimented with class balancing methods such as SMOTE [4] and geometric transformations.

Considering class balancing methods, we experimented with random rotation in ranges [0°, 5°] and [0°, 10°], random horizontal flip, random contrast changes with factor in 0.1, 0.2 or 0.3 and random crop to 360 × 360 and 224 × 224. Further, the combination of random contrast and horizontal flip lead to promising results. Horizontal flip or random contrast alone does not provide any improvement in resulting classification. As well as other methods such as rotation or cropping the image instead of resizing - these methods even lead to accuracy drop. SMOTE provided images with unrealistic colours and artefacts and was not used due to that.

Shadows are a big issue in scene understanding. We took advantage of the method proposed by [7]. According to the authors, it achieves impressive results on pavements, and their model is publicly available[3]. The model is based on a

---

[2] https://github.com/lenoch0d/road-quality-classification.

[3] https://github.com/vinthony/ghost-free-shadow-removal.

Fig. 1. This picture shows examples of the labelled dataset. Each row depicts images given class, starting with first class at the top end sixth at the bottom.

CNN called Dual Hierarchical Aggregation Network and trained on data partially synthesised using their shadow matting generative adversarial network.

We used CLAHE [32] to improve contrast in images - an adaptive histogram equalization with the ability to clip the output limit range. It is essential to choose a proper clip limit, which would help identify defects and not highlight the surface noise. In our case, we empirically found the best clip limit to be 1.

3.2   Classification Models Comparison

Building a machine learning model means experimenting with many different architectures, hyper-parameters, and in our case, also with differently preprocessed datasets. We implemented custom CNNs, but better results were achieved by transfer learning of pre-trained networks: MobileNet [13], MobileNetV2 [23], DenseNet-121 [14], InceptionResNetV2 [26], EfficientNet [28] and EfficientNetV2 [29].

MobileNet [13] is a lightweight and efficient CNN designed with low resource demands and a balanced accuracy-latency trade-off. The architecture uses depthwise separable convolutions [5] with a total 4.24 M parameters[4] and was trained on the ImageNet dataset [22] with resolution of 224×224. However, we primarily trained on images with the resolution of 360×360 to avoid significant detail loss on road surface needed for effective quality classification.

MobileNetV2 [23] is even lighter than the version described above (3.5M neurons), while achieving higher accuracy on the ImageNet dataset. The input preprocessing is the same as for MobileNet.

DenseNet-121 [14] is a deep CNN with 121 layers and 8M neurons. The novelty in this architecture was connecting each layer to every other subsequent layer. Compared to MobileNet, it achieves 5% higher accuracy on the ImageNet dataset. Additionally, we resized the images to the same resolution as with MobileNet.

InceptionResNetV2 [26] is a hybrid of ResNet [11] and Inception [27] architectures. ResNet introduced so-called residual connections, which are direct connections from each layer to a subsequent layer 2–3 hops away. Inception architectures use different kernel sizes in parallel to decrease sensitivity to the size of an object in an image. Therefore, the hybrid network is both deep and wide (572 layers, 55M neurons). Also, it assumes preprocessed data in the same way as for MobileNet networks.

EfficientNet [28] is a CNN built on an observation that balancing network depth, width and resolution can lead to better performance. The authors proposed a new scaling method, which uniformly scales these dimensions using a compound coefficient. There are several versions of the network. The base, EfficientNet-B0, is built on the inverted bottleneck residual blocks of MobileNetV2 and has 5.3M neurons. It expects image inputs with the resolution of 224×224. We experimented with versions B0, B1, B2 and B3, which are scaled for higher image resolutions.

---

[4] Including output classification layers for the ImageNet dataset.

EfficientNetV2 [29] generally follows the same principles as EfficientNet. However, there are some architectural changes, which makes the network even more efficient, such as smaller kernel sizes and expansion ratios. Same as EfficientNet, there are several versions. We picked EfficientNetV2-B3 with the same input expectations and EfficientNetV2-S, which is scaled to a similar resolution as EfficientNet-B4.

3.3   Training

We split all three variations (original, CLAHE, shadow-free) of our dataset into a train set (65%), validation set (20%) and a test set (15%). All datasets were split using random state parameter, which guarantees a reproducibility of a problem the same every time it is run. It means that mentioned experiments were performed on the same data using the same training strategy and so the results are comparable.

Firstly, we used the original dataset to train four networks based on different aforementioned architectures. Each training had two phases. Firstly, we attached a fully connected layer with six neurons to the frozen pre-trained networks. We ran experiments for 50 epochs but we adopted the early stop functionality for cases when our evaluation metric did not improve for 10 epochs.

We used model accuracy as a common evaluation metric to measure models' performance. Then, we picked the best-performing models for the second phase, where we fine-tuned them by continuing training with unfrozen neurons of the pre-trained base network.

The results of the models trained on the original dataset and evaluated on the test set can be seen in Table 2. Based on these results the experiments with preprocessed dataset variations presented further are performed using MobileNet (72.42%).

Table 2. Classification accuracies (Acc.) of pre-trained networks on the original (unchanged) dataset. EfficientNet result is shown for version B1 and EfficientNetV2 result is shown for version S (other versions achieved worse results).

Model	MobileNet	MobileNet V2	DenseNet-121	Inception ResNetV2	EfficientNet	EfficientNetV2
Acc.	**72.42%**	70.39%	70.11%	68.73%	70.11%	69.28%

To understand how well the model classifies in each class, we calculated the F1-scores per class. The results can be seen in Table 3. A visualisation of the model output (labelled roads) can be seen in Fig. 2, where the labels are represented via coloured segments. Although the segments change their colours frequently, the quality trend is visible.

Table 3. Comparison of (test data) accuracy (Acc.) and per-class F1-scores of MobileNet model trained with different augmentation techniques against the baseline model (i.e. trained on unchanged/original data).

Augmentation	Acc.	F1-score					
		1	2	3	4	5	6
Baseline (i.e. none)	72.42%	77.48%	59.77%	73.07%	78.53%	82.54%	71.11%
Using CLAHE	70.48%	76.54%	56.81%	69.76%	78.28%	78.79%	68.29%
Shadow-free	69.28%	77.13%	51.03%	66.41%	77.59%	80.65%	69.77%
Horizontal flip	73.71%	78.65%	62.24%	73.29%	79.59%	82.35%	80.00%
Contrast (factor = 0.1)	71.13%	77.78%	55.44%	69.05%	79.28%	80.56%	79.17%
Contrast (factor = 0.2)	70.20%	76.00%	55.49%	68.31%	79.42%	79.45%	71.11%
Contrast (0.1), hor. flip	71.49%	78.03%	58.82%	70.38%	76.70%	80.60%	66.67%
Contrast (0.2), hor. flip	**75.55%**	**79.07%**	**64.96%**	**75.85%**	**81.86%**	**86.15%**	**81.82%**

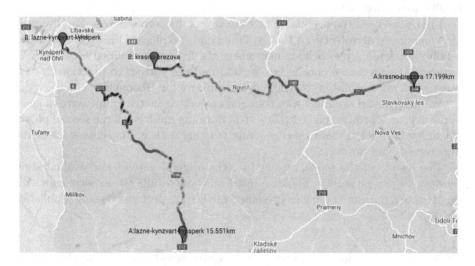

Fig. 2. Our proposed model's predictions on a map using coloured segments. The colours represent classes as follows: green = 1, green-yellow = 2, orange = 3, red = 4, purple = 5, blue = 6. Map base: Google Maps. (Color figure online)

3.4   Discussion

The MobileNet model trained on data augmented with random flips and random contrast with its 75.55% accuracy outperformed the baseline model by more then 3%. F1-scores of the classes 1–6 were increased by $1.59\%, 5.20\%, 2.78\%, 3.33\%, 3.61\%$ and 10.71%. The improvement in the second class is especially valuable since the second class had the lowest F1-score in the baseline model. The second class classification is the most challenging since it contains features from both the first and the third classes (such as perfect surface) and cracks/repairs in a minimal amount.

An explanation of the lower performance of the model trained with a CLAHE preprocessed dataset could be that the higher contrast highlights cracks (noise), making the (neighbouring) classes less separable.

The model using the shadow-free dataset performed even worse than CLAHE. We attribute it to the imperfect shadow removal process. Sometimes, it added artificial defect-looking objects to images without shadows and sometimes the shadow was only partially removed, causing the removed part to look more like a patch repair. As a result, classes became less separable.

In future work, we would like to focus on improving current classifiers. Also, there is potential to use other data augmentation methods, since it turned out to be helpful. The final model could be wrapped into an application taking waypoints as an input and producing a visualisation as an output. Furthermore, this application could automatically suggest directions between two places using only high-quality roads.

4 Conclusion

In this work, we propose a new publicly available RQ Dataset focused on driving pleasure and safety. The dataset contains 7247 road images collected from Google Street View and manually annotated into six classes. We adopted several pre-trained networks to classify road quality. Additionally, we experimented with preprocessing techniques. The best performance was reached by MobileNet network (75.55%) on the augmented dataset using combination of horizontal flip and random contrast. The results showed that the proposed dataset provides several challenges which can be tackled in future work.

Acknowledgements. This research has been supported by SGS grant No. SGS20/ 213/OHK3/3T/18 and by GACR grant No. GA18-18080S.

References

1. Anaconda software distribution (2020). https://docs.anaconda.com/
2. Afenika, A., Gunawan, P.H., Tarwidi, D.: Classification of road surface quality based on SVM method. J. Phys. Conf. Ser. **1641**, 012064 (2020). https://doi.org/ 10.1088/1742-6596/1641/1/012064
3. Arya, D., et al.: Transfer learning-based road damage detection for multiple countries (2020)
4. Chawla, N.V., Bowyer, K.W., Hall, L.O., Kegelmeyer, W.P.: Smote: synthetic minority over-sampling technique. J. Artif. Intell. Res. **16**, 321–357 (6 2002). https://doi.org/10.1613/jair.953, http://dx.doi.org/10.1613/jair.953
5. Chollet, F.: Xception: deep learning with depthwise separable convolutions. In: Proceedings of the IEEE Conference on Computer Vision and Pattern Recognition (CVPR) (July 2017)
6. Cimpoi, M., Maji, S., Kokkinos, I., Vedaldi, A.: Deep filter banks for texture recognition, description, and segmentation (2015)

7. Cun, X., Pun, C.M., Shi, C.: Towards ghost-free shadow removal via dual hierarchical aggregation network and shadow matting gan (2019)
8. Doshi, K., Yilmaz, Y.: Road damage detection using deep ensemble learning (2020)
9. Geiger, A., Lenz, P., Stiller, C., Urtasun, R.: Vision meets robotics: the kitti dataset. Int. J. Robot. Res. (IJRR) (2013). http://www.cvlibs.net/datasets/kitti/raw_data.php
10. Goodfellow, I., Bengio, Y., Courville, A.: Deep Learning. MIT Press, Cambridge (2016)
11. He, K., Zhang, X., Ren, S., Sun, J.: Deep residual learning for image recognition (2015)
12. Hoffmann, M., Mock, M., May, M.: Road-quality classification and bump detection with bicycle-mounted smartphones. In: CEUR Workshop Proceedings, vol. 1088, pp. 39–43 (01 2013)
13. Howard, A.G., et al.: Mobilenets: Efficient convolutional neural networks for mobile vision applications. CoRR abs/1704.04861 (2017)
14. Huang, G., Liu, Z., van der Maaten, L., Weinberger, K.Q.: Densely connected convolutional networks (2018)
15. Lei, X., Liu, C., Li, L., Wang, G.: Automated pavement distress detection and deterioration analysis using street view map. IEEE Access 8, 76163–76172 (2020). https://doi.org/10.1109/ACCESS.2020.2989028
16. Ma, K., Hoai, M., Samaras, D.: Large-scale continual road inspection: visual infrastructure assessment in the wild. In: Proceedings of British Machine Vision Conference (2017). https://www3.cs.stonybrook.edu/~cvl/pavement.html
17. Maddern, W., Pascoe, G., Linegar, C., Newman, P.: 1 Year, 1000 km: the oxford robotcar dataset. Int. J. Robot. Res. (IJRR) 36(1), 3–15 (2017)
18. Nolte, M., Kister, N., Maurer, M.: Assessment of deep convolutional neural networks for road surface classification. In: 2018 21st International Conference on Intelligent Transportation Systems (ITSC), pp. 381–386 (2018). https://doi.org/10.1109/ITSC.2018.8569396
19. Pei, Z., Zhang, X., Lin, R., Shen, H., Tang, J., Yang, Y.: Submission of dd-vision team in global road damage detection 2020 (2020). password: xzc6, https://pan.baidu.com/s/1VjLuNBVJGS34mMMpDkDRGQ,
20. Pezzementi, Z., et al.: Comparing apples and oranges: off-road pedestrian detection on the NREC agricultural person-detection dataset (2017)
21. Rateke, T., Justen, K.A., von Wangenheim, A.: Road surface classification with images captured from low-cost cameras - road traversing knowledge (RTK) dataset. Revista de Informática Teórica e Aplicada (RITA) (2019)
22. Russakovsky, O., et al.: ImageNet large scale visual recognition challenge. Int. J. Comput. Vis. 115(3), 211–252 (2015). https://doi.org/10.1007/s11263-015-0816-y
23. Sandler, M., Howard, A., Zhu, M., Zhmoginov, A., Chen, L.C.: Mobilenetv 2: inverted residuals and linear bottlenecks (2019)
24. Shinzato, P.Y., et al.: Carina dataset: an emerging-country urban scenario benchmark for road detection systems. In: 2016 IEEE 19th International Conference on Intelligent Transportation Systems (ITSC) (2016). https://doi.org/10.1109/ITSC.2016.7795529, dataset available from: http://www.lrm.icmc.usp.br/web/index.php?n=DataSet.Home
25. Smith, M., Baldwin, I., Churchill, W., Paul, R., Newman, P.: The new college vision and laser data set. Int. J. Robot. Res. 28, 595–599 (2009). https://doi.org/10.1177/0278364909103911
26. Szegedy, C., Ioffe, S., Vanhoucke, V., Alemi, A.: Inception-v4, inception-resnet and the impact of residual connections on learning (2016)

27. Szegedy, C., Vanhoucke, V., Ioffe, S., Shlens, J., Wojna, Z.: Rethinking the inception architecture for computer vision (June 2016). https://doi.org/10.1109/CVPR.2016.308

28. Tan, M., Le, Q.V.: Efficientnet: rethinking model scaling for convolutional neural networks (2020)

29. Tan, M., Le, Q.V.: Efficientnetv2: smaller models and faster training (2021)

30. Tiwari, S., Bhandari, R., Raman, B.: Roadcare: a deep-learning based approach to quantifying road surface quality. In: Proceedings of the 3rd ACM SIGCAS Conference on Computing and Sustainable Societies, pp. 231–242. COMPASS 2020, Association for Computing Machinery, New York, NY, USA (2020). https://doi.org/10.1145/3378393.3402284

31. Zanin, M., Messelodi, S., Modena, C.: Diplodoc road stereo sequence (April 2013). https://doi.org/10.13140/RG.2.1.3681.9929, dataset available from: https://tev.fbk.eu/databases/diplodoc-road-stereo-sequence

32. Zuiderveld, K.: Contrast limited adaptive histogram equalization. Graph. Gems IV, 474–485 (1994)

Towards Open Zero-Shot Learning

Federico Marmoreo[1], Julio Ivan Davila Carrazco[1,2(✉)] ⓘ, Jacopo Cavazza[1] ⓘ,
and Vittorio Murino[1,3] ⓘ

[1] Pattern Analysis and Computer Vision (PAVIS), Italian Institute of Technology,
Genoa, Italy
{federico.marmoreo,julio.davila,jacopo.cavazza,vittorio.murino}@iit.it
[2] Department of Marine, Electrical, Electronic and Telecommunications Engineering,
University of Genoa, Genoa, Italy
[3] Department of Computer Science, University of Verona, Verona, Italy

Abstract. In Generalized Zero-Shot Learning (GZSL), unseen categories (for which no visual data are available at training time) can be predicted by leveraging their class embeddings (e.g., a list of attributes describing them) together with a complementary pool of seen classes (paired with both visual data and class embeddings). Despite GZSL is arguably challenging, we posit that knowing in advance the class embeddings, especially for unseen categories, is an actual limit of the applicability of GZSL towards real-world scenarios. To relax this assumption, we propose Open Zero-Shot Learning (OZSL) as the problem of recognizing seen and unseen classes (as in GZSL) while also rejecting instances from unknown categories, for which neither visual data nor class embeddings are provided. We formalize the OZSL problem introducing evaluation protocols, error metrics and benchmark datasets. We also tackle the OZSL problem by proposing and evaluating the idea of performing unknown feature generation.

Keywords: Zero-Shot Learning · Open-Set

1 Introduction

After the advent of deep learning, computer vision has reached near human-level performance on a variety of tasks. However, the main operative assumption behind this outstanding performance is the availability of a large corpus of annotated data and this clearly limits the applicability in a real-world scenario. Generalized Zero-Shot Learning (GZSL) [3] considers the extreme case in which for some of the classes, i.e., the *unseen* classes, no training examples are available. The goal is to correctly classify them at inference time, together with test instances from the seen classes, and this is typically achieved relying on auxiliary semantic (e.g., textual) information describing the classes, the so-called *class embeddings* [11].

For instance, class embeddings can either consist of side information such as manually-defined attributes [11], text embeddings extracted from computational pipelines such as word2vec [15], or CNN+LSTM models trained on Wikipedia

© The Author(s), under exclusive license to Springer Nature Switzerland AG 2022
S. Sclaroff et al. (Eds.): ICIAP 2022, LNCS 13232, pp. 564–575, 2022.
https://doi.org/10.1007/978-3-031-06430-2_47

articles [26]. Desirable features of class embeddings consist in being 1) shared among classes and, at the same time, 2) discriminative. This is how one can transfer knowledge from the classes for which we have annotated visual data, i.e. the *seen* classes, to the unseen ones.

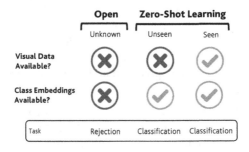

Fig. 1. Open Zero-Shot Learning, a framework where we aim at classifying seen and unseen classes (for which no visual data of the latter is given) while also rejecting (i.e., refusing to take any decision on) unknown classes. Neither visual data nor class embeddings are available for unknown classes.

In GZSL, the challenge is to overcome the bias of the model towards predicting the classes on which it has been directly trained on. To solve the extreme imbalance of the GZSL framework, much effort has been exerted to perform synthetic feature augmentation for the unseen classes [2,4,5,7,9,13,16,19,21, 25,26,29]. By exploiting deep generative models, as Generative Adversarial Networks (GANs) or Variational Auto-Encoders (VAEs), it is indeed possible to take advantage of the class embeddings to generate class consistent features for the unseen classes by training on the seen ones, leading to remarkable performances in GZSL.

However, we claim that the assumption of knowing in advance the full set of classes, the closed-set assumption, and their class embeddings is still a strong limitation for GZSL in real world applications. In fact, while it is reasonable to assume that we can describe all the seen classes with the class embeddings, it seems less reasonable not only to know, but also to describe with the rich semantic content of the class embeddings, all the classes for which we have no visual training data.

We Introduce a New Paradigm, Open Zero-Shot Learning. (Fig. 1). *Open Zero-Shot learning* (OZSL) overcomes the closed-set assumption and goes to the open-set scenario by considering a possible infinite set of classes at inference time. As a consequence, we have three types of classes: 1) *the seen*, for which we have visual data and class semantic descriptors, 2) *the unseen*, for which we have only class embeddings, and 3) *the unknown*, for which we have neither the visual data nor the (semantic) class embeddings. Thus, OZSL extends GZSL with the possibility of performing recognition in the open-set regime [20] where inference has to be jointly performed over seen, unseen and unknown classes in order to classify seen and unseen, and reject unknown ones.

We Build OZSL as the Open-set Generalization of GZSL. We design evaluation protocols, extracting unknown classes as a subpart of unseen classes from typical GZSL benchmark datasets used in the related state of the art [2,4,5,7,9,13,16, 19,21,25,26,29]. We also propose error metrics to allow fair and reproducible comparison across different algorithmic solutions tackling OZSL. We also extend prior GZSL error metrics (harmonic mean of the per-class average accuracy [24]) to better handle the open set scenario. In particular, we consider F1-score between seen and unseen average precision and/or recall scores to better account for successful rejections.

We Approach OZSL by Synthesizing the Unknown. In GZSL, GANs or alternative generative methods [2,4,5,7,9,13,16,19,21,25,26,29] generate visual features conditioned on class embeddings of the unseen classes and train a softmax classifier on top of them as well as of real seen features. We provide a preliminary exploratory analysis on unknown visual features generation by directly using GZSL baseline methods and by evaluating our novel idea to synthesize unknown class embeddings and using them to generate unknown visual features, which we implemented through a variation of Wasserstein GANs [6,17,25], which we term VAcWGAN (variationally-conditioned Wasserstein GAN).

2   Related Work

Initial works in the Zero-Shot Learning [10,12] proposes direct attribute prediction. However, directly predicting the attribute can be unreliable [10], so following works are based on direct class prediction through a compatibility function between visual features and class embeddings (reader can refer to [24] for an overview). Even if these methods showed compelling performance in the conventional Zero-Shot Learning setup, they struggle in the more competitive GZSL setup being biased towards predicting the seen classes.

To address this issue generative approaches for Zero-Shot Learning have been proposed [2,4,5,7,9,13,16,19,21,25,26,29]. As proposed by [26] and [28] almost independently, a (Wasserstein) GAN, conditioned on class embeddings, is paired with a classification loss in order to generate sufficiently discriminative CNN features, which are then fed to a softmax classifier for the final inference stage. Recently, several modifications have been adopted to improve feature generation for ZSL, for instance, by either replacing the GAN with a VAE [2,16] or using the latter two models in parallel [5,26], cycle consistency loss [4,9] or contrastive loss [7]. In [13], class embeddings are regressed from visual features, while semantic-to-visual generation is inverted with another generative, yet opposite, visual-to-semantic stream [21,22].

Differently to all these methods, our GAN-based architecture is different in the way it synthesizes class embeddings for the unknown classes. *To the best of our knowledge, we are the first to synthesize class embeddings.* As a concurrent work to ours, [14] seems to approach the open-set scenario as well but, it approaches the "compositional setup". That is, seen classes are defined as combinations of tags and inference has to be done on unknown combinations. Differently to [14], we put no prior on the classes we need to generalize onto (unseen and unknown mainly).

3   Problem Formulation

In this Section, we relax the closed-set assumption that constraints GZSL methods in knowing class embeddings for all categories (both seen $\mathcal{S}$ and unseen ones $\mathcal{U}$): we therefore attempt to reject unknown categories while not forgetting seen and unseen ones. We do so by proposing OZSL, in which we augment $\mathcal{S}$ and $\mathcal{U}$ with a third set of classes, dubbed *unknown*, and denoted by Ω. Unknown classes are deprived of both visual data and class embeddings (see Fig. 1). We formalize the OZSL problem by instantiating evaluation protocols, datasets and error metrics. We root these in GZSL to ease the transfer of the zero-shot learning community towards the new OZSL paradigm.

3.1   OZSL Evaluation Protocol

In GZSL, seen classes $\mathcal{S}$ are provided of data which are triplets $[\mathbf{x}, y, \mathcal{C}_y]$: $\mathbf{x}$ are vectorial visual embeddings extracted from a deep convnet (usually, ResNet101 [24]) fed by related images, y is the class label and $\mathcal{C}_y$ is a class embeddings (e.g., a list of manually-defined attributes describing the class). Unseen classes $\mathcal{U}$ are instead only given of class embeddings (and labels) $[y, \mathcal{C}_y]$ at training time, hence totally missing visual data.

In OZSL, together with the recognition of seen and unseen classes, we encompass potentially infinitely many classes at inference time. In fact, in addition to classify examples from $\mathcal{S}$ and $\mathcal{U}$, we also consider examples to be rejected since belonging to *unknown* categories we never observed before (no visual data available) and without class embeddings disclosed to the learner. Thus, unknown classes, denoted by Ω, are totally deprived of any visual or semantic information.

Therefore, the task is to train a zero-shot learner to handle the open-set scenario where, not only it has to recognize any unobserved test instance for which visual patterns are apparently matching semantic information of class embeddings, but it has also to avoid to take any decision on instances that seem to have a visual content that is not compatible with any prior semantic knowledge encapsulated in seen and unseen class embeddings.

3.2   OZSL Datasets

In order to allow practitioners to provide experimental results in both the closed-set, i.e., GZSL, and the open-set, the proposed OZSL, we build OZSL benchmark datasets rearranging GZSL ones. Specifically, we consider Animals with

Attributes (AWA) [11], Caltech-UCSD Birds 200 2011 (CUB) [23], Scene Under-standing (SUN) [27], and Oxford Flowers 102 (FLO) [18] since they are, by far, ubiquitous in GZSL literature [2,4,5,9,13,19,21,25,26,29]. We leverage the "Proposed Splits" [24] to be still enabled to use ImageNet pre-trained models to obtain visual descriptors and we stick to their proposed subdivision into seen and unseen classes. We select unknown categories by sampling from unseen classes. In short, we propose to sample 50% of the unseen classes of [24] and transform them to unknown classes, keeping the remaining 50% as unseen categories in OZSL. A complete list of seen, unseen and unknown classes for the selected four benchmark datasets will be publicly available[1].

3.3 Error Metrics

In GZSL, the performance is usually [24] evaluated using the harmonic mean

$$H_{\text{GZSL}} = \frac{2R_{\mathcal{S}}R_{\mathcal{U}}}{R_{\mathcal{S}} + R_{\mathcal{U}}}, \tag{1}$$

between each per-class accuracy $R_{\mathcal{S}}$ and $R_{\mathcal{U}}$, computed over seen and unseen classes, respectively. $R_{\mathcal{S}}$ and $R_{\mathcal{U}}$ are defined as:

$$R_{\mathcal{S}} = \frac{1}{|\mathcal{S}|} \sum_{s \in \mathcal{S}} R_s = \frac{1}{|\mathcal{S}|} \sum_{s \in \mathcal{S}} \frac{TP_s}{TP_s + FN_s}, \tag{2}$$

$$R_{\mathcal{U}} = \frac{1}{|\mathcal{U}|} \sum_{u \in \mathcal{U}} R_u = \frac{1}{|\mathcal{U}|} \sum_{u \in \mathcal{U}} \frac{TP_u}{TP_u + FN_u}. \tag{3}$$

where, in Eq. (2), we compute R_s, for the fixed seen class $s \in \mathcal{S}$, as the ratio between true positives TP_s and the total test examples of the class s, that is the sum of TP_s and the false negatives FN_s for that class. To obtain $R_{\mathcal{S}}$ from R_s, $s \in \mathcal{S}$, we average R_s over the whole list of seen classes (having cardinality $|\mathcal{S}|$). Analogous operations are carried out in Eq. (3) to compute $R_{\mathcal{U}}$, but applied to unseen classes in $\mathcal{U}$, instead. The metrics H_{GZSL}, $R_{\mathcal{S}}$ and $R_{\mathcal{U}}$ were proposed in [24] and adopted by state-of-the-art methods for their experimental validation [2,4,5,9,13,16,19,21,25,26,29].

   In GZSL, given that both seen and unseen classes have to be reliably classified, it makes sense to have error metrics depending upon true positives and false negatives which are computed independently over seen and unseen classes and (harmonically) averaged in order to balance performance over these two sets of categories [24].

   In OZSL, in order to break the closed-set assumption, we need to take into account also false positives FP. In fact, FP simulates cases where examples are predicted as if they belong to that class, albeit their actual ground-truth class is different. Please note that, since we cannot write explicit multi-class classification accuracy scores for the unknown classes Ω - since we do not have anything

---

[1] https://github.com/FMarmoreo/OpenZeroShot.

describing them - we have to rely on false positives, for both seen and unseen classes (FP_s, for every $s \in \mathcal{S}$, and FP_u, for every $u \in \mathcal{U}$), in order to indirectly control the rejection performance. In other words, in order to quantitatively measure the performance of a predictor of seen and unseen classes $\mathcal{S}$ and $\mathcal{U}$, which is also a rejector of unknown classes Ω, we need to control FP_s and FP_u, for every $s \in \mathcal{S}$ and $u \in \mathcal{U}$. This will reduce the possibility of wrongly associating generic unknown instances to any of the seen/unseen classes.

Obviously, the prior control on seen/unseen false positives has to be paired with penalization of "traditional" mis-classifications in a GZSL sense, since we do not want to gain in robustness towards unknown categories while forgetting how to predict seen or unseen classes. Therefore, we propose to measure performance in OZSL through the harmonic mean

$$\mathcal{H}_{\text{OZSL}} = \frac{2F1_\mathcal{S} F1_\mathcal{U}}{F1_\mathcal{S} + F1_\mathcal{U}} \tag{4}$$

of the $F1$ scores $F1_\mathcal{S}$ and $F1_\mathcal{U}$, over seen and unseen classes, defined as

$$F1_\mathcal{S} = \frac{1}{|\mathcal{S}|} \sum_{s \in \mathcal{S}} F1_s = \frac{1}{|\mathcal{S}|} \sum_{s \in \mathcal{S}} \frac{2R_s P_s}{R_s + P_s}, \tag{5}$$

$$F1_\mathcal{U} = \frac{1}{|\mathcal{U}|} \sum_{u \in \mathcal{U}} F1_u = \frac{1}{|\mathcal{U}|} \sum_{u \in \mathcal{U}} \frac{2R_u P_u}{R_u + P_u}. \tag{6}$$

In Eq. (5), for each seen class $s \in \mathcal{S}$, we compute the harmonic mean $F1_s$ of R_s, defined as in Eq. (2), and the precision P_s relative to s. We have that $P_s = \frac{TP_s}{TP_s + FP_s}$, being defined as the ratio of the true positives TP_s for that class and the total test examples classified as belonging to that class, that is the sum of TP_s and false positives FP_s. We repeat the analogous operations over unseen classes to obtain $F1_\mathcal{U}$, as in Eq. (6).

We claim that $\mathcal{H}_{\text{OZSL}}$, as defined in Eq. (4) extends the prior metric H_{GZSL} (in Eq. (1)) by preserving its property of evaluating a correct classification of seen and unseen categories. Concurrently, with $\mathcal{H}_{\text{OZSL}}$, we also inject false positives, formalizing their addition using $F1$ scores, for the sake of controlling any mis-classification involving unknown classes: this is a computable proxy to evaluate performance on unknown classes.

4   Unknown Feature Generation

Feature generators for GZSL, such as [25] or [17], leverage the operative assumption of knowing the class embeddings even for the categories which are unseen at training time. Class embeddings are, in fact, adopted as conditioning factors inside GAN- [7,25], VAE- [2,16] or GAN+VAE-based methods [17,26] to synthesize visual descriptors for the unseen classes. We cannot repeat the very same operation for unknown classes Ω since we have no class embeddings, but we still need to generate visual features because we do not have them as well.

4.1 Direct Unknown Generation (DUG)

Exploiting the generative methods [7,17,25], is possible to train a generative method, trained only on seen categories, to be conditioned on semantic embeddings to generate corresponding visual features. Thus, once the training is complete, is possible to condition on the class embeddings of the unseen classes to generate unseen visual features. Once both the seen and unseen visual features are available, inspired by [8], we take advantage of the MixUp approach to directly generate visual features for the unknown categories.

That is, given two visual features $\mathbf{x}_1$ and $\mathbf{x}_2$, representative of different classes, we mix them with

$$\mathbf{x}_k = \lambda\mathbf{x}_1 + (1 - \lambda)\mathbf{x}_2, \tag{7}$$

where $\lambda \in [0, 1]$ is sampled from a distribution $\mathrm{Beta}(\alpha, \beta)$, and the unknown label is assigned to $\mathbf{x}_k$.

The mixed features present mixed traits of the seen and unseen categories, belonging to any of them, and lie in regions of the feature space in between different classes. By labeling them as unknown, we can heuristically build borders for the classification regions for the seen and unseen classes and create a prior knowledge for classifying the unknown features.

4.2 Semantic Based Unknown Generation (SBUG)

Instead of directly using the visual feature space to generate the unknown features, a different approach that we investigate is to take advantage of the semantic embeddings to generate them. To this end, we propose to adopt a generative process to learn the distribution of the semantic space, as to learn the region of influence of seen and unseen class embeddings (blue and yellow balls in Fig. 2). So doing, we can map class embeddings into a transformed semantic space, and we claim that, inside it, we can generate class embeddings for the unknown classes by performing a mixing approach similar to the one presented in Sect. 4.1. Specifically, we sample the transformed semantic space "in between" the region of interest of seen and unseen classes, obtaining synthetic unknown class embeddings. Using them, we generate unknown visual features which help a classifier in rejecting unknown classes while still reliably classifying seen and unseen ones (from real seen and synthetic unseen visual features, respectively).

Thus, differently from DUG, where we can apply the unknown feature generation over an existing methodology, with SBUG we perform an end-to-end training together with the generative process to learn the mapping of the semantic embeddings in a new, more controllable, semantic space.

VAcWGAN. We introduce a semantic sampler S which is responsible of learning first and second order statistics (μ and Σ) for each of the classes y whose semantic embedding is given (seen and unseen). Once trained, we sample a vector $\mathbf{s}$ from a Gaussian distribution of mean μ and covariance matrix $\Sigma\Sigma^{\top}$. The role of S

is to transform the semantic space through a generative process, as the result of which, seen class embeddings $\mathcal{C}_1, \mathcal{C}_2, \ldots, \mathcal{C}_k$, and unseen ones $\mathcal{C}_{k+1}, \mathcal{C}_{k+2}, \ldots, \mathcal{C}_{k+u}$ are mapped into regions of influence. That is, they are mapped into $\mathcal{N}_1, \mathcal{N}_2, \ldots, \mathcal{N}_k$ (light blue balls in Fig. 2) and $\mathcal{N}_{k+1}, \mathcal{N}_{k+2}, \ldots, \mathcal{N}_{k+u}$ (yellow balls in Fig. 2). We model $\mathcal{N}_1, \mathcal{N}_2, \ldots, \mathcal{N}_k, \mathcal{N}_{k+1}, \mathcal{N}_{k+2}, \ldots, \mathcal{N}_{k+u}$ as Gaussian distributions and we use them to sample the conditioning factor $\mathbf{s}$ which, paired to a random noise vector $\mathbf{z}$ is passed to a Wasserstein GAN. This GAN is trained to generate synthetic visual features $\tilde{\mathbf{x}}$ by making them indistinguishable from the real seen features $\mathbf{x}$ extracted by an ImageNet pre-trained ResNet-101 model.

We call the aforementioned architecture variationally-conditioned Wasserstein GAN (VAcWGAN), which is built over the following optimization: $\min_{G,S} \max_{D} \mathcal{L}$, where

$$\mathcal{L}(\mathbf{x}, \tilde{\mathbf{x}}, \mathbf{s}) = L^{\text{real}}(\mathbf{x}, \mathbf{s}) - L^{\text{fake}}(\tilde{\mathbf{x}}, \mathbf{s})$$
$$= \mathbb{E}_{\mathbf{x} \sim \text{real}}\big[D(\mathbf{x}, \mathbf{s})\big] - \mathbb{E}_{\tilde{\mathbf{x}} \sim \text{gen}}\big[D(\tilde{\mathbf{x}}, \mathbf{s})\big]. \tag{8}$$

In Eq. (8), $\mathcal{L}(\mathbf{x}, \tilde{\mathbf{x}}, \mathbf{s})$ attempts to align the Wasserstein (Earth Mover) distance [1] between the distributions of synthesized features $\tilde{\mathbf{x}}$ over the distribution of the real ones $\mathbf{x}$. We introduce two auxiliary losses for VAcWGAN by jointly considering a standard gradient penalty term [6]

$$\mathcal{R}(\mathbf{x}, \tilde{\mathbf{x}}, \mathbf{s}) = \mathbb{E}_{t \in [0,1]}\big[(\|\nabla D(t\mathbf{x} + (1-t)\tilde{\mathbf{x}}, \mathbf{s})\|_2 - 1)^2\big]$$

which is commonly acknowledged to regularize the whole generation process, increasing computational stability [6]. We used a cross-entropy classification loss [25]

$$C(\tilde{\mathbf{x}}) = -\mathbb{E}_{\tilde{x} \sim \text{gen}}\big[\log p(y|\tilde{\mathbf{x}})\big] \tag{9}$$

which constraints the softmax probability p of classifying $\tilde{\mathbf{x}}$ to belong to the class y: it has to match the prediction done on $\tilde{\mathbf{x}}$ when generated from the class embedding $\mathcal{C}_y$ relative to the class y.

Unknown Generation. We train VAcWGAN using *seen data only*. In addition to generating unseen visual features (as commonly done in GZSL, see Sect. 2), we can also generate the unknown with a two-stages process. Given the generative process that VAcWGAN endow on class embeddings, we estimate the region of interest $\mathcal{N}_1 \cup \mathcal{N}_2 \cup \cdots \cup \mathcal{N}_k \cup \mathcal{N}_{k+1} \cup \mathcal{N}_{k+2} \cup \cdots \cup \mathcal{N}_{k+u}$ of both seen and unseen classes (in a transformed semantic space). We can exploit the complementary of it (i.e., the pink region in Fig. 2) to sample class embeddings that lie in the new semantic space in the regions in between the seen and unseen classes by mixing samples of seen and unseen class embeddings.

Specifically, we sample two semantic embeddings for two different classes $\mathcal{C}_i$ and $\mathcal{C}_j$, sample accordingly to the regions of interest $\mathbf{s}_i \sim \mathcal{N}_i$ and $\mathbf{s}_j \sim \mathcal{N}_j$, and than we mix them with

$$\mathbf{s}_k = \lambda \mathbf{s}_i + (1 - \lambda)\mathbf{s}_j, \tag{10}$$

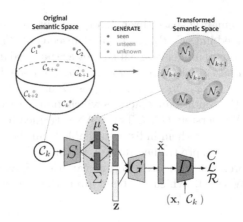

Fig. 2. Using VAcWGAN, we generate unknown class embeddings (in a transformed semantic space) from which, in turn synthetic unknown visual features can be generated. (Color figure online)

where $\lambda \in [0,1]$ is sampled from a distribution $\text{Beta}(\alpha, \beta)$, and the unknown label is assigned to $\mathbf{s}_k$. Once unknown class embeddings are sampled, they can be used as a conditioning factor to generate visual features that can be ascribed to the unknown classes.

5 Experiments

In Table 1 we perform an experimental evaluation between the two strategies of unknown feature generation we presented: DUG (Direct Unknown Generation, Sect. 4.1) and SBUG (Semantic Based Unknown Generation, Sect. 4.2). We combined DUG and SBUG separately and/or jointly to the VAcWGAN architecture (Sect. 4). In the sharp majority of the cases, doing unknown feature generation (with either DUG, SBUG or DUG+SBUG) is better than not doing it. We deem this result highly non-trivial: by attempting to learn how to synthesize unknown descriptors, we are simultaneously better shaping the region of interest of seen and unseen classes, so that the $F1_{\mathcal{U}}$ and $F1_{\mathcal{S}}$ metrics often improve (and so happens for their harmonic mean $\mathcal{H}_{\text{OZSL}}$ as well). For instance, the unknown feature generation improves by $+1.48\%$, $+1.02\%$ and $+0.18\%$ the performance of

Table 1. Direct and semantic unknown generation (DUG and SBUG) for the WAcW-GAN model (check Sect. 4.2).

	AWA			CUB			FLO			SUN		
	$F1_{\mathcal{U}}$	$F1_{\mathcal{S}}$	$\mathcal{H}_{\text{OZSL}}$	$F1_{\mathcal{U}}$	$F1_{\mathcal{S}}$	$\mathcal{H}_{\text{OZSL}}$	$F1_{\mathcal{U}}$	$F1_{\mathcal{S}}$	$\mathcal{H}_{\text{OZSL}}$	$F1_{\mathcal{U}}$	$F1_{\mathcal{S}}$	$\mathcal{H}_{\text{OZSL}}$
VAcWGAN	50.31%	64.84%	56.66%	45.08%	49.23%	47.34%	**47.68%**	72.69%	**57.59%**	**38.05%**	37.33%	37.68%
VAcWGAN + DUG	**51.83%**	65.18%	57.74%	45.48%	49.98%	47.63%	46.25%	**73.30%**	56.71%	33.87%	**38.37%**	**37.68%**
VAcWGAN + SBUG	50.62%	**68.28%**	**58.14%**	45.59%	**51.42%**	48.04%	46.01%	69.40%	55.33%	37.65%	38.07%	35.98%
VAcWGAN + DUG + SBUG	51.63%	66.20%	58.01%	**45.76%**	51.28%	**48.36%**	47.24%	72.00%	57.05%	34.83%	38.53%	36.59%

Table 2. The effect of direct unknown generation (DUG) applied on three benchmark methods: CLSWGAN [25], tf-VAEGAN [17] and CEZSL [7] on the proposed OZSL splits for AWA [11], CUB [23], FLO [18] and SUN [27] datasets.

	AWA			CUB			FLO			SUN		
	$F1_u$	$F1_s$	$\mathcal{H}_{OZSL}$	$F1_u$	$F1_s$	$\mathcal{H}_{OZSL}$	$F1_u$	$F1_s$	$\mathcal{H}_{OZSL}$	$F1_u$	$F1_s$	$\mathcal{H}_{OZSL}$
CLSWGAN	52.37%	65.52%	58.21%	47.01%	52.77%	49.73%	48.47%	76.08%	59.22%	36.31%	38.77%	37.50%
CLSWGAN + DUG	**57.34%**	**67.41%**	**61.97%**	**47.76%**	**53.29%**	**50.37%**	**49.34%**	**76.48%**	**59.98%**	**36.92%**	**39.64%**	**38.23%**
tf-VAEGAN	55.49%	71.47%	62.48%	**52.24%**	56.62%	54.34%	**54.78%**	80.00%	65.03%	43.00%	**41.09%**	42.02%
tf-VAEGAN + DUG	**60.56%**	**71.73%**	**65.68%**	51.60%	**57.92%**	**54.58%**	54.35%	**81.33%**	**65.15%**	**46.53%**	40.76%	**42.10%**
CEZSL	51.70%	**71.66%**	60.07%	5 39.43%	56.83%	46.56%	39.21%	**85.36%**	53.74%	33.35%	**30.83%**	32.04%
CEZSL + DUG	**55.76%**	71.30%	**62.58%**	**40.52%**	**57.54%**	**47.55%**	**40.26%**	85.25%	**54.69%**	**35.17%**	30.15%	**32.47%**

WAcWGAN on AWA, CUB and SUN respectively, while considering the $\mathcal{H}_{OZSL}$ metric and the SBUG, DUG + SBUG and DUG techniques, respectively.

Over DUG and SBUG, the former is advantageous over the latter because it acts directly on the visual space (so that the feature generator has not to be re-trained for unknown synthesis, but can be adapted to it). In view of this consideration, we can apply unknown feature generation to three state-of-the approaches for ZSL, better tailoring them towards the open ZSL regime. Namely, we consider the following methods: the Wasserstein generative adversarial network conditioned on class embeddings (CLSWGAN) [25] and its extension tf-VAEGAN [17] in which this architecture is paired with a variational auto-encoder to boost the generation stage. We also considered the usage of contrastive learning as adopted in CEZSL [7] in tandem with adversarial training. We combine CLSWGAN, tf-VAEGAN and CEZSL with the direct unknown generation that we presented in Sect. 4.1 and dubbed here "DUG" for brevity. As we show in Table 2, the adoption of DUG is always able to improve in performance all the considered baseline approaches with respect to the $\mathcal{H}_{OZSL}$ metric (e.g., +3.76% on AWA for CLSWGAN, +0.73% on SUN for CLSWGAN, +0.99% on CUB for CEZSL and +3.2% on AWA for tf-VAEGAN). Again, we interpret the consistent improvements that we registered as evidence for the effectiveness of performing unknown feature generation for OZSL.

6 Conclusions and Future Work

In this paper, we proposed a novel paradigm, called OZSL, extending traditional ZSL frameworks towards the additional rejection of unknown categories (neither visually nor semantically described) while still recognizing seen and unseen classes. Using the experimental protocols and error metrics that we proposed, our experimental findings suggest that attempting to synthesize unknown descriptors (to be rejected) seems a viable solution for OZSL.

Future works will be aimed at adopting techniques from out-of-domain generalization to better achieve the way we explore the semantic/visual spaces while seeking better strategies to generate the unknown.

References

1. Arjovsky, M., Chintala, S., Bottou, L.: Wasserstein GAN. ArXiv abs/1701.07875 (2017)
2. Arora, G., Verma, V.K., Mishra, A., Rai, P.: Generalized zero-shot learning via synthesized examples. In: The IEEE Conference on Computer Vision and Pattern Recognition (CVPR) (2018)
3. Chao, W.-L., Changpinyo, S., Gong, B., Sha, F.: An empirical study and analysis of generalized zero-shot learning for object recognition in the wild. In: Leibe, B., Matas, J., Sebe, N., Welling, M. (eds.) ECCV 2016. LNCS, vol. 9906, pp. 52–68. Springer, Cham (2016). https://doi.org/10.1007/978-3-319-46475-6_4
4. Felix, R., Vijay Kumar, B.G., Reid, I., Carneiro, G.: Multi-modal cycle-consistent generalized zero-shot learning. In: Ferrari, V., Hebert, M., Sminchisescu, C., Weiss, Y. (eds.) ECCV 2018. LNCS, vol. 11210, pp. 21–37. Springer, Cham (2018). https://doi.org/10.1007/978-3-030-01231-1_2
5. Gao, R., et al.: Zero-VAE-GAN: generating unseen features for generalized and transductive zero-shot learning. IEEE Trans. Image Process. **29**, 3665–3680 (2020)
6. Gulrajani, I., Ahmed, F., Arjovsky, M., Dumoulin, V., Courville, A.C.: Improved training of Wasserstein GANs. In: NIPS (2017)
7. Han, Z., Fu, Z., Chen, S., Yang, J.: Contrastive embedding for generalized zero-shot learning. In: CVPR (2021)
8. Zhang, H., Cisse, M., Dauphin, Y.N., Lopez-Paz, D..: Mixup: beyond empirical risk minimization. In: International Conference on Learning Representations (2018)
9. Huang, H., Wang, C., Yu, P.S., Wang, C.D.: Generative dual adversarial network for generalized zero-shot learning. In: The IEEE Conference on Computer Vision and Pattern Recognition (CVPR) (2019)
10. Jayaraman, D., Grauman, K.: Zero-shot recognition with unreliable attributes. In: NIPS (2014)
11. Lampert, C., Nickisch, H., Harmeling, S.: Learning to detect unseen object classes by between-class attribute transfer. In: The IEEE Conference on Computer Vision and Pattern Recognition (CVPR) (2009)
12. Lampert, C.H., Nickisch, H., Harmeling, S.: Attribute-based classification for zero-shot visual object categorization. IEEE Trans. Pattern Anal. Mach. Intell. **36**(3), 453–465 (2014). https://doi.org/10.1109/TPAMI.2013.140
13. Li, J., Jing, M., Lu, K., Ding, Z., Zhu, L., Huang, Z.: Leveraging the invariant side of generative zero-shot learning. In: The IEEE Conference on Computer Vision and Pattern Recognition (CVPR) (2019)
14. Mancini, M., Naeem, M.F., Xian, Y., Akata, Z.: Open world compositional zero-shot learning. In: CVPR (2021)
15. Mikolov, T., Sutskever, I., Chen, K., Corrado, G.S., Dean, J.: Distributed representations of words and phrases and their compositionality. In: Advances in Neural Information Processing Systems, pp. 3111–3119 (2013)
16. Mishra, A., Reddy, M.S.K., Mittal, A., Murthy, H.A.: A generative model for zero shot learning using conditional variational autoencoders. In: Proceedings of the 2018 IEEE/CVF Conference on Computer Vision and Pattern Recognition Workshops (CVPRW), pp. 2269–22698 (2018)
17. Narayan, S., Gupta, A., Khan, F.S., Snoek, C.G., Shao, L.: Latent embedding feedback and discriminative features for zero-shot classification. In: The European Conference on Computer Vision (ECCV) (2020)

18. Nilsback, M.E., Zisserman, A.: A visual vocabulary for flower classification. In: The IEEE Conference on Computer Vision and Pattern Recognition (CVPR) (2006)

19. Sariyildiz, M.B., Cinbis, R.G.: Gradient matching generative networks for zero-shot learning. In: Proceedings of the 2019 IEEE/CVF Conference on Computer Vision and Pattern Recognition (CVPR), pp. 2163–2173 (2019)

20. Scheirer, W., Rocha, A., Sapkota, A., Boult, T.: Towards open set recognition. IEEE Trans. Pattern Anal. Mach. Intell. (2012)

21. Schonfeld, E., Ebrahimi, S., Sinha, S., Darrell, T., Akata, Z.: Generalized zero- and few-shot learning via aligned variational autoencoders. In: The IEEE Conference on Computer Vision and Pattern Recognition (CVPR), pp. 8247–8255 (June 2019)

22. Shen, Y., Qin, J., Huang, L., Liu, L., Zhu, F., Shao, L.: Invertible zero-shot recognition flows. In: Vedaldi, A., Bischof, H., Brox, T., Frahm, J.-M. (eds.) ECCV 2020. LNCS, vol. 12361, pp. 614–631. Springer, Cham (2020). https://doi.org/10.1007/978-3-030-58517-4_36

23. Welinder, P., Branson, S., Mita, T., Wah, C., Schroff, F., Belongie, S., Perona, P.: Caltech-UCSD Birds 200. Technical report CNS-TR-2010-001, California Institute of Technology (2010)

24. Xian, Y., Lampert, C.H., Schiele, B., Akata, Z.: Zero-shot learning-a comprehensive evaluation of the good, the bad and the ugly. In: Proceedings of the IEEE Transactions on Pattern Analysis and Machine Intelligence (2018)

25. Xian, Y., Lorenz, T., Schiele, B., Akata, Z.: Feature generating networks for zero-shot learning. In: The IEEE Conference on Computer Vision and Pattern Recognition (CVPR) (June 2018)

26. Xian, Y., Sharma, S., Schiele, B., Akata, Z.: F-VAEGAN-D2: a feature generating framework for any-shot learning. In: The IEEE Conference on Computer Vision and Pattern Recognition (CVPR) (June 2019)

27. Xiao, J., Hays, J., Ehinger, K.A., Oliva, A., Torralba, A.: Sun database: large-scale scene recognition from abbey to zoo. In: The IEEE Conference on Computer Vision and Pattern Recognition (CVPR) (2010)

28. Zhu, Y., Elhoseiny, M., Liu, B., Peng, X., Elgammal, A.: A generative adversarial approach for zero-shot learning from noisy texts. In: Proceedings of the IEEE Conference on Computer Vision and Pattern Recognition, pp. 1004–1013 (2018)

29. Zhu, Y., Xie, J., Liu, B., Elgammal, A.: Learning feature-to-feature translator by alternating back-propagation for generative zero-shot learning. In: The IEEE International Conference on Computer Vision (ICCV) (2019)

Spatial-UNet: Deep Learning-Based Lane Detection Using Fisheye Cameras for Autonomous Driving

Salma Moujtahid[1]([✉]), Rachid Benmokhtar[1], Amaury Breheret[2], and Saif-Eddine Boukhdhir[1]

[1] Valeo Vision - Driving Assistance Research (DAR), Paris, France
salma.moujtahid@valeo.com
[2] Mines ParisTech - Center for Robotics, Paris, France

Abstract. This paper addresses the problem of ego and adjacent lanes detection for real-life autonomous driving. Lane detection is a central part in vehicle automation applications, requiring accurate and stable detection in diverse difficult situations with heavy obstructions, erased lane markings or night. To overcome these challenges, we propose a predictive instance segmentation CNN model named S-UNet, combining spatial layers and U-Net style architecture to efficiently detect and infer lines formed by lane markings from non-corrected lateral fisheye cameras. Experimental results show that S-UNet is able to learn the spatial relationships and the continuous prior of lanes and achieves high accuracy and robustness under various conditions in real-life autonomous driving scenarios. It outperforms state-of-the-art model SCNN [17] by +5.57% mean IoU with significant improvement of the line marking detection accuracy.

Keywords: Lane detection · Instance segmentation · Deep learning · Autonomous driving

1 Introduction

In recent years, research has been very active in working towards increasing traffic safety, and providing assistance and comfort automotive and mobility solutions. For these, two tasks are needed: surrounding objects detection to avoid obstacles and lane detection to navigate the road. So road environment perception is an essential and key component for a safe autonomous driving system. It usually requires clear road markings and a high-level quality of infrastructure in order to detect ego and adjacent lanes and thus have stable lateral control.

Cameras are very suitable sensors for wide perception of the environment, especially for lane detection, increasingly being used as low-cost assistance solutions. They are generally small and easy to integrate on a vehicle without changing its shape. In order to manage intelligent lane centering, lane keeping and lane departure warning functionalities, most of autonomous vehicles use frontal cameras to detect only road markings painted on the ground (dashed, continuous markings, stop lines, *etc..*). However, for lane change, lane merge and side impact avoidance collision, 360° surround-view cameras are preferred as they offer a larger view of ego and adjacent lanes. In our proposed system, we employ lateral cameras - positioned on the sides of the vehicle - offering a larger field of view.

Most traditional methods of road marking detection extract low-level line features using various primitives such as color [3], edge [2,11] and texture coupled with Hough transform [9] and particle or Kalman filters [15]. After identifying the lane features, post-processing techniques are employed to filter and group relevant segments together to form the final lanes. These approaches are more adapted to simple environments with significant lighting and line quality.

For more challenging scenarios, deep learning as an emerging technology has demonstrated state-of-the-art performance in solving many computer vision problems. A number of CNN-based approaches have been applied to lane detection and evaluated on different benchmarks [7,8,12,13]. For instance, VPGNet [8] proposes a multi-task network guided by vanishing points for lane and road marking detection. Self Attention Distillation (SAD) was used in [5] for lane detection by performing top-down and hierarchical attention distillation networks to explore light weight methods. However, most of these methods limit their solutions to detecting road lanes based on one frontal camera in controlled environments without taking into account real life scenarios such as bad weather conditions, road mark degradation and severe occlusions. Some recent approaches such as [10] work specifically on a difficult scenario with low lighting, using GANs for style-transfer-based data enhancement and generating images in low-light conditions. Nevertheless, real life scenarios and data is required to train models to be used in real life automated driving systems for optimal safety.

Although classic CNNs for lane detection concentrate on pixel level segmentation of lane markings painted on the road, a predictive model of a line - based on lane markings - is a more interesting approach. A scene understanding and predictive type CNN would help infer lane position even with an obstructed view or with damaged lane marking. To specifically address the issue of line detection and its continuity, Pan and al. [17] proposed the Spatial-CNN (SCNN). They have demonstrated great results on TuSimple benchmark [1] in scene understanding with an effective information propagation of the lane shape. They introduced spatial layers, specifically designed this task, which propagate information in 4 directions to capture the long continuous shape prior of lane markings and poles. SCNN effectively learns the spatial relationship and continuous prior of road line markings and is able to fix the discontinuities which can appear in lane segmentation.

Inspired by this work, we propose the Spatial-UNet model (S-UNet), integrating spatial convolution layers and an optimized U-Net [14] style architecture to efficiently detect ego and adjacent lanes. The network architecture is optimized to preserve the spatial relationships of lane pixels throughout the network and give a better segmentation accuracy. Moreover, S-Unet is applied to the specific sensor context of two lateral fisheye cameras (mounted on the left and

Fig. 1. Images from left and right surround viewing cameras (SVC) system.

right side mirrors of the ego vehicle) rather than the traditional front pinhole camera. Our proposed model is directly applied to the images from the fisheye cameras without correcting the fisheye distortion as to not add computational complexity and keep the field of view. Additionally, one single trained model for both left and right cameras. To meet our specific application needs in terms of sensor type, *i.e.*. lateral fisheye cameras, a hand-labelled dataset for training and evaluation with real world situations is constructed. It covers various challenging use cases such as: erased or damaged road markings, severe occlusions by vehicles or other obstacles, low illumination with day and night use cases, traffics jams, road works and more.

The rest of the paper is organized as follows: in Sect. 2, the used camera sensors are introduced. The proposed S-UNet model architecture is then described in Sect. 3. Experimental and visual results are presented in Sect. 4. Finally, we conclude with a summary of most important results and perspectives.

2   From Surrounding to Lateral View Cameras

In our project, a low-cost architecture of sensors is observed, capable of performing autonomous driving with constrained cost, memory capacity and processing power. In the aim of environment understanding, different sensors technologies exist to make detections robust to different kind of obstacles and road infrastructures. Among these sensors, cameras are ideal sensors for the perception of objects with shapes such as vehicles or lanes.

Our camera system, Surround View Cameras (SVC), combines four fisheye cameras set up on the front and rear car bumpers, left and right side mirrors. They each have 30 m range and 190° horizontal field of view (FoV), making a total 360° detection FoV. However, SVC requires a complex and heavy computational process. Therefore, only left and right cameras are considered instead of four. The high side positioning of the left and right cameras offer better range of detection especially for lane markings, in comparison to the low position of the front camera. As we can see in Fig. 1, the lateral cameras offer a large field of view of ego and adjacent lanes. Moreover, only one model can be trained for both lateral cameras since the views are symmetrical.

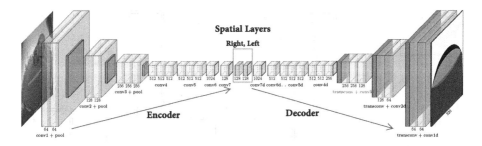

Fig. 2. Proposed S-UNet architecture: (orange) Each convolution is followed with batch normalisation and ReLu; (red): max pooling; (green) spatial Layers; (blue): transverse convolutions. (Color figure online)

With regards to the distortion of the fisheye cameras, we chose to not correct it, as it can be computationally expensive and reduces the range of detection.

3 Proposed S-UNet Architecture

The overall architecture of the proposed S-UNet system is presented in Fig. 2 and can be broken down to 3 main blocs: the encoder, the spatial layers and the decoder (or expansive path).

For the encoder, also called contracting path, we employed a VGG16 [6] architecture as it has demonstrated is performance in feature extraction. Pretrained weights from ImageNet [4] are used to speed up the network convergence.

3.1 Spatial SCNN Layers

The notion of spatial layers is introduced by Pan et al. [17] in order to learn more efficiently the spatial relationship and continuous prior of lane markings in the driving context. The 'spatial' term refers to the spatial information within the image rather than the spatial convolution.

Four spatial convolution layers are proposed corresponding to each spatial direction $\{up, down, left, right\}$. Figure 3 illustrates the left and right spatial layers. The feature map from the previous layer with height H, channels C and width W is divided into W slices following the specific direction, and the convolution is performed in a slice by slice manner. Where in classical CNN, the convolution output feeds the next layer, here the output is added to the next slice and this process continue until the last slice. Thus enabling the passing between pixels across rows and columns inside a layer.

This aggregation of information via processing sliced features and adding them together one by one is similar to recurrent neural networks. Moreover, the propagation of spatial information between neurons throughout the same layer helps to preserve the smoothness and continuity of structured long objects such as lanes.

Since the lateral cameras used are mounted on the left and right sides of the vehicle, the visualized lanes form horizontal lines on the camera as we can

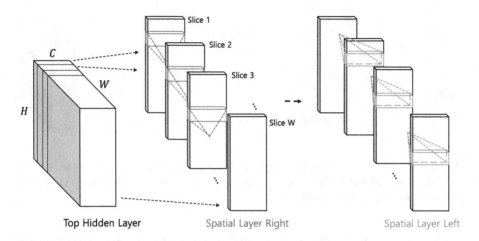

Fig. 3. Spatial left and right layers.

observe in Fig. 1. Thus, the most useful spatial convolution layers are the left and right ones as they propagate pixel information from the left to the right, in a horizontal manner in the feature maps as illustrated in Fig. 3. This will be further confirmed in the experimental evaluation.

3.2   Decoder

In the lane detection context, spatial information in the image is crucial and the spatial relationships of pixels throughout the network need to be preserved. In SCNN [17], the decoder consists of a bi-linear interpolate up-sampling operation. The main drawback of this decoder is the loss of spatial relationship between pixels. It reduces the localization accuracy with the down-sampling operations during encoding, which is not recovered with the up-sampling in the decoding.

In order to keep the spatial information learned throughout the encoder and spatial layers, a decoder inspired from U-Net [14] architecture is introduced. This proposed architecture named S-UNet is designed with a decoder symmetrical to the contracting path. The goal is to allow the network to propagate context information to higher resolution layers during decoding in an optimal manner.

Among various alternative decoders explored, two specific decoders are presented in this paper. The first is employing max-unpooling for the up-scaling operations and the same number of convolutions as in the encoder. The indexes of the maximum value are saved for every max-pooling layer in the encoder and then used in the decoder to map the pixels during unpooling. This has the advantage of better mapping pixels from pooling to unpooling operations. In this architecture (denoted as model 2.1 in Table 1), the expansion path consists of a series of convolutions, with ReLU non-linearities, batch normalisation and max-unpooling layers to obtain the output feature maps with segmented classes.

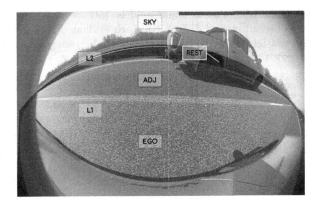

Fig. 4. Description of segmentation classes, detailed in Sect. 4.1.

The second alternative consists of using transposed convolutions for upscaling and the same number of convolutions as in the encoder. Transposed convolutional layers carry out a regular convolution but revert its spatial transformation. Since they have learnable weights, they provide a better generalized upsampling of abstract representations. In this architecture (denoted as model 2.2 in Table 1), the expansion path consists a series of convolutions, with ReLU non-linearities, batch normalisation, and trans-convolutions as illustrated in Fig. 2. As our evaluations will prove, using trans-convolutions outperforms other standard unpooling methods.

4   Evaluation

4.1   Dataset

As our goal is to apply S-UNet to a real life autonomous driving environment, our dataset is constructed from mining real driving recordings using SVC left and right fisheye cameras. These recordings are collected from different vehicles driving in a variety of highway and urban road environments.

A total of 19.000 images were collected including different challenging scenarios such as traffic jams with heavy obstructions, working zones, day and night lighting, erased or damaged lane markings, *etc.*. The model is trained on 16.000 images and evaluated on 3.000 images from recordings independent of the training set.

Six classes {EGO, L_1, ADJ, L_2, SKY, REST} are manually annotated using pixel segmentation to describe the scene, as illustrated in Fig. 4. In our annotation specification, we distinguish each line and lane instead of viewing all lines as one class. L_1 corresponds to the line formed by the closest lane markings and EGO is the ego-lane on which the vehicle is driving. L_2 is the line formed by the lane markings from the adjacent lane and ADJ is the (left or right depending on the camera) adjacent lane. SKY is the area above the horizon line, useful for pitch estimation, and finally, REST corresponds to the remaining areas.

Table 1. Experimental results for different models. IoU scores are presented for each class as well as the overall mean IoU (mIoU). Each evaluated architecture is denoted by its name, the spatial layers (Up, Down, Left, Right) integrated and when data augmentation is added.

Architecture	Spatial layers	Data augment	L_1	L_2	EGO	ADJ	SKY	REST	mIoU
1.0 - CNN	–	–	70.35	42.23	94.99	88.59	97.70	77.16	78.50
1.1 - SCNN	LR	–	73.72	46.83	94.70	89.70	97.71	75.83	79.75
1.2 - SCNN	UDLR	–	75.52	46.11	95.47	90.13	98.20	79.89	80.89
1.3 - SCNN	LR	✓	77.40	51.86	95.19	92.03	98.52	80.31	82.55
2.1 - S-UNet MU	LR	✓	78.16	55.92	95.65	93.05	99.14	83.98	84.32
2.2 - S-UNet TC	LR	✓	**78.35**	**59.49**	95.86	93.46	99.26	85.52	**85.32**

The images are resized to 320×200 from original resolution 1280×800 as a compromise between precision and real time execution.

4.2 Data Augmentation

Data augmentation is a common technique to avoid over-fitting and improve the overall performance. A variety of illumination and noise techniques including affine transformations and local deformations are employed for better generalisation of the model. It includes brightness, blur, contrast and Gaussian noise. A vertical flip is also used to augment the data as our model is trained on images from left and right cameras, thus giving an even wider representation of the scene independently of the camera.

A guided version of the cutout method [16] with a spatial prior of targeted regions is introduced. This technique designates the process of randomly obstructing a region of an image to encourage the network to better use the full context in the image. It can be interpreted as a spatial version of the dropout and helps to give a better network regularisation. In our case, it is motivated by frequent lane markings occlusions caused by surrounding vehicles.

4.3 Experimental Results

The Intersection over Union (IoU) metric is used to quantify the overlap between the target mask and our predictions for each class. IoU measures the number of pixels in common between the target and prediction masks divided by the total number of pixels present across both masks. The results for each class as well as the mean over all classes (mIoU) are presented in Table 1. It is important to note that the values that interest us the most are those of classes L_1 and L_2 (Figure 4). In autonomous driving, these classes present a high interest as they represent the line markings for ego and adjacent lanes. Both are used to generate the geometrical road model and feed the lateral control functionalities (e.g. lane centering, lane change, lateral offset, etc.).

Table 1 regroups the evaluations and comparison of 6 models trained and evaluated on our dataset. Using the baseline model SCNN [17], 3 models are trained with different spatial layers: model 1.0 without spatial layers (giving a simple CNN with a VGG16 encoder), model 1.1 with only left and right spatial layers and model 1.2 using all of them. According to these results, we can confirm the contribution of the spatial layers in better learning the linear shapes as L_1 and L_2 are improved from {70.35% and 42.23%} to {75.72% and 46.83%}.

We can also notice that model 1.1 and 1.2 provide close results {79.75% and 80.89% mIoU}, where the second introduces 4 spatial layers. Since left and right cameras are used, the line-markings are in a horizontal manner, rendering the up and down spatial layers not necessary. For the rest, the spatial LR (Left/Right) layers approach is adopted to save computational power.

In model 1.3, data augmentation (DA) is added. We observe an improvement of performance, especially on the adjacent line class L_2 as DA helps the model to generalise.

The models 2.1 and 2.2 show the behavior of S-UNet approaches proposed in Sect. 3.2 and both prove to outperform the baseline. Model 2.2 with transposed convolutions achieves the best results with 78.35% IoU for L_1 and 59.86% IoU for L_2. The trans-convolution is able to better generalise compared to max-unpooling or the simple up-scale interpolation (in baseline SCNN model) since these techniques are predefined and do not learn from data. Our proposed S-UNet model 2.2 outperforms state-of-the-art SCNN [17] model. We can note that the improvement of the decoder (and DA) is especially beneficial to L_2 since it is the class which suffers the most from occlusions and under-representation.

To further see the performance of our proposed model S-UNet 2.2, visual results of predictions on challenging cases from the evaluation dataset are presented in Fig. 5. In the firsts scenarios (a) and (b), S-UNet is able to infer the adjacent line L2 even with a vehicle occluding a considerable part of it. Scenarios (b) and (c) presents different illuminations situations with a low one in a tunnel and then a high illumination intensity change exiting the tunnel which is usually challenging for cameras. Our model manages to detect lanes in stable manner throughout the light changes.

In the case of scenario (d), there is no adjacent lane, showing that the model does not overfit and successfully detects the ego line marking from the context although it is erased. S-UNet also efficiently predicts the lanes in the difficult case (e) of night lighting.

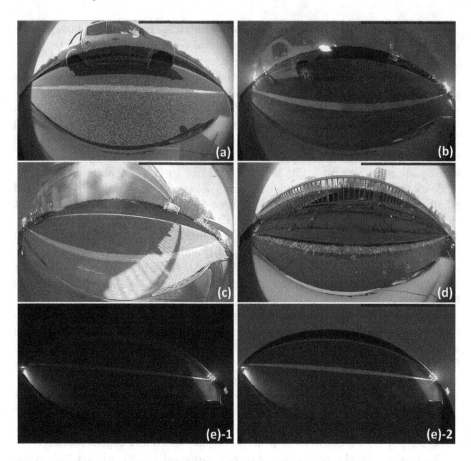

Fig. 5. Visual results of lane prediction based on S-UNet model (2.2) in 5 challenging situations, from various vehicles: (a) severe occlusion, (b) low illumination in tunnel; (c) high illumination change while exiting a tunnel, (d) absent lane makings and no adjacent lane, (e) –1 original image of low illumination at night, (e) –2 prediction result.

5   Conclusions

In this paper, we propose an efficient multi-lanes instance segmentation and detection method combining spatial and U-Net convolutional neural networks principles. Our predictive approach can be applied to lateral fisheye cameras, without distortion correction. The model is also generic as one trained model is used for cameras on both cameras, reducing the computational cost compared to surround view cameras. Extensive experiments demonstrate the performance and robustness of our Spatial-UNet to diverse real life scenarios of autonomous driving. S-UNet outperforms state-of-the-art model SCNN [17] by +4.63% IoU for the ego line (L_1) and a significant improvement of +12.66% IoU for adjacent (L_2) line marking.

The next step for our lane detection module is the geometric modelisation of the segmented lanes. The inferred segmentation masks from the left and right views will be combined to create a complete representation of the road using camera calibration.

Our future work will also focus on introducing temporality with multiple frames for more stable lane line detection, and exploring end-to-end geometrical road model prediction.

References

1. Tusimple benchmark (2017). https://github.com/TuSimple/tusimple-benchmark
2. Bounini, F., Gingras, D., Lapointe, V., Pollart, H.: Autonomous vehicle and real time road lanes detection and tracking. In: IEEE Vehicle Power and Propulsion Conference (VPPC), pp. 1–6 (2015)
3. Chiu, K.Y., Lin, S.F.: Lane detection using color-based segmentation. In: IEEE Intelligent Vehicles Symposium (IV), pp. 706–711 (2005)
4. Deng, J., Dong, W., Socher, R., Li, L.J., Li, K., Fei-Fei, L.: Imagenet: a large-scale hierarchical image database. In: IEEE Conference on Computer Vision and Pattern Recognition (CVPR), pp. 248–255 (2009)
5. Hou, Y., Ma, Z., Liu, C., Loy, C.C.: Learning lightweight lane detection CNNs by self attention distillation. In: Proceedings of the IEEE/CVF International Conference on Computer Vision, pp. 1013–1021 (2019)
6. Karen, S., Andrew, Z.: Very deep convolutional networks for large-scale image recognition. In: International Conference on Learning Representations (ICLR) (2015)
7. Kim, J., Lee, M.: Robust lane detection based on convolutional neural network and random sample consensus. In: Loo, C.K., Yap, K.S., Wong, K.W., Teoh, A., Huang, K. (eds.) ICONIP 2014. LNCS, vol. 8834, pp. 454–461. Springer, Cham (2014). https://doi.org/10.1007/978-3-319-12637-1_57
8. Lee, S., et al.: VPGNet: vanishing point guided network for lane and road marking detection and recognition. In: IEEE International Conference on Computer Vision (ICCV), pp. 1947–1955 (2017)
9. Liu, G., Wörgötter, F., Markelić, I.: Combining statistical hough transform and particle filter for robust lane detection and tracking. In: IEEE Intelligent Vehicles Symposium (IV), pp. 993–997 (2010)
10. Liu, T., Chen, Z., Yang, Y., Wu, Z., Li, H.: Lane detection in low-light conditions using an efficient data enhancement: light conditions style transfer. In: 2020 IEEE Intelligent Vehicles Symposium (IV), pp. 1394–1399. IEEE (2020)
11. Lopez, A., Serrat, J., Canera, C., Lumbreras, F., Graf, T.: Robust lane markings detection and road geometry computation. Int. J. Autom. Technol. **11**, 395–407 (2010)
12. Lou, L., Liu, J., Zhang, Q., Huang, D., Zhou, C., Han, J.: Fast detection of lane based on convolutional neural networks and connected components constraints. In: IEEE Data Driven Control and Learning Systems Conference (DDCLS), pp. 1033–1038 (2019)
13. Neven, D., Brabandere, B.D., Georgoulis, S., Proesmans, M., Gool, L.V.: Towards end-to-end lane detection: an instance segmentation approach. In: IEEE Intelligent Vehicles Symposium (IV), pp. 286–291 (2018)

14. Ronneberger, O., Fischer, P., Brox, T.: U-Net: convolutional networks for biomedical image segmentation. In: Navab, N., Hornegger, J., Wells, W.M., Frangi, A.F. (eds.) MICCAI 2015. LNCS, vol. 9351, pp. 234–241. Springer, Cham (2015). https://doi.org/10.1007/978-3-319-24574-4_28

15. Teng, Z., Kim, J., Kang, D.: Real-time lane detection by using multiple cues. In: IEEE International Conference on Control, Automation and Systems (ICCAS), pp. 2334–2337 (2010)

16. DeVries, T., Taylor, G.W.: Improved regularization of convolutional neural networks with cutout. Computing Research Repository (CoRR) arXiv:1708.04552 (2017)

17. Pan, X., Shi, J., Luo, P., Wang, X., Tang, X.: Spatial as deep: spatial CNN for traffic scene understanding. Computing Research Repository (CoRR) arXiv:1712.06080 (2017)

FourierMask: Instance Segmentation Using Fourier Mapping in Implicit Neural Networks

Hamd ul Moqeet Riaz[✉], Nuri Benbarka, Timon Hoefer, and Andreas Zell

Department of Computer Science (WSI), University of Tuebingen,
Tuebingen, Germany
{hamd.riaz,nuri.benbarka,timon.hoefer,andreas.zell}@uni-tuebingen.de
https://uni-tuebingen.de/de/118829

Abstract. We present FourierMask, which employs Fourier series combined with implicit neural representations to generate instance segmentation masks. We apply a Fourier mapping (FM) to the coordinate locations and utilize the mapped features as inputs to an implicit representation (coordinate based multi layer perceptron (MLP)). FourierMask learns to predict the coefficients of the FM for a particular instance, and therefore adapts the FM to a specific object. This allows FourierMask to be generalized to predict instance segmentation masks from natural images. Since implicit functions are continuous in the domain of input coordinates, we illustrate that by sub-sampling the input pixel coordinates, we can generate higher resolution masks during inference. Furthermore, we train a renderer MLP (*FourierRend*) on the uncertain predictions of FourierMask and illustrate that it significantly improves the quality of the masks. FourierMask shows competitive results on the MS COCO dataset compared to the baseline Mask R-CNN at the same output resolution and surpasses it on higher resolution.

Keywords: Instance segmentation · Implicit representations

1 Introduction

In the past decade, we have witnessed a shift from classical approaches towards deep learning methods for a variety of real world tasks. Due to an ample amount of real and synthetic datasets and high computation power, it has been possible to reliably use these 'black box' models on highly complex and critical use cases. For the field of autonomous driving, perceiving and understanding the environment is crucial. Instance segmentation is one of those tasks which allows autonomous systems to semantically separate different regions in their percepts (images) and at the same time separate objects from each other.

Supplementary Information The online version contains supplementary material available at https://doi.org/10.1007/978-3-031-06430-2_49.

In recent years, the majority of the methods have employed CNNs for the task of instance segmentation. There are methods that generate instance segmentation masks by classifying each pixel of a region of interest as either foreground or background e.g. Mask R-CNN [4]. These methods generally show the best performance but suffer from high computation needs and slow speed. There are methods that predict the contour points around the boundary of the object [21,24]. Although being faster, these methods fail to match the performance of pixel-wise classification methods. Alternatively, there are methods that try to encode the mask contours in a compressed representation [16,23]. These mask representations are compact and meaningful but lack the superior capabilities of pixel-wise methods. Our approach employs a pixel-wise mask representation and additionally we generate the mask using a compact Fourier representation. In the case of segmentation masks, the Fourier series' low-frequency components hold the general shape and high-frequency components hold the edges of the mask. Therefore, our representation is meaningful and can be compressed according to the use case.

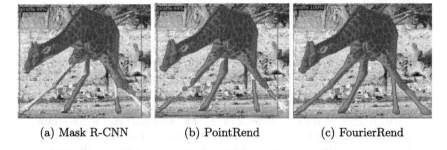

(a) Mask R-CNN   (b) PointRend   (c) FourierRend

Fig. 1. Comparison between Mask R-CNN [4], PointRend [7] and FourierMask.

FourierMask is an *implicit neural representation*. For the image regression task, *implicit representations* learn to predict the RGB values of a particular image given a pixel coordinate. Tancik et al. [18] showed that using a Fourier Mapping of coordinates instead of actual coordinates locations as inputs, allows the implicit networks to learn high-frequency details in images and 3D scenes. Our work draws inspiration from this work and applies their findings to the task of generating masks for instance segmentation. Implicit representations have the advantage of learning and reconstructing fine details, which traditional representations cannot do as effectively in such compact models. As implicit functions are continuous in the domain of input coordinates, we can sub-sample the pixel coordinates to generate higher resolution masks during inference. However, these representations are inherently trained on a single image and have not yet been adopted to a general task of instance segmentation on natural images. Our contributions are as follows:

1. We developed FourierMask, which can replace any mask predictor that uses a region of interest (ROI) to predict a binary mask. It is fully differentiable and end-to-end trainable.

2. We show that implicit representations can be applied to the task of instance segmentation. We achieve this by learning the coefficients of the Fourier mapping of a particular object.

3. As implicit functions are continuous in the domain of input coordinates, we show that we can sub-sample the pixel coordinates to generate higher resolution masks during inference. These higher resolution masks are smoother and improve the performance on MS COCO.

4. We verify and illustrate that the rendering strategy from PointRend[7] brings significant qualitative gains for FourierMask. Our renderer MLP *FourierRend* improves the mask boundary of FourierMask significantly.

2   Related Work

In *two-stage* instance segmentation, the network first detects (proposes) the objects and then predicts a segmentation mask from the detected region. The baseline method for many two stage methods is Mask R-CNN [4]. Mask R-CNN added a mask branch to Faster R-CNN [15], which generated a binary mask that separated the foreground and background pixels in a region of interest. Mask Scoring R-CNN [6] had a network block to learn the quality of the predicted instance masks and regressed the mask IoU. ShapeMask [8] estimated the shape from bounding box detections using shape priors and refined it into a mask by learning instance embeddings. CenterMask [9] added spatial attention-guided mask on top FCOS [19] object detector, which helped to focus on important pixels and diminished noise. PointRend [7] tackled instance segmentation as rendering problem. By sampling unsure points from the feature map and its fine grained features from higher resolution feature map, it was able to predict really crisp object boundaries using a fully connected MLP. Rather than employing binary grid representation of masks, PolyTransform [10] used a polygon representation. These methods accomplish state-of-the-art accuracy, but they are generally slower than one stage methods.

One stage instance segmentation methods predict the instance masks in a single shot, without using any proposed regions/bounding boxes as an intermediate step. YOLACT [1] linearly combined prototype masks and mask coefficients for each instance, to predict masks in real-time speeds. Likewise, Embedmask [25] employed embedding modules for pixels and proposals. ExtremeNet [26] predicted the contour (octagon) around an object using keypoints of the object. Similary, Polarmask [21] used the polar representation to predict a contour from a center point (centerness from FCOS [19]). Dense RepPoints [24] used a large set of points to represent the boundary of objects. FourierNet [16] employed inverse fast Fourier transform (IFFT) to generate a contour around an object represented by polar coordinates. The network learned the coefficients of the Fourier series to predict those contours.

Implicit representations learn to encode a signal as a continuous function. Mescheder et al. [13] proposed occupancy networks, which implicitly represented the 3D surface as the continuous decision boundary of a deep neural network classifier. Mildenhall et al. [14] used implicit neural networks to learn 3D scenes and synthesize novel-views. Tancik et al. [18] showed that mapping input coordinates using a Fourier feature mapping, allows MLPs to learn high frequency function in low dimensional domains. Sirens [17] showed that using periodic activation functions (such as sine) in implicit representations, lets the MLPs learn the information of natural signals and their derivatives better than when using other activation functions such as ReLU.

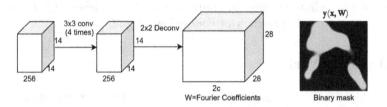

Fig. 2. FourierMask head architecture for a ROI Align size of 14×14. The network predicts Fourier coefficients **W** for each location in the feature map.

3　Method

3.1　FourierMask Head Architecture

This section explains the network architecture of FourierMask. We employ Mask R-CNN [4] as our baseline model. We use a ResNet [5] backbone pre-trained on ImageNet dataset [3], with a feature pyramid network (FPN) [11] architecture. Following from Mask R-CNN, we use a small region proposal network (RPN), which generates k proposal candidates from all feature levels from FPN. To generate fixed size feature maps from these proposal candidates, we use a ROI Align [4] operation. This produces a (k, d, m, m) sized feature map for the mask head, where m is the fixed spatial size after ROI Align operation and d is the number of channels. The structure of the head is shown in the Fig. 2. We apply four convolutions consecutively, each with a kernel size 3×3 and a stride of 1. Then we apply a transposed convolution layer with $2c$ number of filters, which generates a spatial volume of size $2c \times 2m \times 2m$. We call this feature volume **W**, which holds $2c$ Fourier coefficients for each spatial location.

3.2　Fourier Features

In this section, we explain how to obtain the Fourier features from the coefficients **W**. First, we define the Fourier mapping as

$$\gamma(\mathbf{x}) = [cos(2\pi\mathbf{x} \cdot \mathbf{B}), sin(2\pi\mathbf{x} \cdot \mathbf{B})]. \tag{1}$$

Here $\mathbf{x} \in \mathbb{R}^{p \times 2}$ are the pixel coordinates (i, j) normalized to a value in range $[0, 1]$ and p are the total number of pixels in the image. Since sine and cosine have a period of 2π, by normalizing the pixel coordinate $\mathbf{x}$ to a range $[0, 1]$, we ensure one complete image lies in period of 2π. Images are not periodic signals, but since they are bounded by an image resolution, we can safely apply our method to predict 2D binary masks. $\mathbf{B} \in \mathbb{Z}^{2 \times c}$ is the integer lattice matrix which holds the possible combinations of harmonic frequency integers of Fourier series for both dimensions in the image. $\mathbf{B}$ contains the elements of the set S, defined by

$$S = \{0, \ldots, f\} \times \{-f, \ldots, f\} \setminus \{0\} \times \{-f, \ldots, -1\}, \tag{2}$$

where f is the total number of frequencies. $\mathbf{B}$ has non-negative integers in its first row and integers in its second row, except that the second row wont have negative integers for a zero in the first row. This way of defining $\mathbf{B}$ is motivated by the 2D Fourier representation. If we do not limit $\mathbf{B}$ by the total number of frequencies f, we would obtain the original 'sine plus cosine' form of the Fourier series. This property is shown in the supplementary material. However, we have to limit f to a number which fits to our memory constraints and speed requirements. For the case of images (2D input), the possible permutations c can be calculated by:

$$c = (f + 1)(2f + 1) - f \tag{3}$$

Fourier Features are generated as follows:

$$\mathbf{FF}(\mathbf{x}, \mathbf{W}) = \gamma(\mathbf{x}) \circ \mathbf{W}, \tag{4}$$

where $\circ$ is the element-wise product, $\mathbf{W} \in \mathbb{R}^{p \times 2c}$ is the weight matrix predicted by the FourierMask. Note that we flatten the spatial dimension of the prediction beforehand ($p = 2\,\mathrm{m} \times 2\,\mathrm{m}$). Let $\boldsymbol{ff}_i$ be the i_{th} column of $\mathbf{FF}$; then the binary mask $\mathbf{y}$ is defined as

$$\mathbf{y}(\mathbf{x}, \mathbf{W}) = \phi\left(\sum_{i=1}^{2c} \boldsymbol{ff}_i\right). \tag{5}$$

Here ϕ is the sigmoid activation function, which we use to bound the output between 0 and 1. Note that $\mathbf{y}(\mathbf{x}, \mathbf{W})$ can be interpreted as an implicit representation with a single perceptron because it is a linear combination of Fourier features followed by a non linear activation function.

3.3   Fourier Features Based MLP

As shown by [17,18], implicit representations can learn to generate shapes, images etc. from input coordinates very effectively. Following the work from [18], we saw that Fourier mapping of input coordinates lets the MLP learn higher frequencies and consequently generate images with finer detail compared to MLPs without Fourier mapping. Furthermore, [17] showed that using periodic activation functions works better compared to ReLU in implicit neural networks. We employ an MLP with sine activation functions, in which Fourier features (FF) are the input and mask $\mathbf{y}'$ is the output. We have 3 hidden layers (Siren layers), each with 256 neurons. The MLP has a single output neuron, on which we apply a sigmoid function to bound it between 0 and 1. The Fourier features (FF) are generated by the Eq. (4) and they are parameterized by coefficients $\mathbf{W}$ learned by the network and therefore adapted for a specific input ROI. Coordinate based MLPs encode the information of one particular image or shape, but by parameterizing them with learned Fourier coefficients $\mathbf{W}$ of each object, we can generalize them to generate a binary mask of any object.

3.4   MLP as a Renderer - FourierRend

For generating boundary-aligned masks, we used a renderer MLP (FourierRend) which specialized only on the uncertain regions of the mask predicted by Eq. (5). We adopted the rendering strategy from PointRend [7] and made the following modification in the point head (Fig. 3). Rather than sampling coarse mask features in the mask head, we sample the Fourier features ($\mathbf{FF}$) from Eq. (4) at uncertain mask prediction coordinates (the locations where the predictions are near 0.5). Fourier features $\mathbf{FF}(\mathbf{x}, \mathbf{W})$ makes FourierRend an implicit MLP since its input is a function of input coordinates $\mathbf{x}$ and therefore we can take leverage from its implicit nature as discussed before. We concatenate these Fourier features and fine-grained features (from the 'p2' level FPN feature map). We replace the mask predictions from Eq. (5) (coarse predictions), with the fine-detailed predictions from FourierRend. Consequently, we replace uncertain predictions at the boundary, with more accurate predictions of the renderer, resulting in crisp and boundary-aligned masks.

3.5   Training and Loss Function

We concatenate the output $\mathbf{y}$ from Eq. (5) and $\mathbf{y}'$ from the MLP and train both masks in parallel. By training the output $\mathbf{y}$, we learn the coefficients $\mathbf{W}$ of a Fourier series in their true sense. We need these coefficients because we assume that the input for the MLP are Fourier features. We use IoU loss for training the binary masks defined as:

$$\mathbf{IoU\ loss} = \frac{\sum_{i=0}^{N} \min(y_{p_i}, y_{t_i})}{\sum_{j=0}^{N} \max(y_{p_j}, y_{t_j})} \tag{6}$$

y_{p_i} is the predicted value of the pixel i, y_{t_i} is the ground truth value of the pixel i and N is the total number of pixels in the predicted mask.

4   Experiments

For all our experiments we employ a Resnet 50 backbone with feature pyramid network pre-trained on ImageNet [3] unless otherwise stated. We use the Mask R-CNN default settings from detectron2 [20]. We train on the MS COCO [12] training set and show the results on its validation set. We predict class agnostic masks, i.e. rather than predicting a mask for each class in MS COCO, we predict only one mask per ROI. For the baseline, we trained a Mask R-CNN and PointRend [7] with class agnostic masks.

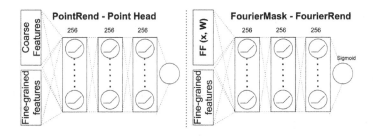

Fig. 3. Difference between point head from PointRend and FourierRend.

4.1 Spectrum Analysis MS COCO

To validate that the Fourier Mapping (Eq. (5)) works for instance mask prediction, we performed a spectrum analysis on the MS COCO training dataset. Along with verifying our method, this analysis gave us an insight on an optimal number of frequencies for the dataset. We performed this experiment by applying a fast Fourier transform on all the target object masks in the COCO training dataset. This Fourier transform gave us the coefficients of a Fourier series, which hold the same meaning as the coefficients prediction **W** of FourierMask. Firstly, we sampled only the lower frequency coefficients of the Fourier series and reconstructed the object's mask by applying Eq. (5). We did this for all the objects' masks in COCO training set and evaluated the IoU loss of the reconstruction compared to the target. Then we incrementally added higher frequency coefficients and repeated the above mentioned procedure until we reached the maximum number of frequencies. The Fig. 4 shows the mean IoU loss at various frequencies. It can be seen that the loss decreases exponentially. We choose the maximum frequency as 12 since it has a low enough reconstruction loss and fits comfortably in our GPU memory. Figure 5 illustrates a visual comparison between the ground truth and reconstructions using varying number of frequencies.

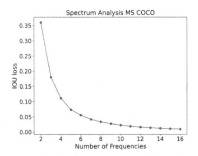

Fig. 4. IOU vs Frequencies.

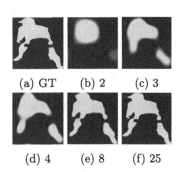

(a) GT   (b) 2   (c) 3

(d) 4   (e) 8   (f) 25

Fig. 5. The ground truth vs its reconstructions at various frequencies.

4.2 Number of Frequencies

To validate our experiment from the previous section, we trained a FourierMask with a similar configuration. Rather than prediciting a set of coefficients for each pixel, we modified the architecture to predict a single vector for the whole image. We applied strided (stride $= 2$) 3×3 convolutions 2 times (on the ROI) to reduce the feature size by 1/4th, and then used a fully connected layer to predict the coefficients. The network architecture is shown in the supplementary material. We applied the Eq. (4) and (5) for generating the mask. We copied the predicted Fourier coefficients p times to match the dimensions for matrix multiplication in Eq. (4). We trained the network with 12 frequencies and an output resolution of 56×56 using IoU loss. In this experiment, we did not add an MLP branch and trained only the Eq. (5). We evaluated the mAP precision of the network on the COCO validation dataset when using a subset of Fourier component frequencies. The network was trained on 12 frequencies, but during inference we incrementally added the higher frequency components starting from the first component. Figure 6 shows the result of this test. The mAP shows a similar trend as seen in Fig. 4 and therefore validates the spectrum analysis and the choice of 12 maximum frequencies. The Fig. 7 shows an example how the masks change when different number of frequencies are used.

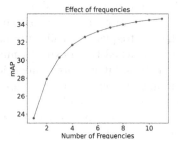

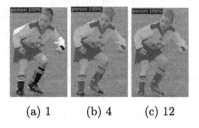

(a) 1   (b) 4   (c) 12

Fig. 6. The mAP when using a subset of trained frequencies.

Fig. 7. Mask predictions using various frequencies

4.3 Fourier Features Based MLP

To validate that the Fourier Feature based MLP improves the performance of FourierMask, we trained 2 networks with the architecture shown in Fig. 2. In this architecture, the network predicts separate Fourier coefficients for each spatial location. The first network was trained on the masks obtained using $\mathbf{y}$ in Eq. (5) and output $\mathbf{y}'$ of MLP (FF + MLP). The second network was trained only using Eq. (5) (FF). Both networks used 12 Fourier frequencies and had an output resolution of 28×28 pixels. We used class agnostic masks and therefore predicted only one class for each region of interest, rather than a mask for each class in the COCO dataset. We had two hidden layers (both with sine activations and

256 neurons) and a single output neuron with sigmoid activation. For the first network (FF + MLP), we took the mean of the masks predicted by $\mathbf{y}$ (Eq. (5)) and output $\mathbf{y}'$ of MLP during inference. The results are shown in the Table 1. As can been seen in the table, the network with an MLP shows the best performance among the models with ResNet-50 backbone. We also trained the same network with a larger ResNeXt-101 [22] backbone. The improvement of more than 4 mAP over Resnet-50 model shows that our model scales well to bigger backbones.

Table 1. Comparison of various FourierMask architectures with Mask R-CNN.

Model	Backbone	mAP
Mask R-CNN	ResNet-50	34.86
FF	ResNet-50	34.89
FF + MLP	ResNet-50	34.97
FF + MLP	ResNeXt-101	39.09

4.4   Higher Resolution Using Pixel Sub-sampling

One of the advantages of our method is that it can predict masks at sub-pixel resolution because implicit representations are continuous in the input domain. We analyzed this by evaluating both the trained networks in Sect. 4.3 on the MS COCO validation set on various pixel steps. For the input $\mathbf{x}$ in the Eq. (1), rather than using integer values of pixels (pixel step of 1), we used a pixel step of $1/2^{s-1}$, where $s \in \mathbb{Z}^+$ is the scaling factor. This effectively scaled both the height and width of the input $\mathbf{x}$ by a factor of s. To match the size of input pixels $\mathbf{x}$, we upsampled the coefficients $\mathbf{W}$ in Eq. (4) in the spatial dimension using bilinear interpolation, by a scaling factor 2^{s-1}. Table 2 shows the evaluation using the two networks explained in Sect. 4.3. We can observe that using a lower pixel step improves the mAP. Figure 8 shows how the mask boundary smooths out when sub sampling the pixels. Note that we trained the network on 28 × 28 output resolution, but we can generate higher resolution output during inference, which is a considerable advantage over other methods. (see supplementary material for analysis with different number of frequencies at various pixel steps.)

4.5   Higher Resolution Using FourierRend

To generate higher resolution masks, we used FourierRend (Sect. 3.4) along with the subdivision strategy from Pointrend [7]. We replaced the predictions from Eq. (5) (coarse predictions), with the fine detailed predictions from the Fourier-Rend. This resulted in masks which were more crisp and boundary aligned. For training FourierRend, we employ the default settings of point selection strategy along with the point loss from PointRend. The results are shown in the Table 3. Here, we also evaluate the mask quality using the *Boundary IOU* [2]

Table 2. Sub-sampling performance and speed (GTX 2080Ti).

Model	Pixel step	Resolution	mAP	Speed (ms)
Mask R-CNN	1	28×28	34.86	48.7
FF	1	28×28	34.89	50.3
FF	1/2	56×56	35.13	59.1
FF	1/4	112×112	**35.18**	68.3
FF + MLP	1	28×28	34.97	52.1
FF + MLP	1/2	56×56	**35.18**	67.0

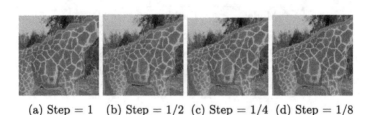

(a) Step = 1   (b) Step = 1/2   (c) Step = 1/4   (d) Step = 1/8

Fig. 8. Subsampling the pixels smooths out the boundaries of the mask.

metric (mAP$_{bound}$), which penalizes the boundary quality more than overall correct pixels. Compared to Mask R-CNN, we see a decent improvement of more than 0.7 mAP$_{mask}$ and 1.6 mAP$_{bound}$ with comparable speeds. We can clearly see visual improvements specially in boundary quality (see Fig. 1 and supplementary material). Compared to PointRend, we observe that the masks are more complete (see Fig. 1) with a reasonably lower inference speed. Note that FourierRend achieves 224×224 in 3 sub-division steps compared to 5 steps of PointRend because FourierRend's initial resolution is 28×28 compared to 7×7 of PointRend.

Table 3. The effect of subdivision inference.

Model	Sub. steps	Resolution	mAP$_{mask}$	mAP$_{bound}$	Speed (ms)
Mask R-CNN	0	28×28	34.86	21.2	48.7
FourierRend	0	28×28	35.01	21.0	48.7
FourierRend	1	56×56	35.63	22.8	52.4
FourierRend	2	112×112	35.64	22.8	55.7
FourierRend	3	224×224	35.64	22.9	59.4
PointRend	5	224×224	36.12	23.5	81.6

5 Conclusion

In this paper, we show how implicit representations combined with the Fourier series can be applied to the task of instance segmentation. We illustrate that the masks generated using our Fourier mapping are compact. The lower Fourier frequencies hold the shape and higher frequencies hold the sharp edges. Furthermore, by sub-sampling the pixel coordinates in our implicit MLP, we can generate higher resolution masks during inference, which are visually smoother and improve the mAP over our baseline Mask R-CNN. We also show that our renderer MLP FourierRend improves the boundary quality of FourierMask significantly. See supplementary materials for more details.

References

1. Bolya, D., Zhou, C., Xiao, F., Lee, Y.J.: Yolact: real-time instance segmentation. In: Proceedings of the IEEE/CVF International Conference on Computer Vision, pp. 9157–9166 (2019)
2. Cheng, B., Girshick, R., Dollár, P., Berg, A.C., Kirillov, A.: Boundary IoU: improving object-centric image segmentation evaluation. In: CVPR (2021)
3. Deng, J., Dong, W., Socher, R., Li, L.J., Li, K., Fei-Fei, L.: Imagenet: a large-scale hierarchical image database. In: 2009 IEEE Conference on Computer Vision and Pattern Recognition, pp. 248–255. IEEE (2009)
4. He, K., Gkioxari, G., Dollár, P., Girshick, R.: Mask r-cnn. In: Proceedings of the IEEE International Conference on Computer Vision, pp. 2961–2969 (2017)
5. He, K., Zhang, X., Ren, S., Sun, J.: Deep residual learning for image recognition. In: Proceedings of the IEEE Conference on Computer Vision and Pattern Recognition (CVPR) (2016)
6. Huang, Z., Huang, L., Gong, Y., Huang, C., Wang, X.: Mask scoring r-cnn. In: Proceedings of the IEEE Conference on Computer Vision and Pattern Recognition, pp. 6409–6418 (2019)
7. Kirillov, A., Wu, Y., He, K., Girshick, R.: Pointrend: Image segmentation as rendering. In: Proceedings of the IEEE/CVF Conference on Computer Vision and Pattern Recognition (CVPR) (2020)
8. Kuo, W., Angelova, A., Malik, J., Lin, T.Y.: Shapemask: learning to segment novel objects by refining shape priors. arXiv preprint arXiv:1904.03239 (2019)
9. Lee, Y., Park, J.: Centermask: real-time anchor-free instance segmentation. In: Proceedings of the IEEE/CVF Conference on Computer Vision and Pattern Recognition, pp. 13906–13915 (2020)
10. Liang, J., Homayounfar, N., Ma, W.C., Xiong, Y., Hu, R., Urtasun, R.: Polytransform: deep polygon transformer for instance segmentation. In: Proceedings of the IEEE/CVF Conference on Computer Vision and Pattern Recognition, pp. 9131–9140 (2020)
11. Lin, T.Y., Dollar, P., Girshick, R., He, K., Hariharan, B., Belongie, S.: Feature pyramid networks for object detection. In: Proceedings of the IEEE Conference on Computer Vision and Pattern Recognition (CVPR) (2017)
12. Lin, T.-Y., et al.: Microsoft COCO: common objects in context. In: Fleet, D., Pajdla, T., Schiele, B., Tuytelaars, T. (eds.) ECCV 2014. LNCS, vol. 8693, pp. 740–755. Springer, Cham (2014). https://doi.org/10.1007/978-3-319-10602-1_48

13. Mescheder, L., Oechsle, M., Niemeyer, M., Nowozin, S., Geiger, A.: Occupancy networks: learning 3D reconstruction in function space. In: Proceedings of the IEEE/CVF Conference on Computer Vision and Pattern Recognition, pp. 4460–4470 (2019)

14. Mildenhall, B., Srinivasan, P.P., Tancik, M., Barron, J.T., Ramamoorthi, R., Ng, R.: NeRF: representing scenes as neural radiance fields for view synthesis. In: Vedaldi, A., Bischof, H., Brox, T., Frahm, J.-M. (eds.) ECCV 2020. LNCS, vol. 12346, pp. 405–421. Springer, Cham (2020). https://doi.org/10.1007/978-3-030-58452-8_24

15. Ren, S., He, K., Girshick, R., Sun, J.: Faster r-cnn: towards real-time object detection with region proposal networks. In: Advances in Neural Information Processing Systems, pp. 91–99 (2015)

16. Riaz, M., Benbarka, N., Zell, A., et al.: Fouriernet: compact mask representation for instance segmentation using differentiable shape decoders. arXiv e-prints arXiv-2002 (2020)

17. Sitzmann, V., Martel, J.N., Bergman, A.W., Lindell, D.B., Wetzstein, G.: Implicit neural representations with periodic activation functions. In: Proceedings of NeurIPS (2020)

18. Tancik, M., et al.: Fourier features let networks learn high frequency functions in low dimensional domains. arXiv preprint arXiv:2006.10739 (2020)

19. Tian, Z., Shen, C., Chen, H., He, T.: FCOS: fully convolutional one-stage object detection. In: Proceedings of the IEEE/CVF International Conference on Computer Vision, pp. 9627–9636 (2019)

20. Wu, Y., Kirillov, A., Massa, F., Lo, W.Y., Girshick, R.: Detectron2 (2019). https://github.com/facebookresearch/detectron2

21. Xie, E., et al.: Polarmask: single shot instance segmentation with polar representation. In: Proceedings of the IEEE/CVF Conference on Computer Vision and Pattern Recognition, pp. 12193–12202 (2020)

22. Xie, S., Girshick, R., Dollár, P., Tu, Z., He, K.: Aggregated residual transformations for deep neural networks. In: Proceedings of the IEEE Conference on Computer Vision and Pattern Recognition, pp. 1492–1500 (2017)

23. Xu, W., Wang, H., Qi, F., Lu, C.: Explicit shape encoding for real-time instance segmentation. In: The IEEE International Conference on Computer Vision (ICCV) (2019)

24. Yang, Z., et al.: Dense reppoints: representing visual objects with dense point sets. arXiv preprint arXiv:1912.11473 2 (2019)

25. Ying, H., Huang, Z., Liu, S., Shao, T., Zhou, K.: Embedmask: embedding coupling for one-stage instance segmentation. arXiv preprint arXiv:1912.01954 (2019)

26. Zhou, X., Zhuo, J., Krahenbuhl, P.: Bottom-up object detection by grouping extreme and center points. In: Proceedings of the IEEE/CVF Conference on Computer Vision and Pattern Recognition, pp. 850–859 (2019)

An Anomaly Detection Approach for Plankton Species Discovery

Vito Paolo Pastore[1,2,3(✉)], Nimrod Megiddo[2], and Simone Bianco[2,3,4P]

[1] MaLGa - DIBRIS, Universitá degli Studi di Genova, Genoa, Italy
Vito.Paolo.Pastore@unige.it
[2] IBM Almaden Research Center, San Jose, CA, USA
megiddo@us.ibm.com
[3] Center for Cellular Construction, San Francisco, CA, USA
[4] Altos Labs, Redwood City, CA, USA
sbianco@altoslabs.com

Abstract. Plankton is one of the most abundant and diverse class of microscopic organisms inhabiting the Earth. Their enormous intra- and inter-species genetic and phenotypic diversity, coupled with the limited amount of large survey data, makes it hard to obtain a complete representation of this important class of organisms. Hence, the classification accuracy of novel supervised machine learning algorithms is bound to be limited by the incompleteness of the training data. In this work we introduce an efficient pipeline centered around a novel anomaly detection algorithm to discover and classify new plankton species, in situ, with the aim of automatically populating a plankton database in an unsupervised fashion. Our pipeline utilizes the concept of anomaly detection to separate a novel species from the ones contained in an initial existing database. Our results show that the implemented algorithm outperforms four state-of-the-art methods for outlier detection on the plankton dataset used in our analysis. Finally, using a leave-one-out approach, we prove that our pipeline is able to identify unknown plankton species with high-accuracy.

Keywords: Anomaly detection · Plankton image analysis

1 Introduction

Plankton is a class of aquatic microorganisms showing significant variability in size and features. Plankton is at the bottom of the aquatic food chain. Phytoplankton microorganisms are responsible for over 50% of all oxygen global primary production [1] and play a fundamental role in climate regulation. Thus, changes in plankton ecology have rippling effects on global climate, as well as deep social and economic consequences [13]. Collecting and analyzing plankton population composition and behavior in the field will enhance our understanding and protect their vital role in a healthy ecosystem [11]. Establishing a morphological and behavioral baseline for healthy plankton allows for such microorganisms to be used as biosensor to detect environmental perturbations

S. Sclaroff et al. (Eds.): ICIAP 2022, LNCS 13232, pp. 599–609, 2022.
https://doi.org/10.1007/978-3-031-06430-2_50

and provide important and timely information about the overall health of the aquatic ecosystem [12]. The exact number of plankton species is not known, but an estimation of oceanic plankton puts the number around 4000 [18]. Indeed, a supervised methodology for classification is limited by the incompleteness of the data, because, while the availability of high-quality annotated plankton data is certainly increasing, it is not yet sufficient for the task at hand. Even if novel computational paradigms are available for fast annotation of large datasets [9], algorithms that require minimal supervision could be a compromise between accuracy and data availability, allowing for the discovery of novel species of plankton not included in the training set. With the development of automatic acquisition system, able to provide a huge amount of plankton images [6], machine learning is becoming the main tool for the characterization of plankton data [2,17,20]. Deep learning and convolutional neural networks have been extensively applied for plankton classification [5,11], in a supervised fashion. However, due to the high number of plankton species and the cost of annotations both in terms of time and resources, unsupervised learning approaches have been proposed, with high accuracy on testing plankton datasets [13,16].

In this paper, we introduce an efficient pipeline centered around a novel anomaly detection method, named TailDeTect (TDT) algorithm. The TDT algorithm builds the boundaries of an average sample set in an assigned multi-dimensional features space. Any deviation from this baseline is labeled as an anomaly. Assuming an initial database is available (i.e., a training set), we propose to train one TDT detector for each of the training species. A sample is considered as a global anomaly if all the trained detectors identify it as an anomaly. Each global anomaly is stored for further investigation. Once a sufficient number of anomalous samples is collected, the anomalies can be used as input to a partitioning algorithm for unsupervised clustering and subsequently assigned to the correct species by an expert of the field. In this way, a TDT detector can be trained for each of the discovered species and added to the initial set.

In this paper, first we schematically describe the proposed pipeline. Then, we provide few details about each step, focusing on the description of the TDT algorithm. We validate our TDT anomaly detector on a plankton dataset [13], showing that it outperforms four available state-of-the-art outlier detection methods. Finally, we use a leave-one-out approach to prove that our pipeline can identify novel species with high-accuracy (92%).

2   Methods

In this section we first describe the proposed pipeline (see Fig. 1 for a schematic representation). Then, we provide details about the main steps, focusing on the description of the TDT algorithm.

2.1   Pipeline

Our pipeline is meant to provide in situ recognition of plankton species not included in the training set. It can either be an actual new species, or a species

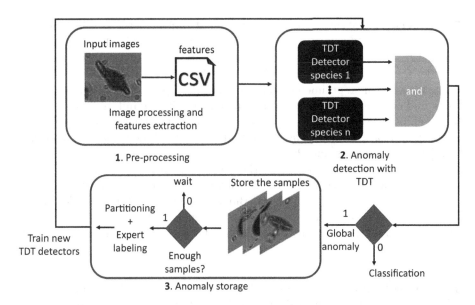

Fig. 1. Schematic representation of the pipeline proposed to identify unseen plankton species.

whose morphology drastically changed as a consequence to environmental perturbations [12].

We suppose that an initial annotated training set is available. This training set can be either labeled by an expert, or obtained with unsupervised methods, as proposed in [13]. Thus, one TDT anomaly detector is trained and stored, for each of the species included in the initial training set. At this stage, we can assume we have a set of trained TDT detectors and an in-situ acquisition system ready to capture plankton images, on top of which our pipeline is meant to be executed. The proposed approach is composed of three main steps:

1. **Pre-processing**: acquired images are binarized so that the plankton cell is identified, and a set of descriptors is computed and used for further steps.
2. **TDT algorithm**: The set of extracted features is fed to each of the trained TDT detectors for testing. A sample is identified as anomalous if all the trained TDT detectors detect it as out-of-class. We refer to such sample as a global anomaly.
3. **Anomaly storage**: The global anomaly is stored and made available for further steps. The detected anomalies correspond to potential new species that can be added to initial set of known classes to automatically populate a plankton database in an unsupervised (or semi-supervised) fashion.

In the next paragraphs, we will provide further information about the three steps composing our pipeline.

Pre-processing. We extract a set of 131 hand crafted features, using the same pre-processing and features extraction procedure described in [13]. To maximize segregation within the same species and separation among different species, we consider features belonging to five different classes: 14 geometric features (first image moment, area, perimeter, width and height, aspect ratio, roundness, circularity and equivalent diameter, rectangularity, minimum and maximum axes of fitting ellipse, eccentricity, and a shape factor (i.e., the foreground's percentage of occupancy of the correspondent bounding box)), 32 moments-based features [8,19] (7 Hu moments and 25 features resulting from Zernike moments up to order 5), 8 features extracted from the gray levels histogram (mean value and standard deviation, skewness, kurtosis, entropy, blue/red, blue/green, green/red) [13], 13 Haralick features [7], 54 Local Binary Pattern and 10 Fourier descriptors [12]. A total number of 131 features have been used to train our TDT algorithm, and to establish the average baseline for each of the species included in the training set.

TDT Algorithm. In this paragraph we provide a detailed description of the TDT algorithm. The proposed method learns the boundaries of an empirical distribution of every feature extracted in experimental conditions that can be assumed to define a baseline. It further classifies a sample as an anomaly when it falls out of the computed boundaries for Max dimensions, where Max is a parameter inferred by the algorithm during training. The TDT algorithm needs only in-class samples for training, providing both classification and anomaly detection with high testing accuracy. See Algorithm 1 and Algorithm 2 for a pseudo-code description of the TDT algorithm training and the testing phase, respectively.

TDT Training Phase. Let us suppose to have a dataset $(x_i, y_i) : x_i \in R^M, y_i = 1, i = 0, ..., N$, where x_i is an m-dimensional real vector, and $y_i = 1$ simply means that all the samples belong to a specific class. A feature-selection method can be used to reduce the dimensionality of the data, as a pre-processing step. An example of this pre-processing step is removing all features which have a distribution with a variance lower than a set threshold. An example of this threshold is the mean value of the distribution of all the variances of the features. The resulting dataset is x_{ij}, with $j = 1, ..., M^* < M$.

A bootstrap approach is used to estimate, with an acceptable confidence level, the mean and standard deviation of the distribution for each feature of our multidimensional dataset, obtaining the mean values $\mu_1, \mu_2, ..., \mu_{M^*}$ and standard deviations $\sigma_1, \sigma_2, ..., \sigma_{M^*}$. We compute the boundaries for anomalous samples using a threshold defined as $\mu \pm n \cdot \sigma$, where n is a parameter yet to be determined. Our plankton training set contains the distribution of each of the 131 features extracted from the acquired sample images for the training class, with a physiological variability within the training population. The TDT algorithm makes no assumptions about the probability distribution from which the data is drawn, except that it has a finite variance.

Using Chebyshev's inequality, the probability that the value of a random variable lies more than n standard deviations away from the mean is less than $\frac{1}{n^2}$. For instance, regardless the distribution of our data, we can say that 75% of values are contained within $\mu \pm 2 \cdot \sigma$. As a comparison, if the distribution is gaussian, about 95% of the samples are contained within the same range.

The TDT algorithm determines for each dimension j the value of n_j so that approximately Z% of the samples are contained within $\mu \pm n_j \cdot \sigma_j$, where Z is a tunable hyperparameter (e.g., 95%). This step produces the values $n_1, n_2, ..., n_{M^*}$, and $t_j = n_j \cdot \sigma_j, j = 1, ..., M^*$, with each dimension of the multi-dimensional dataset x_{ij} having its own boundaries defined by the thresholds $\mu_1 \pm t_1, ..., \mu_{M^*} \pm t_{M^*}$. The values t_j can be estimated directly if there are more than $\frac{1}{Z}$ examples. Otherwise, t_j can be estimated based on σ_j together with Chebyshev's inequality or some knowledge of the probability distribution. Each of the training samples x_{ij} can be out of the boundaries for a certain number of features. A sample will be classified as an anomaly if it is out of the boundaries for more than a defined maximum number of features. Let us denote this value by Max. A simple linear search can be used to describe the value of Max. We can consider $1 - Z$ as the maximum number of anomalies that we allow in the training set, to take into consideration the physiological variability within each feature extracted from the sample plankton image. Starting with Max = 0 (representing the case of all the samples classified as anomaly), the TDT algorithm increases this value with step 1 until the number of detected anomalies is less or equal to the accepted tolerance $1 - Z$.

Algorithm 1. TDT algorithm training

1: input = Num_surr,size_surr,x_train,confidence
2: **for** i in Num_surr **do**
3:     subset = extract _with _replacement(size _surr)
4:     **for** j in features **do**
5:         μ_{ij}, σ_{ij} = compute(subset)
6: **for** j in features **do**
7:     μ_j*, σ_j* = compute_bootstrap(μ_{ij}, σ_{ij})
8:     n_j = fit($\mu_j*, \sigma_j*, x\_train[:, j]$,confidence)
9:     $t_j = n_j \cdot \sigma_j*$
10: Max = fit(x_train,μ_j*, t_j,confidence)
11: output = threshold, Max

TDT Testing Phase. The output of the TDT training is the set of thresholds $\mu_1 \pm t_1, ..., \mu_{M^*} \pm t_{M^*}$, where $t_j = n_j \cdot \sigma_j$ and a value for Max. At this stage, as described in Algorithm 2, a testing sample **x** is considered as an anomaly if falls out of the computed boundaries (i.e., $x_j > \mu_j + t_j \lor x_j < \mu_j + t_j, j = 0, ..., M^*$) a number of times greater than or equal to Max.

Algorithm 2. TDT algorithm testing

1: input = x_test,μ_j*,t_j, Max
2: **for** i in x_test **do**
3:     $count \leftarrow 0$
4:     **for** j in features **do**
5:         **if** x_test[i,j] $> \mu_j * +t_j \vee$ x_test[i,j] $< \mu_j * -t_j$ **then**
6:             $count \leftarrow count + 1$
7: **if** $count > Max$ **then**
8:     $i \leftarrow anomaly$

Anomaly Storage. The features set correspondent to a global anomaly is collected and stored for further steps. We identify a possible approach, consisting in the adoption of a semi-supervised method. First, a partitioning algorithm is used, for instance, to identify candidate clusters that group the entire set of detected global anomaly, according to the extracted morphological features. Then, an expert in the field can assign labels to the computed clusters, identifying if the detected new class is an actual species not included in the training set, or a species whose morphology significantly changed due to perturbations and modifications of the aquatic environment. Finally, for each of the new identified species, a new TDT detector is trained, and add to the collection of trained detectors, so that the novel species can actually be classified as known in further acquisitions.

3   Experiments

3.1   Dataset

The lensless dataset was released in [13]. It includes images extracted from 1-minutes videos of ten species of freshwater plankton, acquired with a lensless microscope[21] (see Fig. 2 for representative examples). For each species, the dataset includes 500 training images and 140 testing images.

3.2   Experiment Details

In our experiments, we use the entire set of 131 features described in Sect. 2.1, without any features selection procedure. We use a number of surrogates equal to 1000 with a level of confidence equal to 98%. Each surrogate is randomly constructed using 100 training samples. The parameter Z, referring to the contamination in the training set (i.e., the percentage of allowed out-of-class training samples) is set to 95%, and is equal for all the species included in the lensless dataset. We compare the TDT algorithm with four state-of-the-art outlier detection methods on the lensless dataset: (i) One-class SVM [15]; (ii) Isolation Forest [10]; (iii) Robust covariance [14]; (iv) Local outlier factor [3]. We use the methods implementation provided in [4]. The methods hyper-parameters are tuned

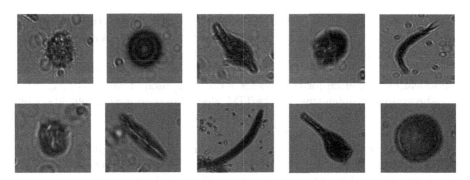

Fig. 2. Sample images from the ten species included in the lensless dataset. From top left to bottom right, the species are the following: (1) Actinosphaerium Nucleofilum; (2) Arcella Vulgaris; (3) Blaepharisma Americanum; (4) Didinium nasutum; (5) Dileptus; (6) Euplotes Eurystomus; (7) Paramecium Bursaria; (8) Spirostomum ambiguum (9) Stentor Coeruleous; (10) Volvox.

with an heuristic procedure to ensure the highest testing accuracy on the lensless dataset, with the *contamination* parameter set accordingly to the analogous parameter for the TDT algorithm.

3.3 Results

In this section, we test and present the results of the proposed pipeline on the lensless dataset. First, we validate our TDT anomaly detection algorithm, on the available test set, for each of the ten plankton species. Then, we apply the whole pipeline using a leave-one-out approach, to evaluate the accuracy of the proposed method to identify an unknown species, i.e., a species not included in the training set.

TDT Algorithm Performances. We extract the features described in Sect. 2.1 to train one TDT detector for each of the ten available species, using the training samples included in the lensless dataset (500 images per species). As described in Sect. 3.1, the lensless test set includes 140 images for each species (with a total of 1400 images). An anomaly detection algorithm provides a binary output information, namely, the test sample is either anomalous or in-class. We validate the TDT algorithm on both anomaly and in-class classification, in terms of testing accuracy. For each species, We use the 140 testing features samples correspondent to the training class to test the in-class classification performances, while all the testing samples belonging to the other species (the remaining 1260 images) are used to compute the anomaly detection accuracy. As a benchmark, we compare the TDT algorithm with four state-of-the-art outlier detection methods: one-class SVM, isolation forest, robust covariance and local outlier factor. Table 1 shows the results averaged over the ten lenless species in terms of testing accuracy. As we can observe, the TDT algorithm outperforms the state-of-the-art benchmarking algorithms both in terms of in-class and anomaly detection accuracy. For completeness, Table 2 and 3 reports the in-class and anomaly detection accuracy separately for each of the 10 lensless species. In these tables, the numerical identification for the species is the same used in caption of Fig. 2.

The average training time for the TDT algorithm is approximately 30 s, while prediction are provided at 60 Hz, on a notebook with Intel Core i7, 2 GHz with 16 GB of RAM.

As a further comparison with deep-learning based approaches, in [13], the authors introduce a neural network to perform anomaly detection. Using the 10 species included in the lensless dataset, they reach an average in-class classification testing accuracy of 0.987, an average anomaly detection accuracy of 0.990 with an overall accuracy of 0.989. The TDT algorithm is able to achieve the same anomaly detection accuracy, with a slightly lower in-class accuracy. However, the final aim of our work is to provide an efficient and computational light pipeline, that can be executed on an embedded device (e.g., a Rasperry-Pi) in order to provide in-situ real time identification of new species of plankton. Thus, the TDT algorithm still represents the best trade-off between complexity, computational time and overall accuracy.

Table 1. Average in-class, anomaly and overall testing accuracy for the proposed algorithm and the 4 benchmark algorithms included in our analysis on the lensless dataset.

Algorithm/Accuracy	In class	Anomaly	Overall
TDT algorithm (Ours)	**0.961**	**0.990**	**0.976**
One-class SVM	0.945	0.901	0.923
Isolation forest	0.939	0.986	0.963
Robust covariance	0.950	0.988	0.969
Local outlier factor	0.941	0.942	0.941

Table 2. In-class testing accuracy for the proposed algorithm and the 4 benchmark algorithms included in our analysis on the lensless dataset.

Algorithm/Species	1	2	3	4	5	6	7	8	9	10
TDT algorithm (Ours)	0.936	0.986	0.964	0.950	0.986	0.936	0.950	0.971	0.993	0.936
One-class SVM	0.921	0.936	0.950	0.936	0.964	0.921	0.936	0.943	0.979	0.964
Isolation forest	0.907	0.950	0.957	0.900	0.971	0.957	0.921	0.971	0.914	0.943
Robust covariance	0.964	0.971	0.957	0.929	0.943	0.929	0.936	0.957	0.957	0.957
Local outlier factor	0.914	0.950	0.993	0.943	0.943	0.936	0.964	0.964	0.921	0.886

Table 3. Anomaly detection testing accuracy for the proposed algorithm and the 4 benchmark algorithms included in our analysis on the lensless dataset.

Algorithm/Species	1	2	3	4	5	6	7	8	9	10
TDT algorithm (Ours)	1.000	0.948	1.000	0.994	1.000	0.999	0.962	0.999	0.996	1.000
One-class SVM	0.992	0.779	0.822	0.881	1.000	0.965	0.917	0.705	0.978	0.975
Isolation forest	1.000	0.994	0.986	0.983	0.998	0.983	0.922	1.000	0.998	1.000
Robust covariance	1.000	0.942	0.994	0.982	1.000	0.998	0.972	1.000	1.000	0.999
Local outlier factor	1.000	0.948	0.988	0.875	0.999	0.827	0.921	0.871	0.996	0.999

The Proposed Pipeline Can Reveal Novel Species. In this section, we test how the pipeline proposed in this paper performs in revealing previously unclassified plankton species, exploiting the lensless dataset. As described in Sect. 2.1, we suppose that a training plankton dataset (e.g., the lensless dataset for freshwater plankton) with a given number of species, is available. Hence, one TDT anomaly detector is trained for each training species. When a new plankton image is acquired in situ, the set of TDT detectors is used in a parallel architecture (see Fig. 1).

A sample is classified as previously unseen (i.e., a global anomaly) if all the trained detectors detect it as an anomaly. Ideally, a sample belonging to a previously unknown species, imaged by a system powered with the proposed pipeline, is recognized as a global anomaly and stored for further steps. To support this hypothesis, we perform a leave-one-out experiment on the lensless dataset. We use one of the ten species as unknown class, while training one TDT algorithm for each of the remaining nine species. We repeat this operation ten times, each time considering a different species as previously undetected. The accuracy averaged with respect to the ten leave-one-out tests is 0.920, that is, in our experiments, the 92.0 % of the times a sample belonging to an unseen species is rejected by all the trained TDT classifiers and identified as a potential new species.

4 Conclusion

In this paper we introduce an efficient pipeline to reveal unknown species of plankton. Our approach is centered around a customized algorithm for anomaly detection, based on features extracted from high-resolution images of aquatic microorganisms. We show that the TDT algorithm reaches a high accuracy in anomaly detection, outperforming four state-of-the-art outlier detection methods, on the plankton dataset used in our analysis. In the proposed pipeline, a TDT detector is trained for each of the species included into the training set. Once the complete set of TDT detectors is available, it is possible to assemble them into a parallel architecture capable of detecting unknown species. Exploiting a leave-one-out approach on the lensless dataset, we show that the engineered pipeline is able to identify a species excluded from the training set, with high accuracy (92.0%).

A high-resolution acquisition system powered with an artificial intelligence based on the described pipeline and our TDT algorithm, has the potential to be useful in field studies where a continuous monitoring of the aquatic environment is necessary. In fact, we propose that detected global anomalies could be stored and then analyzed by expert investigators or undergo an unsupervised process of partitioning. In this way, new TDT detectors could be trained, obtaining a dynamical extension of the available plankton dataset.

It is worth noting that, while we apply our TDT architecture to plankton data, the technique is general and can be applied to other types of biological images (cells, tissues, etc.).

All the operations described in this paper are computationally inexpensive and compatible with low-power computing environments, like a Raspberry-Pi. Even if further work is necessary, our algorithms support the field deployment of AI-powered microscopes continuously screening the aquatic environment. We imagine this system could uncover novel species of plankton, as well as critical modifications to the morphology of the ones already included into the training set, thus enabling the use of plankton as biosensor.

Acknowledgements. We thank all faculty and students in the National Science Foundation Center for Cellular Construction and the Machine Learning Genoa Center (MaLGa) for discussion and critical feedback on the general idea and pipeline. This material is partially based upon work supported by the National Science Foundation under Grant No. DBI-1548297.

References

1. Behrenfeld, M.J., et al.: Biospheric primary production during an ENSO transition. Science **291**(5513), 2594–2597 (2001)
2. Blaschko, M.B., et al.: Automatic in situ identification of plankton. In: 2005 Seventh IEEE Workshops on Applications of Computer Vision (WACV/MOTION 2005), vol. 1, pp. 79–86 (2005). https://doi.org/10.1109/ACVMOT.2005.29
3. Breunig, M.M., Kriegel, H.P., Ng, R.T., Sander, J.: Lof: identifying density-based local outliers. SIGMOD Rec. **29**(2), 93–104 (2000). https://doi.org/10.1145/335191.335388
4. Buitinck, L., et al.: API design for machine learning software: experiences from the scikit-learn project. In: ECML PKDD Workshop: Languages for Data Mining and Machine Learning, pp. 108–122 (2013)
5. Cheng, K., Cheng, X., Wang, Y., Bi, H., Benfield, M.C.: Enhanced convolutional neural network for plankton identification and enumeration. PLOS ONE **14**(7), 1–17 (2019). https://doi.org/10.1371/journal.pone.0219570
6. Fossum, T.O., et al.: Toward adaptive robotic sampling of phytoplankton in the coastal ocean. Sci. Rob. **4**(27), eaav3041 (2019). https://doi.org/10.1126/scirobotics.aav3041
7. Haralick, R.M., Shanmugam, K., Dinstein, I.: Textural features for image classification. IEEE Trans. Syst. Man Cybern. SMC **3**(6), 610–621 (1973). https://doi.org/10.1109/TSMC.1973.4309314
8. Huang, Z., Leng, J.: Analysis of hu's moment invariants on image scaling and rotation, vol. 7, pp. V7–476 (2010). https://doi.org/10.1109/ICCET.2010.5485542

9. Hughes, A.J., et al.: Quanti.us: a tool for rapid, flexible, crowd-based annotation of images. Nature **15**(8), 587–590 (2018). https://doi.org/10.1038/s41592-018-0069-0

10. Liu, F.T., Ting, K.M., Zhou, Z.H.: Isolation forest. In: 2008 Eighth IEEE International Conference on Data Mining, pp. 413–422 (2008). https://doi.org/10.1109/ICDM.2008.17

11. Lumini, A., Nanni, L.: Deep learning and transfer learning features for plankton classification. Ecol. Inf. **51**, 33–43 (2019)

12. Pastore, V.P., Zimmerman, T., Biswas, S.K., Bianco, S.: Establishing the baseline for using plankton as biosensor. In: Imaging, Manipulation, and Analysis of Biomolecules, Cells, and Tissues XVII, vol. 10881, p. 108810H. International Society for Optics and Photonics (2019)

13. Pastore, V.P., Zimmerman, T.G., Biswas, S.K., Bianco, S.: Annotation-free learning of plankton for classification and anomaly detection. Sci. Rep. **10**(1), 12142 (2020). https://doi.org/10.1038/s41598-020-68662-3

14. Rousseeuw, P., Driessen, K.: A fast algorithm for the minimum covariance determinant estimator. Technometrics **41**, 212–223 (1999). https://doi.org/10.1080/00401706.1999.10485670

15. Schölkopf, B., Williamson, R., Smola, A., Shawe-Taylor, J., Platt, J.: Support vector method for novelty detection. In: Proceedings of the 12th International Conference on Neural Information Processing Systems, NIPS 1999, pp. 582–588. MIT Press, Cambridge (1999)

16. Schröder, S.M., Kiko, R., Koch, R.: Morphocluster: efficient annotation of plankton images by clustering. Sensors **20**(11), 3060 (2020)

17. Sosik, H.M., Olson, R.J.: Automated taxonomic classification of phytoplankton sampled with imaging-in-flow cytometry. Limnol. Oceanogr. Methods **5**(6), 204–216 (2007)

18. Sournia, A., Chrdtiennot-Dinet, M.J., Ricard, M.: Marine phytoplankton: how many species in the world ocean? J. Plankton Res. **13**(5), 1093–1099 (1991). https://doi.org/10.1093/plankt/13.5.1093

19. Yang, Z., Fang, T.: On the accuracy of image normalization by zernike moments. Image Vision Comput. **28**(3), 403–413 (2010). https://doi.org/10.1016/j.imavis.2009.06.010

20. Zheng, H., Wang, R., Yu, Z., Wang, N., Gu, Z., Zheng, B.: Automatic plankton image classification combining multiple view features via multiple kernel learning. BMC Bioinf. **18**(16), 570 (2017). https://doi.org/10.1186/s12859-017-1954-8

21. Zimmerman, T., Smith, B.: Lensless stereo microscopic imaging. In: ACM SIGGRAPH 2007: Emerging Technologies, SIGGRAPH 2007, p. 15 (2007). https://doi.org/10.1145/1278280.1278296

Mixing Zero-Shot Learning Up: Learning Unseen Classes from Mixed Features

Julio Ivan Davila Carrazco[1,2](✉) ⓘ, Pietro Morerio[1] ⓘ, Alessio Del Bue[1] ⓘ, and Vittorio Murino[1,3] ⓘ

[1] Pattern Analysis and Computer Vision (PAVIS), Italian Institute of Technology, Genoa, Italy
`pietro.morerio@iit.it, alessio.delbue@iit.it, vittorio.murino@iit.it`
[2] Department of Marine, Electrical, Electronic and Telecommunications Engineering, University of Genoa, Genoa, Italy
`julio.davila@iit.it`
[3] Department of Computer Science, University of Verona, Verona, Italy

Abstract. Zero-Shot Learning (ZSL) objective is to classify instances of classes that were not seen during the training phase. ZSL methods take advantage of side information, i.e., class attributes, to leverage information between the seen and unseen classes. Lately, generative methods have been used to synthesize unseen features in order to train a classifier for the unseen classes. Although generative methods obtain high performance, the learned distribution may not properly represent the real distribution of the unseen classes. We propose an approach to alleviate this issue by creating a new set of mixed features. These mixed features provided a new distribution for the generative method to learn from. By using these mixed features we obtained an +2.2% improvement over tf-VAEGAN in the Oxford Flowers (FLO) dataset.

Keywords: Generative adversarial networks · Mixup · Cutmix

1 Introduction

The outstanding capabilities of today's computational resources together with innovative deep learning models have allowed us to solve complex data driven tasks. In particular, the image classification task was considered extremely hard in the past. Now, models focusing on solving this task have shown performance even greater than those obtained from humans. These remarkable results are mainly because the vast amount of annotated information available for training the models. Such supervision is not always available in the real world making such models unreliable and inefficient. To tackle the unavailability of annotated information, new methodologies had to use either non-annotated information or external sources of information. Zero-Shot Learning (ZSL) applies the latter by exploiting external information in the form of attributes.

© The Author(s), under exclusive license to Springer Nature Switzerland AG 2022
S. Sclaroff et al. (Eds.): ICIAP 2022, LNCS 13232, pp. 610–620, 2022.
https://doi.org/10.1007/978-3-031-06430-2_51

In ZSL, the objective is to classify instances of classes that were not seen during the training phase. ZSL has access to annotated images from certain classes, i.e., the seen classes, and to a set of attributes for all the classes (seen and unseen classes). The idea behind ZSL is to leverage information from seen classes to the unseen classes by using the attributes. ZSL literature have focused on diverse methodologies to solve the problem [1, 6, 8, 12, 13, 15, 16]. An overview on early ZSL methods is presented in [14]. Recently, research has focused on generative methods. In [15], they presented f-CLSWGAN a method that uses a conditional Generative Adversarial Network (GAN) [4] to learn the feature distribution conditioned on the seen classes' attributes. In [8], they integrated a feedback loop to refine the generated features. However, the main issue with the ZSL models is that they are biased towards the seen classes. Especially, generative models learn a distribution based only on the visual features from seen classes. In consequence, this learned distribution could be completely different from the distribution of the unseen classes. This could lead to an inferior performance when classifying only on instances from the unseen classes which is how ZSL evaluates its accuracy.

To improve the learned distribution and to mitigate the bias issue, we present an approach that allows the ZSL models to learn from a new distribution. This new distribution attempts to cover the space where the unseen classes reside. We achieve this by creating new training features for the models to use. Given that seen and unseen classes share common attributes. These new features are based on the available features from the seen classes (seen features). Although the new features are based on the seen features, they seek to resemble the characteristics of the unseen classes.

We take advantage of the available seen features to create the new features. We generate a new feature by combining two seen features. Depending on how we select and combine these seen features, the resulting feature could share attributes similar to the unseen classes. Our approach focus on how to mix the features up by applying two types of mixing techniques. The resulting mixed features will share similarities with seen classes but also they will display characteristics from unseen classes. In the next sections, we expand our approach and present the experimental results obtained from testing our idea.

2 Related Work

2.1 Generative Methods for Zero-Shot Learning

In this section, we review few of the works focusing on solving Zero-Shot Learning by applying generative methods. The generative methods focus on learning a distribution of the seen classes conditioned on the corresponding class attributes. The class attributes are a set of features that describe each class. The idea behind conditioning is to generate unseen synthetic samples by conditioning the generator on the unseen class attributes. The unseen synthetic features are then used to train a classifier. The f-GAN, f-WGAN, and f-CLSWGAN methods [15] apply a GAN to generate visual features instead of generating image pixels.

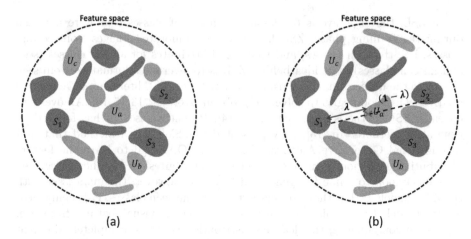

Fig. 1. (a) A representation of the distribution of the seen classes (S) and unseen classes (U) in the feature space. (b) Illustration of how applying Mixup to two features from seen class could generate a new feature with characteristics similar to an unseen class.

f-GAN aims to learn a generator $G : Z \times C \rightarrow X$ where Z is sampled from a Gaussian distribution and C is the attribute embedding of class y in which the GAN is conditioned on, and outputs a CNN image feature $\tilde{x} \in X$ of class y. f-WGAN instead uses a conditional Wasserstein GAN [2] together with a discriminator $D : X \times C \rightarrow \mathbb{R}$ that unlike f-GAN, it is also conditioned on the attribute embedding C. Finally, f-CLSWGAN integrates a classifier that is trained on the input data to guarantee that the generated features are suited for training a classifier. f-VAEGAN [16] uses a Variational Auntoenconder (VAE) [5] combined with a Generative Adversarial Network (GAN) [4]. This model has this setting because it combines the strength of both networks as it has been proved to have a better image generation [7]. tf-VAEGAN [8] expands the work done for f-VAEGAN by introducing a Semantic Embedding Decoder (SED) and a cycle consistency. SED reconstructs the semantic embeddings, i.e., the attributes, from the generated visual features. The SED module is used to apply a cycle consistency which serves to improve the feature synthesis. Finally, their last contribution was the use of the latent embedding from SED as a source of information during the classification phase.

Now, generative methods trained on data from seen classes develop a bias towards such classes. This bias influences the distribution of synthetic features reducing its resemblance to the real distribution of the unseen classes. In Fig. 1(a), we illustrate the distribution of seen and unseen classes in the feature space. Although countermeasures have been taken to mitigate this problem, it still persists. In a latter section, we will present our approach to reduce this issue.

2.2 Generalized Zero-Shot Learning

Zero-Shot Learning is tested on a set containing only instances of unseen classes which makes it unrealistic when applied to a real-world scenario. In the real world, data belonging to both types of classes, i.e., the seen and the unseen, is more likely to be found. So, it would be beneficial to have a model capable of classifying instances coming from both set of classes. Generalized Zero-Shot Learning (GZSL) [3] focuses on solving this problem. The features and labels from the seen classes, as well as, the whole set of class attributes for both the seen and unseen classes are available in the training phase. A relaxation of the problem allows to have access to features from the unseen classes during the training phase. This relaxation of the problem settings is called Transductive GZSL. In literature, several methods that solved ZSL have been used for solving GZSL by training the models on the transductive setting. Generative methods have solved GZSL by adding a second discriminator to their architecture [8,16]. The second discriminator takes as input real unseen instances, as well as, synthetic unseen instances created by the generator conditioned on attributes from the unseen classes. In this paper, we focused mainly on testing our proposed approach only in the unseen classes, i.e., in the ZSL settings. However, we provided experimental results for GZSL as reference for future works.

3 Method

As mentioned earlier, ZSL generative methods are trained using only visual features from seen classes which makes the learned model biased towards such classes. In this section, we present our approach to alleviate this issue. First, we describe the Zero-Shot Learning problem. Let $x \in X$ denote the visual features and $y \in Y^s$ the corresponding labels from a set of M seen classes $Y^s = \{y_1, y_2, ..., y_M\}$. Let $Y^u = \{U_1, U_2, ..., U_N\}$ denote the set of N unseen classes which are disjoint from seen classes Y^s. A set of attributes $a(k) \in A$ is available for each of the k seen Y^s and unseen classes Y^u. The attributes describe the characteristics of each class. The attributes are represented by a set of vectors containing float values. Such values were obtained by performing a series surveys and extracting the statistics of the obtained results [10] or by using fine-grained visual descriptions of image to extract embeddings from a character-based CNN-RNN [11]. The objective in ZSL is to learn a classifier $f_{zsl} : X \rightarrow Y^u$. During training, only the seen visual features X_s, their corresponding class labels Y^s, and all the class attributes $a(k)$ are available. This means that our approach is tested on the inductive setting for ZSL. The attributes are used to leverage the learned information from seen classes to the unseen classes.

As explained, generative models learn the distribution of the seen classes. This creates the bias issue that we try to solve with our approach. To tackle the problem, we trained the generative method with a new set of features and attributes that were created by combining pairs of features and attributes from the seen classes, respectively. To prove our approach, We applied two different mixing techniques. The mixing techniques are Mixup [19] and Cutmix [18] which

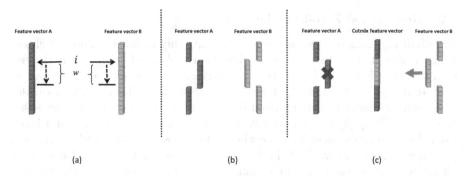

Fig. 2. Steps to perform Cutmix between two feature vectors: (a) **Selecting the segments to cut and to replace.** The segments are selected based on the parameter **i** which represents the starting point in the feature vector and the parameter **w** which is the length of the segment to cut. (b) **Extracting the segments.** The selected segments are remove from the feature vectors. (c) **Replacing the extracted segment.** The extracted segment of feature vector B is inserted in the feature vector A where the segment was cut.

apply different methodologies to obtain new samples (in our case, new features) to train a model. Mixup applies a linear interpolation

$$\mathbf{x}_m = \lambda \mathbf{x}_i + (1 - \lambda)\mathbf{x}_j, \tag{1}$$

where $\mathbf{x}_i$ and $\mathbf{x}_j$ are visual features from the seen classes, $\lambda \in [0, 1]$ is sampled from a distribution $\text{Beta}(\alpha, \beta)$, and $\mathbf{x}_m$ is the resulting mixed visual feature. The intuition behind applying Mixup is to create features that may share similarities with features from the unseen classes. Therefore, the distribution of the mixed features may be similar to the unseen classes. In Fig. 1(b), we illustrate how the linear interpolation of two visual features from seen classes may create features that share similarities with the unseen classes. Now, given that the model is conditioned on the class attributes, we need to create attributes for the mixed features. We decided to create the mixed attributes using the same procedure as with the features by applying Eq. 1 and using the same λ value that was used to mix the visual features. All the mixed features and mixed attributes are created during training phase to guarantee that each batch has a different set of features. The pairs of visual features selected for mixing up are chosen randomly in each iteration. We use the mixed features and their corresponding mixed attributes to train the model.

The second mixing technique Cutmix, mixes a pair of images by cutting a section of an image A and pasting it on an image B. In our case, we are working with visual features instead of images but the methodology can be applied in the same manner. In Fig. 2, we illustrate how Cutmix works for visual features. The first step is to select the segment that would be replaced or cut from the visual features. This is done by selecting a starting point in the feature vector that is sampled from a Uniform distribution and by selecting the length of the segment

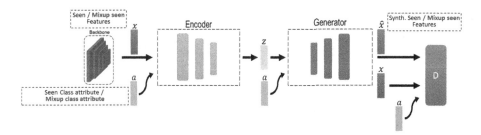

Fig. 3. A simplified representation of a VAEGAN network to solve ZSL. The network is composed of a variational autoencoder (VAE) and a Generative Adversarial Network (GAN) [4]. The two networks share the encoder/generator. The VAEGAN is conditioned on the attribute vector to leverage knowledge from seen classes to unseen classes.

whose size depends on the λ value sampled from a Beta distribution. Then we extract the segments defined by the selected starting point and the segment's length. Finally, we replace the segment in the feature A with the extracted segment from the feature B to create the mixed feature. When applying Cutmix, the mixed attributes are obtained in similar manner as with Mixup by applying an linear interpolation between the corresponding attributes of the selected features. To train the model using Cutmix, we perform the same procedure as with Mixup only by replacing how the features are mixed.

Now, since our approach focuses on creating mixed training features, any model that uses similar features is also able to test them. In particular, we decide to test our approach with two ZSL methods: f-CLSWGAN [15] and tf-VAEGAN [8]. We selected these two methods mainly because 1) tf-VAEGAN is one of the best performing generative methods for ZSL and 2) we wanted to verify if the improvement on the accuracy is mainly because our mixed features or if it also depends on the complexity of the used method. Both methods use a generator but the difference between them is that tf-VAEGAN applies a more complex generative architecture. tf-VAEGAN uses a Variational Autoenconder combined with a Generative Adversarial Network (VAEGAN) to train its generator. In Fig. 3, a simplified version of the used VAEGAN is illustrated. In the VAEGAN, the class attributes are used to condition the encoder, generator, and the discriminator. In our case, we replace the visual features and attributes from seen classes and used exclusively our mixed features and attributes. Next, we show the experimental results obtained from training the ZSL methods using our approach.

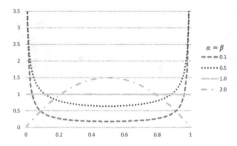

Fig. 4. Probability density function of the Beta distribution. The Beta distribution behaves different depending on the selected values for the alpha and beta parameters. Here, we present four different configurations were the alpha and beta have the same value.

4 Experiments

In Table 1, we present the experimental results for ZSL obtained from training two generative methods: f-CLSWGAN [15] and tf-VAEGAN [8] using three different types of features. The features used in the experiments consisted of the unmixed original visual features of the seen classes and two sets of mixed features that were created by applying Mixup [19] and Cutmix [18] to the available visual features, attributes, and labels from the seen classes training set. A value of 1.0 was used for the Beta distribution's parameters alpha and beta. This configuration provides us with a λ that behaves as an uniform distribution as show in Fig. 4. This means that there is an equal probability of having features that were not mixed and having features that were mixed with a 50%–50% ratio. The methods were tested in two well known datasets: Scene Understanding (SUN) [17], and Oxford Flowers 102 (FLO) [9]. For FLO, the results show an improvement in the accuracy of +0.6% when Cutmix was used for creating the mixed features. Instead, when using Mixup to create the features, FLO obtained an

Table 1. Comparison of the average Top-1 accuracy per class obtained by using three different feature types (**FT**): Original (Orig.) represents the original unmixed visual features used by the baseline, Mixup and Cutmix represent the features created by using the respective methodologies.

	FT	FLO	SUN
f-CLSWGAN	(Baseline) Orig	67.2	60.8
	(Our) Mixup	60.2	59.1
	(Our) Cutmix	61.7	56.8
tf-VAEGAN	(Baseline) Orig	70.8	**66.0**
	(Our) Mixup	**73.0**	65.1
	(Our) Cutmix	71.4	64.3

improvement of +2.2%. However on SUN, the results show that using mixed features does not provide an advantage over using the original features. The difference between FLO and SUN is the way in which their collection of attributes were obtained. FLO uses a text embedding obtained from training a deep learning model with single-sentence visual descriptions of the dataset's images [11]. Instead, SUN uses attributes that were generated after performing a series of surveys to identify common characteristics of the dataset's images [10]. Each of SUN's 102 attributes represent how likely a concept (e.g. fire, diving, medical activity) will appear in the scene. Although doing a linear interpolation might work for certain concepts (e.g. fire, ice), for other concepts (e.g. constructing/building, medical activity, spectating) the same logic might not apply. Especially, when applying Cutmix, the mixed features may not contain the elements that represent certain attributes. This suggests that FLO's attribute embeddings are more suitable for mixing up than SUN's individual attributes. The results obtained using Mixup + tf-VAEGAN are similar or better than its corresponding baseline on contrary to the Mixup + tf-CLSWGAN results that are for most of the cases lower than its baseline. This could indicate that a more complex generative method such as the one used on tf-VAEGAN is more suited for learning from mixed features. The majority of results obtained by using Mixup are better than the results obtained by using Cutmix. The reason could be that Mixup performs the mixing over the whole feature vector while Cutmix only does it on a segment.

Table 2. Average Top-1 accuracy of seen (s) and unseen (u) classes together with their harmonic mean (**H**). To train the models, three different features types (**FT**) were used for the experiments: Original (Orig.) represents the original unmixed visual features used by the baseline, Mixup and Cutmix represent the mixed features created by their respective methodologies.

	FT	FLO			SUN		
		u	s	H	u	s	H
f-CLSWGAN	(Baseline) Orig	59.0	73.8	65.6	42.6	36.6	39.4
	(Our) Mixup	50.7	73.5	60.0	42.9	36.0	39.1
	(Our) Cutmix	53.8	76.3	63.1	42.8	33.1	37.3
tf-VAEGAN	(Baseline) Orig.	62.5	84.1	71.7	45.6	**40.7**	**43.0**
	(Our) Mixup	**65.2**	83.1	**73.1**	**45.8**	38.9	42.1
	(Our) Cutmix	63.5	**84.6**	72.6	45.7	36.7	40.7

By using mixed features, we make available a distribution that may share the feature space with the unseen classes. Given that ZSL is tested only on unseen classes, we expected an increase on accuracy by learning from a similar distribution to the one from unseen classes. An increase that we did observe in our experiments. Now, there is another problem that shares characteristics with

ZSL, so we decided to test on it our approach. In Generalized Zero-Shot Learning (GZSL), the objective is to classify instances from either seen and unseen classes while having only access to the visual features and labels from seen classes and the whole set of attributes during the training phase. In Table 2, we show the results from testing our approach in GZSL. We noticed that integrating the seen classes in the classification impacts the accuracy obtained by using the mixed features of our approach. Although we obtained a higher accuracy (+2.7%) in the unseen classes for FLO, there is trade off with the seen classes' accuracy which is reduced by 1.0%. A similar result is obtained for SUN when applying Mixup. Furthermore, when applying Cutmix, FLO obtained overall better results that the baseline. However, the improvement in the accuracy was lower than the values obtained with Mixup. Although the results for GZSL were not as good as for ZSL, they demonstrated that Mixup is better than Cutmix for mixing features.

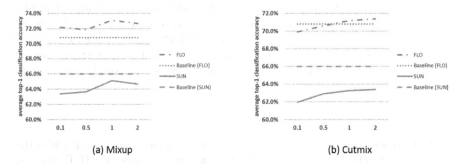

(a) Mixup   (b) Cutmix

Fig. 5. The effect on the Top-1 accuracy in tf-VAEGAN [8] by changing the parameters of the beta distribution that dictates the mixing percentage for each feature. The horizontal axis represent the value used for the alpha and beta parameters of the beta distribution.

Finally, to study the influence of the Beta distribution over the model's accuracy, we perform a series of ablations. We selected four values (0.1, 0.5, 1.0, and 2.0) for the Beta distribution's parameters alpha and beta. The alpha and beta parameters have always the same value for all the experiments. We chose these values because of their probability density functions (PDF). The behavior of each PDF will allow us to test four types of different scenarios. In the first scenario, most of the mixed features will be closely similar to the original features. In the second scenario, the similarity between mixed features and the original ones will be reduce by a small quantity. On the third scenario, the mixup percentages of the features will be equally distributed among the 0% and 50%. Finally in the last scenario, most of the mixed features will have a high mixup percentage between 40% and 50%. In Fig. 4, we illustrate the PDFs of the selected values. By applying different values, we have control over the percentage of each feature that it is mixed. Low values generate a high probability of having a mixed feature that is closer to the original. Instead, high values generate a high probability of having a mixed feature that contains a high percentage of both original features. In Fig. 5, we present the results obtained from running the ablations. For Mixup, both datasets show an improvement in their performance when the values of alpha and beta are higher. Obtaining their highest accuracy when alpha and beta equal to 1.0. Sampling from a Beta distribution with parameters equal to 1.0 is similar to sampling from a Uniform distribution. This means that there is an equal probability between having features almost identical to the original features and having features that are highly mixed. In the case of Cutmix, the higher the probability of cutting out a large segment, the higher is the accuracy obtained by the model. Although the improvement between different alpha and beta configurations is small, they are more stable compare to the ablations' results obtained for Mixup.

5   Conclusions

In this paper, we proposed an approach to alleviate the bias problem presented in ZSL methods. We show how to create mixed features by applying Mixup and Cutmix. We reviewed two generative models to solve Zero-Shot Learning and tested our approach on them. We show the results from testing our approach on two well known datasets Oxford Flowers 102 (FLO) and Scene Understanding (SUN). We demonstrated how by applying our approach an improvement on the accuracy can be obtained under certain conditions. We explained why our approach may not reach the desired results depending on the type of attributes that are available for training. Finally, we performed a set of ablations that showed how given the right configuration our approach can obtained higher results than the state of the art approaches.

References

1. Akata, Z., Perronnin, F., Harchaoui, Z., Schmid, C.: Label-embedding for image classification. IEEE Trans. Pattern Anal. Mach. Intell. **38**(7), 1425–1438 (2016). https://doi.org/10.1109/TPAMI.2015.2487986
2. Arjovsky, M., Chintala, S., Bottou, L.: Wasserstein generative adversarial networks. In: International Conference on Machine Learning, pp. 214–223. PMLR (2017)
3. Chao, W.-L., Changpinyo, S., Gong, B., Sha, F.: An empirical study and analysis of generalized zero-shot learning for object recognition in the wild. In: Leibe, B., Matas, J., Sebe, N., Welling, M. (eds.) ECCV 2016. LNCS, vol. 9906, pp. 52–68. Springer, Cham (2016). https://doi.org/10.1007/978-3-319-46475-6_4
4. Goodfellow, I., et al.: Generative adversarial nets. In: Advances in Neural Information Processing Systems 27 (2014)
5. Kingma, D.P., Welling, M.: Auto-encoding variational Bayes. In: Bengio, Y., LeCun, Y. (eds.) 2nd International Conference on Learning Representations, ICLR 2014, Banff, 14–16 April 2014, Conference Track Proceedings (2014). http://arxiv.org/abs/1312.6114

6. Lampert, C.H., Nickisch, H., Harmeling, S.: Attribute-based classification for zero-shot visual object categorization. IEEE Trans. Pattern Anal. Mach. Intell. **36**(3), 453–465 (2014). https://doi.org/10.1109/TPAMI.2013.140

7. Larsen, A.B.L., Sønderby, S.K., Larochelle, H., Winther, O.: Autoencoding beyond pixels using a learned similarity metric. In: International Conference on Machine Learning, pp. 1558–1566. PMLR (2016)

8. Narayan, S., Gupta, A., Khan, F.S., Snoek, C.G.M., Shao, L.: Latent embedding feedback and discriminative features for zero-shot classification. In: Vedaldi, A., Bischof, H., Brox, T., Frahm, J.-M. (eds.) ECCV 2020. LNCS, vol. 12367, pp. 479–495. Springer, Cham (2020). https://doi.org/10.1007/978-3-030-58542-6_29

9. Nilsback, M.E., Zisserman, A.: A visual vocabulary for flower classification. In: The IEEE Conference on Computer Vision and Pattern Recognition (CVPR) (2006)

10. Patterson, G., Hays, J.: Sun attribute database: discovering, annotating, and recognizing scene attributes. In: 2012 IEEE Conference on Computer Vision and Pattern Recognition, pp. 2751–2758. IEEE (2012)

11. Reed, S., Akata, Z., Lee, H., Schiele, B.: Learning deep representations of fine-grained visual descriptions. In: Proceedings of the IEEE Conference on Computer Vision and Pattern Recognition, pp. 49–58 (2016)

12. Socher, R., Ganjoo, M., Manning, C.D., Ng, A.: Zero-shot learning through cross-modal transfer. In: Advances in Neural Information Processing Systems 26 (2013)

13. Xian, Y., Akata, Z., Sharma, G., Nguyen, Q., Hein, M., Schiele, B.: Latent embeddings for zero-shot classification. In: Proceedings of the IEEE Conference on Computer Vision and Pattern Recognition, pp. 69–77 (2016)

14. Xian, Y., Lampert, H.C., Schiele, B., Akata, Z.: Zero-shot learning - a comprehensive evaluation of the good, the bad and the ugly. TPAMI (2018)

15. Xian, Y., Lorenz, T., Schiele, B., Akata, Z.: Feature generating networks for zero-shot learning. In: IEEE Computer Vision and Pattern Recognition (CVPR) (2018)

16. Xian, Y., Sharma, S., Schiele, B., Akata, Z.: F-VAEGAN-D2: a feature generating framework for any-shot learning. In: IEEE Computer Vision and Pattern Recognition (CVPR) (2019)

17. Xiao, J., Hays, J., Ehinger, K.A., Oliva, A., Torralba, A.: Sun database: large-scale scene recognition from abbey to zoo. In: The IEEE Conference on Computer Vision and Pattern Recognition (CVPR) (2010)

18. Yun, S., Han, D., Oh, S.J., Chun, S., Choe, J., Yoo, Y.: CutMix: regularization strategy to train strong classifiers with localizable features. In: Proceedings of the IEEE/CVF International Conference on Computer Vision, pp. 6023–6032 (2019)

19. Zhang, H., Cisse, M., Dauphin, Y.N., Lopez-Paz, D.: mixup: Beyond empirical risk minimization. arXiv preprint arXiv:1710.09412 (2017)

Self-supervised Learning Through Colorization for Microscopy Images

Vaidehi Pandey, Christoph Brune, and Nicola Strisciuglio$^{(\boxtimes)}$

Faculty of Electrical Engineering, Mathematics and Computer Science,
University of Twente, Enschede, The Netherlands
n.strisciuglio@utwente.nl

Abstract. Training effective models for segmentation or classification of microscopy images is a hard task, complicated by the scarcity of adequately labeled data sets. In this context, self-supervised learning strategies can be deployed to learn suitable image representations from the available large quantity of unlabeled data, e.g. the 500k electron microscopy images that compose the CEM500k data sets.

In this work, we investigate a self-supervised strategy for representation learning based on a colorization pre-text task on microscopy images. We integrate the colorization task into the BYOL (Bootstrap your own latent) self-supervised contrastive pre-training strategy. We train the self-supervised architecture on the CEM500k data set of electron microscopy images. As backbone of the BYOL framework, we investigate the use of Resnet50 and a Stand-alone Self-Attention network, and subsequently test them as feature extractors for downstream classification and segmentation tasks.

The Self-Attention encoders pre-trained with the colorization-based BYOL method are able to learn effective features for segmentation of microscopy images, achieving higher results than those of encoders, both Resnet- and Self-Attention-based, trained with the original BYOL. This shows the effectiveness of colorization as pre-text for a downstream segmentation task on microscopy images. We release the code at https://github.com/nis-research/selfsup-byol-colorization.

Keywords: BYOL · Colorization · Microscopy images · Pre-training · Self-supervised learning

1 Introduction

Deep learning and convolutional networks achieved outstanding results in various computer vision and image processing tasks, such as image classification [11], object detection [1,24], semantic segmentation [3,25], place recognition [9,17], image and video generation [27,30], optical flow [12] and depth estimation [20], among others. In many cases, these models are trained using labeled samples in a supervised learning setting. For instance, semantic segmentation models [3,18, 25,29] require images with pixel-wise labels: they are usually trained on data sets of natural images which contain large amounts of high quality accurately labelled

© The Author(s), under exclusive license to Springer Nature Switzerland AG 2022
S. Sclaroff et al. (Eds.): ICIAP 2022, LNCS 13232, pp. 621–632, 2022.
https://doi.org/10.1007/978-3-031-06430-2_52

images. Examples are Cityscapes [7] or Mapillary Vistas [21], which contain images taken in cities. Collecting and labeling these images is time consuming, but it does not require particular expertise. In the case of medical or microscopy images, instead, acquiring a large number of images is prohibitive and labeling them requires expert knowledge. In [2], authors reported that experts spent 32 to 36 h annotating a microscopy data-set consisting of 165 images.

Recently, self-supervised learning demonstrated to be able to learn effective image representations from unlabeled data [4,5]. The self-supervision is created by defining an artificial pre-text task that exploits intrinsic structures in large amounts of unlabeled data, e.g. classification of image rotation/orientation, reconstruction from mosaique-images, etc. Encoder networks pre-trained with these techniques are then deployed as backbones for various computer vision tasks, and are either fine-tuned on a small amount of application-specific labeled data samples or directly used as feature extractors. In some cases, self-supervised pre-trained networks have achieved results comparable or higher than those of supervised networks [8,10].

In this work, we investigate using a colorization pre-text task in the BYOL pre-training framework to learn representations for microscopy cell image classification and segmentation in a self-supervised fashion. We exploit a large data set of unlabeled microscopy images, namely the CEM500k data set [6]. The use of colorization as pre-text task is motivated by the fact that it relates with shapes and regions of rather uniform color, that are also at the basis of image segmentation. It is thus expected that in the context of microscopy image analysis, this task can help learning some shape priors that could support further segmentation or classification tasks.

The rest of the paper is organized as follows. In Sect. 2, we provide a brief overview of related works while, in Sect. 3, we present our approach and model training strategy. In Sect. 4, we report the results that we achieved and finally draw conclusions in Sect. 5.

2   Related Works

Self-supervised learning methods leverage the data itself to disentangle data representation with no need of labels. The self-supervision is guaranteed by the design of pre-text tasks, which are artificial tasks to be solved by the network. In order to evaluate the quality of the learned representations, downstream tasks such as image classification or semantic segmentation are employed [13]. A good pre-text task is fundamental for self-supervised learning. The choice of the task determines the performance of the model on the downstream tasks. Some of the popular pre-text tasks are colorization [16,28], context prediction via image in-painting [23], jigsaw puzzle [15,22], image generation [31], among others.

The most powerful self-supervised learning methods are based on a pre-text task formulated as a contrastive learning problem, which consists of training two networks by forcing the representation of similar input image-pairs to be close in the latent space, and that of dissimilar input image-pairs to be distant in the latent space. SimCLR [4] (Simple framework for Contrastive Learning) learns self-supervised visual representations by maximizing the loss between dissimilar images (negative pairs) and minimizing the loss between similar images (positive pairs). For each image in a training batch, two augmented versions are generated, which are considered as positive examples. The negative examples are the $2(N-1)$ images in the batch. In [8], it was observed that the performance of SimCLR is influenced by the choice of the augmentation pool for the pre-text task, and that removing the color distortion would result in a considerable drop of results. BYOL [8] discards dissimilar images pairs, making the training process more efficient. The representations learned in the contrastive architecture are processed through two different MLP networks, namely the online and the target network. The encoder in the online network branch is updated via stochastic gradient descent, while the decoder in the target network branch is updated using the exponential moving average (EMA) of the weights of the online network. A ResNet50 pre-trained with BYOL achieved 74% accuracy on ImageNet. Momentum Contrast (MoCo) [10] constructs a dynamic dictionary on-the-fly with a queue and average-moving encoder to support the learning of contrastive representations. It achieved competitive results on various computer vision tasks, namely image classification, detection and segmentation, substantially narrowing the gap with supervised methods. In [5], the authors explore a simplification of the siamese learning framework, called SimSiam, that does not rely on negative sample pairs, large training batches or momentum encoders. They propose a stop-gradient techniques to avoid collapsing solutions.

Self-supervised pre-trained models were deployed in semantic segmentation downstream tasks. Representations learned with BYOL were demonstrated to outperform other pre-trained ones by SimCLR and MoCo in semantic segmentation on the Cityscapes data set [7]. A modification of the in-paining pre-text task was proposed in [26], to overcome some of the limitations of the plain in-painting, which modifies the overall intensity of the input image by removing one or more patches. The use of an adversarial network to produce hard patches to in-paint demonstrated effective for pre-training of good representations for semantic segmentation, achieving higher performance than other methods on the Potsdam, SpaceNet and DG Roads datasets.

3   Data and Methods

3.1   Datasets

The CEM500k data set consists of about 500k electron-microscopy images containing structures at cellular-level, taken from different organisms and with different kinds of microscope. In Fig. 2, we show example images of cells from the organism classes c.elegans, human and mouse. We use images from these three classes to evaluate the performance of the pre-trained encoders on a downstream classification task. In Fig. 2, we show the distribution of the images in the data set, organized according to the type of organisms they are taken from. In total there are eight known types of organism, while a small portion of the data set

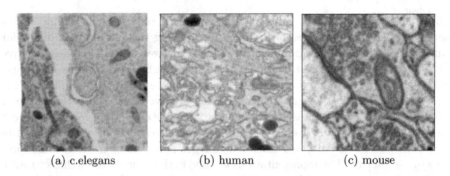

(a) c.elegans   (b) human   (c) mouse

Fig. 1. Example images from the CEM500k data set, taken from the classes of organism (a) c.elegans, (b) human and (c) mouse. We use a subset of these classes to test the downstream classification task.

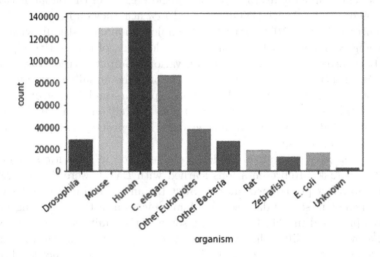

Fig. 2. Distribution of the images in different classes of organism in the CEM500k data set: images from eight known organism classes are present, plus a small portion of images for which the type of organism is unknown.

contains organisms of an unknown type. We use about 200k images for the self-supervised learning stage. The exploit the diversity of the images in the data set a to learn robust visual representations. For the downstream segmentation task, we use two benchmark data sets, namely the Kasthuri++ and Lucchi++ [2,14,19] data sets. They contain cellular-level images of the mouse brain, with labeled mitochondria regions. The Lucchi++ data set (a version of the EPFL Hippocampus dataset reannotated in [2]) contains 165 images with pixel-wise mitochondria annotations, while the Kasthuri++ data set contains 85 training and 75 testing images, also with mitochondria annotation. In Fig. 3a, we show one example image from the Lucchi++ data set, while in Fig. 3b we show the manually-made available ground truth mask of the same image.

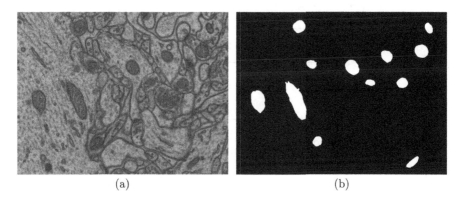

(a)   (b)

Fig. 3. An (a) image from the Lucchi++ data set, together with its (b) manual segmentation ground truth map.

3.2 Self-supervised Training

We use the BYOL framework [8] to train an encoder network for classification and semantic segmentation of microscopy images. We deploy the original BYOL architecture, see Fig. 4. It consists of 1) two encoders that share their weights, 2) an augmentation procedure and 3) a loss function. In BYOL we modify the generation of the augmented images for the target network, replacing the set of augmentation with an image colorization algorithm [28]. We make the code for experiments publicly available[1].

Encoder. The choice of the encoder is an important aspect of this method. In principle, one can choose any type of encoder architecture. In this work, we use a Resnet50 network and a Stand-alone Self-Attention network. While Resnet50 is a well-known convolutional network, the Stand-alone Self-Attention network is a custom modification of Resnet50. We substituted all the convolutional layers, except the first one, with self-attention layers. We thus investigate whether a different type of network, based on the self-attention principle, can be effectively used for self-supervised pre-training of methods for semantic segmentation and classification of structures in microscopy images.

Augmentation/Colorization. We use colorization as a pre-text task for the BYOL self-supervised learning framework. It converts a single-channel gray-scale image to a three-channel Lab image. We use the pre-trained colorization model proposed in [28] to convert the images in the CEM500k dataset from gray-scale to the Lab colorspace. A gray-scale and its corresponding color-augmented image form an input pair for the encoders as shown in Fig. 5. We call BYOL-colorization the method that we design using the colorization pre-text task, and BYOL-original the original version of BYOL, with an extended set of augmentations.

---

[1] Github repository: https://github.com/nis-research/selfsup-byol-colorization.

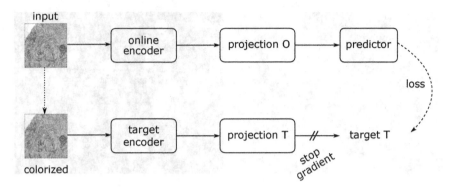

Fig. 4. Architecture of the BYOL framework that we used for self-supervised learning based on colorization pre-text. The online and target encoders have the same architecture but different weights. Only the online encoder and projection network O are trained via back-propagation (see stop gradient on the target network branch).

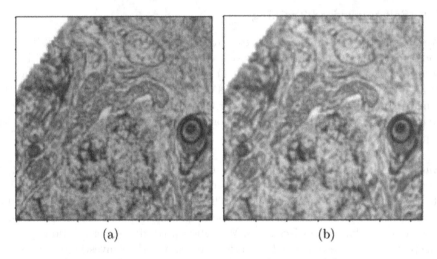

Fig. 5. An (a) example image from the CEM500k data set and (b) its colorized version obtained by using the model proposed in [28]

Loss Function and Training. The online and target encoder network in the BYOL architecture take as input the original gray-scale image and its colorized version, respectively. They have the same architecture but do not share weights. The target network provides the regression target to train the online network. The weights of the online network are optimized by back-propagating the gradient of an L2 regression loss function that compares the representations computed by the online and target network. The weights of the target network are updated as the exponential weighted average of the weights of the online network, according to the training scheme proposed in [8].

3.3   Downstream Tasks

We evaluate the image representation learned in the self-supervised pre-training on two downstream tasks, namely image classification and segmentation.

Classification. We use the pre-trained encoder as feature extractor in combination with a logistic regression model, to classify the images into three classes, namely *human cells, mouse cells, and c.elegans cells.* We train only the logistic regression model using a subset of the CEM500k dataset, which was not used for pre-training. In Fig. 6a, we depict the flow diagram of the classification downstream task.

Semantic Segmentation. We considered semantic segmentation, namely the task of pixel-wise labeling image regions to belong to one out of a number of classes of interest, as another interesting downstream task to investigate for microscopy images. We use the pre-trained encoder as backbone for a UNet-like architecture, which contains a decoder network that computes a segmentation map of the same size of the input image. We compared the representation power of different backbones, pre-trained using BYOL-original, our BYOL-colorization, U-Net encoder and ResNet-50 pre-trained on ImageNet. We perform a fine-tuning stage, where the weights of the encoder stay unchanged while the weights of the decoder only are updated by back-propagation. In Fig. 6b, we show the flow diagram of the segmentation downstream task.

4   Experiments and Results

4.1   Experiments

We use the encoders pre-trained on the CEM500k data set as feature extractors for a classification and a semantic segmentation downstream task. For the classification task, we deploy our BYOL-colorization pre-trained encoders, namely the ResNet50 and Self-Attention networks, to extract features from images of a subset of the CEM500k data set. We then use these features together with a logistic regression classifier. We compare the performance of our encoders with that of similar encoders pre-trained with the original BYOL algorithm on the CEM500k dataset, and with a ResNet50 and a Self-Attention network pre-trained on ImageNet. Finally, we also use the encoder trained in [2]. While training the logistic regression classifier, we freeze the weights of the pre-trained encoders, so that we can test the effectiveness and quality of the pre-trained representations without adapting them to the downstream task.

Similarly, we compare the representation capabilities of our pre-trained encoders with those of BYOl-original pre-trained encoders on the task of semantic segmentation. Also for this experiment, we deploy the Resnet50 and the Self-Attention encoders as backbones. For evaluation purpose, we use the data sets proposed in [2,14] and [2,19], which contain segmentation labels. We freeze the encoder weights, and embed them into a U-Net architecture for segmentation, of which we fine-tune only the decoder part.

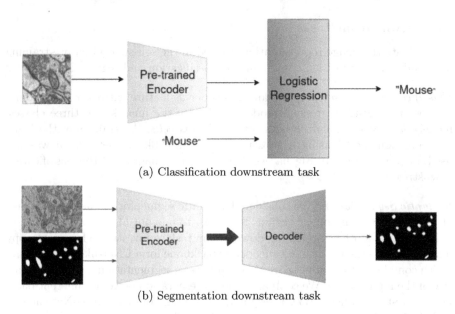

(a) Classification downstream task

(b) Segmentation downstream task

Fig. 6. Sketch diagram of the (a) classification and (b) segmentation downstream task. The weights of the pre-trained encoders (yellow boxes) are not updated while tuning the classifiers (red and blue boxes for classification and segmentation, respectiely) for the downstream tasks. (Color figure online)

4.2   Metrics

To assess and compare the performance of the pre-trained encoders on the classification downstream task, we computed the accuracy of classification. For the semantic segmentation task, instead, we measure the performance in terms of mean Intersection-over-Union (IoU) over the considered classes.

Table 1. Classification Results on cem500k dataset: The table shows the results of classification on cem500k dataset.

Encoder	Pre-training	Accuracy (%)
Resnet50	BYOL-colorization	71.75
Resnet50	BYOL-original	72.3
Resnet50	ImageNet	70.715
Self-Attention	BYOL-colorization	59.03
Self-Attention	BYOL-original	**75.67**
Self-Attention	ImageNet	55.32
[2]	Segmentation	36.83

4.3   Results

In Table 1, we report the results that we achieved on the downstream classification task. The Stand-alone Self-Attention encoder pre-trained with our BYOL-colorization achieved an accuracy of 59.03% while that pre-trained with the original BYOL achieved an accuracy of 75.67%. Our encoder improves upon the performance of the ImageNet Self-attention encoder by 3.71%. The Resnet50 encoder derived from our BYOL-colorization pre-training achieved an accuracy of 71.75%, which is slightly less than the performance of the Resnet50 derived from the BYOL-original pre-training by 0.8%. Our encoder performs better than the ImageNet pre-trained Resnet50 by 1%. The encoder of the network proposed in [2] achieved an accuracy of 36.83%, which is much lower than that of our BYOL-colorization pre-trained ResNet50 by 36.14%.

The BYOL-colorization pre-training allows to learn representations from unlabeled microscopy images which are more effective for classification than image representations learned on natural images from ImageNet. However, the diversity of data augmentation used in the original BYOL self-supervised pre-training approach allows to disentangle better features that are more effective for the classification task.

We report the results achieved on the downstream segmentation task in Table 2. We froze the weights of the pre-trained encoders and only trained the decoders for semantic segmentation. Self-Attention and Resnet50 encoders trained with our proposed BYOL-colorization pre-training achieved an mIoU score equal to 0.7034 and 0.6593 on the Lucchi++ data set, and equal to 0.7167 and 0.6839 on the Kasthuri++ data set. For both data sets, our BYOL-colorization pre-trained Self-Attention encoders achieved higher results than those of the BYOL-original pre-trained encoders. The results demonstrate that the use of colorization in our pre-training strategy contributes to learn suitable features for semantic segmentation of microscopy images. This is attributable

Table 2. Comparison of our pre-training strategy with BYOL pre-training: The table shows the results of semantic segmentation when the encoders pre-trained with our pre-training strategy is compared against the BYOL. The weights of the encoders are not updated during the training on semantic segmentation dataset.

Dataset	Encoder	Pre-training	mIoU
Lucchi++	Resnet50	BYOL-colorization	0.6593
	Self-Attention	BYOL-colorization	**0.7034**
	Resnet50	BYOL-original	0.6743
	Self-Attention	BYOL-original	0.6530
Kasthuri++	Resnet50	BYOL-colorization	0.6839
	Self-Attention	BYOL-colorization	**0.7167**
	Resnet50	BYOL-original	0.7036
	Self-Attention	BYOL-original	0.6849

to the fact that the colorization task induces the network to learn shape and color-region specific characteristics of the images, which better relate to the segmentation task. The wider range of augmentations learned in the original BYOL pre-training are not effectively tuned for segmentation.

The performance gap of pre-trained encoders for semantic segmentation of microscopy images with respect to supervised models is still large. The U-Net model adapted to the Lucchi++ and Kasthuri++ data sets proposed in [2] achieved an mIoU score (0.946 and 0.92) higher than that of BYOL-pre-trained encoder. In [2] the U-Net was trained for 1000 epochs, which is ten times larger than our 100 epochs fine-tuning of the decoder only, on the very few training images in the data sets, which may incur in overfitting, indicating that further investigation in the direction of evaluating the generalization properties of these networks is needed.

5   Conclusions

We investigated the feasibility of learning microscopy image representations from a large amount of unlabeled data in a self-supervised fashion. We thus address the problem of scarcity of unlabeled images, by training several Resnet50 and Self-Attention encoders using the BYOL self-supervised learning framework.

We demonstrated that using colorization as a pre-text task is effective to learn robust representations for semantic segmentation, and achieved better segmentation results than those obtained by encoders pre-trained using the set of augmentations designed for the original BYOL. For a classification downstream task, instead, the representation learned by the original BYOL showed slightly superior performance. The promising insights gained from the experiments open possibilities for further investigations in the direction of filling the performance gap between self-supervised and supervised methods for microscopy images, the latter of which may incur in overfitting caused by the long training schedules on very few labeled images.

References

1. Carion, N., Massa, F., Synnaeve, G., Usunier, N., Kirillov, A., Zagoruyko, S.: End-to-end object detection with transformers. cite arxiv:2005.12872 (2020)
2. Casser, V., Kang, K., Pfister, H., Haehn, D.: Fast mitochondria detection for connectomics. Nat. Methods **16**(12), 1247–1253 (2019)
3. Chen, L., Papandreou, G., Schroff, F., Adam, H.: Rethinking atrous convolution for semantic image segmentation. CoRR abs/1706.05587 (2017). http://arxiv.org/abs/1706.05587
4. Chen, T., Kornblith, S., Norouzi, M., Hinton, G.: A simple framework for contrastive learning of visual representations (2020)
5. Chen, X., He, K.: Exploring simple Siamese representation learning. In: 2021 IEEE/CVF Conference on Computer Vision and Pattern Recognition (CVPR), pp. 15745–15753 (2021). https://doi.org/10.1109/CVPR46437.2021.01549

6. Conrad, R., Narayan, K.: CEM500k, a large-scale heterogeneous unlabeled cellular electron microscopy image dataset for deep learning. eLife **10**, e65894 (2021). https://doi.org/10.7554/eLife.65894

7. Cordts, M., et al.: The cityscapes dataset for semantic urban scene understanding. In: Proceedings of the IEEE Conference on Computer Vision and Pattern Recognition (CVPR) (2016)

8. Grill, J.B., et al.: Bootstrap your own latent: A new approach to self-supervised learning (2020)

9. Hausler, S., Garg, S., Xu, M., Milford, M., Fischer, T.: Patch-NetVLAD: multiscale fusion of locally-global descriptors for place recognition. In: CVPR, pp. 14141–14152 (2021)

10. He, K., Fan, H., Wu, Y., Xie, S., Girshick, R.: Momentum contrast for unsupervised visual representation learning. In: 2020 IEEE/CVF Conference on Computer Vision and Pattern Recognition (CVPR), pp. 9726–9735 (2020). https://doi.org/10.1109/CVPR42600.2020.00975

11. He, K., Zhang, X., Ren, S., Sun, J.: Deep residual learning for image recognition. In: CVPR, pp. 770–778 (2016)

12. Ilg, E., Mayer, N., Saikia, T., Keuper, M., Dosovitskiy, A., Brox, T.: FlowNet 2.0: evolution of optical flow estimation with deep networks. In: IEEE Conference on Computer Vision and Pattern Recognition (CVPR), July 2017

13. Jing, L., Tian, Y.: Self-supervised visual feature learning with deep neural networks: A survey. CoRR abs/1902.06162 (2019). http://arxiv.org/abs/1902.06162

14. Kasthuri, N., et al.: Saturated reconstruction of a volume of neocortex. Cell **162**(3), 648–661 (2015). https://doi.org/10.1016/j.cell.2015.06.054

15. Kim, D., Cho, D., Yoo, D., Kweon, I.S.: Learning image representations by completing damaged jigsaw puzzles (2018)

16. Larsson, G., Maire, M., Shakhnarovich, G.: Learning representations for automatic colorization. CoRR abs/1603.06668 (2016). http://arxiv.org/abs/1603.06668

17. Leyva-Vallina, M., Strisciuglio, N., Petkov, N.: Generalized contrastive optimization of Siamese networks for place recognition. CoRR abs/2103.06638 (2021). https://arxiv.org/abs/2103.06638

18. Long, J., Shelhamer, E., Darrell, T.: Fully convolutional networks for semantic segmentation. CoRR abs/1411.4038 (2014). http://arxiv.org/abs/1411.4038

19. Lucchi, A., Smith, K., Achanta, R., Knott, G., Fua, P.: Supervoxel-based segmentation of mitochondria in EM image stacks with learned shape features. IEEE Trans. Med. Imag. **31**(2), 474–486 (2012). https://doi.org/10.1109/TMI.2011.2171705

20. Mayer, N., et al.: A large dataset to train convolutional networks for disparity, optical flow, and scene flow estimation. In: IEEE CVPR, pp. 4040–4048. arXiv:1512.02134 (2016)

21. Neuhold, G., Ollmann, T., Rota Bulo, S., Kontschieder, P.: The Mapillary Vistas dataset for semantic understanding of street scenes. In: Proceedings of the IEEE International Conference on Computer Vision (ICCV), October 2017

22. Noroozi, M., Favaro, P.: Unsupervised learning of visual representations by solving jigsaw puzzles (2017)

23. Pathak, D., Krähenbühl, P., Donahue, J., Darrell, T., Efros, A.A.: Context encoders: Feature learning by inpainting. CoRR abs/1604.07379 (2016). http://arxiv.org/abs/1604.07379

24. Redmon, J., Divvala, S., Girshick, R., Farhadi, A.: You only look once: unified, realtime object detection. In: 2016 IEEE Conference on Computer Vision and Pattern Recognition (CVPR), pp. 779–788 (2016). https://doi.org/10.1109/CVPR.2016.91

25. Ronneberger, O., Fischer, P., Brox, T.: U-Net: convolutional networks for biomedical image segmentation. In: Navab, N., Hornegger, J., Wells, W.M., Frangi, A.F. (eds.) MICCAI 2015. LNCS, vol. 9351, pp. 234–241. Springer, Cham (2015). https://doi.org/10.1007/978-3-319-24574-4_28

26. Singh, S., et al.: Self-supervised feature learning for semantic segmentation of overhead imagery. In: BMVC (2018)

27. Tulyakov, S., Liu, M.Y., Yang, X., Kautz, J.: MoCoGAN: decomposing motion and content for video generation. In: IEEE Conference on Computer Vision and Pattern Recognition (CVPR), pp. 1526–1535 (2018)

28. Zhang, R., Isola, P., Efros, A.A.: Colorful image colorization. CoRR abs/1603.08511 (2016). http://arxiv.org/abs/1603.08511

29. Zhao, H., Shi, J., Qi, X., Wang, X., Jia, J.: Pyramid scene parsing network. CoRR abs/1612.01105 (2016). http://arxiv.org/abs/1612.01105

30. Zhu, J.Y., Park, T., Isola, P., Efros, A.A.: Unpaired image-to-image translation using cycle-consistent adversarial networks. In: Proceedings of the IEEE International Conference on Computer Vision, pp. 2223–2232 (2017)

31. Zhu, J., Park, T., Isola, P., Efros, A.A.: Unpaired image-to-image translation using cycle-consistent adversarial networks. CoRR abs/1703.10593 (2017). http://arxiv.org/abs/1703.10593

MOBDrone: A Drone Video Dataset for Man OverBoard Rescue

Donato Cafarelli[1], Luca Ciampi[1], Lucia Vadicamo[1(✉)], Claudio Gennaro[1], Andrea Berton[3], Marco Paterni[2], Chiara Benvenuti[2], Mirko Passera[2], and Fabrizio Falchi[1]

[1] Institute of Information Science and Technologies, CNR, Pisa, Italy
`lucia.vadicamo@isti.cnr.it`
[2] Institute of Clinical Physiology, CNR, Pisa, Italy
[3] Institute of Geosciences and Earth Resources, CNR,
Via G. Moruzzi 1, 56124 Pisa, Italy

Abstract. Modern Unmanned Aerial Vehicles (UAV) equipped with cameras can play an essential role in speeding up the identification and rescue of people who have fallen overboard, i.e., man overboard (MOB). To this end, Artificial Intelligence techniques can be leveraged for the automatic understanding of visual data acquired from drones. However, detecting people at sea in aerial imagery is challenging primarily due to the lack of specialized annotated datasets for training and testing detectors for this task. To fill this gap, we introduce and publicly release the MOBDrone benchmark, a collection of more than 125K drone-view images in a marine environment under several conditions, such as different altitudes, camera shooting angles, and illumination. We manually annotated more than 180K objects, of which about 113K man overboard, precisely localizing them with bounding boxes. Moreover, we conduct a thorough performance analysis of several state-of-the-art object detectors on the MOBDrone data, serving as baselines for further research.

Keywords: Man overboard · Object detection · Unmanned Aerial Vehicles · Drone · Benchmark

1 Introduction

The 2021 Annual Overview of Marine Casualties and Incidents [9] reported that $22,532$ marine casualties and incidents were occurred between 2014 and 2020 in the waters of EU Member States or involving EU ships. $7,051$ of these events involved people, with 550 lives lost and $6,921$ injured. The main events that resulted in fatalities were ship collisions and people slipping/falling into the

Supported by NAUSICAA - "NAUtical Safety by means of Integrated Computer-Assistance Appliances 4.0", a project funded by the Tuscany region (CUP D44E20003410009).
D. Cafarelli, L. Ciampi and L. Vadicamo—Co-first authors.

water. Of the falls, 9.8% were falling overboard, resulting in 84 lives lost. Survival chances in a Man Overboard (MOB) incident depend on many variables, including the height of the fall, the water temperature, the sea state, and the weather conditions, along with the rescue operation time, the person's state of consciousness and ability to swim, to name but a few. Unfortunately, in most cases (estimated between 85–90%), it ends in death [12]. Indeed, the rescue operations are usually long and complicated. If the person falls overboard while the boat is navigating (e.g., at a speed of 18 knots), the time that elapses from when the alarm is given to when the boat can slow down and turn 180° to return to the MOB point is several minutes. Not to mention that, for safety reasons, the boat cannot turn back rapidly as it would risk running over the victim. Moreover, since the exact rescue point is not always detectable due to sea currents and alarm delays, it is clear that the MOB scenario is very critical and dangerous.

Quick and effective search and rescue operations (SAR) are crucial to increasing the victim's chances of survival. To this end, it is essential to determine a limited search area and plan paths for rescue boats [20]. Unmanned Aerial Vehicles (UAVs) equipped with thermal and/or video cameras can be profitably used to localize and track people overboard, thus expediting rescue operations and increasing their probability of success. In this regard, the "NAUtical Safety by means of Integrated Computer-Assistance Appliances 4.0" (NAUSICAA) project aims at creating a system for medium and large boats in which the conventional control, propulsion, and thrust systems are integrated with a series of latest generation sensors (including lidar systems, cameras, radar, drones) for assistance during the navigation and mooring phases. Specifically, within the project, we will use commercial aerial drones (equipped with a video camera) and Artificial Intelligence (AI) techniques to search for people overboard automatically.

Many AI techniques have achieved outstanding results in localizing and recognizing people and objects in images and video frames in recent years [21,22,26]. However, evaluating these approaches (or developing new ones) in a MOB scenario is difficult due to the lack of labeled data. Although many annotated datasets containing people and objects in everyday scenarios are publicly available, to the best of our knowledge, the same cannot be said for the case of aerial footage of people and objects in marine environments. To fill this gap, we collected and publicly released [3] a large-scale dataset of aerial footage of people who, being in the water, simulated the need to be rescued. Our dataset, named *MOBDrone*, contains 66 video clips with 126,170 frames manually annotated with more than 180K bounding boxes (of which more than 113K belonging to the *person* category). The videos were gathered from one UAV flying at an altitude of 10 to 60 m above the mean sea level.

This paper introduces our dataset and describes the data collection and annotation processes. Moreover, it presents an in-depth experimental analysis of the performance of several state-of-the-art object detectors on this newly established MOB scenario, serving as baselines. We hope that this benchmark and the preliminary results may become a reference point for the scientific community concerning the localization of MOBs from UAV imagery.

Evaluation code and all other resources for reproducing the results are available at http://aimh.isti.cnr.it/dataset/MOBDrone.

2   Related Work

In the last years, many annotated datasets have been released for supporting the supervised learning of modern detectors based on deep neural networks [2,6,7, 17]. However, only a few include images or videos taken from UAVs, and most are not focused on the marine environment. This section briefly reviews some of these drone-view datasets suitable for object detection.

VisDrone [27] is the largest object detection and tracking dataset in this category. It consists of 179,264 frames extracted from 263 video clips captured by various drone-mounted cameras covering different urban and suburban areas under various weather and lighting condition. Frames are manually annotated with more than 2.6 million bounding boxes localizing targets such as pedestrians, vehicles, and bicycles. Another remarkable dataset is UAVTD [8], suitable for vehicle detection. It consists of 80K images gathered from a UAV platform in different urban scenarios and contains 2,700 vehicles annotated with bounding boxes. Another annotated dataset for car detection is the MOR-UAV [19], which comprises more than 10K drone-view images. Finally, CAPRK [15] is a view drone dataset exploited for detecting and counting parked vehicles [1].

Few works have been done to date on creating datasets of images taken by drones of people in marine environments. Lygouras et al. [18] addressed the problem of open water human detection by conducting real-time recognition onboard a rescue hexacopter. They gathered a swimmers dataset composed of images collected from the internet and frames recorded from a drone. In total, the dataset consists of just 4,500 full HD images. Recently, Varga et al. [24] released the SeaDroneDataset that contains over 54K annotated frames captured from various altitudes and viewing angles. The dataset mainly contains people swimming in open water, and the frames are annotated using six classes: swimmer, floater (swimmer with life jacket), swimmer† (a person on a boat not wearing a life jacket), floater† (a person on a boat wearing a life jacket), life jacket, and boat. The main difference with our dataset is that we focus on people at sea without a life jacket (since in the fall into the water from a large ship it is unlikely that the person was previously wearing a life jacket), and we also consider different scenarios of the person's state of consciousness. Nevertheless, the SeaDronesSee dataset is an excellent reference for the task of human detection and tracking in the marine environment that we plan to use in the future, at least for some classes, in conjunction with our MOBDrone for training and testing deep neural networks. Finally, Ferau et al. in [11] faced the problem of assisting SAR operations in MOB incidents using autonomous UAV-based systems. Unlike our work, they aim to locate people in the water by analyzing images recorded with thermal instead of video cameras. The automatic detection and classification of objects in water from thermal images acquired using UAVs was also explored in [16].

3   The MOBDrone Dataset

Our *MOBDrone Dataset*, which we publicly released at [3], aims to overcome the lack of large public datasets of drone-based imagery for overboard human detection. Its realization required nearly 80 h of work between data acquisition, post-processing, and annotation, involving, among others, a certified pilot of the Fly&Sense Service of the CNR of Pisa for UAV flight operations and two professional divers for in-water activities. In the following, we detail the processes of data collection and curation.

Data Collection. We carried out the drone shooting activities in the Gombo beach of the Migliarino, San Rossore, and Massaciuccoli Park (Pisa, Italy). This choice was dictated for privacy reasons and to ensure compliance with the safety protocols of UAV flight operations. Indeed, the Gombo beach is a segregated area that can be accessed only after obtaining the appropriate authorization from the Park Authority.

To guarantee variability of data, we identified several dimensions of interest, including (i) *subjects/objects to be filmed* (people, lifebuoys, boats, rocks, pieces of wood, parts of the land, and whatever else there is naturally), (ii) *person's state of consciousness* (conscious, semiconscious or unconscious), (iii) *person's visual appearance* (man, woman, persons in light, dark or colored clothing, persons in a bathing suit, etc.), (iv) *light changes* (different shooting times), (v) *altitude and camera directions* (e.g., try to fly at high altitudes to see a more significant portion of the sea, and at lower altitudes better to see the objects and a possible man overboard, also changing the camera shooting angle).

We gathered a total of 49 videos at high resolution (4K) exploiting the DJI FC6310 camera of the Phantom 4 Pro V2 drone. The camera angle was perpendicular to the water (90°), except for a small set of shots where a 45° angle was used. Two professional divers (one male and one female) simulated various scenarios of a person overboard, including a conscious person (swimming, floating, or waving their arms to attract attention) and an unconscious person (floating body in a supine or prone position, or partially floating, i.e., part of the body is below the water surface). Some videos incidentally captured people close to the portion of the sea where our divers were positioned. We split these videos into multiple video clips to remove portions where people were identifiable for privacy concerns. The final dataset contains 66 videos that we post-processed, as described in the following section.

Data Curation. First, we converted the 66 video clips captured in the data acquisition campaign from 4K to 1080p resolution. Then, we extracted the frames from the videos at a rate of 30 FPS, obtaining a total of 126,170 images (see Table 1 for summary statistics). Finally, a human expert annotator manually annotated them. Specifically, the annotation process took approximately 60 h, and the Computer Vision Annotation Tool (CVAT) [23] was used. Although our work focuses on localizing and recognizing people, we also annotated other

Table 1. Dataset details. The MOBDrone benchmark contains 126,170 drone-view images at six different heights in MOB scenarios.

Altitude	# Images	# Video clips
10 m	958	1
20 m	10,053	6
30 m	29,404	15
40 m	33,046	13
50 m	29,183	16
60 m	23,526	15
tot	**126,170**	**66**

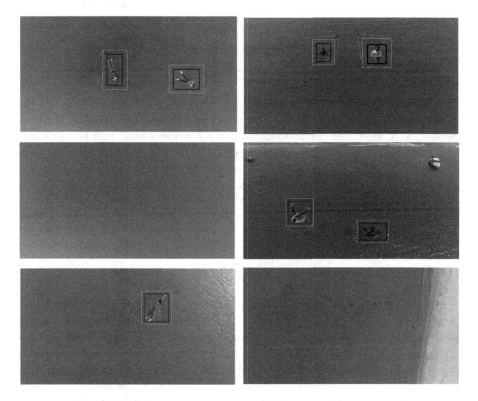

Fig. 1. Samples of the MOBDrone Dataset. Examples of images captured at different altitudes, light conditions, and camera directions. The bounding box annotations localizing the labeled objects are also shown. Objects belonging to the *person* category, which is the one of paramount interest in MOB scenarios, are outlined with red bounding boxes and zoomed. Note that 27.72% of the images do not contain objects (i.e., images of clear water) and that interfering objects in the background, such as rocks, often trigger false positive detections. (Color figure online)

Table 2. Annotation statistics. We labeled with bounding boxes 181,689 objects belonging to 5 categories.

Class	#Annotations	#Images	Samples
person	113, 408	77, 365	
boat	39, 967	31, 238	
wood	15, 980	9, 040	
life buoy	10, 401	10, 386	
surfboard	1, 933	1, 933	
no object		34, 976	
total	181,689		

objects present in the scenes. In particular, we considered a total of 5 classes (*person, boat, surfboard, wood, life_buoy*). We provide a bounding box precisely localizing each instance of the objects of interest. The total number of annotations is $181, 689$, of which the ones related to the *person* class, which is of primary interest in the MOB scenario, is $113, 408$. However, note that about 27.72% of the images do not contain any objects (i.e., images of clear water). We report some statistics concerning the annotations in Table 2, while we show some samples of our dataset in Fig. 1.

4   Detection Performance Analysis

In this section, we evaluate several state-of-the-art object detectors on our MOB-Drone dataset[1], focusing on the detection of the overboard people, i.e., on the localization of the object instances belonging to the class *person*. In the first part of our performance analysis, we compare 9 of the most popular and performing object detectors present in the literature. Then, we look upon the best three ones, performing a more in-depth analysis of the obtained results.

The detection methods considered in our analysis can be roughly grouped into three categories, i.e., anchor-based Convolutional Neural Network (CNN) methods, anchor-free CNN methods, and Transformer-based methods. We briefly summarize them below. We refer the reader to the papers describing the specific detectors for more details.

---

[1] Although in this work we exploited the whole dataset as a test benchmark, in [3] we provide training and test splits.

Table 3. Comparison of the considered detectors. mAP@[0.50:0.95] is the AP averaged over 10 IoU thresholds in the range [0.50 : 0.95] with a step size of 0.05, while AP50 is the AP computed at the single IoU threshold value of 0.50.

Method	AP50 ↑	mAP@[0.50:0.95] ↑
VarifocalNet [25]	**0.378**	**0.144**
TOOD [10]	0.314	0.116
Deformable DETR [28]	0.199	0.075
YOLOX [13]	0.126	0.049
Faster R-CNN [22]	0.126	0.041
CenterNet [26]	0.124	0.041
DETR [4]	0.128	0.040
Mask R-CNN [14]	0.109	0.033
YOLOv3 [21]	0.011	0.009

Anchor-based CNN methods compute bounding box locations and class labels of object instances exploiting CNN-based architectures that rely on anchors, i.e., prior bounding boxes with various scales and aspect ratios. They can be divided into two groups: i) the two-stage approach, where a first module is responsible for generating a sparse set of object proposals and a second module is in charge of refining these predictions and classifying the objects; and ii) the one-stage approach that directly regresses to bounding boxes by sampling over regular and dense locations, skipping the region proposal stage. Here, we use Faster R-CNN [22] and Mask R-CNN [14] regarding the first group, and YOLOv3 [21], TOOD [10] and VarifocalNet (VfNet) [25] concerning the second one.

Anchor-free CNN methods rely on the prediction of key-points, such as corner or center points, to predict the objects, instead of using anchor boxes and their inherent limitations. In this work, we exploit CenterNet [26] and YOLOX [13].

Transformer-based methods rely on the recently introduced Transformer attention modules in processing image feature maps, removing the need for hand-designed components like a non-maximum suppression procedure or anchor generation. In this paper, we consider DEtection TRansformer (DETR) [4] and one of its evolution, Deformable DETR [28].

   We evaluate and compare the above-described detectors over our MOBDrone dataset following the golden standard Average Precision (AP), i.e., the average precision value for recall values over 0 to 1. Specifically, we consider the MS COCO mAP@[0.50:0.95] [17], i.e., the AP averaged over 10 IoU thresholds in the range [0.50, 0.95] with a step size of 0.05, and the AP50, i.e., the AP computed at the single IoU threshold value of 0.50. We refer the reader to [17] for more details. All the detection techniques that we employed were pre-trained[2] on the COCO dataset [17], a popular collection of images in everyday contexts compris-

---

[2] Pre-trained models are available, e.g., in the model zoo of MMDetection project [5].

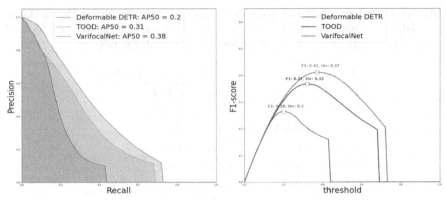

(a) Precision vs. Recall curves (IoU=0.5). (b) F_1-score vs. detection threshold curves
Areas under curves correspond to AP50.   (IoU=0.5).

Fig. 2. Comparison of the three best detectors. We report Precision-Recall (a), and F_1-detection threshold (b) curves of the three best models (VfNet, TOOD, and Deformable DETR). VfNet shows best performances.

Fig. 3. Detections produced by VarifocalNet. We indicate false positives in green, false negatives in yellow, true positive in blue and gt in red. (Color figure online)

ing objects belonging to 80 different categories, of which is present the *person* class. To evaluate the performance of the detectors, we filtered the obtained detections considering only the ones classified as *person*. We report the obtained results in Table 3. The model which turns out to be the most performing is VarifocalNet, considering both the metrics, followed by TOOD and Deformable DETR. However, in general, all the detectors exhibit moderate performance, indicating the difficulties in localizing persons in this challenging scenario. We deem that the most significant metrics in our scenario is the AP50, since i) the dataset is manually labeled by humans and therefore is accurate in terms of classification and inaccurate in terms of boundaries, ii) it is *not* crucial to precisely localize instances, i.e., it is critical to detect overboard persons but the quality of the localization is less important. With this in mind, in the following, we show an in-depth analysis of the three AP50 best models, i.e., VarifocalNet, TOOD, and Deformable DETR.

Table 4. Comparison of the three best detectors at different altitudes. AP50 and F_1 are the AP and the F_1-score computed with IoU set to 0.50.

Altitude	VarifocalNet [25] AP50 ↑	F_1 ↑	TOOD [10] AP50 ↑	F_1 ↑	Deformable DETR [28] AP50 ↑	F_1 ↑
10 m	0.973	0.444	**0.989**	0.363	0.959	**0.636**
20 m	**0.771**	**0.318**	0.681	0.308	0.514	0.279
30 m	0.400	0.199	**0.407**	**0.223**	0.240	0.210
40 m	**0.540**	**0.226**	0.406	0.203	0.314	0.209
50 m	**0.241**	**0.161**	0.187	0.140	0.107	0.08
60 m	**0.205**	**0.223**	0.171	0.196	0.063	0.131

In Fig. 2a, we report the Precision-Recall curves, i.e., precision and recall values for different detection confidence thresholds, of these three best detectors while setting the IoU threshold at 0.50. Areas under these curves correspond to AP50 values. As can be seen, the VarifocalNet detector exhibits the best performance at all confidence thresholds. The same trend is confirmed in Fig. 2b, where we show F_1-score values (where $F_1 = 2 \times \frac{Precision \times Recall}{Precision + Recall}$) at different detection confidence thresholds, again setting the IoU threshold at 0.50. Still, VarifocalNet shows superior performance compared to the other two detectors. Please note that the maximum values of these curves indicate the detection confidence score that may be used by potential users, enclosing a trade-off between the resulting Precision and Recall values. Figure 3 shows some qualitative outputs produced by VarifocalNet using this confidence score.

In Table 4, we show a comparison of the best three detectors at different altitudes in terms of AP50 and F_1-score. As expected, in general, performances decrease with increasing altitude. However, it is interesting to note that TOOD and Deformable DETR particularly struggle to detect small objects, i.e., when the altitude is above 40 m, while achieving comparable or even better results than VarifocalNet at altitudes below 30 m.

Finally, in Table 5, we report a classwise analysis of the obtained detections, i.e., we consider the detections belonging to all the 80 classes and not only the detections classified as *person*. Specifically, we take into account also errors due to misclassified objects, i.e., detected objects that matched with *person* annotations but that were classified as objects belonging to another category. We define the *True Positive Rate* (TPR) as the ratio between the number of correctly detected and classified person instances (TP) and the total number of person instances in the ground-truth (P). On the other hand, we define $dTPR(c) = \frac{dTP(c)}{P}$ as the *detection True Positive Rate* for the output class c with respect to the target *person* class, that is the number $dTP(c)$ of person instances detected correctly (i.e., considering only the IoU of predicted and target bounding boxes) but classified with category c divided by the total number of person instances in the ground-truth. In other words, $dTPR(c)$ gives us the proportion of person

Table 5. Classwise Analysis. We consider the detections of all 80 COCO classes, accounting for errors due to misclassified objects, i.e., detected objects that matched with *person* annotations but that were classified as objects of another category. TPR is the *True Positive Rate* with respect to the target *person* class. dTPR is the ratio of person instances correctly detected but misclassified. The overall *detection Recall* (dR) is the proportion of detected person instances considering also misclassified objects; the overall *detection Miss Rate* is the proportion of person instances that were not detected at all. We set IoU to 0.5.

Method	Person	Bird	Airplane	Kite	Other	Overall	
	TPR ↑	dTPR ↑	dTPR ↑	dTPR ↑	dTPR ↑	dR ↑	dMR ↓
VfNet [25]	0.285	0.266	0.190	0.067	0.012	**0.818**	**0.182**
TOOD [10]	**0.326**	0.118	0.212	0.125	0.017	0.799	0.201
Def. DETR [28]	0.206	0.311	0.072	0.041	0.026	0.657	0.343

instances that were detected correctly but misclassified with category c. The sum of the TPR and all the dTPR(c) gives the overall *detection Recall* (dR), i.e., the ratio of person instances detected correctly without considering the output classification. Similarly, the overall *detection Miss Rate*, defined as dMR=1-dR, is the portion of person instances that were not detected at all. For example, from Table 5, we can observe that the pre-trained VarifocalNet correctly detected 81.8% of the ground-truth person instances even if, in most cases, it misclassified them. This may suggest that the same model fine-tuned on MOB data may have room for growth in localizing *person* instances.

5   Conclusion and Future Directions

This paper presents the MOBDrone benchmark, a large-scale drone-view dataset suitable for detecting persons overboard. It is part of the NAUSICAA project aiming at creating a control system that, for the first time, uses aerial and marine drones and augmented and virtual reality to provide increased safety to medium and large vessels. Specifically, we collected more than 125K images, and we manually annotated more than 180K objects in a marine environment under several conditions, like different altitudes and camera shooting angles. Furthermore, we report an in-depth experimental evaluation of several state-of-the-art object detectors, serving as baselines for further research on this topic. Our analysis shows that detectors pre-trained on standard datasets of everyday objects exhibit moderate performance in localizing and recognizing people at sea in aerial images acquired at mid-high altitudes. The classification stage is the primary source of error for the best of the tested models, i.e., VarifocalNet, as about 82% of the ground-truth persons were correctly detected but misclassified, thus suggesting that the same model fine-tuned on MOB data may have room for growth. To this end, as a future direction, we plan to extend our dataset with additional annotated data for the supervised training procedure, also considering synthetic images coming from virtual worlds.

References

1. Amato, G., Ciampi, L., Falchi, F., Gennaro, C.: Counting vehicles with deep learning in onboard UAV imagery. In: 2019 IEEE Symposium on Computers and Communications (ISCC). IEEE, June 2019. https://doi.org/10.1109/iscc47284.2019.8969620
2. Amato, G., Ciampi, L., Falchi, F., Gennaro, C., Messina, N.: Learning pedestrian detection from virtual worlds. In: Ricci, E., Rota Bulò, S., Snoek, C., Lanz, O., Messelodi, S., Sebe, N. (eds.) ICIAP 2019. LNCS, vol. 11751, pp. 302–312. Springer, Cham (2019). https://doi.org/10.1007/978-3-030-30642-7_27
3. Cafarelli, D., et al.: MOBDrone: a large-scale drone-view dataset for man overboard detection, February 2022. https://doi.org/10.5281/zenodo.5996890
4. Carion, N., Massa, F., Synnaeve, G., Usunier, N., Kirillov, A., Zagoruyko, S.: End-to-end object detection with transformers. In: Vedaldi, A., Bischof, H., Brox, T., Frahm, J.-M. (eds.) ECCV 2020. LNCS, vol. 12346, pp. 213–229. Springer, Cham (2020). https://doi.org/10.1007/978-3-030-58452-8_13
5. Chen, K., et al.: MMDetection: Open MMLab detection toolbox and benchmark. arXiv preprint arXiv:1906.07155 (2019)
6. Ciampi, L., Messina, N., Falchi, F., Gennaro, C., Amato, G.: Virtual to real adaptation of pedestrian detectors. Sensors **20**(18), 5250 (2020). https://doi.org/10.3390/s20185250
7. Ciampi, L., Santiago, C., Costeira, J., Gennaro, C., Amato, G.: Domain adaptation for traffic density estimation. In: Proceedings of the 16th International Joint Conference on Computer Vision, Imaging and Computer Graphics Theory and Applications. SCITEPRESS - Science and Technology Publications (2021). https://doi.org/10.5220/0010303401850195
8. Du, D., et al.: The unmanned aerial vehicle benchmark: object detection and tracking. In: Ferrari, V., Hebert, M., Sminchisescu, C., Weiss, Y. (eds.) ECCV 2018. LNCS, vol. 11214, pp. 375–391. Springer, Cham (2018). https://doi.org/10.1007/978-3-030-01249-6_23
9. European Maritime Safety Agency: Annual overview of marine casualties and incidents 2021 (2021)
10. Feng, C., Zhong, Y., Gao, Y., Scott, M.R., Huang, W.: TOOD: task-aligned one-stage object detection. In: Proceedings of the IEEE/CVF International Conference on Computer Vision (ICCV), pp. 3510–3519, October 2021
11. Feraru, V.A., Andersen, R.E., Boukas, E.: Towards an autonomous UAV-based system to assist search and rescue operations in man overboard incidents. In: 2020 IEEE International Symposium on Safety, Security, and Rescue Robotics (SSRR), pp. 57–64. IEEE (2020). https://doi.org/10.1109/SSRR50563.2020.9292632
12. Garay, E.: What Happens When Someone Falls Off a Cruise Ship (2017). https://www.cntraveler.com/story/what-happens-when-someone-falls-off-a-cruise-ship. Accessed 25 Jan 2022
13. Ge, Z., Liu, S., Wang, F., Li, Z., Sun, J.: YOLOX: exceeding YOLO series in 2021. arXiv preprint arXiv:2107.08430 (2021)
14. He, K., Gkioxari, G., Dollár, P., Girshick, R.B.: Mask R-CNN. In: IEEE International Conference on Computer Vision, ICCV 2017, pp. 2980–2988. IEEE Computer Society (2017). https://doi.org/10.1109/ICCV.2017.322
15. Hsieh, M.R., Lin, Y.L., Hsu, W.H.: Drone-based object counting by spatially regularized regional proposal network. In: Proceedings of the IEEE International Conference on Computer Vision, pp. 4145–4153 (2017)

16. Leira, F.S., Johansen, T.A., Fossen, T.I.: Automatic detection, classification and tracking of objects in the ocean surface from UAVs using a thermal camera. In: 2015 IEEE Aerospace Conference, pp. 1–10. IEEE (2015). https://doi.org/10.1109/AERO.2015.7119238

17. Lin, T.-Y., et al.: Microsoft COCO: common objects in context. In: Fleet, D., Pajdla, T., Schiele, B., Tuytelaars, T. (eds.) ECCV 2014. LNCS, vol. 8693, pp. 740–755. Springer, Cham (2014). https://doi.org/10.1007/978-3-319-10602-1_48

18. Lygouras, E., Santavas, N., Taitzoglou, A., Tarchanidis, K., Mitropoulos, A., Gasteratos, A.: Unsupervised human detection with an embedded vision system on a fully autonomous UAV for search and rescue operations. Sensors 19(16), 3542 (2019). https://doi.org/10.3390/s19163542

19. Mandal, M., Kumar, L.K., Vipparthi, S.K.: MOR-UAV: a benchmark dataset and baselines for moving object recognition in UAV videos. In: Proceedings of the 28th ACM International Conference on Multimedia, pp. 2626–2635 (2020). https://doi.org/10.1145/3394171.3413934

20. Mou, J., Hu, T., Chen, P., Chen, L.: Cooperative MASS path planning for marine man overboard search. Ocean Eng. 235, 109376 (2021). https://doi.org/10.1016/j.oceaneng.2021.109376

21. Redmon, J., Farhadi, A.: Yolov3: An incremental improvement. arXiv preprint arXiv:1804.02767 (2018)

22. Ren, S., He, K., Girshick, R., Sun, J.: Faster r-CNN: towards real-time object detection with region proposal networks. IEEE Trans. Pattern Anal. Mach. Intell. 39(6), 1137–1149 (2017). https://doi.org/10.1109/tpami.2016.2577031

23. Sekachev, B. et al.: Computer Vision Annotation Tool (CVAT) (2020). https://github.com/openvinotoolkit/cvat

24. Varga, L.A., Kiefer, B., Messmer, M., Zell, A.: SeaDronesSee: a maritime benchmark for detecting humans in open water. In: Proceedings of the IEEE/CVF Winter Conference on Applications of Computer Vision (WACV), pp. 2260–2270, January 2022

25. Zhang, H., Wang, Y., Dayoub, F., Sunderhauf, N.: VarifocalNet: an IoU-aware dense object detector. In: 2021 IEEE/CVF Conference on Computer Vision and Pattern Recognition (CVPR). IEEE, June 2021. https://doi.org/10.1109/cvpr46437.2021.00841

26. Zhou, X., Wang, D., Krähenbühl, P.: Objects as points. arXiv preprint arXiv:1904.07850 (2019)

27. Zhu, P., et al.: Detection and tracking meet drones challenge. IEEE Trans. Pattern Anal. Mach. Intell. 01, 1 (2021). https://doi.org/10.1109/TPAMI.2021.3119563

28. Zhu, X., Su, W., Lu, L., Li, B., Wang, X., Dai, J.: Deformable DETR: deformable transformers for end-to-end object detection. In: 9th International Conference on Learning Representations, ICLR 2021. OpenReview.net (2021)

6 DoF Pose Regression via Differentiable Rendering

Andrea Simpsi$^{(\boxtimes)}$, Marco Roggerini, Marco Cannici ,
and Matteo Matteucci

Politecnico di Milano, 20133 Milan, Italy
{andrea.simpsi,marco.cannici,matteo.matteucci}@polimi.it,
marco.roggerini@mail.polimi.it

Abstract. Six Degrees of Freedom (6DoF) pose estimation is a crucial task in computer vision. It consists in identifying the 3D translation and rotation of an object with respect to the observer system of coordinates. When this is obtained from a single image, we name it Monocular 6 DoF Regression and it is a very prominent task in several fields such as robot manipulation, autonomous driving, scene reconstruction, augmented reality as well as aerospace. The prevailing methods used to tackle this task, according to the literature, are the direct regression of the object's pose from the input image and the regression of the objects' keypoints followed by a Perspective-n-Point algorithm to obtain its pose.

While the former requires a lot of data to train the deep neural networks used to accomplish the task, the latter requires costly annotations of keypoints for the objects to be regressed. In this work, we propose a new method to address 6DoF pose estimation using differentiable rendering along the entire pipeline. First, we reconstruct the 3D model of an object with a differentiable rendering technique. Then, we use this information to enrich our dataset with new images and useful annotations and regress a first estimation of the 6DoF pose. Finally, we refine this coarse pose with a render-and-compare approach using differentiable rendering. We tested our method on ESA's Pose Estimation Challenge using the SPEED dataset. Our approach achieves competitive results on the benchmark challenge, and the render-and-compare step is shown to be able to enhance the performance of existing state-of-the-art algorithms.

Keywords: 6-DoF pose estimation · Pose refinement · Differentiable rendering

1 Introduction

Six Degrees of Freedom (6DoF) pose estimation from monocular images is a crucial task in computer vision: it consists in obtaining the translation and rotation

---

Supplementary Information The online version contains supplementary material available at https://doi.org/10.1007/978-3-031-06430-2_54.

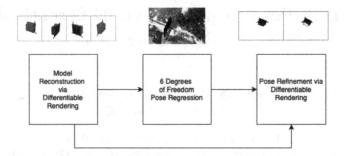

Fig. 1. A schematic version of the proposed pipeline.

of an object with respect to a fixed system of coordinates from a single image. This task is very challenging since a single 2D image does not directly carry the depth information, and some form of a priori knowledge is required. In the last years, this topic has obtained a growing interest in different fields, as investigated by Sahin et al. in their review [30]. Among these, the two most known are surely robotics manipulation, where an autonomous device can pick up different objects and place them in the desired target location, and autonomous driving, where pose estimation of other vehicles and pedestrians is the base for collision avoidance in autonomous navigation. The space domain is another field interested in this task, as highlighted in the survey of Opromolla et al. [21]. Indeed, spacecraft pose estimation is a relevant problem in different scenarios, such as formation flying [5], comet and asteroid exploration [7,14], on-orbit servicing [9,33], and active debris removal [1,8]. This interest is further shown by the recent ESA's Pose Estimation Challenges [2,4], where machine learning enthusiasts from around the world joined together to tackle the 6DoF pose estimation problem of spacecrafts.

The solutions proposed by the participants to the ESA's Pose Estimation Challenge [2] had fallen into the two typical solutions in the literature for pose retrieval from monocular images. Either the pose of the object is directly regressed with a deep neural network from the available image (from now on *Direct Regression*), or some key points of the object are detected and then used by a Perspective-n-Point (PnP) algorithm for pose estimation (from now on *Keypoints Regression*). Both Direct Regression and Keypoints Regression achieved important results in different fields. The former has shown to be robust and easy to apply to different objects, but it is usually reported to be less accurate than Keypoints Regression. Moreover, Direct Regression requires a huge amount of data for training the deep neural network regressor. On the other hand, Keypoints Regression, while obtaining excellent results in the accuracy of the estimated pose, requires a costly annotation of the ground truth position of the key points and the 3D model of the object to be recognized.

In this work, we present a novel method to obtain the 6 DoF pose of an object entirely based on differentiable rendering, overcoming Direct Regression's huge data requirement and Keypoints Regression's need for time consuming 3D model annotations. As from the pipeline in Fig. 1, at first an accurate 3D model of an object from a small set of RGBA images is obtained. This model, obtained by minimizing the visual loss through a differentiable renderer, can then be used to generate new data and annotations. Then, we regress a coarse estimation of the object's pose, which is finally refined with a render-and-compare technique using again differentiable rendering. Experimental results on real data show that our method is competitive in ESA's challenge [2], and that the render-and-compare technique is effective in improving further the already good estimates made by state-of-the-art methods. Furthermore, we have reconstructed an extremely accurate model of the Tango satellite used in the challenge [2], that can be used to generate a virtually unlimited number of new images and thus improve the results even further.

The code of the pipeline is publicly available for experimenting with the proposed algorithm[1] on different domains as the method can work with virtually any kind of object. We show, indeed, that it is possible to reconstruct the 3D model of other objects, and from that reconstruction to apply the entire pipeline we presented, as it does not depend on any specific 3D model.

2   Related Works

6DoF Pose Estimation with Neural Networks. Deep neural networks for 6DoF pose regression can be categorized by the representation they use. *Classification methods* discretize the continuous space of rotations and positions [20,32,34] and frame the pose regression as a classification problem. Despite being easy to implement, this approach lead two different issues. Firstly, classes are independent one another, while they should be tied together by their proximity. Secondly, discretization in classes may lead to a loss of performance.

In *regression methods*, the neural network regresses directly the position and rotation of the object in a continuous way [13,19,35]. Generally, in these models, the rotation is represented as a quaternion, but in the study of Zhou et al. [38] the strength of a 6D vector representation was proved. Su et al. [32] have shown that rendered images can be used to train a neural network for pose regression effectively, thus solving the issue of data requirements. Xu et al. [36] propose a regressor leveraging the use of a differentiable renderer for the estimation of the human 3D pose and obtain superior performance against state-of-the-art, despite being only applicable to humanoid shapes. Other works have used differentiable renderer for 6DoF pose estimation [22,23,37], but, unlike in our work, they do not obtain an accurate mesh of the target object.

Finally, *keypoints methods* do not regress directly the 6DoF pose of the object from the image. Instead, some keypoints of the object are regressed first, and then the position and rotation of the object are obtained from them with a PnP algorithm [24,26]. Despite these methods are generally the most accurate, they often require costly annotations.

---

[1] Code can be found at https://github.com/Simpands/diff-6dof-regression.

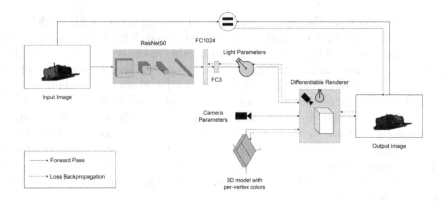

Fig. 2. A scheme of the network used in the model reconstruction (Sect. 3.1) step.

Differentiable Rendering. Differentiable rendering represents a class of techniques that enable the integration of a rendering procedure in an end-to-end optimization pipeline by obtaining useful gradients of the rendering process. One of the main discriminators among methods for differentiable rendering is the type of data representation they use. We focus here on reviewing the literature that uses a mesh-based representation, as this work relies on the same. Among mesh-based methods, there are two families which can be distinguished by the part of the rendering pipeline they approximate: the first family approximates the backward pass of the rendering process (approximated gradients), while the second family approximates the forward pass (approximated rendering). Regarding the *approximated gradients* family, Loper et al. [18] developed the first general-purpose differentiable renderer, called OpenDR. However, because of its general-purpose nature, OpenDR is often unsuitable for usage with neural networks. Kato et al. [12] proposed the Neural 3D Mesh Renderer (N3MR) and designed its gradients specifically for neural networks training. In particular, they propose to approximate the color's transition across pixels, which is typically discrete, with a smooth transition function with improved differentiability. Rhodin et al. [29] focused on a different solution, *approximated rendering*, that approximates the rasterization process by making objects' edges semi-transparent and thus ensuring differentiability. The Soft Rasterizer of Liu et al. [17] expanded this idea by modeling the contribution of each triangle to a pixel's color as a probability function. Finally, Chen et al. [6] proposed the DIB-R system, a novel approach that deals independently with foreground and background pixels to approximate the rasterization process. The foreground pixels are determined only by one face with a weighted interpolation of its local proprieties, while the background pixels are a distance-related function of global geometry.

3   Method

In this section, we describe in details our method to obtain the 6 degrees of freedom pose of an object using differentiable rendering. Our pipeline, shown in

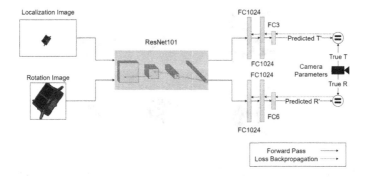

Fig. 3. A scheme of the network used in the pose regression (Sect. 3.2) step.

Fig. 1, is composed by three different steps. In the first step, *model reconstruction*, we obtain an accurate 3D model of an object from a small set of RGBA images. This model, then, can be used to generate new data and annotations via rendering. In the second step, *pose regression*, we regress a coarse estimation of the object's pose which we then refine in the last step, *pose refinement*, that is a render-and-compare technique using again differentiable rendering.

3.1 Model Reconstruction

The network used for model reconstruction is shown in Fig. 2. The goal of this neural network is to regress the position and color of the vertices of a 3D object. This allows us to obtain a good approximation of the true shape of the 3D object which can then be used for model refinement and data generation. Our network is composed by a convolutional neural network used to detect the light direction in the image and a 3D model (initialized as a icosphere with 10242 vertices and 20480 faces). This information (together with the pose of the object, provided by the dataset) is fed into a differentiable renderer to create an image which is then compared to the input image. This comparison is done with an Intersection over Union loss [27] on the alpha channel of the two images and a Mean Absolute Error on the color channels of the two images. From this comparison we optimize both the shape and the colors of the vertices of the 3D object and the parameters of the light detection network. In the Fig. 4, images generated with the model reconstructed in this process are shown.

3.2 Pose Regression

The second step of the pipeline is addressed via direct regression both on the translation and on the rotation of the object, which we perform through the network shown in Fig. 3. This network is composed by two separate branches: a *Rotation Branch*, used to regresses the rotation of the camera, and a *Localization Branch*, used to regress the position of the camera with respect to the object. Both branches use a feature extractor with the same architecture.

Fig. 4. The top row shows some of the input images used during the model reconstruction step. The bottom row shows some of the images generated from the model obtained during the model reconstruction.

The translation branch takes in input the full image (before any cropping) of size 384×240, which we call *Localization Image*. This image is processed in the feature extractor and then in two fully connected layers with 1024 neurons, followed by a fully connected layer with 3 neurons. This branch outputs the predicted position of the object, encoded as a 3D vector. The rotation branch takes in input only the region of interest representing the object, re-scaled to a fixed dimension of 240×240, and processes it with the feature extractor. We call this image *Rotation Image* as it is used by the network to regress the object rotation. Then, these features are passed trough 2 fully-connected layers with 1024 neurons and finally trough a fully-connected layer with 6 neurons. The feature extractor's output changes between the two branches as the dimension of the input image differs between the Translation Branch and the Rotation Branch (square 240×240 image for the Rotation Branch, w.r.t. a rectangular 384×240 image for the Translation Branch).

The loss used to train this network is given by the sum $L = L_T + \lambda L_R$, where L_T and L_R are the translation and rotation losses defined as it follows:

$$L_T\left(\widehat{T}, T\right) = \frac{1}{3}\left|\widehat{T} - T\right|_1, \quad L_R\left(\widehat{R}, R\right) = \sum_{e \in E}\left|\widehat{R}(e) - R(e)\right|, \qquad (1)$$

where T is the ground-truth translation vector and $\widehat{T}$ the predicted one. E is the set of all elements of a 3×3 matrix, $\widehat{R}$ is the predicted rotation matrix and R is the ground-truth one. In this work we set $\lambda = 1$.

The 6D vector predicted as output of the network is converted to the rotation matrix R as follows:

$$c_1 = \frac{v_1}{|v_1|}, \quad c_2 = \frac{u_2}{|u_2|}, \quad u_2 = v_2 - \langle c_1, v_2 \rangle\, c_1, \quad R = \begin{bmatrix} c_1 & c_2 & c_1 \times c_2 \end{bmatrix}, \quad (2)$$

where v_1 is the vector composed by the first 3 values of the 6D output vector, while v_2 is the vector composed by the last 3 values. This matrix is orthonormal by construction, and thus a valid 3D rotation matrix. During training the 3D model extracted by the model reconstruction process is used for data generation and augmentation.

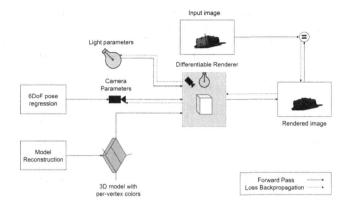

Fig. 5. A scheme of the network used in the pose refinement (Sect. 3.3) step.

3.3 Pose Refinement

In the final step of the pipeline, we use a render-and-compare approach exploiting the differentiable renderer for pose refinement. This technique consists in rendering with a differentiable renderer the image corresponding to a starting guess, e.g., the output of the 6DoF pose estimation process, compare this generated image to the source RGBA image, and then improve the predicted position based on this comparison. The comparison is given by the sum $L_{tot} = L_{IoU} + \lambda L_{col}$ of the Intersection Over Union loss L_{IoU} proposed in [27] and the color loss $\mathcal{L}_{col}$ defined as the Mean Absolute Error between each channel of the two images:

$$L_{col} = \frac{1}{|C|} \sum_{c \in C} \frac{1}{|P|} \sum_{p \in P} |I_1(c,p) - I_2(c,p)| \tag{3}$$

where C is the set of the color channels of the image, P is the set of all pixels of the image and I_1, I_2 are the two images to compare. The optimization procedure is applied with an iterative approach. In particular, as we can notice from Fig. 5, the differentiable renderer takes in input:

- A 3D model of the object in the image;
- A rotation matrix R_m, which can be obtained either from 3 Euler angles in case of the Euler regression, 4 un-normalized values in case of the quaternion regression, or 6 un-normalized values in case of the 6D vector rotation regression;
- A translation vector $T = [X_t, Y_t, Z_t]$, representing the translation of the camera from the object;
- A directional light object, comprehensive of the direction of the light $[X_l, Y_l, Z_l]$ and the light color (ambient color, diffuse color and specular color).

During the iterative refinement process, the differentiable renderer synthesizes the image given the current input parameters and compares it with the

Table 1. The improvement obtained with the pose refining step on the score calculated on the Pose Estimation Challenge [2]. The value of UrsoNet in the *competition score* is the value obtained reported in the leaderboard. Instead, UrsoNet *before refinement* is the value obtained using the pre-trained network given by the authors.

Method	Competition score	Pose error	
		Before refinement	After refinement
6D vector	–	0.38221	0.23191
Quaternion	–	0.47944	0.29602
Euler angles	–	0.83889	0.74775
Classification	–	1.20432	1.00247
UrsoNet	0.05546	0.07311	0.05279
UrsoNet Real	0.14763	–	0.14263

one observed, from which the object pose has to be estimated. The loss L_{tot} is then computed and its gradient backpropagated to update all renderer's inputs, except for the 3D model, in such a way to make the two images closer and thus improve the estimated pose. Regarding the rotation matrix, we decide not to optimize it directly, or to predict a normalized quaternion, as they both have to lay on a manifold and this cannot be enforced by direct gradient descent methods. Alternatively, manifold-aware gradient descend methods can be used and they can replace the method used in this work directly.

4   Experimental Evaluation

In this section we evaluate the proposed pipeline in the ESA's Pose Estimation Challenge [2]. We use a PyTorch3D's [28] renderer to implement all architectures that require a rendering step and ResNet [11] based backbones as feature extractors. Implementation details are provided in the supplemental materials, along with additional visualizations of the pipeline's outputs.

4.1   Dataset

The dataset used to test our proposed pipeline is the SPEED dataset [31] provided for the ESA's Pose Estimation Challenge [2]. This dataset is composed by in 4 different subsets. The *train* set consists of 12000 synthetic annotated images created by merging real images of the Earth with renderings of the Tango spacecraft at different positions and orientations. A small set of 5 *real* images is also provided with hand-made annotations of the pose, created by taking a photo of a replica of the satellite in a studio. The test set is divided in 2998 synthetic images generated similarly to the training ones, and a set of 300 real images. The dataset provides the annotations on the pose of the satellite: the satellite's position is described by a 3D vector, while the orientation of the satellite is described as a quaternion. Moreover, the satellite provides also the characteristics of the camera used to obtain the images.

For each image of this dataset, we add an alpha channel representing the binary segmentation of the satellite. To obtain the segmentation on the train set, we use the model estimated in the reconstruction step (Sect. 3.1), render it at the position and orientation provided by the ground truth annotations, and use the rendered silhouette as a segmentation mask. Instead, to obtain the segmentation of the image of the test set, we use the Mask R-CNN [10] segmentation network with ResNet-101 [11] and Feature Pyramid Networks [15] as a backbone, pre-trained of COCO [16] 2017, and implemented in the Detectron2 framework [3]. We fine-tune this network using a dataset obtained by combining the train set augmented with the alpha channel annotations just discussed, a set of 5950 images created by segmenting the satellite from the previous set and performing additional data augmentation operations (addition of random images of the Earth as background, random rotation and resize of Tango, and darkening of the image), and a set of 100000 synthetic images rendered using the 3D model we estimated in Sect. 3.1.

4.2 Results

In this section, we report the result obtained by this pipeline in the ESA pose estimation challenge [2] just introduced. We do not provide any quantitative evaluation of the 3D model reconstruction as the true model of the Tango satellite is not available. A qualitative evaluation of the reconstruction is given by the rendering reported in Fig. 4, and the effectiveness of the pose refinement can be considered as an indirect assessment of its quality too.

In Table 1, we report the quantitative results for the whole pipeline, before and after the pose refinement step. In particular, we have evaluated our model with different methods for representing 3D rotations. In the *6D vector* case the model shown in Sect. 3.2 regresses the 6D rotation vector and the pose refinement optimization is performed on the elements of the 6D vector prior to transform them into a rotation matrix. In the *Quaternion* case the regression model output is a 4D unconstrained quaternion orientation, and the pose refinement optimization is performed on such a 4D vector, from which we obtain a real quaternion trough a normalization prior to convert it into the rotation matrix R. In the *Euler Angles* case we regress a 3D vector of Euler angles and the optimization is performed directly on these 3 Euler angles, which are applied in the order $Z \rightarrow X \rightarrow Y$ to obtain the final rotation matrix R. We have also tested a *Classification approach* for the angle regression by discretizing each Euler angle into 18 classes. Each class is converted to its corresponding Euler angle prior to optimize them directly in the pose refinement step. These are then applied in the $Z \rightarrow X \rightarrow Y$ order to obtain the final rotation matrix R. As we can see from the Table 1 the process of pose refinement using differentiable rendering improves the score in all situations. The 6D vector representations is the one performing the best, since, being continuous, it is easier to regress by a neural network, as also reported in [38]. The Euler angles representation and its classification simplification are those the worse as they both suffer from gimbal lock.

We also notice that the quality of the final estimate strongly depends on the quality of the starting estimate. To verify to what extent this improvement is possible we evaluated the effectiveness of pose refinement using the pose estimate obtained by one of the best competitors in the ESA's challenge, i.e., UrsoNet [25]. *UrsoNet* is a neural network that competed for the first tranche of the Pose Estimation Challenge, achieving third position. The authors of UrsoNet made their neural network architecture and the weights used for the competition available for the public, so we can perform a direct comparison with their approach. Their neural network is composed by a pre-trained feature extractor followed by two branches: the localization branch that directly regresses the translation vector T, and a rotation branch that obtains the quaternion q via probabilistic fitting. We used this neural network with the provided pre-trained weights, scored it on the challenge website and use this result as a baseline. Then, we applied our proposed pose refinement step to UrsoNet predictions. The order of Euler angles follows the convention $Z \rightarrow X \rightarrow Y$ to obtain the final rotation matrix R. The results on UrsoNet are reported both for the test set of rendered images and for the test set of real images. We can do this only for this scenario: ESA's scoring page only shows the score of the rendered test set and not of the real test set. This score can be only obtained from the leaderboard (showing the best score both on the real images and on the rendered images). Once again, the proposed pose refinement approach is able to improve the results, proving that it can be used to enhance any pose regression estimation, even of high quality.

5   Conclusions

In this work we presented a new method to address the 6 degrees of freedom pose estimation using differentiable rendering. With this method, first we reconstruct the 3D model of an object with a differentiable rendering technique, then we use this information to enrich our dataset with new images and useful annotations, and regress a first estimation of the six degrees of freedom. Finally, we refine this coarse pose with a render-and-compare approach using differentiable rendering.

The use of differentiable rendering in the pose refinement scenario proved itself to be very solid and consistent, enhancing the prediction in each different scenario we tested it in, even enhancing the score of one of the best competitors of the first tranche of the challenge held by ESA [2]. As for the reduction in data requirement the method we propose has been able to reconstruct the 3D model of the object to localize from few images, nevertheless the pose regression part has required data augmentation via rendering.

To improve the performance of the pose regressor we are currently working on a end-to-end architecture for training which uses the differentiable renderer to compute an additional unsupervised loss beside the supervised one currently used.

References

1. Active debris removal: recent progress and current trends. Acta Astronaut. **85**, 51–60 (2013). https://doi.org/10.1016/j.actaastro.2012.11.009
2. Pose estimation challenge 2019 post mortem (2020). https://kelvins.esa.int/pose-estimation-challenge-post-mortem/
3. Detectron2: a PyTorch-based modular object detection library (2021). https://github.com/facebookresearch/detectron2
4. Pose estimation challenge 2021 (2022). https://kelvins.esa.int/pose-estimation-2021/
5. Bauer, F., Hartman, K., How, J., Bristow, J., Weidow, D., Busse, F.: Enabling spacecraft formation flying through spaceborne GPS and enhanced automation technologies (2002)
6. Chen, W., et al.: Learning to predict 3D objects with an interpolation-based differentiable renderer (2019)
7. Cheng, A.F.: Near Earth Asteroid Rendezvous: Mission Summary, pp. 351–366 (2002)
8. Clerc, X., Retat, I.: Astrium vision on space debris removal. In: Proceeding of the 63rd International Astronautical Congress (IAC 2012), Napoli, Italy, vol. 15 (2012)
9. Flores-Abad, A., Ma, O., Pham, K., Ulrich, S.: A review of space robotics technologies for on-orbit servicing. Prog. Aerosp. Sci. **68**, 1–26 (2014)
10. He, K., Gkioxari, G., Dollar, P., Girshick, R.: Mask R-CNN. In: Proceedings of the IEEE International Conference on Computer Vision (ICCV), October 2017
11. He, K., Zhang, X., Ren, S., Sun, J.: Deep residual learning for image recognition. CoRR arXiv:1512.03385 (2015)
12. Kato, H., Ushiku, Y., Harada, T.: Neural 3D mesh renderer (2017)
13. Kendall, A., Grimes, M., Cipolla, R.: Convolutional networks for real-time 6-DOF camera relocalization. CoRR arXiv:1505.07427 (2015)
14. Kubota, T., Sawai, S., Hashimoto, T., Kawaguchi, J.: Robotics and autonomous technology for asteroid sample return mission. In: ICAR 2005. Proceedings, 12th International Conference on Advanced Robotics, pp. 31–38 (2005). https://doi.org/10.1109/ICAR.2005.1507387
15. Lin, T.Y., Dollár, P., Girshick, R., He, K., Hariharan, B., Belongie, S.: Feature pyramid networks for object detection. In: Proceedings of the IEEE Conference on Computer Vision and Pattern Recognition, pp. 2117–2125 (2017)
16. Lin, T., et al.: Microsoft COCO: common objects in context. CoRR arXiv:1405.0312 (2014)
17. Liu, S., Li, T., Chen, W., Li, H.: Soft rasterizer: a differentiable renderer for image-based 3D reasoning (2019)
18. Loper, M.M., Black, M.J.: OpenDR: an approximate differentiable renderer. In: Fleet, D., Pajdla, T., Schiele, B., Tuytelaars, T. (eds.) ECCV 2014. LNCS, vol. 8695, pp. 154–169. Springer, Cham (2014). https://doi.org/10.1007/978-3-319-10584-0_11
19. Mahendran, S., Ali, H., Vidal, R.: 3D pose regression using convolutional neural networks. CoRR arXiv:1708.05628 (2017)
20. Massa, F., Aubry, M., Marlet, R.: Convolutional neural networks for joint object detection and pose estimation: a comparative study. CoRR arXiv:1412.7190 (2014)
21. Opromolla, R., Fasano, G., Rufino, G., Grassi, M.: A review of cooperative and uncooperative spacecraft pose determination techniques for close-proximity operations. Prog. Aerosp. Sci. **93**, 53–72 (2017). https://doi.org/10.1016/j.paerosci.2017.07.001

22. Palazzi, A., Bergamini, L., Calderara, S., Cucchiara, R.: End-to-end 6-DoF object pose estimation through differentiable rasterization. In: Proceedings, Part III, Munich, Germany, 8–14 September 2018, pp. 702–715 (2019). https://doi.org/10.1007/978-3-030-11015-4_53

23. Park, K., Mousavian, A., Xiang, Y., Fox, D.: LatentFusion: end-to-end differentiable reconstruction and rendering for unseen object pose estimation. CoRR arXiv:1912.00416 (2019)

24. Pavlakos, G., Zhou, X., Chan, A., Derpanis, K.G., Daniilidis, K.: 6-DoF object pose from semantic keypoints. CoRR arXiv:1703.04670 (2017)

25. Proenca, P.F., Gao, Y.: Deep learning for spacecraft pose estimation from photo-realistic rendering (2019)

26. Rad, M., Lepetit, V.: BB8: a scalable, accurate, robust to partial occlusion method for predicting the 3D poses of challenging objects without using depth. CoRR arXiv:1703.10896 (2017)

27. Rahman, M.A., Wang, Y.: Optimizing intersection-over-union in deep neural networks for image segmentation. In: Bebis, G., et al. (eds.) ISVC 2016. LNCS, vol. 10072, pp. 234–244. Springer, Cham (2016). https://doi.org/10.1007/978-3-319-50835-1_22

28. Ravi, N., et al.: Accelerating 3D deep learning with PyTorch3D. arXiv:2007.08501 (2020)

29. Rhodin, H., Robertini, N., Richardt, C., Seidel, H.P., Theobalt, C.: A versatile scene model with differentiable visibility applied to generative pose estimation (2016)

30. Sahin, C., Garcia-Hernando, G., Sock, J., Kim, T.K.: A review on object pose recovery: from 3D bounding box detectors to full 6D pose estimators (2020)

31. Sharma, S., D'Amico, S.: Pose estimation for non-cooperative rendezvous using neural networks. arXiv preprint arXiv:1906.09868 (2019)

32. Su, H., Qi, C.R., Li, Y., Guibas, L.J.: Render for CNN: viewpoint estimation in images using CNNs trained with rendered 3D model views. CoRR arXiv:1505.05641 (2015)

33. Tatsch, A., Fitz-Coy, N., Gladun, S.: On-orbit servicing: a brief survey. In: Proceedings of the IEEE International Workshop on Safety, Security, and Rescue Robotics (SSRR 2006), pp. 276–281 (2006)

34. Tulsiani, S., Malik, J.: Viewpoints and keypoints. CoRR arXiv:1411.6067 (2014)

35. Xiang, Y., Schmidt, T., Narayanan, V., Fox, D.: PoseCNN: a convolutional neural network for 6d object pose estimation in cluttered scenes. CoRR arXiv:1711.00199 (2017)

36. Xu, Y., Zhu, S.C., Tung, T.: DenseRaC: joint 3D pose and shape estimation by dense render-and-compare (2019)

37. Yang, Z., Yu, X., Yang, Y.: DSC-PoseNet: learning 6DoF object pose estimation via dual-scale consistency (2021)

38. Zhou, Y., Barnes, C., Lu, J., Yang, J., Li, H.: On the continuity of rotation representations in neural networks. CoRR arXiv:1812.07035 (2018)

Accelerating Video Object Detection by Exploiting Prior Object Locations

Berk Ulker[1(✉)], Sander Stuijk[1], Henk Corporaal[1], and Rob Wijnhoven[2]

[1] Eindhoven University of Technology, Eindhoven, Netherlands
{b.ulker,s.stuijk,h.corporaal}@tue.nl
[2] ViNotion BV, Eindhoven, Netherlands
rob.wijnhoven@vinotion.nl

Abstract. We provide a set of generic modifications to improve the execution efficiency of single-shot object detectors by exploiting prior object locations in video sequences. We propose a crop-based method to accelerate object detection tasks. It dynamically generates crop regions based on prior information and exploits scene sparsity enabling focused use of computational resources. In contrast to prior work, smaller input resolutions for processing crop regions are used to further reduce computational load. The execution efficiency is increased by avoiding multiple executions of the detector in full resolution. Data augmentations are used to successfully train these lower-resolution networks and maintain their accuracy at the baseline level while reducing inference time. Experiments with two public datasets, UA-DETRAC [13] and UAVDT [2], using the SSD-ML [19] object detection architecture with 128×128, 64×64 and 32×32 input resolutions show that we can achieve a maximum speedup by a factor of 1.7 on the UA-DETRAC dataset, and 1.6 on the UAVDT dataset while delivering the same level of accuracy as the base method. An extensive set of experiments demonstrates the speed-accuracy trade-off and shows that our method can achieve accuracy comparable to state-of-the-art methods at lower execution time.

1 Introduction

In the domain of video surveillance, scene monitoring is often performed on embedded devices that impose stringent resource constraints requiring efficient implementation of the analytics pipeline. Recent research in object detection introduced several improvements and new concepts, pushing both the accuracy and complexity of state-of-the-art methods higher. While such improvements can conveniently be employed in generic object detection, the use of such methods in resource-limited domains with real-time requirements poses a significant challenge. In this work, we aim to tackle this challenge by proposing a method that exploits prior object locations to accelerate object detection for video sequences.

Traditional algorithms for video-based object detection include background subtraction and motion detection [16]. Although these methods are computationally efficient, they are surpassed by deep learning-based methods in terms of

This work is funded by the NWO Perspectief program ZERO.

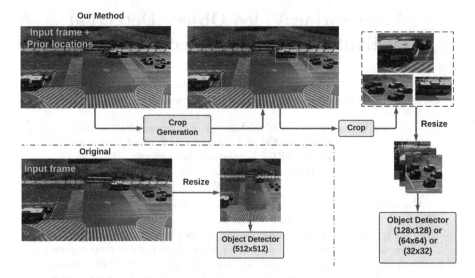

Fig. 1. Pre-processing procedure. Object prior information is grouped into crop regions after iterative merging. Next regions are resized and used for detection.

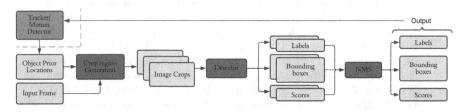

Fig. 2. Our crop based method in video processing pipeline. A set of crop regions covering prior object locations are generated. Cropped images are scaled to the input network resolution, and fed into the detector. Results from multiple detections are fused using non-maxima suppression (NMS).

accuracy. Proposed solutions are inspired by two main approaches to the problem. The first focuses on the inspection of the relationship of several images over time using video-based CNN detectors. Such video-CNN solutions, designed to use temporal information in the detector, are often applied in the domain of action recognition and perform accurate detection in the presence of video artifacts, but are computationally expensive. The second approach applies object detection on still images followed by a tracking stage to exploit the temporal information. Popular detection architectures employed in such solutions are based on YOLO [8], SSD [6] and Faster-RCNN [11]. Most solutions perform object detection independently per frame, without using temporal information. In contrast, we aim to increase execution efficiency by exploiting prior object location information that is present in video sequences.

We assume that prior object locations are available as input for object detection. This information can originate from a tracker that provides a location from previously detected objects, or from a motion detector that detects objects entering the scene. In this work, our scope is to investigate possible gains in run-time efficiency that can be obtained with a still-image-based object detector and prior location information. Our assumption is valid for use cases including real-life systems where the object detection is part of a video processing pipeline in which a tracking and/or motion detection component is commonly present.

The proposed detection system is shown in Fig. 2. We seek to achieve more efficient execution of the object detector by two key improvements to traditional detector pipelines. First, instead of processing the input image directly with an object detector, we perform detection on several smaller image crops focused around the prior object locations. As suggested in earlier works, selective cropping enables to exploit the sparse and uneven distribution of objects in the scene by removing image regions without objects from the processing pipeline. Second, we introduce detector networks with smaller input resolutions, which can obtain similar levels of accuracy at a lower computational cost. Our analysis focuses on the combined effect of both enhancements, as suggested gains are achieved only in combination.

Our contributions can be summarized as follows:

We provide a generic modification of object detectors to increase execution efficiency and maintain satisfactory accuracy levels by exploiting object sparsity through prior object location information in video sequences. We accelerate detection by modification of the object detector architecture for smaller input resolutions (from 512×512 to 32×32 pixels). To maintain accuracy levels, we provide a data augmentation technique for training with smaller input images. Our method achieves accuracy levels obtained by a detector that uses a higher image resolution, while reducing the execution time.

An empirical evaluation of the algorithm on two real-world datasets: UA-DETRAC (road-side traffic surveillance) and UAVDT (traffic surveillance from a drone), a scan of different detector network input resolutions and configuration parameters, analyzing the speed-accuracy trade-off characteristics of the proposed method.

2   Related Work

Research effort has focused on increasing the detection accuracy in recent years. As a general improvement to the popular SSD and Faster-RCNN models, the additional feature pyramid structure used in Mask-RCNN [1] and RetinaNet [4] partially address this problem at the cost of an increased execution time. The two-stage approach in Faster-RCNN already partially reduces this run-time increase by first generating region proposals and then classifying each proposal. Although this results in a higher detection quality when compared to single-shot object detectors such as YOLO and SSD, the computational complexity is also higher. In contrast to the two-stage approaches which generate region proposals from the current image only, we focus on limiting the number of proposals by exploiting prior object locations obtained analysing previous video frames.

Several other works consider explicit crop regions that are independently processed. Ružičk and Franchetti [10] propose a two-stage method, which uses two instances of YOLO-v2 [9], for generating an attention map on the downsampled image, and detecting objects in higher-resolution crops. Unel *et al.* [7] and Zhang *et al.* [18] propose the use of static tiles, where the input image is divided into fixed overlapping crops. Together with the original image, these crops are used for training and inference. While the results show improvement in accuracy, static tiling methods are inefficient as they perform multiple detections on all image pixels, completely ignoring the sparse distribution of the objects in the image. Wang *et al.* [12] propose an adaptive tiling approach for video sequences, which selects the crop size from a predefined set, using an iterative approach. The optimal set of predefined crops is independently searched in each consecutive image.

To address the inefficiency of static methods and enable focused resource allocation, several methods have been proposed to generate a suitable set of crop regions. Zhang *et al.* [17] propose an adaptive cropping method, which enhances detection accuracy for difficult targets, using a difficult region estimation network. DMNet [3] is another multi-stage method where an estimated density map is used to generate crops. AdaZoom [14] proposes a dynamic zooming strategy for selecting crop regions using reinforcement learning. Yang *et al.* [15] propose a method to reduce the number of crops using a cluster proposal sub-network. While the proposed method shares our goal of increasing detection efficiency, a separate network is used for object cluster generation. Furthermore, the number of crops are hard-limited by an empirically set value per dataset, which is significantly lower than the average number of crops we generate per image. This limits the granularity of the generated crops by enforcing larger regions. Above methods focus on improving the overall accuracy but are computationally more complex than the baseline detectors. This increase in complexity is caused by the introduced attention mechanisms that select crops, plus the fact that the same detector is executed multiple times for each input image (once for each crop). Note that the same internal detector resolution is used for processing each crop, effectively increasing the resolution of the image for detection, leading to an inherently increased detection accuracy. As main goal of this work is acceleration of the detection, we exploit the sparse distribution of objects to generate a minimal set of crops, and use detector networks with smaller input resolutions.

3   Proposed Method

As shown in Fig. 2, our method consists of three main stages for inference, which are crop generation, object detector and non-maxima suppression (NMS). We first generate a set of regions using the available object prior locations (boxes). We assume that these prior locations are available from a prior tracking or motion detection component. We follow an iterative approach to merge the initial prior boxes assigned to each individual object, resulting in crop regions. The resulting crops are then resized and individually processed by the detector. Finally, we use a class-agnostic NMS to fuse the results from the individual crops into the final set of detected object boxes. While our approach is generically applicable for object detectors, we chose SSD-ML [19] based on SSD [6] considering speed and accuracy characteristics.

Generation of Crop Regions: In this stage, we take inspiration from the dynamic crop region generation proposed in earlier works [15,17]. The goal of crop generation is the fusion of the initial prior object locations into a set of crop regions that will be processed by the object detector. We try to limit the redundant area enclosed in the regions, while ensuring coverage for all objects of interest. Since we are working with video streams, it is valid to assume that prior object locations are available as boxes in the area of the actual object in the scene (vehicles or people). Quality of the prior locations bounds overall performance by changing number of regions processed, and ratio of the objects covered by at least one detection window. We try to find a good trade-off between the effective capturing of the prior locations with many crop regions and reducing the number of crop regions to limit the computational load.

We present an effective three step merging procedure. First, input prior locations are expanded with an empirically set margin E_{prior} before merge. Subsequently, an upper-triangular boolean merge candidates matrix is built. Finally, this matrix is traversed and updated starting from top-left entry, merging boxes. To prevent loss of the parts of the objects due to errors in prior information, expansion operation in the first step is executed. In the second step, a 2-D square candidate matrix is constructed where each input box is represented by a row and column. Each cell of the matrix represents validity of a merge between corresponding pair of input boxes. For every cell in upper triangular part of the matrix, we calculate the size of the new region after merging, as the maximum value of the new width and height. If the new dimensions are smaller than the merge size threshold C_s, the two regions are designated as candidates for merging, and corresponding cell is set to True. In third step, this matrix is traversed row-wise starting from the top-left entry of the upper triangular part. For each cell with a valid merge, corresponding pair is merged, and an element-wise AND operation between the two rows representing the merged boxes is performed to update the corresponding rows and the relationship between the newly-merged box and the other boxes. Rows corresponding to a previously merged box are skipped. To compensate for possible errors in the object prior locations or size estimates, we extend the set of final merged boxes with an additional margin E_{final}. Although this method does not find the optimal solution, it provides adequate results with fast execution.

Object Detection: We use the SSD-ML object detector architecture for our experiments. Note that our proposed cropping method works with basically any detection architecture, but we found SSD-ML a good representation of single-shot detectors and easy to train. SSD-ML is an extension of the original SSD detector that enables hierarchical classification. The main difference with SSD is the decoupling of detection and classification tasks in the detection head. We use the SSD-ML extension because of two benefits: negative sampling per batch instead of per image during training, and the calculation of the objectness score, which we found to give improved results when used as input for NMS. We did not use the hierarchical classification feature and kept our network architecture similar to the original SSD, apart from the objectness calculation.

We use the VGG-16 as backbone network. We trained SSD-ML object detectors for resolutions of 128×128, 64×64 and 32×32 pixels. For smaller input resolutions we pruned network when a feature layer size of $1 \times 1 \times Channels$ is reached. The number of feature layers for input resolutions 128×128, 64×64 and 32×32 are 5, 4, 3 respectively. For each resolution, we used equally spaced object prior sizes, with default SSD configuration for aspect ratios.

Training: Modifications in the training procedure are implemented to adapt the training for our crop-based approach and the smaller input resolutions. Similarly to earlier crop-based methods, we use crops for training the object detector. Considering the presence of significant object sparsity in many training images, we did not employ a random crop selection to avoid using training images with limited or no positive samples. Instead, we ensure that at least one object is present by first selecting a random ground truth box and building the crop region around it (see Fig. 3). In order to control cropping behaviour during training, we define training crop size limit T_s, and initialize window with this size with a random offset R_m between -25% and $+25\%$. We remove the random extension via padding as present in SSD training pipeline from the preprocessing stage, since we obtain a similar effect via offset R_m.

Fig. 3. Visualization of the training sample generation. First, a ground truth box is randomly selected based on pre-defined probabilities. Initial region including selected box, with dimensions randomized around T_s with margin R_m is randomly placed. Resulting crop is resized and objects with less than 50% area coverage are discarded to form final training sample.

The traffic surveillance datasets UAVDT and UA-DETRAC suffer from a strong class imbalance. The number of samples from classes other than 'car' are significantly lower in most of the sequences, leading to poor precision for these classes. Furthermore, our random sampling approach does not improve this situation, as it is not guaranteed for every object in a training image batch to be present in the final crop used in training. To alleviate this, we first increased the number of samples belonging to the underrepresented classes, by simply duplicating images containing objects other than 'car' class in each epoch. Second, we introduced a bias for the random initial box selection. Instead of random selection from all ground truth boxes, we prioritize boxes from underrepresented classes. For both the UAVDT and UA-DETRAC datasets, we boost the probability of selection of classes other than 'car' to 0.9. Improvements result in an increase in per-class AP for bus (1.2%) and truck (3.9%) classes and an improved overall AP (1.0%), while causing a limited accuracy loss for car class (−1.5%).

Inference: Inference is executed on a batch built of resized image crops belonging to the same frame for increased efficiency. While we use the same cropping strategy and parameters, the resize factor is different for each input resolution (see Fig. 1). Crop region size limit parameter C_s controls generated crops, independent of the input resolution of the object detector selected. Use of different input resolutions in such configurations results in different speed-accuracy characteristics, caused by the detector network performance.

An additional post-processing step is required for merging results from multiple detections. In order to eliminate false positives consisting of duplicated and partial detections near crop region boundaries, we employed NMS. However, instead of using class-based NMS as in SSD [6], we execute class-agnostic NMS by using objectness scores calculated in SSD-ML [19] as score input.

4   Experimental Setup

We conduct our experiments on two different hardware platforms, Jetson TX2 and GTX Titan X. The measured execution times include the individual crop generation, pre-processing, inference, and post-processing stages. All network implementations are performed on the default Pytorch[1] software framework. In the scope of this work, we did neither use additional optimization techniques such as quantization and pruning, nor optimization tools such as TensorRT. Our proposed method is complementary to such acceleration techniques.

Datasets: To demonstrate the effectiveness of our method, we run performance evaluations on two different public datasets. **UA-DETRAC**, is composed of more than 100 video sequences recorded at 24 locations from static cameras, and contains over 140,000 frames with a resolution of 960 × 540 pixels. Typical sequences contain real-world traffic scenes captured from a birds-eye view. The dataset consists of 'car', 'van', 'bus' and 'other' classes, with majority of the samples being 'car'. **UAVDT**, object detection dataset contains 50 video sequences of 23,256 training images, and 15,069 test images with a resolution of 1080 × 540 pixels. Videos are acquired using a UAV, providing a birds-eye view of urban locations. The dataset consists of 'car', 'bus' and 'truck' classes, with the

---

[1] Experiments performed with PyTorch v1.8, CUDA and cuDNN 10.2.

majority of the samples belonging to the 'car' class. Compared to UA-DETRAC, the number of objects with a small size compared to the image size is higher.

Evaluation: To compare with related work we use the MS-COCO [5] AP (average precision), AP_{50} and AP_{75} metrics. While AP_{50} and AP_{75} use a single Intersection over Union (IoU) threshold, AP is the average precision under multiple thresholds, ranging from 0.5 to 0.95 with incremental steps of 0.05. It must be noted that evaluation under higher IoU threshold often yields lower average precision, as the localization errors are penalized more strict. In addition to the AP metrics, we also extract per-class precision which is useful to explore the effect of class imbalance as present in the datasets.

Inference Parameters: In our experiments we set E_{prior} to 10 pixels for both datasets, and E_{final} to 10 and 15 for UA-DETRAC and UAVDT respectively. To obtain different points in the speed-accuracy trade-off, we used multiple values for the threshold C_s ranging from 32 to 256 with steps of 32.

Table 1. Results obtained on UA-DETRAC(top) and UAVDT(bottom) datasets. 'Avg Inferences' indicate how many inferences per frame are executed and is equal to the number of crop regions extracted. For the second group, the crop size parameter is tuned to obtain similar accuracy to the base model. Base method SSD-ML 512×512 executes inference on the complete image.

Model	UA-DETRAC						UAVDT					
	Avg inferences	Avg precision			Execution time (ms)		Avg inferences	Avg precision			Execution time (ms)	
		AP	AP_50	AP_75	TX2	Titan X		AP	AP_50	AP_75	TX2	Titan X
512×512	1	41.4	56.0	49.7	531.6	35.0	1	15.6	27.1	16.7	531.6	35.0
128×128	5.9	42.3	63.5	49.8	234.5	28.7	3.9	16.1	27.3	17.3	163.0	22.9
64×64	5.9	41.1	64.1	48.4	108.4	20.4	7.6	15.9	28.9	16.2	124.1	21.5
32×32	10.3	42.4	65.2	53.1	146.2	26.3	9.9	15.9	29.5	16.5	132.6	28.2
128×128	7.8	44.9	66.4	54.1	259.2	32.4	13.9	19.3	32.9	20.7	278.6	32.7
64×64	12.0	52.4	81.5	64.8	167.6	32.9	13.9	21.5	36.5	23.4	184.0	36.4
32×32	11.5	45.6	73.0	53.6	128.8	29.55	13.9	17.4	31.2	18.4	174.1	36.5

Training: We trained SSD-ML models with input resolutions of 512×512 for the base network, 128×128, 64×64 and 32×32 pixels for our crop based method. For both datasets, we updated the training parameters from [19]. For base resolution of 512×512, we trained for 240k iterations with a batch size of 16, initial learning rate of 5×10^{-4} while decreasing with a factor of 10 after 160k and 200k. We increased initial learning rate to 1×10^{-3} for other resolutions. For all networks, we used a momentum of 0.9, and warm-up period of 5,000 iterations. For training our models, we used augmentations introduced in Sect. 3.

5   Results

We examine speed and accuracy operating characteristics of our method, under various configurations of training and inference parameters. Table 1 shows the comparison of base method SSD-ML 512 with our method in available input resolutions and two different parameter configurations: minimum execution time at matching accuracy and maximum accuracy at matching execution time. Our method can accelerate detection with a factor of 1.7 with minimal accuracy loss of 0.3 AP, with 64×64 input size on UA-DETRAC set. A speedup factor of 1.6 can be achieved in combination with an accuracy boost of 0.3 AP on UAVDT with same configuration. Performance comparison under sweep of inference crop size parameter C_s is shown in Fig. 4. For both datasets and all resolution levels, our method achieves more feasible operating points.

While for UA-DETRAC, best performance is obtained using 64×64 network, the Pareto front obtained on UAVDT in Fig. 4 involves operating points from all three resolution levels. One reason for this behavior is the larger change in number of inferences per frame and total execution times on UAVDT dataset, for different C_s values. For values of $C_s = 32$ and $C_s = 256$, average number of inferences are 5.2 and 12 for UA-DETRAC, 3.9 and 20.3 for UAVDT.

Inference execution time is the largest part of the overall execution time, followed by NMS. Execution times are mostly smaller for smaller input resolutions when other parameters are fixed. A notable exception is the higher execution time with $C - s = 32$ and 32×32 input size. Due to large number of detections,

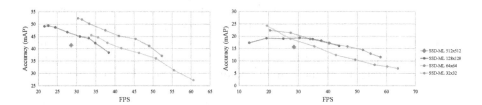

Fig. 4. Speed-Accuracy characteristics of our method, compared to SSD-ML 512. Obtained using UA-DETRAC (left) and UAVDT (right) sets, on Titan X. Each point is collected using a different C_s, ranging from 32 to 256 with step of 32.

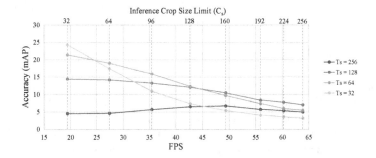

Fig. 5. Comparison of SSD-ML 32×32, performance for different T_s and C_s values, obtained from UAVDT dataset. Accuracy decreases with increasing values of C_s parameter and with increasing gap between T_s and C_s.

execution time for NMS jumps to 8.6, from 4.2 with 64×64 input size, causing a higher overall execution time for 32×32 input size.

Effect of Crop Size Limit Parameters C_s and T_s: C_s controls how many crop regions are generated, and how large is the down-size factor when resizing crops for the detection for inference. Tuning this parameter closer to zero causes a crop region to be assigned to almost every prior object location, while closer to image dimensions causes method generating very large crops.

We measured system performance by sweeping this parameter, from 32 to 256 with step size of 32. Our results show that increasing C_s yields a higher processing rate for all cases, as number of inferences per frame decreases. For most cases, AP decreases with increasing C_s. While this effect is more dominant as seen in Fig. 4, difference between training crop size T_s and C_s can also affect the accuracy negatively. An example is shown in Fig. 4, where SSD-ML 128×128 model trained using $T_s = 128$ delivers higher accuracy when C_s is increased from 32 to 64, decreasing the gap between two parameters.

For resolutions of 64×64 and 32×32, smaller values of C_s yields a higher ratio of increase in AP_{75} than the increase in AP_{50}, indicating better localization of the detected objects as shown in Table 1. This is mainly because of the higher spatial resolution of the detector inputs generated using smaller C_s.

The effect of training crop size T_s on network performance is shown on Fig. 5. T_s has a significant effect as, for the same cost of processing, the accuracy can be boosted significantly by tuning this parameter. However, there is no single optimal T_s value yielding best performance for every inference configuration. For both sets, the optimal operating points are achieved using different hyperparameter values, leading to an increase in training complexity due to search step for optimal configuration.

6   Conclusions

We presented a novel crop-based object detection method for use in video-based systems, such as road-side or drone-based traffic surveillance. We exploited prior object location information to focus the crop regions on relevant object regions in the scene. Each crop region is processed independently by a lower-resolution detector network for execution efficiency. A crop size limit parameter is used to control the amount of generated crops and their size, which shifts the operating point on the speed/accuracy trade-off. We implemented our crop-based system with the SSD-ML object detector using 128×128, 64×64 and 32×32 input resolutions, and used data augmentations during training.

We experimented with the UA-DETRAC traffic surveillance and the UAVDT drone datasets to measure execution efficiency and accuracy of our method under different training and inference crop size limits. Compared to the base SSD-ML detector, we achieved several Pareto efficient operating points with different input resolutions. Our method achieved up to 1.7× and 1.6× speedup on UA-DETRAC and UAVDT sets, respectively. By tuning inference crop size, accuracy is improved with 11% AP at 1.1× speed for UA-DETRAC and with 8.6% AP at 0.7× speed for UAVDT.

Inference and training crop sizes have significant impact on speed and accuracy. While decreasing the crop size limit increases the number of crops and improves the detection rate, the resulting difference in the image crop size compared to the crops used for training crop size, reduces the detection rate. Our analysis showed that, depending on the preferred speed/accuracy operating point, the optimal set of parameters and input resolutions are different. For UA-DETRAC, the network resolution of 64 × 64 is preferred image size for the complete speed/accuracy range. For UAVDT, the optimal parameters include operating points from all three input resolutions. This is caused by reduced processing rate for smaller inference crop size limit values. While reducing this parameter increases the number of inferences per frame for both datasets, increase is more significant for UAVDT where ratio of smaller objects are higher, and objects are more sparsely distributed due to recording setting.

References

1. Abdulla, W.: Mask R-CNN for object detection and instance segmentation on Keras and TensorFlow (2017). https://github.com/matterport/Mask_RCNN
2. Du, D., et al.: The unmanned aerial vehicle benchmark: object detection and tracking. In: Proceedings of the European Conference on Computer Vision (ECCV), pp. 370–386 (2018)
3. Li, C., Yang, T., Zhu, S., Chen, C., Guan, S.: Density map guided object detection in aerial images. In: Proceedings of the IEEE/CVF Conference on Computer Vision and Pattern Recognition Workshops, pp. 190–191 (2020)
4. Lin, T.Y., Dollár, P., Girshick, R., He, K., Hariharan, B., Belongie, S.: Feature pyramid networks for object detection. In: Proceedings of the IEEE Conference on Computer Vision and Pattern Recognition, pp. 2117–2125 (2017)
5. Lin, T.Y., et al.: Microsoft COCO: common objects in context. In: Fleet, D., Pajdla, T., Schiele, B., Tuytelaars, T. (eds.) ECCV 2014. LNCS, vol. 8693, pp. 740–755. Springer, Cham (2014). https://doi.org/10.1007/978-3-319-10602-1_48
6. Liu, W., et al.: SSD: single shot MultiBox detector. In: Leibe, B., Matas, J., Sebe, N., Welling, M. (eds.) ECCV 2016. LNCS, vol. 9905, pp. 21–37. Springer, Cham (2016). https://doi.org/10.1007/978-3-319-46448-0_2
7. Ozge Unel, F., Ozkalayci, B.O., Cigla, C.: The power of tiling for small object detection. In: Proceedings of the IEEE/CVF Conference on Computer Vision and Pattern Recognition Workshops (2019)
8. Redmon, J., Divvala, S., Girshick, R., Farhadi, A.: You only look once: unified, real-time object detection. In: Proceedings of the IEEE Conference on Computer Vision and Pattern Recognition, pp. 779–788 (2016)
9. Redmon, J., Farhadi, A.: YOLO9000: better, faster, stronger. In: Proceedings of the IEEE Conference on Computer Vision and Pattern Recognition, pp. 7263–7271 (2017)
10. Růžička, V., Franchetti, F.: Fast and accurate object detection in high resolution 4K and 8K video using GPUs. In: 2018 IEEE High Performance extreme Computing Conference (HPEC), pp. 1–7. IEEE (2018)

11. Sun, X., Wu, P., Hoi, S.C.: Face detection using deep learning: an improved faster RCNN approach. Neurocomputing **299**, 42–50 (2018)
12. Wang, Y., Mao, K., Chen, T., Yin, Y., He, S., Chen, G.: Accelerating real-time object detection in high-resolution video surveillance. Concurr. Comput. Pract. Exp., e6307 (2021)
13. Wen, L., et al.: UA-DETRAC: a new benchmark and protocol for multi-object detection and tracking. Comput. Vis. Image Underst. **193**, 102907 (2020)
14. Xu, J., Li, Y., Wang, S.: AdaZoom: adaptive zoom network for multi-scale object detection in large scenes. arXiv preprint arXiv:2106.10409 (2021)
15. Yang, F., Fan, H., Chu, P., Blasch, E., Ling, H.: Clustered object detection in aerial images. In: Proceedings of the IEEE/CVF International Conference on Computer Vision, pp. 8311–8320 (2019)
16. Yilmaz, A., Javed, O., Shah, M.: Object tracking: a survey. ACM Comput. Surv. (CSUR) **38**(4), 13-es (2006)
17. Zhang, J., Huang, J., Chen, X., Zhang, D.: How to fully exploit the abilities of aerial image detectors. In: Proceedings of the IEEE/CVF International Conference on Computer Vision Workshops (2019)
18. Zhang, X., Izquierdo, E., Chandramouli, K.: Dense and small object detection in UAV vision based on cascade network. In: Proceedings of the IEEE/CVF International Conference on Computer Vision Workshops (2019)
19. Zwemer, M., Wijnhoven, R.G., et al.: SSD-ML: hierarchical object classification for traffic surveillance. In: 15th International Conference on Computer Vision. Imaging and Computer Graphics Theory and Applications (VISAPP2020), pp. 250–259. SCITEPRESS-Science and Technology Publications, LDA (2020)

Deep Autoencoders for Anomaly Detection in Textured Images Using CW-SSIM

Andrea Bionda, Luca Frittoli[(✉)], and Giacomo Boracchi

DEIB, Politecnico di Milano, via Ponzio 34/5, Milan, Italy
andrea.bionda@mail.polimi.it,
{luca.frittoli,giacomo.boracchi}@polimi.it

Abstract. Detecting anomalous regions in images is a frequently encountered problem in industrial monitoring. A relevant example is the analysis of tissues and other products that in normal conditions conform to a specific texture, while defects introduce changes in the normal pattern. We address the anomaly detection problem by training a deep autoencoder, and we show that adopting a loss function based on Complex Wavelet Structural Similarity (CW-SSIM) yields superior detection performance on this type of images compared to traditional autoencoder loss functions. Our experiments on well-known anomaly detection benchmarks show that a simple model trained with this loss function can achieve comparable or superior performance to state-of-the-art methods leveraging deeper, larger and more computationally demanding neural networks.

Keywords: Anomaly detection · Unsupervised learning · Segmentation · Autoencoder · Texture similarity

1 Introduction

The automatic detection and localization of anomalous regions within images is a crucial problem in many industrial scenarios where the production volume prevents human supervision. In most cases, anomaly detection solutions are required to identify regions that deviate either in appearance, structure or contrast from the rest of the image, which instead follows a pattern or texture representing normal conditions. As an example, consider images of nanofiber tissues [5], namely polymer fibers with diameters down to one hundred nanometers, which look like a texture of randomly overlapping filaments. In this context, it is crucial to automatically locate defects, i.e., regions that deviate from this texture, a problem that was first addressed in [6]. Figure 1 represents the textures from the Nanofiber [6] and MVTec dataset [2], and different kinds of anomalies.

Nanofiber images are a relevant example of many industrial monitoring scenarios where normal images are characterized by specific textures or repetitive patterns that can differ in orientation, scale and shape from image to image. For this reason, it is rarely possible to detect anomalies in test images by direct

© The Author(s), under exclusive license to Springer Nature Switzerland AG 2022
S. Sclaroff et al. (Eds.): ICIAP 2022, LNCS 13232, pp. 669–680, 2022.
https://doi.org/10.1007/978-3-031-06430-2_56

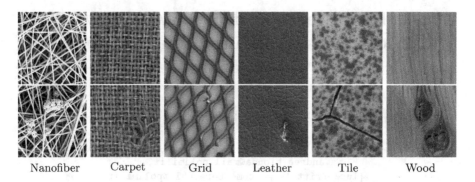

Nanofiber   Carpet   Grid   Leather   Tile   Wood

Fig. 1. Examples of normal (above) and anomalous images (below) from the Nanofiber [6] and MVTec [2] datasets.

comparison against a template or a reference. Moreover, anomaly detection is performed by assigning an anomaly score to each pixel of the image, while the vast majority of anomaly detection methods is designed to classify an entire image as normal or anomalous [17]. In our case, anomaly detection is seen as a segmentation problem where images are to be segmented into normal and anomalous regions. Since anomalies are typically rare, and their annotation is expensive and time-consuming, supervised learning is impractical in most industrial scenarios. Thus, anomaly detection is usually addressed by unsupervised models trained exclusively on normal images, which are usually abundant and do not require annotation [2].

A mainstream approach to detect anomalous regions consists in processing the image in small patches either by sparse representations [6] or deep neural networks [3,14,25]. Here, the patch size determines a trade-off between the resolution of the predictions, which increases when using smaller patches, and the detection accuracy, which improves when using larger patches since each patch contains more information. In this respect, deep neural networks have the advantage that they can adjust their *effective receptive fields* during training, and learn what is the most informative area inside a fixed-size patch. Thus, the patch size is perhaps a less influential parameter in deep neural networks than in traditional machine learning methods for anomaly detection.

Convolutional autoencoders, namely convolutional neural networks (CNNs) trained to extract a latent representation and then reconstruct the original image, have been widely employed for anomaly detection purposes [4,25]. The rationale is that an autoencoder trained exclusively on normal images will accurately reconstruct only the normal regions of a test image. Thus, it is possible to detect as anomalous those regions of a test image that substantially differ from their reconstruction. Autoencoders are usually trained to minimize the *Mean Square Error* (MSE) between the input and reconstructed images. However, the MSE is not well suited for textured images since it measures the average pixelwise difference, while repetitive patterns are substantially more relevant than individual pixel values. For this reason, a loss function based on the *Structural Similarity* index (SSIM) [22] can improve the anomaly detection performance [4].

We propose to address anomaly detection in textured images using convolutional autoencoders trained and tested using similarity metrics specifically designed to compare textures. In particular, we employ *Complex Wavelet Structural Similarity* (CW-SSIM) [23] to train an autoencoder, and also to compare the original and reconstructed images at test time. Using CW-SSIM substantially improves the reconstruction quality compared to MSE, and yields better robustness to differences in scale and orientation compared to SSIM. Our experiments on the Nanofiber [6] and MVTec Texture [2] datasets show that a simple autoencoder can outperform comparably more complicated and computationally expensive models such as those in [3,14]. The code is available at https://github.com/AndreaBiondaPolimi/Anomaly-Detection-Autoencoders-using-CW-SSIM.

Our solution is currently designed to process grayscale images, since CW-SSIM extracts structural and contrast information from single-channel images. This type of images can be encountered in many circumstances, e.g. where quality control is performed by X-ray imaging or electronic microscopes [6]. Our solution yields excellent performance even on some MVTec datasets containing RGB images, which we convert to grayscale before feeding them to our autoencoder. However, it turns out that, in some MVTec images, anomalies only affect the color of certain regions. Thus, color information is crucial for anomaly detection, which encourages us to investigate how to better handle RGB images.

2   Problem Formulation

Let us denote by $I : \mathcal{X} \to \mathbb{N}^c$ the image to be analyzed, where $\mathcal{X} \subset \mathbb{Z}^2$ is a pixel grid and c is the number of color channels, whose intensity can range from 0 to $2^d - 1$, where d is the color depth. We primarily consider grayscale images (i.e., $c = 1$), since in several industrial monitoring scenarios images are acquired by X-ray sensors or electronic microscopes [6]. Our goal is to locate anomalous regions in I, i.e., to estimate the unknown anomaly mask $\Omega : \mathcal{X} \to \{0, 1\}$:

$$\Omega(i, j) = \begin{cases} 1 & \text{if pixel } (i, j) \text{ falls inside an anomalous region,} \\ 0 & \text{otherwise.} \end{cases} \tag{1}$$

In particular, we are interested in estimating an anomaly mask $\hat{\Omega} : \mathcal{X} \to \{0, 1\}$ that approximates Ω as well as possible. Typically, this is done by combining anomaly scores computed on image patches, which we indicate by x.

We assume that a training set TR containing only normal (i.e., anomaly-free) images is provided, so we address anomaly detection in an unsupervised manner. These are reasonable settings in industrial monitoring, where normal images are abundant, while anomalies are rare and expensive to annotate. Moreover, anomalies can vary in shape, size and other characteristics, and an annotated training set might not encompass all the possible anomalies that will be encountered during testing. Our goal is to detect as anomalous all the regions of I that do not comply with the structure of normal images provided for training.

3   Related Work

Anomaly detection in images is a widely studied problem. The vast majority of the existing solutions follow a one-class classification approach, where the entire image is classified as normal or anomalous. In this paper we address the more challenging problem of detecting anomalous regions within the image, thus we refer to [17] for a recent survey of the one-class classification literature.

The first solutions proposed to detect anomalous regions in images apply unsupervised learning methods, such as *One-Class Support Vector Machine* (OC-SVM) [12] or *Kernel Density Estimation* (KDE) [15], to features extracted from image patches. The extracted features can either be hand-crafted features, such as the response of the image to Gabor filters [20] or regularity measures [7], or data-driven features such as dictionaries learned to yield *sparse representations* [6]. Another approach models the texture by an autoregressive model and detects anomalies using the residuals [9]. In general, data-driven models are more effective than those based on hand-crafted features, thanks to their greater flexibility [6].

Nowadays, the vast majority of anomaly detection methods leverage features extracted by *Convolutional Neural Networks* (CNNs), trained to address a different classification problem. A relevant example is the work by *Napoletano et al.* [14], where a pre-trained ResNet-18 [10] is used to extract feature vectors from normal patches. Then, these features are reduced in dimension by PCA and clustered by K-Means to form a dictionary of normal features. The anomaly score on each test patch is computed as the Euclidean distance between its projected feature vector and the most similar dictionary atom. Another way to exploit pre-trained CNNs is the *Student-Teacher* approach [3], which is based on a feature-extracting network (pre-trained on natural images), referred to as the Teacher, and a set of Student networks trained on anomaly-free patches to estimate the Teacher's output in a regression problem. During testing, the anomaly score of each pixel is computed by combining the regression error and the prediction variance of the Student networks. The intuition is that the Students can extract similar features compared to the Teacher only from normal patches.

The most popular deep learning methods for anomaly detection in images are autoencoders [25], namely CNNs trained to extract a compact *latent representation* from the input image and then to reconstruct the input image from the *latent representation*. The most common loss function is simply the MSE between the input and reconstructed images, even though *Bergmann et al.* [4] show that using SSIM [22] as training loss instead of the MSE substantially improves the reconstruction quality, as it takes into account local structures instead of differences in individual pixel values. During testing, the autoencoder can accurately reconstruct only patches that are similar to the anomaly-free training data, thus anomalous regions can be detected by analyzing the pixelwise reconstruction error. In this work, we extend [4] and demonstrate that a loss function based on CW-SSIM [23] further improves the anomaly detection performance.

It is also worth mentioning some more sophisticated anomaly detectors based on *Generative Adversarial Networks* (GANs) [1,18] and *reconstruction by inpainting* [24], which are becoming increasingly popular. However, we do not consider these methods since we primarily focus on autoencoders, and investigate the impact of using different loss functions. Moreover, so far these solutions based on GANs have been applied to produce a single anomaly score for the whole input image, and cannot be easily adapted to localize anomalies within images [17].

4   Proposed Solution

We address anomaly detection using an autoencoder [25] trained on small image patches. Autoencoders consist of two sub-networks: an *encoder* $\mathcal{E}$ and a *decoder* $\mathcal{D}$. The encoder extracts a feature vector $z = \mathcal{E}(x)$, called *latent representation*, from a patch x of the input image. The decoder is an upsampling network taking as input z and returning a patch $y = \mathcal{D}(z)$ having the same size as x. The two sub-networks are jointly trained on patches of images from TR, which do not contain anomalies, to minimize a loss function measuring the pixelwise reconstruction error, namely the difference between y and x. This loss is typically defined as the MSE, but in [4] it has been shown that replacing the MSE with structural similarity metrics such as SSIM [22] improves the reconstruction and anomaly detection performance.

Structural similarity metrics were initially designed to classify textured images [26] and to assess their quality [22] by detecting differences with respect to a template. We focus on metrics based on the *steerable filter decomposition* [8], which can extract informative features describing a texture. For this reason, we propose to train the autoencoder to maximize the Complex Wavelet Structural Similarity (CW-SSIM) index [23] between input and reconstructed images. We also employ CW-SSIM at test time to estimate the anomaly map by comparing the input and reconstructed images. Compared to other structural similarity metrics (including SSIM), CW-SSIM is more robust to scaling, translations and rotations, and we show that this further improves the reconstruction quality and the anomaly detection ability compared to the autoencoder in [4].

In what follows we first illustrate the steerable filter decomposition (Sect. 4.1), then we introduce the CW-SSIM index and illustrate how we employ it in the loss function of the autoencoder (Sect. 4.2). Finally, we present our anomaly detection procedure, which is also based on CW-SSIM (Section 4.3).

4.1   Steerable Filter Decomposition

Steerable filter decomposition is based on recursive application of filtering and subsampling operations, stacked in the *steerable pyramid algorithm*. It has been demonstrated that the statistics obtained by this procedure can characterize well a broad class of textures [16]. This algorithm decomposes the input patch x into M different *subbands* x^m in the frequency domain, where $m \in \{1, \ldots, M\}$, at S different scales. The first subband is obtained by applying the *Fast Fourier Transform* (FFT) to the original patch. The patch is then downsampled by 2×2

average pooling and convolved with O oriented kernels $S-2$ times to obtain the subbands at different scales. The patch is downsampled one more time to obtain the last subband. In the end we have

$$M = O(S - 2) + 2 \tag{2}$$

subbands, where the term "+2" takes into account the first and the last subbands. The design of the filters and more implementation details can be found in [8,19].

4.2   CW-SSIM as Loss Function

As in [4], we train the autoencoder over randomly cropped patches $\{x\}$ from anomaly-free images in TR. During training, the loss function $\mathcal{L}$ compares each patch x with the corresponding $y = \mathcal{D}(\mathcal{E}(x))$ reconstructed by the autoencoder:

$$\mathcal{L}(x,y) = 1 - \frac{1}{M} \sum_{m=1}^{M} \frac{1}{L} \sum_{l=1}^{L} \text{CW-SSIM}(x_l^m, y_l^m), \tag{3}$$

where x^m and y^m indicate the subbands in which x and y are decomposed by the steerable filter algorithm, for $m \in \{1, \ldots, M\}$. The CW-SSIM index is computed by dividing each subband into L small windows x_l^m and y_l^m of size $R \times R$ following a regular grid sampling with stride 1, where $l \in \{1, \ldots, L\}$. The term $\text{CW-SSIM}(x_l^m, y_l^m)$, which compares two windows at the same spatial location l and subband m, is defined as:

$$\text{CW-SSIM}(x_l^m, y_l^m) = \frac{2|\langle x_l^m, y_l^m \rangle| + K}{\|x_l^m\|_2^2 + \|y_l^m\|_2^2 + K}, \tag{4}$$

where $\langle \cdot, \cdot \rangle$ and $\| \cdot \|_2^2$ indicate the inner product and norm of complex vectors, and K is a small constant preventing the denominator from vanishing [23]. In (4) the windows x_l^m, y_l^m are assumed to be flattened to column vectors. The loss function in (3) represents a dissimilarity score between x and y ranging in $[0, 2]$.

Implementation Details. We empirically chose the window size $R = 7$ for CW-SSIM and the number of orientations $O = 6$ for the steerable filter decomposition. Since we observed that subbands with a spatial dimension lower than 16×16 pixels in the last subband do not contain useful information for anomaly detection, we set the number of scales $S = 5$, starting from patches of size 256×256.

Our autoencoder features a simple architecture, with an encoder $\mathcal{E}$ made of 5 convolutional layers each, deployed with a kernel size of 4 and stride 2, and a symmetric decoder $\mathcal{D}$. We construct the training set by randomly cropping 50,000 overlapping patches from normal images. We train the autoencoder for 400 epochs, using the ADAM optimizer [11] with an initial learning rate of $1 \cdot 10^{-3}$ and a weight decay set to 0.5 every 20 epochs.

4.3 CW-SSIM for Anomaly Detection

We employ CW-SSIM also at inference time, to estimate the anomaly masks by comparing the original and reconstructed test images. To do so, we first divide each test image I into patches, following a regular grid sampling with stride 16. Then, we reconstruct these patches through our autoencoder, and merge them to form a full-size reconstructed image J, averaging the overlapping regions. To identify anomalous regions in I, we first apply the subband decomposition on both I and J, divide the two images into L' $R \times R$ windows I_l^m, J_l^m, where $m \in \{1, \ldots, M\}$ and $l \in \{1, \ldots, L'\}$, and then compare these windows by the CW-SSIM index (4). We define the anomaly score $\mathcal{S}(l)$ for each pair I_l^m, J_l^m as the average CW-SSIM index over all the subbands:

$$\mathcal{S}(l) = 1 - \frac{1}{M} \sum_{m=1}^{M} \text{CW-SSIM}(I_l^m, J_l^m). \tag{5}$$

Then, we compute the anomaly score $\hat{\omega}(i, j)$ for each pixel (i, j) by averaging the scores $\mathcal{S}(l)$ of the windows containing the pixel (i, j). The CW-SSIM window size R and number of orientations O in the subband decomposition are the same used during training ($R = 7$ and $O = 6$), in order to extract the same type of features. To produce more robust results, we average the anomaly maps obtained computing the CW-SSIM index several times, changing the number of scales $S \in \{7, 8, 9\}$. Using a larger number of scales increases the dimensionality reduction factor, thus compensating for the fact that, at test time, we compare full-size images, while during training the CW-SSIM index is computed only on small patches. Then, we generate the estimated anomaly mask $\hat{\Omega}$ by detecting as anomalous those pixels (i, j) whose anomaly score exceeds a threshold γ, i.e., $\hat{\omega}(i, j) \geq \gamma$. We define the threshold γ to obtain a false positive rate (FPR) of 0.05 over the validation images, which are defective images that will be no used during test. Finally, we apply circular erosion to $\hat{\Omega}$ over a disk of radius 10 pixels to remove outliers and to reduce jagged edges, thus improving the coverage of anomalous regions, as shown in [6].

5 Experiments

Here we present the experiments we performed to evaluate the anomaly detection performance of our solution. First, we describe the datasets (Sect. 5.1) and the figures of merit we employ to test our solution (Sect. 5.2). Then we briefly present the methods against which we compare our solution (Sect. 5.3), and finally we illustrate and discuss our experimental results (Sect. 5.4).

5.1 Benchmarking Datasets

We evaluate the performance of our methods on two anomaly detection datasets: the *Nanofiber* and the *MVTec Texture* datasets (see Fig. 1 for examples of normal and anomalous images). The Nanofiber dataset [6] contains images of nanofibrous material acquired by a *Scanning Electron Microscope* (SEM), with a pixel size of few tens nanometers. In normal conditions, the material presents a non-periodic continuous texture of filamentous elements with a diameter of about 10 pixels. The anomalous regions consist of agglomerates of filamentous material, and can appear in any location within the image. The annotated defects have a diameter that ranges between 20 to 350 pixels. The images are grayscale (i.e., $c = 1$) with 8-bit color depth ($d = 8$), and have size 696×1024 pixels.

The MVTec Texture dataset is the portion of the MVTec anomaly detection dataset [2] that contains only textured images. We exclude images depicting objects, which are out of the scope of this paper. There are 5 different categories of either regular (*Carpet, Grid*) or random textures (*Leather, Tile, Wood*). Test images can contain a variety of defects, such as surface anomalies (e.g., scratches, cuts), or missing parts, with a diameter between 40 to 390 pixels. The dataset is mainly composed of RGB images (i.e., $c = 3$), apart from the Grid images, which are grayscale. All the images have 8-bit color depth and spatial dimension ranging between 700×700 and 1024×1024. Due to the design of CW-SSIM, we convert RGB images to grayscale before presenting them to our autoencoder.

5.2   Figures of Merit

We assess the anomaly detection performance by the area under the ROC curve (AUC), a well-known threshold-independent performance metric, computed from the anomaly scores of each pixel in test images. To enable a direct comparison with the results in [2], in our experiments on the MVTec dataset we compute the AUC up to FPR = 0.3 (normalized so that the maximum attainable value is 1). On the Nanofiber dataset, we also evaluate the *defect coverage*, which compares the connected components of the ground-truth and the estimated anomaly mask obtained by a detector yielding FPR = 0.05. This metric was proposed in [6] to assess the detection performance independently from the size of the anomalous regions, which influences the AUC since larger anomalies contain more pixels [6].

5.3   Considered Methods

In our experiments we compare our solution with autoencoders sharing the same architecture but trained with different loss functions: MSE, SSIM as in [4], and MS-SSIM [21]. For a fair comparison, we re-train our SSIM autoencoder rather than reporting the results from [4], whose autoencoder has a simpler architecture than ours. We also evaluate the method presented in [14], which we indicate by *Feature Dictionary*, and the Student-Teacher method [3], which represents the state of the art on the MVTec datasets. On the Nanofiber dataset, we also evaluate the sparse representation-based method proposed in [6], which we refer to as *Sparse Coding*. The SSIM and MS-SSIM autoencoders and Sparse Coding are designed to process grayscale images, thus it is necessary to convert RGB images to grayscale before applying these methods.

Table 1. AUC and defect coverage of the considered methods on the Nanofiber dataset. Our solution is the best method, outperforming the state of the art on this dataset.

Metrics	Sparse Coding [6]	Feature Dict. [14]	MSE Autoencoder	SSIM Autoencoder	MS-SSIM Autoencoder	CW-SSIM Autoencoder
AUC	0.926	0.974	0.665	0.961	0.962	**0.977**
Def. Cov.	0.650	0.850	0.000	0.880	0.870	**0.960**

Table 2. Normalized AUC of the considered methods of the MVTec Texture dataset. We indicate by (G) grayscale textures and by (C) color textures. Our solution outperforms the state of the art in three out of five textures from this dataset.

Dataset	Feature Dict. [14]	Student-Teacher [3]	MSE Autoencoder	SSIM Autoencoder	MS-SSIM Autoencoder	CW-SSIM Autoencoder
Carpet (C)	0.943	0.927	0.543	0.754	0.753	**0.947**
Grid (G)	0.872	0.974	0.662	0.852	0.927	**0.978**
Leather (C)	0.819	0.976	0.658	0.800	0.696	**0.979**
Tile (C)	0.854	**0.946**	0.552	0.650	0.679	0.847
Wood (C)	0.720	**0.895**	0.574	0.706	0.640	0.753

We report the anomaly detection performance on the Nanofiber dataset in [14] for Feature Dictionary and in [6] for Sparse Coding, and the performance on the MVTec Texture datasets of Feature Dictionary and Student-Teacher as reported in [2]. Since the Student-Teacher implementation is not publicly available, we were not able to test it on the Nanofiber dataset.

5.4 Results and Discussion

In Table 1 we report the AUC and median defect coverage of the considered methods on the Nanofiber dataset [6]. These results show that our CW-SSIM autoencoder is the best at detecting anomalous regions, slightly outperforming the state-of-the-art Feature Dictionary [14] in terms of AUC. The difference is even more substantial in defect coverage, meaning that our solution is superior in detecting also small anomalous regions.

Table 2 reports the normalized AUC (up to FPR = 0.3) of the considered methods on the different categories in the MVTec Texture dataset [2]. Our CW-SSIM autoencoder represents the best-performing method on three out of five categories of textures, namely Grid (which is grayscale), Carpet and Leather (which are color). We remark that our solution, which is based on a simple architecture with $5.57 \cdot 10^6$ trainable parameters, outperforms the state-of-the-art Feature Dictionary and Student-Teacher approaches, which leverage comparably more complicated pre-trained models with respectively $11.46 \cdot 10^6$ and $26.07 \cdot 10^6$ parameters. Most remarkably, our solution yields better AUC than competing methods also on color textures like Carpet and Leather, even though we convert RGB images to grayscale before feeding them to our autoencoder, while Feature Dictionary and Student-Teacher take as input the original RGB images. In

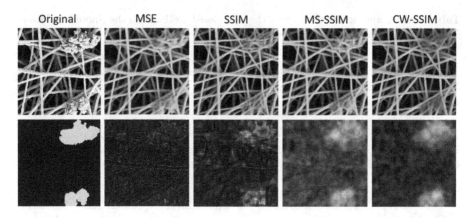

Fig. 2. An example of reconstruction and anomaly scores produced by autoencoders trained with different loss functions from a Nanofiber image. This image shows that autoencoders trained with structural similarity metrics, and CW-SSIM in particular, yield better reconstruction quality and superior anomaly detection performance than a traditional MSE autoencoder.

contrast, on Tile and Wood, our solution performs similarly to Feature Dictionary but worse than Student-Teacher. We speculate that this is due to the fact that color information is crucial to detect anomalies in these textures, and this motivates us to further investigate how to manage colors in our solution.

The results in Tables 1 and 2 confirm that a loss function based on SSIM substantially improve the anomaly detection performance of autoencoders compared to using the MSE, as shown in [4]. Our experiments show that using MS-SSIM yields similar performance to SSIM, while our loss function and anomaly detection procedure based on CW-SSIM further improve the results, making our solution comparable or superior to the state of the art. This is due to the fact that autoencoders based on structural similarity metrics, and CW-SSIM in particular, achieve higher reconstruction quality of normal images than MSE autoencoders, and this yields superior anomaly detection performance. We qualitatively illustrate this in Fig. 2, which shows an image from the Nanofiber dataset, its reconstruction produced by autoencoders based on MSE, SSIM, MS-SSIM, and CW-SSIM, and the corresponding anomaly maps $\hat{\omega}$ compared to the ground truth.

6   Conclusions and Future Work

We propose an effective training strategy for autoencoders to detect anomalous regions in textured images. In particular, we employ a custom loss function based on CW-SSIM [23] for both training and testing. Our experiments show that our solution improves the state of the art on several anomaly detection benchmarks, outperforming deep neural networks with more sophisticated and computationally demanding architectures. Although CW-SSIM is designed for grayscale images, thus it cannot exploit color information, our solution can effectively detect anomalous regions also from RGB images converted to grayscale. Future work will investigate how to combine CW-SSIM and *Optimal Color Composition Distance* (OCCD) [13] in our custom loss function for autoencoders, as suggested in [26], since in some cases color information might be crucial for anomaly detection. Moreover, we will make our autoencoder fully convolutional to achieve more efficient training and testing.

Acknowledgement. We gratefully acknowledge the support of NVIDIA Corporation with the four RTX A6000 GPUs granted through the Applied Research Accelerator Program to Politecnico di Milano.

References

1. Akcay, S., Atapour-Abarghouei, A., Breckon, T.P.: GANomaly: semi-supervised anomaly detection via adversarial training. In: Jawahar, C.V., Li, H., Mori, G., Schindler, K. (eds.) ACCV 2018. LNCS, vol. 11363, pp. 622–637. Springer, Cham (2019). https://doi.org/10.1007/978-3-030-20893-6_39
2. Bergmann, P., Batzner, K., Fauser, M., Sattlegger, D., Steger, C.: The MVTec anomaly detection dataset: a comprehensive real-world dataset for unsupervised anomaly detection. Int. J. Comput. Vis. **129**(4), 1038–1059 (2021)
3. Bergmann, P., Fauser, M., Sattlegger, D., Steger, C.: Uninformed students: student-teacher anomaly detection with discriminative latent embeddings. In: Proceedings of the IEEE/CVF Conference on Computer Vision and Pattern Recognition, pp. 4183–4192 (2020)
4. Bergmann., P., Löwe., S., Fauser., M., Sattlegger., D., Steger., C.: Improving unsupervised defect segmentation by applying structural similarity to autoencoders. In: International Joint Conference on Computer Vision, Imaging and Computer Graphics Theory and Applications (VISAPP), pp. 372–380. SciTePress (2019)
5. Bölgen, N., Vaseashta, A.: Nanofibers for tissue engineering and regenerative medicine. In: Sontea, V., Tiginyanu, I. (eds.) 3rd International Conference on Nanotechnologies and Biomedical Engineering. IP, vol. 55, pp. 319–322. Springer, Singapore (2016). https://doi.org/10.1007/978-981-287-736-9_77
6. Carrera, D., Manganini, F., Boracchi, G., Lanzarone, E.: Defect detection in SEM images of nanofibrous materials. IEEE Trans. Ind. Inf. **13**(2), 551–561 (2016)
7. Chetverikov, D., Hanbury, A.: Finding defects in texture using regularity and local orientation. Pattern Recogn. **35**(10), 2165–2180 (2002)
8. Freeman, W.T., Adelson, E.H., et al.: The design and use of steerable filters. IEEE Trans. Pattern Anal. Mach. Intell. **13**, 891–906 (1991)
9. Haindl, M., Grim, J., Mikeš, S.: Texture defect detection. In: Kropatsch, W.G., Kampel, M., Hanbury, A. (eds.) CAIP 2007. LNCS, vol. 4673, pp. 987–994. Springer, Heidelberg (2007). https://doi.org/10.1007/978-3-540-74272-2_122
10. He, K., Zhang, X., Ren, S., Sun, J.: Deep residual learning for image recognition. In: Proceedings of the IEEE Conference on Computer Vision and Pattern Recognition, pp. 770–778 (2016)
11. Kingma, D.P., Ba, J.: Adam: a method for stochastic optimization. In: 3rd International Conference on Learning Representations (2015)

12. Li, K.L., Huang, H.K., Tian, S.F., Xu, W.: Improving one-class SVM for anomaly detection. In: Proceedings of the 2003 International Conference on Machine Learning and Cybernetics, vol. 5, pp. 3077–3081. IEEE (2003)

13. Mojsilovic, A., Hu, H., Soljanin, E.: Extraction of perceptually important colors and similarity measurement for image matching, retrieval and analysis. IEEE Trans. Image Process. **11**(11), 1238–1248 (2002)

14. Napoletano, P., Piccoli, F., Schettini, R.: Anomaly detection in nanofibrous materials by CNN-based self-similarity. Sensors **18**(1), 209 (2018)

15. Parzen, E.: On estimation of a probability density function and mode. Ann. Math. Stat. **33**, 1065–1076 (1962)

16. Portilla, J., Simoncelli, E.P.: A parametric texture model based on joint statistics of complex wavelet coefficients. Int. J. Comput. Vis. **40**, 49–70 (2000)

17. Ruff, L., Kauffmann, J.R., Vandermeulen, R.A., Montavon, G., Samek, W., Kloft, M., Dietterich, T.G., Müller, K.R.: A unifying review of deep and shallow anomaly detection. Proc. IEEE **109**(5), 756–795 (2021)

18. Schlegl, T., Seeböck, P., Waldstein, S.M., Langs, G., Schmidt-Erfurth, U.: f-AnoGAN: fast unsupervised anomaly detection with generative adversarial networks. Med. Image Anal. **54**, 30–44 (2019)

19. Simoncelli, E.P., Freeman, W.T., Adelson, E.H., Heeger, D.J.: Shiftable multiscale transforms. IEEE Trans. Inf. Theory **38**, 587–607 (1992)

20. Tsa, D.M., Wu, S.K.: Automated surface inspection using Gabor filters. Int. J. Adv. Manuf. Technol. **16**(7), 474–482 (2000)

21. Wang, Z., Simoncelli, E., Bovik, A.: Multiscale structural similarity for image quality assessment. In: Conference Record of the Asilomar Conference on Signals, Systems and Computers, vol. 2, pp. 1398–1402 (2003)

22. Wang, Z., Bovik, A.C., Sheikh, H.R., Simoncelli, E.P.: Image quality assessment: from error visibility to structural similarity. IEEE Trans. Image Process. **13**(4), 600–612 (2004)

23. Wang, Z., Simoncelli, E.P.: Translation insensitive image similarity in complex wavelet domain. In: IEEE International Conference on Acoustics, Speech, and Signal Processing (ICASSP), 2005. vol. 2, pp. 573–576. IEEE (2005)

24. Zavrtanik, V., Kristan, M., Skočaj, D.: Reconstruction by inpainting for visual anomaly detection. Pattern Recogn. **112**, 107706 (2021)

25. Zhou, C., Paffenroth, R.C.: Anomaly detection with robust deep autoencoders. In: Proceedings of the 23rd ACM SIGKDD International Conference on Knowledge Discovery and Data Mining, pp. 665–674 (2017)

26. Zujovic, J., Pappas, T.N., Neuhoff, D.L.: Structural similarity metrics for texture analysis and retrieval. In: IEEE International Conference on Image Processing (ICIP), pp. 2225–2228. IEEE (2009)

Self-supervised Prototype Conditional Few-Shot Object Detection

Daisuke Kobayashi[(✉)]

Corporate Research and Development Center, Toshiba Corporation, Kanagawa, Japan
daisuke32.kobayashi@toshiba.co.jp

Abstract. Traditional deep learning-based object detection methods require a large amount of annotation for training, and creating such a dataset is expensive. Few-shot object detection which detects a new category of objects with a small amount of data is a difficult task. Many previous methods have been applied to detect new categories with few data by fine-tuning. However, fine-tuning methods cannot be sufficiently adapted quickly to new environments because of time-consuming training. In this study, we develop methods that can be adapted to new categories without fine-tuning. A prototype conditional detection module built on prototypes for each category is intended to be conditioned and detected by prototypes. Moreover, a self-supervised conditional detection module improves the representational capability for detecting new categories by a self-supervised task that uses embeddings outside the annotation region to detect the same regions before and after image transformation. The results of extensive experiments on the PASCAL VOC and MS COCO show the effectiveness of the proposed method.

Keywords: Few-shot object detection · Few-shot learning · Object detection

1 Introduction

In recent years, deep learning has achieved excellent performance in a variety of computer vision tasks and has been used in embedded systems, such as automatic robot control and automated driving. In such systems, new environments need to be explored and understood rapidly. Therefore, it is desirable to detect new categories on the fly with only a small amount of data. Few-shot object detection (FSOD) is one method to solving these problems. FSOD uses generalized object features learned from rich data containing objects in base categories to make detection possible even for sparse data containing objects in novel categories. Early work on FSOD used meta-learning to improve performance by dividing base categories and simulating detection of new categories [8,31,32]. The challenge with these methods is that they require a complex training process and data partitioning. In contrast, because fine-tuning-based methods are

S. Sclaroff et al. (Eds.): ICIAP 2022, LNCS 13232, pp. 681–692, 2022.
https://doi.org/10.1007/978-3-031-06430-2_57

simple and effective in freezing the parameters of the first layer of the network and fine-tuning only the final layer, many related methods have been proposed in recent years [23, 28].

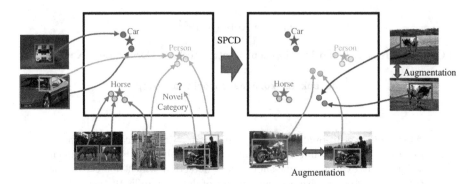

Fig. 1. Illustration of the embedding space in proposal regions. The circles in the figure represent the embedding of each object, and the stars represent the prototypes of base categories. As shown in the left figure, it is difficult to obtain embeddings for detecting new categories because regions outside base categories are uniformly treated as negative during base learning. As shown in the figure on the right, our SPCD learns embeddings to detect objects in the background region generated in a self-supervised manner.

However, there are still some issues with conventional FSOD methods. Many previous methods require fine-tuning, including meta-learning-based methods. In embedded systems, computer resources are not sufficient to perform computationally intensive learning, which is a challenge for immediate adaptation. In addition, when training a large amount of data in advance, identification of objects in base categories is learned, but objects outside base categories are learned as negatives, so the representational ability to identify objects in novel categories is insufficient.

To solve this problem, we develop a self-supervised prototype conditional detection approach (SPCD) that can be adapted to novel categories without fine-tuning. As shown in Fig. 2, the overall structure consists of a Prototype Conditional Prediction Module (PCPM) and a Self-Supervised Conditional Prediction Module (SCPM). In the PCPM, we add a layer that outputs embedding to a RCNN in Faster-RCNN [21], and create prototypes for each object category. In the RCNN, we use the dynamic parameters generated from the prototypes for each category to conditionally perform box regression and classification. This makes it easy to add new categories simply by preparing prototypes. However, as shown Fig. 1, since categories other than base categories are treated as negative, the expressive power is insufficient to identify novel categories. Therefore, we improved the representation capability with the SCPM to extract rectangular regions from outside the base category regions and taught the network to detect the same regions before and after adding image transformation in a self-supervised manner.

The contributions in this paper can be summarized as follows.

- We propose a simple and effective prototype-based FSOD architecture that does not require fine-tuning.
- To cope with data-scarce scenarios, we improve detection performance by using a self-supervised approach that extracts features even from background regions.
- Experiments showed that the proposed method can achieve state-of-the-art results.

2 Related Work

2.1 Object Detection

Object detection is a fundamental task in computer vision, which involves classifying categories of objects and estimating their locations. Object detection can be broadly classified into one-stage detection methods [2,12,14,18–20] and two-stage detection methods [3,5,9,21].

One-stage detection methods estimate category scores and bounding box coordinates on a dense grid simultaneously. Two-stage detection methods predict a category-agnostic object-ness region and improve the proposed region such as Faster RCNN [21]. These methods rely on predefined anchor boxes, but recently, anchor-free methods have been proposed that do not require anchor design [4, 25,33].

2.2 Few-Shot Object Detection

Previous methods of FSOD can be mainly divided into transfer learning based and meta learning-based methods. Transfer learning-based methods employ general object detection methods and focus on the fine-tuning process to effectively transfer features learned in base categories to novel categories. TFA [28] shows performance improvement by fine-tuning only the final layer of a Faster RCNN. MPSR [30] extracts features from multi-resolution images to accommodate objects of various sizes. FSCE [23] introduces contrastive learning to the detection had to improve performance. UP-FSOD [29] introduces universal prototype augmented features to learn invariant object characteristics, different from the prototypes that are separately learned from each category. DeFRCN [16] introduces a layer to separate RPN and RCNN and calibrate a classification score. Meta learning based methods adapt efficiently to new categories by building and learning an auxiliary task in which a target category is conditioned on the support set. These methods still require fine-tuning. FSRW [11] and Meta RCNN [32] extract meta-functions from base categories and adjust them by reweighting functions in new categories. FSIW [31] performs the same process as Meta RCNN, but more complexly aggregates features from a support set. AttentionRPN [8] detects features by mapping a query image to a support image by contrastive learning.

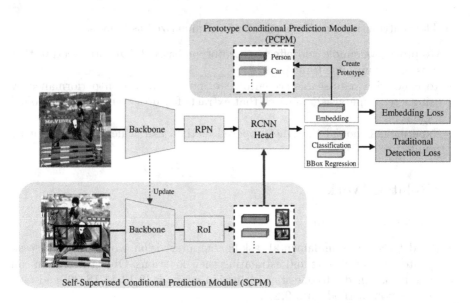

Fig. 2. The architecture of our SPCD. Our SPCD is based on Faster RCNN. PCPM generates prototypes for each category and performs classifications and box regressions conditioned by the prototypes. SCPM performs classification and box regression conditioned by the embeddings of self-supervised generated regions.

2.3   Prototype-Based Related Tasks

Prototypes have been shown to be effective for few-shot image classification [22,27]. In incremental learning, when training additional new data, prototypes of each category are stored in memory and retrained to prevent catastrophic forgetting [17,34]. In open world recognition, a prototype-based classifier is used to recognize unknown categories as unknown and adaptively improve the model when new labels are provided [1,15]. However, most of these existing methods are proposed to address image classification and cannot be easily adapted to FSOD. Our simple prototype-based FSOD approach is structured to be extendable to incremental learning and open world recognition settings.

3   Approach

This section describes how we developed a self-supervised prototype conditional detection approach (SPCD) for detecting novel categories without fine-tuning.

3.1   Problem Definition

As with previous methods, this study followed the standard problem setup for FSOD. First, we acquire object detection knowledge on a large, rich base data set D_{base} with annotation, and then quickly adapted the network to novel categories with a small set of novel support data D_{novel}. Base categories C_{base} and novel categories C_{novel} in the support data set do not overlap.

3.2   Prototype Conditional Prediction Module

To support novel categories without fine-tuning, AttentionRPN inputs a support image and a query image, which is learned by contrastive learning to recognize objects of the same category as the support image from the query image. This method has difficulties in the way the support image is selected. In contrast, the PCPM has a simple structure in which prototypes of features for each category are kept, and detection is conditioned on the prototypes for each category.

We add an embedding head to generate prototypes. It is connected through ROIAlign to obtain embeddings for ROI generated by RPN. Prototypes for each category are reserved in memory, and embeddings of rectangles corresponding to ground truth rectangles are updated with an exponential moving average (EMA) as prototypes for that category. The loss function for embeddings is computed in cross-entropy as follows so that the embeddings and outputs of the same categories are close. x are network outputs from embedding head and p represents the prototype of the category corresponding to x. P is a softmax function with temperature.

$$\mathcal{L}_{embed} = \frac{1}{N} \sum_{i=1}^{N} L(P(x), P(p)) \tag{1}$$

$$L(a, b) = -a \log b \tag{2}$$

The box classifier of RCNN is calculated based on cosine similarity between embeddings and the prototypes as follows. α is a learnable scaling factor.

$$S(a, b) = \alpha \frac{a^\top b}{\|a\| \cdot \|b\|} \tag{3}$$

Box regression in RCNN are conditioned on the prototypes by PCPM motivated by CondInst [24]. It learns the parameters of convolution for box regression from prototypes by fully connected layers. The overall supervised loss when training is,

$$\mathcal{L}_{sup} = \mathcal{L}_{rpn} + \mathcal{L}_{cls} + \mathcal{L}_{reg} + \lambda \mathcal{L}_{embed} \tag{4}$$

where $\mathcal{L}_{rpn}$ is loss of foreground and background classification and box regression for each anchor, $\mathcal{L}_{cls}$ is a focal loss for the box classifier, $\mathcal{L}_{reg}$ is a smoothed $L1$ loss for the box regressor, $\mathcal{L}_{embed}$ is embedding loss for the embedding branch. λ is set as 1.0 by default to balance the scale of the losses. If fine-tuning is not used, adaptation to novel categories is prototyped by embeddings of annotation region averaged over each category by forward processing. On the other hand, for fine-tuning, backbone is frozen and only the other parameters are updated.

3.3   Self-supervised Conditional Prediction Module

In training with a large base data set, only base categories are annotated and other regions are treated as negative. Therefore, the representational capability of the model depends on base categories. The SCPM improves the representational capability of the model by using regions other than base categories to deal with unknown categories. Specifically, for image transformations such as horizontal flipping or color transformation, Selective Search [26] generates a rectangle from regions other than the base categories, and embeddings are obtained using ROIAlign. We consider these as prototypes, and learn to detect the same region before the image transformation by conditioning on prototypes as in the PCPM.

To ensure accurate and stable embeddings, we introduce the teacher-student mutual learning. There are two models with the same architecture, a student model and a teacher model. At the beginning of training, both the student model and the teacher model are randomly initialized. The overall loss of the student model is,

$$\mathcal{L}_{all} = \mathcal{L}_{sup} + \mathcal{L}_{selfsup\_cls} + \mathcal{L}_{selfsup\_reg} + \lambda \mathcal{L}_{selfsup\_embed} \qquad (5)$$

where $\mathcal{L}_{sup}$ is loss for annotated data, $\mathcal{L}_{selfsup\_cls}$ is a focal loss and $\mathcal{L}_{selfsup\_reg}$ is a smoothed $L1$ loss and $\mathcal{L}_{selfsup\_embed}$ is a embedding loss conditioned on the embedding of the teacher model for data generated in a self-supervised manner. Furthermore, the embedding loss in the SPCM can be easily implemented in other FSOD methods to improve performance.

4   Experiments

4.1   Experimental Setting

We evaluate our method with PASCAL VOC [7] and MS COCO [13]. We use conventional settings to construct few-shot datasets for fair comparison. For PASCAL VOC, we divide the 20 categories into 15 base categories and 5 new categories. We use three types of splitting according to the traditional settings. K object annotations for each novel category are available and K is set to 1, 2, 3, 5, and 10. For MS COCO, we use 20 categories that overlap with PASCAL VOC out of 80 categories as novel categories with $K = 10$, and 30 and the remaining 60 categories are used as base categories.

As evaluation metrics for object detection, AP_{50} is used for PASCAL VOC and AP for MS COCO. All results are averages of multiple runs.

4.2   Implementation Details

Our FSOD framework is implemented based on Faster RCNN with ResNet-50 [10] as the backbone. The channel number of prototypes is set as 256. We adopt SGD to optimize our network end-to-end with a mini-batch size of 16, momentum of 0.9 and weight decay of $1e^{-5}$. A learning rate is set to 0.02 for base training and 0.001 for fine-tuning.

In the SCPM, a strong data augmentation is used for the student model and a weak data augmentation is used for the teacher model. For the strong data augmentation, color jitter, solarize jitter, brightness jitter, contrast jitter, and sharpness jitter are applied as color transformation, scale jitter, rotate, shift, x-y shear are applied as geometric transformation, and Cutout [6] is applied as other augmentation. For the generation of self-supervised bounding boxes, 10 boxes are randomly selected from the boxes obtained by Selective Search.

4.3   Comparison of Results

Table 1 shows the evaluation results detected without fine-tuning for the three different data partitions of PASCAL VOC. Our proposed method can be fine-tuned, and Table 2 shows the evaluation results with fine-tuning. Basically, as the number of novel category instances increases, the detection performance improves. However, without fine-tuning, it may not contribute to the improvement of accuracy. This indicates that the quality of the selected novel category instances affects the accuracy more than the number of novel category instances. The detection accuracy of the proposed method in the case without fine-tuning is consistently better than that of AttentionRPN. Table 2 shows that the proposed method can be further improved by fine-tuning when the number of novel category instances is large. The backbone of our proposed method is ResNet50, but the results are close to the state-of-the-art methods with ResNet101 backbone.

Table 3 shows the evaluation results of detection without fine-tuning for the MS COCO. Table 4 shows the evaluation results for the case with fine-tuning. Our proposed method consistently outperforms Attention RPN, and results are 1.9 and 3.1 points higher for 10 and 30 shots in the case without fine-tuning.

For a small number of shots, the proposed method without fine-tuning even outperforms several methods with fine-tuning. Without fine-tuning is superior when the number of shots is small and immediate adaptation is required without training, while fine-tuning is superior when the number of shots is large and training time is allowd.

In Fig. 3, we show the detection results of AttentionRPN and our method based on the 1 shot case without fine-tuning. 'bird', 'bus', 'cow', 'motorbike', and 'sofa' are included in the novel category, we can see that the proposed method can detect the novel category properly.

Table 1. Few-shot object detection evaluation results on PASCAL VOC without fine-tuning.

Method	Backbone	Novel Set1					Novel Set2					Novel Set3				
		1	2	3	5	10	1	2	3	5	10	1	2	3	5	10
AttentionRPN [8]	Res50	21.2	16.5	19.7	21.8	22.3	13.8	14.3	15.8	16.8	15.2	19.1	15.0	24.9	25.7	21.6
SPCD(Ours)	Res50	**46.0**	**37.0**	**45.3**	**51.4**	**55.0**	**29.6**	**26.7**	**37.0**	**30.2**	**34.6**	**41.0**	**32.4**	**30.7**	**44.0**	**43.4**

Table 2. Few-shot object detection evaluation results on PASCAL VOC with fine-tuning.

Method	Backbone	Novel Set1					Novel Set2					Novel Set3				
		1	2	3	5	10	1	2	3	5	10	1	2	3	5	10
FRCN-ft [32]	Res101	13.8	19.6	32.8	41.5	45.6	7.9	15.3	26.2	31.6	39.1	9.8	11.3	19.1	35.0	45.1
Meta R-CNN [32]	Res101	19.9	25.5	35.0	45.7	51.5	10.4	19.4	29.6	34.8	45.4	14.3	18.2	27.5	41.2	48.1
TFA [28]	Res101	39.8	36.1	44.7	55.7	56.0	23.5	26.9	34.1	35.1	39.1	30.8	34.8	42.8	49.5	49.8
FSIW [31]	Res101	24.2	35.3	42.2	49.1	57.3	21.6	24.6	31.9	37.0	45.7	21.2	30.0	37.2	43.8	49.6
MPSR [30]	Res101	41.7	–	51.4	55.2	61.8	24.4	–	39.2	39.9	47.8	35.6	–	42.3	48.0	49.7
FSCE [23]	Res101	44.2	43.8	51.4	61.9	**63.4**	27.3	29.5	43.5	44.2	**50.2**	37.2	41.9	47.5	54.6	**58.5**
UP-FSOD [29]	Res101	43.8	47.8	50.3	55.4	61.7	**31.2**	30.5	41.2	42.2	48.3	35.5	39.7	43.9	50.6	53.5
DeFRCN [16]	Res101	**53.6**	**57.5**	**61.5**	**64.1**	60.8	30.1	**38.1**	**47.0**	**53.3**	47.9	**48.4**	**50.9**	**52.3**	**54.9**	57.4
AttentionRPN [8]	Res50	35.0	36.0	39.1	51.7	55.7	20.8	23.4	35.9	37.0	43.3	31.9	30.8	38.2	48.9	51.6
SPCD(Ours)	Res50	**43.9**	**48.0**	**49.1**	**52.7**	**61.6**	**26.2**	**28.7**	**41.0**	**41.0**	**44.4**	**42.2**	**44.3**	**47.7**	**52.0**	**52.8**

Table 3. Few-shot detection evaluation results on MS COCO without fine-tuning.

Method	Backbone	10-shot	30-shot
AttentionRPN [8]	Res50	5.9	6.4
SPCD(Ours)	Res50	**7.8**	**9.5**

Table 4. Few-shot detection evaluation results on MS COCO with fine-tuning.

Method	Backbone	10-shot	30-shot
TFA [28]	Res101	10.0	13.7
MSPR [30]	Res101	9.8	14.1
FSCE [23]	Res101	11.9	15.3
UP-FSOD [29]	Res101	11.0	15.6
DeFRCN [16]	Res101	**18.5**	**22.6**
FRCN-ft [32]	Res50	6.5	11.1
Meta R-CNN [32]	Res50	8.7	12.4
AttentionRPN [8]	Res50	9.2	14.8
FSIW [31]	Res50	**12.5**	14.7
SPCD(Ours)	Res50	11.5	**16.4**

Fig. 3. Detection results based on the 1 shot case without fine-tuning on Novel Set 1 of PASCAL VOC. The first row shows the results of AttentionRPN. The second row shows the results of our SPCD.

4.4   Ablation Study

This section describes the ablation study on 1 shot and 10 shots of a novel set from PASCAL VOC without fine-tuning and analyzes the contribution of each module.

Table 5 shows the comparison results of different modules. The first row shows the results without SCPM. The second rows shows the results without $\mathcal{L}_{selfsup\_cls}$ and $\mathcal{L}_{selfsup\_reg}$ in the SCPM. It shows that the performance is greatly improved by the SCPM.

Table 6 show the method using random and Selective Search as the self-supervised bounding box generation methods in SCPM, respectively. We show that using Selective Search as the self-supervised bounding box generation method results in a significant improvement of 5.1 points for 10-shot.

Table 7 shows the results of the comparison for different numbers of self-supervised bounding box generation by SCPM. The performance improved as the number of generated bounding boxes increased, and the best result is obtained with 10 boxes.

Table 5. Effectiveness of different modules on PASCAL VOC.

Method	Shot number	
	1	10
PCPM	22.9	41.9
PCPM+SCPM (w/o $L_{selfsup\_cls}$ and $L_{selfsup\_reg}$)	38.2	43.1
PCPM+SCPM	**46.0**	**55.0**

Table 6. Effectiveness of different self-supervised bounding box generation methods in SCPM.

Method	Shot number	
	1	10
PCPM+SCPM (Random)	38.6	49.9
PCPM+SCPM (Selective Search)	**46.0**	**55.0**

Table 7. Comparison of the number of self-supervised bounding boxes generated by SCPM.

Number of self-supervised bounding boxes	Shot number	
	1	10
1	40.0	47.7
5	44.1	52.3
10	**46.0**	**55.0**
20	44.0	53.3

5 Conclusion

In this paper, we proposed a new architecture SPCD that can be learned in a self-supervised manner with respect to FSOD, that can be adapted to new categories without fine-tuning. The PCPM, built on prototypes for each category is intended to be conditioned and detected by prototypes. Moreover, the SCPM improves the representational capability for detecting new categories by a self-supervised task that uses embeddings outside the annotation region to detect the same regions before and after image transformation. The extensive results of the PASCAL VOC and MS COCO demonstrate the effectiveness of our method.

References

1. Bendale, A., Boult, T.: Towards open world recognition. In: Proceedings of the IEEE Conference on Computer Vision and Pattern Recognition, pp. 1893–1902 (2015)
2. Bochkovskiy, A., Wang, C.Y., Liao, H.Y.M.: Yolov4: optimal speed and accuracy of object detection. arXiv preprint arXiv:2004.10934 (2020)
3. Cai, Z., Vasconcelos, N.: Cascade R-CNN: delving into high quality object detection. In: Proceedings of the IEEE Conference on Computer Vision and Pattern Recognition, pp. 6154–6162 (2018)
4. Carion, N., Massa, F., Synnaeve, G., Usunier, N., Kirillov, A., Zagoruyko, S.: End-to-end object detection with transformers. In: Vedaldi, A., Bischof, H., Brox, T., Frahm, J.-M. (eds.) ECCV 2020. LNCS, vol. 12346, pp. 213–229. Springer, Cham (2020). https://doi.org/10.1007/978-3-030-58452-8_13
5. Dai, J., Li, Y., He, K., Sun, J.: R-FCN: object detection via region-based fully convolutional networks. In: Advances in Neural Information Processing Systems, vol. 29 (2016)

6. DeVries, T., Taylor, G.W.: Improved regularization of convolutional neural networks with cutout. arXiv preprint arXiv:1708.04552 (2017)
7. Everingham, M., Gool, L.V., Williams, C.K., Winn, J., Zisserman, A.: The pascal visual object classes (VOC) challenge. Int. J. Comput. Vis. **88**(2), 303–338 (2010)
8. Fan, Q., Zhuo, W., Chi-Keung, T., Tai, Y.W.: Few-shot object detection with attention-RPN and multi-relation detector. In: Proceedings of the IEEE/CVF Conference on Computer Vision and Pattern Recognition, pp. 4013–4022 (2020)
9. He, K., Dollár, G.G.P., Girshick, R.: Mask R-CNN. In: Proceedings of the IEEE International Conference on Computer Vision, pp. 2980–2988 (2017)
10. He, K., Zhang, X., Ren, S., Sun, J.: Deep residual learning for image recognition. In: Proceedings of the IEEE Conference on Computer Vision and Pattern Recognition, pp. 770–778 (2016)
11. Kang, B., Liu, Z., Wang, X., Yu, F., Feng, J., Darrell, T.: Few-shot object detection via feature reweighting. In: Proceedings of the IEEE/CVF International Conference on Computer Vision, pp. 8420–8429 (2019)
12. Lin, T.Y., Goyal, P., Girshick, R., He, K., Dollár, P.: Focal loss for dense object detection. In: Proceedings of the IEEE International Conference on Computer Vision, pp. 2980–2988 (2017)
13. Lin, T-Y., et al.: Microsoft COCO: common objects in context. In: Fleet, D., Pajdla, T., Schiele, B., Tuytelaars, T. (eds.) ECCV 2014. LNCS, vol. 8693, pp. 740–755. Springer, Cham (2014). https://doi.org/10.1007/978-3-319-10602-1_48
14. Liu, W., et al.: SSD: single shot multiBox detector. In: Leibe, B., Matas, J., Sebe, N., Welling, M. (eds.) ECCV 2016. LNCS, vol. 9905, pp. 21–37. Springer, Cham (2016). https://doi.org/10.1007/978-3-319-46448-0_2
15. Pernici, F., Bartoli, F., Bruni, M., Bimbo, A.D.: Memory based online learning of deep representations from video streams. In: Proceedings of the IEEE Conference on Computer Vision and Pattern Recognition, pp. 2324–2334 (2018)
16. Qiao, L., Zhao, Y., Li, Z., Qiu, X., Wu, J., Zhang, C.: DEFRCN: Decoupled faster R-CNN for few-shot object detection. In: Proceedings of the IEEE/CVF International Conference on Computer Vision, pp. 8681–8690 (2021)
17. Rebuffi, S.A., Kolesnikov, A., Sperl, G., Lampert, C.H.: ICARL: incremental classifier and representation learning. In: Proceedings of the IEEE Conference on Computer Vision and Pattern Recognition, pp. 2001–2010 (2017)
18. Redmon, J., Divvala, S., Girshick, R., Farhadi, A.: You only look once: unified, real-time object detection. In: Proceedings of the IEEE Conference on Computer Vision and Pattern Recognition, pp. 779–788 (2016)
19. Redmon, J., Farhadi, A.: Yolo9000: better, faster, stronger. In: Proceedings of the IEEE Conference on Computer Vision and Pattern Recognition, pp. 7263–7271 (2017)
20. Redmon, J., Farhadi, A.: Yolov3: an incremental improvement. arXiv preprint arXiv:1804.02767 (2018)
21. Ren, S., He, K., Girshick, R., Sun, J.: Faster R-CNN: towards real-time object detection with region proposal networks. In: Advances in Neural Information Processing Systems, pp. 91–99 (2015)
22. Snell, J., Swersky, K., Zemel, R.: Prototypical networks for few-shot learning. In: Advances in Neural Information Processing Systems, vol. 30 (2017)
23. Sun, B., Li, B., Cai, S., Yuan, Y., Zhang, C.: FSCE: few-shot object detection via contrastive proposal encoding. In: Proceedings of the IEEE/CVF Conference on Computer Vision and Pattern Recognition, pp. 7352–7362 (2021)

24. Tian, Z., Shen, C., Chen, H.: Conditional convolutions for instance segmentation. In: Vedaldi, A., Bischof, H., Brox, T., Frahm, J.-M. (eds.) ECCV 2020. LNCS, vol. 12346, pp. 282–298. Springer, Cham (2020). https://doi.org/10.1007/978-3-030-58452-8_17

25. Tian, Z., Shen, C., Chen, H., He, T.: FCOS: Fully convolutional one-stage object detection. In: Proceedings of the IEEE/CVF International Conference on Computer Vision, pp. 9627–9636 (2019)

26. Uijlings, J.R., Sande, K.E.V.D., Gevers, T., Smeulders, A.W.: Selective search for object recognition. In. J. Comput. Vis. **104**(2), 154–171 (2013)

27. Vinyals, O., Blundell, C., Lillicrap, T., Koray, K., Wierstra, D.: Matching networks for one shot learning. In: Advances in Neural Information Processing Systems, vol. 29 (2016)

28. Wang, X., Huang, T.E., Darrell, T., E, J.G., Yu, F.: Frustratingly simple few-shot object detection. In: International Conference on Machine Learning (2020)

29. Wu, A., Han, Y., Zhu, L., Yang, Y.: Universal-prototype enhancing for few-shot object detection. In: Proceedings of the IEEE/CVF International Conference on Computer Vision, pp. 9567–9576 (2021)

30. Wu, J., Liu, S., Huang, D., Wang, Y.: Multi-scale positive sample refinement for few-shot object detection. In: Vedaldi, A., Bischof, H., Brox, T., Frahm, J.-M. (eds.) ECCV 2020. LNCS, vol. 12361, pp. 456–472. Springer, Cham (2020). https://doi.org/10.1007/978-3-030-58517-4_27

31. Xiao, Y., Marlet, R.: Few-shot object detection and viewpoint estimation for objects in the wild. In: Vedaldi, A., Bischof, H., Brox, T., Frahm, J.-M. (eds.) ECCV 2020. LNCS, vol. 12362, pp. 192–210. Springer, Cham (2020). https://doi.org/10.1007/978-3-030-58520-4_12

32. Yan, X., Chen, Z., Xu, A., Wang, X., Liang, X., Lin, L.: Meta R-CNN: towards general solver for instance-level low-shot learning. In: Proceedings of the IEEE/CVF International Conference on Computer Vision, pp. 9577–9586 (2019)

33. Zhou, X., Wang, D., Krähenbühl, P.: Objects as points. arXiv preprint arXiv:1904.07850 (2019)

34. Zhu, F., Zhang, X.Y., Wang, C., Yin, F., Liu, C.L.: Prototype augmentation and self-supervision for incremental learning. In: Proceedings of the IEEE/CVF Conference on Computer Vision and Pattern Recognition, pp. 5871–5880 (2021)

Low-Light Image Enhancement Using Image-to-Frequency Filter Learning

Rayan Al Sobbahi and Joe Tekli[(⊠)]

Department of Electrical and Computer Engineering, Lebanese American University (LAU),
Byblos 36, Lebanon
rayan.alsobbahi@lau.edu, joe.tekli@lau.edu.lb

Abstract. Low-light image (LLI) enhancement techniques have recently demonstrated remarkable progress especially with the use of deep learning (DL) approaches. Yet most existing techniques adopt an image-to-image learning paradigm where DL model architectures are constrained due to latent image feature reconstruction. In this paper, we propose a new LLI enhancement solution titled LLHFNet (Low-light Homomorphic Filtering Network) which performs image-to-frequency filter learning. It is designed independently from custom DL architectures and can be seamlessly coupled with existing feature extractors like ResNet50 and VGG16. We have conducted a large battery of experiments using SICE and Pascal VOC datasets to evaluate LLHFNet's enhancement quality. Our solution consistently ranks among the best existing image enhancement techniques and is able to robustly handle LLIs and normal-light images (NLIs).

Keywords: Image enhancement · Low-light conditions · Deep learning · Homomorphic filtering

1 Introduction

Modern artificial intelligence-based applications like autonomous spacecrafts, drones, autopilot car systems, robots, and security surveillance systems, among others, rely on visualizing and understanding outdoor environments. While these systems show good performance during normal and clear outdoor conditions, yet varying weather conditions and poor illumination might challenge their visual perception and compromise their performance [1, 2]. Low-light conditions account to a considerable time of our daily lives and can significantly affect the robustness of such systems and hinder their market deployment [1]. Hence, low-light image (LLI) enhancement has emerged as an image processing task that aims at illuminating LLIs and improving their visual quality.

LLI enhancement techniques have been largely investigated recently. Many traditional approaches use gamma correction methods [3], some rely on histogram equalization methods [4], while others follow the Retinex theory model [5]. More recently, Deep learning (DL) techniques have demonstrated better performance and efficiency compared with traditional methods [6, 7]. Yet most of these models the image-to-image learning paradigm where the deep network architecture is constrained to produce an output image through latent feature reconstruction.

In this work, we introduce a novel LLI enhancement model titled LLHFNet (Low-light Homomorphic Filtering Network) based on image-to-frequency filter learning. Our approach aims at learning two homomorphic filter parameters, which are consequently applied on the input LLI to perform enhancement. This removes the constraints of performing full image reconstruction, while designing the DL model to solely focus on the image enhancement task. LLHFNet is independent from any specific architecture and can be seamlessly coupled with typical feature extractors utilized with existing classification models including ResNet50 [8], VGG16 [9], and MobileNetv2 [10], among others. We perform a battery of experiments to evaluate the performance of our approach. Results show improved results compared with recent LLI enhancement models, where our solution is able to robustly handle LLIs and normal-light images (NLIs).

Section 2 provides a review of the related works. Section 3 explains preliminaries on homomorphic filtering, Sect. 4 describes our LLI enhancement model. Section 5 describes our experimental evaluation and results, before concluding in Sect. 6.

2 Related Works

Traditional LLI enhancement techniques rely on mathematical or algorithmic models to perform the enhancement task. Many approaches use gamma correction methods [3], some rely on histogram equalization methods [4], while others follow the Retinex theory model [5] or utilize homomorphic filtering (HF) to perform image enhancement [11, 12]. In contrast with traditional approaches, Deep Learning (DL) techniques are essentially data-driven, where training datasets of LLIs and NLIs are used to drive the learning process. They have gained great attention in the past few years as the most effective solutions to perform LLI enhancement, outperforming many traditional methods based on histogram equalization, e.g., [4, 13, 14] and Retinex theory, e.g., [5, 15–17]. LLNet [6] is one of the first DL approaches for LLI enhancement. Its architecture is based on a stacked-sparse denoising autoencoder (SSDA) made of three denoising autoencoder layers comprising hidden units with no use of convolutional layers. In [18] authors introduce RetinexNet consisting of two subnetworks: i) DecomNet which learns the Retinex decomposition of the image into its reflectance and illumination components, and ii) EnhanceNet which uses a dedicated encoder-decoder structure to perform illumination adjustment and enhancement. In [19], the authors introduce GLADNET made of a global illumination estimation step, using an encoder-decoder structure followed by a reconstruction step through a series of convolutional layers. In [20], the authors propose MBLLEN, a multi-branch network which extracts the LLI features at each of its 10 convolutional layers through a special feature extraction module, and then enhances the features at each layer using an encoder-decoder network. The authors in [21] propose a DeepUPE to perform an image-to-illumination map learning. It consists of an encoder network based on a pre-trained VGG16 model [9], followed by a bilateral grid based up-sampling step to produce the image's full resolution illumination map, which is used to enhance the image based on the Retinex model. The authors in [22] introduce EnlightenGAN, an unsupervised generative adversarial network (GAN) approach based on attention guided U-Net [23] as its generator backbone, in addition to a global relativistic discriminator [24], and a local discriminator to handle spatially varying light conditions in the image. A recent approach in [25] proposes ZeroDCE, a zero reference deep curve estimation model which does not require any paired or unpaired training data. The authors reformulate the LLI enhancement task: from image-to-image learning into image-to-light curve learning. The light enhancement curves are estimated for each pixel using a lightweight deep curve estimation network (DCE-Net) thus resulting in an output with the same size of the input image.

DL LLI enhancement techniques are designed based on carefully curated architectures usually embodied within encoder-decoder networks to reconstruct latent features back to the image domain. In contrast, we introduce a novel LLI enhancement solution which performs image-to-frequency filter learning using Homomorphic Filtering (HF), focusing solely on the enhancement task independently of any specific DL architecture.

3   Preliminaries on Homomorphic Filtering

LLI enhancement models based on HF adopt the Retinex model representation of an image as a combination of illumination and reflective components. HF aims at converting the illumination and reflectance components which combine multiplicatively, into an additive form in the logarithmic domain [26]. The additive components are separated linearly in the Fourier transform frequency domain in which high-frequency components are associated with reflectance while low-frequency components correspond to illumination. A high-pass filter is used to suppress low frequencies and amplify high frequencies [26]. Figure 1 depicts the flow of the HF algorithm adopted in our approach.

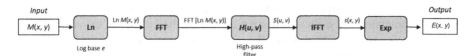

Fig. 1. HF algorithm flow (adapted based on [26]).

Step 0. The algorithm accepts as input an image following the Retinex Model:

$$M(x, y) = I(x, y) \times R(x, y) \tag{1}$$

where $M(x, y)$ is the original image, $I(x, y)$ is the illumination component, and $R(x, y)$ is the reflectance component.

Step 1. The logarithm of both sides of the Retinex model is taken to convert the illumination and reflective components from multiplicative form to additive form:

$$\ln M(x, y) = \ln I(x, y) + \ln R(x, y) \tag{2}$$

Step 2. The fast Fourier transform is applied to convert the image from the spatial domain to the frequency domain:

$$M(u, v) = I(u, v) + R(u, v) \tag{3}$$

where $M(u, v)$, $I(u, v)$ and $R(u, v)$ are the Fourier transforms of $M(x, y)$, $I(x, y)$ and $R(x, y)$ respectively. Note that $I(u, v)$ is mainly concentrated in the low frequency range while $R(u, v)$ is concentrated in the high frequency range.

Step 3. An appropriate high-pass filter with transfer function $H(u, v)$ is applied to perform the enhancement:

$$S(u, v) = H(u, v) \times M(u, v) = H(u, v)I(u, v) + H(u, v)R(u, v) \qquad (4)$$

Step 4. The inverse Fourier transform is applied to transform the image from the frequency domain to the spatial domain. Let $s(x, y)$ be the inverse Fourier transform of $S(u, v)$, then the inverse Fourier transform of Formula 4 becomes:

$$s(x, y) = IFFT(H(u, v)I(u, v)) + IFFT(H(u, v)R(u, v)) = h_I(x, y) + h_R(x, y) \quad (5)$$

Step 5. Finally, the exponential operation is applied on Formula 5 to obtain the enhanced image denoted by $E(x, y)$:

$$E(x, y) = \exp[s(x, y)] = \exp\left[h_I(x, y)\right]\exp[h_R(x, y)] \qquad (6)$$

4   LLI Enhancement Model

We design a new model titled LLHFNet (Low-light Homomorphic Filtering Network) which performs image-to-frequency filter learning instead of the typical image-to-image learning paradigm adopted by most existing solutions. The overall model architecture is depicted in Fig. 2. It is based on HF where a special filter of two parameters is devised to filter the image frequency components in the Fourier transform domain. The two parameters are estimated using a typical DL-based feature extractor utilized in classification models. We describe the main components of our model including: i) enhancement filter design, ii) DL network architecture, and iii) loss function.

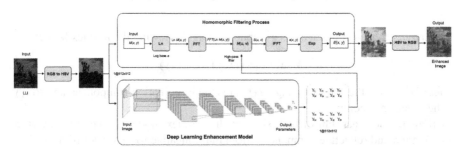

Fig. 2. LLHFNet image enhancement framework.

4.1 Enhancement Filter Design

A core part of the HF algorithm is the frequency filtering transform $H(u, v)$. In our design, we aim to produce a simple and effective filter transform that can be easily learned by the enhancement network. Here, the Fourier transform of the original image, i.e., $M(u, v)$ at $(0,0)$, represents its DC-term[1] which corresponds to its average brightness in the spatial domain [27]. We make two interesting observations: i) $M(0, 0)$ with LLIs is a large negative value reflecting the low brightness of these images, whereas ii) $M(0, 0)$ for NLIs is either a small negative value or a positive value reflecting the normal brightness of these images. Based on the latter observations, we assume that brightness can be enhanced by increasing $M(0, 0)$. We define our enhancement filter:

$$H(u, v) = \begin{cases} \gamma_L & (0, 0) \\ \gamma_H & otherwise \end{cases} \tag{7}$$

where $\gamma_L \in [0, 1]$ denotes the lightness parameter associated with low-frequency components and is placed at $H(0, 0)$, and $\gamma_H \in [0, 1]$ denotes the sharpness parameter associated with the remaining higher-frequency components of $M(u, v)$ corresponding to the image variations. The filter's behavior can be described as follows: i) the smaller (larger) the value of parameter γ_L, the higher (lower) the brightness level of the image, ii) the larger (smaller) the value of γ_H, the sharper (blurrier) the contents of the image.

We run the HF algorithm by applying our enhancement filter on the Value channel of the HSV (Hue-Saturation-Value) color domain, instead of using the Red, Green, and Blue channels of the RGB domain. We make this choice for the following reasons: i) it is more efficient to apply the Fourier transform and its inverse on one channel only instead of three, ii) the Value channel in HSV describes the lightness of the image which we aim to improve; while Hue and Saturation remain unchanged, and iii) HSV allows more simplicity with only two required parameters, compared with the RGB domain which may require two parameters for each of its channels to achieve good quality.

LLI	Enhanced Image ($\gamma_L = 0.35, \gamma_H = 0.45$)

a. LLI with low exposure and the enhanced image

LLI	Enhanced Image ($\gamma_L = 0.60, \gamma_H = 0.70$)

b. LLI with medium exposure and the enhanced image

Fig. 3. LLIs with from the SICE dataset [28] and their enhanced counterparts.

---

[1] The DC-term is the 0 Hz term and is equivalent to the average of all the samples in the sampling window.

Figure 3 provides two examples highlighting the behavior of our enhancement filter with different exposure levels. On the one hand, Fig. 3a presents a LLI with a low exposure level, requiring parameter values <0.5 ($\gamma_L = 0.35$, $\gamma_H = 0.45$, cf. Formula 7) to produce a visually pleasing enhanced image with minimal artifacts. On the other hand, Fig. 3b presents a LLI with a medium exposure level, requiring relatively higher parameter values ($\gamma_L = 0.60$, $\gamma_H = 0.70$) to perform a minimal enhancement while avoiding overexposure. Here, there is a need to identify and fine-tune the parametric values of the filter function in order to maximize image enhancement quality. So, we develop a DL network model which can powerfully and efficiently extract high-level features from input images and allow estimating the values of parameters γ_L and γ_H while handling different input exposure levels.

4.2 Deep Learner Network Architecture

Our DL network architecture is depicted in Fig. 4. It consists of two main parts: i) feature extractor, and ii) enhancement head. The *feature extractor* is responsible for extracting high-level features from the input images. Our solution allows the usage of any feature extractor network (e.g., VGG16 [9], ResNet50 [8], MobileNetv2 [10], SqueezeNet [29], among others) to perform the image-to-filter mapping, which comes down to estimating filter parameters γ_L and γ_H. We modify the first layer of the extractor to accept as input the Value channel of the image represented in the HSV domain.

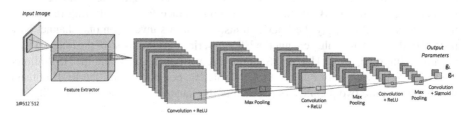

Fig. 4. DL enhancement model network architecture.

The *enhancement head* consists of four convolutional layers followed by ReLU activation and max pooling layers, allowing to downsize the feature maps obtained from the feature extractor. The last convolutional layer is followed by an adaptive average pooling layer to resize the network output to size $1 \times 2 \times 1$, and then a Sigmoid activation function to limit the 2 output values representing γ_L and γ_H to the range [0,1], following our enhancement filter definition described in the previous section.

4.3 Enhancement Loss Function

The loss function is a major element of the LLI enhancement model and drives the entire learning process. In our approach, we adopt a supervised training setting in which reference-based loss functions are needed. We rely on Multi-scale Structural Similarity Index Measure (MS-SSIM) [30] for our loss function. MS-SSIM is an advanced version of SSIM which conducts assessment over multiple scales of the image. SSIM is widely used for image quality assessment as it can capture image contrast, structure, and illumination, e.g., [21, 25, 31], and is adopted as a loss function in many recent studies, e.g., [32–34]. Yet a recent empirical evaluation in [35] shows that quantitative image quality assessment metrics do not always correlate with the human perception of visual quality, due to the disparity between computational enhancement (done by the machine) and enhancement quality (perceived by humans). While the latter miscorrelation is difficult to evaluate through the loss function with existing image-to-image learning models, yet it is easier to monitor with our image-to-filter enhancement model (which seeks to learn two filter parameters only, rather than learning the image as a whole). In this context, a preliminary evaluation of our enhancement model shows two contradictory observations. On the one hand, an MS-SSIM based loss function may show a tendency to generate values for the lightness parameter γ_L which are greater than those of the sharpness parameter γ_H. This tends to produce enhanced images which are smoothed with distinctive color deviations, making them perceptually unpleasing. On the other hand, this tendency is encouraged by lower MS-SSIM loss values indicating that the metric is failing to properly quantify the quality of these enhanced images. To minimize the impact of this miscorrelation between qualitative perception and quantitative measure, we add a regularization term to the loss function, encouraging the learner model to generate values for γ_H which are greater than γ_L while reducing the overall loss value. More formally:

$$\text{enhLoss}(I_{\text{Enhanced}}, I_{\text{NLI}}) = 1 - \text{MS\_SSIM}(I_{\text{Enhanced}}, I_{\text{NLI}}) + \alpha \times \ell \quad (8)$$

where *enhLoss* designates the enhancement loss function, $I_{Enhanced}$ is the enhanced image, I_{NLI} is the normal light image, $\ell = \gamma_L - \gamma_H$ is the regularization term, $\alpha >= 0$ is a linear weight parameter highlighting the impact of regularization on overall loss. Our empirical evaluation shows that values of α ranging between [0.05, 0.1] produce satisfactory LLI enhancement results (in our experiments, we use $\alpha = 0.08$).

5   Experimental Evaluation

We perform an image quality assessment that aims at evaluating whether an image is visually pleasing and how it is visually perceived. We conduct both quantitative and qualitative evaluations, by evaluating the visual quality achieved by 5 prominent enhancement models (2 traditional solutions: SRIE [17] and LIME [15], and 3 DL-based solutions: ZeroDCE[2] [25], EnlightenGAN[3] [22] and DeepUPE[4] [21]). We compare the models with LLHFNet[5] implemented using PyTorch on a P100 Tesla Nvidia GPU, with a batch size of 8. We utilize an Adam optimizer with default parameters and a reduce-on-plateau decay-learning rate with an initial value of 1e − 4 for network optimization. Our prototype implementation and experimental data are available online (See Footnote 5).

---

[2] https://github.com/Li-Chongyi/Zero-DCE.
[3] https://github.com/TAMU-VITA/EnlightenGAN.
[4] https://github.com/wangruixing/DeepUPE.
[5] https://github.com/rayanalsubbahi/LLHFNet.

5.1 Experimental Data

We use the well-known SICE dataset [28] to conduct our training and testing experiments. We adopt two subsets for: i) training and ii) testing. The training subset consists of 2,150 image pairs from Part 1 of SICE, excluding extremely underexposed and overexposed images (which are difficult to handle and may tend to disrupt the training process). We resize all the training images to 512×512, and perform cross validation where 1700 pairs (i.e., 80%) are used for model learning and 450 pairs (i.e., 20%) are used for model evaluation. Although the training dataset seems relatively small, yet our enhancement model does not require huge training data since it relies on powerful pre-trained feature extractors for its backbone. In this experiment, we utilize five pre-trained extractors including VGG16 [9], ResNet50 [8], MobileNetv2 [10], SqueezeNet [29] and DenseNet [36]. The testing subset consists of 767 paired LLIs/NLIs collected from Part 2 of the SICE dataset [28] and resized to $1200 \times 900 \times 3$ following the same approach adopted in [25] to perform our empirical evaluations. We additionally employ 3,000 images for testing from the well-known Pascal VOC 2007 dataset [37] and synthetically generate LLIs considering five different exposure levels using gamma correction with gamma values {4.5, 3.5, 2.5} corresponding to low-exposure levels, and gamma values {0.5, 0.8} corresponding to high-exposure levels. The subset is divided equally among the used (Υ corrected) exposure levels and all images are resized to 512×512.

5.2 Quantitative Evaluation Results

We run the enhancement models against three objective metrics commonly used in the literature: Structural Similarity index (SSIM) [38], Peak Signal to Noise Ratio (PSNR), and Mean Absolute Error (MAE). Table 1 shows quantitative IQA results comparing LLHFNet with its prominent counterparts, applied on SICE and Pascal VOC 2007 testing subsets. Our solution produces the best PSNR and MAE average scores, and the second best scores following SSIM on SICE while it ranks as the best following all three metrics on Pascal VOC 2007. Table 2 provides the average scores of LLHFNet using different feature extractors, including: ResNet50, MobileNetv2, VGG16, DenseNet, and SqueezeNet. MobileNetv2 and VGG16 produce some of the best average scores across all evaluation metrics using SICE subset. This is probably due to their dense architectures. ResNet50 shows second best result for all metrics on Pascal VOC 2007 indicating the effectiveness of the residual architecture in learning the filter parameters. SqueezeNet produces the worst results across all evaluation metrics, which is probably due to its lightweight architecture. Yet all LLHFNet variants show consistently competitive results when compared with the enhancement solutions in Table 1.

Table 1. Comparing the quality of existing LLI enhancement models. LLHFNet uses MobileNetv2 [10] as its feature extractor.

	a. SICE dataset			**b. Pascal VOC dataset**		
Model	SSIM ↑	PSNR ↑	MAE ↓	SSIM ↑	PSNR ↑	MAE ↓
LLHFNet	**0.58**	**16.89**	**94.99**	**0.734**	**15.77**	**117.07**
ZeroDCE	**0.59**	16.57	98.78	0.67	14.96	139.05
EnlightenGAN	**0.59**	16.21	102.78	0.6284	13.63	152.32
DeepUPE	0.49	13.52	142.01	**0.730**	14.30	143.89
LIME	0.57	16.17	108.12	0.6286	13.33	159.68
SRIE	0.54	14.41	127.08	0.629	13.50	154.69

Table 2. Comparing different feature extractors used with LLHFNet.

	a. SICE dataset			**b. Pascal VOC dataset**		
Feature Extractor	SSIM ↑	PSNR ↑	MAE ↓	SSIM ↑	PSNR ↑	MAE ↓
MobileNetv2	**0.583**	16.896	94.992	**0.734**	**15.775**	**117.078**
VGG16	**0.582**	**16.897**	**94.064**	0.728	15.590	121.214
ResNet50	0.577	16.686	96.152	**0.731**	**15.696**	**119.922**
DenseNet	0.576	16.716	97.253	**0.730**	15.563	120.855
SqueezeNet	0.575	16.593	99.129	0.726	15.442	122.063

Results in this experiment show that LLHFNet can be effectively used with different feature extractors, making it independent of any specific architecture.

5.3 Qualitative Evaluation Results

In addition to the quantitative evaluation, we also perform a qualitative evaluation to assess the human visual perception of images enhanced by our model and its five counterparts considered in this experiment. We randomly select 20 images from the SICE testing subset, and display the reference input LLI and the enhanced image side by side in a dedicated online survey[6]. Responders are asked to rate each image considering three visual IQA criteria including: i) level of exposure (over/under-exposed regions), ii) color deviations, and iii) overall beauty of the image. A total of 76 testers (senior computer engineering and master's students) were invited to contribute in the experiment, and independently rate every enhancement model on an integer scale from 1 to 10 (i.e., worst to best). We also deal with inconsistencies in image ratings by computing the average score for every image, and then eliminating ratings which have an extreme deviation from the average (e.g., ratings which are extremely low/high for images deemed visually pleasing/unpleasing). A total of at least 1200 responses were collected for each model, with every image receiving 60 rating scores. The ratings are aggregated for every enhancement model to evaluate its overall perceptual quality. Results are provided in Fig. 5, and sample LLIs and enhanced images are shown in Fig. 6.

---

[6] https://forms.gle/FrjzGAZXpyKqGRnw9

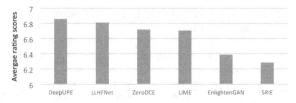

Fig. 5. Average user ratings for the enhancement models ranked from best to worst.

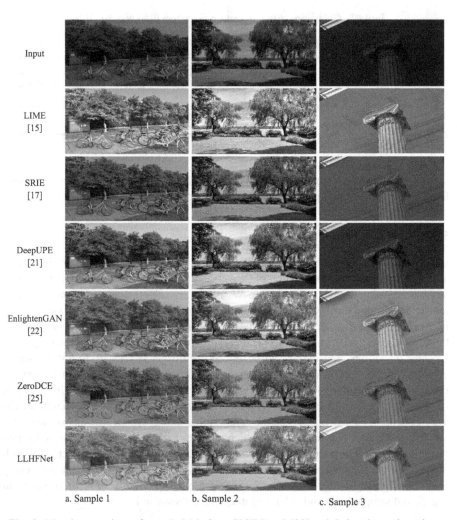

a. Sample 1   b. Sample 2   c. Sample 3

Fig. 6. Visual comparison of sample LLIs from SICE Part 2 [28] and their enhanced versions.

Results in Fig. 5 show that LLHFNet ranks second best among the five compared models, and is thus favored by human testers. Sample LLIs in Fig. 6 show that LLHFNet produces visually pleasing enhanced images with minimal artifacts. In the first image (Fig. 6a), our model is able to uncover the dark regions of the fence and is able to effectively restore the green colors of the trees. In the second image (Fig. 6b), our model properly restores the colors of the trees, grass, and white clouds without overexposing them (compared with EnlightenGAN where the clouds are overexposed, and ZeroDCE and SRIE where the cloud colors and overall image colors deviate into blue). In the third image (Fig. 6c), our model shows a good illumination level and produces results comparable with to ZeroDCE and SRIE. The reader can refer to [39] for a detailed description of the experimental results, as well as the whole framework.

6   Conclusion

In this paper, we introduce a new LLI enhancement solution titled LLHFNet (Low-light Homomorphic Filtering Network) based on image-to-frequency filter learning. The network is designed independently from a custom architecture and can use many feature extractors commonly adopted in object classification. Experimental results show improved enhancement quality on LLIs, and is ranked among the best enhancement models compared with recent solutions. We are currently conducting an empirical study to evaluate the performance of our solution on extremely LLIs. In the near future, we aim to integrate and evaluate our enhancement model in high-level tasks like object detection [40], image semantization [41], localization and tracking [42, 43], and multi-label image recognition [44].

References

1. Yang, W., et al.: Advancing image understanding in poor visibility environments: a collective benchmark study. IEEE Trans. Image Process. (IEEE TIP) **29**, 5737–5752 (2020)
2. Scheirer, W., et al.: Bridging the gap between computational photography and visual recognition. IEEE TPAMI, p. 1 (2020)
3. Zhi, N., et al.: An enhancement algorithm for coal mine low illumination images based on Bi-gamma function. J. Liaoning Tech. Univ. **37**(1), 191–197 (2018)
4. Kansal, S., et al.: Image contrast enhancement using unsharp masking and histogram equalization. Multim. Tools Appl. **77**(20), 26919–26938 (2018)
5. Ren, X., et al.: LR3M: robust low-light enhancement via low-rank regularized retinex model. IEEE Trans. Image Process. **29**, 5862–5876 (2020)
6. Lore, K.G., et al.: LLNet: a deep autoencoder approach to natural LLI enhancement. Pattern Recognit. **61**, 650–662 (2017)
7. Li, C., et al.: LightenNet: a convolutional neural network for weakly illuminated image enhancement. Pattern Recognit. Lett. **104**, 15–22 (2018)
8. He, K., et al.: Deep residual learning for image recognition. IEEE CVPR Conference, pp. 770–778. IEEE (2016)
9. Simonyan, K., et al.: Very deep convolutional networks for large-scale image recognition. arXiv preprint arXiv:1409.1556 (2015)
10. Sandler, M., et al.: MobileNetV2: inverted residuals and linear bottlenecks. In: IEEE CVPR, pp. 4510–4520 (2018)

11. Zhang, C., et al.: Color image enhancement based on local spatial homomorphic filtering and gradient domain variance guided image filtering. J. Electron. Imaging. **27**(6), 1 (2018)
12. Han, L., et al.: Using HSV space real-color image enhanced by homomorphic filter in two channels. Comput. Eng. Appl. **45**(27), 18–20 (2009)
13. Al-Wadud, A., et al.: A dynamic histogram equalization for image contrast enhancement. IEEE Trans. Consum. Electron. **53**(2), 593–600 (2007)
14. Gu, K.: Automatic contrast enhancement technology with saliency preservation. IEEE TCSVT **25**(9), 1480–1494 (2015)
15. Li, L., et al.: A low-light image enhancement method for both denoising and contrast enlarging. In: IEEE International Conference on Image Processing (ICIP), pp. 3730–3734. IEEE (2015)
16. Li, M., et al.: Structure-revealing low-light image enhancement via robust retinex model. IEEE Trans. Image Process. **27**(6), 2828–2841 (2018)
17. Fu, X., et al.: A weighted variational model for simultaneous reflectance and illumination estimation. In: 2016 IEEE Conference on Computer Vision and Pattern Recognition, pp. 2782–2790 (2016)
18. Wei, C., et al.: Deep retinex decomposition for low-light enhancement. In: International BMVC Conference, pp. 1–12 (2018)
19. Wang, W., et al.: GLADNet: low-light enhancement network with global awareness. In: IEEE Automatic Face and Gesture Recognition, pp. 751–755. IEEE, New York (2018)
20. Lv, F., et al.: MBLLEN: low-light image/video enhancement using CNNs. In: Proceedings of the British Machine Vision Conference, pp. 1–13 (2018)
21. Wang, R., et al.: Underexposed photo enhancement using deep illumination estimation. In: IEEE CVPR Conference, pp. 6842–6850 (2019)
22. Jiang, Y., et al.: EnlightenGAN: deep light enhancement without paired supervision. IEEE TIP **30**, 2340–2349 (2021)
23. Ronneberger, O., Fischer, P., Brox, T.: U-Net: convolutional networks for biomedical image segmentation. In: Navab, N., Hornegger, J., Wells, W.M., Frangi, A.F. (eds.) MICCAI 2015. LNCS, vol. 9351, pp. 234–241. Springer, Cham (2015). https://doi.org/10.1007/978-3-319-24574-4_28
24. Jolicoeur-Martineau, A.: The Relativistic Discriminator: A Key Element Missing from Standard GAN. arXiv preprint arXiv:1807.00734 (2018)
25. Guo, C., et al.: Zero-reference deep curve estimation for low-light image enhancement. In: IEEE CVPR, pp. 1777–1786 (2020)
26. Wang, W., et al.: An experiment-based review of LLI enhancement methods. IEEE Access **8**, 87884–87917 (2020)
27. Gonzalez, C., Woods, E.: Digital Image Processing, 4th edn. Pearson, New York (2018)
28. Cai, J., et al.: Learning a deep single image contrast enhancer from multi-exposure images. IEEE TIP **27**(4), 2049–2062 (2018)
29. Iandola, F.N.: Squeezenet: Alexnet-level Accuracy with 50x Fewer Parameters and <0.5 mb Model Size. arXiv preprint arXiv:1602.07360 (2016)
30. Wang, Z., et al.: Multiscale structural similarity for image quality assessment. In: 37th Asilomar Conference on Signals, Systems, and Computers, pp. 1398–1402. IEEE (2003)
31. Ren, W., et al.: Low-light image enhancement via a deep hybrid network. IEEE Trans. Image Process. **28**, 4364–4375 (2019)
32. Xiang, Y., Fu, Y., Zhang, L., Huang, H.: An effective network with ConvLSTM for low-light image enhancement. In: Lin, Z., et al. (eds.) PRCV 2019. LNCS, vol. 11858, pp. 221–233. Springer, Cham (2019). https://doi.org/10.1007/978-3-030-31723-2_19
33. Zhang, Y.: Kindling the darkness: a practical low-light image enhancer. In: ACM MM Conference, pp. 1632–1640 (2019)

34. Shi, Y., et al.: Low-light Image Enhancement Algorithm Based on Retinex and Generative Adversarial Network. Computing Research Repository. arXiv preprint arXiv:1906.06027 (2019)
35. Al Sobbahi, R., Tekli, J.: Comparing deep learning models for low-light image enhancement and their impact on object detection and classification. Signal Process. Image Comm. J. (2022)
36. Huang, G., et al.: Densely connected convolutional networks. In: IEEE CVPR Conference, pp. 4700–4708. IEEE (2017)
37. Everingham, M., et al.: Pascal visual object classes (VOC) challenge. Int. J. Comput. Vis. **88**(2), 303–338 (2010)
38. Wang, Z., et al.: Image quality assessment: from error visibility to structural similarity. IEEE Trans. Image Process. **13**(4), 600–612 (2004)
39. Al Sobbahi, R., Tekli, J.: Low-light homomorphic filtering network for integrating image enhancement and classification. Signal Process. Image Comm. **100**, 116527 (2022)
40. Salem, C., et al.: An image processing and genetic algorithm-based approach for the detection of melanoma in patients. Method. Inf. Med. **57**(1), 74–80 (2018)
41. Salameh, K., Tekli, J., Chbeir, R.: SVG-to-RDF image semantization. In: Traina, A.J.M., Traina, C., Cordeiro, R.L.F. (eds.) SISAP 2014. LNCS, vol. 8821, pp. 214–228. Springer, Cham (2014). https://doi.org/10.1007/978-3-319-11988-5_20
42. Ebrahimi, D., et al.: Autonomous UAV trajectory for localizing ground objects: a reinforcement learning approach. IEEE Trans. Mobile Comput. **20**(4), 1312–1324 (2021)
43. Samir, M., et al.: Age of information aware trajectory planning of UAVs in intelligent transportation systems: a deep learning approach. IEEE Trans. Veh. Technol. **69**(11), 12382–12395 (2020)
44. Laib, L., et al.: A probabilistic topic model for event-based image classification and multi-label annotation. Signal Process. Image Comm. **76**, 283–294 (2019)

Synthetic Data of Randomly Piled, Similar Objects for Deep Learning-Based Object Detection

Janis Arents[1]([✉])[iD], Bernd Lesser[2][iD], Andis Bizuns[1][iD], Roberts Kadikis[1][iD],
Elvijs Buls[1][iD], and Modris Greitans[1][iD]

[1] Institute of Electronics and Computer Science, 14 Dzerbenes St., Riga 1006, Latvia
{janis.arents,andis.bizuns,roberts.kadikis,elvijs.buls,
modris_greitans}@edi.lv
[2] Virtual Vehicle Research GmbH, Inffeldgasse 21a, 8010 Graz, Austria
bernd.lesser@v2c2.at
http://www.edi.lv/en, https://www.v2c2.at

Abstract. Currently, the best object detection results are achieved by supervised deep learning methods, however, these methods depend on annotated training data. With the synthetic data generation approach, we intend to mimic the real data characteristics and diversify the dataset by a systematic rendering of highly realistic synthetic pictures. We systematically explore how different combinations and portions of real and synthetic datasets affect object detectors performance. The developed synthetic data generation framework shows promising results in deep learning-based object detection tasks and can supplement real data when the variety of real training data is insufficient. However, when synthetic data ratio increases over real data ratio, a decrease in average precision can be observed, which has the most affect on 0.75-0.95 IoU threshold range.

Keywords: Synthetic data · Deep learning · Object detection

1 Introduction

Even though we have almost spent a decade since the term Industry 4.0 was introduced [16], industrial robots by themselves still mostly have limited intelligence. One of the most popular ways to make robots smarter is by attaching cameras [4] that can acquire information on a specific region of interest or perceive the environment around the robot and adjust their movements respectively [19]. Therefore, by giving the robots an ability to *see*, comprehend and act accordingly, we can automate tasks that traditionally require human intelligence or complex and very spacious machines [5].

Specific computer vision problems - object detection and instance segmentation - are two of the main prerequisites to automate a number of tasks where industrial robots must proceed in an unstructured environment. The detection indicates where in the camera's frame an object is located, and which class does this object belong to. Whereas segmentation determines which class does every pixel of an image belongs to. Instance segmentation is a type of segmentation that differentiates among pixels belonging to different instances of the same class. With this information, one can acquire a visible shape of a specific object and use it to determine objects pose [12], which in turn is handy for picking up and manipulating the object. However, the segmentation task typically is computationally more expensive than object detection or classification tasks [9], therefore in the case of randomly piled objects it is rather inefficient to segment all of the objects in the region of interest. In this article, we focus on detecting objects that have the highest possibility to result in a successful grasp.

Object detection is hard in the case of randomly piled objects. The objects are often only partially visible, and when the pile consists of similar or even the same objects, the similar features that could be used to detect the unobstructed objects are scattered all over the pile. There are still challenges to retaining low false-positive rates in cluttered environments [14]. In such environments robotic grasping is hardly competing with human performance, therefore many manipulations still require manual work or rather big and expensive machines that are hard to adjust if product assortment changes.

Currently, in case of known and finite amount of instances, the best object detection results are achieved by supervised deep learning methods, for example EfficientDet [25] and HRNet-OCR [29]. However, these methods depend on annotated training data. Each new object requires numerous new training examples of pile images, and labelling of such images is a time consuming manual labour, especially in the case of image segmentation tasks. To alleviate the training data acquisition process and simplify the use of modern computer vision methods in industry, we turn to data synthesis.

Image synthesis or rendering is a process of generating digital images from virtual scenes. The photo-realism of rendered images, videos, and computer games keeps increasing. Also, the tools for creating virtual environments with included physics simulation are becoming more user-friendly and affordable. For example, such tools as Blender, Unity, and Unreal Engine can be used free of charge. Therefore AI and computer vision research community increasingly use such tools to generate data, train systems in virtual environments directly, and to adjust and develop more vision-oriented tools, for example, an open-source plugin UnrealCV [20].

The main goal of this article is to explore and develop a synthetic data generation framework for object detection tasks. We compare manually labelled data with synthetically generated data and analyze how does the real data, synthetic data and different combinations of both affect the precision of object detection models. Taking into account the above mentioned the rest of the paper is structured as follows: Sect. 2 describes state-of-the-art developments in the fields of object detection and data synthesis. Section 3 describes our proposed method. In Sect. 4 several different tests are performed and the obtained results are analyzed. Section 5 concludes this article.

2   Related work

Object detection in images has improved significantly in the last years [31]. Current best detectors are based on deep neural networks and trained in a supervised fashion. Notable detectors include types of Region-Based Convolutional Neural Networks (Faster R-CNN [24], Mask R-CNN [13], Beta R-CNN [28]). Even though the mean average precision (mAP) is an important aspect to compare between object detectors, other aspects such as computational efficiency, memory consumption and inference time also play an important role in selecting an appropriate object detector for the intended use-case. In this sense, the YOLO [22,23] variants provide better leverage between the precision and speed when compared to others [31].

Detection and segmentation of objects that are randomly piled combined with an industrial robot pick and place operation [3] are commonly referred to as bin-picking. Even though this problem has been addressed by the research community for several decades, it remains one of the most challenging tasks in robotics [2,6]. Multiple instances of the same type of objects that are randomly stacked in a pile introduce a variety of challenging conditions, such as the similarity between foreground objects and the background, occlusions, and clutter. That in a combination with sensor noise contribute to the complexity of object detection [4].

Most of the research is focused on applications that have publicly available large data-sets on commonly used objects, whereas in industrial applications object types can be specific to the respective product. Model-free grasping techniques partly addresses data absence [17] and can proceed without having prior knowledge of the objects, but this method complicates the post-gripping [33] process and can introduce additional steps for precise positioning. The latest achievements in the field of computer-generated imagery widen opportunities in synthetic real-life like data generation for object detection tasks in this particular scenario.

Synthetic data for object detection tasks typically has been composed by placing foreground objects on background scenes with different parameters that can be varied. Some approaches proceed with 2D images that are placed on a set of background images [7,11], with set of rules or physical simulation for piling the objects realistically. The level of complexity for such and similar methods is lower, however they are typically restricted to 2D nature that contributes to the lack of realism, which therefore decreases detectors performance. 3D graphical engines have also been utilized in synthetic data generation, however, these are typically dedicated for household situations or usage with every day objects [26,27].

A comprehensive overview of the existing state of the art datasets relevant for object pose estimation for industrial bin-picking is given in [15]. Even though we don't address pose estimation task in this article, this overview gives a valuable insights in the current progress and the applicability of the available data sets. It shows that most previous approaches are not well suited to be directly used for deep learning methods in bin-picking task due to either a lack in the amount of data, its variety or incomplete ground truth information.

Even though modern object detection methods are effective at detecting of traffic signs [32], pedestrians [10], household objects [8] etc., the methods fall short in industrial settings due to lack of training data. Our proposed method utilizes modern graphics engines to generate realistic data for training neural networks for robotic grasping in industrial specific scenarios, where data can be use-case specific and change over time. We focus on generating versatile data by a vast amount of adjustable parameters. Thus the parameters can be adjusted by specific needs. In this case, we extract the data of unobstructed objects in corresponding orientation to train AI models only on the objects that have the highest probability to result in a successful grasp when the model is deployed. The object detection described in this article serves as the first step in the whole pick and place process, and only information about the most promising objects that could be grasped will be processed further.

With respect to the synthetically generated data, in this work, we create a new dataset fully suitable for deep learning by the means of the amount of data and sufficient ground truth annotations. Although we currently feature only one 3D model, we aim at synthesizing photo-realistic images by means of higher resolutions, materials, textures, lighting, reflections and indirect light bounces instead to reduce the reality gap.

3   Proposed Method

3.1   Synthetic Generation of Realistic Pile Images and Annotations

We aim at automating the systematic rendering of highly realistic synthetic pictures to generate data sets for training the object detection algorithm. The image generator obtains images by arranging any kind of objects that have 3D models within a virtual 3D scene from which it renders highly realistic images. In this case, it is a box with white bottles. By tuning parameters of the 3D scene such as an object, camera and light positions, object colour or texture and surface properties, brightness, contrast, and saturation, a large image diversity can be generated in resolution and levels of realism depending on the user needs. By further exporting relevant ground truth data from the 3D scene including the location and orientation of objects within a rendered image, the generated data is fully annotated.

By sampling all possible configurations of objects and image parameters within the 3D scene (in arbitrary, user-defined granularity), a modification space is defined allowing for the automated generation of large synthetic data sets, for which the diversity of images is controlled by the user. By systematically defining appropriate scenes and modification spaces, the image generator can be used to generate not only training and validation data sets in sufficient quantity (overcoming a lack of training data, which is often a limiting problem), but in general, also allows for generating data sets specifically designed for specific experiments

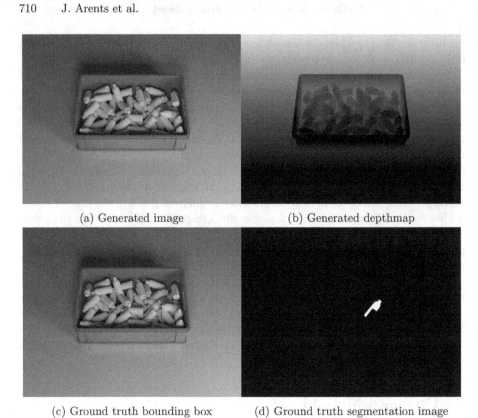

(a) Generated image   (b) Generated depthmap

(c) Ground truth bounding box   (d) Ground truth segmentation image

Fig. 1. Synthetic data rendered with Blender

Fig. 2. Different perspectives and light conditions

later on (e.g. to compare across different classification models or to extract specific insights of the algorithm).

In this work, we focus on automating the generation of realistic data sets for training our object detection algorithm, i.e. highly realistic images of piled objects together with their corresponding depth-map, for which we annotate every single object visible in the scene with location and orientation ground truth. Corresponding instance-segmented images are also generated as depicted in Fig. 1d. For image generation, we use the free open-source tool, for which we fully automated the rendering of parametrizable 3D scenes (as they have been defined within the scope of this work) in *python* by making use of blender's python API.

We implemented a basic set of functions in python for setting up virtual 3D scenes and controlling the image rendering process, where we kept the configuration of objects and image parameters for the defined configuration space as variables. These functions have been implemented in such a way, that the whole process of image generation iterates across all possible configurations of varying image parameters within the given configuration space (i.e. the placement of the 3D objects into different positions of the 3D scene, applying different textures and shaders to the objects, different positioning of light sources and the camera, running the rendering process) is being automated and executed in headless mode and no manual interaction via Blender's GUI is required.

To additionally allow for a more realistic object placement within the considered 3D scenes, especially for piling up any number of objects more realistically and naturally, we further make use of Blender's integrated physics engine, which enables us to optionally apply simple rigid body simulations based on convex hull collision detection. For postprocessing and annotation of images, we further make use of OpenCV, which we make available to blender's internal python interpreter via pip.

3.2 Training Setup

A wide variety of machine learning algorithms are currently available to detect objects in images. At the beginning of the study, YOLOv3 [23] was one of the most popular real-time, single-stage object detection algorithms with the best AP scores and FPS trade-offs [31]. In addition, YOLOv3 can be used with a variety of frameworks [1,30] and has a large community to discuss important issues. We used YOLOv3 for all following experiments

To train and evaluate the YOLOv3 models, we used the Darknet [1] framework and already pre-trained convolutional weights [21] *darknet53.conv.74* on Imagenet. This framework is a branch of the original Darknet open-source framework [21] with various improvements such as anchor calculation or charts. Darknet framework is implemented using the *c++* programming language and is efficient at training other popular neural networks as well. Training and evaluation was performed on two separate *Linux* OS workstations with *Nvidias RTX 3090* and *A100* GPUs.

4   Tests and Results

The performed tests are structured in a way to evaluate how object detector trained on synthetic datasets performs in real-life scenarios compared to object detector trained on real datasets. Even though the main focus is concentrated on synthetic data usage, we also systematically explore how different combinations and portions of real and synthetic datasets affect object detectors performance. In respect to the use-case, the main goal is to detect objects that are most likely to result in a successful grasp, therefore we also analyze in how many scenes at least one object has been found above a certain Intersection over Union (IoU) threshold.

4.1   Datasets

Real Datasets. The real data was gathered by randomly distributing bottles in a box. For each of the acquired images, the positions of the bottles were altered. The intensity of lightning and camera exposure time was systematically modified to acquire a high diversity of different lighting conditions. In total for training purposes, 2200 real images were acquired and manually labelled from which 1760 images for training and 440 images for validation purposes. Furthermore, the training and validation datasets were rotated four times by 90°, in total resulting in 7040 training images and 1760 validation images.

Two test datasets were gathered and manually labelled, first dataset *Test 1* was captured in similar conditions as the real training dataset, however *Test 2* was captured with different camera and in different conditions.

Synthetic Datasets. For the experiments considered in this work, we have generated a synthetic dataset in the same amount as the real dataset, consisting of 8800 photo-realistic high-resolution scenes. For every individual scene we fill an initially empty box with randomly placed bottles by making use of Blender's physics simulation engine to achieve realistic positioning and orientation. We use realistic textures and Blender's principled BSDF shader nodes to achieve realistic renderings of the scenes including lighting, reflections and indirect light bounces.

After filling the box with the bottles, we create a series of renderings for which we vary power levels of 4 different light sources and the orientation of the camera, which orbits around the box and renders the scene from 16 different angles as depicted in Fig. 2. For every camera orientation, we also generate a depth image and the segmentation images of the individual bottles as seen by the camera and labelled by the object ID as illustrated in Fig. 1. We further generate a separate annotation file for every camera perspective which contains the individual object's orientation and rotation in-camera coordinates together with the object's visibility percentage.

4.2 Evaluation Metrics

The most common metric used to measure the accuracy of the object detection in the images is average precision (AP) [18], which is also utilized in this article to evaluate the performance of object detection in our experiments. In our case the Darknet framework with variously set (0.5–0.95) IoU thresholds is used to perform AP measurements. With the IoU we measure the overlapping area between the ground truth and the predicted bounding box and evaluate the precision over different thresholds. True positive (TP), false positive (FP), and false negative (FN) estimates for each detected object are used to calculate precision and recall to perform AP measurements.

4.3 Comparisons

The performance and precision of the proposed synthetic data generation approach were evaluated by various aspects. First, the object detection performance was investigated with deep learning models trained using different ratios of synthetic and real data combinations and by utilizing the maximum amounts of the acquired data. Starting with 100% of real data, the real data ratio was incrementally decreased by 10% at the time by substituting it with the synthetic data as depicted in Table 1.

Table 1. Evaluation of object detectors precision

Data distribution			Test 1			Test 2		
Real	Synthetic	Real/Synthetic Ratio %	Step	AP @0.5:0.95	OD %	Step	AP @0.5:0.95	OD %
8800	0	100/0	9100	88.61	100.00	9100	69.22	96.95
7920	880	90/10	7900	88.61	100.00	9200	71.04	98.47
7040	1760	80/20	6000	88.36	100.00	8000	73.23	100.00
6160	2640	70/30	6300	**88.65**	100.00	7600	72.83	100.00
5280	3520	40/60	6900	88.22	100.00	7400	72.50	100.00
4400	4400	50/50	4400	85.82	100.00	5000	**73.84**	100.00
3520	5280	40/60	8000	85.59	100.00	8700	70.27	100.00
2640	6160	30/70	7200	84.39	100.00	5900	64.57	100.00
1760	7040	20/80	7200	84.23	100.00	4500	63.62	100.00
880	7920	10/90	8200	82.59	100.00	7000	63.25	100.00
0	8800	0/100	7700	70.01	100.00	5000	38.66	100.00

The evaluation was performed on the described datasets - *Test 1* and *Test 2*. Object detector evaluated on data close to real training data *(Test 1)* scored similar average precision results when real data amount was higher than synthetic data as depicted in Fig. 3a. This also holds true to higher IoU threshold values from 0.85 to 0.95. All of the trained models showed similar average precision results in the IoU threshold region from 0.5–0.8. The main difference can be seen in the case when the model is trained on purely synthetic data, as the precision remarkably decreases.

A different situation can be seen when the object detector is evaluated on a test dataset that contains different environmental parameters - *Test 2*. In this case the synthetic data supplements real data and increases average precision, whilst achieving the highest precision on a 50/50 ratio. Similarly as with the evaluation results on *Test 1* dataset, also in this case object detector trained on purely synthetic datasets showed the least precision.

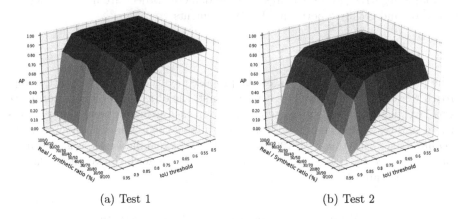

(a) Test 1   (b) Test 2

Fig. 3. Average precision of object detection models on real images over different IoU thresholds, viewed by the ratio of real to synthetic data in the training datesets

Even though the object detector trained on real data or different combinations outperforms the detector trained on purely synthetic data, the main precision aspects in this article are connected to the specific use-case. Respectively, the goal is to detect at least one object in the scene with an IoU threshold above 0.95. Obtained results on this aspect are depicted in Table 1 under columns Object Detected (OD). On both test datasets, the trained models could meet this requirement, except in Test 2 case, when the model was trained on purely real data and in following 90/10 ratio.

5   Conclusion

In dynamic environments, especially in the case of randomly piled objects, a lot of uncertainties and different environmental conditions can be present. Ideally, these different conditions should be covered by the training data set in a deep learning-based object detection task to satisfy the precision requirements. However, gathering and labelling the real data is a tedious task and requires a certain amount of human resources and in some cases, it is complicated to recreate all the possible configurations. With the synthetic data generation approach, we intend to mimic the real data characteristics and diversify the dataset by a systematic rendering of highly realistic synthetic pictures. By tuning image parameters such as an object, camera and light positions, object colour or texture and surface properties, brightness, contrast, saturation, a large image diversity can be generated in resolution and level of realism depending on the requirements. In this article we explore the usage of synthetic data in this particular scenario with one object type, however, the generator can arrange any kind of objects that have 3D models.

The generated dataset, real dataset and different combinations of both were used to train an object detector. The trained models were evaluated on two different test datasets. In most of the cases when the real data ratio was higher than synthetic data, the model achieved higher precision ratings. Even though the models trained on purely synthetic images has lower average precision on real images, the achieved precision is sufficient in our use-case. Thus, by diversifying the training dataset with synthetic images a precision increase can be observed, however when the synthetic ratio is over 50%, the precision decreases.

The developed synthetic data generation framework shows promising results in deep learning-based object detection tasks and can supplement real data when the variety of real training data is insufficient. However, adding the synthetic data to real data requires testing to find the peak of precision, as adding more synthetic images results in lower precision. In respect to future work, the synthetic data generator will be further improved and utilized in object segmentation tasks.

Acknowledgements. The work has been performed in the project AI4DI: Artificial Intelligence for Digitizing Industry, under grant agreement No 826060. The project is co-funded by grants from Germany, Austria, Finland, France, Norway, Latvia, Belgium, Italy, Switzerland, and Czech Republic and - Electronic Component Systems for European Leadership Joint Undertaking (ECSEL JU). In Austria the project was also funded by the program "IKT der Zukunft" of the Austrian Federal Ministry for Climate Action (BMK). Parts of the publication was written at Virtual Vehicle Research GmbH in Graz and partially funded within the COMET K2 Competence Centers for Excellent Technologies from the Austrian Federal Ministry for Climate Action (BMK), the Austrian Federal Ministry for Digital and Economic Affairs (BMDW), the Province of Styria (Dept. 12) and the Styrian Business Promotion Agency (SFG). The Austrian Research Promotion Agency (FFG) has been authorised for the programme management.

References

1. AlexeyAB: darknet. https://github.com/AlexeyAB/darknet. Accessed 20 Dec 2021
2. Alonso, M., Izaguirre, A., Graña, M.: Current research trends in robot grasping and bin picking. In: Graña, M., López-Guede, J.M., Etxaniz, O., Herrero, Á., Sáez, J.A., Quintián, H., Corchado, E. (eds.) SOCO'18-CISIS'18-ICEUTE'18 2018. AISC, vol. 771, pp. 367–376. Springer, Cham (2019). https://doi.org/10.1007/978-3-319-94120-2_35

3. Arents, J., Cacurs, R., Greitans, M.: Integration of computervision and artificial intelligence subsystems with robot operating system based motion planning for industrial robots. Autom. Control Comput. Sci. **52**(5), 392–401 (2018)

4. Arents, J., Greitans, M.: Smart industrial robot control trends, challenges and opportunities within manufacturing. Appl. Sci. **12**(2), 937 (2022). https://doi.org/10.3390/app12020937

5. Arents, J., Greitans, M., Lesser, B.: Artificial Intelligence for Digitising Industry, Applications, chap. Construction of a Smart Vision-Guided Robot System for Manipulation in a Dynamic Environment, pp. 205–220. https://www.riverpublishers.com/book_details.php?book_id=967, https://doi.org/10.13052/rp-9788770226639 (2021)

6. Buchholz, D.: Bin-Picking - New Approaches for a Classical Problem. Ph.D. thesis, (Jul 2015), https://publikationsserver.tu-braunschweig.de/receive/dbbs_mods_00060699

7. Buls, E., Kadikis, R., Cacurs, R., Ārents, J.: Generation of synthetic training data for object detection in piles. In: Eleventh International Conference on Machine Vision (ICMV 2018). vol. 11041, pp. 533–540. International Society for Optics and Photonics, SPIE (2019). https://doi.org/10.1117/12.2523203, https://doi.org/10.1117/12.2523203

8. Chu, F.J., Xu, R., Vela, P.A.: Real-world multiobject, multigrasp detection. IEEE Robot. Autom. Lett. **3**(4), 3355–3362 (2018)

9. Das, A., Kandan, S., Yogamani, S., Krizek, P.: Design of real-time semantic segmentation decoder for automated driving (2019)

10. Dollar, P., Wojek, C., Schiele, B., Perona, P.: Pedestrian detection: An evaluation of the state of the art. IEEE Trans. Pattern Anal. Mach. Intell. **34**(4), 743–761 (2011)

11. Dwibedi, D., Misra, I., Hebert, M.: Cut, paste and learn: Surprisingly easy synthesis for instance detection. In: 2017 IEEE International Conference on Computer Vision (ICCV), pp. 1310–1319 (2017). https://doi.org/10.1109/ICCV.2017.146

12. Gupta, S., Girshick, R., Arbeláez, P., Malik, J.: Learning rich features from RGB-D images for object detection and segmentation. In: Fleet, D., Pajdla, T., Schiele, B., Tuytelaars, T. (eds.) ECCV 2014. LNCS, vol. 8695, pp. 345–360. Springer, Cham (2014). https://doi.org/10.1007/978-3-319-10584-0_23

13. He, K., Gkioxari, G., Dollár, P., Girshick, R.: Mask R-CNN. In: Proceedings of the IEEE international conference on computer vision, pp. 2961–2969 (2017)

14. He, R., Rojas, J., Guan, Y.: A 3D object detection and pose estimation pipeline using RGB-D images. In: 2017 IEEE International Conference on Robotics and Biomimetics (ROBIO), pp. 1527–1532. IEEE (2017)

15. Kleeberger, K., Landgraf, C., Huber, M.F.: Large-scale 6D object pose estimation dataset for industrial bin-picking (2019)

16. Luenendonk, M.: Industry 4.0: definition, design principles, challenges, and the future of employment (2019). Accessed 24 2020

17. Mousavian, A., Eppner, C., Fox, D.: 6-dof graspnet: variational grasp generation for object manipulation (2019)

18. Padilla, R., Netto, S.L., da Silva, E.A.B.: A survey on performance metrics for object-detection algorithms. In: 2020 International Conference on Systems, Signals and Image Processing (IWSSIP), pp. 237–242 (2020). https://doi.org/10.1109/IWSSIP48289.2020.9145130

19. Pérez, L., Rodríguez, Í., Rodríguez, N., Usamentiaga, R., García, D.F.: Robot guidance using machine vision techniques in industrial environments: a comparative review. Sensors **16**(3), 335 (2016)

20. Qiu, W., Yuille, A.: UnrealCV: connecting computer vision to unreal engine. In: Hua, G., Jégou, H. (eds.) ECCV 2016. LNCS, vol. 9915, pp. 909–916. Springer, Cham (2016). https://doi.org/10.1007/978-3-319-49409-8_75

21. Redmon, J.: Darknet: open source neural networks in c. http://pjreddie.com/darknet/ (2013–2016)

22. Redmon, J., Farhadi, A.: Yolo9000: better, faster, stronger. In: Proceedings of the IEEE Conference on Computer Vision and Pattern Recognition, pp. 7263–7271 (2017)

23. Redmon, J., Farhadi, A.: Yolov3: An incremental improvement. arXiv preprint arXiv:1804.02767 (2018)

24. Ren, S., He, K., Girshick, R., Sun, J.: Faster r-cnn: Towards real-time object detection with region proposal networks. IEEE Transactions on Pattern Analysis and Machine Intelligence 39 (06 2015). https://doi.org/10.1109/TPAMI.2016.2577031

25. Tan, M., Pang, R., Le, Q.V.: Efficientdet: scalable and efficient object detection. In: Proceedings of the IEEE/CVF Conference on Computer Vision and Pattern Recognition, pp. 10781–10790 (2020)

26. Tremblay, J., To, T., Sundaralingam, B., Xiang, Y., Fox, D., Birchfield, S.: Deep object pose estimation for semantic robotic grasping of household objects. In: CoRL (2018)

27. Wang, K., Shi, F., Wang, W., Nan, Y., Lian, S.: Synthetic data generation and adaption for object detection in smart vending machines. CoRR abs/1904.12294 http://arxiv.org/abs/1904.12294 (2019)

28. Xu, Z., Li, B., Yuan, Y., Dang, A.: Beta R-CNN: Looking into pedestrian detection from another perspective. In: NeurIPS (2020)

29. Yuan, Y., Chen, X., Wang, J.: Object-contextual representations for semantic segmentation. In: Vedaldi, A., Bischof, H., Brox, T., Frahm, J.-M. (eds.) ECCV 2020. LNCS, vol. 12351, pp. 173–190. Springer, Cham (2020). https://doi.org/10.1007/978-3-030-58539-6_11

30. YunYang1994: tensorflow-yolov3. https://github.com/YunYang1994/tensorflow-yolov3 Accessed Dec 20 2021

31. Zaidi, S.S.A., Ansari, M.S., Aslam, A., Kanwal, N., Asghar, M., Lee, B.: A survey of modern deep learning based object detection models. In: Digital Signal Processing, p. 103514 (2022)

32. Zhu, Z., Liang, D., Zhang, S., Huang, X., Li, B., Hu, S.: Traffic-sign detection and classification in the wild. In: Proceedings of the IEEE conference on computer vision and pattern recognition. pp. 2110–2118 (2016)

33. Zoghlami, F., Kurrek, P., Jocas, M., Masala, G., Salehi, V.: Design of a deep post gripping perception framework for industrial robots. J. Comput. Inf. Sci. Eng. **21**, 1–14 (2020). https://doi.org/10.1115/1.4048204

Key Landmarks Detection of Cleft Lip-Repaired Partially Occluded Facial Images for Aesthetics Outcome Assessment

Paul Bakaki[1,4], Bruce Richard[2], Ella Pereira[1], Aristides Tagalakis[1], Andy Ness[3], Ardhendu Behera[1], and Yonghuai Liu[1(✉)]

[1] Faculty of Arts and Science, Edge Hill University, Lancashire L39 4QP, UK
{bakakip,pereirae,Aristides.Tagalakis,Beheraa,
yonghuai.liu}@edgehill.ac.uk
[2] Birmingham Children's Hospital, Steelhouse Lane, Birmingham B4 6NH, UK
brucerichard@blueyonder.co.uk
[3] British Dental School, University of Bristol, Bristol BS1 2LY, UK
Andy.Ness@bristol.ac.uk
[4] Department of Computer Science, Makerere University,
P.O. Box 7062, Kampala, Uganda

Abstract. This paper proposes a novel method for the detection of the symmetrical axis of the cropped face required for the aesthetic outcome estimation from the facial images of patients after their cleft treatment. It firstly applies the Gaussian filter to smooth the images in order to compress noise on the subsequent tasks, then the bilateral semantic segmentation network is applied to segment the facial components out and each region is assigned a distinct colour, thirdly the Canny edge detector is applied to detect the facial feature points and all the contours are further detected and classified into three thirds according to their height. Fourthly, the centres of mass of detected feature points on the contours and the average of all these centres are used to estimate four potential symmetrical axes of the face, the one with minimum Manhattan distance from all the detected feature points is finally selected as the optimal one and used to estimate the aesthetic numerical score through the shape analysis in structural similarity measure. The experimental results based on a publicly accessible dataset shows that it performs well and better than one existing method.

Keywords: Cleft · Facial image · Aesthetic outcome estimation · Symmetrical axis · Shape analysis

1 Introduction

Cleft lip (CL) is one of the most common maxillofacial congenital malformations with high surgical treatment costs [1]. Computational studies aim to demonstrate the potential for objective outcome of aesthetic assessment following cleft

S. Sclaroff et al. (Eds.): ICIAP 2022, LNCS 13232, pp. 718–729, 2022.
https://doi.org/10.1007/978-3-031-06430-2_60

surgical treatment. The overall intention is to aid audit of the various surgical practices, by encouraging only those with better aesthetic outcome to surgically treat cleft lip. This has the potential to influence most practitioners to adhere to the set standard surgical guidelines for cleft-related treatment [2]. The mouth lip beauty is a targeted outcome measure. The obvious distortion of the lip morphology hinders detection and identification of key features, considered essential for beauty. The features depicted from the facial aesthetic outcome significantly aid towards categorization as success or failure of a cleft repair. Eventually, this aids any audit of different cleft repair practices by assessing the beauty of the mouth lips [3,4].

Determining facial features in images/videos is predominantly premised on a detected face [5,6]. Therefore, face detection is a major component of facial feature identification studies. Facial anonymization of aesthetic outcomes is a convention for cleft lip related studies [7]. It is logically commendable and ethically a best practice for unbiased outcome assessment audit of different practices. Anonymization obstructs biased human assessment from any eye colour, ears shape, hair etc., unlike computer-based assessment [7,8]. Consequently, the images used during outcome assessment bear significant partial occlusion.

Occlusion in computer vision started in the 1960s when Guzman proposed to detect faint lines in polyhedral drawings [9]. Consequently, it has been a subject of study in computer vision for detection of hidden facial features using convolutional neural networks with an attention mechanism [10]; facial appearance and shape learning to robustly detect facial features using an occlusion-adaptive deep network [11]; and cascaded pose regression (CPR) [12,13].

Some of the facial image features of significant importance include inner eye corners (i.e. inner canthus, lacrimal punctum and inner canthal distance), nose features (tip, ala, root, and nasal base) and mouth features (upper/lower lip vermillion, oral commissure, vermillion border) [14,15]. Presence of these features in the aesthetics symbolizes beauty. Therefore, computer vision tools aim to detect and locate these features, hence assess beauty using symmetry and other suitable shape defining parameters [16].

Most facial features occur in group or pairwise classification and can be used to determine the symmetric or asymmetric nature of a facial aesthetic. Key mouth features are used in [17] to determine the symmetry axis from which shape analysis is applied for facial aesthetic assessment.

Scars and other skin residues from surgical repair and photography effects can naturally cause features occlusion and influence aesthetic outcome assessment. Given this fact, deep learning techniques and regression studies have registered success regarding feature detection. Our approach disregards face detection because all facial aesthetics in use are anonymized. This study proposes a deep learning-based approach for the detection of facial features from cropped images for the analysis of their aesthetic outcome.

Dollár, Welinder and Perona [13] studied occluded facial landmarks detection using a statistical regression analysis framework. A facial image is partitioned into nine equal portions with anticipated landmarks positions. This study has been applied to normal facial images in several datasets such as WFLF [18] and COFW [12]. A more robust approach (RCPR) introduced in [12] operates under difficult occlusion with the intention to improve the performance in [13].

So far, all related studies reviewed assume that occlusion is casually created using external objects such as spectacles, caps, hair styling etc. This study introduces and investigates a unique case of CL aesthetics where occlusion originates from the surgical treatment procedure and ethical norms.

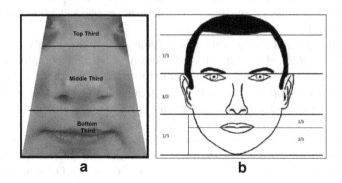

Fig. 1. Illustration and adaptation of horizontal thirds under occlusion of a cropped face (left) and full face (right).

2   Methodology

The approach taken aims to detect as many feature points as possible. Additionally, the objective is to classify and group the detected feature points in the three apparent segments of the facial aesthetics: upper third, middle third and bottom third, represented as the periorbital region, nose region and lips/oral region respectively [3,19] (Fig. 1).

Successful categorization of these features plays a crucial role in the determination of the most befitting symmetric axis of the face. To this end, a deep learning-based method is proposed to detect the facial feature points of interest in the three regions: pre-processing, feature detection, symmetrical axis estimation and numerical score estimation that are detailed below.

1. Pre-process the aesthetic image using an appropriate filter. Filters have an enhancement and smoothing effect to facilitate generation of better segmentation results [20]. Several filters such as Gaussian, Laplacian of Gaussian, median and the others [20] can be used for this purpose. However, given the nature of our dataset's aesthetics, a 3×3 Gaussian filter was the best choice because image pixels are evenly distributed despite any image degenerative conditions with each element set to 1. Without the Gaussian filtering, less features are detected. Other filters can be designed. Examples of other filters are: mean $f = \frac{1}{9}\left(\begin{smallmatrix} 1 & 1 & 1 \\ 1 & 1 & 1 \\ 1 & 1 & 1 \end{smallmatrix}\right)$ and a vertical Sobel filter: $f = \left(\begin{smallmatrix} -1 & 0 & 1 \\ -2 & 0 & 2 \\ -1 & 0 & 1 \end{smallmatrix}\right)$ etc. Some of the outputs from these filters are presented in Fig. 2.

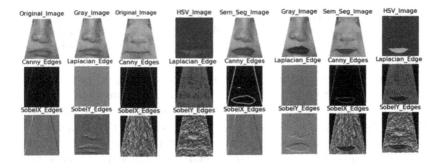

Fig. 2. Different filters used to visualize features for potential classification. Left four columns show that facial features are not clearly localized. Right four columns show clearer features from the same filters after segmentation using an ML approach.

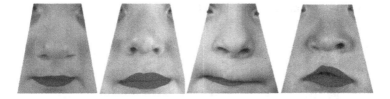

Fig. 3. Segmentation results. Mouth region properly detected in all. From Left to Right: first - right eye corner not detected, fourth - all eye corners not detected.

2. Separate the salient regions through semantic segmentation because facial images present segmentation challenges using ordinary techniques due to low contrast [21]. Inner canthus and oral region features are most salient. Figure 3 shows that not all the key regions will be detected due to poor anonymization procedures. Further, skin colour tone and scars from surgical treatment complicates the detection of any features [22]. The nose region is not segmented semantically but through any edges that may be detected. Consequently, the head orientation during photo-taking (either looking straight or downwards) influences the detection of the nose region edges and feature points following luminance contrast. Both the bilateral semantic network segmentation algorithm [23] and high-resolution network segmentation [24] produce appropriate segmentation results. The former is a faster and less resource intensive approach. We utilize the detailed module and semantic module of the bilateral segmentation network to acquire the image low/high level features and the semantics of each pixel, respectively. The two modules are combined through a real-time fusion module. The outcome is a clearly segmented mouth region, and the eye canthus, where possible. Figure 3 shows the red-segmented mouth region and red-segmented eye canthi (*middle two*). For the different scenarios, as discussed in the results section, different features are therefore considered as inputs to the top network layers.

Segmentation therefore aids the detection of the mouth region and the inner canthi, but it usually completely missed the nose region. The nose was not considered a key CL surgical outcome, hence excluded from training the segmentation network. Besides, the baseline dataset for the bilateral network is annotated to incorporate the nose as a whole, [32], without the two nostrils. Eventually, the two nostrils would be considered significant to determine the symmetry of the partially occluded faces. To this end, we propose to apply Canny edge detection [25]. This further results into more feature points with higher accuracy for detection of the mouth and eye regions, Fig. 4, *middle*. Each of the regions should have feature points to aid with symmetry detection. For eyes, the interest lies with the closest inner canthus distance and the median distance while for the mouth region, the philtrum, vermillion borders, and oral commissures are desirable. Within the nose region, the tip, nostrils and their base are of interest. Figure 4, *left* shows that classification of feature points can be more complicated before segmentation. To find as many feature points as possible in each region, distinct colours are assigned for improved contrast, as illustrated in Fig. 4, *right*.

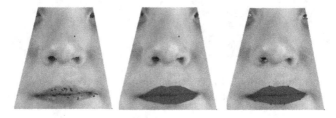

Fig. 4. Features identified per horizontal partition. Left: Largely disorderly without segmentation. Middle: Shows improved features mapping and detection after segmentation. Right: Classified per horizontal partition. (Color figure online)

3. Because of the partially occluded and anonymised nature of dataset, skin residues etc., some feature points within the key regions are disconnected. Implying there is need to identify connected components for stability to aid the determination of the symmetry. This is aided by filtering the detected feature points using Canny edge detector with the lower and upper thresholds set as decrease and increase by 0.33 of the median of pixel intensities of the whole image. The integral features lie along the eyes corner/inner canthus, nose tips/base/root/nostrils and mouth boundary/ philtrum/oral commissures. Once the feature points are detected, different colours, other than red *(for the predicted set, PS)* and green for the *ground truth sets (GT1, GT2 and GT3)*, are assigned to the feature points per different horizontal third for visualization (Fig. 4, *right*).

4. Link the respective feature points as contours. Successfully determining the feature point's perimeter suggests presence of a closed area/contour. At this stage, all possible contours have been identified as re-sampled contours from fully connected shapes without self-intersection, following the library implementation of [26], Fig. 4, *Left*. Some contours may be very small but neces-

sary for the location of the position of the features of interest. For example, a detailed execution shows that the green feature points representing identification of the nose region features (Fig. 4, *right*) are actually more than visibly displayed.

5. Partition the feature points on the detected contours into the three horizontal thirds based on their heights and classify them into the respective horizontal third segment. Figure 4, *Center and Right* illustrate the features' locations using a colour coding for each of the three horizontal thirds.

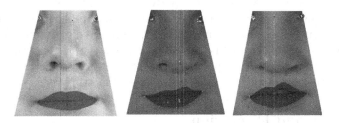

Fig. 5. Potential symmetric axes plotted based on component positions and their averages.

6. Determine the centres of mass of the contours in each of the three horizontal thirds. Draw a vertical line through each of the centres of the mass and also the average of these centres from the top to bottom boundary, yielding four potential symmetrical axes (Fig. 5).

7. Given four plotted potential axes of symmetry (Fig. 5), only one axis should be considered as an optimal one. Determine the Manhattan distance of each of the feature points from each of the potential symmetric axes through *Eq.* 1. Due to aggressive features detection, it has been experimentally proven that there are enough points from which to determine the symmetric axis.

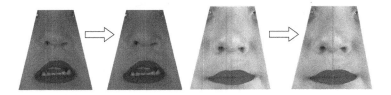

Fig. 6. Most suitable symmetric axis selected using average Manhattan distance.

$$dist(axis_k) = \sum_j \sum_{i=0}^{n} \mid axis_k - p_{ij} \mid, k = 1, 2, 3, 4 \qquad (1)$$

where n is the number of the detected contours in the image, $axis_k$ is a potential vertical axis of symmetry, p_{ij} is a feature point j in the ith contour. The symmetric axis is finally determined as the one with the minimum distance (please see the green line in Fig. 5 and used in Fig. 6).

8. The determined symmetric axis is the basis for dividing the mouth/lip region into two sections to aid lip shape analysis [17] under different appropriate scenarios. Lip shape analysis aims to determine how evenly/unevenly shaped the lip region is on either side of the symmetric axis. Generally, shape analysis has been applied to describe human perception features in medical images using contrast improvement ratio (CIR) [27] for example. A study by Loncaric [28] reviews other shape analysis techniques applied to images. In this study, a structural similarity measure [29] is preferred to automatically determine how agreeable the human visual perception is of the mouth lips, nose or a combination of both. Shape analysis is reduced to a structural comparison between the two mouth lip sides using the symmetric axis as a basis.

3   Results and Discussion

The distribution of the number of key feature points per horizontal third of the aesthetics from the public CCUK dataset is shown in Fig. 7. The four subcategories of the public dataset are: (i) Predicted Set (PS) - a set of aesthetics obtained through the proposed algorithm above, and (ii) three expert-generated Ground Truth sets $GT1$, $GT2$, and $GT3$. However, $GT3$ has not been considered in this study because it offers only a single-feature, the mouth/lip boundary. Ground truth sets are generated by manual annotation of the lip region using ImageJ, an open source software.

There exists more features in the upper and bottom horizontal thirds (blue and green, seen in Fig. 7 *left*), probably as expected. Figure 7 *right* presents a more detailed breakdown of features points distribution per dataset subcategory per horizontal third.

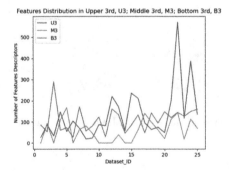 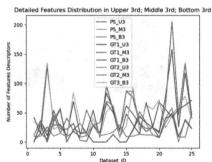

Fig. 7. The number of features detected across the 3 different horizontal upper, middle and bottom thirds: U3, M3 and B3 of each of the 3 considered sub-datasets. (Color figure online)

Fig. 8. Visualization of mouth region in Scenario 1 (left), upper lip in Scenario 2 (middle) and both nose and mouth regions in Scenario 3 (right).

Mouth lip and/or nose shape analysis was performed through three scenarios to assess the beauty of the aesthetic outcome of the surgical repair.

Scenario 1: Mouth region only (Fig. 8, *Left*). The physical surgical repair to the cleft on the upper lip is usually taken, in consideration of its alignment with the lower lip. Hence, the consideration of the whole mouth region is a natural occurrence when performing aesthetic outcome assessment. Similarity on either side of the symmetric axis through the mouth region is expected for features such as commissures and philtrum.

Scenario 2: The Upper lip, (Fig. 8, *Middle*) is the actual region of the mouth that is surgically repaired. Therefore, it is a trivial and fair choice to investigate its structural feature (dis)similarity.

Scenario 3: Combination of the nose region and mouth/lips region (Fig. 8, *Right*). Whereas cleft surgical repair is usually performed on the upper lip, human aesthetic outcome assessment naturally occurs with the awareness of other neighbouring features [30]. The closest feature available and applicable to our dataset is the nose region. It is almost trivial to observe any (mal)alignment between the nose and mouth.

These scenarios facilitate shape and structural computation and comparison using colour images, as presented before human assessors [31]. This also implies that the efficacy of our method can be determined by comparing the shape and structural computation with the human-generated numeric score *(HNS)* by human assessors. In [17], binary images for the lip/mouth region were used. After computing the structural similarity measure, s, it is converted to a numeric score between 1 and 5. Three (3) models, designed for conversion of s into a numeric score, are defined as follows: Model 1 $(M1)$: $f(s) = 5(1 - s^2) + s$, Model 2 $(M2)$: $f(s) = \exp((1 - s) \ln 5)$, and Model 3 $(M3)$: $f(s) = 5 - \frac{4s}{(1+s)^{\frac{1}{100}}}$ where $0 \leq s \leq 1$ and $1 \leq f(s) \leq 5$.

Therefore, s is computed for each of the three subsets of the main dataset *(PS, GT1, GT2)* and then automatically converted into their respective numeric scores *(PS_ AENS, GT1_AENS* and *GT2_AENS)*, from which correlation coefficients were calculated against *HNS*. The higher the coefficient, the more accurate the estimated aesthetic numeric score.

Since *PS* is determined automatically, its respective numeric scores *PS_AENS* are also automatically obtained. The correlation coefficient between *HNS* and *PS_AENS* is the most significant correlation (MSC) because it compares computer-generated numeric scores with human-generated numeric scores as

Table 1. Correlation coefficients between the aesthetics scores estimated using different methods for different scenarios (S1, S2 and S3) and different models (M1, M2 and M3).

	PS_AENS vs HNS			GT1 vs HNS			GT2 vs HNS		
	S1	S2	S3	S1	S2	S3	S1	S2	S3
M1	0.236	−0.247	0.205	0.079	0.062	0.039	0.055	0.181	0.024
M2	0.200	−0.192	0.151	0.007	0.072	−0.033	−0.056	0.567	−0.102
M3	0.219	−0.226	0.176	0.041	0.052	−0.003	−0.011	0.457	0.056

Table 2. Correlation coefficients between the aesthetics scores estimated using different methods for different scenarios (S1, S2 and S3) and different models (M1, M2 and M3).

	PS_AENS vs GT1			PS_AENS vs GT2			GT1 vs GT2		
	S1	S2	S3	S1	S2	S3	S1	S2	S3
M1	0.809	0.560	0.811	0.854	0.653	0.850	0.903	0.888	0.906
M2	0.827	0.356	0.825	0.834	0.560	0.842	0.940	0.910	0.940
M3	0.836	0.654	0.833	0.856	0.456	0.856	0.924	0.908	0.928

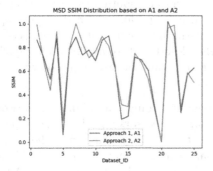
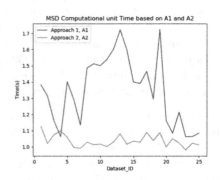

Fig. 9. SSIM (*left*) and computation time (*right*) of different approaches. (Color figure online)

seen in Table 1. The highest MSC is 23.6% (Model 1, Scenario 1). Overall, Scenario 1 also presents the best MSC. Further reduction of the region of interest (RoI) to study only the upper lip (Scenario 2) produces negative correlation results. It is a potential indication that this is not consistent with the practice and how to determine the region of interest requires further investigation.

Besides, the proposed method has produced a highest correlation result of 94.0% between *GT1_AENS* and *GT2_AENS*, Model 2, Scenarios 1 and 3, as seen in Table 2. This is also a significant outcome, implying potential higher similarity in the datasets generated by the human experts (*GT1* and *GT2*).

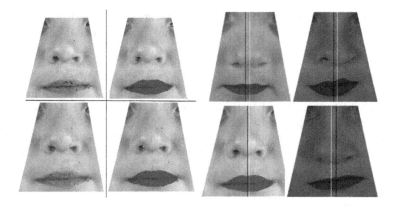

Fig. 10. Features (left two columns) detected using SIFT (Left column) and proposed approach (Second column). Then, symmetrical axis detection from some example cleft images (right two columns) using approach 1 (A1 - Black axis by [17]) and approach 2 (A2 - White axis by the proposed method)

Figure 9 *(right, orange)* presents a comparative study between the proposed method (referred to as approach 2, A2) and the existing one (referred to as approach 1, A1) [17]. It shows that the latter usually generates lower structural similarity across the dataset than the former. This is because it uses more feature points and thus generates more accurate symmetric axes. Figure 9 *(right, orange)* also shows that A2 takes shorter time than A1. It is an indicator that the whole face feature detection is faster than partial feature detection. Basically, it is easier to perceive and assess a whole face than its portion. Figure 10 compares feature detection using SIFT algorithm and A2, and shows how A1 and A2 map symmetric axes.

4   Conclusion and Future Works

Detecting key feature positions requires an aggressive approach that combines deep learning and traditional approaches. For instance, a deep learning approach for segmentation combined with traditional edge detection, led to detection of more features (nose region especially) and feature points within the various regions of interest. Automatic structural comparison and analysis of colour aesthetics outcomes is more in harmony with human visual perception and judgement due to inclusion of luminosity and contrast features. This is represented by the consistent MSC results across the scenarios for the three models, M1, M2 and M3. Finally, anonymised and occluded facial aesthetics in our dataset have more features in the upper and lower horizontal third segments, implying that they may provide more potential for the estimation of the aesthetics from the cleft images. Other shape analysis techniques applicable to biomedical images as reviewed in [28] will be tested in future studies. The next step will be investigating purely deep learning techniques to extract, detect and, where necessary predict, facial aesthetic features specific to the public CCUK dataset.

Acknowledgments. The facial images are the cropped and anonymised anteroposterior (A/P) photos of 5-year-old children from the Cleft Care UK (CCUK). This publication presents data derived from the Cleft Care UK Resource (an independent study funded by the National Institute for Health Research (NIHR) under its Programme Grants for Applied Research scheme RP-PG-0707-10034). PB was funded by Graduate Teaching Assistantship, Edge Hill University; YL was partially funded by Shaanxi Province Key Research and Development Plan General Project-Industrial Field (2021GY-171).

References

1. Zhang, Q., Yue, Y., Shi, B., Yuan, Z.: A bibliometric analysis of Cleft lip and palate-related publication trends from 2000 to 2017. Cleft Palate-Craniofacial J. **56**, 658–669 (2019)
2. de Ladeira, P.R.S., Alonso, N.: Protocols in Cleft lip and palate treatment: systematic review. Plast. Surg. Int. **2012**, 1–9 (2012)
3. Hashim, P.W., Nia, J.K., Taliercio, M., Goldenberg, G.: Ideals of facial beauty. Cutis. **100**, 222–224 (2017)
4. Kar, M., Muluk, N.B., Bafaqeeh, S.A., Cingi, C.: È Possibile Definire Le Labbra Ideali? Acta Otorhinolaryngol. Ital. **38**, 67–72 (2018)
5. Hassaballah, M., Bekhet, S., Rashed, A.A.M., Zhang, G.: Facial features detection and localization. In: Hassaballah, M., Hosny, K.M. (eds.) Recent Advances in Computer Vision. SCI, vol. 804, pp. 33–59. Springer, Cham (2019). https://doi.org/10.1007/978-3-030-03000-1_2
6. Reisfeld, D., Yeshurun, Y.: Robust detection of facial features by generalized symmetry (1992)
7. Lee, T.V.N., et al.: Is there a correlation between Nasolabial appearance and dentoalveolar relationships in patients with repaired unilateral Cleft lip and palate? Cleft Palate-Craniofacial J. **57**, 21–28 (2019)
8. Shkoukani, M.A., Chen, M., Vong, A.: Cleft lip - a comprehensive review. Front. Pediatr. **1**, 1–10 (2013)
9. Hoiem, D., Efros, A.A., Hebert, M.: Recovering occlusion boundaries from an image. Int. J. Comput. Vis. **91**, 328–346 (2011)
10. Li, Y., Zeng, J., Shan, S., Chen, X.: Occlusion aware facial expression recognition using CNN with attention mechanism. IEEE Trans. Image Process. **28**, 2439–2450 (2019)
11. Zhu, M., Shi, D., Zheng, M., Sadiq, M.: Robust facial landmark detection via occlusion-adaptive deep networks. In: IEEE/CVF Conference on Computer Vision and Pattern Recognition (CVPR), June 2019, pp. 3481–3491 (2019)
12. Burgos-Artizzu, X.P., Perona, P., Dollar, P.: Robust face landmark estimation under occlusion. In: Proceedings of the IEEE International Conference on Computer Vision, pp. 1513–1520 (2013)
13. Dollár, P., Welinder, P., Perona, P.: Cascaded pose regression. In: 2010 IEEE Computer Society Conference on Computer Vision and Pattern Recognition, pp. 1078–1085 (2010)

14. Hennekam, R.C.M., Cormier-Daire, V., Hall, J.G., Méhes, K., Patton, M., Stevenson, R.E.: Elements of morphology: standard terminology for the nose and philtrum. Am. J. Med. Genet. Part A. **149**, 61–76 (2009)
15. Hall, B.D., Graham, J.M., Cassidy, S.B., Opitz, J.M.: Elements of morphology: standard terminology for the periorbital region. Am. J. Med. Genet. Part A **149**, 29–39 (2009)
16. Sharma, V.P., Bella, H., Cadier, M.M., Pigott, R.W., Goodacre, T.E.E., Richard, B.M.: Outcomes in facial aesthetics in cleft lip and palate surgery: a systematic review. J. Plast. Reconstr. Aesthetic Surg. **65**, 1233–1245 (2012)
17. Bakaki, P., Richard, B., Pereira, E., Tagalakis, A., Ness, A., Liu, Y.: Shape analysis approach towards assessment of Cleft lip repair outcome. In: Tsapatsoulis, N., Panayides, A., Theocharides, T., Lanitis, A., Pattichis, C., Vento, M. (eds.) CAIP 2021. LNCS, vol. 13052, pp. 165–174. Springer, Cham (2021). https://doi.org/10.1007/978-3-030-89128-2_16
18. Sagonas, C., Tzimiropoulos, G., Zafeiriou, S., Pantic, M.: 300 faces in-the-wild challenge: the first facial landmark localization challenge. In: 2013 IEEE International Conference on Computer Vision Workshops, pp. 397–403 (2013)
19. Erian, A., Shiffman, M.A.: Advanced Surgical Facial Rejuvenation: Art and Clinical Practice, pp. 1–740. Springer, Heidelberg (2010). https://doi.org/10.1007/978-3-642-17838-2
20. Frery, A.C.: Image filtering. In: de Mello, C.A.B. (ed.) Digital Document Analysis and Processing, pp. 55–70. Nova Science Pub Inc, New York (2013). https://doi.org/10.1201/b10797-8
21. Oliveira, R.B., Filho, M.E., Ma, Z., Papa, J.P., Pereira, A.S., Tavares, J.M.R.S.: Computational methods for the image segmentation of pigmented skin lesions: A review. Comput. Methods Programs Biomed. **131**, 127–141 (2016)
22. Sandy, J., Kilpatrick, N., Ireland, A.: Treatment outcome for children born with cleft lip and palate. Front. Oral Biol. **16**, 91–100 (2012)
23. Yu, C., Gao, C., Wang, J., Yu, G., Shen, C., Sang, N.: BiSeNet V2: bilateral network with guided aggregation for real-time semantic segmentation. arXiv (2020)
24. Wang, J., et al.: Deep high-resolution representation learning for visual recognition. IEEE Trans. Pattern Anal. Mach. Intell. **43**, 3349–3364 (2020)
25. Canny, J.: A computational approach to edge detection. IEEE Trans. Pattern Anal. Mach. Intell. **PAMI-8**, 679–698 (1986)
26. Wu, S.-T., da Silva, A.C.G., Márquez, M.R.G.: The Douglas-Peucker algorithm: sufficiency conditions for non-self-intersections. J. Brazilian Comput. Soc. **9**, 67–84 (2004)
27. Kimori, Y.: Morphological image processing for quantitative shape analysis of biomedical structures: effective contrast enhancement. J. Synchrotron Radiat. **20**, 848–853 (2013)
28. Loncaric, S.: A survey of shape analysis techniques. Pattern Recogn. **31**(8), 983–1001 (1998)
29. Wang, Z., Bovik, A.C., Sheikh, H.R., Simoncelli, E.P.: Image quality assessment: from error visibility to structural similarity. IEEE Trans. Image Process. **13**, 600–612 (2004)
30. Deall, C.E., et al.: Facial aesthetic outcomes of Cleft surgery: assessment of discrete lip and nose images compared with digital symmetry analysis. Plast. Reconstr. Surg. **138**, 855–862 (2016)
31. Mosmuller, D.G.M., et al.: Scoring systems of cleft-related facial deformities: a review of literature. Cleft Palate-Craniofacial J. **50**, 286–296 (2013)
32. Liu, Z., Luo, P., Wang, X., Tang, X.: Deep learning face attributes in the wild. In: IEEE International Conference on Computer Vision (ICCV) (2015)

Exploring Fusion Strategies in Deep Multimodal Affect Prediction

Sabrina Patania[(✉)] , Alessandro D'Amelio , and Raffaella Lanzarotti

Univerità degli Studi di Milano, Milan, Italy
{sabrina.patania,alessandro.damelio,raffaella.lanzarotti}@unimi.it

Abstract. In this work, we explore the effectiveness of multimodal models for estimating the emotional state expressed continuously in the Valence/Arousal space. We consider four modalities typically adopted for the emotion recognition, namely audio (voice), video (face expression), electrocardiogram (ECG), and electrodermal activity (EDA), investigating different mixtures of them. To this aim, a CNN-based feature extraction module is adopted for each of the considered modalities, and an RNN-based module for modelling the dynamics of the affective behaviour. The fusion is performed in three different ways: at feature-level (after the CNN feature extraction), at model-level (combining the RNN layer's outputs) and at prediction-level (late fusion). Results obtained on the publicly available RECOLA dataset, demonstrate that the use of multiple modalities improves the prediction performance. The best results are achieved exploiting the contribution of all the considered modalities, and employing the late fusion, but even mixtures of two modalities (especially audio and video) bring significant benefits.

Keywords: Multimodal emotion recognition · Deep learning · Multimodal fusion

1 Introduction

Emotions have a massive impact in humans life: a positive emotional state can help to perform better in many fields and also positively affect the health state; conversely, negative emotions can degenerate into many psychological diseases, such as depression, especially if it is accumulated and repressed for a very long time [1]. Thus, detecting (automatically) negative emotions can play a crucial role, allowing to promptly intervene. Emotions have also a leading role in both verbal and non-verbal communication, being one of the most important aspects of human-human interaction. Let think to the e-learning [2] or to the game domain [3]: knowing the emotion of the user allows to tune the program to the need conveyed by the emotion (e.g. lesson/game too boring or too difficult). As a consequence, it earned a fundamental role in the development of human-computer interaction (HCI) systems [4].

© The Author(s), under exclusive license to Springer Nature Switzerland AG 2022
S. Sclaroff et al. (Eds.): ICIAP 2022, LNCS 13232, pp. 730–741, 2022.
https://doi.org/10.1007/978-3-031-06430-2_61

As a matter of fact the expression of an emotional state relies on many pathways or "modalities" [5]. Some of these are easily accessible (e.g. facial expression [6–8], body gesture [9], and prosody [10]), other are hidden to the observer during the emotional interaction (e.g. physiological signals such as EEG [11], ECG [12], and EDA [13]). It is no surprising that the vast majority of models for the detection of the emotional state of a subject rely on such "visible" signals, often taken individually.

Nonetheless, the interpretation of each single modality presents a certain degree of uncertainty that can be disambiguate or at least reduced observing two or more signals [5]. Furthermore, to guarantee a certain versatility, we should take into account the event that some signals are not available (e.g. corrupted video, noisy audio, ...) [14]. In this vein, as stigmatized in many studies ([4,14–16]), considering the joint occurrence of many modalities allows to build more robust models of affect prediction. As shown in Sect. 2, several mixtures of modalities have been proposed, but to our knowledge a systematic evaluation in a common framework is still lacking.

In this work we develop a deep architecture, suitable to evaluate several mixture modalities and to employ different fusion strategies. The goal is the prediction of the affective state defined as a continuous trajectory in the 2-dimensional Valence/Arousal space, combining in different ways both visible and non-visible signals, namely video, audio, ECG, and EDA. Dedicated Convolutional Neural Networks (CNNs) are employed to characterize each single modality, then combining the outputs employing different fusion strategies. The dynamics of the prediction is modelled through Recurrent Neural Networks.

2 Related Work

In 2015 Soleymani *et al.* [17] published a work about the integration of electroencephalograms (EEG) and facial features extracted from video recordings into an Emotion Recognition (ER) system. The purpose of their work was to test the correlation between EEG signal and facial features deformations to see if this can bring to better performances in a multimodal system compared to single modality ones. As a principal result, the authors found that feature level fusion with EEG does not lead to relevant performance improvements, being facial features a lot more informative than EEG in emotion recognition tasks.

In 2017 Tzirakis *et al.* [16] proposed a multimodal end-to-end emotion recognition system based on audio-video analysis. Emotions are described as a continuous signal based on valence and arousal concepts. This paper is developed and tested on RECOLA dataset, evaluating the multimodality on video and audio. First, audio segments and video-frames are processed with a convolutional neural network (CNN): audio is fed into two different layers with 1D convolutional filters; video frames are passed into pre-trained and fine-tuned residual Network, ResNet50 [18]; then, both modalities are fed as input to a 2-layers LSTM network to learn the temporal correlation between consequent samples. The output of the recurrent neural network (LSTM in this case) is then passed into a 2-neuron fully connected network (FCN), to output 2 different values per sample, respectively predicted valence and arousal.

Zhang et al. [19] analysed different ways to perform a fusion between modalities in an ER task and propose a novel method to fuse audio and video inputs based on a Deep Belief Network (DBN), which is formed by multiple Restricted Boltzmann Machines (RBMs). At first, they computed 3D Mel-Spectrograms from audio and cluster faces extracted from contiguous frames; then, audio and video tracks are subdivided into equally long sequences and single modalities are analysed through the fine-tuning of two different deep networks, to extract deep features: AlexNet for audio and C3D-Sports-1M for video. The features extracted are then fused using a DBN: the two feature-vector are concatenated, then the fused vector is fed into the DBN, composed of two different layers of RBMs. At last, a global utterance prediction is performed on the result of average pooling of the sequences final fused vector (the output of the DBN).

Du et al. [20] proposed a specific application of a multimodal ER system, the real-time monitoring of a gamer emotional status. Authors had to use only non-contact solutions to not interfere with the gaming process, so they opted for a system based on video input and heart rate value and variability, inferring the heartbeat from variations in video red channel. To perform the emotion recognition task, authors have used a sort of feature-level fusion of the two modalities. Feature vectors are firstly concatenated and then fed as input to a particular self-organizing map called SOM-BP, which overcomes SOM 'dead-neurons' problem by back-propagating the error when misclassification occurs.

In [14] Boccignone et al. pursued a simulation-based approach in which the multimodal nature of the affective interplay between humans involved in dyadic interactions is exploited. In this work an affective deep latent space is built from the combination of multiple modalities, namely facial expression and physiological signals (Heart Rate Variability and Electrodermal Activity). Once trained, the model is able to simulate affective behaviour (facial expression, variation of physiological signals) via sampling trajectories in the learned latent space representation.

At last, Ho et al. [21] performed a multimodal ER system based on speech and text. As feature representation, they computed Mel-Spectrograms from audio and embedded words using BERT [22]. Extracted features are both singularly fed into a batch normalisation layer and then into a gated recurrent unit (GRU). Next, both the outputs of GRUs are put as input to a multi-level multi-head fusion attention (MMFA) module. Global average pooling (GAP) is applied to the fused output of MMFA, to minimize overfitting and reduce dimension over time.

3   Proposed Model

As highlighted in Sect. 2, the key choices in designing a multimodal ER system concern the modalities to take into account, the fusion strategy, and of course the learning method. In this work we investigate deep architectures, adopting fusion at different levels (feature-, model-, and prediction-level), and four modalities, namely video, audio, ECG and EDA, evaluating several combinations of them. Inspired by [16], the proposed model consists of a dedicated CNN architecture for each considered modality. A shallow 1D CNN architecture has been adopted to learn representations for Audio and Physiological signals. For what concerns the visual modality, the well known pre-trained VGGFace CNN has been employed. Dynamics is modelled through a RNN architecture. These models are described more in depth in the following.

3.1 Audio/Physiological Network Architecture

The neural network used to extract deep features from one-dimensional data (either audio or physiological signals) is quite shallow, as it includes only two convolutional layers. This network aims at processing the input raw data and extracting their representation in the feature space. The structure of this network is shown in Fig. 1.

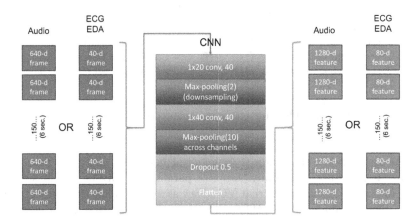

Fig. 1. 1D feature extraction architecture: this model processes groups of one-dimensional data (audio, ECG and EDA) with a 2-layers convolutional network, to extract their representations.

As can be noted, such network ingests the raw data and produces a 1280-dimensional feature representation. Depending on the modality (audio or physiological signals), the input to the network may have different length depending on the different sampling frequency.

3.2   Video Network Architecture

The video network architecture is in charge of extracting the relevant information from face images in order to perform prediction of affect. Given the wide availability of pre-trained models of facial descriptors, we decided to exploit the feature characterization of one of these architectures. The chosen model is VGGFace, a CNN network trained in 2015 by Parkhi *et al.* [23]. The VGGFace structure, is shown in Fig. 2. Each frame is first pre-processed by employing a face detector module [24]. The detected face is cropped and then fed as input to the VGGFace architecture which yields a 2622-dimensional feature vector.

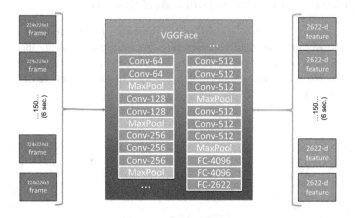

Fig. 2. 2D feature extraction architecture (VGGFace)

Note that, unlike the Audio/Physiological network, VGGFace has a quite complex architecture with many trainable parameters. Given the limited availability of data, we decided to treat VGGFace as a stand-alone feature extractor. Hence, during training, its weights will be frozen in order to avoid overfitting.

3.3   Temporal Modelling

The temporal modelling is a crucial step in the overall process. Here, the model extracts information about temporal relations between data that are useful to predict a much more accurate signal.

In this work, we adopted a recurrent neural network to model the dynamics of the emotional experience of a given subject; in particular, Peephole LSTM cells have been employed (cfr. Fig. 3). In order to define the best performing architecture, a series of experiments with different number of RNN layers (1 or 2) and units (cells) per layer ([8, 16, 32, 64, 28, 256] have been performed.

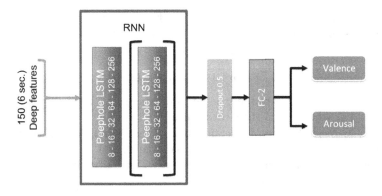

Fig. 3. RNN and final prediction structure: note that the second Peephole LSTM layer is optional.

4   Experimental Results

4.1   Dataset

The dataset employed for experiments is the REmote COLaborative and Affective interactions (RECOLA) dataset [25]. It is based on spontaneous interactions collected from a collaborative task performed remotely and contains multimodal data (audio, video, ECG and EDA) with detailed annotation of emotional and social primitives, evaluated from both internal (by the subjects) and external (by annotators) views. The emotional state has been annotated continuously in time and scale in two dimensions, valence and arousal. Data has been recorded from 46 French-speaking participants, who had to achieve a survival task in a group, remotely: they had to survive in a disaster scenario, i.e. a plane crash, ranking 15 objects by their importance. The emotional state has been annotated by 6 external annotators: they evaluated continuously valence and arousal with values ranging from -1 to 1 with steps of 0.01. In order to both capture the most interesting part of the discussion tasks and to limit the data to be annotated, the external evaluators only noted down the first 5 min of every discussion session.

4.2   Model Training

Model training has been performed by first training each unimodal network separately. In order to do so, 1 or two layers of Peephole LSTM have been stacked on top of each network (cfr. Fig. 3). Training is then carried out via back-propagation using the Adam optimizer in order to minimize the loss defined as the arithmetic mean of the Lin's Concordance Correlation Coefficient (CCC) [26] (ρ_c) as in [16].

CCC measures the agreement between two variables x and y representing in our case, the ground truth (continuous V/A annotations) and the predictions (predicted V/A trajectories), respectively. As the standard Pearson's Correlation coefficient ρ, it ranges between -1 and 1, where -1 means total disagreement and 1 means total agreement. The CCC is defined as:

$$\rho_c = \frac{2\rho\sigma_{xy}}{\sigma_x^2 + \sigma_y^2 + (\mu_x - \mu_y)^2} \tag{1}$$

where σ_x, σ_y and μ_x, μ_y are the standard deviations and means of the two variables x and y, while σ_{xy} is their covariance.

The CCC-based loss can be defined as $L = 1 - \rho_c$; call $L_{Valence}$ and $L_{Arousal}$ the two loss functions derived from the comparison of the predicted Valence and Arousal with the respective annotations. The global loss is given by the arithmetic mean of the single losses:

$$L_{Tot} = \frac{L_{Valence} + L_{Arousal}}{2} \tag{2}$$

Once each of the "unimodal" network has been trained separately (with the exception of the "video" network, that acts as standalone feature extractor), the results can be combined in order to obtain a multimodal prediction of affect. To this end, we employ three different fusion methods: feature-level fusion, model-level fusion and late fusion.

4.3   Feature-Level Fusion

In feature-level fusion the last layers of the unimodal networks to be considered for fusion are used as the feature characterizations of those specific modalities at the given time step. The feature vectors coming from all the considered modalities are joined together so to yield the multimodal characterization of affect at that time step. This last feature vector is then fed as input to the RNN architecture (Fig. 4).

4.4   Model-Level Fusion

In model-level fusion, the feature vector is obtained by taking the last layer of the RNN of each considered modality. As with the feature-level approach, such vectors are joined in order to obtain a multimodal summary of affect. Note that in this case the feature vector is able to capture the dynamics of the affective interaction.

As the fusion is located in the network after the modelling of the temporal features (with RNN), these multimodal networks only need to train the Fully-Connected network (FCN) that outputs the final prediction (Fig. 5).

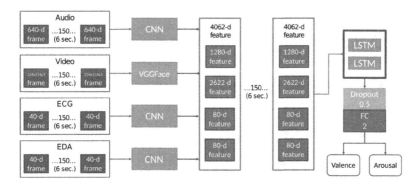

Fig. 4. Feature-level fusion, taking into account all the modalities

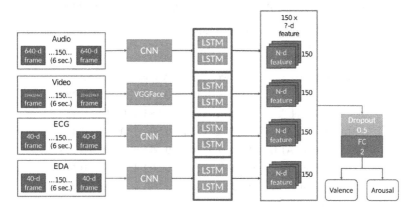

Fig. 5. Model-level fusion, taking into account all the modalities

4.5 Late Fusion

In the late fusion approach, predictions from the uni-modal models are combined and weighted by the average CCC value achieved in the training phase.

This way, modalities that achieve better results are more influential for the final prediction. The late fusion approach is depicted in Fig. 6.

4.6 Results

We trained and tested the architecture adopting the three fusion strategies, and referring to different combinations of modalities, namely Audio + Video, Audio + ECG, Audio + EDA, Video + ECG, Video + EDA, and All. In Table 1 we report the performance obtained employing the three fusion strategies, in the cases of the two configurations that achieved the best performances: Audio + Video and All. In all the reported results the 2-layer LSTM architecture with 256 units has been selected for the temporal modelling step, as it gave the highest performances.

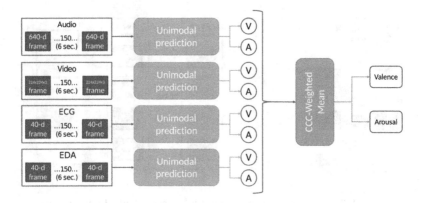

Fig. 6. Late Fusion approach, taking into account all the modalities

Table 1. Average CCC values (on the evaluation set) of the multimodal prediction employing different fusion strategies

Fusion type	Valence	Arousal	Avg
Feature-level AV	0.354 ± 0.152	**0.610 ± 0.124**	0.482
Feature-level ALL	0.332 ± 0.150	0.586 ± 0.113	0.459
Model-level AV	0.359 ± 0.163	0.560 ± 0.123	0.459
Model-level ALL	0.368 ± 0.166	0.563 ± 0.128	0.465
Late fusion AV	0.399 ± 0.150	0.602 ± 0.105	**0.501**
Late fusion ALL	**0.424 ± 0.203**	0.585 ± 0.114	**0.505**

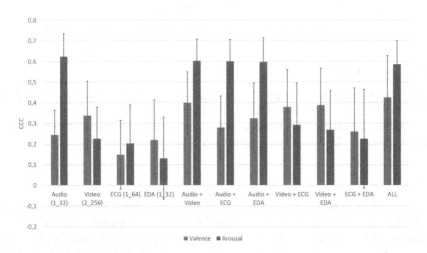

Fig. 7. CCC values on the evaluation set using late fusion with different modalities.

In Fig. 7 we plot the results for all the configurations, focusing on the most effective approach (late fusion); the unimodal performances are shown, too. It's worth noticing that most combinations of multiple modalities outperforms the unimodal performances, audio + video being the most successful.

5   Discussion and Conclusions

This work dealt with the development and analysis of a multimodal emotion recognition system based on deep neural network architectures. The proposed model comprises, for each modality, three main modules: a feature extraction module (based on deep convolutional neural networks), a dynamics modelling module (implemented as an LSTM architecture) and a final prediction module (composed of a 2-neurons fully connected layer). The model has been trained and validated on the popular and freely available RECOLA dataset. Different configurations of RNN structures have been explored, varying the number of layers and units per layer. Notably, the deeper (2-layers) and bigger (256 units) RNN architecture yielded the best performances. In addition, the contribution of each modality to the prediction of either Valence or Arousal has been inspected.

To this end, different combinations of modalities have been examined in a multimodal training and testing perspective: Audio + Video, Audio + ECG, Audio + EDA, Video + ECG, Video + EDA, and all modalities together (ALL). We attempted three different fusion strategies: in the first, named feature-level fusion, we fused modalities at feature-extraction level, that is by joining CNNs feature representations and re-training the RNN and Fully-Connected modules; the second method (model-level fusion) involves the fusion at the RNN-output-level, thus re-training only the Fully-Connected module; in the last method, named late fusion, we used the predictions from the "unimodal" models and combined them via weighted averaging, each weight being given by their respective average training CCC score.

The late fusion seemed to be the most effective, followed by feature-level fusion. Notably, any combination of modalities improves predictions or - at worst - achieves similar results, if compared to the respective "unimodal" performances. In particular, it emerges that the most effective pair of modalities is audio + video, that significantly improves Valence prediction compared to the video-based "unimodal" model, while achieving similar CCC test performances in predicting Arousal w.r.t. the audio-based one. In summary, the fusion of more modalities allows to create a more stable and reliable model, even if the addition of physiological signals led only to slight improvements in performances. These results encourage the adoption of multimodal systems of affect recognition as delivering more robust and stable predictive models.

Acknowledgement. This work was part of the project n. 2018-0858 title "Stairway to elders: bridging space, time and emotions in their social environment for wellbeing" supported by Fondazione CARIPLO.

References

1. Grossi, G., Lanzarotti, R., Napoletano, P., Noceti, N., Odone, F.: Positive technology for elderly well-being: a review. Pattern Recogn. Lett. **137**, 61–70 (2020)
2. Sun, A., Li, Y.-J., Huang, Y.-M., Li, Q.: Using facial expression to detect emotion in e-learning system: a deep learning method. In: Huang, T.-C., Lau, R., Huang, Y.-M., Spaniol, M., Yuen, C.-H. (eds.) SETE 2017. LNCS, vol. 10676, pp. 446–455. Springer, Cham (2017). https://doi.org/10.1007/978-3-319-71084-6_52
3. Du, G., Zhou, W., Li, C., Li, D., Liu, P.X.: An emotion recognition method for game evaluation based on electroencephalogram. IEEE Trans. Affect. Comput. 1 (2020)
4. Sebe, N., Cohen, I., Gevers, T., Huang, T.S.: Multimodal approaches for emotion recognition: a survey. In: Internet Imaging VI, vol. 5670, pp. 56–67. International Society for Optics and Photonics (2005)
5. Nguyen, D., Nguyen, K., Sridharan, S., Dean, D., Fookes, C.: Deep spatio-temporal feature fusion with compact bilinear pooling for multimodal emotion recognition. Comput. Vis. Image Underst. **174**, 33–42 (2018)
6. Jain, N., Kumar, S., Kumar, A., Shamsolmoali, P., Zareapoor, M.: Hybrid deep neural networks for face emotion recognition. Pattern Recogn. Lett. **115**, 101–106 (2018)
7. Bursic, S., Boccignone, G., Ferrara, A., D'Amelio, A., Lanzarotti, R.: Improving the accuracy of automatic facial expression recognition in speaking subjects with deep learning. Appl. Sci. **10**(11), 4002 (2020)
8. Cuculo, V., D'Amelio, A.: OpenFACS: an open source FACS-based 3D face animation system. In: Zhao, Y., Barnes, N., Chen, B., Westermann, R., Kong, X., Lin, C. (eds.) ICIG 2019. LNCS, vol. 11902, pp. 232–242. Springer, Cham (2019). https://doi.org/10.1007/978-3-030-34110-7_20
9. Noroozi, F., Kaminska, D., Corneanu, C., Sapinski, T., Escalera, S., Anbarjafari, G.: Survey on emotional body gesture recognition. IEEE Trans. Affect. Comput. **12**, 505–523 (2018)
10. Albanie, S., Nagrani, A., Vedaldi, A., Zisserman, A.: Emotion recognition in speech using cross-modal transfer in the wild. In: Proceedings of the 26th ACM International Conference on Multimedia, pp. 292–301 (2018)
11. Song, T., Zheng, W., Song, P., Cui, Z.: EEG emotion recognition using dynamical graph convolutional neural networks. IEEE Trans. Affect. Comput. **11**(3), 532–541 (2018)
12. Sarkar, P., Etemad, A.: Self-supervised ECG representation learning for emotion recognition. IEEE Trans. Affect. Comput. 1 (2020)
13. Shukla, J., Barreda-Angeles, M., Oliver, J., Nandi, G.C., Puig, D.: Feature extraction and selection for emotion recognition from electrodermal activity. IEEE Trans. Affect. Comput. **12**(4), 857–869 (2019)
14. Boccignone, G., Conte, D., Cuculo, V., D'Amelio, A., Grossi, G., Lanzarotti, R.: Deep construction of an affective latent space via multimodal enactment. IEEE Trans. Cogn. Dev. Syst. **10**(4), 865–880 (2018)
15. Schuller, B., Valstar, M., Cowie, R., Pantic, M.: The first audio/visual emotion challenge and workshop – an introduction. In: D'Mello, S., Graesser, A., Schuller, B., Martin, J.-C. (eds.) ACII 2011. LNCS, vol. 6975, p. 322. Springer, Heidelberg (2011). https://doi.org/10.1007/978-3-642-24571-8_42
16. Tzirakis, P., Trigeorgis, G., Nicolaou, M.A., Schuller, B.W., Zafeiriou, S.: End-to-end multimodal emotion recognition using deep neural networks. IEEE J. Sel. Topics Signal Process. **11**(8), 1301–1309 (2017)

17. Soleymani, M., Asghari-Esfeden, S., Fu, Y., Pantic, M.: Analysis of EEG signals and facial expressions for continuous emotion detection. IEEE Trans. Affect. Comput. **7**(1), 17–28 (2015)
18. He, K., Zhang, X., Ren, S., Sun, J.: Deep residual learning for image recognition. In: Proceedings of the IEEE Conference on Computer Vision and Pattern Recognition, pp. 770–778 (2016)
19. Zhang, S., Zhang, S., Huang, T., Gao, W., Tian, Q.: Learning affective features with a hybrid deep model for audio-visual emotion recognition. IEEE Trans. Circ. Syst. Video Technol. **28**(10), 3030–3043 (2017)
20. Du, G., Long, S., Yuan, H.: Non-contact emotion recognition combining heart rate and facial expression for interactive gaming environments. IEEE Access **8**, 11896–11906 (2020)
21. Ho, N.-H., Yang, H.-J., Kim, S.-H., Lee, G.: Multimodal approach of speech emotion recognition using multi-level multi-head fusion attention-based recurrent neural network. IEEE Access **8**, 61672–61686 (2020)
22. Devlin, J., Chang, M.-W., Lee, K., Toutanova, K.: BERT: pre-training of deep bidirectional transformers for language understanding. arXiv preprint arXiv:1810.04805 (2018)
23. Parkhi, O.M., Vedaldi, A., Zisserman, A.: Deep face recognition (2015)
24. Zhang, K., Zhang, Z., Li, Z., Qiao, Y.: Joint face detection and alignment using multitask cascaded convolutional networks. IEEE Signal Process. Lett. **23**(10), 1499–1503 (2016)
25. Ringeval, F., Sonderegger, A., Sauer, J., Lalanne, D.: Introducing the recola multimodal corpus of remote collaborative and affective interactions. In: 2013 10th IEEE International Conference and Workshops on Automatic Face and Gesture Recognition (FG), pp. 1–8. IEEE (2013)
26. Lawrence, I., Lin, K.: A concordance correlation coefficient to evaluate reproducibility. Biometrics **45**, 255–268 (1989)

A Contrastive Distillation Approach for Incremental Semantic Segmentation in Aerial Images

Edoardo Arnaudo[1,2]([✉]) [iD], Fabio Cermelli[1] [iD], Antonio Tavera[1] [iD],
Claudio Rossi[2] [iD], and Barbara Caputo[1] [iD]

[1] Politecnico di Torino, Torino, Italy
{edoardo.arnaudo,fabio.cermelli,antonio.tavera,barbara.caputo}@polito.it
[2] LINKS Foundation, Torino, Italy
{edoardo.arnaudo,claudio.rossi}@linksfoundation.com

Abstract. Incremental learning represents a crucial task in aerial image processing, especially given the limited availability of large-scale annotated datasets. A major issue concerning current deep neural architectures is known as catastrophic forgetting, namely the inability to faithfully maintain past knowledge once a new set of data is provided for retraining. Over the years, several techniques have been proposed to mitigate this problem for image classification and object detection. However, only recently the focus has shifted towards more complex downstream tasks such as instance or semantic segmentation. Starting from incremental-class learning for semantic segmentation tasks, our goal is to adapt this strategy to the aerial domain, exploiting a peculiar feature that differentiates it from natural images, namely the orientation. In addition to the standard knowledge distillation approach, we propose a contrastive regularization, where any given input is compared with its augmented version (i.e. flipping and rotations) in order to minimize the difference between the segmentation features produced by both inputs. We show the effectiveness of our solution on the Potsdam dataset, outperforming the incremental baseline in every test (Code available at: https://github.com/edornd/contrastive-distillation).

Keywords: Semantic segmentation · Incremental learning · Aerial images

1 Introduction

Semantic Segmentation represents a key task in aerial image processing, given its wide range of applications, from urban contexts [21], land cover and monitoring [33] or agricultural settings [32].

However, the majority of state-of-art solutions are designed to perform on a static set of categories by means of a full end-to-end training, with no option to integrate new knowledge. Without precautions, deep neural networks tend in fact

© The Author(s), under exclusive license to Springer Nature Switzerland AG 2022
S. Sclaroff et al. (Eds.): ICIAP 2022, LNCS 13232, pp. 742–754, 2022.
https://doi.org/10.1007/978-3-031-06430-2_62

to forget previously acquired information when a new training set is provided, resulting in poor performance on the old classes.

This phenomenon, known as *catastrophic forgetting* [16], has been addressed and successfully mitigated through a range of different methods [12,15,25], mostly considering image classification or object detection. In recent years, a greater deal of effort has been put on specific downstream tasks such as semantic segmentation, with solutions involving representation consistency [9], replay-based methods [29], or knowledge distillation [3]. The problem of incremental learning is extremely relevant also in aerial settings where, despite the growth in resources and data, the scarcity of large-scale annotated aerial datasets remains a crucial drawback for practical applications. In fact, it is often the case that images are collected in the same geographical area [21], or that the data itself is not immediately available, but rather acquired and processed periodically.

In this work, we propose to tackle the problem of incremental-class learning (ICL) in the context of semantic segmentation, focusing on aerial imagery. Leveraging on the MiB framework [3], a distillation-based method specifically designed for semantic segmentation tasks, we introduce an additional regularisation based on contrastive distillation, with the aim of exploiting a distinctive feature of such images, namely their invariance to orientation. We explicitly model this feature by comparing the activations produced by the framework on the input and its transformed version, minimising their difference. A first step involves the student network, comparing pairs of augmented inputs, then activations are also compared with the teacher from the previous incremental step, to improve the knowledge distillation. We evaluate our solution on the Potsdam benchmark dataset [21], where it consistently outperforms the robust incremental baseline in every setting. In summary, our contributions can be listed as follows:

- We address the problem of ICL in semantic segmentation of aerial images, providing benchmark results on a popular dataset.
- We propose a new regularization and distillation approach based on contrastive representation learning, addressing the arbitrary orientation of the inputs, one of the key aspects of aerial images.

2   Related Work

Aerial Semantic Segmentation. Thanks to the recent advancements in deep learning, many semantic segmentation approaches have been proposed over the years [5,8,14,35], focusing mostly on natural images. Most common methods revolve around encoder-decoder, fully-convolutional architectures [6]. These techniques have been successfully applied to the field of aerial images in wide range of contexts, such as semantic labelling in urban [1,8,18] or agricultural scenarios [18,32], or land cover tasks [23]. Despite the strong similarities with the natural counterparts, aerial images present some peculiar differences that have been addressed with varying approaches: first, satellite imagery are seldom limited to the visible spectrum and often include additional frequencies [32]. Common solutions to this problem include simpler solutions such as the duplication of input weights [20] or finer multi-modal approaches comprising the fusion of different modalities [23,33]. Last, a peculiar aspect of aerial and satellite images is represented by the top-down view, in which the orientation becomes arbitrary. In our work we propose to leverage on this peculiar feature, already successfully exploited in classification tasks [24,31], by applying a contrastive regularization to both the segmentation task and the incremental tasks, to further improve the knowledge distillation between steps.

Incremental Learning. Catastrophic forgetting [16], meaning the inability to remember past knowledge upon learning new information, represents a major issue concerning current deep learning solutions. Several techniques have been proposed to mitigate this issue, with different approaches: replay-based methods [25,29], exploiting exemplars from old classes parameter isolation parameter-based methods [15], involving a selective pruning so that the weights representing old labels are maintained through the learning steps, and memory-based approaches [34], where important parameters from previous steps are consolidated, forcing the model to maintain a robust representation for old classes. Last, one of the most effective techniques focuses on data and exploits knowledge distillation [3,12]. The latter is usually carried out with a *teacher-student* approach. Considering Semantic Segmentation on aerial imagery, a first proposal is represented by [29]: here, an hybrid approach comprising both knowledge distillation and additional supporting exemplars is employed. Similarly, in [9] the distillation approach is improved by strengthening the internal representations throughout the learning steps. Compared to image classification, semantic segmentation presents peculiarities that may lead to poor performances when not addressed, such as the presence of a common *background* label. In MiB [3], this issue is tackled by taking into consideration this distributional shift, by means of unbiased losses and regularizations with respect to the background label.

Contrastive Learning. Contrastive learning has become one of the most promising recent techniques in deep learning, closing the gap between supervised and self-supervised settings [2,7,10,17], or even improving the former by learning more robust representations [11]. The objective of Contrastive Representation Learning (CRL) is to cluster together latent representations of similar samples (i.e. *positive examples*), while at the same time increasing the distances between instance representations of different categories *(i.e. negative examples)*. CRL is often applied exploiting pretext tasks (i.e. manually devised tasks solely based on the image itself), including: geometric or color transformations [7,17], image reconstruction from its parts [19,28], or cross-modal techniques [13,30]. These additional tasks can also be paired with more traditional supervised settings such as semantic segmentation, in order to improve the results on the main task [28,30], deal with low resource datasets [4], or integrate additional modalities [22,30]. Here, we propose a similar approach where the same inputs are augmented twice, however we exploit the resulting representations as a further regularization to induce further invariance with respect to the applied transformations, during both standard training and knowledge distillation.

3 Methodology

3.1 Problem Statement

We address the problem of Incremental-Class Learning (ICL) for Semantic Segmentation on aerial images, where we suppose that different portions of data are provided sequentially, each one with a different set of labels.

First, we can define Semantic Segmentation as a pixel-wise classification, where each pixel x_i composing a generic image $x \in X$ with constant dimensions $H \times W$, is associated with a label $y_i \in Y$ representing its category, or eventually associated with a generic and comprehensive *background* class $b \in Y$. The training can be defined as learning a model f_θ with parameters θ, mapping from the image space X to the pixel-wise label space Y, namely: $f_\theta : X \mapsto \mathbb{R}^{|H \times W \times Y|}$.

Considering now the ICL setting, we require multiple sequential training phases named *learning steps*, in which we provide a different set of data samples and labels every time. Specifically, at each step t, we expand the previous set of labels Y^{t-1} with the additional ground truth Y^t, obtaining a new set of labels $C^t = Y^{t-1} \cup Y^t$. At each phase, we are also provided with a new training set D^t, such that each pixel-wise label y_i belongs to one of the current categories Y^t or the generic background class b. We then train a new model f_θ^t on the whole set of categories C^t, deriving the old labels from the outputs of the previous model $f_\theta^{t-1} : X \mapsto \mathbb{R}^{|H \times W \times Y^{t-1}|}$ and the new labels via standard training, exploiting the dataset for the current step. The final goal is to obtain a single model, able to perform well on both and new classes, namely $f_\theta^t : X \mapsto \mathbb{R}^{|H \times W \times C^t|}$.

3.2 Baseline

As previously mentioned, we adopt MiB as robust incremental baseline [3]. In ICL applied in the context of image classification, a standard approach involves a two-way training, combining a supervised loss on the dataset at the current step D_t with an additional term to maintain the old knowledge. In the case of the selected framework, the latter is carried out through distillation of the old model's outputs. Specifically, the final loss at each learning step becomes:

$$L(\theta^t) = L_{CE}(\theta^t) + \lambda L_{KD}(\theta^t) \tag{1}$$

where $L_{CE}(\theta^t)$ represents a supervised Cross-Entropy loss, while $L_{KD}(\theta^t)$ represents the Knowledge Distillation term at step t from the previous model $f_{\theta^{t-1}}$, weighted by a factor λ.

As briefly stated in Sect. 2, it is common that two sets of categories, namely Y_i and Y_j share the common background class b, however the semantic regions of the image are assigned to such label is often different in every set. This aspect of semantic segmentation needs to be dealt with during the incremental steps, taking into account that a pixel labeled as background in the dataset D_t might instead belong to one of the previous classes from step 0 to $t-1$. Thus, for each pixel i of a generic image x, the predicted probability $q(i, b)$ for the background class is substituted with:

$$q(i,b) = \sum_{k \in Y^{t-1}} q_x^t(i,k) \qquad (2)$$

In other words, the background is not considered as a category on its own, but rather a probability of having an old class *or* actual background.

A similar concept is adopted for the distillation component, where the following distillation loss is applied:

$$L_{KD}^{\theta^t}(x,y) = \frac{1}{N} \sum_{i \in x} \sum_{c \in Y^{t-1}} q_x^{t-1}(i,c) log(q_x^t(i,c)) \qquad (3)$$

Where the last term refers to the predicted probabilities for the new model with respect to the old classes. Given that the contribution for the new labels is provided by the Cross-Entropy loss, we require that $q_x^t(i,c) = 0, \forall c \in Y^t \setminus \{b\}$. In every other case, the term represents the predicted probability for the new model of having a label c for a pixel i, normalized across old classes as reported in [3]. Again, the distributional shift of the background class needs to be addressed for the incremental learning as well. Consequently, the predicted probability $q_x^t(i,b)$ for this special class is rewritten as:

$$q_x^t(i,b) = \sum_{k \in Y^t} q_x^t(i,k) \qquad (4)$$

In other terms, the predicted probability for the background class of the new model is substituted with the probability of having a new class *or* the background. In fact, we expect regions belonging to the new classes to be ignored by the previous model, thus labelled as generic background.

Moreover, excluding similarities among categories, it is extremely likely that predictions for $f_{\theta^{t-1}}$ for the current classes Y^t will fall under the background class. For this reason, we perform the same weight initialization for the final classifier as proposed in [3], so that its outputs for the new classes are uniformly distributed around the background from the very beginning to ease convergence.

3.3   Contrastive Distillation

As stated in Sect. 2, a major difference between natural and aerial images is represented by their orientation: in the former case, the point of view is fundamental to the correct detection of an entity. In fact, in a common scenario we expect to find background and foreground entities in a specific part of the image (e.g. animals in a specific pose, sky on top, ground on the bottom). In the latter case instead, given that the orientation is often arbitrary and simply given by the direction of the observation mean, image rotations around the top-down axis become meaningless for the correct classification or detection.

Therefore, we explicitly model this orientation bias by introducing an additional regularization, both to the supervised training and the incremental knowledge distillation, using a contrastive-based approach. Specifically, given a generic

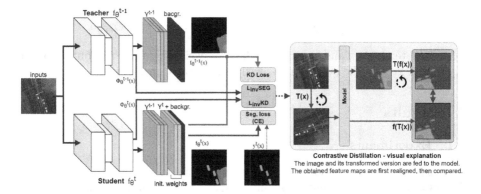

Fig. 1. Overview of the ICL setting on aerial images. For each step t, both the image x and its augmented version $T(x)$ are provided to the old (top) and new (bottom) models. New classes are trained with supervised training on the available ground truth (blue), while old categories are learned through KD (green). Last, features of the augmented inputs are confronted with the augmented features of the normal input, on both distillation and supervised training (red). (Color figure online)

input image x, we can obtain the output features of the current model $\phi_\theta(x)$ (thus excluding the final classifier). At the same time, given the same image transformed with augmentation T, the model should output a new activation, namely $\phi_\theta(T(x))$. Given the invariance to rotation, we can assume that both outputs are comparable, minus a transformation, which can be directly applied to the first activation. Formally, we can therefore introduce a regularization term at each learning step t, namely L_{inv}^{SEG} as:

$$L_{inv}^{SEG} = MSE(\phi_{\theta^t}(T(x)), \ T(\phi_{\theta^t}(x))) \qquad (5)$$

In other words, the additional term minimizes the differences between *the features of the model on the transformed image and the transformed features of the same model on the original image*, exploiting a Mean Squared Error between the features.

In an ICL setting, we are also interested in transferring the knowledge between $f_{\theta^{t-1}}$ and f_{θ^t}, so that the previous outputs are maintained as unaltered as possible. Together with the standard KD loss from Eq. (3), we can apply the same invariance principle between old and new models. More formally, at each step $t > 0$ we can introduce a further regularization as:

$$L_{inv}^{KD} = MSE(\phi_{\theta^t}(T(x)), \ T(\phi_{\theta^{t-1}}(x))) \qquad (6)$$

Simply put, this term minimizes the difference between the features of the new model derived from the transformed image and the transformed features of the old model, obtained from the non-augmented version of the input.

In summary our method comprises three regularizations, therefore the final loss to be minimized can be expressed as:

$$L(\theta^t) = L_{CE}(\theta^t) + \lambda L_{KD}(\theta^t) + \eta L_{inv}^{SEG}(\theta^t) + \rho L_{inv}^{KD}(\theta^t), \tag{7}$$

where the terms λ, η and ρ are scalar factors, weighting the contribution of the additional losses. The overall framework is illustrated in Fig. 1.

4   Results

4.1   Experiments

As described in the previous section, we build our method on top of the MiB framework, which represents a strong baseline for ICL in segmentation tasks. We perform all our experiments on the Potsdam dataset [21], a well known benchmark on aerial imagery providing an urban land cover subdivided into six classes: *impervious surfaces, building, low vegetation, trees, cars* and *clutter*. The dataset contains 38 large patches taken from the namesake city, where each patch has a fixed size of 6000×6000. Each patch comes with a sampling resolution of $5cm$ and provides five different modalities, namely: red (R), blue (B), green (G), infrared (IR) and a normalized digital surface map (DSM), all encoded as TIFF files. Given our focus on ICL, we only include in our tests inputs composed of RGB and RGBIR, discarding the additional surface map.

Every incremental set of labels is assumed to be disjoint from the previous ones. However, given the aerial setting, it is quite common that each image contains many of the available labels. For this reason, we first split the set into disjoint partitions, such that each split only contains a single label. Formally, considering a full dataset $D \subset X \times Y^{|H \times W|}$, we subdivide the available data into $|Y|$ disjoint partitions D_y such that $D_i \cap D_j = \emptyset \; \forall i, j \in Y$ where $i \neq j$, and each partition only contains a set of labels $Y_i = \{i, b\}$, i.e. the set of images is unique for each partition and each split only contributes to the whole training with a single label, or a generic background. Every incremental step will then include a variable number of classes, which will in turn require all the partitions corresponding to the involved categories.

4.2   Implementation Details

For all the experiments we adopted an encoder-decoder architecture with residual connections, based on the Res-UNet model [8]. Since memory requirements are crucial for the incremental setting, we introduce two optimizations: first, we swap the standard ResNet backbone with an equivalent yet more efficient TResNet with ImageNet pretraining [26]. The latter applies a series of optimizations aimed at maximizing the data throughput on GPU, while at the same time improving the performance over the classical residual architectures. For the experiments concerning four input channels, namely RGBIR, we expand the input layers duplicating the weights of the red channels, with a similar approach to [20]. Second, we apply in-place activated batch normalization also on the decoder, as proposed in [27], further reducing the memory footprint of the architecture.

We train the model for 80 epochs for each step, using AdamW as optimizer with learning rate of 10^{-3} and a cosine annealing scheduler, while reducing to 10^{-4} for the last steps. We adopt a batch size of 8, with effective size equal to 16 given that the pairs generated via contrastive augmentation are also exploited for the supervised training. Given the large size of the inputs, we tile the 6000×6000 images of the Potsdam dataset into patches with size 512×512 with overlap of 12 pixels, which is the minimum amount required to avoid partial tiles while also minimizing the replication of the image content. We perform robust data augmentation as in [8] in every setting, focusing on elastic transformations. Considering the contrastive regularization, we maintain the setting provided in [3]. We set the factors $\eta = \rho = 0.1$ in every test and evaluate as transformation random vertical and horizontal flipping, with rotations by 90-degree angles. In order to monitor the performances, we select 15% of the training set as validation. The final results are reported as F1 scores on the benchmark test set.

4.3   Potsdam Dataset

Given the high similarities among image patches and the uniform distribution of the labels among the tiles, the overlapped setting [3, 29], (i.e. images are kept even if they contain future classes), is not complex enough for a robust evaluation of the proposed regularizations. For this reason, we implement the *split* protocol described in Sect. 4.1: we first tile the original patches to obtain fixed-size input images, then we partition the dataset into 5 different disjoint sets, where each one is associated with a single label. Then, we randomly assign each tile to the smallest set among the labels present in the current tile, obtaining a uniform allocation of the data samples among the classes. This configuration can be seen as having 5 different datasets, where each one only contains a single type of annotation. The disjoint splits ensure that the model will work on unseen images at each step, further increasing the robustness of the tests.

We perform tests for two different configurations: first we replicate the testing scenario proposed in [29] where we suppose to receive, for the initial step, the labels for *building* and *trees*, then *impervious surfaces* and *low vegetation*, and as last step *car* (3-2-1). Second, we perform a more challenging test with the same order of labels, but provided sequentially (5S). For this last configuration, we exclude the *clutter* category, since it is not included in the official benchmarks [21]. Results for both configurations are shown in Table 1 and Table 2. Given the framework explicitly designed for segmentation, the MiB baseline performs reasonably well, even considering the fully sequential setting. However, the contrastive distillation approach consistently improves the performances in every experiment and every step, as reported in Sect. 4.3, even in the multi-spectral tests. We note that in the simpler 3-2-1 setting the RGB baseline performs on par with the regularized version. We argue because of both the effectiveness of the standard approach and the robust backbone pretrained on RGB images. However, in more challenging scenarios such as 5S, the contribution of the additional regularization is far more prominent, with a total increase over MiB of around 4% (Fig. 2).

Table 1. Class-wise and average results (F1 Score) obtained after 3 incremental steps (3-2-1). Double vertical lines indicate label groups for each step.

Method	Building	Tree	Clutter	Surf.	Low veg.	Car	Avg.
MiB (RGB)	0.9116	0.8217	0.2766	0.8918	0.7589	0.8500	0.7517
MiB + CD (RGB)	0.9209	0.8085	0.3119	0.9021	0.7619	0.8541	**0.7599**
MiB (RGBIR)	0.8708	0.8062	0.2682	0.8773	0.7414	0.8176	0.7303
MiB + CD (RGBIR)	0.9178	0.8190	0.3128	0.8950	0.7635	0.8515	**0.7598**

Table 2. Class-wise and avg. results (F1 Score) obtained after 5 incremental steps (5S).

Method	Building	Tree	Surfaces	Low veg.	Car	Avg.
MiB (RGB)	0.8451	0.7449	0.7912	0.7011	0.6759	0.7810
MiB + CD (RGB)	0.9015	0.7515	0.8848	0.7313	0.8287	**0.8195**
MiB (RGBIR)	0.8564	0.7007	0.8575	0.6862	0.8228	0.7847
MiB + CD (RGBIR)	0.8770	0.7740	0.8755	0.7343	0.8437	**0.8209**

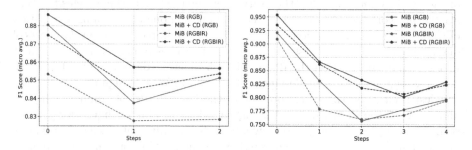

Fig. 2. Micro-averaged F1 scores over the incremental steps in the 3-2-1 configuration (left) and 5S (right). Blue indicates Contrastive Distillation (CD), dashed lines the RGBIR version. (Color figure online)

4.4   Ablation Study

In Table 3, we report an ablation study highlighting the contribution of our proposals, on the *split 5S* configuration with RGB input. We first start from a simple finetuning (FT): as expected, a new training without considering previous knowledge is detrimental for every step but the last. We then test on the MiB framework that already provides excellent results, with an average increment of more than 60% over the simple FT baseline. Naively introducing the single L_{inv}^{SEG}, (i.e. acting on the current step only) results in better scores for the last class, as expected. However, this negatively affects the performance on previously seen categories, which are not taken into consideration. On the other hand, applying the single L_{inv}^{KD} between current and old model allows for higher scores for previous categories, increasing the average score of 2%, though without obtaining any boost on the labels for the current step. Combining the two regularizations, it is possible to both improve over the current step and increase the performance over old classes, with a significant boost of around 4% over the strong MiB baseline and close to the theoretical upper bound of the *offline* test, representing a static multi-class learning over the whole set at once. As additional test for the distillation capabilities of the regularization, in the second row of Table 3 we report results for finetuning, using both the unbiased cross-entropy from [3] and CD, without actual distillation loss. The scores confirm that the additional losses actively contribute in maintaining previous knowledge (Fig. 3).

Table 3. Ablation study applied to the $5S$ setting, as class-wise F1 Scores of the last incremental step and averaged across classes.

Method	Building	Tree	Imp. surf.	Low veg.	Car	Average
FT	0.0	0.0	0.0	0.0	0.8708	0.1742
FT, Unb. CE + CD	0.6118	0.4927	0.6924	0.2909	0.5275	0.5231
MiB	0.8491	0.7625	0.8480	0.6751	0.7703	0.7810
MiB + L_{inv}^{SEG}	0.8178	0.7452	0.8514	0.6781	0.8186	0.7822
MiB + L_{inv}^{KD}	0.9079	0.7522	0.8815	0.7011	0.7895	0.8064
MiB + L_{inv}^{SEG} + L_{inv}^{KD}	0.9015	0.7515	0.8848	0.7313	0.8287	**0.8196**
Offline	0.9510	0.8535	0.9063	0.8415	0.8942	0.8893

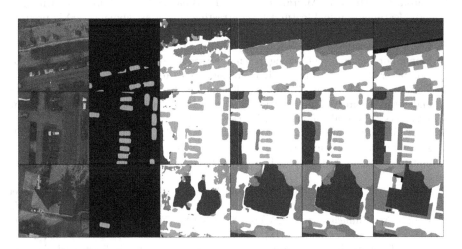

Fig. 3. From left to right: input, finetuning (FT), Finetuning with unbiased CE and Contrastive Distillation (FT + CD), Modelling the Background (MiB), MiB with Contrastive Distillation (MiB + CD), ground truth.

5   Conclusions

We addressed the problem of incremental learning in the context of semantic segmentation of aerial imagery, proposing a new regularization based on contrastive distillation to explicitly model the orientation invariance of such top-down images. In our experiments, we first provide benchmark results for the current state-of-the-art technique on natural images, already displaying excellent performances. We then demonstrate the effectiveness of our simple additional solution leveraging on the same framework, that consistently outperforms the strong baseline leading to a more stable sequential training. Nevertheless, incremental learning remains a challenging problem, especially considering different data sources and domains. Future works could provide more insight on this technique with additional datasets and explore more diverse scenarios, where datasets not only come with different annotations, but also from different domains.

Acknowledgements. This work was developed in the context of the Horizon 2020 projects SHELTER (grant agreement n.821282) and SAFERS (grant agreement n.869353).

References

1. Audebert, N., Le Saux, B., Lefèvre, S.: Beyond RGB: very high resolution urban remote sensing with multimodal deep networks. ISPRS J. Phot. Rem. Sens. **140**, 20–32 (2018)
2. Caron, M., Misra, I., Mairal, J., Goyal, P., Bojanowski, P., Joulin, A.: Unsupervised learning of visual features by contrasting cluster assignments. arXiv preprint arXiv:2006.09882 (2020)
3. Cermelli, F., Mancini, M., Rota Bulò, S., Ricci, E., Caputo, B.: Modeling the background for incremental learning in semantic segmentation. In: IEEE Conference on Computer Vision and Pattern Recognition, June 2020 (2020)
4. Chaitanya, K., Erdil, E., Karani, N., Konukoglu, E.: Contrastive learning of global and local features for medical image segmentation with limited annotations. In: Advances in Neural Information Processing System (2020)
5. Chen, L.C., Papandreou, G., Kokkinos, I., Murphy, K., Yuille, A.L.: DeepLab: semantic image segmentation with deep convolutional nets, atrous convolution, and fully connected CRFs. IEEE Trans. Pattern Anal. Mach. Intell. **40**(4), 834–848 (2017)
6. Chen, L.-C., Zhu, Y., Papandreou, G., Schroff, F., Adam, H.: Encoder-decoder with atrous separable convolution for semantic image segmentation. In: Ferrari, V., Hebert, M., Sminchisescu, C., Weiss, Y. (eds.) ECCV 2018. LNCS, vol. 11211, pp. 833–851. Springer, Cham (2018). https://doi.org/10.1007/978-3-030-01234-2_49
7. Chen, T., Kornblith, S., Norouzi, M., Hinton, G.: A simple framework for contrastive learning of visual representations. In: International Conference on Machine Learning, pp. 1597–1607. PMLR (2020)
8. Diakogiannis, F.I., Waldner, F., Caccetta, P., Wu, C.: ResUNet-a: a deep learning framework for semantic segmentation of remotely sensed data. ISPRS J. Photogram. Rem. Sens. **162**, 94–114 (2020)

9. Feng, Y., Sun, X., Diao, W., Li, J., Gao, X., Fu, K.: Continual learning with structured inheritance for semantic segmentation in aerial imagery. IEEE Trans. Geosci. Rem. Sens. **60**, 1–17 (2021)

10. He, K., Fan, H., Wu, Y., Xie, S., Girshick, R.: Momentum contrast for unsupervised visual representation learning. In: IEEE Conference on Computer Vision and Pattern Recognition, pp. 9729–9738 (2020)

11. Khosla, P., et al.: Supervised contrastive learning. In: Advances in Neural Information Processing System, vol. 33, pp. 18661–18673 (2020)

12. Li, Z., Hoiem, D.: Learning without forgetting. IEEE Trans. Pattern Anal. Mach. Intell. **40**(12), 2935–2947 (2017)

13. Loghmani, M.R., Robbiano, L., Planamente, M., Park, K., Caputo, B., Vincze, M.: Unsupervised domain adaptation through inter-modal rotation for RGB-D object recognition. IEEE Robot. Autom. Lett. **5**(4), 6631–6638 (2020). https://doi.org/10.1109/LRA.2020.3007092

14. Long, J., Shelhamer, E., Darrell, T.: Fully convolutional networks for semantic segmentation. In: IEEE Conference on Computer Vision and Pattern Recognition, pp. 3431–3440 (2015)

15. Mallya, A., Lazebnik, S.: PackNet: adding multiple tasks to a single network by iterative pruning. In: IEEE/CVF Conference on Computer Vision and Pattern Recognition (CVPR), June 2018, pp. 7765–7773 (2018). https://doi.org/10.1109/CVPR.2018.00810

16. McCloskey, M., Cohen, N.J.: Catastrophic interference in connectionist networks: the sequential learning problem. Psych. Learn. Motiv. **24**, 109–165 (1989)

17. Misra, I., van der Maaten, L.: Self-supervised learning of pretext-invariant representations. In: IEEE Conference on Computer Vision and Pattern Recognition, June 2020 (2020)

18. Nogueira, K., Dalla Mura, M., Chanussot, J., Schwartz, W.R., dos Santos, J.A.: Learning to semantically segment high-resolution remote sensing images. In: International Conference on Pattern Recognition, pp. 3566–3571 (2016)

19. Noroozi, M., Favaro, P.: Unsupervised learning of visual representations by solving jigsaw puzzles. In: Leibe, B., Matas, J., Sebe, N., Welling, M. (eds.) ECCV 2016. LNCS, vol. 9910, pp. 69–84. Springer, Cham (2016). https://doi.org/10.1007/978-3-319-46466-4_5

20. Pan, B., Shi, Z., Xu, X., Shi, T., Zhang, N., Zhu, X.: CoinNet: copy initialization network for multispectral imagery semantic segmentation. IEEE Geos. Rem. Sens. Lett. **16**(5), 816–820 (2019). https://doi.org/10.1109/LGRS.2018.2880756

21. The International Society for Photogrammetry and Remote Sensing: Potsdam dataset (2018)

22. Pielawski, N., et al.: CoMIR: contrastive multimodal image representation for registration. In: Advances in Neural Information Processing Systems, vol. 33, pp. 18433–18444 (2020)

23. Piramanayagam, S., Saber, E., Schwartzkopf, W., Koehler, F.W.: Supervised classification of multisensor remotely sensed images using a deep learning framework. Rem. Sens. **10**(9) (2018). https://doi.org/10.3390/rs10091429

24. Qi, K., Yang, C., Hu, C., Shen, Y., Shen, S., Wu, H.: Rotation invariance regularization for remote sensing image scene classification with convolutional neural networks. Rem. Sens. **13**(4) (2021). https://doi.org/10.3390/rs13040569

25. Rebuffi, S.A., Kolesnikov, A., Sperl, G., Lampert, C.H.: iCaRL: incremental classifier and representation learning. In: IEEE Conference on Computer Vision and Pattern Recognition, pp. 2001–2010 (2017)

26. Ridnik, T., Lawen, H., Noy, A., Friedman, I.: TResNet: high performance GPU-dedicated architecture. In: Winter Conference on Applications of Computer Vision, pp. 1399–1408 (2021)
27. Rota Bulò, S., Porzi, L., Kontschieder, P.: In-place activated batchnorm for memory-optimized training of DNNs. In: IEEE Conference on Computer Vision and Pattern Recognition (2018)
28. Singh, S., et al.: Self-supervised feature learning for semantic segmentation of overhead imagery. In: The British Machine Vision Conference, vol. 1, p. 4 (2018)
29. Tasar, O., Tarabalka, Y., Alliez, P.: Incremental learning for semantic segmentation of large-scale remote sensing data. IEEE J. Sel. Top. App. Earth Observ. Rem. Sens. **12**(9), 3524–3537 (2019)
30. Valada, A., Mohan, R., Burgard, W.: Self-supervised model adaptation for multimodal semantic segmentation. Int. J. Comput. Vis. **128**(5), 1239–1285 (2020)
31. Wang, G., Wang, X., Fan, B., Pan, C.: Feature extraction by rotation-invariant matrix representation for object detection in aerial image. IEEE Geos. Rem. Sens. Lett. **14**(6), 851–855 (2017). https://doi.org/10.1109/LGRS.2017.2683495
32. Yang, S., Yu, S., Zhao, B., Wang, Y.: Reducing the feature divergence of RGB and near-infrared images using switchable normalization. In: IEEE Conference on Computer Vision and Pattern Recognition Workshop, June 2020, pp. 206–211 (2020). https://doi.org/10.1109/CVPRW50498.2020.00031
33. Yuan, Q., Shafri, H.Z.M., Alias, A.H., Hashim, S.J.: Multiscale semantic feature optimization and fusion network for building extraction using high-resolution aerial images and LiDAR data. Rem. Sens. **13**(13), 2473 (2021). https://doi.org/10.3390/rs13132473
34. Zenke, F., Poole, B., Ganguli, S.: Continual learning through synaptic intelligence. In: International Conference on Machine Learning, ICML 2017, vol. 70, pp. 3987–3995 (2017)
35. Zhao, H., Shi, J., Qi, X., Wang, X., Jia, J.: Pyramid scene parsing network. In: IEEE Conference on Computer Vision and Pattern Recognition, July 2017 (2017)

Image Novelty Detection Based on Mean-Shift and Typical Set Size

Matthias Hermann[1($\boxtimes$)], Bastian Goldlücke[2], and Matthias O. Franz[1]

[1] HTWG Konstanz, Konstanz, Germany
mhermann@htwg-konstanz.de
[2] Universität Konstanz, Konstanz, Germany

Abstract. The detection of anomalous or novel images given a training dataset of only clean reference data (inliers) is an important task in computer vision. We propose a new shallow approach that represents both inlier and outlier images as ensembles of patches, which allows us to effectively detect novelties as mean shifts between reference data and outliers with the Hotelling T^2 test. Since mean-shift can only be detected when the outlier ensemble is sufficiently separate from the typical set of the inlier distribution, this typical set acts as a *blind spot* for novelty detection. We therefore minimize its estimated size as our selection rule for critical hyperparameters, such as, e.g., the size of the patches is crucial. To showcase the capabilities of our approach, we compare results with classical and deep learning methods on the popular datasets MNIST and CIFAR-10, and demonstrate its real-world applicability in a large-scale industrial inspection scenario.

Keywords: Image novelty detection · Independent component analysis · Mean-shift

1 Introduction

Novelty detection is a semi-supervised approach to anomaly detection where all available training data belongs to a single class. The task is to learn the class boundary of the reference class such that the model can classify test data into known (inliers) and novel examples (outliers). Such models output a score based on a single input example that can be used for classification. As a consequence, the decision process is not robust against noise contamination or overlapping distributions between inliers and outliers.

This motivates the main idea of our approach to novelty detection: representing both training and test images as ensembles of image patches. Instead of establishing a relation of a single test data point to an inlier distribution, this allows us to compare the training and test ensembles against each other which is inherently more robust. The literature provides a broad range of statistics for testing whether two datasets originate from the same distribution. Here, we use

Supported by Bundesministerium für Bildung und Forschung (BMBF, 01IS19083A).

S. Sclaroff et al. (Eds.): ICIAP 2022, LNCS 13232, pp. 755–766, 2022.
https://doi.org/10.1007/978-3-031-06430-2_63

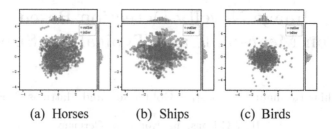

(a) Horses   (b) Ships   (c) Birds

Fig. 1. Patch distribution of the two largest principal components of CIFAR-10 showcasing different mean shifts between reference data (blue) and anomalies (yellow). Note the considerable overlap of both distributions. (Color figure online)

the Hotelling T^2 test [6] for this purpose. It is based on computing the mean shift between the training and test ensembles which is particularly simple to compute and therefore suitable for large datasets.

In our experiments, we found that the success of this approach critically depends on the details of how the patch ensemble is extracted from the input images. The most important parameters are the number and the size of the patches and the image basis in which the patches are represented. These parameters depend on the specific dataset and thus have to be found by an automatic model selection procedure. Our approach is based on the observation that a mean shift can only be detected when the outlier ensemble is sufficiently separate from the *typical set* of the inlier distribution. The typical set has a total probability close to one which is a consequence of the asymptotic equipartition property [AEP; 3]. Any outlier which falls inside the volume occupied by the typical set has no chance of being detected (see Fig. 1). Thus, the typical set acts as a *blind spot* for novelty detection which needs to be kept as small as possible. By estimating the volume of the typical set, an optimal parameter set can be found for each dataset.

In this work, we contribute a shallow algorithm based on the Hotelling T^2 test and independent component analysis (ICA) for image novelty detection with overlapping distributions. Notably, the hyperparameters of our model are selected using typical set theory for finding a patch ensemble which optimally represents the input images. We show in extensive experiments on raw pixel data that our approach not only achieves comparable results to Deep Learning approaches on MNIST and CIFAR-10, but is also applicable to a large-scale industrial inspection scenario, due to its simple architecture and fast predictions[1].

2   Related Work

We refer the reader to [14] for a good overview over the general techniques for novelty detection and focus on images in the following. In terms of image novelty

---

[1] https://github.com/matherm/mean-shift.

detection, current approaches rely on compression [1], generative models [4], feature engineering [18], or known statistics of images [7]. Well-known instances of the compression class are variational Autoencoders [9] or probabilistic PCA [17]. Besides, there exist autoencoders that are specialized to novelty detection, such as Latent Space Autoregression [1]. Density-based approaches model the training data with a statistical parametric model $\mathbf{x} \sim \mathbf{p}_\theta(\mathbf{x})$, where the parameters θ are learned by maximizing the likelihood function [4]. Here, the densities of input examples $\mathbf{p}_\theta(\mathbf{x})$ are directly available, and anomalies can be classified by using this density as a scoring rule. Kernel space methods project the data into a feature space $\mathcal{F}$ by computing a nonlinear mapping $\phi(\mathbf{x})$. Anomaly detection is conducted in feature space, where typically the norm of $||\phi(\mathbf{x})||$ is used as scoring function [15]. Natural image statistics provide image priors that can be used to derive suitable feature spaces for novelty detection [7]. Our approach is related in the sense that we also optimize within the independent components (ICA) framework, although our starting point is quite different. In contrast to existing methods, we propose to transform the input images into patches and to compute statistics of patch ensembles. A similar approach was proposed by [11] in an out-of-distribution scenario, where they used a hypothesis testing framework to test for typicality. However, they analyzed ensembles of multiple input images, instead of ensembles of patches of a single input image, and they did not optimize the typical set size.

3   Data Preparation

Our method first transforms a given data set of reference input images $\mathbf{I}_0, \cdots, \mathbf{I}_N$ into an ensemble $\mathbf{X}$ of patches. We describe these preparatory steps in this section. For a given test image $\mathbf{I}^*$, we apply the same preprocessing and extract an ensemble $\mathbf{x}^*$ of patches. We then measure the mean-shift of the ensemble mean $\mu(\mathbf{x}^*)$ with respect to the mean of all given training patches $\mu(\mathbf{X})$. The anomaly score $\mu\text{shift}(\mathbf{x})$ is based on the Hotelling T^2 test [6] and derived in Sect. 4. Finally, we describe how to automatically select optimal hyperparameters for the method in Sect. 5.

We process the input examples $\mathbf{I}_i(x, y)$ on a common scale and apply contrast normalization as preprocessing [7]. The normalization step centers and projects all given input examples onto the unit sphere with respect to the L_2-norm. This is achieved by first removing the pixel-wise mean and rescaling afterwards,

$$\mathbf{I}' = \frac{\mathbf{I} - \frac{1}{D}\sum_{x,y}\mathbf{I}(x,y)}{\left\|\mathbf{I} - \frac{1}{D}\sum_{x,y}\mathbf{I}(x,y)\right\|_{L_2}}, \tag{1}$$

where D is the number of pixels - treating colors as additional pixels - in the image. To ensure the numerical stability of the algorithm, we globally divide all examples by the standard deviation $\text{std}(\mathbf{I}')$ over all of the N training examples. Having a set of preprocessed images $\mathcal{I} = \{\mathbf{I}'_0/s, \ldots, \mathbf{I}'_N/s\}$, the important part of our algorithm is to generate patch ensembles, instead of processing the full image. There are several possible strategies for cropping square image patches from an image and we tested different sampling strategies without noticing significant differences. Therefore, we propose to simply crop the patches in a sliding window fashion and extract all *valid* patches inside the image without crossing the border. The horizontal and vertical stride τ of the sliding window helps controlling the total number of cropped patches. We did not notice a performance-critical impact of this parameter and keep it fixed to $\tau = 2$. This means that the maximum number S of distinct image patches per input image is only limited by the size D of the image and the patch size P, i.e., the larger the patch size, the fewer patches can be extracted. To distinguish between an input data point and ensembles of patches, we denote a single patch of the i-th example by $\mathbf{x}_i$ and use $\mathbf{x}_i(s)$ for indexing the ensemble where necessary. We flatten the extracted patches with c color channels and organize the vectors of all computed reference patches in a long design matrix $\mathbf{X} \in \mathbb{R}^{NS \times M}$, where $M = cP^2$.

4 Mean-Shift Detection

We perform mean-shift detection with the Hotelling T^2 test [6]. This is a multivariate generalization of Student's t-test and allows for computing the significance of mean-shifts between two populations. First, we present the test in its classical form. In the second part, we derive a feature space interpretation that reveals relevant hyperparameters that are needed for model selection in Sect. 5.

In our case, we compare the pixel-wise mean

$$\mu = \frac{1}{NS} \sum_{i}^{N} \sum_{s}^{S} \mathbf{x}_i(s) \tag{2}$$

computed over all training patches with the pixel-wise mean

$$\mu^* = \frac{1}{S} \sum_{s}^{S} \mathbf{x}^*(s). \tag{3}$$

of the patches $\mathbf{x}^*$ extracted from a single test example. The unnormalized Hotelling T^2 test statistic for a dependent test sample is given by

$$\tilde{T}^2 = (\mu^* - \mu)^T \Sigma^{-1} (\mu^* - \mu), \tag{4}$$

where

$$\Sigma = \frac{1}{NS - 1} (\mathbf{X} - \mu)^T (\mathbf{X} - \mu) \tag{5}$$

is the covariance matrix of the training dataset $\mathbf{X}$. In other words, the Hotelling $\tilde{T}^2$ statistic is the Mahalanobis distance between the two mean vectors. Note that because our samples are equally sized, we neglect the constant normalization factor $\frac{NS^2}{NS+S}$ for simplicity.

To obtain a feature space interpretation, the Mahalanobis distance can be computed by first whitening the data with a whitening transformation $\mathbf{A}$ and then computing the standard L_2 distance of the whitened mean vectors. This allows us to reveal relevant hyperparameters, such as the noise floor and the rotation freedom. Due to the linearity of $\mathbf{A}$, this is equivalent to applying $\mathbf{A}$ to the mean difference vector in the original pixel space:

$$\tilde{T}^2 = \|\mathbf{A}(\boldsymbol{\mu}^* - \boldsymbol{\mu})\|_{L_2} \tag{6}$$

The whitening transformation $\mathbf{A}$ can be decomposed into an orthogonal matrix $\mathbf{W}$ containing the Eigenvectors of the covariance matrix $\boldsymbol{\Sigma}$ as columns, a diagonal scaling matrix $\mathbf{S}^{-1/2}$, and an arbitrary rotation matrix $\mathbf{R}$, such that

$$\mathbf{A} = \mathbf{R}\mathbf{S}^{-1/2}\mathbf{W}, \tag{7}$$

with $\boldsymbol{\Sigma} = \mathbf{W}^{\mathbf{T}}\mathbf{S}\mathbf{W}$, $\mathbf{W}\mathbf{W}^{\mathbf{T}} = \mathbf{I}$, and $\mathbf{R} \in SO(M)$. The matrix $\mathbf{S}$ consists of the variances $s_i, \cdots, s_M$ along the components in $\mathbf{W}$.

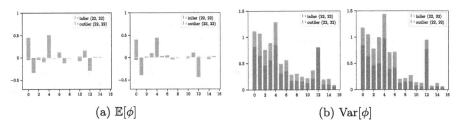

(a) $\mathbb{E}[\phi]$ (b) $\mathrm{Var}[\phi]$

Fig. 2. Mean-shifts and variances of the first 16 principal components of ϕ for the deer class of CIFAR-10 with patch size $P = 22$. The left half of (a) and (b) shows the statistics with unoptimized rotation matrix, the right half after optimization.

In this transformation, we also reduce the data dimension M to k by removing the noise floor. This is done by truncating the matrices $\mathbf{W}$ and $\mathbf{S}$ by removing the dimensions with smallest variance. Across experiments, we found it helpful to control the number of informative variables k by a rule, instead of fixing the number of retained features. We retain all components up to a fixed threshold of explained variance [7], in our case 90%. We denote this number by $k = k_{90}$. Furthermore, it is known that whitened data stay whitened under rotation, so we can apply an arbitrary rotation matrix $\mathbf{R}$ without changing the $\tilde{T}^2$ statistic. We will cover the choice of this rotation freedom in more detail in Sect. 5 and show why it is crucial for model selection.

For visualization, it is useful to decompose the $\tilde{T}^2$ statistic into the feature vector

$$\phi(\mathbf{x}) = \mathbf{A}\boldsymbol{\mu}(\mathbf{x}) \tag{8}$$

and the derived anomaly score

$$\mu\mathrm{shift}(\mathbf{x}) = \|\phi(\mathbf{x}) - \phi(\mathbf{X})\|_{L_2}. \tag{9}$$

Figure 1 depicts the two largest components of $\phi(\mathbf{x})$ and $\phi(\mathbf{x}^*)$ of the CIFAR-10 dataset showing the large overlap between both distribution. Nevertheless, a mean shift is still detectable in some components despite of the large variance of the feature vectors $\phi(\mathbf{x})$, see Fig. 2.

5   Typical Set Size Minimization

For learning the model, we need to find a suitable whitening matrix $\mathbf{A}$ of the image patches. The matrices $\mathbf{S}^{-1/2}$ and $\mathbf{W}$ can be computed by standard Eigenvector decomposition of the covariance matrix $\boldsymbol{\Sigma}$. However, during this procedure, also the remaining hyperparameters $\Theta = \{P, \mathbf{R}\}$ need to be set. Again, P is the patch size, and $\mathbf{R}$ the arbitrary rotation. Figure 3 shows the visual impact of the patch size P on the ensemble statistics. Depending on the size, the appearance of the classes changes drastically, particularly in terms of dissimilarity between neighboring image patches. This observation motivates the use of an entropy-related measure of dissimilarity or disorder for hyperparameter selection. Since we do not observe the outliers, we can only manipulate the statistics of the transformed reference data. In the introduction, we argued that a good strategy for model selection is to keep the size $|\mathcal{A}(\cdot)|$ of the typical set as small as possible as this limits the blind spot of the mean shift detection mechanism. A central relationship between the size of the typical set and the entropy of the feature distribution [3,11] is

$$\log |\mathcal{A}(\phi)| \leq f(\mathcal{H}[\phi]), \tag{10}$$

where $f(\cdot)$ is a monotonically increasing function which satisfies certain constraints. This means that in order to keep the *blind spot* small, we need to minimize the entropy of ϕ.

Directly minimizing entropy of stochastic variables is heavily studied in the field of sparse coding and Independent Component Analysis [ICA; 2]. A central measure in that field is the so-called negentropy, which is the negative of entropy. Negentropy has an appealing feature that arises from the maximum entropy principle, i.e., given a fixed variance the maximum entropy distribution is a Gaussian [3]. This relation can be utilized by the construction of a negentropy approximation [2] that uses the Gaussian distribution as contrast

$$\mathcal{J}[\phi] \propto \sum_i^k (g(\phi_i) - g(\gamma))^2, \tag{11}$$

where $g = \log \cosh(\cdot)$, $\gamma \sim \mathcal{N}(0,1)$, and ϕ is centered. As a consequence, the model selection rule simplifies to a linear search over the patch size P and a non-convex optimization of the rotation matrix $\mathbf{R}$,

$$\underset{P,\mathbf{R}}{\operatorname{argmax}} \mathcal{J}[\phi], \tag{12}$$

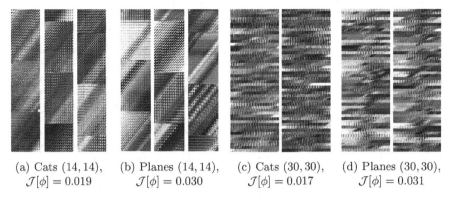

(a) Cats $(14, 14)$,   (b) Planes $(14, 14)$,   (c) Cats $(30, 30)$,   (d) Planes $(30, 30)$,
$\mathcal{J}[\phi] = 0.019$ $\mathcal{J}[\phi] = 0.030$ $\mathcal{J}[\phi] = 0.017$ $\mathcal{J}[\phi] = 0.031$

Fig. 3. Patch ensembles from two classes of CIFAR-10, cats and planes, with two different patch sizes. The plane class has a more homogeneous appearance (e.g. blue sky), while the cat class is more chaotic (e.g. cat pose) yielding a smaller negentropy $\mathcal{J}$. (Color figure online)

with $P \in [14, \sqrt{D/c} - \tau]$ and $\mathbf{R} \in \mathrm{SO}(k)$. We chose 14 as minimum patch size to avoid the pathological case of selecting too small patches containing zero image content, such as black spots in MNIST. Again, the rotation matrix $\mathbf{R}$ needs only to be optimized as this model freedom highly impacts the negentropy measure, but does not change the Mahalanobis distance (Eq. 4). While P is found by a grid search, optimizing the rotation matrix $\mathbf{R}$ is a non-convex problem, in particular, the solution is constrained to be an orthogonal matrix. As a first step, we decompose the gradient via the chain rule,

$$\nabla_{\mathbf{R}} \mathcal{J} = \nabla_\phi \mathcal{J} \nabla_{\mathbf{R}} \phi, \tag{13}$$

where $\nabla_\phi \mathcal{J}$ is a $1 \times k$ row vector and $\nabla_{\mathbf{R}} \phi$ is a $k \times k^2$ matrix. We organize the resulting $1 \times k^2$ gradient $\nabla_{\mathbf{R}} \mathcal{J}$ as $k \times k$ matrix for further processing. Note that our problem is different to standard ICA, as we are computing the gradient w.r.t. to the average across multiple patches (cf. Eq. 8), instead of a single patch.

The orthogonality constraint can be enforced by performing gradient ascent only inside the Lie group $\mathrm{SO}(k)$ [*gradient flow*; 12]. Hence, the gradient $\nabla_{\mathbf{R}} \mathcal{J}$ of the loss function represents an infinitesimal rotation which has the form of a skew-symmetric matrix. The set of all skew-symmetric matrices is called the Lie algebra $\mathfrak{so}(k)$ associated with the Lie group $\mathrm{SO}(k)$. Every skew-symmetric matrix $\mathbf{\Theta}$ can be uniquely parameterized by a vector $\mathbf{r}$ of dimension $k(k-1)/2$, denoted as the Plücker coordinates. The rotation matrix $\mathbf{R}$ associated to its generating skew-symmetric matrix $\mathbf{\Theta}$ is $\mathbf{R} = \exp(\mathbf{\Theta})$.

In order to compute a valid gradient step beyond the neighborhood of $\mathbf{R}$, the gradient direction needs to be expressed by the Lie bracket. We refer to [12] for the full derivation of the relation between the two gradient expressions using the commutator

$$\nabla_{\mathbf{\Theta}} \mathcal{J} = (\nabla_{\mathbf{R}} \mathcal{J})^T \mathbf{R} - \mathbf{R}^T (\nabla_{\mathbf{R}} \mathcal{J}). \tag{14}$$

	SVM	KDE	VAE	LSA	DSVD	μshift
Plane	0.630	0.658	0.688	**0.735**	0.617	0.731
Car	0.440	0.520	0.403	0.580	0.659	**0.711**
Bird	0.649	0.657	0.679	**0.690**	0.508	0.498
Cat	0.487	0.497	0.528	0.542	0.591	**0.609**
Deer	0.735	0.727	0.748	**0.761**	0.609	0.582
Dog	0.500	0.496	0.519	0.546	**0.657**	0.620
Frog	0.725	**0.758**	0.695	0.751	0.677	0.724
Horse	0.533	0.564	0.500	0.535	0.673	**0.718**
Ship	0.649	0.680	0.700	0.717	0.759	**0.805**
Truck	0.508	0.540	0.398	0.548	0.730	**0.751**
	0.586	0.610	0.586	0.641	0.648	**0.675**

Table 1. AUC on CIFAR-10.

	SVM	KDE	VAE	LSA	DSVD	μshift
0	0.988	0.885	**0.998**	0.993	0.980	0.997
1	0.999	0.996	**0.999**	0.999	0.997	0.993
2	0.902	0.710	0.962	0.959	0.917	**0.986**
3	0.950	0.693	0.947	0.966	0.919	**0.979**
4	0.955	0.844	0.965	0.956	0.949	**0.971**
5	0.968	0.776	0.963	0.964	0.885	**0.981**
6	0.978	0.861	0.995	0.994	0.983	**0.995**
7	0.965	0.884	0.974	**0.980**	0.946	0.973
8	0.853	0.669	0.905	0.953	0.939	**0.969**
9	0.955	0.825	0.978	**0.981**	0.965	0.977
	0.951	0.814	0.969	0.975	0.948	**0.982**

Table 2. AUC on MNIST.

We initially set $\mathbf{R}_0 \sim \mathcal{N}(0, \mathbf{I})$ and compute the feature vectors ϕ_i using Eq. 8. Afterwards we compute the negentropy $\mathcal{J}$ using Eq. 11 and the gradient w.r.t. $\mathbf{R}$ using Eq. 14. From there, we compute the corresponding parameter vector $\mathbf{r}$ for gradient ascent by taking the upper triangular matrix of $\nabla_{\boldsymbol{\Theta}}\mathcal{J}$ corresponding to the Plücker coordinates. Unfortunately, rotation matrices for $k > 2$ are not commutative. Hence we cannot make additive steps of ascent in $\mathfrak{so}(k)$ and need to map between the Lie algebra and $\mathrm{SO}(k)$ in every iteration by using $\exp(\boldsymbol{\Theta})$. The *gradient flow* update rule is then given by

$$\mathbf{R}_{i+1}^T = \exp(\eta\,\boldsymbol{\Theta}_{\mathbf{r}_i})\,\mathbf{R}_i^T, \tag{15}$$

with step size η and the skew-symmetric matrix $\boldsymbol{\Theta}_{\mathbf{r}_i}$ parametrized by $\mathbf{r}_i$ at the i-th iteration step. For acceleration, we use the ADAM optimizer [8] with $\eta = 1e^{-2}$ and optimize for 20 epochs at maximum (Table 1).

6   Evaluation

We compare our approach to the following standard algorithms: OC-SVM [10], Kernel Density Estimator (KDE), and Variational Autoencoder (VAE) [9]. Furthermore, we compare two algorithms especially designed for one-class classification and not rely on prior-knowledge or transfer learning: Latent Space Autoregression (LSA) [1][2] and Deep SVD (DSVD) [15][3]. We use the standard measure of area under the receiver operating characteristic curve (AUC) for evaluating performance [16].

First, we test the method on the CIFAR-10 challenge. For training, we use all examples from the training split of a single class. Inlier data is the corresponding test split of the particular class, consisting of 1000 examples. For the outlier data, the same number of examples is sampled from the other classes of the

---

[2] https://github.com/ChangYungHua/Latent-space-AR/.

[3] https://github.com/lukasruff/Deep-SVDD-PyTorch.

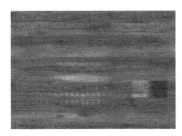

Fig. 4. Example of a scanned wooden texture with nozzle faults and color drifts from the digital printing dataset. Note that the nozzle faults are only 2px and therefore difficult to see.

Table 3. AUC on the industrial digital printing dataset.

	SVM	KDE	VAE	LSA	DSVD	μshift
0	0.812	0.785	0.802	0.717	0.910	**0.924**
1	0.513	0.542	0.484	0.614	**0.676**	0.652
2	0.591	0.501	0.606	0.499	0.648	**0.678**
3	0.534	**0.820**	0.584	0.700	0.775	0.609
4	0.615	0.563	0.487	0.523	**0.692**	0.616
5	0.646	0.545	0.645	0.541	0.594	**0.686**
6	0.573	**0.703**	0.573	0.551	0.671	0.646
7	0.441	0.518	0.572	0.529	**0.673**	0.571
8	0.838	0.559	0.802	0.433	0.545	**0.844**
9	0.609	0.676	0.743	0.647	0.736	**0.758**
	0.617	0.621	0.629	0.575	0.692	**0.698**

test split. Table 2 shows the results of the experiment. We see that on average μshift yields the best results on CIFAR-10. It is interesting to note that the bird class is difficult for DSVD and ours. We attribute this result to the almost orthogonal feature spaces of inliers and outliers in this case (see further discussion in Sect. 6.2). As a second example, we evaluate the methods on the MNIST challenge. Again, for training, we use all examples from the training split of a single class, consisting of 6000 examples. Inlier data is the corresponding test split from the same class, consisting of 1000 examples. For the outlier data the same number of examples is sampled from the other classes of the test split. Table 2 summarizes the results. The image statistics of MNIST are completely different from CIFAR-10. As a result, we see that a different set of algorithms performs better on this dataset. However, μshift consistently achieves the best results on the average.

Finally, we also test performance on an industrial digital printing dataset. The dataset consists of ten different wooden textures that are printed and scanned by an industrial inspection system at a resolution of 300 dpi. Figure 4 shows the clean reference, an error-free scan, and a scan with printing anomalies, such as nozzle faults and color drifts. For training, the reference scan was divided into 7100 square non-overlapping 96 × 96 RGB images. The 3308 inlier examples were gathered in the same way from the additional error-free scan, the 3308 outlier examples from the anomalous scan. The results in Table 3 show a similar trend to CIFAR-10 and MNIST, confirming the comparatively good performance of μshift in a large-scale industrial inspection scenario with large patch sizes.

6.1   Runtime

Due to the minimalistic design of the algorithm, the algorithm can operate very efficiently on unseen test data. This is of particular importance for industrial use cases. The computation of the test scores for the digital printing dataset took about 800 ms (8270 FPS) on average using a single GPU 1080 GTX. The most expensive operations are the extraction of the S patches and the single $k \times M$ matrix multiplication. Note, that both operations can be implemented efficiently with a single 2D convolution plus a spatial averaging operation.

6.2   Limitations

The way we construct our features before applying the Hotelling T^2 test assumes that the feature spaces of the reference and outlier examples are largely overlapping and that there is a single mode. If this is not the case, it might well happen that many outliers are located in orthogonal directions with respect to the feature space of the inliers. As a result, many features of these examples are mapped to the null space of the whitening transform such that they cannot contribute to a detectable mean shift. We already suspected that this effect might be responsible for the poor performance of μshift on the bird class example in Table 2 since the bird class in CIFAR-10 is very different from the other classes. In order to confirm this intuition, we created an artificial test problem where we took our references and inliers from the cats class of CIFAR-10 whereas the outliers came from the three class of MNIST. Here, the outliers occupy a very low-dimensional feature subspace that accounts for only a small proportion of the variance in the reference class. Table 4 clearly shows that a mean-shift detection does not work in this case. Interestingly, when reversing directions and using MNIST as the reference class, mean-shift detection becomes possible again, because the relatively rich feature space of the cats class has enough variance also in the small subspace occupied by the MNIST features.

Table 4. AUC of mean-shift detection for non-overlapping feature spaces.

	Cats vs. 3 ($\mathcal{J}[\phi]$)	3 vs. Cats ($\mathcal{J}[\phi]$)
$P = 14$	0.36 (0.020)	1.00 (0.005)
$P = 18$	0.15 (0.016)	0.99 (0.005)
$P = 22$	0.12 (0.024)	1.00 (0.008)
$P = 26$	0.11 (0.031)	1.00 (0.008)

7   Conclusion

In this work, we propose a new algorithm for the task of image novelty detection based on mean-shifts between the reference and the overlapping outlier patch distributions. The decision function is based on comparing patch ensembles instead of entire images which leads to an increased robustness and sensitivity of the algorithm. The chosen patch size and representation turns out to be the critical parameters for the success of this approach. A choice of these parameters based on minimizing the size of the critical set leads to satisfactory results that can even surpass deep learning methods. Our experiments show consistent applicability over multiple datasets, both in standard novelty detection tasks and in an industrial application scenario. Moreover, the computational load of the proposed algorithm is relatively small which recommends its application to large scale problems. A limitation of our method arises when the feature spaces of inliers and outliers do not overlap or multiple modes appear. This causes the outlier features to fall into the null-space of the whitening transform which makes mean-shift detection impossible. For practical applications, domain knowledge needs to be used to decide whether this special case prevails in the problem at hand. If so, mean-shift detection on raw pixels is not desirable and utilizing prior knowledge, feature extractors (e.g., [5,13]) or models based on other loss functions, such as reconstruction loss, are better suited.

References

1. Abati, D., Porrello, A., Calderara, S., Cucchiara, R.: Latent space autoregression for novelty detection. In: Proceedings of the IEEE Conference on Computer Vision and Pattern Recognition, pp. 481–490 (2019)
2. Comon, P.: Independent component analysis, a new concept? Signal Process. **36**(3), 287–314 (1994)
3. Cover, T.M.: Elements of Information Theory. Wiley, New York (1999)
4. Dinh, L., Sohl-Dickstein, J., Bengio, S.: Density estimation using real NVP. In: ICLR (Poster) (2016)
5. Hendrycks, D., Mu, N., Cubuk, E.D., Zoph, B., Gilmer, J., Lakshminarayanan, B.: Augmix: a simple data processing method to improve robustness and uncertainty. arXiv preprint arXiv:1912.02781 (2019)
6. Hotelling, H.: The generalization of student's ratio. In: Kotz, S., Johnson, N.L. (eds.) Breakthroughs in Statistics, pp. 54–65. Springer, Heidelberg (1992). https://doi.org/10.1007/978-1-4612-0919-5_4
7. Hyvärinen, A., Hurri, J., Hoyer, P.O.: Natural Image Statistics: A Probabilistic Approach to Early Computational Vision, vol. 39. Springer, Heidelberg (2009). https://doi.org/10.1007/978-1-84882-491-1
8. Kingma, D.P., Ba, J.L.: Adam: a method for stochastic optimization. In: ICLR 2015: International Conference on Learning Representations (2015)
9. Kingma, D.P., Welling, M.: Stochastic gradient VB and the variational auto-encoder. In: Second International Conference on Learning Representations, ICLR, vol. 19 (2014)
10. Manevitz, L.M., Yousef, M.: One-class SVMs for document classification. J. Mach. Learn. Res. **2**, 139–154 (2001)

11. Nalisnick, E., Matsukawa, A., Teh, Y.W., Lakshminarayanan, B.: Detecting out-of-distribution inputs to deep generative models using a test for typicality. arXiv preprint arXiv:1906.02994, vol. 5, p. 5 (2019)
12. Plumbley, M.D.: Geometrical methods for non-negative ICA: manifolds, lie groups and toral subalgebras. Neurocomputing **67**, 161–197 (2005)
13. Rippel, O., Mertens, P., Merhof, D.: Modeling the distribution of normal data in pre-trained deep features for anomaly detection. In: 2020 25th International Conference on Pattern Recognition (ICPR), pp. 6726–6733. IEEE (2021)
14. Ruff, L., et al.: A unifying review of deep and shallow anomaly detection. In: Proceedings of the IEEE (2021)
15. Ruff, L., et al.: Deep one-class classification. In: International Conference on Machine Learning, pp. 4393–4402. PMLR (2018)
16. Schubert, E., Wojdanowski, R., Zimek, A., Kriegel, H.P.: On evaluation of outlier rankings and outlier scores. In: Proceedings of the 2012 SIAM International Conference on Data Mining, pp. 1047–1058. SIAM (2012)
17. Tipping, M.E., Bishop, C.M.: Probabilistic principal component analysis. J. R. Stat. Soc. Ser. B (Stat. Methodol.) **61**(3), 611–622 (1999)
18. Xie, X.: A review of recent advances in surface defect detection using texture analysis techniques. In: ELCVIA: Electronic Letters on Computer Vision and Image Analysis, pp. 1–22 (2008)

Author Index

Printed in the United States
by Baker & Taylor Publisher Services